4th International Conference on Sustainable Agriculture and Environment 2017

IOP Conference Series: Earth and Environmental Science
Volume 142

Surakarta, Indonesia
10 – 12 August 2017

ISBN: 978-1-5108-6109-1
ISSN: 1755-1307

Printed from e-media with permission by:

Curran Associates, Inc.
57 Morehouse Lane
Red Hook, NY 12571

Some format issues inherent in the e-media version may also appear in this print version.

Printed by Curran Associates, Inc. (2018)

For permission requests, please contact the Institute of Physics at the address below.

Institute of Physics
Dirac House, Temple Back
Bristol BS1 6BE UK

Phone: 44 1 17 929 7481
Fax: 44 1 17 920 0979

techtracking@iop.org

Additional copies of this publication are available from:

Curran Associates, Inc.
57 Morehouse Lane
Red Hook, NY 12571 USA
Phone: 845-758-0400
Fax: 845-758-2633
Email: curran@proceedings.com
Web: www.proceedings.com

Table of contents

Volume 142

The 4th International Conference on Sustainable Agriculture and Environment (4th ICSAE)

10–12 August 2017, Surakarta, Indonesia

Accepted papers received: 19 March 2018
Published online: 9 April 2018

Preface

The 4th International Conference on Sustainable Agriculture and Environment (4th ICSAE)

Peer review statement

Papers

Feed efficiency of diets with different energy and protein concentrations supplemented with methionine in laying quails

A Ratriyanto, R Indreswari, A M P Nuhriawangsa and E Purwanti.....1

Egg quality of quails fed low methionine diet supplemented with betaine

A Ratriyanto, R Indreswari, R Dewanti and S Wahyuningsih.....6

Genetic parameters on Bali cattle progeny test population

A R Hariansyah, A Raharjo, A Zainuri, Y Parwoto, D Prasetiyo, S Prastowo and N Widyas.....10

The number of service per conception of Indonesian Friesian Holstein with artificial insemination in Selo, Boyolali, Central Java, Indonesia

A M Wicaksono, A Pramono, A Susilowati, Sutarno, N Widyas and S Prastowo.....15

Persistency of milk yield in Indonesian Holstein cows

N Widyas, F Y Putra, T Nugroho, A Pramono, A Susilowati, Sutarno and S Prastowo.....19

Predicting Madura cattle growth curve using non-linear model

N Widyas, S Prastowo, T S M Widi and E Baliarti.....23

Age effect on post freezing sperm viability of Bali cattle (*Bos javanicus*)

R D Hapsari, Y Khalifah, N Widyas, A Pramono and S Prastowo.....27

Physical quality of Simental Ongole crossbred silverside meat at various boiling times

J Riyanto, M Cahyadi and W S Guntari.....31

In silico study of protein to protein interaction analysis of AMP-activated protein kinase and mitochondrial activity in three different farm animal species

S Prastowo and N Widyas.....35

Birth and weaning weight of kids from different Boer goat crosses

T Nugroho, A Nurhidayati, A I Ayuningtyas, C Kustiyani, S Prastowo and N Widyas.....40

Effect of benzyl amino purine and indole-3-acetic acid on propagation of *Sterculia foetida in vitro*

E Yuniastuti, C E Widodo, Samanhudi and M N I Delfianti.....44

The effect of dosages of microbial consortia formulation and synthetic fertilizer on the growth and yield of field-grown chili

N Istifadah, D Sapta, H Krestini, B Natalie, P Suryatmana, A Nurbaity and R Hidersah.....49

Isolation of Mycorrhizal *Rhizoctonia* as resistance inducer of *Dendrobium macrophyllum* to drought

R Soelistijono, Daryanti and M T Handayani.....56

Dynamics of N-NH_4^+, N-NO_3^-, and total soil nitrogen in paddy field with azolla and biochar

W S Dewi, G I Wahyuningsih, J Syamsiyah and Mujiyo.....61

Study of buffer substrate and Arenga wood fiber size on hydroponic Kailan (*Brassica alboglabra*)

D Harjoko, M D Anggraheny and R B Arniputri.....68

The evaluation of water hyacinth (*Eichhornia crassiper*) control program in Rawapening Lake, Central Java Indonesia

N Hidayati, T R Soeprobowati and M Helmi.....77

Wastewater treatment by nanofiltration membranes

R Mulyanti and H Susanto.....82

The competitiveness of medicinal plants in Central Java Indonesia

E W Riptanti, R A Qonita and R U Fajarningsih.....88

The growth response of *Artemisia annua* L. to organic fertilizer type in lowland

L Prasetyo, Y Widyastuti, A Yunus and Samanhudi.....97

Polyploid response of *Artemisia annua* L. to colchicine treatment

A Yunus, Parjanto, Samanhudi, M P Hikam and Y Widyastuti.....101

Artemisia annua respon to various types of organic fertilizer and dose in lowland

A Yunus, Samanhudi, N Brahmanto and Y Widyastuti.....108

Shading and vermicompost effect on growth and flavonoid content of Tapak Liman (*Elephantopus scaber* L.)

R Y A Dawiyah, A Yunus, Samanhudi, Y Widiyastuti and Widodo.....114

A review of phytoremediation technology: heavy metals uptake by plants

A Sumiahadi and R Acar.....123

The effect of protected sardine fish oil as feed supplement on ruminal fermentation

A Pramono, D T Widayati and E Handayanta.....132

Nutrient digestibility of vegetables waste flour on male quail (*Coturnix coturnix japonica*)

A Pramono, M S Primadhani, W Swastike and J Sutrisno.....136

Study of antidiarrheal and hematology profile of laboratory rat fed with yogurt containing local probiotic and purple sweet potato extract

A I N Tari, C B Handayani and S Hartati.....140

The effect of CMC and arabic gum stabilizer combination on the characteristics of soursop velva (*Annona muricata L.*)

N H R Parnanto, B Yudhistira, S R Pertiwi and A Pangestika.....147

Characteristic study of chitosan addition in Tilapia (*Oreochromis niloticus*) bone based gelatin film

W Atmaka, B Yudhistira and M I S Putro.....153

Local durian (*Durio zibethinus* murr.) exploration for potentially superior tree as parents in Ngrambe District, Ngawi

E Yuniastuti, A Anggita, Nandariyah and Sukaya.....162

Weed diversity identification on growth phases of twelve maize varieties

M Rahayu, P Yudono, D Indradewa and E Hanudin.....171

Effectiveness and efficiency of potassium fertilizer application to increase the production and quality of rice in entisols

S Hartati, Suryono and D Purnomo.....180

Application of Azolla and intermittent irrigation to improve the productivity and nutrient contents of local black rice variety

Sulandjari and M B Yunindanova.....188

Stomata character and chlorophyll content of tomato in response to Zn application under drought condition

A T Sakya, E Sulistyaningsih, D Indradewa and B H Purwanto.....194

The potential of turmeric (*Curcuma xanthorrhiza*) in agroforestry system based on silk tree (*Albizia chinensis*)

D Purnomo, M S Budiastuti, A T Sakya and M I Cholid.....200

Influence of spacing and depth of planting to growth and yield of arrowroot (*Marantha arundinacea*)

M Qodliyati, Supriyono and S Nyoto.....205

The development of the food barn in achieving food security of poor household in Wonogiri Regency

Suminah.....211

The yield and quality of black rice varieties in different altitude

E Purwanto, W Hidayati and Nandariyah.....219

Effect of manure and plants spacing on yield and flavonoid content of *Elephantopus scaber* L.

D Riyana, Y Widiyastuti, H Widodo, E Purwanto and Samanhudi.....226

Analysis of rice purchase decision on rice consumer in Bandung city

K Kusno, A Imannurdin, N Syamsiyah and E Djuwendah.....234

First record of entomopathogenic fungi on autumn leaf Caterpillar (*Doleschallia bisaltide*)

A K Dayanti, Sholahuddin, A Yunus and D Subositi.....241

The utilization of microbes as a fermentation agent to reduce saponin in Trembesi leaves (*Sammanea saman*)

A K Sariri, A M W Mulyono and A I N Tari.....246

Development of small and medium business (SMES) of mango dodol processing to increase the added value (a case study in Ujungjaya Village, Indramayu District, West Java)

L Sulistyowati, P Pardian, N Syamsyiah and Y Deliana.....251

Response of *Artemisia annua* L. to shade and manure fertilizer application in lowland altitude

H H Permana, Y Widyastuti, Samanhudi and A Yunus.....259

Yields and content of ursolic acid in pearl grass (*Hedyotis corymbosa*) when utilizing cow manure fertilizer in different shades

L D Jayanti, A Yunus, B Pujiasmanto and Y Widyastuti.....266

The potential of arbuscular mycorrhizal fungi application on aggregrate stability in alfisol soil

J Syamsiyah, A Herawati and Mujiyo.....275

Effect of utilization of tomato extract and foliar fertilizer as media on shoots multiplication of banana cv Ambon in vitro

L A Bidhari, E Purwanto and A Yunus.....281

Analysis of the effectiveness of fertilizer subsidy policy and its effect on rice production in Karanganyar Regency

A T Mulyadiana, S Marwanti and W Rahayu.....285

Performance of Arrowroot (*Marantha arundinacea*) in various light intensities

M B Oktafani, Supriyono, MTh S Budiastuti and D Purnomo.....291

Effect aquadest-extracted *Gloriosa superba* seed as mutagen on morphology of *Artemisia annua*

S I Rahmawati, A Susilowati, A Yunus and Y Widyastuti.....295

Performance of Mentik Wangi rice (*Oryza sativa*, L.) M2 generation from gamma ray irradiation

A Yunus, Parjanto, Nandariyah and S Wulandari.....299

Dynamic models of farmers adaptation to climate change (case of rice farmers in Cemoro Watershed, Central Java, Indonesia)

Sugihardjo, J Sutrisno, P Setyono and Suntoro.....307

Morphology, production, and chemical content performance of black rice Matesih accession with several comparisons

Nandariyah, E Purwanto and A N Meidini.....313

Determination colchicine content in aquadest-extracted *Gloriosa superba* seed from Sukoharjo and Gunung Kidul

S I Rahmawati, A Susilowati, A Yunus and Y Widyastuti.....321

Farmers's perception and strategies for the development of sustainable livelihoods in disaster prone areas

S Anantanyu, Suwarto and Suminah.....326

The effect of shade and vermicompost application on yield and flavonoid levels of Tempuyung (*Sonchus arvensis*)

D P Putri, Y Widyastuti, W S Dewi and A Yunus.....333

The effect of cutting origin and organic plant growth regulator on the growth of Daun Ungu (*Graptophyllum pictum*) through stem cutting method

S P Pratama, A Yunus, E Purwanto and Y Widyastuti.....344

The impact of seasonal rice price changes on rice self-consumption in farm household of rural Java

S W Ani and E Antriyandarti.....352

Drought resistant of bacteria producing exopolysaccharide and IAA in rhizosphere of soybean plant (*Glycine max*) in Wonogiri Regency Central Java Indonesia

A Susilowati, A A Puspita and A Yunus.....358

Analysis of soybean production and import trends and its import factors in Indonesia

I H Ningrum, H Irianto and E W Riptanti.....365

Paddy farmer households' participation and food security level in special effort program in Seputih Raman sub-district of Central Lampung Regency

K K Rangga and Y A Syarief.....373

The analysis of export commodity competitiveness in Central Java Province at period 2011-2015

M G Elysi, Darsono and E W Riptanti.....378

Building agribusiness model of LEISA to achieve sustainable agriculture in Surian Subdistrict of Sumedang Regency West Java Indonesia

E Djuwendah, T Priyatna, K Kusno, Y Deliana and E Wulandari.....385

The performance of M2 generation of *Mentik Wangi Susu* rice resulted from gamma ray iradiation

A Yunus, Parjanto and I Y Pratama.....393

Antagonism of rice phylloplane fungi against *Cercospora oryzae*

A Mardani and Hadiwiyono.....403

Effectiveness of cow manure and mycorrhiza on the growth of soybean

M N A Muktiyanta, Samanhudi, A Yunus, B Pujiasmanto and S Minardi.....409

Aplication of giberelins on flowering and yield of two varieties of shallot in lowland

E Triharyanto, S Nyoto and I Yusrifani.....416

Adaptation of six shallots varieties to phosphate solubilizing bacteria on the flower formation, seeds fromation, and yields on the lowland

E Triharyanto, Sudadi and S Rawandari.....426

Defense reactions of bean genotypes to bacterial pathogens in controlled conditions

B Uysal and K K Bastas.....433

Farmer readiness for adopting stevia cultivation (a case study at District of Pasir Jambu, Regency of Bandung)

D Supyandi, Y Sukayat and H Hapsari.....438

Utilizing Bacillus to inhibit the growth and infection by sheath blight pathogen, *Rhizoctoniasolani* in rice

R Margani, Hadiwiyono and S Widadi.....444

Sabuk Janur: tools to move community participation in reducing natural disasters and environment (case study at Lawu mount slope in Indonesia)

E Lestari, S H Purnomo, Komariah and A Wibowo.....451

The land use potential of flood-prone rice fields using floating rice system in Bojonegoro regency in East Java

H Irianto, Mujiyo, E W Riptanti and A Qonita.....459

The fficiency of Mycorrhiza biofertilizer treatment to the growth and yield of soybean

Samanhudi, B Pujiasmanto, Sudadi, I H Putra and H M Mumtazah.....465

The influence of ihsan attitude and economic condition to farmer household food security

H Hendrarini, E S Rahayu, Kusnandar, R J Sunarsono and T Soedarto.....472

Effect of CMC and arabic gum in the manufacture of jackfruit velva (*Artocarpus heterophyllus*)

B Yudhistira, N H Riyadi, A D Pangestika and S R Pertiwi.....477

Sustainable development of smallholder crop-livestock farming in developing countries

S Ates, H Cicek, L W Bell, H C Norman, D E Mayberry, S Kassam, D B. Hannaway and M Louhaichi.....485

Economic activities results and problems of the farm engaged in production of stick tomatoes in Erdemli (Mersin) Province

M Direk and S Topkara.....496

The effect of mixed liming and NPK fertilizer to yield of some rice varieties on new openings of acid sulfate tidal swamp land

A Akhmad, W S Dewi, S Sagiman and Suntoro.....504

Organic pollutant removal from edible oil process wastewater using electrocoagulation

S Sharma, O T Can, M Hammed, D Nawarathna and H Simsek.....511

IOP Conf. Series: Earth and Environmental Science **142** (2018) 011001 doi:10.1088/1755-1315/142/1/011001

PREFACE

On behalf of organizing committee, I would like to express our sincere gratitude for all of your contributions and participation in the 4th International Conference of Sustainable Agriculture (ICSAE) which held in Surakarta Indonesia, 10-12 August 2017. This conference belongs to ICSAE conference series that already held in Surakarta Indonesia (2013), Konya Turkey (2015) and Warsaw Poland (2016). International Conference on Sustainable Agriculture and Environment was aimed to provide a platform for researchers and academics as well as practicing professionals from all over the world, to present their research and professional development activities in agriculture, environment, food and other relevant subjects. This conference series was an effort to identify the ideas, practices and policies that constitute our concept of sustainable agriculture. We are doing this activity for two reasons: 1) to clarify the research agenda and priorities of our goal, and 2) to suggest practical approaches which may be appropriate in moving toward a better concept of sustainable agriculture. The concept of sustainable agriculture itself is still evolving and thus, we published not as a definitive or final statement, but as an invitation to continue our dialogue.

At this conference, there were 170 papers submitted from 15 different countries. Each paper proposal was evaluated by two reviewers and 120 papers were presented. However, only papers which meet the certain criteria were selected for published in this *IOP Conference Proceeding Series*. We would like to thank to all of our conference committees as well as to all participants without whom the conference won't gain the current success. We are sure that, 4th ICSAE will be the flagship conference for researchers, students and professionals in the areas related to Agriculture & Environment. Their contributions have benefited our countries and the world by disseminating their research advancements and discoveries. Our ideas were network building and knowledge exchange in order to straighten existing partnership and foster new collaborations.

New policies are needed to simultaneously promote environmental health, economic profitability, and social and economic equity. Collaborations must be created to address these policies concerning local, regional, and national level. In addition to strategies for preserving natural resources and changing production practices, sustainable agriculture requires a commitment to change the public policies, economic institutions, and social values. Strategies for the changes must take into account the complex, reciprocal and ever-changing relationship between agricultural production and the broader society.

Critical discussion on sustainable agriculture concept will and should continue. Understanding will deepen and answers will continue to come. On-going dialog is important for another reason: intense communication among sustainable agriculture practitioners and contributors which would enable us to synchrone our agenda, focus and issues. This measure was taken in order to avoid "sustainable agriculture" main frame from becoming so all-encompassing as to become meaningless.

Finally, it is important to point out that reaching toward the goal of sustainable agriculture is the responsibility of all of the system's stake holders, including farmers, laborers, policymakers, researchers, retailers, and consumers. Each group has its own part to play, its own unique contribution to make, to strengthen the sustainable agriculture community. See you soon at the coming conference of 5th ICSAE that will be held in Tunisia.

Ahmad Yunus, Prof. Dr.
Universitas Sebelas Maret, Surakarta Indonesia
Chairman of 4th ICSAE Conference Committee
10-12 August 2017, Surakarta Indonesia.

IOP Conf. Series: Earth and Environmental Science **142** (2018) 011002 doi:10.1088/1755-1315/142/1/011002

Peer review statement

All papers published in this volume of *IOP Conference Series: Earth and Environmental Science* have been peer reviewed through processes administered by the proceedings Editors. Reviews were conducted by expert referees to the professional and scientific standards expected of a proceedings journal published by IOP Publishing.

Feed efficiency of diets with different energy and protein concentrations supplemented with methionine in laying quails

A Ratriyanto, R Indreswari, A M P Nuhriawangsa and E Purwanti
Department of Animal Science, Faculty of Agriculture, Universitas Sebelas Maret, Surakarta, Indonesia

E-mail: ratriyanto@staff.uns.ac.id

Abstract. The study was conducted to evaluate the feed efficiency of quail diets containing different concentrations of metabolizable energy (ME) and crude protein (CP) with constant ratio and supplemented with methionine. Four hundred laying quails (*Coturnix coturnix japonica*) were randomly assigned to four experimental diets in a 2×2 factorial arrangement. Each dietary treatment used 5 replicates of 20 quails. Two basal diets were formulated to contain 2,800 kcal kg^{-1} ME and 18.7% CP (High ME-CP) and 2,600 kcal kg^{-1} ME and 17.3% CP (Low ME-CP). Each basal diet was supplemented with 0 and 0.12% methionine. The High ME-CP diets generated lower feed consumption but higher egg mass and feed efficiency ($P<0.01$) compared with the Low ME-CP. Furthermore, supplementation of methionine increased egg mass, feed efficiency, energy efficiency ratio and protein efficiency ratio ($P<0.01$). The High ME-CP supplemented with methionine resulted the highest feed efficiency followed by the Low ME-CP supplemented with methionine, while both High ME-CP and Low ME-CP without methionine supplementation resulted the lowest feed efficiency ($P<0.05$). In addition, ME and CP consumption of the birds were not influenced by the treatments. Thus, feeding High ME-CP supplemented with 0.12% methionine provided benefit to improve the feed efficiency in laying quails.

1. Introduction

Energy and protein concentrations in the diet play an important role in livestock productivity and are critical in the evaluation of poultry performance [1]. It is generally accepted that the poultry diet should be formulated precisely to obtain an optimal performance [2]. The nutrient concentration is important both in the nutritional aspect and practical application in terms of its economics, thus, feed formulation with lower protein or energy concentration enables the decrease of feed cost [3].

Feed consumption in poultry is regulated among others by nutrient density in the diet [4] and more specifically poultry consume the feed to fulfill the requirement of energy and protein [5]. The too high nutrients content in quail's diet may induce heat stress as a result of metabolic processes [2] and leading to inefficiency [6]. Therefore, feeding quails with low energy and protein diet is expected to minimize the heat production, especially in tropical climates [7]. However, the quail's performance may decrease if the energy and protein are provided below the requirement. In addition, adjustment of energy and protein concentration in the diet is necessary to avoid excessive feed consumption and fat deposition in laying poultry [8]. The recommendation for crude protein (CP) content is 20% and metabolizable energy (ME) is 2,900 kcal kg^{-1} for laying quail raised at the temperature 21°C [9]. However, in higher temperature condition, some researchers evaluated the nutrient requirement for laying quails to be 18% CP and 2,700 kcal kg^{-1} ME [10].

The requirement of energy and protein is correlated with the methionine level in the diet. Methionine requirement increases as ME level increases [11]. Methionine lessened the energy excretion, thus, the energy absorbed and digested by the animals can be improved [12]. Methionine supplementation also improved the immunity in chickens fed diet containing low ME [13], and may raise the laying performance [14]. Moreover, methionine serves as a methyl group donor which plays an important function in metabolism of energy and protein [15]. Thus, this study evaluated the effects of ME and CP concentrations with constant ratio and methionine supplementation to the diet on feed efficiency of quails.

2. Methods

In total, 400 twenty-five-days laying quails (*Coturnix coturnix japonica*) with average body weight of 80.83 ± 2.47 g were randomly assigned to four experimental diets in a 2 × 2 factorial arrangement. Two basal diets were formulated to contain 2,800 kcal kg^{-1} ME and 18.7% CP (High ME-CP) and 2,600 kcal kg^{-1} ME and 17.3% CP (Low ME-CP). Each basal diet was supplemented with 0 and 0.12% methionine. Each dietary treatment used 5 replicates of 20 quails. Nutrient composition of the basal diets is presented in Table 1. The experimental diets were obtained by supplementing 0.12% methionine to the basal diets in the expense of rice bran [16].

Table 1. Nutrient composition of the basal diets

Nutrients	High ME-CP	Low ME-CP
Metabolizable energy (kcal/kg)	2,800	2,600
Crude protein (%)	18.70	17.30
Calcium (%)	3.30	3.30
Nonphytate phosphorus (%)	0.43	0.43
Lysine (%)	1.18	1.18
Methionine (%)	0.40	0.40

The birds were randomly distributed to 20 pens with the size 120 × 30 × 25 cm. During experiment, water and feed were provided *ad libitum*. Commercial grower diet was fed during pre-experimental period in growing phase. Then, the quails were fed a basal diet containing 2,800 kcal kg^{-1} ME and 18.7% CP from the age of 42 days. The experimental diets were fed for 2 periods of 28 days after egg production reached 50%. The observed data included the feed, ME and CP consumption, egg mass, feed efficiency as well as energy and protein efficiency ratio (EER and PER). The EER was calculated as grams of egg mass per 100 kcal kg^{-1} ME consumption, and PER was calculated as grams of egg mass per gram of CP consumption [17].

The obtained data were submitted to analysis of variance using R statistic to determine the effect of treatment and it was continued to Duncan's test if the treatment indicated significant effect [18].

3. Results and Discussion

3.1. Nutrient consumption

There was no interaction between dietary ME-CP concentration and methionine supplementation on feed, energy and protein consumption (Table 2). This indicated that the quails consumed the same amount of energy and protein both fed Low ME-CP and High ME-CP with or without methionine supplementation. Furthermore, feed consumption in quails fed Low ME-CP was higher ($P<0.01$) than quails fed High ME-CP resulting in the equal amount of ME and CP consumption. This phenomenon confirms glucostatic and amino static theoris which described that the birds consume the feed to meet the requirement of energy and protein/amino acids [5]. In agreement with this finding, other authors observed in quails that feeding 2,700 kcal kg^{-1} ME generated higher feed consumption than feeding 2,900 kcal kg^{-1} [19]. However, feeding 16.5 to 19.5% CP for laying quails did not affect feed consumption [17].

Supplementation of methionine did not affect feed, ME and CP consumption of the quails (Table 2). Feed consumption is affected by amino acids concentration in the blood [5]. It is hypothesized that methionine suplementation did not alter amino acid profils in the blood since the basal diets contained adequate total sulphur amino acids. In support with this observation, methionine supplementation did not affect feed consumption in broiler chickens [20].

Table 2. Feed, metabolizable energy and protein consumption in quails fed diets containing different ME and CP concentration and supplemented with methionine

Treatments		Feed Consumption (g day^{-1})	Metabolizable Energy Consump-tion (g day^{-1})	Crude Protein Consumption (g day^{-1})
Interaction ME-CP × methionine				
High ME-CP	0.00%	24.04	67.32	4.50
High ME-CP	0.12%	24.36	68.20	4.55
Low ME-CP	0.00%	25.57	66.49	4.42
Low ME-CP	0.12%	25.93	67.41	4.49
Significance		NS	NS	NS
Effect of ME-CP				
High ME-CP		24.20^{b}	67.76	4.53
Low ME-CP		25.75^{a}	66.95	4.45
Significance		**	NS	NS
Effect of methionine				
0.00%		24.81	66.90	4.46
0.12%		25.14	67.81	4.52
Significance		NS	NS	NS

^{a,b} Means within a treatment and column with different superscripts differ significantly (P<0.01)
NS: Not significant
** P<0.01

3.2. Feed and nutrient efficiency

An interaction was observed between ME-CP concentration and methionine supplementation on feed efficiency of quails (Table 3). Methionine supplementation, independently ME-CP concentration, increased feed efficiency, with the highest feed efficiency was observed for feeding High ME-CP supplemented with methionine (P<0.05). This finding may be attributed to the combined effects of improvement in feed efficiency of feeding High ME-CP and methionine supplemented diets. Similar result was also observed in previous study in quails [21].

Feeding High ME-CP improved (P<0.01) egg mass and feed efficiency by 1.9 and 8.9%, respectively, compared with feeding Low ME-CP (Table 3) which is in accordance with previous observations in laying quails [17, 21]. Feed efficiency is affected among others by feed consumption. Since the quails fed High ME-CP consumed less feed but generated higher egg mass than the quails fed High ME-CP, the feed efficiency was higher for feeding High ME-CP. Furthermore, energy and protein efficiency ratio were not affected by ME-CP concentration, which was in accordance with the ME and CP consumption in this study.

Methionine supplementation enhanced (P<0.01) egg mass by 15.7% and feed efficiency, energy and protein efficiency ratio by 14.4, 14.4, and 14.6%, respectively (Table 3). The improvement in feed efficiency obtained with the dietary methionine supplementation may be attributed to its role as a methyl group donor [16, 22]. Its suggests that methionine as a methyl group donor is used in protein and energy metabolisms, thus the higher egg mas and feed efficiency can be obtained due to methionine supplementation [21]. Previous observation indicated that dietary methionine increased egg mass and feed efficiency in quails [23]. Similar to this finding, supplementation of betaine, other methyl group donor, improved energy and protein efficiency ratio in quails [17].

Table 3. Egg mass, feed efficiency, energy, and protein efficiency ratio in quails fed diets containing different ME and CP concentration and supplemented with methionine

Treatments		Egg Mass ($g\ day^{-1}$)	Feed Efficiency (%)	Energy Efficiency Ratio	Protein Efficiency Ratio
Interaction ME-CP × methionine					
High ME-CP	0.00%	6.38	26.53^{c}	9.48	1.42
High ME-CP	0.12%	7.62	31.30^{a}	11.18	1.67
Low ME-CP	0.00%	6.47	25.30^{c}	9.73	1.46
Low ME-CP	0.12%	7.26	28.01^{b}	10.77	1.62
Significance		NS	*	NS	NS
Effect of ME-CP					
High ME-CP		7.00^{a}	28.92^{a}	10.33	1.55
Low ME-CP		6.87^{b}	26.66^{b}	10.25	1.54
Significance		**	**	NS	NS
Effect of methionine					
0.00%		6.43^{b}	25.92^{b}	9.60^{b}	1.44^{b}
0.12%		7.44^{a}	29.66^{a}	10.98^{a}	1.65^{a}
Significance		**	**	**	**

^{a,b} Means within a treatment and column with different superscripts differ significantly ($P<0.05$)
NS: Not significant
* $P<0.05$
** $P<0.01$

4. Conclusion

Dietary methionine supplementation independently metabolizable energy and crude protein concentration improved feed efficiency in quails. Feeding quails with 2,800 kcal kg^{-1} metabolizable energy and 18.7% crude protein generated higher feed consumption, egg mass, and feed efficiency than feeding 2,600 kcal kg^{-1} metabolizable energy and 17.3% crude protein. Moreover, supplementation of methionine enhanced egg mass, feed efficiency, energy and protein efficiency ratio in quails.

References

[1] Dairo F A S, Adesehinwa A O K, Oluwasola T A and Oluyemi J A 2010 High and low dietary energy and protein levels for broiler chickens *Afr. J. Agric. Res.* **5** 2030–8
[2] Li Y X, Wang Y Q, Pang Y Z, Li J X, Xie X H, Guo T J and Li W Q 2011 The effect of crude protein level in diets on laying performance, nutrient digestibility of yellow quails *Int. J. Poult. Sci.* **10** 110–2
[3] Kamran Z, Sarwar M, Nisa M, Nadeem M A, Mahmood S, Babar M E and Ahmed S 2008 Effect of low-protein diets having constant energy-to-protein ratio on performance and carcass characteristics of broiler chickens from one to thirty-five days of age *Poult. Sci.* **87** 468–74
[4] Moura G de S, Barreto S L de T and Lanna E A T 2010 Efeito da redução da densidade energética de dietas sobre as características do ovo de codorna japonesa *Rev. Bras. Zootec.* **39** 1266–71
[5] Ferket P R and Gernat A G 2006 Factors that affect feed intake of meat birds: A review. *Int. J.*

Poult. Sci. **5** 905–11
[6] De Faria Filho D E, Campos D M B, Alfonso-Torres K A, Vieira B S, Rosa P S, Vaz A M, Macari M and Furlan R L 2007 Protein levels for heat-exposed broilers: Performance, nutrients digestibility, and energy and protein metabolism *Int. J. Poult. Sci.* **6** 187–94
[7] Furlan R L, Faria Filho D E de, Rosa P S and Macari M 2004 Does low-protein diet improve broiler performance under heat stress conditions? *Braz. J. Poult. Sci.* **6** 71–9
[8] Kiiskinen T and Helander E 1998 Effects of restricted methionine and energy intake on egg weight and shell quality *Agric. Food Sci. Finl.* **7** 513–21
[9] Nutrition Research Council 1994 *Nutrient Requirements of Poultry* (Washington DC: National Academic Press)
[10] Ghazvinian K, Irani M, Jamshidi R and Mirzaei-aghsaghali A 2011 The effect of energy to protein ratio on production performance and characteristics of Japanese quail eggs **2** 122–8
[11] Harms R H, Russell G B, Harlow H and Ivey F J 1998 The influence of methionine on commercial laying hens *J. Appl. Poult. Res.* **7** 45–52
[12] Piliang G W and Djojosoebagio S 2006 *Fisiologi Nutrisi* (Bogor: Percetakan Institut Pertanian Bogor)
[13] Mirzaaghatabar F, Saki A A, Zamani P, Aliarabi H and Matin H R H 2011 Effect of different levels of diet methionine and metabolisable energy on broiler performance and immune system *Food Agric. Immunol.* **22** 93–103
[14] Koreleski J and Świątkiewic S 2010 Effect of methionine and energy level in high protein *Ann. Anim. Sci.* **10** 83–91
[15] Ratriyanto A, Mosenthin R, Bauer E and Eklund M 2009 Metabolic, osmoregulatory and nutritional functions of betaine in monogastric animals *Asian-Australasian J. Anim. Sci.* **22** 1461–76
[16] Ratriyanto A, Mosenthin R, Jezierny D, Sauer N and Eklund M 2009 Betaine, organic acids and inulin do not affect ileal and total tract nutrient digestibility or microbial fermentation in piglets *J. Anim. Feed Sci.* **18** 453–64
[17] Ratriyanto A, Indreswari R, Nuhriawangsa A, Ratriyanto A, Indreswari R and Nuhriawangsa A 2017 Effects of dietary protein level and betaine supplementation on nutrient digestibility and performance of Japanese quails *Rev. Bras. Ciênc. Avíc.* **19** 445–54
[18] R Core Team 2015 *R: A Language and Environment for Statistical Computing* (Vienna: R Foundation for Statistical Computing)
[19] Gheisari A, Halaji H A, Maghsoudinegad G and Toghyani M 2011 Effect of different dietary levels of energy and protein on performance of Japanese quails (*Coturnix coturnix japonica*) *Proc. 2nd Int. Conf. on Agricultural and Animal Science* IPCBEE vol 22 (Singapore: IACSIT Press) pp 156–9
[20] Ratriyanto A, Indreswari R and Sunarto 2014 Effects of protein levels and supplementation of methyl group donor on nutrient digestibility and performance of broiler chickens in the tropics *Int. J. Poult. Sci.* **13** 575–81
[21] Ratriyanto A, Indreswari R, Nuhriawangsa A M P and Arifin A A 2017 Dietary metabolizable energy and methionine affect performance of quails *Proceeding of the 1st International Conference on Tropical Agriculture* (Cham: Springer International Publishing) pp 329–35
[22] Ezzat W and Shoeib M S 2011 Impact of betaine, vitamin c and folic acid supplementations to the diet on productive and reproductive performance of Matrouh poultry *Egypt. Poult. Sci.* 521–37
[23] Khairani K, Sumiati S and Wiryawan K G 2015 Egg production and quality of quails fed diets with varying levels of methionine and choline chloride *Med. Pet.* **39** 34–9

Egg quality of quails fed low methionine diet supplemented with betaine

A Ratriyanto, R Indreswari, R Dewanti and S Wahyuningsih
Department of Animal Science, Faculty of Agriculture, Universitas Sebelas Maret, Surakarta, Indonesia

E-mail: ratriyanto@staff.uns.ac.id

Abstract. This experiment investigated the effect of betaine supplementation to low methionine diet on egg quality of quails. A total of 340 laying quails (*Coturnix coturnix japonica)* was divided into 4 dietary treatments with 5 replicates of 17 quails each. The experiment was assigned in a completely randomized design. The four dietary treatments were the low methionine diet (0.3% methionine) without betaine supplementation and the low methionine diet supplemented with 0.07, 0.14, and 0.21% betaine. The experimental diets were applied for 8 weeks and the egg quality traits were measured at the age of 16 and 20 weeks. The data were subjected to analysis of variance, and when the treatment indicated significant effect, it was continued to orthogonal polynomial test to determine the optimum level of betaine. Increasing dietary levels of betaine increased the fat content of the egg with the linear regression of $y = 11.0949 + 4.1914x$ ($R^2 = 0.18$). However, supplementation of betaine did not affect protein content, yolk, albumen, and eggshell percentage. It can be concluded that betaine supplementation up to 0.21% to low methionine diet only had little effect in improving the quality traits of quail eggs.

1. Introduction

Methionine and betaine are the potential feed additives as methyl goup (CH_3) donor [1]. Methionine is an essential amino acid whose metabolites are used in a variety of fundamental biological processes including protein deposition and synthesis of S-adenosyl methionine [2]. Providing other methyl group donor (*e.g.* betaine) may spare methionine as methyl group donor or provide the methyl group necessary to convert homocysteine to methionine, thus the requirement for methionine in the diet can be reduced [3].

Betaine (trimethyl glycine) has two main biological functions, *i.e.* as a methyl group donor and an organic osmolyte. As a methyl group donor, betaine can spare the use of other methyl group donors such as methionine and choline. Thus, betaine is expected to increase the availability of methionine for protein synthesis, resulting in optimal performance [1]. As an organic osmolyte, betaine protects cells from various osmotic pressures, such as cell hydration [4].

Methyl group cannot be syntesized in the body of poultry, thus it must be supplied in the diet [1]. Previous studies showed that betaine supplementation to methionine-deficient diet increased betaine homocysteine-methyltransferase (BHMT) activity in chickens, indicating that poultry have specific requirements for methyl groups [5]. Observation in poultry revealed that betaine maintained performance of laying hens fed diets reduced in methionine and without choline chloride [6]. Supplementation of 0.1% betaine increased eggshell thickness, haugh unit and eggshell percentage but did not affect yolk percentage in laying hens [7]. Studies on the potential of betaine as a methyl group donor in laying poultry are still limited and the results are inconsistent [6, 8]. Betaine supplementation to low methionine diet is expected to provide methyl groups which have the potential to improve egg quality. The objective of this study was to investigate the effect and optimum level of betaine supplementation in low methionine diet egg quality of quails.

2. Methods

A total of 340 laying Japanese quails (*Coturnix coturnix japonica*) aged 4 weeks with an average body weight of 98.31 ± 8.67 grams was used in this study. The study was designed as a completely randomized design with 4 dietary treatments, each treatment with 5 replicates containing 17 quails. The basal diet was formulated to meet nutrient requirement of laying quails according to the recommendation of Indonesian National Standard [9] except for the amino acid methionine which was set at 0.3%, lower than the methionine requirement standard of 0.4 (Table 1).

Table 1. Composition and nutrient content of the basal diet

Ingredients	Proportion (%)
Yellow corn	25.00
Rice bran	21.00
Soybean meal	25.00
Fish meal	5.50
Cassava by product meal	15.50
Dicalcium phosphate	1.75
Limestone	5.50
Premix	0.50
NaCl	0.25
Nutrient content	
Metabolizable energy (kcal/kg)	2,887.61
Crude protein (%)	18.51
Calcium (%)	3.41
Available phosporus (%)	0.70
Lysine (%)	1,02
Methionine (%)	0,30

The diet was fed without betaine supplementation or supplemented with 0.07, 0.14, and 0.21% betaine. Betaine was supplemented at the expense of rice bran [10]. During the experiment, the quails had free access to water and feed. The experimental diets were applied for 8 weeks after the egg production has reached 10%. The egg quality traits were measured for 3 consecutive days at 16 and 20 weeks. In addition, standard managemental practice was applied during this study.

The observed variables were yolk, albumen, and eggshell percentage, and protein and fat content of the eggs. Data were subjected to variance analysis and if it showed an effect of treatment, then it was continued with orthogonal polynomial test to determine the optimum level of treatment.

3. Results and Discussion

Betaine supplementation to low methionine diet did not affect yolk, albumen, and eggshell percentage as well as protein content of the eggs (Table 2). Egg size is mostly determined by the yolk and albumen. The results of this study were in accordance with previous study in Lohmann Brown laying hens, in which supplementation of 0.035% betaine did not affect yolk and eggshell percentage [6]. Similarly, Ezzat et al. [7] did not observed any effect of 0.1% betaine supplementation on yolk, albumen, and eggshell percentage in Matrouh laying hens.

Table 2. Effect of betaine supplementation to low methionine diet on egg quality traits of quails

Variables	Betaine Levels (%)				P
	0	0.07	0.14	0.21	Value
Yolk percentage (%)	38.94	33.21	35.05	34.46	0.10
Albumen percentage (%)	53.33	58.94	55.40	57.36	0.14
Eggshell percentage (%)	8.73	8.27	7.37	8.47	0.29
Protein content (%)	11.76	11.80	11.59	11.41	0.21
Fat content (%)	10.94	11.33	12.26	11.61	0.03

Previous studies revealed that betaine supplementation increased protein synthesis and decreased protein degradation as indicated by increased in serum protein, globulin and albumin levels [11]. However, increasing dietary betaine levels in this study could not optimize the availability of methionine for protein synthesis. Presumably methionine is used for immune systems, where the requirement of methionine for immunity is greater than for protein synthesis [1]. This results was in support with previous observation, in which betaine supplementation did not affect egg quality traits in laying hens [12]. Different to this result, Ratriyanto et al. [13] observed that supplementation of 0.06 and 0.12% betaine to methionine adequate diet enhanced egg quality traits of quails including yolk, albumen, and eggshell weight.

Furthermore, the homeostasis mechanism regulates the amount of calcium absorbed according to the requirement of mineral [14], thus the eggshell percentage was not affected by the treatment. In agreement with this study, supplementation of 0.035% betaine did not affect eggshell percentage in Lohmann Brown laying hens [6]. Meanwhile, according to Ryu et al. [15] betaine supplementation at levels of 0.05% to 0.2% increased eggshell strength in laying hens.

Increasing dietary betaine supplementation to low methionine diet increased linearly ($P<0.05$) fat content of the eggs, with the regression $y = 11.0949 + 4.1914x$ ($R^2 = 0.18$) (Figure 1). Betaine donates the methyl group during the transmetylation reaction to synthesize important metabolically substances such as carnitine and creatine that play an important role in fat metabolism [1]. Betaine is also a lipotropic agent and involved in the synthesis of chylomicron that play a role in fat absorption [16]. In addition, betaine donates its methyl groups to synthesis of lecithin which is required in the transportation of fat in the body, thus providing the fat for egg formation [17].

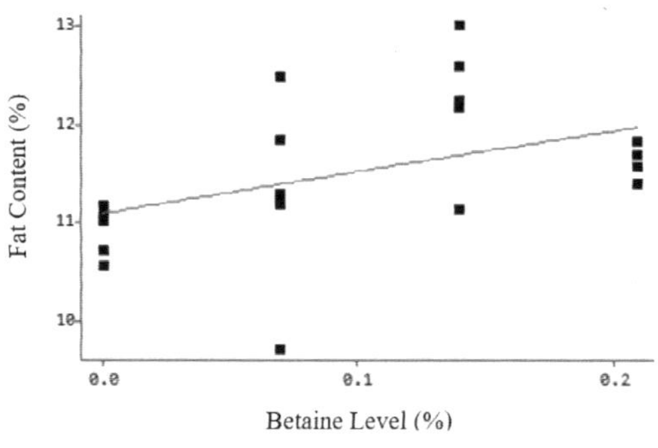

Figure 1. Effect of betaine supplementation on fat content of the eggs

4. Conclusion

Betaine supplementation to low methionine diet increased fat content but did not affect the percentage of yolk, albumen, eggshell, and protein content of quail eggs. Furthermore, increasing dietary betaine supplementation linearly increased fat content of the eggs with the regression $y = 11.0949 + 4.1914x$ ($R^2 = 0.18$).

References

[1] Ratriyanto A, Mosenthin R, Bauer E and Eklund M 2009 Metabolic, osmoregulatory and nutritional functions of betaine in monogastric animals *Asian-Austral. J. Anim. Sci.* **22** 1461–76

[2] Finkelstein J D, Martin J J and Harris B J 1988 Methionine metabolism in mammals *J. Biol. Chem.* **263** 11750–4

[3] Sun H, Yang W R, Yang Z B, Wang Y, Jiang S Z and Zhang G G 2008 Effects of betaine supplementation to methionine deficient diet on growth performance and carcass characteristics of broilers *Am. J. Anim. Vet. Sci.* **3** 78–84

[4] Siljander-Rasi H, Peuranen S, Tiihonen K, Virtanen E, Kettunen H, Alaviuhkola T and Simmins P 2003 Effect of equi-molar dietary betaine and choline addition on performance, carcass quality and physiological parameters of pigs *J. Anim. Sci.* **76** 55–62

[5] Emmert J L, Garrow T A and Baker D H 1996 Hepatic betaine-homocysteine methyltransferase activity in the chicken is influenced by dietary intake of sulfur amino acids, choline and betaine. *J. Nutr.* **126** 2050–8

[6] Hruby M, Ombabi A and Schlagheck A 2005 Natural betaine maintains layer performance in methionine/choline chloride reduced diets *Proc. 15th Eur. Symp. on Poultry Nutrition* (Balatonfured, Hungary) pp 507–9

[7] Ezzat W and Shoeib M S 2011 Impact of betaine, vitamin c and folic acid supplementations to the diet on productive and reproductive performance of Matrouh poultry *Egypt. Poult. Sci.* 521–37

[8] Lu J and Zou X 2006 Effects of adding betaine on laying performance and contents of serum yolk precursors VLDL and VTG in laying hen *J. Zhejiang Univ.* **32** 287–291

[9] Indonesian National Standard 2006 *Pakan puyuh bertelur* (quail layer) (Jakarta: Badan Standardisasi Nasional)

[10] Ratriyanto A, Mosenthin R, Jezierny D, Sauer N and Eklund M 2009 Betaine, organic acids and inulin do not affect ileal and total tract nutrient digestibility or microbial fermentation in piglets *J. Anim. Feed Sci.* **18** 453–64

[11] Rao R S V., Raju M V L N, Panda A K, Saharia P and Sunder S G 2011 Effect of supplementing betaine on performance, carcass traits and immune responses in broiler chicken fed diets containing different concentrations of methionine *Asian-Austral. J. Anim. Sci.* **24** 662–9

[12] Gudev D, Popova-Ralcheva S, Yanchev I, Moneva P, Petkov E and Ignatova M 2011 Effect of betaine on egg performance and some blood constituents in laying hens reared indoor under natural summer temperatures and varying levels of air ammonia *Bulg. J. Agric. Sci.* **17** 859–66

[13] Ratriyanto A, Indreswari R, Nuhriawangsa A, Ratriyanto A, Indreswari R and Nuhriawangsa A 2017 Effects of dietary protein level and betaine supplementation on nutrient digestibility and performance of Japanese quails *Rev. Bras. Ciênc. Avíc.* **19** 445–54

[14] Underwood E J and Suttle N F 1999 *The Mineral Nutrition of Livestock* (Oxon: CABI Publishing)

[15] Ryu M S, Cho K H, Shin W J and Ryu K S 2002 Influence of dietary supplemental betaine on performance and egg quality of laying hens during the heat stress *Korean J. Poult. Sci.* **29** 117–23

[16] Yao Z and McLeod R S 1994 Synthesis and secretion of hepatic apolipoprotein B-containing lipoproteins *Biochim. Biophys. Acta - Lipids Lipid Metab.* **1212** 152–66

[17] Saunderson C L and Mackinlay J 1990 Changes in body-weight, composition and hepatic enzyme activities in response to dietary methionine, betaine and choline levels in growing chicks. *Br. J. Nutr.* **63** 339–49

Genetic parameters on Bali cattle progeny test population

A R Hariansyah[1], A Raharjo[1], A Zainuri[1], Y Parwoto[2], D Prasetiyo[2], S Prastowo[1], N Widyas[1]

[1] Animal Science Department, Faculty of Agriculture, Universitas Sebelas Maret, Surakarta Indonesia

[2] Bali Cattle Breeding Center Pulukan, Indonesia

e-mail: nwidyas@gmail.com

Abstract. Bali cattle (*Bos javanicus*) is Indonesian indigenous cattle with having superior genetics potential on fitness traits in tropical environment and low feed quality. Bali Cattle Breeding Center Pulukan Indonesia conducted progeny test per annum in order to select bulls using offspring's phenotype. This paper aimed to estimate the genetic parameters of yearling weight in Bali cattle progeny test populations and to observe the variation between periods in the above breeding center. Data were collected from the year of 2013 to 2014. There were four bulls (3 tests, 1 AI control) in 2013 and five bulls (4 tests, 1 AI) in 2014. Thirty breeding females were allocated per paddock per bull and allowed to mate naturally. In total 80 and 104 offspring's records were obtained from 2013 and 2014 data, respectively. We built half-sib family model to estimate the additive genetic variance due to the sire and later estimate the breeding value (EBV) of each sire. Results showed that in 2013 the heritability (h^2) for yearling weight was 0.19 while in 2014 was 0.79. In both years, tested bulls had higher EBV compared to the control bulls. The remarkable difference of heritability between years was due to the variations among bull candidates which might differ every year with regards to their origins. The fact that the EBV of tested bulls were higher than the control bulls gave us insight that despite the conservation policy and the continuous departure of Bali cattle bulls outside the Island, the population could still maintain its genetic quality.

1. Introduction

Bali cattle (*Bos javanicus*) is Indonesian indegenous cattle that was domesticated in Bali Island. Bali catttle widely spread over the Indonesian region, however its developed in eastern part of Indonesia such as South Sulawesi, West and East Nusa Tenggara, Central Sulawesi, South Kalimantan and West Kalimantan [1]. This cattle is highly adaptable and robust towards low feed quality [2], high fertility, good reproduction ability and excellent carcass quality [3]. Compared to exotic breed, Bali cattle has smaller body weight, 211-484 kg when mature [1,3], and slow growth rate [3].

Improving Bali cattle productivity can be done by improving feed quality [4–6], rearing management [6] and genetic quality by crossbreeding. In crossbreeding method, careful management is a must to avoid the decrease of Bali cattle genetic quality [5]. A best way to improve genetic quality is through male selection then applied in breeding program, this expected to enhance productive traits of Bali Cattle [7].

Currently, Indonesian government running Bali cattle breeding facilities in Bali which perform selection program using its phenotype offspring call progeny test. The program will result estimate breeding value (EBV) of specific parents, particularly male, which relatively have better performance compared to the population average. Doing selection by phenotypic trait should pay attention in the h^2 of the trait. In this regard h^2 defined as a degree of specific trait that inherited from parent to its offspring [8]. Higher h^2 represent strong resemblance between parent and offspring [9] as well as in the contrary. Those two values, h^2 and EBV, named as genetic parameters are important value to guide the selection program. For that, this paper aimed to estimate the genetic parameters of Bali cattle in Bali Cattle Breeding Center Pulukan Indonesia. We will use weaning weight (WW) and yearling weight (YW) data from offspring during progeny test program for data analysis.

2. Methods

2.1 Data

Data were collected from Bali Cattle Breeding Center Pulukan Indonesia. The data consists of YW records in 2013 and 2014. Tested bulls (n = 7) were mated naturally with female and 2 AI bulls we used as control bulls. Each bull was mated with 30 cows in every paddock. Yearling weight were adjusted to 365-day of age using the following formula:

$$YW_A = \left(\frac{\text{YW-WW}}{\text{rearing period (days)}} \times 160 + \text{WW}_A\right)$$

YW_A = adjusted yearling weight (365 days)
YW = actual yearling weight
WW = weaning weight
WW_A = adjusted weaning weight (205) [10]

2.2 Estimation of heritability

A linear model was built to calculate h^2 of YW as follows:

$$y_{ijk} = \mu + sex_i + S_{ij} + e_{ijk}$$

y_{ijk} = performance of individual
μ = mean
sex_i = sex of the offspring
S_{ij} = sire
e_{ijk} = error

Followed with analysis of variance and separation of variance components as presented in Table 1

Table 1. Analysis of variance and separation of variance components

Source	Df[1]	SS[2]	MS[3]
Sire	s-1	SSs[4]	EMs[5]
Offspring	n-s	SSp[6]	KTw
Total	n-1	JKt	

Note :
[1]degree of freedom
[2]sum of square
[3]means of square
[4]sum of square sire
[5]expected means square
[6]sum of square progeny-sire [11]

Following the variance analysis, h^2 value was calculated using the following formula:

$$h^2 = \frac{4\times\sigma_S^2}{\sigma_S^2+\sigma_e^2}$$

Where:

σ_S^2 = additives sire
h^2 = heritability
σ_e^2 = error [12]

2.3 Estimated of Breeding Value

The calculation of bulls breeding value was based on the performance of his half-sib family according to the following formula:

$$EBV = \frac{\frac{1}{4}nh^2}{1+(n-1)h^2}\left(\bar{P}_{offspring}-\bar{P}_{population}\right)$$

Where:

EBV = estimated breeding value
$\bar{P}_{offspring}$ = mean of the offspring
$\bar{P}_{populasi}$ = mean of population
h^2 = heritability
n = offspring number

3. Results and Discussion

The h^2 of YW in Bali cattle at 2013 and 2014 were 0.14 and 0.79, respectively (Table 2). In 2014 YW h^2 was higher compared to 2013, this represent genetic improvement means a success of selection program [13]. High h^2 indicate large number of genetic variants are inherited by the parent to offspring [16]. Moreover, the value of h^2 will vary in different population [17].

Table 2. Heritability value of yearling weight

Year	Heritability	σ_e^2	σ_S^2
2013	0.19	194.7	82.2
2014	0.79	16.5	21.25

Estimated breeding value is defined as relative value based on position in the population, according to the performance derived from its offspring. High EBV means more genetic factors will affect offspring's performance. Therefore, EBV can be used as reference in the selection program. Results shows different EBV of YW in different bulls in every year of progeny test program (Figure 1). We can see positive and negative value of EBV, negative means that the bull has EBV below the population's average [16].

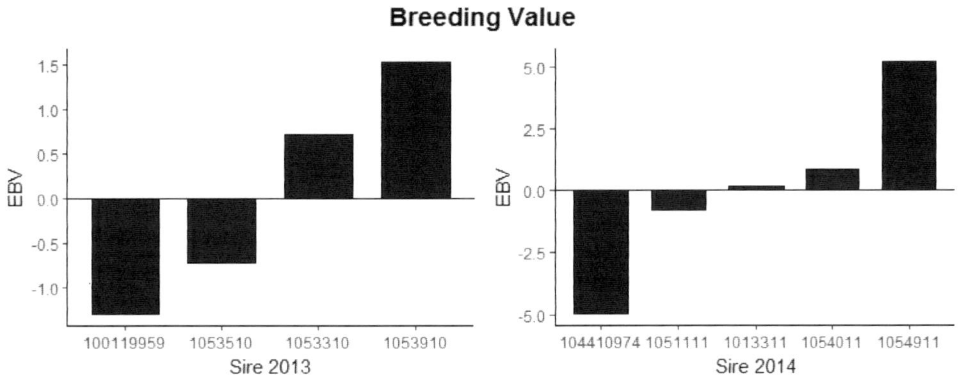

Figure 1. Bali cattle bull EBV of yearling weight

To select bull for breeding program, bulls can be ranked according to his EBV value. Bulls with positive EBV may select as elite bulls for genetic improvement of specific trait in the next generation [17]. As explained in the previous report, it is still rooms to improve Bali cattle genetic quality through selection process [18], following carefully consideration to use that genetic potential for pure or crossbred purposes [19].

4. Conclusions

According to the result of h^2 and EBV in this paper, we show that there is variation of bulls in Bali Cattle Breeding Centre Pulukan Indonesia. It is giving us an insight that the population could still maintain its genetic quality. A sustainable selection process must be considered well to improve specific trait in Bali Cattle, while keeping its variation for future generation.

Acknowledgment

This study was funded by the Maintenance Research Group Grant, Ruminant Animal Research Group, Provided by Universitas Sebelas Maret Surakarta Indonesia at fiscal year 2017.

References

[1] Talib C, Entwistle K, Siregar A, Budiarti-Tunner S and Lindsay D 2002 Survey of Population and Production Dynamics of Bali Cattle and Existing Breeding Programs in Indonesia *Strategies to Improve Bali Cattle in Eastern Indonesia* ed K Entwistle dan D R Lindsay (Australian Centre for International Agricultural Research) Pages 3–9

[2] Copland J 1997 Bali Cattle: Origins in Indonesia *Jembrana Disease and the Bovine Lentivirus* ed G E Wilcox, S Soeharsono, D M N Dharma and J W Copland (Australian Centre for International Agricultural Research) Pages 29–33

[3] Wiryosuhanto S 1997 Bali Cattle–Their Economic Importance in Indonesia *Jembrana Disease and the Bovine Lentivirus* ed G E Wilcox, S Soeharsono, D M N Dharma and J W Copland (Australian Centre for International Agricultural Research) Pages 34–42

[4] Oka L 2002 Performance of Bali Cattle Heifers and Calves prior to Weaning Weight n a Feedlot System *Strategies to Improve Bali Cattle in Eastern Indonesia* ed K Entwistle dan D R Lindsay (Australian Centre for International Agricultural Research) Pages 14–16

[5] Mastika I M 2002 Feeding Strategies to Improve the Production Performance and Meat Quality of Bali Cattle (Bos sondaicus) *Strategies to Improve Bali Cattle in Eastern Indonesia* ed K Entwistle dan D R Lindsay (Australian Centre for International Agricultural Research) Pages 10–13

[6] Bamualim A and Wirdahayati R B 2002 Nutrition and Management Strategies to Improve Bali Cattle Productivity in Nusa Tenggara *Strategies to Improve Bali Cattle in Eastern Indonesia*

ed K Entwistle dan D R Lindsay (Australian Centre for International Agricultural Research) Pages 17–22
[7] Miller S 2010 Genetic improvement of beef cattle through opportunities in genomics *Rev. Bras. Zootec.* **39** 247–255
[8] Cassel B 2009 Using Heritability for Genetic Improvement *Virginia Cooperative Extension* **404-084** 1–4
[9] Wray N R and Visscher P M 2008 Estimating trait heritability *Nature Education* **1** 1–4
[10] Warwick E J, Astuti J M and Hardjosubroto W 1990 *Pemulian Ternak* (Yogyakarta: Gadjah Mada Univ. Press)
[11] Hardjosubroto W 1994 *Aplikasi Pemuliaan Ternak di Lapangan* (Jakarta: Gramedia Widiasarana Indonesia)
[12] Nagy I 2011 *Quantitative genetics* (Kaposvár University)
[13] Kaswati, Sumadi and Ngadiyono N 2013 Estimasi nilai heritabilitas berat lahir, sapih, dan umur satu tahun pada sapi bali di balai pembibitan ternak unggul sapi bali *Bul. Pet.* **37** 74–8
[14] Falconer D S and Mackay T F C 1996 *Introduction to Quantitave Genetics* (London: Longman Group Ltd)
[15] Fuerst C 2012 Effect of inbreeding depression on survival of Austrian Brown Swiss calves and heifers *J. Dairy Sci.* **95** 6086–92
[16] Hardjosubroto W 1994 *Aplikasi Pemuliaan Ternak di Lapangan* (Jakarta: Gramedia Widiasarana Indonesia)
[17] Goddard M G and Smith C 1990 Optimum number of bull sires in dairy cattle breeding *J. Dairy Sci.* **73** 1113–22
[18] Widyas N, Nugroho T and Prastowo S 2017 Rooms for genetic improvement in Indonesian Bali cattle population. IOP Conference Series: Materials Science and Engineering **193** 12037
[19] Prastowo S, Widi T and Widyas N 2017 Preliminary analysis on hybrid vigor in Indonesian indigenous and crossbred cattle population using data from published studies. IOP Conference Series: Materials Science and Engineering **193** 12028

IOP Conf. Series: Earth and Environmental Science **142** (2018) 012004 doi:10.1088/1755-1315/142/1/012004

The number of service per conception of Indonesian Friesian Holstein with artificial insemination in Selo, Boyolali, Central Java, Indonesia

A M Wicaksono[1], A Pramono[1], A Susilowati[2], Sutarno[2], N Widyas[1], S Prastowo[1]
[1] Animal Science Departement, Faculty of Agriculture, Sebelas Maret University, Surakarta, Indonesia
[2] Department of Bioscience, Graduate Program, Sebelas Maret University, Surakarta, Indonesia

E-mail: prastowo@staff.uns.ac.id

Abstract. Boyolali is an area in Central Java Indonesia, it has large number of Indonesian Friesian Holstein (IFH; dairy cattle). To improve its population as well as genetic quality of milk production, artificial insemination (AI) is widely applied as mating program. The success of AI can be evaluated from the number of service per conception (S/C), represent a number of service using AI to achieve one pregnancy. Its mirroring mating management and reproductive efficiency in dairy cattle, estimated in herd during specific time and location. For that, this study aims to estimate S/C in Selo, Boyolali during October 2016 to January 2017. Data were gathered with 95% confidence level. Sample size were 367 IFH, visited and selected purposively based on criteria one-time partus, 3 y.o and have complete AI record. Animal data were collected in reproduction and mating management. In addition, 124 dairy farmer who have minimum 5 years experiences in rearing IFH cow were interviewed as respondent in estrus detection, followed with 2 skilled inseminators for AI performing time data. Result shows that S/C is 1.71, this mean one pregnancy need 1.71 times AI services. In the estrus detection, most of dairy farmers were able to observe estrus sign in vulva color, size and the present of mucus by visual. Moreover, AI was performed in 9 to 12 hours after the sign of estrus observed. It is concluded that AI of IFH in Selo, Boyolali has been successfully applied, however there are still rooms to improve the reproduction efficiency through mating management in regard to lower S/C.

1. Introduction

The application of AI in Indonesian dairy cattle has brought a huge impact in genetic and economic perspective. Genetically, AI supporting the dissemination of high genetic quality of selected bull and in economic view, its increase milk production capability. Milk production depend on the ability of cow to pregnant then initiate lactation cycle, however decline in reproduction efficiency of high milk production become a main issue [1]. In dairy cattle, reproductive efficiency can be estimated using S/C data. Previous study shows decline of S/C represent reducing herd fertility [2,3]. Moreover, fertility is highly influenced by management and environmental factors [4] for example mating management, estrus detection, insemination time, feed and season [3].

Selo Boyolali is one of area in Central Java Indonesia which has big number of IFH (8,192 heads; Boyolali statistical bureau 2016) [5]. It has mild climate, 1,200 – 1,500 m above sea level, ambient temperature 17-20°C and rainfall intensity 2,448 mm (https://id.climate-data.org/location/26748/) therefore suitable for IFH rearing. Since years ago, mating system in this area has been rely on AI which performed by skilled and certified local inseminators according to the government policy. In the other side semen used for AI is provided by the government, therefore the genetic of IFH could be maintained. It is widely known that reproduction is key component in enhance population as well as milk production. However, the efficiency must be measure time to time for evaluation purpose. For that, this paper aims to evaluate IFH reproductive efficiency by S/C, following the explanation in the related factor in Selo Boyolali. Result could be beneficial for strategic planning to enhance dairy cattle herd fertility.

2. Methods

A ground survey was done during October 2016 – January 2017 in Selo Boyolali using confidence level 95%. According to that, 367 heads of IFH were selected as sample size. Three locations in Selo Boyolali named Samiran (latitude, longitude, altitude) -7.498938, 110.470918, 1,474 m, Jeruk at -7.481306, 110.490056, 1257 m and Tarubatang at -7.493889, 10.481361, 1,426 m (http://www.mapcoordinates.net/en) were selected. In addition, 124 dairy farmers were visited to gather estrus observation data, as well as 2 skilled inseminators for AI performing time data. To estimated S/C, number of total AI performed during January to October 2016 were calculate then divided with total number of pregnant cows in the same period in the same cows used as sample in this study. Estrus data were presented in percentage and AI performing time was in hours after estrus signs were observed.

3. Result and discussions

S/C in Selo Boyolali was estimated as 1.7, this mean to achieve 1 pregnancy need 1.7 times AI services. Best S/C is closer to 1 while higher represent more inefficient of herd fertility [2]. S/C depends largely on the breeding system used. Lower S/C can be found in AI compared to natural mating. According to the previous study, different locations resulted different S/C. For example in Bogor West Java S/C was 1.81 in Kebon Peder, 1.97 in Tajur Halang and 1.82 in Cibeureum [6], 1.6 in Mojosongo Boyolali Central Java [7], 2.93 in Malang East Java [8] and 1.87 in Bengkulu Sumatra [9]. These data represent different in rearing management and mating system used. This result is in line with the previous review where explaining the fertility is influenced by management and environment [4].

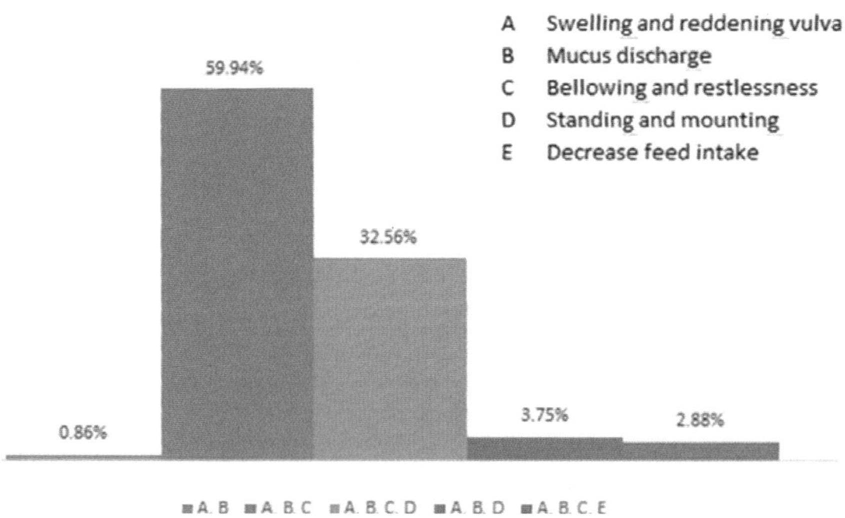

Figure 1. Estrus sign observed visually

S/C belongs to variables of cattle reproductive performance which represent mating management efficiency and involve intensive estrus detection. As resulted from interview, dairy farmer knowledge in estrus detection need to be improved more intense. The fact dairy farmer (Figure 1) are able to observe the estrus visually. Fail in observe estrus increase the fall of determine best time for insemination followed with reducing the possibility of fertilization [3,5]. When cattle are fail to pregnant at the first time of AI, the more AI service is needed during next estrus cycle which means increasing S/C.

Fertilization is depending on the service. To get pregnant, ovum need to be fertilized by sperm after ovulation and before that cow must be come into estrus. In this regards, estrus is the key player in determining S/C. A fail to detect estrus result in fail of conception [5]. Estrus is the most important element of cattle herd management, inadequate of estrus detection affect to the herd fertility. The fact, estrus can only observed visually by the dairy farmer [3,5]. However, some obstacles such as silent heat and in appropriate farm visit became factors belongs to failure of estrus detection. In this study, dairy farmers are able to detect the present of estrus sign by visual observation of vulva (redder, bigger and the present of cervical mucus) compared to non-estrus period. Taking into account, the dairy cattle farmer in Selo Boyolali has been experience in rearing for more than 5 years.

In dairy cattle, normally estrus last about 18 hours, and best time to inseminate is at 9-16 hours after the beginning of observed the signs of estrus as reported in the previous study [10]. In practical, dairy farmer and the inseminator in Selo Boyolali already know the best time to perform AI. If heat observed in the morning then the latest time for AI is in the afternoon. Meanwhile, if observed in the afternoon then latest time is the next morning. The range of AI time in this study was 9-12 hours. According to the inseminators, frozen semen was thawed in air temperature for 20-30 seconds before AI and deposit at cervix ring number 4.

All in all, reproductive performance of dairy cattle in Selo Boyolali categorized as good. Improving dairy farmer estrus detection could be a way to enhance the herd fertility. Moreover, the other factor such as mating management, nutrition and other factor which support good reproduction performance is needed.

4. Conclusion

It is concluded that, the implementation of AI in Selo Boyolali have been successfully. However, factors to improve reproduction efficiency are still open to manage. Improving fertility is not a single work, many factors interact. Enhancing estrus detection method, nutrition and recording would be a tool in reducing S/C.

References

[1] Lucy M C 2001 Reproductive loss in high-producing dairy cattle: Where will it end? *J. Dairy Sci.* **84** 1277–93

[2] Jones R I, Stewart P G, Dairymen T, Box P O and Africa S 1992 Estimating true and apparent number *of* services per conception, estrus detection intensity and calving interval in dairy herds *Theriogenology* **37** 1327–39

[3] De Kruif A 1978 Factors influencing the fertility of a cattle population *J. Reprod. Fertil.* **54** 507–18

[4] Weigel K A 2004 Improving the reproductive efficiency of dairy cattle through genetic *selection J. Dairy Sci.* **87** E86–92

[5] Barr H L 1975 Influence of estrus detection on days open in dairy herds *J. Dairy Sci.* **58** 246–7

[6] Agustina D, Mansjoer S S and Purwanto B P 2001 Performa reproduksi sapi perah pada tiga zona *klimatik* di Bogor *J.II.Pert.Indon* **10** 50–7

[7] Fitrianti A 2003 *Penampilan Reproduksi Sapi Perah Di Peternakan Sapi Perah Rakyat Wilayah Kerja KUD Mojosongo Kabupaten Boyolali Jawa Tengah* (Skripsi: Fakultas Kedokteran Hewan Institut *Pertanian* Bogor)

IOP Conf. Series: Earth and Environmental Science **142** (2018) 012004 doi:10.1088/1755-1315/142/1/012004

[8] Wahyuningsih S 2013 Tampilan reproduksi sapi perah pada berbagai paritas di Desa Kemiri Kecamatan Jabung Kabupaten Malang *J. Ternak Tropika* **14** 13–22

[9] Sulistyowati E, Kuswadi E, Sutarno L and Tampubolon G 2015 Penampilan reproduksi sapi perah FH (Fries Holland) dan pertumbuhan pedetnya pada umur 1- 3 bulan (Studi kasus di Desa Air Duku dan Desa Air Putih Kali Bandung, Selupu Rejang, Rejang Lebong, Bengkulu) *J. Sain Pet. Indon.* **4 21–26**

[10] Foote R H 1979 Time of artificial insemination and fertility in dairy cattle *J.Dairy Sci.* **62** *355*–8

Persistency of milk yield in Indonesian Holstein cows

N Widyas[1], F Y Putra[1], T Nugroho[1], A Pramono[1], A Susilowati[2], Sutarno[2] and S Prastowo[1]
[1] Animal Science Department, Agriculture Faculty, Universitas Sebelas Maret, Surakarta, Indonesia
[2] Department of Bioscience, Graduate Program, Universitas Sebelas Maret, Surakarta, Indonesia

E-mail: nwidyas@gmail.com

Abstract. Milk yield is an important trait in dairy industry; thus, information regarding this phenotype is essential to measure the productivity of a farm. Total milk yield in one lactation period was often predicted using information from samples collected within certain time intervals. The rate of change of milk production between two-time intervals is defined as persistency. This article aims to estimate the persistency of milk yield between lactation 1, 2 and 3 in Indonesian Friesian Holstein (IFH) cows. Data was collected from Limpakuwus stable, Baturraden Dairy Cattle Breeding Centre, Central Java Indonesia. Records were obtained from cows which started lactating on 2013 until the end of third lactation around the beginning of 2016. Milk yield from the first (L1), second (L2) and third (L3) lactations of 21 cows were recorded in kilograms. Samples were collected in 30 days basis interval started from the 10th day of lactation up to the 10th month. In this population, the cows first calving was around February – April 2013; while the second and third calving occurred all over the relevant year. The mean of milk yield for L1, L2 and L3 were 17.77±3.70, 16.09±5.17 and 13.73±4.02 Kg respectively. The peak of milk yields was achieved at the second month of the lactation for L1, L2 and L3. The persistency from the second to the tenth test days were 97, 93 and 94% for L1, L2 and L3, respectively. Milk yield persistency is representing ability of cow in maintain milk production after peak during lactation period. The more persistent shows better performance of dairy cattle as well as farm management. For that, persistency value could be used as valuable information in evaluating the management in Indonesian dairy farms.

1. Introduction

Milk yield is an important trait in dairy industry as a measure of a farm's productivity. However, to keep milk production and farm's records daily is a tedious and labor-costly operation. In order to acquire data on their productivity, dairy farms conducted observations on certain days, on regular interval basis during the cow's lactation period. This procedure of obtaining dairy data is called the test day records. Test days data has been extensively used in dairy researches and proven to yielded reliable results [1–3].

Milk persistency is a measure of the rate of change of milk production between test days. It is defined as the milk yield at one test expressed as a percentage milk yield at an earlier test [4]. High persistency shows slow rate of milk decline, whereas low persistency is associated with rapid decline in milk yield [5]. In the economic importance, persistency will determine the shape of lactation curve. Persistency provides valuable information regarding the amount of milk produced during one lactation period which represents biological aspects of the cow, farm management and cost efficiency. Factors influenced persistency are reproduction performance [6], energy in the diet [7,8], mastitis, metabolic disorder and also herd social interaction [4].

Holstein breed dominates the dairy cattle population in Indonesia (Indonesian Friesian Holstein = IFH). A government facility hosted this breed, the Baturraden Dairy Cattle Breeding Centre, Central Java, Indonesia which also served as test station. Within which dairy cows undergone various observations and treatments. Records from this station are used as standards for Indonesian Holstein population in the country. This paper aims to estimate milk yield persistency of IFH in the Baturraden breeding center for standard purpose. The result could be beneficial to evaluate the efficiency of IFH milk production under tropical management practice of dairy farm in Indonesia.

2. Methods

Data was collected from Limpakuwus stable, Baturraden Dairy Cattle Breeding Centre, Central Java Indonesia. Records were obtained from cows which started lactating on 2013 until the end of third lactation around the beginning of 2016. Milk yield from the first (L1), second (L2) and third (L3) lactations of 21 cows were recorded in kilograms per test day. Samples were collected in 30 days basis interval during the 10 months of a lactation period. Records for the first moth of lactation were taken from the 10th day.

The average of milk yield across lactation months and every lactation period were plotted to visualize the trajectory of milk yield. A linear model was built to account for the possible effects of different lactation periods and calving months. Persistency is defined as how well the cows maintain their milk yield after the peak was reached. It is estimated in percent according to the following formula:

$$PERSISTENCY = \frac{milk\ Kg\ in\ later\ test}{milk\ Kg\ in\ earlier\ test} \times 100\%$$

3. Results and discussions

Baturaden Dairy Cattle Breeding Centre is an Indonesian government facility located in Central Java within which Holstein cows test herds were kept and observed. Indonesian Holstein were originated either as imported cows from Australia, Canada, New Zealand and USA [9,10] or Indonesian born which dams and sires were the aforementioned imported cattle. Recording from twenty-one cows which have recording from three lactation periods housed in the same stable were used in the analysis. Records were obtained on test day's basis, with 30 days interval spanning for 10 months of each lactation period. The mean of daily milk yield for L1, L2 and L3 were 17.77±3.70, 16.09±5.17 and 13.73±4.02 kg respectively. These findings were in agreement with previous study [10] whom observed that on average, test day milk yield in Indonesian Holstein cows were between 11.9 – 17.3 kg.

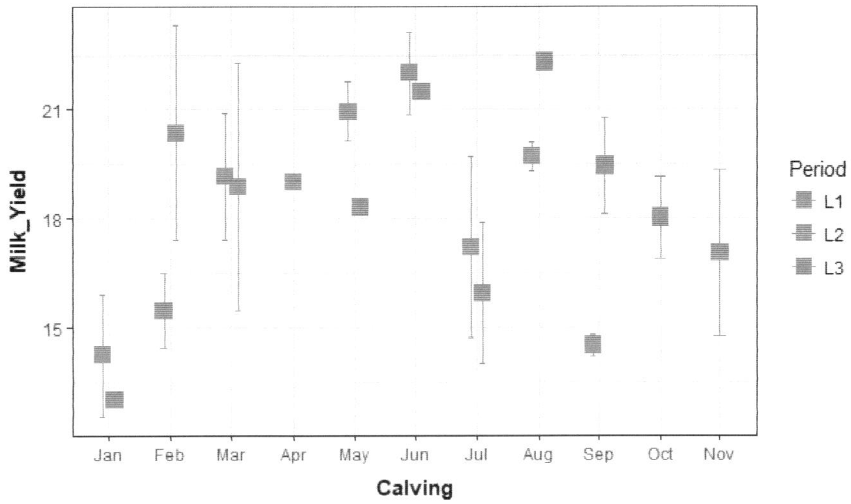

Figure 1. The effect of lactation period and calving month on milk yield

A linear model accounting for the effects of lactation period, calving month and their interaction on daily milk yield was build and presented in Figure 1. In this population, the cows first calving (and first lactation = L1) was around January – March 2013; while the L2 calving occurred on February – September 2014 and L3 calving occurred from May 2015 - January 2016. Analysis of variance (ANOVA) showed that both lactation period and calving month gave significant effect on milk yield whilst there was no interaction effect detected ($p>0.05$). Despite the ANOVA results, there were no clear trends of milk yield across months as we would expect due to the different seasons. Our hypotheses were that 1) The weather variation in Indonesia is not differed greatly among seasons except for the rainfall; 2) the management and feed supply in this test station remained stable throughout the year hence environmental factors were minimum.

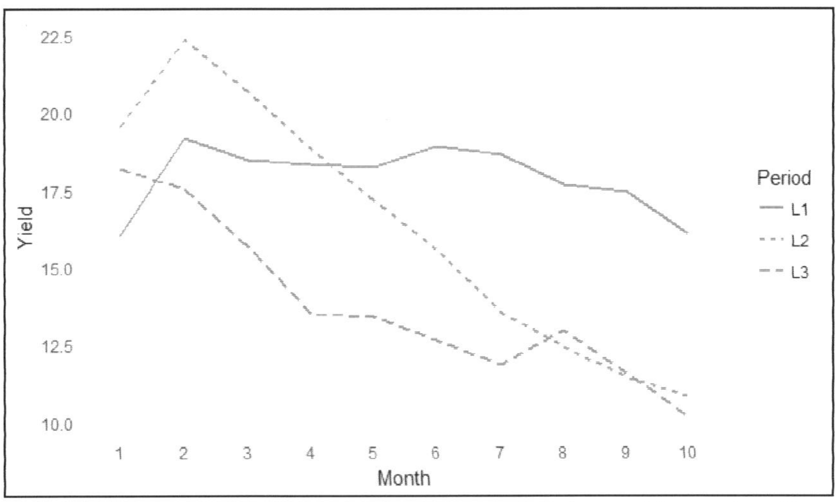

Figure 2. The trajectory of milk yield (lactation curve) during 10 months of lactation

Milk yield over 10 months of lactation period is presented in Figure 2. On X axes were the lactation months and on Y axis were the milk yield. It is seen that the peak of milk yield happened during the second month of lactation which values were 19.19±4.04, 22.39±4.63 and 17.54±3.32 kg per day for L1, L2 and L3 respectively. Whereas the persistency from the second to the tenth months were 97, 93 and 94% for L1, L2 and L3, respectively. Different persistency value across lactation period could be interpreted as the influence of many factors, however in this study we can't account whether specific factor is the key player due to lack of data.

4. Conclusions

It is concluded that milk yield persistency of IFH in this study shows different value in different lactation periods. Numerous factor associated with environment and management might be interact each other in this regard. Moreover, the persistency value from this test station could be used as the standard in evaluation the productivity and efficiency of dairy farm in Indonesia.

Acknowledgment

The data presentation of this paper was funded by the International Collaboration Grant provided by Universitas Sebelas Maret Surakarta Indonesia in the fiscal year 2017. The grant tittle is *"Comparative genetic diversity analysis of local zebu, Fogera cattle of western Amhara, Ethiopia and Indonesian Friesian Holstein by using microsatellite markers for enhancing milk production and climate change adaptation in selected production and model sites"*.

References

[1] Jensen J 2001 Genetic evaluation of dairy cattle using test-day models *J. Dairy Sci.* **84** 2803–12
[2] Machado S G, Freitas M A R and Gadini C H 1999 Genetic parameters of test day milk yields of Holstein cows *Genet. Mol. Biol.* **22** 383–6
[3] Costa C N, De Melo C M R, Packer I U, De Freitas A F, Teixeira N M and Cobuci J A 2008 Genetic parameters for test day milk yield of first lactation Holstein cows estimated by random regression using Legendre polynomials *Rev. Bras. Zootec.* **37** 602–8
[4] DHI 2017 *Persistency of milk production*
[5] Jingar S, Mehla R K, Singh M and Roy A K 2014 Lactation Curve Pattern and Prediction of Milk Production Performance in Crossbred Cows *J. Vet. Med.* **2014** 1–6
[6] Dobson H, Smith R F, Knight C H, Sheldon I M and Glasgow G 2007 The high producing dairy cow and its reproductive performance *Reprod. Domestic Anim.* **42** 17–23
[7] Nørgaard J, Sørensen A, Sørensen M T, Andersen J B and Sejrsen K 2005 Mammary cell turnover and enzyme activity in dairy cows: effects of milking frequency and diet energy density *J. Dairy Sci.* **88** 975–82
[8] Linde R Van Der, Groen A and Jong G De 1998 Estimation of genetic parameters for persistency of milk production in dairy cattle *Analysis* 1–4
[9] Tabany A, Purwanto B P, Toharmat T and Anggraeni A 2011 hubungan masa kosong dengan produktivitas pada sapi perah Friesian Holstein di Baturraden, Indonesia *Med. Pet.* **34** 77–82
[10] Anggraeni A 2012 Genetic Improvement of milk quantity and milk quality in Holstein Friesian cattle by selection *Wartazoa* **22** 1–11

IOP Conf. Series: Earth and Environmental Science **142** (2018) 012006 doi:10.1088/1755-1315/142/1/012006

Predicting Madura cattle growth curve using non-linear model

N Widyas[1], S Prastowo[1], T S M Widi[2] and E Baliarti[2]
[1] Animal Science Department, Faculty of Agriculture, Universitas Sebelas Maret, Surakarta, Indonesia
[2] Department of Animal Production, Faculty of Animal Science, University of Gadjah Mada, Yogyakarta, Indonesia

E-mail: nwidyas@gmail.com

Abstract. Madura cattle is Indonesian native. It is a composite breed that has undergone hundreds of years of selection and domestication to reach nowadays remarkable uniformity. Crossbreeding has reached the isle of Madura and the Madrasin, a cross between Madura cows and Limousine semen emerged. This paper aimed to compare the growth curve between Madrasin and one type of pure Madura cows, the common Madura cattle (Madura) using non-linear models. Madura cattles are kept traditionally thus reliable records are hardly available. Data were collected from small holder farmers in Madura. Cows from different age classes (<6 months, 6-12 months, 1-2years, 2-3years, 3-5years and >5years) were observed, and body measurements (chest girth, body length and wither height) were taken. In total 63 Madura and 120 Madrasin records obtained. Linear model was built with cattle sub-populations and age as explanatory variables. Body weights were estimated based on the chest girth. Growth curves were built using logistic regression. Results showed that within the same age, Madrasin has significantly larger body compared to Madura ($p<0.05$). The logistic models fit better for Madura and Madrasin cattle data; with the estimated MSE for these models were 39.09 and 759.28 with prediction accuracy of 99 and 92% for Madura and Madrasin, respectively. Prediction of growth curve using logistic regression model performed well in both types of Madura cattle. However, attempts to administer accurate data on Madura cattle are necessary to better characterize and study these cattle.

1. Introduction

Madura isle, located at the northeast of Java coast is the home of a native and unique cattle breed called the Madura cattle. This breed is a product of cross mating between Zebu (*Bos indicus*) and Banteng (*Bos javanicus*) which happened around 1,500 years ago [1]–[3]. The fame of Madura cattle in cultural events, however, started around the 13th century during the Dutch colony, when the *karapan* or bull racing started to take place. The cattle culture in Madura is very intense, until recently, local people refuse the inclusion of other cattle breeds into their region. Around the year of 2000, Indonesian government introduced Limousine breed through artificial insemination (AI) program to the Maduran farmers and it was surprisingly well received in some areas [2], [4], and thus, the Madrasin cattle started to exist. It was revealed that nowadays Madura cattle are divided into four distinct sub populations named common Madura cattle (Madura), Karapan, Sonok and Madrasin (Rudi Haryanto, Personal communication, August 20, 2017). The production system of every sub population differs; while Karapan is the sub population specialized for bull races and Sonok is the body conformation contest cow's population, Madura cattle and Madrasin belong to the commercial meat producer class.

Having the advantages of a tropical breed, Madura cattle excel in heat stress, adaptability, tick resistance and high tolerance towards low quality feed [4]–[6]. However, the genetic and physical characteristic of Madura sub population showed remarkable decrease due to unplanned traditional breeding including the selection and mating system [2]. The introduction of Limousine breed as a new genetic resource was objected in improving this once decreasing quality of Madura cattle.

All sub populations of Madura cattle were kept by smallholder farmers with low number of cattle per family [2]. Nowadays farmers still manage their cattle traditionally with semi intensive farming system and no documented records; thus, reliable information are very limited. As both sub populations objectives were for meat production, growth trait is important in determining their potentials. This paper aimed to observe and compare the growth trajectories of Madura and Madrasin cattle using body measurements variables. This information shall serve as guidelines for farmers in order to keep in tract of the cattle growth as well as to give warning in case of anomalies occurred hence early precaution could be made.

2. Methods

Data were collected from female common Madura cattle and Madrasin sub populations from various ages. We conducted direct observations on farms with cattle classified as common Madura and Madrasin. We measure their body variables including chest girth (CG), wither height (WH) and body length (BL). The ages were then classified in to six classes inclusive: 1 (< 6 months old), 2 (6-12 months), 3 (1-2 years old), 4 (2-3 years old), 5 (3-5 years old) and 6 (>5 years old). Number of observations per age classes were presented in Table 1. In total 63 Madura and 120 Madrasin records were obtained.

Table 1. Number of observations per age classes.

Age	**Sub-population**	
	Madura	Madrasin
1 (< 6 mo)	3	0
2 (6-12 mo)	4	4
3 (1-2 yo)	2	8
4 (2-3 yo)	16	44
5 (3.5 yo)	23	46
6 (>5 yo)	15	18

*no new born calves were available for measurements

A linear model was built to identify the possibility of having significant systematic effects of age classes and sub populations with $\alpha = 0.05$.

$$y_{ijk} = \mu + Age_i + P_j + e_{ijk}$$

where,

y : response variable of body measurements
μ : general mean
Age : Age effect
P : Sub population effect
e : random residuals

We estimated the body weight (kg) in our data based on the chest girth (cm) using the measuring tape developed by FHK Ogawa Seki Co. Ltd, Tokyo, Japan. We built the growth curve by fitting the means of body weight from each age classes into a plot. A logistic regression model was then used to predict the body weights for both sub populations.

$$y(t) = \frac{\alpha}{1 + (\beta \exp(-kt))}$$

where,
y(t) : growth at time t
t : time point
α : upper asymptote
β : growth range
k : growth rate

The estimated and predicted body weight data were then compared to estimate the accuracy of the non-linear prediction model.

3. Results and Discussion

The arithmetic means and standard deviations of the body measurements variables are presented in Table 2. It is clearly visible that Madrasin have significantly larger build in all body dimensions compared to Madura cattle ($p<0.05$). This finding is in agreement with Kustiyah (2005) [7] which mentioned that crossbred individuals between Limousin and Madura cattle yielded heavier body weight compared to pure Madura cattle in all three time frame (birth, weaning and post-weaning). The body measurements growth trends in different age classes can also be seen in Table 2.

Table 2. Mean and standard deviation of cattle's body measurements.

Variables	Body measurements (cm)		
	Chest girth	Wither height	Body length
Sub-populations			
Madura[a]	147.95±16.58	113.95±8.32	107.95±13.72
Madrasin[b]	153.87±22.23	120.24±9.90	115.32±11.01
Age Group			
1	102.00±3.46	89.83±3.75	74.30±6.46
2	124.25±14.87	102.93±7.59	90.44±12.78
3	126.90±16.57	114.80±5.29	97.20±6.02
4	152.20±21.16	119.59±7.99	114.58±8.94
5	157.07±15.79	119.23±8.94	115.86±9.34
6	159.00±14.24	120.14±9.67	116.73±10.96

[a,b] different superscripts showed significant difference with p-value < 0.05

Cattle's body weight were estimated based on the chest girth following the measuring tape developed by FHK Ogawa Seki Co. Ltd, Tokyo, Japan. This approach was taken due to the circumstances in the field where weighing the cattle using standard scale is not feasible.

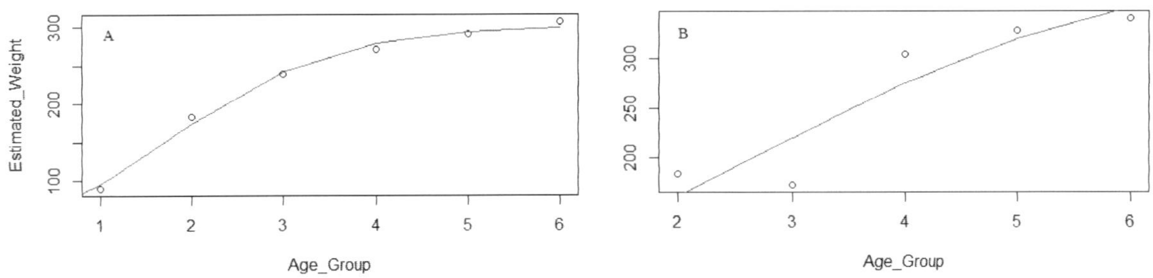

Figure 1. Growth curve based on the estimated body weight A) Madura cattle, B) Madrasin cattle.

The estimated body weights were then plotted as in Figure 1 where in X axis was the age classes and in Y was the estimated body weight. A logistic regression model was built to predict the body weights across sub-populations and age classes and plotted as the lines in Figure 1. It is shown that the model fit the Madura cattle data better than the Madrasin. The mean square error (MSE) for Madura cattle was 39.09 whereas for Madrasin cattle was a lot larger at 759.28. Correlation value between estimated and predicted body weights was used to describe the accuracy of the logistic model prediction. We estimated the accuracy of prediction for Madura cattle was 99% and Madrasin cattle was 92%.

4. Conclusions

We concluded that body dimensions wise, Madrasin cattle are larger compared to common Madura cattle. Non-linear model can be used to visualize the growth trend in both cattle sub-populations with considerably high accuracy.

Acknowledgement

Writers would like to acknowledge Mr. Rudi Haryanto, Head of UPTD Waru, Pamekasan, Madura, East Java for the fruitful discussions and for providing supporting data.

References

[1] Nijman I J, Otsen M, Verkaar E L C, de Ruijter C, Hanekamp E, Ochieng J W, Shamshad S, Rege J E O, Hanotte O, Barwegen M W, Sulawati T, and Lenstra J A 2003 Hybridization of banteng (Bos javanicus) and zebu (Bos indicus) revealed by mitochondrial DNA, satellite DNA, AFLP and microsatellites *Heredity (Edinb)* **90** 10–16

[2] Widi T S M 2015 *Mapping the impact of crossbreeding in smallholder cattle systems in Indonesia* Wageningen University and Research Centre

[3] Mohamad K, Olsson M, van Tol H T A, Mikko S, Vlamings B H, Andersson G, Rodríguez-Martínez H, Purwantara B, Paling R W, Colenbrander B, and Lenstra J A 2009 On the origin of Indonesian cattle *PLoS One* **4** 1–6.

[4] Hartatik T, Mahardika D A, Widi T S M, and Baliarti E 2009 Karakteristik dan kinerja induk sapi silangan Limousin-Madura dan Madura di Kabupaten Sumenep dan Pamekasan *Bul. Pet.* **33** 143–147

[5] Sutarno and Setyawan A D 2015 Review : Genetic diversity of local and exotic cattle and their crossbreeding impact on the quality of Indonesian cattle **16** 327–354

[6] Payne W J and Rollinson D 1976 *Madura cattle J. Z. Tierzücht Zuchtsbiol* **93** 89–100

[7] Kutsiyah F, Kusmartono and T Susilawati T 2003 Studi Komparatif Produktivitas antara Sapi Madura dan Persilangannya dengan Limousin di Pulau Madura *JITV* **8** 98–106

IOP Conf. Series: Earth and Environmental Science **142** (2018) 012007 doi:10.1088/1755-1315/142/1/012007

Age effect on post freezing sperm viability of Bali cattle (*Bos javanicus*)

R D Hapsari[1], Y Khalifah[2], N Widyas[1], A Pramono[1] and S Prastowo[1]
[1] Animal Science Department, Agriculture Faculty, Universitas Sebelas Maret, Surakarta, Indonesia
[2] Singosari Artificial Insemination Center, Malang, East Java, Indonesia

E-mail: prastowo@staff.uns.ac.id

Abstract. Post freezing sperm viability is one of factors which determine artificial insemination success. In the other side, bull's or sire age influences the semen quality through sperm membrane constituent. It is known that freezing process change the sperm membrane during the processing stage. This research aims to know the effect of sire age on post freezing sperm viability of Bali cattle. The samples were collected in Singosari Artificial Insemination Centre, Malang, East Java, Indonesia on September - November 2016. Eight Bali cattle (4 and 7 y.o, 4 heads in each group) were used as semen source. Semen was collected using artificial vagina, 10 times spanning for 5 weeks (2 times per week, interval 3 and 4 days) in a row. The samples were then evaluated at fresh, chill and frozen stage. Fresh semen was diluted in Tris-citrate-egg yolk 20% (v/v) followed with chilling and freezing. Semen qualities were observed as sperm % motility (MOT), % live sperm using eosin-nigrosine staining (EOS) and % sperm membrane integrity using hypoosmotic swelling test (HOS). Variable comparisons between age groups were done using t-test. On the average, 4 y.o bulls showed higher semen quality at fresh, chill and frozen compared to 7 y.o in MOT (68.00±6.39 vs 65.9±7.62; 56.40±3.71 vs 54.33±5.83; 44.25±3.52 vs 40.40±7.06), EOS (72.08±6.63 vs 71.82±7.38; 57.81±3.83 vs 57.41±6.32; 53.16 ±8.41 vs 46.49±9.13) and HOS (60.85±13.91 vs 54.84±13.43; 53.16 ±8.41 vs 46.49±9.13; 44.6±9.39 vs 33.8±10.70) respectively. Statistical analysis results showed that age was significantly ($P<0.05$) affecting HOS at chill stage and MOT and HOS at frozen. In conclusion, younger Bali cattle (4 y.o) have more viable post freezing sperm compared to the older ones (7 y.o).

1. Introduction

Bali cattle is an Indonesian native which has many advantages, like well adapted to the tropics [1] and high carcass percentage and high fertility [2]. Fertility of Bali cattle is higher compared to Europe cattle or another tropical cattle, that is 82–86% [3]. To spread that genetic potential, artificial insemination (AI) is the efficient way compared to other reproductive technology. AI is a first generation reproductive biotechnology that has made a profound contribution to the genetic improvement[4]. To be able to use as superior bull, the fertility of Bali cattle need to be tested through observation of libido ability and semen characteristics both macroscopic and microscopic. In term of semen quality, it is influenced by breed, reproductive health status, technical skills, age and bull genetic constitution [5].

Sperm viability will determine the success of AI. Low sperm viability will decrease the achievement of conception rates. Bull's age is very important factor to determining the sperm viability. Previous study shows that bulls at age 2 to 7 y.o are able to produce high quality semen with high conception rates compared to older [6]. Moreover, in young bull (less than 2 y. o; at puberty), more abnormal sperm produced with account to low fertilization rate[7]. During frozen semen production, temperature change the quality and structure of sperm membrane. It is reported that these condition modify lipids and alter functional state of membrane[8], therefore more or less 50% of sperm will die during freezing process [9]. Based on the previous description, this study aimed to know the effect of age on post freezing sperm viability of Bali cattle.

2. Materials and Methods

2.1. Semen samples

This study was conducted in Singosari Artificial Insemination Centre, Malang, East Java, Indonesia during September - November 2016. Eight Bali cattle (4 and 7 y.o, 4 head in each group) were used as semen source. Semen was collected using artificial vagina, 10 times spanning for 5 weeks (2× per week, interval 3 and 4 days) in a row. Fresh semen was diluted in Tris- citrate -egg yolk 20% (v/v) followed with chilling and freezing. The samples were then evaluated at fresh, chill and frozen stage.

2.2. Semen quality

Semen qualities were observed as:

2.2.1. Sperm motility (MOT) It can be defined as the percentage of sperm which moving in straight forward direction. Sperms were observed by placing the drop of samples semen on the slide glass followed by observation under microscope (40× magnificent). Sperm motility was scored 0–100%.

2.2.2. Livesperm (EOS) Live sperms were evaluated with differential staining technique using eosin-nigrosine dye according to the previous protocol [10]. The death sperm will absorb the dye and the observation was done in 200 sperms followed with percentage calculation as follows:

$$\%\ \text{EOS} = \frac{\text{live spermatozoa}}{\text{total spermatozoa}} \times 100\%$$

2.2.3. Sperm membrane integrity(HOS) The sperm membrane integrity was evaluated using hypoosmotic swelling test. Two microliter of semen sample were mixed with 20 μL hypoosmotic solution (0.90 g D-fructose and 0.490 g Trisodium citrate in 100 ml aquadest with osmolarity 150 m osmoles), followed with homogenization and incubation at 37°C for 30 minutes. The incubated solution were then evaluate with differential staining technique using eosin-nigrosine dye. Sperms with intact membrane will show swelling or coiling tail [10]. Moreover, the percentage of intact sperm membrane was calculated using following formula:

$$\%\ \text{HOS} = \frac{\text{intact spermatozoa}}{\text{total spermatozoa}} \times 100\%$$

2.2.4. Statistical analysis Semen quality comparison between age groups was done using t-test in this study. All the statistical analysis and graph presentation were done using R language and statistical environment.

3. Result and discussion

On the average, 4 y.o bulls shows higher semen quality at fresh, chill and frozen stage compare to 7 y.o (Table 1). The result showed that age is significantly affecting HOS at chill stage and MOT and HOS at frozen stage. This result is similar with the previous study [11] which shows that age gave significant influence to the MOT. The MOT and HOS were affected by bull age because of their viability. Earlier study reported that genetic, age, breed and individual will influence the sperm viability with the thermal shock during the thawing[12]. Moreover, 4 y.o have better sperm viability in the thermal shock compared to. 7 y.o. In the younger bull and has higher semen quality, show lower decrease from fresh to chill or frozen stage (Table 2 and Figure 1).

Table 1. Mean and standard deviation of semen quality based on bull's age

Bull age (y.o)	MOT (%)	EOS (%)	HOS (%)
Fresh stage			
4	68.00±6.39	72.08±6.63	60.85±13.91
7	65.90±7.62	71.82±7.38	54.84±13.43
Chill stage			
4	56.40±3.71	57.81±3.83	53.16 ±8.41[a]
7	54.33±5.83	57.41±6.32	46.49±9.13[b]
Frozen stage			
4	44.25±3.52[a]	39.47±9.57	44.60±9.39[a]
7	40.40±7.06[b]	35.16±11.74	33.80±10.70[b]

[ab]Different superscripts within same row differ significantly ($p<0.05$)

Table 2. Decrease of semen quality based on bull's age

Variables	Processing stage	Bull age	
		4 y.o	7 y.o
MOT (%)	Fresh	68.00±6.39	65.90±7.62
	Chill	56.40±3.71	54.33±5.83
	Frozen	44.25±3.52	40.40±7.06
EOS (%)	Fresh	72.08±6.63	71.82±7.38
	Chill	57.81±3.83	57.41±6.32
	Frozen	39.47±9.57	35.16±11.74
HOS (%)	Fresh	60.85±13.91	54.84±13.43
	Chill	53.16 ±8.41	46.49±9.13
	Frozen	44.6±9.39	33.8±10.70

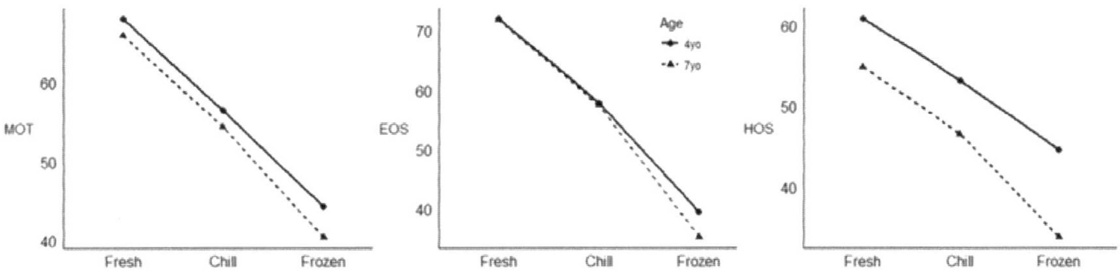

Figure 1. The decrease of semen quality during semen processing

The decrease of semen quality was suggested by the damage of sperm membrane. Cryopreservation process in frozen semen production will induces the damaged of sperm membrane [13]. The cryopreservation has number of potential stress that damaging plasma membrane such as the temperature variation, osmotic and toxic stress, and the formation of ice in the extracellular environment[8]. The damage sperm membrane commonly occurs in the midpiece where mitochondria are located. As reported mitochondria are wrapped around midpiece to provide energy for tail movement [14], therefore any damage of mitochondria resulting low sperm motility.

IOP Conf. Series: Earth and Environmental Science **142** (2018) 012007 doi:10.1088/1755-1315/142/1/012007

4. Conclusion

In this study, it is concluded that younger Bali cattle has more viable sperm compared to the older ones at post freezing stage. That result might be accounted by the condition of sperm membrane in 4 y.o bulls that better compared to 7 y.o.

References

[1] Talib C, Entwistle K, Siregar A, Budiarti-Turner S and Lindsay D 2002 Survey of population and production dynamics of bali cattle and existing breeding programs in Indonesia *Strategies to Improve Bali Cattle in Eastern Indonesia* (ACIAR Proceedings) pp 2–9

[2] Bamualim A and Wirdahayati R B 2002 Nutrition and management strategies to improve Bali cattle productivity in Nusa Tenggara *Strategies to Improve Bali Cattle in Eastern Indonesia* (ACIAR Proccedings) pp 17–22

[3] Gunawan, Pamungkas D and Affandhy L 1998 *Sapi Bali* (Yogyakarta: Kanisius)

[4] Lemma A and Shemsu T 2015 Effect of age and breed on semen quality and breeding soundness evaluation of pre-service young bulls *J. Reprod. Infertil.* **6** 35–40

[5] Bhakat M, Mohanty T K, Raina V S, Gupta A K, Khan H M, Mahapatra R K and Sarkar M 2011 Effect of age and season on semen quality parameters in Sahiwal bulls *Trop. Anim. Health Prod.* **43** 1161–8

[6] Yulianto P and Saparinto C 2014 *Beternak Sapi Limousin* (Jakarta: Penebar Swadaya)

[7] Hafez E S E 1993 *Reproduction in Farm Animals* (Maryland: Lippincott Wiliams and Wilkins)

[8] Watson P F 2000 The causes of reduced fertility with cryopreserved semen *Anim. Reprod. Sci.* 481–92

[9] Lessard C, Parent S, Leclerc P, Bailey J L and Sullivan R 2000 Cryopreservation alters the levels of the bull sperm surface protein P25b *J. Androl.* **21** 700–7

[10] Shukla M K 2011 *Applied Veterinary Andrology and Frozen Semen Technology* (New Delhi, India: New India Publising Agency)

[11] Aminasari P D 2009 *Pengaruh Umur Pejantan terhadap Kualitas Semen Beku Sapi Limousin* (Malang: Universitas Brawijaya)

[12] Chenoweth P J 2005 Genetic sperm defects *Theriogenology* **64** 457–68

[13] Januskauskas A, Johannisson A and Rodriguez-Martinez H 2003 Subtle membrane changes in cryopreserved bull semen in relation with sperm viability, chromatin structure, and field fertility *Theriogenology* **60** 743–58

[14] Connell M O, Mcclure N and Lewis S E M 2002 The effects of cryopreservation on sperm morphology, motility and mitochondrial function *Hum. Reprod.* **17** 704–9

Physical quality of Simental Ongole crossbred silverside meat at various boiling times

J Riyanto, M Cahyadi and W S Guntari
Animal Husbandry Department, Faculty of Agriculture, Universitas Sebelas Maret, Surakarta, Indonesia

E-mail : jokoriyanto@staff.uns.ac.id

Abstract. This study aims to determine the physical quality of silverside beef meat at various boiling times. Samples that have been used are the back thigh or silverside meat. Treatment of boiling meat included TR (meat without boiled), R15 (boiled 15 minutes), and R30 (boiled for 30 minutes). The experimental design using Completely Randomized Design with 3 replications. Each replication was done in triple physical quality test. Determination of physical quality was performed at the Livestock Industry and Processing Laboratory at Sebelas Maret University Surakarta and the Meat Technology Laboratory at the Faculty of Animal Husbandry of Gadjah Mada University. The result of variance analysis showed that boiling affect cooking loss ($P \geq 0.05$) and but did not affect ($P \leq 0,05$) pH, water holding capacity and meat tenderness. The conclusions of the study showed that boiling for 15 minutes and 30 minutes decreased the cooking loss of Simental Ongole Crossbred silverside meat. Meat physical quality of pH, water holding capacity and the value of tenderness is not affected by boiling for 15 and 30 minutes.

1. Introduction

Meat is a high-quality food source of animal protein. Consumers prefer meat with a texture that is not tough. The physical quality of meat is influenced by many factors such as cooking methods. Some of the physical properties of fresh meat such as structure, firmness and texture are difficult to measure objectively. Physical quality is related to the visible external appearance of the flesh, which consists of pH value, water holding capacity, tenderness and cooking loss.

The process of boiling meat improves the digestibility of protein but can also lower nutritional value. The cooking of protein foods that are not well controlled can decrease the nutritional value of the protein [1]. The higher the cooking temperature and the longer the boiling time the greater the level of the lost meat fluid. The amount of fluid obtained during boiling will increase further, it describes some of the protein damage with the damage of amino acids [2].

The method of boiling meat is based on three main factors including surface temperature of meat, external temperature and heat exchange method i.e. direct contact with heat, air or steam sources [3]. Winarso [4] mentioned that boiling meat at a temperature of 80-90^{0}C in 30 to 60 minutes increased pH, cooking loss, tenderness and decreased water holding capacity. The purpose of this research was to know the physical quality of beef with various cooking times related to the improvement of physical quality of meat.

2. Methods

This meat samples of the study were obtained from ongole crossbred (Simpo) which had been fattened for 90 days. The fattening was done in the CV Izzah Sejahtera Multifarm in Village Sambi District Boyolali. Fattening feed consisted of fermented rice straw and basal concentrate with a ratio of 30: 70%. The basal concentrate was composed of 9.10% coffee skin, 14.40% palm cake, 20.86% rice bran, 9.23% pollard, 0.61% mineral, 10.80% molasses, 15.00% formaldehyde-protected soyabean groat. The feed contained crude protein 10.45%, crude fat 6.28%, crude fiber 19.87%, organic material 88.91%, nitrogen-free extract 52.34% and total digestible nutrient 69.74%

All cows are slaughtered in slaughter houses according to Islam Method. Carcass preparation was done according to Australian Meat and Livestock Corporation [5]. The meat samples used in this study were the silverside meat. Sampling on the whole carcass was done by splitting into two halves right along the spine. Cutting between the front and rear carcass is done right between the ribs 12 and 13. The muscle samples were taken from the cross section of the rear carcass. The carcass were put into the cool box containing ice cubes to keep freshness then taken to the laboratory for examination.

Treatment of boiling meat includes TR (meat without boiled), R15 (boiled meat 15 minutes), and R30 (boiled meat 30 minutes). The experiment design using Randomized Complete Design as many as 3 replications of each replication was done triple physical quality test. The average difference in treatment effect was tested further using DMRT Test.

Preparation of the meat sample was done by cutting the meat with the same thickness, part and fiber direction. Fresh meat samples were tested for physical quality. Samples were weighed (180 - 200 g) [6] and included in tightly clipped polyethylene plastic bags. The samples were boiled using an electric stove at 100^0 C for 15 and 30 minutes temperature controlled [4]. Temperature is measured using a thermometer, while time is measured using stopwatch [8].

Meat pH values were tested in accordance with [6] methods in [7], samples 10 g of meat using a digital pH meter. Assessment of water holding capacity of meat using Hamm method in [6] by determining the value of free water content and total water content. Determination of free water content by taking 0.3 g meat then put on filter paper and cover the meat using filter paper on it.

Determination of total water content was done by measuring the water content i.e. weighing the initial sample with 5g digital scales then putting the sample on the filter paper then putting it in the oven at 105^0C for 8 hours and weighing the final weight of the sample [7]. Assessment of cooking loss in accordance with [6] methods in [7] by boiling meat samples measuring ± 100 g at a temperature of 80^0C. Meat tenderness value was determined by shear-force method [7]. Meat tenderness value using warner tool bratzler shear force device.

3. Results and Discussion

The physical quality of beef Simpo of silverside part in various boiling times can be seen in Table 1.

Table 1. Effect of boiling on pH value, water holding capacity, cooking loss, meat tenderness of Silverside Simental Ongole Crossbred meat

Treatment	TR (meat without boiled)	TR 15 (boiled meat 15 minutes)	TR 30 (boiled meat 30 minutes)
pH	6.60 ±0.12	6.12± 0.56	6.22 ± 0.10
Water holding capacity (%)	28.68±8.34	22.06±6.46	16.05±5.55
Cooking loss (%)	31.59^a± 2.41	16.28^b± 0.94	17.96^b± 1.37
Meat tenderness (kg/cm^2)	10.81± 2.95	9.28± 1.83	8.98± 1.61

^{a,b} different superscripts on the same raw showed significant differences (P <0.05)

The results showed no effect of boiling duration on pH value. Boiling at a temperature of 100^0C for 15 and 30 minutes has not yet led to a fast-running glycolysis process. Previous research [9] indicated that beef cooked at 95^0C for 10 minutes has a pH value of 5.7. The pH value of raw meat and meat cooked in the study showed normal pH values. The pH values in raw meat ranged from 5.5 to 7.0 [2] while the range of normal cooking pH values ranged from 5.1 to 6.1 [10].

The boiling time did not affect WHC (P <0.08). Boiling up to 30 minutes has not caused the breakdown of protein structures and the same protein's ability to bind water. Boiling at temperatures over 60°C causes protein denaturation that will lower the value of juice in the meat so that the meat is more tender. The results of this study were consistent with previous study, in which the duration of boiling meat 60 and 90 minutes tended to decrease the water holding capacity shown by 19.76 and 19.07%, respectively [11]. Higher crude protein content decreased the capacity of meat to absorb water [12]. According to Lawrie [2] pH is affected by the final pH of the meat, when the pH value is higher or lower than the isoelectric point of meat (5.0 - 5.1). The results showed that pH values on the meat were above the isoelectric point. The increased pH value will lead to an increase in WHC value.

The length of boiling time affect to cooking loss (P <0.05). Soeparno [7] states that cooking loss is affected by the temperature and duration of boiling. The higher the cooking temperature the greater the fluid content lost from the meat. The results of this study was in accordance with Ridwan [13] who showed that the value of cooking loss decreased by 44.04; 40.21 and 37.23% resulting in improved physical quality of the meat. Soeparno [7] mentioned that the cooking loss varies between 1.5 -54.5%. The value of cooking loss in the study ranged from 17 to 30% which shows the value of normal cooking loss.

The boiling duration up to 30 minutes didi not affect the meat tenderness although the tenderness tends to decrease (P <0.11). The boiling time up to 30 minutes causes the denaturation of myofibril proteins, resulted the tender meat which according to consumer's preference. The results were in accordance with [14] that the value of tenderness of boiled beef at external temperature 100 °C with internal temperature meat 75 and 60 °C resulted the same tenderness value. Suryati and Arief [15] stated that the very soft meat has the power of warner bratzler <4.15 kg cm^{-2}, tender meat 4.15 - <5.86 kg cm^{-2}, slightly tender meat 5.86 - <7.56 kg cm^{-2}, a bit tough meat of 7.56 - <9.27 kg cm^{-2}, tough meat 9.27 - <10.97 kg cm^{-2} and very tough meat ≥ 10.97 kg cm^{-2}.

4. Conclusion

The cooking loss in Simpo Crossbred silverside meat decreased with increasing boiling duration. The meat that were boiled for 15 and 30 minutes did not have different cooking loss. Furthermore, physical qualities in terms of pH, water holding capacity and tenderness in fresh meat and boiled meat 15 and 30 minutes were not different.

Acknowledgments

The authors wish to thank the Ministry of Research and Technology of Indonesia for providing research fund with the PUPT research scheme, 2015.

References

[1] Wyrwiszj J, Połtorak A, Poławska E, Pierzchała M, Jozwik A, Zalewska M, Zaremba R and Wierzbicka A 2012 The impact of heat treatment methods on the physical properties and cooking yield of selected muscles from Limousine crossbred cattle *Anim. Sci. Papers Reports* **30** 339-351

[2] Lawrie R A 2003 *Meat Science* Fifth Edition Aminudin Parakkasi Translation (Jakarta: University of Indonesia)

[3] Janz J A M, Aalhus J L, Dugan M E R and Price M A 2006 Mapping method for the description of warner-bratzler shear force gradients in beef longissimus thoracic et lumborum and semitendinosus *Meat Sci.* **72** 79-90

[4] Winarso D 2003 Changes in physical characteristics due to differences in age, muscle type, time and temperature of boiling in chicken meat *J. Indon. Trop. Anim Agric.* **28** 119-132

[5] Australian Meat and Livestock Corporation (AMLC) 1991 *Study the Australian Feeder Steer Trade to Indonesia Lot Feeding Industry* (Australia: Asian Marketing Division)

IOP Conf. Series: Earth and Environmental Science **142** (2018) 012008 doi:10.1088/1755-1315/142/1/012008

[6] Bouton P E and Harris P V 1972 The effect of cooking temperature and time on some mechanical properties of meat *J.Food Sci.* **37** 140-144
[7] Soeparno 2005 *Meat Science and Technology. 5th Print* (Yogyakarta: Gadjah Mada University Press)
[8] Subagyo W C 2014 *Characteristics of Bali beef protein and Wagyu after boiling.* Thesis. Graduate program. (Denpasar: Udayana University)
[9] Honikel K O 2004 *Water-holding capacity of meat* (Kulmbach, Germany: Prosiding Federal Center for Meat Research) pp 389 - 400
[10] Buckle K A, Edwards R A, Fleet G H 1987 *Food Science* (Jakarta: University of Indonesia Press)
[11] Lukman H 2004 *The Influence of Different Temperature and Cooking Time Against Chemical Composition and Physical Quality of afkirf duck meat* Thesis (Yogyakarta: Gadjah Mada University)
[12] Bailey C R, G C Duff, S R Sanders, J L Treichel, L H Baumgard, J A Marchello, D W Schafer and C P McMurphy 2008 Effect of increasing crude protein concentrations on performance and carcass characteristics of developing and finishing steers and heifers *Anim. Feed Sci. Tech.* **142** 111-120
[13] Ridwan T 2011 *Characteristics of Physical Beef Dara Brahman Cross by Giving Different Concentrate Types* Essay (Bogor: Bogor Agricultural Institute)
[14] Kołczak T, Krzysztoforski K and Palka K 2008 Effect of post-mortem aging, method of heating and reheating on collagen solubility, shear force and texture parameters of bovine muscles *Polish J. Food Nut. Sci.* **58** 27 - 32
[15] Suryati, I and Arief I I 2005 *Testing power of Warner Bratzler, cooking and organoleptik shrinkage as a predictor of the level of tenderness of beef preferred by consumers* Research report (Bogor: Faculty of Animal Husbandry, Bogor Agricultural University)

IOP Conf. Series: Earth and Environmental Science **142** (2018) 012009 doi:10.1088/1755-1315/142/1/012009

In silico study of protein to protein interaction analysis of AMP-activated protein kinase and mitochondrial activity in three different farm animal species

S Prastowo and N Widyas
Animal Science Department, Faculty of Agriculture, Universitas Sebelas Maret, Surakarta, Indonesia

E-mail: prastowo@staff.uns.ac.id

Abstract. AMP-activated protein kinase (AMPK) is cellular energy censor which works based on ATP and AMP concentration. This protein interacts with mitochondria in determine its activity to generate energy for cell metabolism purposes. For that, this paper aims to compare the protein to protein interaction of AMPK and mitochondrial activity genes in the metabolism of known animal farm (domesticated) that are cattle (*Bos taurus*), pig (*Sus scrofa*) and chicken (*Gallus gallus*). In silico study was done using STRING V.10 as prominent protein interaction database, followed with biological function comparison in KEGG PATHWAY database. Set of genes (12 in total) were used as input analysis that are PRKAA1, PRKAA2, PRKAB1, PRKAB2, PRKAG1, PRKAG2, PRKAG3, PPARGC1, ACC, CPT1B, NRF2 and SOD. The first 7 genes belong to gene in AMPK family, while the last 5 belong to mitochondrial activity genes. The protein interaction result shows 11, 8 and 5 metabolism pathways in *Bos taurus*, *Sus scrofa* and *Gallus gallus*, respectively. The top pathway in *Bos taurus* is AMPK signaling pathway (10 genes), *Sus scrofa* is Adipocytokine signaling pathway (8 genes) and *Gallus gallus* is FoxO signaling pathway (5 genes). Moreover, the common pathways found in those 3 species are Adipocytokine signaling pathway, Insulin signaling pathway and FoxO signaling pathway. Genes clustered in Adipocytokine and Insulin signaling pathway are PRKAA2, PPARGC1A, PRKAB1 and PRKAG2. While, in FoxO signaling pathway are PRKAA2, PRKAB1, PRKAG2. According to that, we found PRKAA2, PRKAB1 and PRKAG2 are the common genes. Based on the bioinformatics analysis, we can demonstrate that protein to protein interaction shows distinct different of metabolism in different species. However, further validation is needed to give a clear explanation.

1. Introduction

AMPK is a protein working as energy censor by sensing the balance of cellular ADP and AMP concentration [1,2]. It's known as protein kinase AMP-activated, abbreviated as PRAKA in the NCBI database as the official gene name, a heterotrimetric complex consist of α, β and γ subunit. According to the previous study, AMPK has a close relation with mitochondrial activity in serving cell with energy[3], any defect of its function resulted in metabolic disorder [4,5]. In addition, a wide range of AMPK expression have been found in many living organism such as cattle [6], pig [7], chicken [8] and mice [9].

As widely known, specific gene encode specific protein. In the multicellular organism activity of protein interact each other [10] to support its metabolic function, thus in specific organism/species different protein express different metabolic function. A protein to protein interaction will show the direct link to gene expression and metabolism pattern [11]. Knowing its interaction could be a way to identify and understand the key player in the molecular level [12], such as identify gene or protein function which associated with specific defect of metabolism [13].

Considering the AMPK and mitochondria activity as cellular energy regulator and to know the specific function of interact proteins, therefore this study aim to reveal the AMPK protein to protein interaction using existing protein database in three different animal species. The result will give us a picture of different metabolism occurred in different species.

2. Methods

Three known farm animal species were selected that are cattle (*Bos taurus*), pig (*Sus scrofa*) and chicken (*Gallus gallus*). To compare metabolism in selected species, as the analysis input for this in silico study, set of genes related to AMPK and mitochondrial activity were taken from the previous studies[14,15]. Those genes are in AMPK family (PRKAA1, PRKAA2, PRKAB1, PRKAB2, PRKAG1, PRKAG2 and PRKAG3) and mitochondrial activity (PPARGC1, ACC, CPT1B, NRF2 and SOD), genes abbreviation and accession number based on official name in NCBI database (www.ncbi.nlm.nih.gov). The protein to protein interaction was retrieved from functional protein association network using STRING V.10 (https://string-db.org/) according to the previous study [16], followed with protein biological function classification using KEEG PATHWAY database (http://www.kegg.jp/) to retrieve its metabolism pathway. Common genes in the same metabolism pathway were clustered with the help of Venn diagram (http://bioinformatics.psb.ugent.be/webtools/Venn/).

3. Result and Discussions

Retrieved metabolic pathway form protein to protein interaction in *Bos taurus*, *Sus scrofa* and *Gallus gallus* were listed in Table 1, 2 and 3 respectively. As listed, AMPK signaling pathway, Adipocytokine signaling pathway and Insulin signaling are in the top listed pathways in those three species. In each pathway there are different number of observed genes that match with clustered protein network. For example, there are 10 genes in protein network which belongs to AMPK signaling pathway in *Bos taurus*. This result shows that protein interact each other working together to support specific metabolism pathway [11].

Table 1. Observed metabolism pathways in *Bos taurus*

No	Pathway Description	Observed Gene Count	Matching proteins in your network
1	AMPK signaling pathway	10	ACACA,CPT1B,PPARGC1A,PRKAA1,PRKAA2,PRKAB1,PRKAB2, PRKAG1,PRKAG2,PRKAG3
2	Adipocytokine signaling pathway	9	CPT1B,PPARGC1A,PRKAA1,PRKAA2,PRKAB1,PRKAB2,PRKAG1 ,PRKAG2,PRKAG3
3	Circadian rhythm	7	PRKAA1,PRKAA2,PRKAB1,PRKAB2,PRKAG1,PRKAG2,PRKAG3
4	Insulin signaling pathway	9	ACACA,PPARGC1A,PRKAA1,PRKAA2,PRKAB1,PRKAB2, PRKAG1,PRKAG2,PRKAG3
5	Hypertrophic cardiomyopathy (HCM)	7	PRKAA1,PRKAA2,PRKAB1,PRKAB2,PRKAG1,PRKAG2, PRKAG3
6	FoxO signaling pathway	7	PRKAA1,PRKAA2,PRKAB1,PRKAB2,PRKAG1,PRKAG2,PRKAG3
7	Oxytocin signaling pathway	7	PRKAA1,PRKAA2,PRKAB1,PRKAB2,PRKAG1,PRKAG2,PRKAG3
8	Non-alcoholic fatty liver disease (NAFLD)	7	PRKAA1,PRKAA2,PRKAB1,PRKAB2,PRKAG1,PRKAG2,PRKAG3
9	Regulation of autophagy	2	PRKAA1,PRKAA2
10	Fatty acid metabolism	2	ACACA,CPT1B
11	mTOR signaling pathway	2	PRKAA1,PRKAA2

Table 2. Observed metabolism pathways in *Sus scrofa*

No	Pathway Description	Observed Gene Count	Matching proteins in your network
1	Adipocytokine signaling pathway	8	CPT1B,PPARGC1,PRKAA2,PRKAB1,PRKAB2,PRKAG1,PRKAG2,PRKAG3
2	AMPK signaling pathway	8	CPT1B,PPARGC1,PRKAA2,PRKAB1,PRKAB2,PRKAG1, PRKAG2,PRKAG3
3	Circadian rhythm	6	PRKAA2,PRKAB1,PRKAB2,PRKAG1,PRKAG2,PRKAG3
4	Insulin signaling pathway	7	PPARGC1,PRKAA2,PRKAB1,PRKAB2,PRKAG1,PRKAG,PRKAG3
5	Hypertrophic cardiomyopathy (HCM)	6	PRKAA2,PRKAB1,PRKAB2,PRKAG1,PRKAG2,PRKAG3
6	FoxO signaling pathway	6	PRKAA2,PRKAB1,PRKAB2,PRKAG1,PRKAG2,PRKAG3
7	Oxytocin signaling pathway	6	PRKAA2,PRKAB1,PRKAB2,PRKAG1,PRKAG2,PRKAG3
8	Non-alcoholic fatty liver disease (NAFLD)	6	PRKAA2,PRKAB1,PRKAB2,PRKAG1,PRKAG2,PRKAG3

Table 3. Observed metabolism pathways in *Gallus gallus*

No	Pathway Description	Observed Gene Count	Matching proteins in your network
1	Insulin signaling pathway	7	ENSGALG00000005439,PPARGC1A,PRKAA1,PRKAA2,PRKAB1,PRKAG2,TTLL4
2	Adipocytokine signaling pathway	6	PPARGC1A,PRKAA1,PRKAA2,PRKAB1,PRKAG2,TTLL4
3	FoxO signaling pathway	5	PRKAA1,PRKAA2,PRKAB1,PRKAG2,TTLL4
4	Regulation of autophagy	2	PRKAA1,PRKAA2
5	mTOR signaling pathway	2	PRKAA1,PRKAA2

According to the listed pathway in 3 species in this study, we clustered those pathways using Venn diagram and found 3 common pathways that are Adipocytokine signaling pathway, Insulin signaling pathway, FoxO signaling pathway (Table 4). In the classification of genes responsible in each of those 3 metabolism pathways, we found different set of distinct genes in specific species (Table 5). As we can see here, group of genes interact together in every specific pathway and species which support the previous report [11].

Table 4. Common pathways based on species group

Species	Observed Pathways	Pathway name
Bos taurus, *Sus scrofa* *Gallus gallus*	3	Adipocytokine signaling pathway, Insulin signaling pathway, FoxO signaling pathway
Bos taurus, *Sus scrofa*	5	Hypertrophic cardiomyopathy (HCM), Non-alcoholic fatty liver disease (NAFLD), AMPK signaling pathway, Oxytocin signaling pathway, Circadian rhythm
Bos taurus, *Gallus gallus*	2	mTOR signaling pathway, Regulation of autophagy
Bos Taurus	1	Fatty acid metabolism

Table 5. Distinct genes in the metabolism pathway in each species

Adipocytokine signaling pathway			Insulin signaling pathway			FoxO signaling pathway		
Bt	Ss	Gg	Bt	Ss	Gg	Bt	Ss	Gg
CPT1B	CPT1B	PPARGC1A	ACACA	PPARGC1A	ENSGALG00000005439	PRKAA1	PRKAA2	PRKAA1
PPARGC1A	PPARGC1A	PRKAA1	PPARGC1A	PRKAA2	PPARGC1A	PRKAA2	PRKAB1	PRKAA2
PRKAA1	PRKAA2	PRKAA2	PRKAA1	PRKAB1	PRKAA1	PRKAB1	PRKAB2	PRKAB1
PRKAA2	PRKAB1	PRKAB1	PRKAA2	PRKAB2	PRKAA2	PRKAB2	PRKAG1	PRKAG2
PRKAB1	PRKAB2	PRKAG2	PRKAB1	PRKAG1	PRKAB1	PRKAG1	PRKAG2	TTLL4
PRKAB2	PRKAG1	TTLL4	PRKAB2	PRKAG2	PRKAG2	PRKAG2	PRKAG3	
PRKAG1	PRKAG2		PRKAG1	PRKAG3	TTLL4	PRKAG3		
PRKAG2	PRKAG3		PRKAG2					
PRKAG3			PRKAG3					

Bt: *Bos taurus*, Ss: *Sus scrofa*, Gg: *Gallus gallus*

In the 3 species of this study, we found common genes that interact within Adipocytokine signaling pathway, Insulin signaling pathway and FoxO signaling pathway (Table 6), that are PRKAA2, PRKAB1 and PRKAG2. Whether these genes worked as the key player in the interact proteins was the question to be addressed and validated. As short explanation, those 3 pathways belong to cell signaling which determine intracellular signaling molecules for energy metabolism. Adipocytokine signaling is an important regulator of energy intake and metabolic rate through adipocyte cell and positively correlated with leptin production. This pathway related with glucose utilization and fatty acid oxidation via AMPK activation [17,18]. Insulin signaling pathway work in determine the insulin binding to its receptor results in the tyrosine phosphorylation of insulin receptor substrates by the insulin receptor tyrosine kinase for controlling glucose uptake into the cell [19], as we know glucose used as energy source. The last one is FoxO signaling pathway which regulates the expression of genes in cellular physiological events and act as the integrator of homeostasis maintenance signal [20].

Table 6. Common genes in the same pathway

Pathway	Total gene	Gene name
Adipocytokine signaling pathway	4	PRKAA2, PPARGC1A, PRKAB1, PRKAG2
Insulin signaling pathway	4	PRKAA2, PPARGC1A, PRKAB1, PRKAG2
FoxO signaling pathway	3	PRKAA2, PRKAB1, PRKAG2

The *in silico* work in this study shows the possibility of using common genes in the different species as metabolic marker, for example, according to the gene similarity name. However, the different in specific gene sequence in each species need to be considered. For that, carefully wet lab need to be employed to validate the results to give a distinct function in each species.

4. Conclusion

This study shows that AMPK plays in different metabolism pathway of *Bos taurus*, *Sus scrofa* and *Gallus gallus*. As the cellular metabolic regulator, AMPK may act in different pathway by expressing different protein to protein interaction in specific species, thus its expression need to be characterized individually.

References

[1] Towler M C and Hardie D G 2007 AMP-activated protein kinase in metabolic control and insulin signaling *Circ. Res.* **100** 328–41
[2] Hardie D G and Hawley S A 2001 AMP-activated protein kinase: the energy charge hypothesis revisited *Bioessays* **23** 1112–9
[3] Dugan L L, You Y H, Ali S S, Diamond-Stanic M, Miyamoto S, DeCleves A E, Andreyev A, Quach T, Ly S, Shekhtman G, Nguyen W, Chepetan A, Le T P, Wang L, Xu M, Paik K P, Fogo A, Viollet B, Murphy A, Brosius F, Naviaux R K and Sharma K 2013 AMPK dysregulation promotes diabetes-related reduction of superoxide and mitochondrial function *J. Clin. Invest.* **123** 4888–99
[4] Viollet B, Mounier R, Leclerc J, Yazigi A, Foretz M and Andreelli F 2007 Targeting AMP-activated protein kinase as a novel therapeutic approach for the treatment of metabolic disorders *Diabetes Metab.* **33** 395–402
[5] Viollet B, Horman S, Leclerc J, Lantier L, Foretz M, Billaud M, Giri S and Andreelli F 2010 AMPK inhibition in health and disease *Crit. Rev. Biochem. Mol. Biol.* **45** 276–95
[6] Tosca L, Chabrolle C, Uzbekova S and Dupont J 2007 Effects of metformin on bovine granulosa cells steroidogenesis: possible involvement of adenosine 5' monophosphate-activated protein kinase (AMPK) *Biol. Reprod.* **76** 368–78
[7] Park S K, Sheffler T L, Spurlock M E, Grant A L and Gerrard D E 2009 Chronic activation of

5'-AMP-activated protein kinase changes myosin heavy chain expression in growing pigs *J. Anim. Sci.* **87** 3124–33
[8] Nguyen T M D, Alves S, Grasseau I, Métayer-Coustard S, Praud C, Froment P and Blesbois E 2014 Central role of 5'-AMP-activated protein kinase in chicken sperm functions1 *Biol. Reprod.* **91** 121
[9] Chen J, Hudson E, Chi M M, Chang A S, Moley K H, Hardie D G and Downs S M 2006 AMPK regulation of mouse oocyte meiotic resumption in vitro *Dev. Biol.* **291** 227–38
[10] Phizicky E M and Fields S 1995 Protein-protein interactions: methods for detection and analysis *Microbiol. Rev.* **59** 94–123
[11] Van der Knaap J A and Verrijzer C P 2016 Undercover: gene control by metabolites and metabolic enzymes *Genes Dev.* **30** 2345–69
[12] Xing S, Wallmeroth N, Berendzen K W and Grefen C 2016 Techniques for the analysis of protein-protein interactions in vivo *Plant Physiol.* **171** 727–58
[13] Sevimoglu T and Arga K Y 2014 The role of protein interaction networks in systems biomedicine *Comput. Struct. Biotechnol. J.* **11** 22–7
[14] Prastowo S, Amin A, Rings F, Held E, Wondim D S, Gad A, Neuhoff C, Tholen E, Looft C, Schellander K, Tesfaye D and Hoelker M 2017 Fateful triad of reactive oxygen species, mitochondrial dysfunction and lipid accumulation is associated with expression outline of the AMP-activated protein kinase pathway in bovine blastocysts *Reprod. Fertil. Dev.* **29** 890–905
[15] Amin A, Gad A, Salilew-Wondim D, Prastowo S, Held E, Hoelker M, Rings F, Tholen E, Neuhoff C, Looft C, Schellander K and Tesfaye D 2014 Bovine embryo survival under oxidative-stress conditions is associated with activity of the NRF2-mediated oxidative-stress-response pathway *Mol. Reprod. Dev.* **81** 497–513
[16] Szklarczyk D, Franceschini A, Wyder S, Forslund K, Heller D, Huerta-Cepas J, Simonovic M, Roth A, Santos A, Tsafou K P, Kuhn M, Bork P, Jensen L J and Von Mering C 2015 STRING v10: Protein-protein interaction networks, integrated over the tree of life *Nucleic Acids Res.* **43** D447–52
[17] Yamauchi T, Kamon J, Minokoshi Y, Ito Y, Waki H, Uchida S, Yamashita S, Noda M, Kita S, Ueki K, Eto K, Akanuma Y, Froguel P, Foufelle F, Ferre P, Carling D, Kimura S, Nagai R, Kahn B B and Kadowaki T 2002 Adiponectin stimulates glucose utilization and fatty-acid oxidation by activating AMP-activated protein kinase *Nature Med.* **8** 1288–95
[18] Minokoshi Y, Kim Y-B, Peroni O D, Fryer L G D, Müller C, Carling D and Kahn B B 2002 Leptin stimulates fatty-acid oxidation by activating AMP-activated protein kinase *Nature* **415** 339–43
[19] Bevan P 2001 Insulin signalling *J.Cell Sci.* **114** 1429–30
[20] Eijkelenboom A and Burgering B M T 2013 FOXOs: Signalling integrators for homeostasis maintenance *Nature Rev. Mol. Cell Biol.* **14** 83–97

IOP Conf. Series: Earth and Environmental Science **142** (2018) 012010 doi:10.1088/1755-1315/142/1/012010

Birth and weaning weight of kids from different Boer goat crosses

T Nugroho, A Nurhidayati, A I Ayuningtyas, C Kustiyani, S Prastowo and N Widyas
Animal Science Department, Faculty of Agriculture, Universitas Sebelas Maret, Surakarta, Indonesia

E-mail: nwidyas@gmail.com

Abstract. Crossbreeding in Indonesian goat has been a common practice to improve genetic quality by mating the local does with exotic goat to produce offspring. The offspring, kids, are expected to have higher production performance compared to its parents, but robust towards tropical environment especially the available feed resources. Therefore, the aim of this paper is to characterize the production performance of offspring resulted from Boer crossed with different doe breeds. Birth and weaning weight (adjusted for 77 days of age) data were collected from kid records and their pedigree that born (n = 4,042) between January 2012 – June 2015 in CV. Kambing Burja, East Java, Indonesia. In this study, purebred Boer bucks (n = 23) were mated to three distinct doe breeds: pure Boer (n = 161), local Jawarandu (n = 700) and Boer × Jawarandu cross (Boerja, n = 501) to produce offspring. To analyze effect of weight data, linear model was built using breed (doe), year, and kid sex as fixed effects. According to the linear model, year gave effect to birth and weaning weight. Results showed all those fixed factors were significantly (P<0.01) affecting to birth and weaning weight. Boer offspring shows heaviest birth (3.16±0.60 kg, 2.99±0.63 kg and 2.84±0.58 kg; respectively) and weaning weight (15.02±3.94 kg, 13.67±3.87 kg and 13.48±3.70 kg; respectively); compared to both Boerja and Jawarandu. Moreover, sex also gave significant effect where male was heavier than female. It is concluded goat production performance in this study was affected by the factors of breed, year and sex. To achieve better performance in goat crossbred, breed, composition became a factor that need to be considered.

1. Introduction

Indonesia has a large population of goats, it is about 19 million in 2016 [1]. Local goat in Indonesia consists of various breeds, including Kacang goat, Costa, Jawawandu, Etawa Grade, Gembrong and many others [2]. Indonesian local goat has some advantages such as high adaptability, good reproductive ability, good mothering ability and prolific. However, Indonesian local goats have low meat production capabilities [3].

To solve this problem, Indonesian government and breeders conducted crossbreeding to increase the genetic potential of local goat as meat producers [4]. The most popular goat that have high meat production capability is Boer goat. Boer goat is a meat-type goat originating from South Africa [5] has heavy body weight, rapid growth [6] and highly adaptable to the environment [7]. In addition, Boer has high fertility, high prolificacy and good mothering abilities [8]. The results of this crossbreeding are expected to produce goats that have good production and reproduction ability but also have high adaptability to the environment and feed sources.

Birth weight and weaning weight are among of the parameters to determine the production potential of goat. One of the factors affecting birth weight and weaning weight is the effect of dam [9,10]. The results of crossbreeding between Boer goat with Kacang goat produced kids with birth weight and weaning weight of 32% and 35% higher respectively compared to pure Kacang goat [11]. Meanwhile crossbred between Boer goat with Etawa Grade produce kids with birth weight and weaning weight 15,6% and 10,07% higher than Etawa Grade goat [12].

There is no record found about the product of crossbreeding from Boer buck and Local Jawarandu doe. Jawarandu goat are crossbred from Etawa Grade with Kacang goat. For that the purpose of this study was try to found any differences in kids' birth weight and weaning crossed Boer bucks with three different goat breeds.

2. Methods

Data were collected from CV. Kambing Burja, East Java, Indonesia. Data obtained from January 2012 to June 2015. Boer bucks (n = 23) were mated into three distinct doe breeds, following these mating scenarios:

MS1: Boer Bucks x Boer does (n = 161)
MS2: Boer Bucks x Jawarandu (Boerja; n = 700); and
MS3: Boer Bucks x Boerja (n = 501)

Boer bucks and does are imported from Australia, but some Boer doe were born in Indonesia as the result of MS1. Jawarandu goat are crossbred of Etawa Grade with Kacang goat obtained from farmer around East Java. Boerja goat are product of Boer buck mate with Jawarandu doe in CV. Kambing Burja as the result of MS2. A total of 4,042 offsprings with three different breed compositions was produced. Birth and weaning weight were recorded.

Female goats are kept in colony, containing 20-25 goat each flock. Female goats are not separated by its breed, therefore no different in management. Boer bucks and all of does breed naturally by joining (male : female is 1 : 20-25) in the flock for 45 days.

Birth and weaning (adjusted for 77 days of age) weight data were analyzed by linear model. Year of born, offspring sex and doe breed as fixed effects according to the following basic model:

$$y_{ijkl} = \mu + Year_i + Sex_j + DB_k + \varepsilon_{ijkl}$$

Where
Y_{ijkl} : response (birth and weaning weight)
μ : general mean
Year : year of birth effect (i = 1,2,3,4)
Sex : offspring sex effect (j = 1,2)
DB : doe breed effect (k = 1,2,3)
ε_{ijkl} : residuals

3. Results and discussions

Result showed that year of birth affect ($P < 0,01$) birth weight and weaning weight (Table 1). Birth and weaning weight of kids vary widely each year. The heaviest birth and weaning weight gained in 2014, then lightest birth weight gained in 2012 and weaning weight in 2013. This result is in agreement with several reports [5,10]. The high variation of birth and weaning weight in each years can be explained from changes in management, climate and sample size [10]. Variations every year might be explained partly by differences of rainfall which in influence grass production and feed availability [13].

Result showed that kid sex significantly influence ($P < 0.01$) birth and weaning weight (Table 1). The male kids are always heavier than female. This is because the rate of prenatal growth of male is higher than female [14].The male kids always grew faster than female, this indicates that the male sex more quickly adaptable to the environment [10].

Dam breed has significant effect (P < 0.01) on birth and weaning weight (Table 1). Boer doe produces heaviest birth weight compared to Burja and Jawarandu doe. Jawarandu doe produces lighter kids compared to Boerja. The higher birth weight of the kids from Boer does is not surprising and is consistent with previous reports [15]. Birth weight is determined by extra-chromosomal inheritance and intrauterine environmental or dam effect [9]. Parent as an individual gives great influence to birth weight, in relation to the ability of each parent in providing good uterine environment during pregnancy through fulfillment of nutrient for embryo. Body measurement of the doe related to uterus capacity and nutrient supplying for embryo [16]. Boer doe is the biggest and Jawarandu is lightest. Based on that body measurement, Boer doe has bigger carrying capacity than Burja and Jawarandu.

Table 1. Number of records (n), mean and standard deviation of birth and weaning weight of kids in various dam breed, year and sex.

Factor	Level	n	Birth Weight (kg)	Weaning Weight (kg)
	Boer	561	3.16 ± 0.60^{a}	15.02 ± 3.94^{a}
Dam Breed	Boerja	1,231	2.99 ± 0.63^{b}	13.67 ± 3.87^{b}
	Jawarandu	2,250	2.84 ± 0.58^{c}	13.48 ± 3.70^{b}
		P value	< 0.01	< 0.01
	2012	704	2.64 ± 0.47^{d}	13.42 ± 3.20^{b}
	2013	1,172	2.69 ± 0.48^{c}	12.94 ± 3.74^{c}
Year	2014	1,392	3.23 ± 0.62^{a}	14.86 ± 4.04^{a}
	2015	774	3.00 ± 0.60^{b}	13.29 ± 3.60^{b}
		P value	< 0.01	< 0.01
Sex	Male	2,001	2.99 ± 0.63^{a}	14.19 ± 4.01^{a}
	Female	2,041	2.86 ± 0.58^{b}	13.31 ± 3.56^{b}
		P value	< 0.01	< 0.01

Note: Different superscripts in the same factors and traits were significantly different (P < 0,01).

Postnatal growth was affected by big proportion of environment compared to genetic potency. Then dam effect just affected by supplying milk [17]. Boer goat have a high growth and origin from South Africa [15] which has same tropic climate with Indonesian environment, thus, Boer doe can produce kids with greater weaning weight than others. Boerja and Jawarandu doe produce kids with the same weaning weight. We assumed that this is because Boerja and Jawarandu have almost the same potential. Boerja doe is crossbred of Jawarandu and Boer, so it already has 3 compositions of breed. The addition of Boer breed was not able to increase the weaning weight.

4. Conclusions

The fixed effects such as year, kids sex and dam breed are very important to birth and weaning weight of goats. Crossing between Boer bucks with Boerja and Jawarandu doe resulted in kids with similar weight. Considering blood composition in resulted offspring for better production performance need to be carefully planned along with better recording system to evaluate.

References

[1] BPS-Statistics Indonesia 2017 *Statistical Yearbook of Indonesia* (Jakarta: BPS-Statistics Indonesia)
[2] Periasamy K, Vahidi S M, Silva P, Faruque M, Naqvi A, Basar M, Cao J, Zhao S, Thi L, Pichler R, Garcia M, Shamsuddin M, Boettcher P, Fernando J, Han J, Ajmone P, Diallo A and Viljoen G J 2017 Mapping molecular diversity of indigenous goat genetic resources of Asia

Small Rumin. Res. **148** 2–10
[3] Budisatria I G S, Udo H M J, Eilers C H A M and Van Der Zijpp A J 2007 Dynamics of small ruminant production A case study of Central Java, Indonesia *Outlook Agric.* **36** 145–52
[4] FAO 2003 *National Report on Animal Genetic Resources Indonesia (A Strategic Policy Document)* (Food and Agriculture Organization (FAO))
[5] Schoeman S, Els J and van Niekerk M 1997 Variance components of early growth traits in the Boer goat *Small Rumin. Res.* **26** 15–20
[6] Casey N H and van Niekerk W A 1988 The boer goat. I. Origin, adaptability, performance testing reproduction and milk production *Small Rumin. Res.* **1** 291–302
[7] Malan S W 2000 The improved Boer goat *Small Rumin. Res.* **36** 165–70
[8] Greyling J P 2000 Reproduction traits in the Boer goat doe *Small Rumin. Res.* **36** 171–7
[9] Zhang C, Yang L and Shen Z 2008 Variance components and genetic parameters for weight and size at birth in the Boer goat *Livest. Sci.* **115** 73–9
[10] Zhang C, Zhang Y, Xu D, Li X, Su J and Yang L 2009 Genetic and phenotypic parameter estimates for growth traits in Boer goat *Livest. Sci.* **124** 66–71
[11] Elieser S, Sumadi, Budisatria G S and Subandriyo 2012 Productivity comparison between Boer and Kacang goat dam *J. Indon. Trop. Anim. Agric.* **37** 15–21
[12] Kostaman T and Sutama I 2006 Korelasi bobot badan induk dengan lama bunting, litter size, dan bobot lahir anak kambing Peranakan Etawah *Seminar Nasional Teknologi Peternakan dan Veteriner* hal 522–7
[13] Al-Shorepy S A, Alhadrami G A and Abdulwahab K 2002 Genetic and phenotypic parameters for early growth traits in Emirati goat *Small Rumin. Res.***45** 217–23
[14] Sutama I K, Budiarsana I G M, Setianto H and Priyanto A 1995 Productive and reproductive performances of young Ettawah-cross does *Indon. J. Anim. Vet. Sci.* **1** 81–5
[15] van Niekerk W A and Casey N H 1988 The boer goat. II. Growth, nutrient requirements, carcass and meat quality *Small Rumin. Res.* **1** 355–68
[16] van Niekerk M, Botha M E, Casey N and Schoeman S J 1996 Heritability estimates for pre-weaning growth traits in the Adelaide Boer goat flock *South African J. Anim. Sci.* **26** 6–10
[17] Robinson O W 1981 The influence of maternal effects on the efficiency of selection; A review *Livest. Prod. Sci.* **8** 121–37

IOP Conf. Series: Earth and Environmental Science **142** (2018) 012011 doi:10.1088/1755-1315/142/1/012011

Effect of benzyl amino purine and indole-3-acetic acid on propagation of *Sterculia foetida in vitro*

E Yuniastuti, C E Widodo, Samanhudi and M N I Delfianti
Study Program of Agrotechnology, Faculty of Agriculture, Universitas Sebelas Maret Surakarta, Indonesia

E-mail: yuniastutisibuea@staff.uns.ac.id

Abstract. *Sterculia foetida* is an oval seed plants that can be used as biofuel, which is one of the environmental friendly fuels. This plant is quite hard to find because not many peoples cultivate the plants. An in vitro propagation is one way to preserve the plant. This research aimed to determine optimum concentration of benzyl amino purine (BAP) and indole-3-acetic acid (IAA) to propagate *S. foetida in vitro*. The results showed that woody plant medium (WPM) added by 4 mg L BAP^{-1} and 0.5 mg L IAA^{-1} was able to produce complete plantlet, whereas those added by 4 mg L BAP^{-1} and 1 mg L IAA^{-1} generated the best growth of shoot and leaves.

1. Introduction

Sterculia foetida is a type of potential plant to be developed as a source of biofuel which has not been cultivated [1]. This plant can be grown in the tropics and sub-tropics (30°N-35°S). Oval-shaped seed is white with yellow seed coat [2]. The main composition of dry beans is the fats (51.78%), protein (21.61%), starch (12.1%), sugar (5%), cellulose (5.51%) and ash (3.9%) [3].

In this research, propagation of plant through tissue culture. Propagation by tissue culture is expected to provide seed mass, uniforms and all year round. Tissue culture also produces pathogen-free plant and germplasm storage [4]. Tissue culture media used, are woody plant medium (WPM) with plant growth regulator (PGR) benzyl amino purine (BAP) and indole-3-acetic acid (IAA). The BAP is a class of cytokines with function for cell division in meristematic tissue, stimulates the differentiation of the cells generated in meristem and encourage the growth of side shoots, leaves apical dominance and expansion. The IAA is a commonly used class of auxin in tissue culture serves to spur the process of cell elongation. This research aimed to determine optimum concentration of BAP and IAA to propagate *S. foetida* by *in vitro*.

2. Methods

Research was held at Laboratory of Plant Physiology and Biotechnology, Faculty of Agriculture Universitas Sebelas Maret Surakarta. The explants were maintained in the green house for three months. Materials used in the study were explant from stem of *S. foetida*, BAP (0, 2, 4, and 6 mg L^{-1}), IAA (0, 0.5 and 1 mg L^{-1}) and some nutrients used in the manufacture of WPM medium. The data were analyzed using descriptive methods that describe the results of observational studies.

3. Results and discussion

3.1. Callus and shoot emergence

Callus induction from petiole explant was readily obtained within 2 weeks [5]. Formation of callus culture media is very important to observe because it is hopeful to promote callus differentiation. In this research, callus that first appeared in the initiation medium was white translucent with crumb texture. The results of BAP and IAA with varying concentrations to produce callus growth is uneven (Fig. 2a). The emergence of the fastest callus was found in the treatment of 4 mg L BAP^{-1} and 4 mg L BAP^{-1} + 1 mg L IAA^{-1} are two weeks after planting, while largely the result of interaction of BAP and IAA capable of producing callus on the 4th week after planting. The IAA concentration of 0.5 mg L^{-1} and 1 mg L BAP^{-1}can accelerate emergence of a callus. According Rao and Purohit [6] high concentrations of auxin stimulates callus formation and morphogenesis pressing.

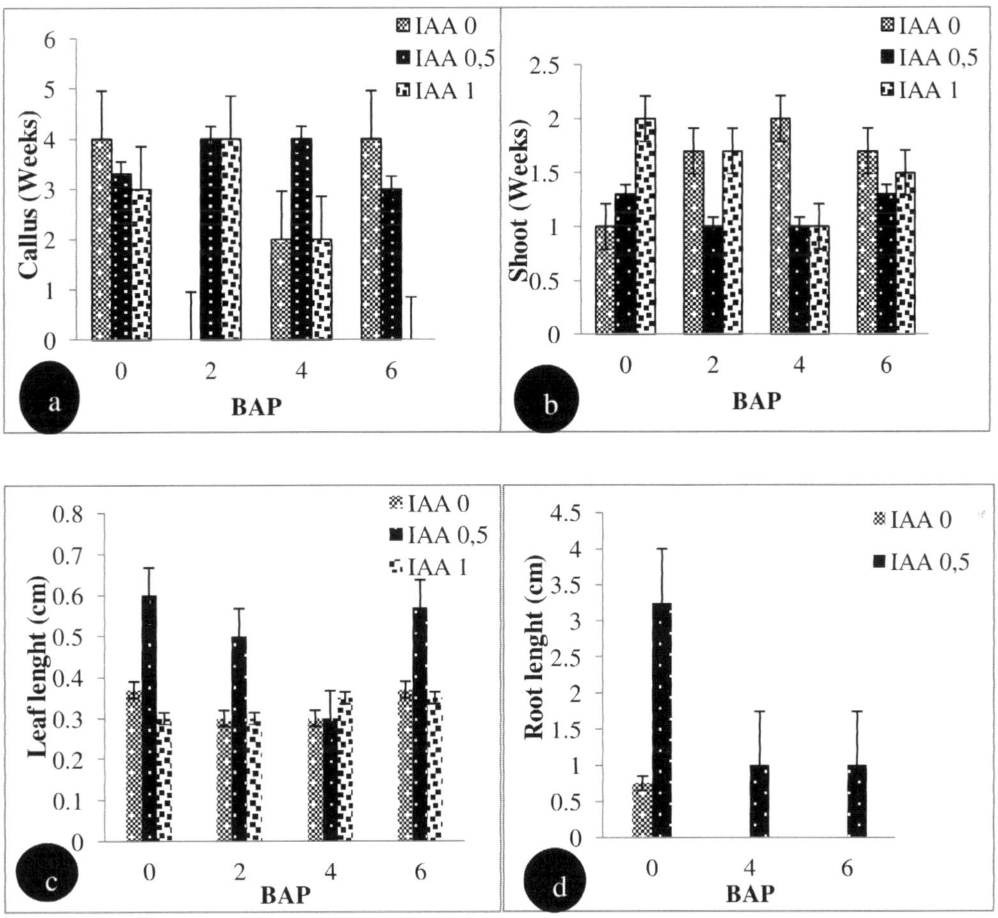

Figure 1. (a) Effect of BAP and IAA on the callus emergence of *Sterculia foetida* explant, (b) shoot emergence, (c) leaf length, (d) root length

The emergence of shoots characterized by their greenish bulge in the armpit leaves. Shoots growing from the meristem tissue composed of components promeristem or the basics of the new organs begin to form. The morphological diversity of plant organs depends on the developmental origin primordia, shoot meristem generate leaves [7]. The highest shoot regeneration potential was observed on stem and leaf segments [5]. In this research, the shoots appeared first in the treatment of 2 mg L BAP^{-1} + 0.5 mg L IAA^{-1}; 4 mg L BAP^{-1} + 0.5 mg L IAA^{-1}; 4 mg L BAP^{-1} + IAA 1 mg L^{-1}; and control (Fig. 2b). The most of the four treatments shoots appear in the first week. The addition of IAA at 6 mg L BAP^{-1} in the culture medium can accelerate the emergence of shoots, because the nature of the IAA at low concentration between 0.5-1 mg L^{-1} combined with high concentrations of cytokines can trigger the growth of shoots [8].

3.2. Leaf Length

Leaves for the plant have an important role because it the center of the leaf photosynthesis. Although the mechanism of formation of a crop that can perform photosynthesis by itself, conditions *in vitro* have not clearly known yet [9]. Number of leaves in this research was calculated based on observations of plantlets at 12 weeks after planting. The number of leaves was affected by the addition of plant growth regulators into the media.

The average number of leaves was ranged from 1 to 3 sheets. (Fig. 2b). Number of leaves that appear in visible explants varied. Variations in the number of leaves is possible because the endogenous hormone levels are not exactly the same so its response to the addition of plant growth regulator also varies [10]. Combination of BAP and IAA can successfully leverage in tissue culture if using specific concentration. Application of auxin with cytokinin (BAP) at the appropriate concentration can spur the growth of explants, especially in the formation of leaf, shoot and intensive segment [11]. The role of auxin in the plant growth of tissue culture, especially arranged leaf expansion [12].

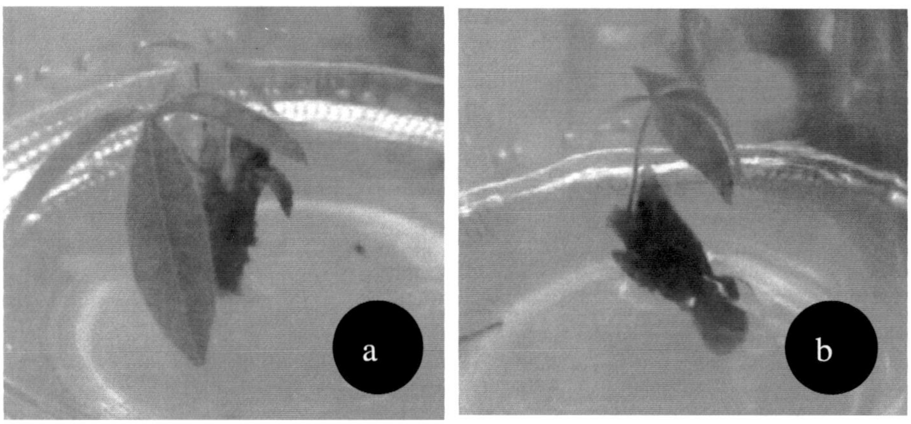

Figure 2. (a) Number of leaves with 0 mg L BAP^{-1} and 0.5 mg L IAA^{-1}. (b) 4 mg L BAP^{-1} and 0.5 mg L IAA^{-1}

In leaf morphogenesis, the control of cell proliferation seems to be related to the control of cell size [13]. Leaf growth is a process of differentiation leaf buds, with the addition of plant growth regulator such as auxin and cytokinin encourage the differentiation process. Leaf growth begins periclinal their division followed by cell growth in children and cause a bulge, is leaf primordia. Leaf surface area will increase caused by the activity of cleavage anticlinal [14].

The average length of the largest leaf obtained at 0.5 mg L IAA^{-1} treatment is 0.6 cm and a combination of 6 mg L BAP^{-1} and 0.5 mg L IAA^{-1} is 0.57 cm (Fig. 1c). High concentration of BAP combined with 0.5 mg L IAA^{-1} had leaves longer, while some combination of BAP and IAA treatment of most other produce leaf length of 0.3 cm. This is because the ability of each cell receiving a different PGR response and have optimal limit [15]. High concentrations of BAP were able to increase leaf area of explant [16].

In most of the higher plants, the root is the part that contained in the soil, the plant mainly serves the cantilever body and for the absorption of water and minerals [17]. Plantlets root formation is one thing that is advantageous, because it can increase the growth during the process of propagation *in vitro*. Many roots that come out so much more facilitate and expedite absorption of nutrients and water to optimal plant growth.

3.3. Root Shoot

Adding BAP and IAA response for explant able to produce a single root. It also occurred in the research [18] that were capable of producing a single root from stem explant. Roots in explant representing 1. The roots can be directly formed on the explants, either from the network or callus, if the media is given with sufficient auxin. Low concentrations of auxin could stimulate root growth [19].

The roots of plants are an important part in the growth. The plant has root length also good position in the growth media. The fact that developmental stages of root cells are roughly correlated with distance from the apical meristem [20]. Root elongation process is accelerated by the presence of auxin in the media. Elongation is caused by enlargement of cells that are formed when the embryo is still developing in the mother plant [14].

The highest root length was obtained from the concentration of 0 mg/LBAP + 0.5 mg/LIAA is 3.25 cm (Fig. 1d). Control also produce roots that is 0.75 cm, but the results were higher than the results of additional treatment BAP 4 mg/Land 6 mg/L. The higher concentration of BAP generate root length. The effect of BAP on root elongation activity, concentrations of cytokines is thought to stimulate cell division effective root [21].

4. Conclusions

It can be concluded that: (a) WPM medium added by 4 ppm BAP and 0.5 ppm IAA is able to produce complete plantlet by forming buds, leaves and roots. (b) WPM medium added by 4 ppm BAP and 1 ppm IAA produce the best growth of shoot and leave. (c) Callus can be formed in almost any combination of BAP and IAA treatment.

Acknowledgements

The authors acknowledged the financial support by INSINAS RISTEK DIKTI 2016.

References

[1] Sudrajat D J, Nurhasybi and Dida S 2011 Technology for improving the germination of seeded seeds (*Sterculia foetida* Linn.) *Forest Plant Res.* **8** 301–14

[2] Yuniastuti E 2008 Intensive research program report on the identification and selection of plant pragmatism (*Sterculia foetida*) as well as plant propagation technology in vitro for the provision of biofuel raw materials (Surakarta: Faculty of Agriculture, Sebelas Maret University)

[3] Ong H C, Silitonga U S, Masjuki H H, Mahlia T M I and Chong W T 2013 Production and comparative fuel properties of biodiesel from non-edible oils: *Jatropha curcas, Sterculia foetida* and *Ceiba pentandra J. Energy. Convers. Manag.* **73** 245–55

[4] Gaspar T, Kevers C, Penel C and Greppin H 1996 Plant hormones and plant growth regulators in plant tissue culture *In Vitro Cell. Dev. Biol-Plant* **32** 272–89

[5] Sujatha N and Mukta N 1996 Morphogenesis and plant regeneration from tissue cultures of *Jatropha curcas PCTOC* **44** 135–41

[6] Rao M S and Purohit S D 2006 In vitro shoot bud differentiation and plantlet regeneration in *Celastrus paniculatus Wild. Biol. Plant.* **50** 501–506

[7] Benkova E, Michniewicz M, Saurer M and Teichmann T 2003 Local, efflux-dependent auxin gradients as a common module for plant organ formation *Cell* **115** 591–602.

[8] George E F and Sherring ton P D 1984 *Plant Propagation by Tissue Culture* (Eversley England: Exegetics Limited)

[9] Wetherell D F 1982 *Introduction to In Vitro Plant Propagation* (New Jersey: Avery Publishing

IOP Conf. Series: Earth and Environmental Science **142** (2018) 012011 doi:10.1088/1755-1315/142/1/012011

Group Inc)
[10] Widyawati G 2010 *Effect of NAA and BAP concentration variation on Jatropha curcas L. induction of jatropha callus* Thesis Faculty of AgricultureUniversity of March Surakarta (ID).
[11] Basri Z 2008 Multiplication of four varieties of chrysanthemum through tissue culture techniques *J Agroland* 15(04): 271-277.
[12] Teale W D, Paponov I A and Palme K 2006 Auxin in action: signalling, transport and the control of plant growth and development *Mol Cell Biol* 7: 846-859.
[13] Tsukaya H 2003 Organ shape and size: a lesson from studies of leaf morphogenesis *Curr Opin Biotechnol* 6: 57-62.
[14] Salisbury F B dan Ross C W 1995 *Plant Physiology III* (Bandung (ID): ITB).
[15] Santoso J, Mathius N T, Sastraprawira U, Suryatmana G dan Saodah D 2004 Propagation of quinine plants (Cinchona ledgeriana moens) and (C. Succirubra Pavon) through doubling of axillary buds *A Tower Plantation* 72(1):11-27.
[16] Romano A, Noronha C and Matins-loucao MA 1992 Influence of growth regulators on shoot proliferation in *Quercus suber* L. Ann. Bot 70: 531-536.
[17] Zulkarnain 2013 *Plant Crop Culure Crop Breeding Solution*. (Jakarta (ID): PT. Earth Script).
[18] Roostika I, Sunarlim N dan Mariska I 2005 Mikropropagasi of mangosteen plant (*Garcinia mangostana*) *J AgroBiogen* 1(1): 20-25.
[19] Deore A C and Johnson T S 2008 High frequency of plant regeneration from leaf-disc cultures of *Jatropha curcas* L.: an important biodiesel plant *Plant Biotechnol Rep* 2: 7-11.
[20] Birnbaum K, Shasha D E, Wang J Y, Jung J W, et al. 2003 A gene expression map of the *Arabidopsis* root. *Science* 302: 1956-1960.
[21] Pisesha P A, Mattjik N A dan Sukma D 2008 The influence of IAA, IBA, BAP, and coconut water concentration on the formation of poinsettium root (Euphorbia pulcherrima Wild Et Klotz) in vitro. Paper seminar of agronomy and horticulture department. Faculty of Agriculture IPB. Bogor (ID). URL: http://repository.ipb.ac.id/handle/ 123456789/2640.

IOP Conf. Series: Earth and Environmental Science **142** (2018) 012012 doi:10.1088/1755-1315/142/1/012012

The effect of dosages of microbial consortia formulation and synthetic fertilizer on the growth and yield of field-grown chili

N Istifadah[1], D Sapta[2], H Krestini[3], B Natalie[2], P Suryatmana[2], A Nurbaity[2] and R Hidersah[2]
[1]Department of Plant Pests and Diseases, Agriculture Faculty, Universitas Padjadjaran, Jl Raya Bandung-Sumedang Km 21, Jatinangor 45363, Jawa Barat; Indonesia
[2]Department Soil Sciences, Agriculture Faculty, Universitas Padjadjaran , Jl Raya Bandung-Sumedang Km 21, Jatinangor 45363, Jawa Barat; Indonesia
[3]Vegetable Research Station, Lembang, Bandung Barat, Jawa Barat, Indonesia

E-mail: n.istifadah@unpad.ac.id

Abstract. Chili (*Capsicum annuum*, L) is one of important horticultural crop in Indonesia. Formulation of microbial consortia containing *Bacillus subtilis, Pseudomonas* sp., *Azotobacter chroococcum* and *Trichoderma harzianum* has been developed. This study evaluated the effects of dosage of the microbial formulation combined with NPK fertilizer on growth and yield of chili plants in the field experiment. The experiment was arranged in completely randomized design of factorial, in which the first factor was dosage of formulation (0, 2.5, 5.0, 7.5, 10 g per plant) and the second factor was NPK fertilizer dosage (0, 25, 50 and 75% of the standard dosage). The treatments were replicated three times. For application, the formulation was mixed with chicken manure 1:10 (w/v). The results showed that application of microbial formulation solely improved the chili growth. There was interaction between dosages of the microbial formulation and NPK fertilizer in improving plant height, nitrogen availability and the chili yield, while there was no interaction between those dosages in improving the root length. Combination between microbial formulation at the dosage of 5.0-7.5 g per plant combined with NPK fertilizer with the dosage 50 or 75% of the standard dosage support relatively better growth and the chili yield.

1. Introduction

Fertilizers are important agricultural input supporting plant production. Conventional agriculture depends on high input of synthetic fertilizers. The long-term use of such input, however, has led to various negative impacts to the environment such as decreasing soil quality and environmental pollution. Concerning the sustainability of food production, currently agriculture development has been directed toward eco-friendly agricultural systems. One of eco-friendly technology supporting such system is the use of plant growth promoting microorganisms (PGPM) to reduce high input of synthetic or inorganic fertilizers.

Plant growth promoting microorganisms are microorganisms, usually bacteria, fungi, protozoa, inhabiting plant rhizosphere that have abilities to improve plant growth [1, 2]. The abilities of the microorganisms to promote plant growth were through direct or indirect mechanisms. The direct mechanisms comprise the production of plant growth hormone, supporting the nutrient availability such as through nitrogen fixation, phosphorous solubilization, production of iron chelating agents (siderophore). The indirect mechanisms encompass the suppression of plant pathogens and deleterious microbes [2, 3, 4].

We have developed microbial formulation containing PGPM consortia which are *Bacillus subtilis, Pseudomonas* sp. (a fluorescent Pseudomonad), *Azotobacter chroococcum* and *Trichoderma harzianum. Azotobacter* has been known as non-symbiotic Nitrogen-fixing bacteria that can also produce plant hormone such as indole acetic acid (IAA), gibberellins (GA) and cytokinins (CK) [5] and also vitamins such as thiamine and riboflavin [6]. Fluorescent *Pseudomonads* are nonpathogenic Pseudomonas that can promote plant growth through production iron-chelating agent, siderophore [7, 8] plant hormones such as auxin (indole acetic acid, IAA), cytokinins [9] and suppression of plant pathogens [7, 9]. Another plant growth promoting microbes is *Bacillus. Bacillus subtilis* has been reported to produce cytokinins [7, 10] and mediating synthesis of other hormones such as IAA [10, 11]. *Trichoderma harzianum* has also ability to increase plant growth through production of phytohomone such as IAA and Giberelic acid [12] and solubilization of phosphate and micronutrients [13].

The abilities of PGPM to support plant growth make them potential to be used as biofertilizer that can reduce the requirement of synthetic or inorganic fertilizers [1, 14]. This paper discussed the effects of dosage of the microbial formulation combined with NPK fertilizer on the growth and yield of chili plants in the field experiment.

2. Methods

The field experiment was conducted in Citespong, Lembang, West Bandung, West Java with andisol soil type. The experiment was arranged in Completely Randomized Design with factorial combination. The first factor was the dosages of microbial consortia (0.0, 2.5, 5.0, 7.5 and 10 g) and the second factor was the dosages of NPK (0, 25, 50 and 75% of standard dosages). Each treatment consisted of 28 plants, which were planted in 60 × 60 cm planting space. Each treatment was replicated three times.

For all treatments, chicken manure was applied with the rate as that usually used by the famers (15 ton per ha including the part used for mixing with the formulation). Formulation was prepared based on the method developed by Istifadah *et al.* [15]. For application, the formulation was mixed with chicken manures 1:10 (v/v) and incubated for 7 days. The mixture was then applied in the planting holes just before transplanting. The NPK fertilizers (the rate depending on the treatment with the standard dosage as recommended for chili) were applied at 3, 6 and 9 weeks after transplanting. The chili cultivar used was Ciko.

The plant height and stem diameter were observed at 2,4, 6 and 8 weeks after transplanting. The chili yield was accumulated based on two harvesting times.

3. Results and Discussions

The results showed that in general, the microbial formulation and NPK fertilizer enhanced the chili plant height, even though in two weeks after transplanting (WAT), there was no significant different on the plant height (Table 1). The experiment was conducted in andisol soil in which in all treatments including the check, composted chicken manure was applied in the soil. In the early growth stage, the nutrients supported by organic matters probable were still sufficient to support the growth of young chili plants.

At 4-8 weeks after transplanting there was interaction between dosage of the formulation and NPK fertilizer on the plant height. At fourth weeks after transplanting, the microbial formulation without NPK fertilizer significantly increased the chili height, whilst the NPK fertilizer had significant effect on chili height (Table 2). In general, there is no effect of combination between the microbial formulation and NPK fertilizer on plant height. This means in this stage, application of microbial formulation solely is enough to support chili growth.

Table 1. The effect of microbial formulation and NPK fertilizer in chili plant height (2 WAT)

Treatments	Plant Height (cm)	Treatments	Plant Height (cm)
ao : Check	9.7	bo : Check	9.5
a1 : Formulation 2.5 g per plant	9.9	b1 : NPK 25% of the standard dosage	9.6
a2 : Formulation 5.0 g / plant	9.5	b2 : NPK 50% of the standard dosage	9.9
a3 : Formulation 7.5 g per plant	9.6	b3 : NPK 75 % of the standard dosage	10.1
a4 : Formulation 10 g per plant	10.2		

Table 2. The effect of microbial formulation and NPK fertilizer in chili plant height (4 WAT)

Dosage of Microbial Formulation	Dosage of NPK fertilizer			
	B0: without NPK fertilizer	B1: 25% of standard dosage	B2: 50 % of standard dosage	B3: 75% of standard dosage
A0: Without Formulation	19.5 (a) A	22.1 (a) A	22.0 (a) A	22.6 (a) A
A1: Formulation, 2.5 g per plant	22.8 (b) B	21.3 (a) AB	21.0 (a) AB	20.1 (a) A
A2: Formulation, 5.0 g per plant	22.7 (b) A	22.1 (a) A	22.3 (a) A	22.2 (a) A
A3: Formulation, 7.5 g per plant	22.4 (b) A	22.0 (a) A	21.1 (a) A	22.0 (a) A
A4: Formulation, 10.0 g per plant	21.7 (b) AB	21.5 (a) A	23.6 (a) B	23.2 (a) AB

Data in the same column followed by the same lower case letter is not significantly different based on Tukey' HSD ($p<0.01$). Data in the same row followed by the same capital letter is not significantly different based on Tukey' HSD ($p<0.01$)

At six weeks after transplanting, the microbial formulation or NPK fertilizer applied solely increased the plant height (Table 3). Combination between them, however, did not increase the plant height, compared to the respective dosage of the formulation or NPK fertilizer, applied solely. Combination between microbial formulation and NPK that supported better plant growth were combination between microbial formulation at dosage of 5.0 g per plant and NPK fertilizer 25% of the standard dosage or combination formulation at the dosage 2.5 g per plant and NPK fertilizer 75% of standard dosages.

Table 3. The effect of microbial formulation and NPK fertilizer in chili plant height (6 WAT)

Dosage of Microbial Formulation	B0: without NPK fertilizer	B1: 25% of standard dosage	B2: 50 % of standard dosage	B3: 75% of standard dosage
A0: Without Formulation	24.1 (a) A	29.2 (ab) B	31.2 (a) B	30.3 (ab) B
A1: Formulation, 2.5 g per plant	30.5 (c) AB	30.3 (b) AB	27.7 (a) A	**34.4 (b) B**
A2: Formulation, 5.0 g per plant	26.8 (ab) A	**31 (b) C**	28.1 (a) AB	29.5 (ab) BC
A3: Formulation, 7.5 g per plant	29.4 (bc) A	27.4 (a) A	28.1 (a) A	26.3 (a) A
A4: Formulation, 10.0 g per plant	28.1 (bc) A	29.9 (ab) A	29.5 (a) A	29.5 (ab) A

Data in the same column followed by the same lower case letter is not significantly different based on Tukey' HSD ($p<0.01$). Data in the same row followed by the same capital letter is not significantly different based on Tukey' HSD ($p<0.01$)

At eight weeks after transplanting, the microbial formulation and NPK fertilizer applied solely also increased the plant height (Table 4). Combination between them generally did not increase in plant height. The combination that resulted in significant increase was combination between microbial formulation at the dosage of 7.5 g per plant and NPK fertilizer at the dosage 50% of standard dosage.

Table 4. The effect of microbial formulation and NPK fertilizer in chili plant height (8 WAT)

Dosage of Microbial formulation	B0: without NPK fertilizer	B1: 25% of standard dosage	B2: 50 % of standard dosage	B3: 75% of standard dosage
A0: Without Formulation	48.7 (a) A	52.3 (ab) B	52.0 (a) B	51.6 (a) B
A1: Formulation, 2.5 g per plant	51.4 (ab) A	51.1 (a) A	52.3 (a) A	53.2 (a) A
A2: Formulation, 5.0 g per plant	52.7(b) A	51.5 (a) A	52.3 (a) A	52.6 (a) A
A3: Formulation, 7.5 g per plant	53.3 (b) AB	52.1 (a) A	**55.8 (b) B**	52.9 (a) A
A4: Formulation, 10.0 g per plant	53.0 (b) A	52.4 (a) A	51.3 (a) A	52.2 (a) A

Data in the same column followed by the same lower case letter is not significantly different based on Tukey' HSD ($p<0.01$). Data in the same row followed by the same capital letter is not significantly different based on Tukey' HSD ($p<0.01$)

Based on other parameters such as stem diameter and root development, there was no interaction between the dosages of microbial formulation and NPK fertilizer, and therefore the analysis was based on the main factors. Based on stem diameter of chili plants, the application of microbial formulation or NPK fertilizer did not provide any significant effects. The diameter in any treatments was not significantly different to that of the control plants (Table 5).

Based on chili root fresh weight, the application of microbial formulation of 5.0 g per plant increased the root growth (Table 5). Meanwhile the application of NPK fertilizer at the dosage 50% of standard dosage increased the fresh root weight (Table 6).

Table 5. The effect of microbial formulation on stem diameter and root development

Treatments	Average of Stem Diameter (cm)	Average of Root Length (cm)
ao : Check	0.76 a	26.98 a
a1 : Formulation 2.5 g per plant	0.76 a	28.22 ab
a2 : Formulation 5.0 g per plant	0.76 a	29.36 b
a3 : Formulation 7.5 g per plant	0.77 a	28.08 ab
a4 : Formulation 10 g per plant	0.77 a	28.27 ab

Data in the same column followed by the same letter is not significantly different based on Tukey' HSD ($p<0.01$).

Table 6. The effect of microbial formulation and NPK fertilizer in chili plant height

Treatments	Average of Stem Diameter (cm)	Average of Root Length (cm)
bo : Check	0.76 a	27.14 a
b1 : NPK Dosage 25% of the standard	0.77 a	28.24 ab
b2 : NPK Dosage 50% of the standard	0.76 a	29.23 b
b3 : NPK Dosage 75 % of the standard	0.76 a	28.11 ab

Data in the same column followed by the same letter is not significantly different based on Tukey' HSD ($p<0.01$).

The application of microbial formulation and NPK fertilizer solely increased the nitrogen availability in the chili rhizosphere. The increase in nitrogen availability in the chili rhizosphere may be due to the existence of nitrogen-fixing bacteria, *A. chroococcum,* in the formulation. There was an interaction between the dosage of formulation and NPK fertilizer. The combination that led to highest nitrogen availability in the soil was combination between microbial formulation at the dosage of 10 g per plant and NPK fertilizer with dosage 25% of standard dosage (Table 7).

Table 7. The effect of microbial formulation and NPK fertilizer on the nitrogen availability in the chili rhizophere

Dosage of Microbial Formulation	B0: without NPK fertilizer	B1: 25% of standard dosage	B2: 50 % of standard dosage	B3: 75% of standard dosage
A0: Without Formulation	15.75 (a) A	21.76 (a) B	21.96 (a) B	24.93 (ab) B
A1: Formulation, 2.5 g per plant	19.62 (ab) A	28.25 (bc) B	22.84 (a) AB	21.02 (a) A
A2: Formulation, 5.0 g per plant	21.35 (bc) A	22.63 (ab) AB	26.53 (a) B	23.80 (ab) AB
A3: Formulation, 7.5 g per plant	26.37 (c) B	23.89 (ab) AB	20.61 (a) A	26.53 (b) B
A4: Formulation, 10.0 g per plant	23.33 (bc) A	**30.85 (c) B**	27.11 (a) AB	24.59 (ab) A

Data in the same column followed by the same lower case letter is not significantly different based on Tukey' HSD (p<0.01). Data in the same row followed by the same capital letter is not significantly different based on Tukey' HSD (p<0.01)

There was interaction between application of microbial formulation and NPK fertilizer on the healthy chili yield (Table 8). Many chili fruits were drop due to lack of water. Many harvested chilies also showed cracked skin, and therefore they were not considered as healthy fruits. The application of microbial formulation at dosage of 7.5 g per plant increased the healthy chili yield. Meanwhile, the application of NPK fertilizer at reduced dosage (25-75% of the standard dosage) also increased the healthy yield (Table 8). Even though in many combinations the weight of healthy chili accumulated from two harvest times was not significantly different to that of the control plants, there were certain combinations such as application of microbial formulation 7.5 g per plant and NPK fertilizer 75% of standard dosage, that significantly higher than that of the check plants. In this treatment, the healthy chili yield increased 61.4%.

Table 8. The effect of microbial formulation and NPK fertilizer on the chili yields per plant

Dosage of Microbial Formulation	B0: without NPK fertilizer	B1: 25% of standard dosage	B2: 50 % of standard dosage	B3: 75% of standard dosage
A0: Without Formulation	144.9 (a) A	191.8 (a) B	191.9 (a) B	161.6(a) AB
A1: Formulation, 2.5 g per plant	174.7 (ab) A	214.9(a) A	208.8 (a) A	178.1 (a) A
A2: Formulation, 5.0 g per plant	133.6 (a) A	143.6 (a) A	206.8 (a) B	189.03 (ab) B
A3: Formulation, 7.5 g per plant	207.1 (b) A	183.0(a) A	207.0 (a) A	**233.8 (b) A**
A4: Formulation, 10.0 g per plant	162.0 (ab) A	152.5 (a) A	207.8(a) A	143.6 (a) A

Data in the same column followed by the same lower case letter is not significantly different based on Tukey' HSD (p<0.01). Data in the same row followed by the same capital letter is not significantly different based on Tukey' HSD (p<0.01)

The overall results showed that in early vegetative growth (four weeks after transplanting), application of microbial formulation or NPK fertilizer solely increased the chili vegetative growth. At 6 – 8 weeks after transplanting, there was an increase in the plant growth if formulation and NPK fertilizer combined at appropriate dosages. There was tendency that the lower dosage of microbial formulation required relatively higher dosages of NPK fertilizer and vice versa. This means that the microbial formulation and NPK fertilizer can be complementary. In this study, the application of microbial formulation could reduce the requirements of the NPK dosage up to 75% reduction. For supporting roots system, the dosage of NPK fertilizer could reduce to 50% of the recommended dosage.

The abilities of microbial formulation to increase the chili growth and reduced the requirement of NPK fertilizer seems due to the abilities of microbes in the formulation to produce plant hormone and support nutrient availability. *Azotobacter chroococcum* used in this study is nitrogen-fixing bacteria that also produced phytohormon [16]. The abilities of *A. chroococcum* to support the nitrogen availability was confirmed by the increase on the nitrogen availability in the chili rhizosphere treated by the microbial formulation. *Pseudomonas* sp. used in this study also can produce phytohormone kinetin, zeatin, IAA [17] and phosphatase enzyme leading to solubilization of soil phosphate [17, 18]. Other microbes, *T. harzianum* [12] and *B. subtilis* [7,10] are also known as plant growth promoting microbes that capable of producing plant hormone.

Even though the microbial formulation could support the early vegetative growth of chili, it should be combined with synthetic fertilizer when the plants were in the end of vegetative stages (7-8 weeks after transplanting). Combination that sufficient to support the growth and yield of chili were application of 7.5 g formulation combined with NPK 50% or 75% of the standard dosage. This means that microbial formulation could reduce the requirement of synthetic fertilizers or improve their efficiency [1,19]. The use of PGPM in plant production system is very beneficial as they are not only support the nutrients management, but they can also play important role in plant protection as many of PGPM are antagonistic to plant pathogens [20].

Acknowledgments
This study was a part of Collaborative Research between University and Industry called 'RAPID' funded by DRPM, Kemenristekdikti.

References
[1] Malusà E, Pinzari F and Canfora L 2016 Efficacy of biofertilizers: Challenges to improve crop production (*Microbial Inoculants in Sustainable Agricultural Productivity*) Eds. Singh, D P. Singh, H B, Prabha, R (India: Springer) pp 17–40
[2] Yadav A N, Priyanka V, Divjot K and Kusam L R 2017 *Int. J. Environ Sci. Nat. Res.* **3** 555601. DOI : 10.19080/IJESNR.2017.03.555601
[3] Whips J M 2001 *J. Exp. Bot.* **52** 487–511
[4] Glick B R 2012 *Scientifica* vol 2012, Article ID 963401, 15 pages
[5] Sethi S K and Adhikary S P 2012 *Dyn. Biochem. Process Biotechnol. Mo.l Biol.* **6** (Special Issue I) 68–74
[6] Revillas J J, Rodelas B, Pozo C, Martinez-Toledo M V, Gonzalez L J 2000 *J. Appl. Microbiol.* **89** 486–493
[7] Santoyoa G, Orozco-Mosquedaa M C and Govindappab M 2012 *Biocontrol Sci. Tech.* **22** 855–872
[8] Dorjey S, Dolkar D, and Sharma R 2017 *Int. J. Curr. Micribiol. App. Sci.* **6** 1335–1344
[9] Preston G M 2004 *Phil. Trans. R. Soc. Lond.* B **359** 907–918
[10] Arkhipova T N, Vaselov S U, Melentiev A I, Martynenko E V, Kudoyorova G R 2005 *Plant Soil* **272** 201-209
[11] Tahir H A S, Gu Q, Wu H, Raza W, Hanif A, Wu L, Colman M V and Gao X 2017 *Front. Microbiol.* **8** 171
[12] Sofo A., Scopa A, Manfra M, De Nisco M, Tenore G, Troisi J, Di Fiori R, Novellino E 2011 *Plant Growth Regul.* **65** 421–425
[13] Li R X, Cai F, Pang G, Shen Q R, Li R, Chen W 2015 *PLoS ONE* **10** e0130081
[14] Adesemoye O and Kloepper J W 2009 *Appl Microbiol. Biotechnol.* **85** 1–12

IOP Conf. Series: Earth and Environmental Science **142** (2018) 012012 doi:10.1088/1755-1315/142/1/012012

[15] Istifadah N, Djaya L and Nurbaity N 2015 *Research Report* LPPM Unpad
[16] Hindersah R. Arief D H, Sumarni Y 2000 Prosiding Seminar Nasional Bioteknologi Pertanian, Yogyakarta pp 6-7
[17] Fitriatin B N, Yuniarti A, Turmuktini T and Saman M 2013 *Soil Water J.* **2** 547–554
[18] Fitriatin B N, Arief D H, Simarmata T Santosa D A and Joy B 2011 *J. Soil. Sci. Environ. Manag.* **2** 299-303
[19] Duarah I, Deka M, Saikia N and Boruah H P D 2011 *Biotech.* **1** 227–238
[20] Fuentes-Ramirez L, Caballero J and Mellado Z 2005 *PGPR: Biocontrol and Biofertilization, A. Siddiqui (ed.)* (Dordrecht: Springer) pp 143–172.

Isolation of Mycorrhizal *Rhizoctonia* as resistance inducer of *Dendrobium macrophyllum* to drought

R Soelistijono, Daryanti and M T Handayani
Agricultural Faculty of Universitas Tunas Pembangunan, Jl. Balekambang Lor No.1, Surakarta (57139), Indonesia

E-mail: sulistyo.utp@gmail.com

Abstract. One of the obstacles encountered in the cultivation of orchids *Dendrobium macrophyllum* is difficult to cultivate in areas with high drought due to the slow absorption of nutrients. Based on previous research, the mycorrhizal binucleate *Rhizoctonia* (BNR) has the ability to increase the resistance of vanilla (*Vanilla planifolia* Andrews) to drought, but it has never been tried on orchid *Dendrobium macrophyllum*. The objectives of this study was to isolate resistance inducer organisms by induced resistance techniques on orchids against drought. It is expected that the administration of mycorrhizal *Rhizoctonia* can increase the absorption of nutrients in *D. macrophyllum* which is exposed to high water stress. Each treatment consisted of 3 replications of 3 potted plants. The characterization of mycorrhizal *Rhizoctonia* isolate from *D. macrophyllum* root from Surakarta, Kopeng, Magelang, and Yogyakarta did not different morphologically. Character equations are in colony color, cell length and number of cores, while character differences are present in cell width and all isolates are capable of forming a *peloton* structure.

1. Introduction

Drought is a major factor in orchid cultivation. Long-term observation results indicated that the long dry season due to the phenomenon of La Nina global climate anomalies in general occur periodically every 5 years. Orchid *D. macrophyllum* is a group of *Dendrobium* sp. which is resistant to drought stress. *D. macrophyllum* is an endemic orchid at the foot of Merbabu Mountain. However, in recent years, this orchid is very hard to find in its habitat. This is due to environmental conditions that are less supportive (drought stress factor). By the Department of Commerce, the orchid *D. macrophyllum* is categorized into Appendix 2 plants, which are rare and non-tradable crops except cultivation.

D. macrophyllum is rare because it is not able to live in environment with drought stress, this because in the flowering phase requires enough water, thus, inhibit the flowering phase (generative phase) in the dry season (Figure 1).

Figure 1. Stressed *Dendrobium macrophyllum* by low water

Lack of nutrients absorbed by the root of epiphytic orchids caused by high water stress factors affect the productivity of flowering *D. macrophyllum*. This because the structure of epiphytic orchids do not intersect with the soil, thereby can not directly absorb nutrients from the soil and rely solely on water uptake in the epiphytic root that causes the lack of nutrients. Water stress factor is a major factor in the cultivation of orchids and influences the flowering phase of the plant. If the minimum water content within the plant roots is fulfilled, then the flowering process will occur quickly. Increased plant resistance to water stress can be done by inoculation using various biological agents due to the increase of several compounds that stimulate the nutrient uptake process and this is one of the effects of induced mechanisms *in vitro*.

The mycorrhizal *Rhizoctonia* is thought to have the ability to induce orchid plants [4]. The resistance mechanism in *D. macrophyllum* inoculated mycorrhizal *Rhizoctonia* presumably due to the formation of *peloton* structures [9], which indicates the existence of intermediate associations between mycorrhizal *Rhizoctonia* and *D. macrophyllum*. Until now there has been no report that definitely mentions the use of mycorrhizal *Rhizoctonia* in improving the resilience of orchid plants in high water stress conditions. It is therefore necessary to conduct research on the role of mycorrhizal *Rhizoctonia* in increasing flowering productivity in *D. macrophyllum* on high water stress by isolating it from the wild.

2. Methods

Isolation of mycorrhizal *Rhizoctonia* was performed from healthy *D. macrophyllum* root from various sites according to Otero *et al.* [7]. Identification of mycorrhizal Rhizoctonia by schlerotium and mycorrhizal identified by Barnett & Hunter [3] included: (1) sclerotium form, (2) hyphae branching forms, and (3) the number of nucleotide.

D. macrophyllum development stage *in vitro* by [8] conducted in the tissue culture laboratory of Agricultural Faculty of UTP included: Stage 1: Observation of embryo development (0-3 weeks after seed was sown); Stage 2: Protocorm observation (3-10 weeks after seed was sown); Stage 3: Observing the growth of seedling (10-20 weeks after the seed is sown).

Observation of *peloton* on transverse slices of orchid root of *D. macrophyllum for* each treatment were done by floroglucinol painting using method of Kutscha [6]. *Peloton* observations were performed 24, 48, and 96 hours after induction of mycorrhizal *Rhizoctonia*.

3. Results and discussion

Figure 2 shows the identification of mycorrhizal according to Barnett & Hunter [3] includes the color of the colony, the shape of the hypha branch, and the number of nuclei seen. Isolation of mycorrhizal *Rhizoctonia* from *D. macrophyllum* root in various places was obtained by several isolates of mycorrhizal *Rhizoctonia*, from Surakarta obtained isolate R1, from Magelang obtained isolate R2, from Kopeng obtained isolate, and from Yogyakarta obtained R4.

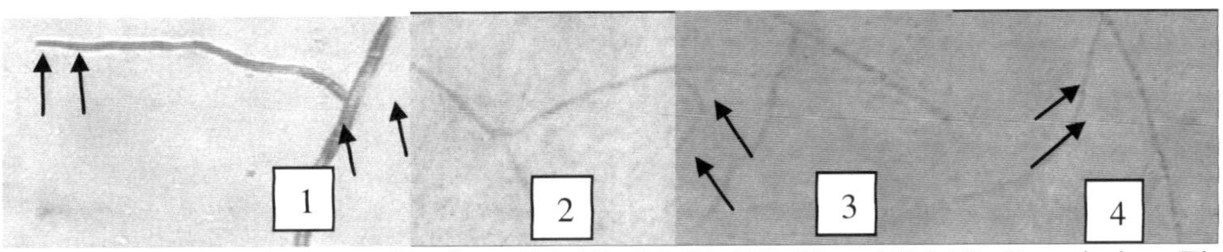

Figure 2. The morphology of isolates R1 (1), R2 (2), R3 (3) and R4 (4). Description: The arrow pointed the cell nucleus.

Figure 3 shows the propagation of culture in vitro obtained exsplant *D. macrophyllum* to be inoculated mycorrhizal *Rhizoctonia* to see the *peloton* in the orchid cortex. From figure 3, we can see the exsplant *D. macrophyllum* to able inoculated mycorrhizal *Rhizoctonia* is 8 weeks after seed planting (WASP), because the root of explant has occurred. Mycorrhizal *Rhizoctonia* is able to

IOP Conf. Series: Earth and Environmental Science **142** (2018) 012013 doi:10.1088/1755-1315/142/1/012013

symbiotize with orchid root tissue and form a hyphae coagulant that clumps on the cortical tissue of the roots. The coagulated hifa coil structure is called a *peloton* [9].

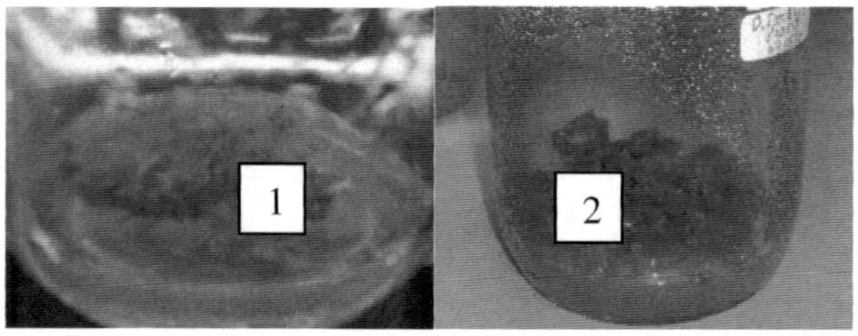

Figure 3. Explant *Dendrobium macrophyllum* inoculated by mycorrhizal *Rhizoctonia*.
Note: 1) on 4 weeks after seed planting (WASP): 2) on 8 WASP

Figure 4 shows the *pelotons* are intracellular hyphae in the cortex network of roots and usually only exist in a certain period before lysis [11] which can be seen the assosiation betwen mychorrhizal *Rhizoctonia* with exsplant *D. macrophyllum*.

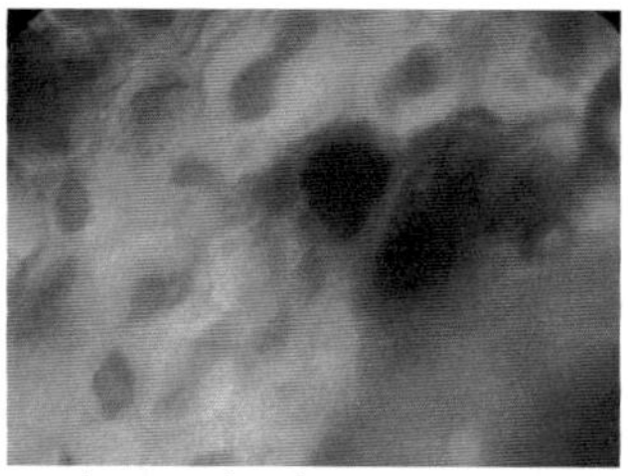

Figure 4. The structure of the *pelotons* in the root cortex *Dendrobium macrophyllum.*

Table 1 shows characteristics of various isolates of mycorrhizal *Rhizoctonia* isolated from Surakarta, Kopeng, Magelang, and Yogyakarta having differences in cell width and length. The isolations were done with 40 sample from each location.

Table 1. The characteristics mycorrhizal *Rhizoctonia* of *D. macrophyllum*

Mychorrhizal *Rhizoctonia*			Colony color	Sclerotium formation
	Cell width	Cell length		
Rl	5,3-9,1	43,0-128,0	Brownish white	+
R2	5,1-9,0	51,0-131,0	Light brown	+
R3	7,2-9,0	40,0-137,3	Light brown	+
R4	5,9-11,5	42,5-171,5	Light brown	+

Note: R1: Mycorrhizal *Rhizoctonia* isolated from Surakarta; R3: mycorrhizal *Rhizoctonia* isolated from Kopeng; R2: Mycorrhizal *Rhizoctonia* isolated from Magelang, and R4: mycorrhizal *Rhizoctonia* isolated from Yogyakarta.
+ = Sclerotium is formed.

Table 2 shows the isolate mycorrhizal *Rhizoctonia* isolated from Surakarta, Kopeng, Magelang, and Yogyakarta also had differences in the number of nucleotide within each cell.

Table 2. The nuclei number of hyphae cell of mycorrhizal *Rhizoctonia*

Rhizoctonia sp. (4 isolate)	The number cells based on the number of nucleotide					*Range*	Information
	1	2	3	4	5		
R1	3	15	7	5	0	2	binucleate
R3	3	16	4	7	0	2	binucleate
R2	7	14	5	5	0	2	binucleate
R4	4	15	8	3	0	2	binucleate

Note: R1: Mycorrhizal *Rhizoctonia* isolated from Surakarta; R3: mycorrhizal *Rhizoctonia* isolated from Kopeng; R2: Mycorrhizal *Rhizoctonia* isolated from Magelang, and R4: mycorrhizal *Rhizoctonia* isolated from Yogyakarta.

Colonies of mycorrhizal *Rhizoctonia* are different depending on their group (isolates grouping). The mycorrhizal *Rhizoctonia* isolates (R2, R3, and R4) obtained, were mostly brownish-brown, while the colonies of *Rhizoctonia* isolate R1 were white colored. [1] found that 9 isolates of mycorrhizal *Rhizoctonia* isolated from *Spathoglottis plicata* from various places in Thailand showed white colonies. Meanwhile Agustini *et al.* [2] found different things in the botanical garden of Cycloops Jayapura, that were 10 orchids. The exsplants of *D. macrophyllum* inoculated by mycorrhizal *Rhizoctonia* were 8 weeks after seed planting (WASP), because the root of explant has occurred. The *pelotons* structure is already formed in the root cortex part of *D. macrophyllum*. It is expected that the *peloton* structure will help to absorb the nutrients needed by *D. macrophyllum* from the surrounding environment.

Mycorrhizal *Rhizoctonia* has cell size ranges from 5.1μ to 11.5μ. Table 1 showed that each of the mycorrhizal *Rhizoctonia* isolates obtained from various places in Surakarta, Kopeng, Magelang, and Yogyakarta have a more uniform cell size. This is in accordance with research on mycorrhizal *Rhizoctonia* in orchids such as *Ceratobasidium*, *Thanatephorus* (Ceratobasidiales), *Sebacina* (Exidiales) and *Tulasnella* (Tulasnellales), in which the width of mycorrhizal *Rhizoctonia* cells can be more than 10 μm and grouped in *Tulasnella* [5].

Isolates Rl, R3, R2, and R4 are mostly hypha cells with average nuclei ranging (range) 2. According to Sneh *et al.* [9], isolates with a nucleotide number of 1-3 cells were binucleic groups, whereas isolates with more than 3 nuclei cells were multinucleated. From the opinion of Sneh *et al.* [9] isolates Rl, R3, R2, and R4 are binucleate. This is in accordance with previous research in orchid plants [1], [5], [10], [12], which concluded most of the isolates of *Rhizoctonia* sp. are mycorrhizal and belong to the binucleate group. Mycorrhizal isolates obtained colony color varied from white to black. Therefore, based on morphological observation of colloidal color of the mycorrhizal *Rhizoctonia* isolate from *D. macrophyllum* orchids, it can be said that colony color can not be used as a differentiator between each isolate.

4. Conclusion

The characterization mycorrhizal *Rhizoctonia* isolates of *D. macrophyllum* root from Kopeng, Magelang, Surakarta, and Yogyakarta did not different morphologically. Character similarity is in colony color, cell length and number of cores, while character differences are in cell width. The association results of *D. macrophyllum* with mycorrhizal *Rhizoctonia*, formed a *peloton* structure in the root cor

Acknowledgments

On this occasion, we are grateful to the Directorate General of Higher Education, whom funded the study in accordance with the Contract of Research Execution Number: 007/K6/KM/BASERISET/ 2016 dated May 4, 2016.

References

[1] Athipunyakom P, Manoch L 2008 Isolation and identification of mycorrhizal fungi from eleven terrestrial orchids *J. Plant Prot.* **15** 195-207

[2] Agustini V, Supeni S and Suharno 2009 Mycorrhiza association of terestrial orchids of cycloops nature reserve, jayapura *Biodiversitas* **10** 175 – 180

[3] Barnett H L, Hunter B B 1972 *Illustrated Genera of Imperfect fungi*, 3rd edition (Minneapolos, Minnesota: Burgess Publishing Company) 241

[4] Dressler R L 1990 *The orchids, natural history and classification* (Massachusetts, Cambridge: Harvard University Press) 332

[5] Garcia V G, Portal Onco M A and Rubio Susan V 2006 Review Biology and systematics of the form genus rhizoctonia *Span. J. Agric. Res.* **4** 55-79

[6] Kutscha, G 1972 The suitability of certain strains or studying lignification in balsam fir, Abies balsamina (L.) MILL Technical Bulletin 53 Life Sciences and Agricultural Experiment Station

[7] Otero J T, Ackerman J D and Bayman P 2002 Diversity and host specificity of endophytic Rhizoctonia-like fungi from tropical orchids *Am. J. Bot.* **89** 1852 – 1858

[8] Semiarti E, Indrianto A, Purwantoro A, Isminingsih S, Suseno N, Ishikawa T, Yoshioka Y, Machida Y and Machida C 2007 Agrobacterium-mediated transformation of the wild orchid species Phalaenopsis amabilis *Plant Biotechnol.* **24** 255-272

[9] Sneh B, Burpee L and Ogoshi A 1991 *Identification of Rhizoctonia Species* St. Paul MN: APS Press

[10] Soelistijono 2013 Utilization of rhizoctonia mycorrhiza for root rot disease control in orchid plant spathoglottis plicata http://etd.ugm.ac.id/index.php?mod=penelitian_detail&sub=PenelitianDetail&act=view&typ=html&buku_id=59657&obyek_4

[11] Taylor D L, Bruns T D, Leake J R and Read D J 2002 *Mycorrhiza Specifity and Function in Myco-heterotrophic Plants* (Berlin: Springer-Verlag) 413

IOP Conf. Series: Earth and Environmental Science **142** (2018) 012014 doi:10.1088/1755-1315/142/1/012014

Dynamics of N-NH_4^+, N-NO_3^-, and total soil nitrogen in paddy field with azolla and biochar

W S Dewi[1], G I Wahyuningsih[2], J Syamsiyah[1] and Mujiyo[1]
[1] Lecturer staff of Agrotechnology, Faculty of Agriculture, Universitas Sebelas Maret, Jl. Ir. Sutami 36A, Surakarta, 57126, Indonesia
[2] Undergraduate student of Agrotechnology, Faculty of Agriculture, Universitas Sebelas Maret, Jl. Ir. Sutami 36A, Surakarta, 57126, Indonesia

E-mail: wsdewi2000@gmail.com

Abstract. Nitrogen (N) is one of macronutrients which is dynamic in the soil and becomes constraint factor for rice crops. The addition of nitrogen fertilizers and its absorption in paddy field causes the dynamics of nitrogen, thus declines of N absorption efficiency. The aim of this research is to know influence Azolla, biochar and different varieties application on N-NH_4^+, N-NO_3^-, and total soil N in paddy field. This research was conducted in a screen house located in Jumantono Laboratory, Faculty of Agriculture, Universitas Sebelas Maret (UNS) with altitude 170 m asl from April to June 2016. Treatment factors that were examined consisted of azolla (0 and 10 tons/ha), biochar (0 and 2 tons/ha), and rice varieties (Cisadane, Memberamo, Ciherang, IR64). The results of this research showed that there was no interaction between azolla, biochar and varieties. Nevertheless, azolla treatment with dose of 10 tons/ha increased soil NH_4^+ content (41 days after planting, DAP) by 13.4% but tend to decrease at 70 and 90 DAP. Biochar treatment with dose of 2 ton/ha increases NO_3^- soil content (70 DAP) by 1.7% but decreases total N soil by 5.8% (41 DAP) and 4.7% (90 DAP). Different rice varieties generated different soil NH_4^+ content (41 DAP) and rice root volume. Cisadane variety can increase soil NH_4^+ content (41 DAP) by 52.08% and root volume by 51.80% (90 DAP) compared with Ciherang variety. Organic rice field management with azolla and biochar affects the availability of N in the soil and increase N absorption efficiency through its role in increasing rice root volume.

1. Introduction

Rice is a staple food and economic resource for Indonesian and becomes strategic value in achieving National Food Security. According to Indonesian Central Bureau of Statistics (BPS) [1], rice production decreased by 0.43 million tons or 0.61%. This decline is caused by the lack of nutrient management system, particularly the management of N, P, K nutrients which are unbalanced in intensified paddy field [2].

Nitrogen is one of the most important nutrients, as it improves grain quality and yield by increasing the number of tillers, leaf area development, grain formation, grain filling and protein synthesis [3]. In general, nitrogen in the soil is mobile and dynamic, it can easily change from one form to another form such as from NH_4^+ to NO_3^-, NO, N_2O and N_2 [4] leads to low absorption efficiency and total absorption of nitrogen in rice crops.

The use of organic substances such as azolla and biochar may provide nitrogen for rice crops [5] and improve environmental conditions as it increases soil CEC [6]. Azolla is one of the most suitable sources of nitrogen developed in paddy fields, since it is able to reduce nitrogen from the atmosphere to ammonia [7], in the amount of 100-170 kg N ha^{-1} per year [8]. The high content of nitrogen produced by azolla biomass must be balanced with maximum management to optimized nitrogen utilization, for example by using biochar and high yielding varieties.

Nitrogen availability in the soil is influenced by several factors such as the addition of nitrogen fertilizer, nutrient absorption and nitrogen cycle occurring in the soil. Because of the complexity of the nitrogen transformation, it is necessary to investigate the dynamics of nitrogen on paddy fields. This study aims to examine the effect of azolla, biochar and varieties on the dynamics of NH_4^+, NO_3^- and total nitrogen soil in paddy soil.

2. Methods

This research was conducted in a screen house located in Jumantono Laboratory, Faculty of Agriculture, Universitas Sebelas Maret (UNS) with altitude 170 m asl from April to June 2016. The characteristics of soil used as planting medium are presented on Table 1. This soil is an organic soil originated from Kebonagung, Sidoharjo, Wonogiri Regency.

Table 1. Soil Chemical Properties before Treatment

Parameter	Value
Organic C	4.19%
NH_4^+	0.004 ppm
NO_3^-	0.02 ppm
Total N	0.84%
pH H_2O	6.58
K_2O	2.31 cmol(+)/kg
P_2O_5 (Olsen)	12.33 ppm
CEC	44.51 cmol(+)/kg

This research used 0 and 10 tons/ha azolla as green manure and 0 and 2 tons/ha biochar as well as four different rice varieties (Cisadane, Memberamo, Ciherang and IR64). Half the dose of azolla and the whole dose of biochar were applied before planting by being immersed. Five days later half the remaining dose of azolla was applied by being spread on the surface. After azolla fully grown, 60% of azolla was embedded into the soil relate to the second weeding activities of azolla. Meanwhile, the rest of azolla (40%) was allowed to continue to grow until the rice crop was harvested.

This experiment used completely randomized block design (CRBD) and the data were analyzed by using ANOVA with F test at the level of 95% and then followed by DMR test at the level of 95% as well as correlation test. Various parameters were measured including NH_4^+ and NO_3^- soil (determining ammonium method with KCl 1 N), total nitrogen soil (Kjedahl) and root volume of paddy.

3. Results and Discussion

3.1 Effect of Azolla on N-NH_4^+, N-NO_3^-, Total N Soil and Rice Root Volume

The application of azolla with dose of 10 tons/ha increased ($p<0.01$) soil NH_4^+ content by 13.4% (41 days after planting, DAP) and rice root volume (41, 70 and 90 DAP; Table 2). It is in line with Mujiyo et al. [5] in which the use of azolla fertilizer increased NH_4^+ content in the soil, since Azolla fertilizers mineralized and transformed into simple components such as NH_4^+ and NO_3^-. Root volume on 41, 70, and 90 DAP also increased by 6.3, 8.3 and 13.3%, respectively (Table 2), compare to those without azolla application. According to Fosu-Mensah et al. [9] azolla increased the whole growth of rice plants as indicated by plant height, total dry weight, and grain yield.

Elements derived from the high N content of azolla will be released into the soil after azolla decomposition [10], followed by increasing total N of soil [11]. Nevertheless, in this research NO_3^- and total nitrogen soil was not affected ($p>0.05$) by azolla application. It is assumed that NH_4^+ released by azolla has not been nitrified to NO_3^- and NO_2^- but only increases NH_4^+ content. Furthermore, the increase in NH_4^+ did not increase the total Nitrogen of soil since total nitrogen consists of organic and inorganic N, by which amount of organic N in the soil is more than 90% [11]. According to Suntoro et al. [12] azolla as a green manure has not been able to increase the total N of soil, because azolla must be decomposed first to provide nitrogen. This process occurs gradually and takes quite a long time to completely decomposed.

Table 2. Effect of Azolla Application on NH_4^+ Soil Content (41 DAP) and Root Volume (41, 70, 90 DAP)

Dose of Azolla (tons/ha)	NH_4^+ Soil Content (ppm)	Root Volume (cm^3) 41 DAP	70 DAP	90 DAP
0	0.575[a]	0.426[a]	2.457[a]	8.054[a]
10	0.652[b]	0.453[b]	2.660[b]	9.126[b]

The number followed by the same letter in the same column shows no significant difference in the DMRT test of 5%

3.2 Effect of Biochar on N-NH_4^+, N-NO_3^-, Total N Soil and Rice Root Volume

The application of biochar with dose of 2 tons/ha enhanced ($p<0.05$) soil NO_3^- content by 1.7% (70 DAP) and total soil nitrogen (41 and 90 DAP; Table 3). The small magnitude in the increase of NO_3^- content was due to the first season of application biochar has a low adsorption capacity compared with long-standing biochar, as has been explained by Cheng et al. [13]. Based on several studies, Clough et al. [14] summerized that biochar is able to absorb NO_3^- longer in the soil, so it is not washed away and can be absorbed by plants. Furthermore, Dempster et al. [15] explained that biochar application increased water holding capacity led to reduce nutrients leaching in the soil, including NO_3^-. However, application of 2 tons/ha of biochar treatment generated lower total nitrogen content than without biochar treatment (Table 3). It is assumed that application of biochar in the soil can improve the physical and chemical properties of the soil, thus, supports the plant roots to absorb nitrogen. In line with this research, Santoso [16] explained that N total in the soil decreased due to absorption by the rice plants.

Table 3. Effect of Biochar Application on NO_3^- Soil Content (70 DAP) and Total N Soil Content (41 and 90 DAP)

Dose of Biochar (ton/ha)	NO_3^- Soil Content (ppm)	Total N Soil Content (%) 41 DAP	90 DAP
0	0.0058[a]	0.580[a]	1.034[a]
2	0.0059[b]	0.548[b]	0.988[b]

The number followed by the same letter in the same column shows no significant difference in the DMRT test of 5%

3.3 Effect of Different Rice Varieties on N-NH_4^+, N-NO_3^-, Total N Soil and Rice Root Volume

The application of different rice varieties resulted in different ($p<0.01$) rice root volume (41, 70 and 90 DAP). Table 4 shows that two varieties (Cisadane and Memberamo) generated the higher root volume than other varieties. The largest increase of rice root volume was found for Memberamo compared to Ciherang variety on 41 and 70 DAP by 80.0% and 52.0%, respectively. While on 90 DAP, the largest increase was found in Cisadane variety 51.8% compared to Ciherang variety. This result indicates that the use of different rice varieties can provide a varied growth response. According to Adisarwanti [17] the growth response of rice plants varies depending on genetic factors and growing environmental conditions. The use of different varieties also increased NH_4^+ soil content (41 DAP) with the highest increase in Cisadane variety which is 52.08% compared with Ciherang variety and 14.77% compared with Memberamo variety and 35.94% compared to IR64 variety.

Table 4. Effect of Varieties Application on NH_4^+ Soil Content (41 DAP) and Root Volume (41, 70, 90 DAP)

Rice Varieties	NH_4^+ Soil Content (ppm)	Root Volume (cm^3) 41 DAP	70 DAP	90 DAP
Cisadane	0.730^c	0.730^c	2.826^c	10.217^c
Memberamo	0.636^b	0.636^d	2.904^c	8.898^b
Ciherang	0.480^a	0.480^a	1.909^a	6.732^a
IR64	0.537^a	0.537^b	2.231^b	7.521^a

The number followed by the same letter in the same column shows no significant difference in the DMRT test of 5%

Nitrogen availability is indispensable for the root growth and development, and the preferred form of Nitrogen for rice plants is NH_4^+ [18]. A good plant's root growth affects the availability of nutrients in the soil including NH_4^+. Santoso [16] explains that the reaction between root plant and soil will affect the environmental conditions of the rhizosphere by releasing CO_2, H^+, HCO_3^- and other organic compounds, which affect the availability of nutrients. Furthermore, Mantelin and Touraine [19] explain that the condition of rhizosphere is influenced by root activity that becomes habitat for soil microorganisms because of organic nutrients availability are beneficial to microorganisms' growth. The microbial colonization that occurs will affect the biology of roots and plants.

3.4 Effect of Azolla, Biochar and Different Rice Varieties on the Dynamics of N-NH_4^+, N-NO_3^- and Total N Soil

Azolla treatment with dose of 10 tons/ha significantly affect the dynamics of N-NH_4^+ in the soil (Fig. 1), whereas 2 tons/ha biochar treatment significantly affected the dynamics of N-NO_3^- (Fig. 2) and total N soil (Fig. 3). The decrease in NH_4^+ soil content from 41 DAP to 90 DAP (Fig. 1) was caused by an increase in the rice root volume (Fig. 4), and supported by the close relationship between the decrease in NH_4^+ soil content and the increase of rice root volume (Fig. 5). This result is in line with Sipahutar and Kasno [20] that the decrease of NH_4^+ content in the soil is caused by faster absorption of NH_4^+ ions by plants and partly lost or transformed into a form which is not available to plants.

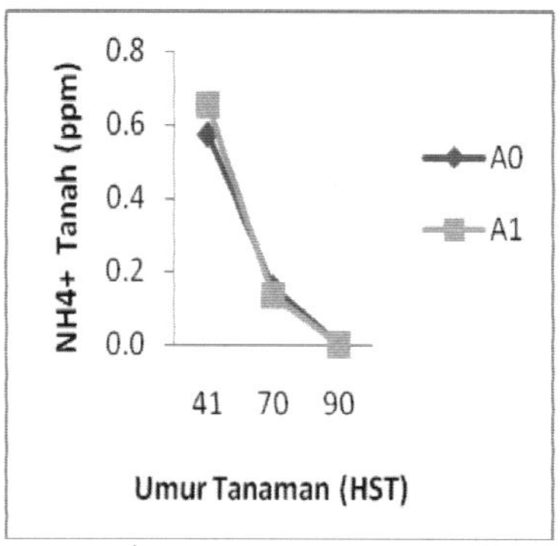

Figure 1. NH_4^+ soil content 41, 70 and 90 DAP

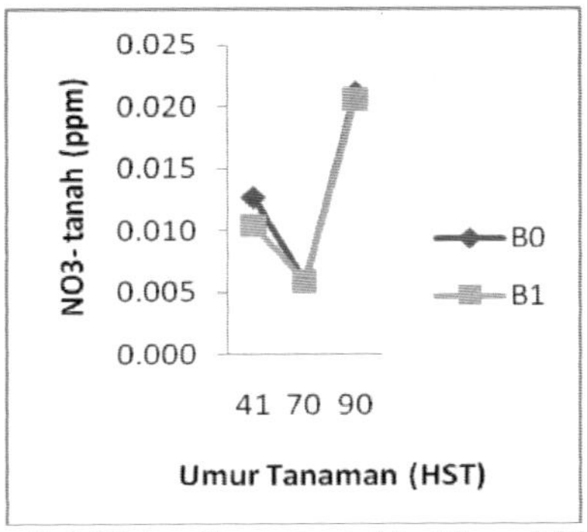

Figure 2. NO_3^- soil content 41, 70 and 90 DAP

The increase of NO_3^- soil content from 70 DAP to 90 DAP (Fig. 2) is inversely proportional to the NH_4^+ content (Fig. 1). This is in line with Purwanto [21] who observed that the decrease of NH_4^+ in soil increased NO_3^- concentration in the soil due to nitrification. The decreased of total N soil content at 70 DAP and increased at 90 DAP (Fig. 3) was influenced by the rapid process of mineralization and immobilization of nitrogen. The increasing of total N is due to inorganic N such as NO_3^- and NH_4^+ used by soil microorganisms for the decomposition process of crop residue which leads to increase N-organic soil [22].

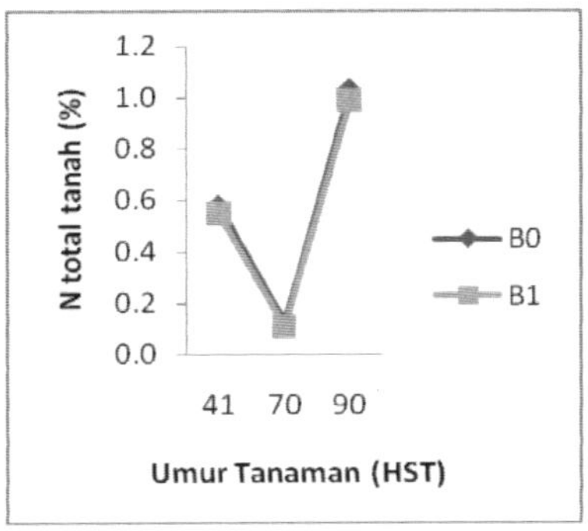

Figure 3. Total N soil content 41, 70 and 90 DAP

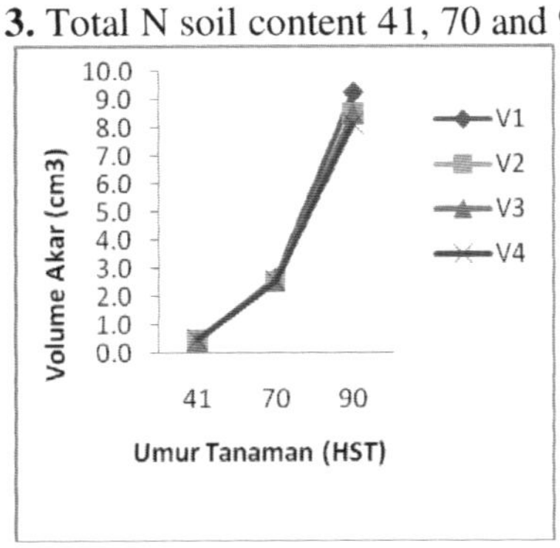

Figure 4. Rice root volume 41, 70 and 90 DAP

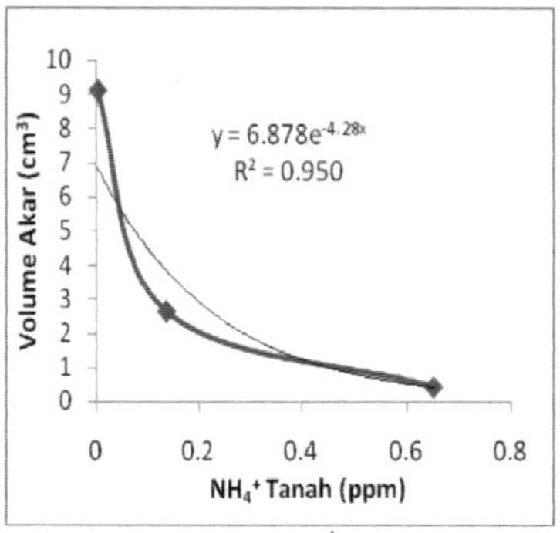

Figure 5. Correlation graph of NH_4^+ soil content and root volume

4. Conclusion

Application of azolla with dose of 10 tons/ha enhanced the soil NH_4^+ content at 41 DAP but tends to decrease at 70 and 90 DAP. Azolla increased rice root volume at 41, 70 and 90 DAP. Biochar with dose of 2 tons/ha increased NO_3^- soil content at 70 DAP but decrease N total of soil content at 41 DAP by and 90 DAP. The use of different varieties caused different NH_4^+ soil content and roots volume. Cisadane variety generated higher NH_4^+ soil content at 41 DAP and root volume at 90 DAP compared with Ciherang variety. Organic rice field management with azolla and biochar affect the availability of N in the soil and increased N absorption efficiency through its role in increasing the root volume of rice crops.

Acknowledgement

This research was supported by INSENTIF RISET SINAS, The Ministry of Research, Technology and Higher Education, Republic of Indonesia, 2016. We also would like to thank "TIM PENELITIAN PADI 2016" (Rizka and Krista) for their participation on paper preparation.

References

[1] Badan Pusat Statistik 2015 *https://www.bps.go.id/Brs/view/id/1157*
[2] Sirappa M P, Rieuwpassa A J, Waas E D 2007 *J. Pengkajian dan Pengembangan Teknologi Pertanian* **10** 48-56
[3] Patty E S, Kaya E, Silahooy C H 2013 *Agrologia* **2** 51-58
[4] Winarso S 2005 *Kesuburan Tanah, Dasar Kesehatan dan Kualitas Tanah* (Yogyakarta: Gava Media)
[5] Mujiyo, Sunarminto B H, Hanudin E and Widada J 2011 *Agronomika* **11** 167-178
[6] Gani A 2010 Multiguna Arang Hayati Biochar *Sinar Tani*
[7] Kuncarawati I L, Husen S and Rukhiyat M 2005 *J. Dedikasi* 310-16
[8] Awodun M A 2008 *World J.Agric. Sci.* **4** 157-160
[9] Fosu-Mensah B Y, Vlek P L G, Manske G, and Mensah M 2015 *J. Agric. Sci.* **7** 118-130
[10] Syamsiyah J, Sunarminto B H, Mujiyo 2016 Sains Tanah*J. Soil Sci. Agroclimat.* **13** 68-73
[11] Yadav R K, Abraham G, Singh Y V, Singh P K 2014 *Proc. of Indian National Science Academy*, India, pp. 301-316
[12] Suntoro, Syamsiyah J, Tiyanto F A A 2013 Sains Tanah*J. Soil Sci. Agroclimat.* **10** 45-54
[13] Cheng C H, Lehmann J, Engelhard M 2008 *Geochim Cosmochim. Acta* **72** 1598-1610
[14] Clough T J, Leo M C, Claudia K, Cristoph M 2013 *Agronomy* **3** 275-293
[15] Dempster D N, Jones D L, Murphy D M 2012 *Soil. Res.* **50** 216–221
[16] Santoso S 2011 *J. Agrivigor* **10** 319-330.
[17] Adisarwanti T 2006 *Budidaya Kedelai dengan Pemupukan yang Efektif dan Pengoptimalan Peran Bintil Akar* (Jakarta: Penebar Swadaya)
[18] Makarim A K, Suhartatik E 2009 Morfologi dan Fisiologi Tanaman Padi, Balai Besar Penelitian Tanaman Padi *http://www.litbang.pertanian.go.id/special/padi/bbpadi_2009_itkp_11.pdf*

[19] Mantelin S and Touraine B 2004 *J. Exp. Bot.* **55** 27-34
[20] Sipahutar I A dan Kasno A 2009 Dinamika Hara N Pada Lahan Sawah Intensifikasi Bermineral Liat Dominan 2:1, Balai Penelitian Tanah
[21] Purwanto, Handayanto E, Suparyogo D, Hairiah K 2017 *Pelita Perkebunan* **23** 35-56
[22] Yeasmin S, Islam A K M, Islam A K M A 2012 *J. Agric. Tech.* **8** 775-793

IOP Conf. Series: Earth and Environmental Science **142** (2018) 012015 doi:10.1088/1755-1315/142/1/012015

Study of buffer substrate and Arenga wood fiber size on hydroponic Kailan (*Brassica alboglabra*)

D Harjoko, M D Anggraheny and R B Arniputri
Agrotechnology Department, Faculty of Agriculture, Universitas Sebelas Maret, Surakarta, Indonesia

E-mail: dwifpuns@yahoo.com

Abstract. Kailan is a kind of vegetable that has high economic value, however its prospect is not well developed. One of obstacles in Kailan cultivation is the limitation of fertile soil, that can be solved by using hydroponic substrate. Considering its amount and potential, the fiber waste of Arenga wood was selected as substrate candidate. For that, this research aims to study the growth and yield of Kailan with different soaking treatment using buffer solution and size of Arenga wood fiber in the hydroponic substrate. Research was conducted at Green House Laboratory, Faculty of Agriculture Sebelas Maret University Surakarta from February to May 2017. The treatments were soaking buffer solution with EC 1.2 $mScm^{-1}$; 1.4 $mScm^{-1}$; and 1.6 $mScm^{-1}$ and the size of Arenga fiber <1 cm, 1-2 cm and 2-3 cm. In this experiment, sand media was used as control. Result show that, soaking in 1.6 $mScm^{-1}$ EC buffer solution with Arenga fiber size lower than 3 cm gives higher root volume compared to other treatments combination.

1. Introduction

Public awareness of health importance of and the fulfillment daily nutrition led to high quality vegetable and fruit production demand. Kailan belongs to a new vegetable and has not been developed although it has economic value and high prospect. The popularity and consumption of Brassica vegetables is increasing due to its nutritional value. Phenolics and other bioactive compounds are distributed differently depending on Brassica [1]. This vegetable also contains anticarcinogenic and antioxidant that are beneficial to the health of the body [2].

The cultivation problem of Kailan is lack of fertile agriculture soil. One effort to solved this, is by improving the culture technique using Hydroponics that land-saving crops. Some of of hydroponics advantages is the efficiency of water consumption and nutrition [3,4] minimize pesticide use and improve yields [5]

Most of the palm trees in the tropics are exploited from the wild plan and still few are cultivated [6]. Sugar is a widely used multipurpose tree species scattered in several regions of Indonesia [7]. Almost all physical parts of Arenga can be used from stems, roots, and leaves, the Arenga industry produces waste that has negative potential to cause problems if not handled properly. Arenga wood fiber waste has large amount and its utilization has not been maximized. Therefore, fiber from Arenga wood can be used as growing media for hydroponic substrate.

Growing media is important in hydroponic substrate cultivation activities. Each media has its own physical and chemical characteristic. It needs special treatment on organic media so that can improve the quality of planting media, one of them is by soaking in buffer solution (buffered substrate) [8] soaking buffer on coco coir medium gives a real effect on the number of flowers and the length of stems Gerbera plant (*Gerbera jamesonii*). Therefore, it is necessary to study the growth and yield of Kailan which is cultivated with soaking treatment in buffer solution and the size of Arenga wood fiber as growing media on hydroponic substrate.

2. Methods

This research was conducted at the Green House Laboratory Faculty of Agriculture Sebelas Maret University Surakarta in February to May 2017. The experimental study was completely randomized design (CRD). The treatments was soaking in buffer solution with EC 1.2 $mScm^{-1}$; 1.4 $mScm^{-1}$; and 1,6 $mScm^{-1}$ and the size of Arenga fiber at <1 cm, 1-2 cm, 2-3 cm and sand media used as control.

The study steps are laboratory media analysis consist of physical analysis (bulk density, particle density, and water holding capacity) and chemical analysis (N, P, K, Mg, Ca, Fe and Mn), followed with implementation of research include seeding, media preparation, planting and harvesting.

The variables observed were fresh weight, consumption weight, stem diameter, plant height, leaf number, leaf area, chlorophyll, root length, root weight, root volume, and biomass. Resulted data were analyzed using ANOVA (*Analysis of Variance*; α =5%), any significant differences will be followed with orthogonal contrast test.

3. Result and discussion

3.1. Weight

Plant weight is one of parameters in the observation variable [9] which is necessary to determine biomass content. The result of variance analysis showed that all treatments did not give real effect to all weight variable of KKailan, that are total weight and consumption weight. Table 1 shows the average of yield on total weight and consumption weight of Kailan, wherevalue are 8.09-11. g and 7.12-10.76 g respectively.

Environmental factors were suspected to be one of factors affecting plant growth especially fluctuating temperatures in the Greenhouse. The average daily temperature is 30°C and the daily average humidity is 66.2%. The ideal temperature and humidity are the main concerns for hydroponic cultivation of Kailan because it is very important on plant growth. The air temperature affected tothe various physiological processes that take place in plant organs as they relate to the growth and production of plants [10]

The soaking buffer solution treatment on Arenga wood fiber media generally gives lower yield than without soaking in the buffer. The soaking buffer on Arenga wood fiber media causes the accumulation of Potassium (K). Known in the Arenga wood fiber has contained K elements of 17000 ppm. The unbalanced composition resulted in stunted growth of generalized KKailan plants. This suggests that all buffer soaking treatments and sizes have not yielded optimum results on total weight and consumption weight.

3.2. Stem

Growth is an important process in plant life. One of the characteristics of growth is the increase in plant stem size due to cell division and enlargement. Observation parameters in bar variable that is stem diameter and plant height presented in Table 2.

Measurement of stem diameter was done at the end of observation at 6 WAP. Each treatment showed average results of diameter Kailan not much different. The data presented shows the stem diameter range of 0.39-0.57 cm and plant height range 10.00-12.53 cm. The growth of diameter is influenced by the activity of photosynthesis, this growth takes place when the photosynthesis results such as respiration, leaf replacement, the growth of roots and height has been fulfilled [11].

Plant height is one of the parameters measured in this study. Plant height is calculated from the base of the stem to the point of growing. Based on High growth chart of Kailan (Figure 1) has a difference every week and continues to grow to 6 WAP. High growth gap of Kailan is less visible. The graph of treatment A (control treatment) showed slower increase in plant height compared to other treatments. Graph of treatment J (size 2-3 cm without soaking buffer solution) showed the most rapid increase of plant height.

Table 1. Average observation of Kailan weight variables

Fiber size (cm)	EC ($mScm^{-1}$)	Observation Variable (g)	
		Total weight	Consumption weight
<1	0	11.65	10.03
	1.2	10.25	9.01
	1.4	9.59	8.44
	1.6	10.32	9.33
1-2	0	8.09	7.12
	1.2	9.55	8.33
	1.4	9.77	7.95
	1.6	9.66	8.20
2-3	0	10.28	10.28
	1.2	8.63	7.18
	1.4	11.13	9.69
	1.6	9.39	7.89
Control		11.93	10.76

Information : EC (*Electrical Conductivity*)

Table 2. Average observation of Kailan stem variables

Fiber size (cm)	EC ($mScm^{-1}$)	Observation Variable	
		Stem diameter (cm)	Plant height 6 WAP (cm)
<1	0	0.50	11.10
	1.2	0.46	11.30
	1.4	0.45	10.87
	1.6	0.49	10.33
1-2	0	0.39	11.83
	1.2	0.53	10.97
	1.4	0.46	11.17
	1.6	0.48	11.03
2-3	0	0.52	12.53
	1.2	0.45	10.60
	1.4	0.49	12.11
	1.6	0,.47	11.27
Control		0.57	10.00

Information : EC (*Electrical Conductivity*), WAP (Week After Planting)

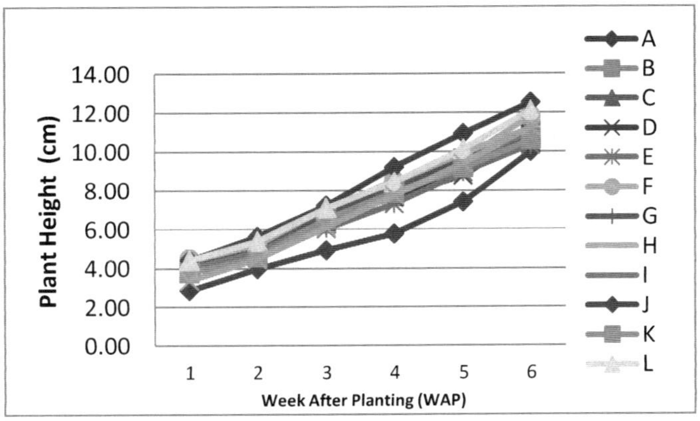

Figure 1. Height growth chart of Kailan

3.3. Leaves

The main organ of the plant where photosynthesis takes place is the leaves. Plants catch light using a pigment called chlorophyll that gives the green color to the plant [12]. The result of variance analysis showed that all treatments did not give any significant effect to all leaf variables is leaf number, leaf area and chlorophyll. The average observation result of Kailan leaf variables is presented in Table 3.

Table 3. Average observation of Kailan leaves variables

Fiber size (cm)	EC ($mScm^{-1}$)	Observation Variable		
		Leaf number (sheet)	Leaf area (cm^2)	Chlorophyll
<1	0	6.87	205.78	45.53
	1.2	6.80	211.89	47.03
	1.4	6.33	187.89	48.33
	1.6	6.93	222.00	48.93
1-2	0	6.73	180.78	45.17
	1.2	6.60	181.89	46.90
	1.4	7.33	207.56	45.03
	1.6	7.07	189.78	46.13
2-3	0	7.07	232.44	46.40
	1.2	6.13	157.56	47.90
	1.4	7.07	218.56	43.20
	1.6	7.07	184.33	47.37
Control		7.87	246.11	50.53

Information : EC (*Electrical Conductivity*)

Based on Table 3, it is known that the average range of Kailan leaf number is between 6.13-7.87. Average range of leaf area between 180.78 cm^2-246.11 cm^2. While the average range of chlorophyll is between 45.03-50.53. Known in the Arenga wood fiber media has contained K elements of 17000 ppm. [13] Mg deficiency may occur when NH^{4+}, K^+, or Ca^{++} concentrations in high rooting medium. The cause is the weakest Mg^{++} in the competition among the cations [14] adds that Magnesium in plants is a central atom in the formation of chlorophyll molecules, so it is very important in relation to photosynthesis.

The result showed a decrease of Magnesium (Mg) element as the value of EC buffer increases. The unbalanced composition resulted in the inhibition of chlorophyll growth in general Kailan plants.

Symptoms of magnesium nutrient deficiency (Mg) seen on the appearance of Kailan leaves. The leaves appearing yellow spots and in the bones of the leaves remain green. Magnesium (Mg) deficient plants show yellowing leaves while venous leaves remain green [14].

3.4. Roots

Root is an important part of a plant, good media greatly influence to the success of new roots growth and formation. Based on TTable 4 it is known that the average range of root length is 9.73-14,5 cm and the average of Kailan root weight result shows 1,11-1,70 g. In general, the treatment of soaking buffer on Arenga wood fiber media gives better result than sand media (control).

The treatment of soaking buffer on the Arenga wood fiber media causes the accumulation of Phosphate (P) and Calcium (Ca). It is shown in the results of chemical analysis of the content of Arenga wood fiber media, soaking buffer tends to increase the elements of P and Ca. Elements of P and Ca according to the previous study [13] play a role in the growing point of root formation.

Table 4. Average observation of Kailan root variables

Fiber size (cm)	EC ($mScm^{-1}$)	Observation Variable		
		Root length (cm)	Root weight (g)	Root volume (mL)
<1	0	12.67	1.57	0.49
	1.2	11.77	1.67	0.44
	1.4	12.67	1.27	0.46
	1.6	14.50	1.76	0.51
1-2	0	13.30	1.19	0.49
	1.2	12.50	1.26	0.40
	1.4	12.63	1.39	0.49
	1.6	12.73	1.43	0.65
2-3	0	13.27	1.47	0.43
	1.2	10.83	1.11	0.46
	1.4	13.13	1.45	0.43
	1.6	12.87	1.70	0.70
Control		9.73	1.13	0.28

Information : EC (*Electrical Conductivity*)

Distribution of roots in the substrate can be affected by the particle size distribution of the substrate [16]. The result of physical analysis on Arenga wood fiber showed that the size treatment caused different values of bulk density, particle density and water holding capacity. The result of physical analysis showed sand medium having bulk density 1,54 g mL^{-1}. The average length of the control treatment root with the sand medium showed the lowest yield. The result of bulk density from low to high is shown consecutively on the medium of aren size 2-3 cm, 1-2 cm size, aren size <1 cm that is with value 0,06 g mL^{-1}; 0.07 g mL^{-1}; and 0.08 g mL^{-1}.

According to [17] by knowing the weight of bulk density and particle density it can be calculated the number of total pores of soil. Particle density sand media showed 2.10 g mL^{-1}; Arenga wood fiber <1 cm 0,50 g mL^{-1}; Size 1-2 cm 0.45 g mL^{-1}; And measures 2-3 cm 0.38 g mL^{-1}. This shows that the higher value of bulk density, the higher value of particle density shown in the sand media. Size of Arenga wood fiber affects the physical character which is indicated by the increasing value of bulk density and particle density as the smaller size of Arenga wood fiber.

The result of variance analysis showed that the treatment give a real effect to the root volume. The results are significantdifferent ($P<0.05$) from this root volume and then continued orthogonal contrast test (Table 5).

There are 12 comparisons presented in Table 5. Comparison based on fiber size grouping and EC buffer. The results show that there is no significant different on the same fiber size groupings and different EC buffers. These results indicate that the treatment of Arenga wood fiber size <1 cm, 1-2 cm and 2-3 cm did not give differences in yield to root volume growth. The difference of physical analysis result on very small palm fiber media causes overall size treatment does not give different result. Based on these results it can be seen that the larger the size of Arenga wood fiber does not affect the growth of Kailan root volume.

The result of orthogonal contrast test on Kailan root volume (Table 5) shows a significant difference in DHL vs EIM and BFJ vs EIM ratio and very significant difference in 3 comparison, A vs Others, EIM vs CDGHKL and CGK vs EIM. Comparison of A vs others compares control (sand media) with all treatments (palm fiber media). The results showed that treatment A (control) give the lowest yield on growth of root volume compared with all treatment using Arenga wood fiber media. This suggests that Kailan planting using Arenga wood fiber media gives better influence from sand media on root volume growth.

Table 5. Test of orthogonal contrast treatment to root volume

Comparison	Information
A vs Others	**
BFJ vs CDEGHIKLM	-
CGK vs DHL	-
C vs GK	-
D vs HL	-
E vs IM	-
EIM vs CDGHKL	**
BFJ vs CGK	-
DHL vs EIM	*
DHL vs CEGIKM	-
CGK vs EIM	**
BFJ vs EIM	*

Information :

*	:	significantly different (p>5%)
**	:	very significantly different (p>1%)
-	:	not significantly different
A vs Others	:	Control compared to all treatments
BFJ vs CDEGHIKLM	:	Arenga wood fiber size <1 cm, 1-2 cm and 2-3 cm without soaking buffer solution compared to B + soaking buffer solution EC 1.2 $mScm^{-1}$, 1.4 $mScm^{-1}$, 1.6 $mScm^{-1}$ and F + soaking buffer solution EC 1.2; 1,4; 1,6 and J + soaking buffer solution of EC 1,2 $mScm^{-1}$; 1,4 $mScm^{-1}$; 1,6 $mScm^{-1}$
CGK vs DHL	:	B + F; J + Soaking buffer solution EC 1.2 $mScm^{-1}$ compared to B; F; J + Soaking buffer solution EC 1.4 $mScm^{-1}$
C vs GK	:	B + Soaking buffer solution EC 1.2 compared with F + Soaking buffer solution EC 1.2 $mScm^{-1}$; J + Soaking buffer solution EC 1.2 $mScm^{-1}$
D vs HL	:	B + Soaking buffer solution EC 1.4 compared with F + Soaking buffer solution EC 1.4 $mScm^{-1}$; J + Soaking buffer solution EC 1.4 $mScm^{-1}$
E vs IM	:	B + Soaking buffer solution EC 1.6 $mScm^{-1}$compared with F + Soaking buffer solution EC 1.6 $mScm^{-1}$; J + Soaking buffer solution EC 1.6 $mScm^{-1}$
EIM vs CDGHKL	:	B; F; J + Soaking buffer solution EC 1.6 $mScm^{-1}$compared with B; F; J + Soaking buffer solution EC 1,2 $mScm^{-1}$and 1,4 $mScm^{-1}$
BFJ vs CGK	:	Arenga wood fiber size <1 cm, 1-2 cm and 2-3 cm without soaking buffer solution compared with B; F; J + soaking buffer solution EC 1.2 $mScm^{-1}$
DHL vs EIM	:	B; F; J + Soaking buffer solution EC 1.4 $mScm^{-1}$ compared with B; F; J + Soaking buffer solution EC 1.6 $mScm^{-1}$
DHL vs CEGIKM	:	B; F; J + Soaking buffer solution EC 1.4 compared with B Soaking buffer solution EC 1,2 $mScm^{-1}$and 1,6 $mScm^{-1}$; F + Soaking EC buffer solution 1,2 $mScm^{-1}$and 1,6 $mScm^{-1}$; J Soaking buffer solution EC 1,2 $mScm^{-1}$and 1,6 $mScm^{-1}$
CGK vs EIM	:	B + F; J + Soaking buffer solution EC 1.2 $mScm^{-1}$compared to B; F; J + Soaking buffer solution EC 1.6 $mScm^{-1}$
BFJ vs EIM	:	Arenga wood fiber size <1 cm, 1-2 cm and 2-3 cm without soaking buffer solution compared with B; F; J + Soaking buffer solution EC 1.6 $mScm^{-1}$

Comparison of DHL vs EIM, BFJ vs EIM, EIM vs CDGHKL and CGK vs EIM are groupings that compare the same size of Arenga wood fiber (<1 cm, 1-2 cm and 2-3 cm) with different EC of 1.2 $mScm^{-1}$; 1.4 $mScm^{-1}$; And 1.6 $mScm^{-1}$. The results showed that EIM treatment (EC immersion buffer 1,6 $mScm^{-1}$ on Arenga wood fiber media size less than 3 cm) gave the best effect to growth of root volume. Based on this result it can be said that the higher giving EC buffer on Arenga wood fiber media can increase root volume in Kailan.

Root volume can be used as a plant growth parameter [18], growth and development in plants is a process of growth of size which includes volume, weight, cell number with cell differentiation which further form the plant organ.

3.5. Biomass

Biomass of Kailan consists of leaf biomass, stems and roots. The results of the variance analysis showed that all treatments had no significant different results. Based on Table 6, it is known that the average range of leaf dry weight is between 0.37-0.54 g. The average stem dry weight range is between 0.33-0.52 g. The more number of leaves and the wider the leaf that captures the light then the CO_2 assimilation increases so that the results of the dry weight of the plant is high.

Table 6. Average observation of Kailan biomass variables

Fiber size (cm)	EC ($mScm^{-1}$)	Observation Variable			
		Biomass (g)			
		Leaf	Stem	Root	Total
<1	0	0.52	0.32	0.36	0.97
	1.2	0.49	0.32	0.35	0.93
	1.4	0.47	0.33	0.34	0.91
	1.6	0.46	0.36	0.37	0.97
1-2	0	0.40	0.30	0.33	0.82
	1.2	0.43	0.28	0.52	1.00
	1.4	0.40	0.30	0.35	0.86
	1.6	0.47	0.30	0.39	0.95
2-3	0	0.50	0.38	0.34	1.00
	1.2	0.37	0.25	0.34	0.74
	1.4	0.50	0.38	0.34	0.99
	1.6	0.42	0.32	0.43	0.93
Control		0.54	0.32	0.38	1.01

Information : EC (*Electrical Conductivity*)

Excess element K according to [19] potentially fix it the absorption of elements Calcium (Ca) and Magnesium (Mg). In addition, that if the excessive K element there is the possibility of an antagonist with elements of Fe and Mn. The results showed an increase in the elemental content of Calcium (Ca) and the decrease of Magnesium (Mg) element as the value of EC buffer increases. On the other hand, the buffer soaking treatment increases the Manganese (Mn) element. The need for plants to the element of Mn [19] is actually not too much, but has an important role in the process of forming chlorophyll, enzyme activator and nitrogen absorption. Biomass results in general show that buffer soaking treatment has not given optimal results.

4. Conclusions and suggestions

4.1. Conclusions

Based on the research results can be conclude:

- Size affects the physical character of Arenga wood fiber, shown by the increment of bulk density, particle density and water holding capacity value in line with smaller size of the fiber.
- Before soaking in buffer solution, Arenga wood fiberhas high content of Potassium (K) which is 17000 ppm.
- Soaking in the buffer solution increases Potassium (K) content and decreases Magnesium (Mg) as the value mirroring in the increment of EC buffer.
- Soaking buffer EC 1.6 $mScm^{-1}$ with size <3 cm gives high root volume.

4.2. Suggestions
Based on this research, it is necessary to modify the size of palm fiber and the reduction of potassium (K) element in the buffer solution.

Acknowledgments
This research was supported by Ministry of Research, Technology and Higher Education Republic of Indonesia who funded this research.

References
[1] Cartea M, Francisco M, Soengas P and Velasco P 2011 Phenolic compounds in Brassica vegetables *J. Molecules.* **16** 251
[2] Jeffery E, Brown A, Kurilich A, Keck A, Matusheski N, Klein B, et al. 2003. Variation in content of bioactive components in broccoli. *J. Food Compos. Anal.* **16** 323
[3] Bradley P and Marulanda C 2000 Simplified hydroponics to reduce global hunger *J. Acta Hort.* **554** 289
[4] Sheikh B 2006 Hydroponics: key to sustain agriculture in water stressed and urban environment. *J. Agric., Agril. Eng, Vet. Sci.* **22** 53
[5] Resh H and Howard M 2012 Hydroponic food production: A definitive Giudebook for the advanced hoe gardener and the commercial hydroponic grower. California : CRC Press
[6] Byg A and Balslev H 2006 Palms in indigenous and settler communities in southeastern Ecuador: farmers' perceptions and cultivation practices *J. Agrofor Syst.* **67** 147
[7] Mogea J, Seibert B and Smits W 1991 Multipurpose palms: the sugar palm (*Arenga pinnata* (Wurmb) Merr.) *J. Agrofor Syst.* **13** 111
[8] Poulter, Rachel, Bors and Aleksandra 2010 *Quantifying differences between treated and untreated coir substrate* Project Report FL08009 (Sydney: Horticulture Australia Limited)
[9] Buntoro H, Rogomulyo R and Trisnowati S 2014 Pengaruh takaran pupuk kandang dan intensitas cahaya terhadap pertumbuhan dan hasil temu putih (Curcuma zedoria L.) *J. Vegetalika.* **3** 29
[10] Laksono and Sugiono 2016 Karakteristik agronomis tanaman Kailan (Brassica oleraceae L. Var. Acephala DC.) kultivar full white 921 akibat jeis media tanam organik dan nilai EC (Electrical Conductivity) pada hidroponik sistem wick. Program studi Agroteknologi Fakultas Pertanian Universitas Singaperbangsa Karawang
[11] Hardjana K 2013. Model hubungan tinggi dan diameter tajuk dengan diameter setinggi dada pada tegakan tengkawang tungkul putih (Shorea macrophylla (de vriese) p.s. ashton) dan tungkul merah (Shorea stenoptera burck.) di Semboja, Kabupaten Sanggau *J Penelitian Dipterokarpa.* **7** 7
[12] Pertamawati 2010 Pengaruh fotosintesis terhadap pertumbuhan tanaman kentang (*Solanum tuberosum* L.) dalam lingkungan fotoautotrof secara invitro. *J. Sains dan Teknologi Indonesia* **12** 31
[13] Sutiyoso Y 2003 Meramu pupuk hidroponik: tanaman sayur, tanaman buah, tanaman bunga (Jakarta: Penebar Swadaya)
[14] Winarso S 2005 *Kesuburan tanah dasar kesehatan dan kualitas tanah* (Yogyakarta: Gava

media)
[15] Ingram P, Zhu, Shariff A, Davis IW, Benfey P and Elich T 2012 High-throughput imaging and analysis of root system architecture in Brachypodium distachyon under differential nutrient availability. *J. Philos. Trans. R. Soc. Lond. B Biol. Sci.* **367** 1559
[16] Hardjowigeno 2003 *Ilmu tanah* (Jakarta: Akademika pressindo)
[17] Mukaromah, Nurhidayati and Nurfadilah 2013 Pengaruh sumber dan konsentrasi nitrogen terhadap pertumubuhan dan perkembangan biji Dendrobium laxiflorum J.J Smith secara in vitro. *J. sains dan seni pomits* **2** 2337
[18] Arifin R 2016 *Bisnis hidroponik ala kebun sayur* (Jakarta: AgroMedia)

IOP Conf. Series: Earth and Environmental Science **142** (2018) 012016 doi:10.1088/1755-1315/142/1/012016

The evaluation of water hyacinth (*Eichhornia crassiper*) control program in Rawapening Lake, Central Java Indonesia

N Hidayati[1], T R Soeprobowati[2,3] and M Helmi[4]
[1] Master Program of Environmental Science, School of Postgraduate Studies, Diponegoro University, Semarang, Indonesia
[2] School of Postgraduate Studies, Diponegoro University, Semarang, Indonesia
[3] Department of Biology, Faculty of Science and Mathematics, Diponegoro University, Semarang, Indonesia
[4] Depertment of Oceanography, Faculty of Fisheries and Marine Science, Diponegoro University, Semarang, Indonesia

Email: trsoeprobowati@live.undip.ac.id

Abstract. The existence of water hyacinths and other aquatic plants have been a major concern in Rawapening Lake for many years. Nutrient input from water catchment area and fish feed residues suspected to leads eutrophication, a condition that induces uncontrolled growth of aquatic plants. In dry season, aquatic plants cover almost 70% of lake area. This problem should be handled properly due to wide range of lake function such as water resources, fish farming, power plants, flood control, irrigation and many other important things. In 2011, Rawapening Lake was appointed as pilot project of Save Indonesian Lake Movement: the Indonesian movement for lakes ecosystem conservation and rehabilitation. This project consists of 6 super priority programs and 11 priority programs. This paper will evaluate the first super priority program which aims to control water hyacinth bloom. Result show that the three indicators in water hyacinth control program was not achieved. The coverage area of Water hyacinth was not reduced, tend to increase during period 2012 to 2016. We suggesting better coordination should be performed in order to avoid policies misinterpretation and to clarify the authority from each institution. We also give a support to the establishment of lake zonation plan and keep using all the three methods of cleaning water hyacinth with a maximum population remained at 20%.

1. Introduction

One of sustainable development goals (SDGs) is to ensure the availability and sustainable management of water and sanitation for all [1]. This goal was in line with one of Indonesian national development agenda during 2014-2019 increasing water security [2]. One of its indicators is improving water quality in 15 priority lakes and 5 priority river areas.

Since 2010, the Indonesian government has set fifteen lakes to prioritize its management. This is based on the urgency of its ecosystem damage due to over capacity [3]–[5]. There are at least four symptoms for a lake that has exceeded its capacity: the diminishing production of fisheries, the rapid growth of aquatic plants, pollution of water sources and the loss of endemic organism, and the silting up and depletion of lake [6].

Uncontrolled growth of aquatic plants is the most recognizable characteristic and most common feature of lakes in the world. This phenomenon occurs due to eutrophication, a condition of increased nutrient content of waters [7,8]. Nutrient input may come from waste and agricultural activity in water catchment area [9] and fish feed residues from floating net cages. For an example, Lake Rawapening has experienced the phenomenon of the floating island since the 1920s [10]. Since that time, endless problem of aquatic plants, especially water hyacinth, is began. In 2005, it was observed that hyacinth cover the water are until 65% [11] even more, until70%, in dry season [5]. This condition threatened lake function as water resources, fish farming, power plants, flood control, irrigation, tourism, etc.

To overcome the problems experienced by those 15 priority lakes, in October 14th, 2011 nine ministries have an agreement to conduct a save lake movement [10] stand Indonesian movement for lakes ecosystem conservation and rehabilitation. Rawapening Lake was appointed as the pilot project for period 2012-2016. The program of Rawapening consists of 6 super-priority programs and 11 priority programs with the first super-priority is to control water hyacinth bloom.

Now, 5i years have passed since the save lake movement program launched. For that this study aims to determine the extent to which water hyacinth control program indicators have been achieved and what efforts have been done by the government to pursue these targets.

2. Methods

There are 3 activities in water hyacinth control program: mechanical, biological and chemical strategies. Those activities have 3 indicators: the percentage of water hyacinth coverage for mechanical, the number of grass carp fish for biological and the amount of eco-friendly herbicide for chemical efforts. The percentage of water hyacinth coverage is measured by geographical information system (GIS) technique, while the number of grass carp fish and eco-friendly herbicide is assessed from the information of the implementing agency.

High resolution satellite imagery (HRSI) was used to measure the coverage of water hyacinth in the beginning of program (2012) and by the end (2016). The satellite imageries were generated from GeoEye-1 and WorldView-2, selected of the adjacent date which has a clear image quality i.e. July 5, 2012 and July 16, 2016. Groundcheck of 15 points was done using global positioning system (GPS) to ensure the suitability of the image with field conditions.

Image data was processed using ArcMap 10.1. Manual digitization was used to create lake's surface boundary. Supervised classification was used to distinguish water vegetation with others. Then, classified aquatic plants were digitized and grouped to obtain the size of the water hyacinth coverage area.

The achievement of the controlling activities of water hyacinth is confirmed to program implementing agency to determine the constraints they have faced.

3. Result and discussion

Rawapening Lake is a semi-natural lake located at 7°4' - 7°30' S and 110°24'46 " - 110°49'06 " E at altitudes between 455 - 465 meters above sea level. Administratively, it is located in Semarang Regency, Jawa Tengah Province with water catchment area covering Semarang regency (89.53%), Salatiga city (9,31%) and Magelang Regency (1,2%) [11]. Rawapening has 9 inlets which are Galeh, Torong, Panjang, Legi, Parat, Sraten, Rengas, Kedung Ringin dan Ringis and 1 outlet: Tuntang River.

Water hyacinth control program involving three government agencies from different level governance that are Ministry of Public Works and Public Housing Directorate General of Water Resources Balai Besar Wilayah Sungai Pemali Juana (BBWSPJ), Jawa Tengah Provincial Agency of Water Resources (Dinas Pekerjaan Umum Sumberdaya Air dan Tata Ruang/Dinas PUSDATARU) and Jawa Tengah Provincial Agency of Marine and Fisheries (Dinas Kelautan dan Perikanan/DKP). BBWSPJ was appointed as leading sector and program coordinator. Based on Ministry of Public Works and Public Housing Regulation No.4/2015, Rawapening Lake is managed by central government (BBWSPJ), while the rivers which cross regency boundaries around Rawapening are managed by regional government (Dinas PUSDATARU).

The first activity in water hyacinth control program is reducing water hyacinth coverage using mechanical equipment conducted by BBWSPJ. From 2011 until 2016, BBWSPJ only using excavator to cleaned water hyacinth combined with human resources by empowering surrounding community. Funded as much as 500 million Indonesian Rupiah/ year, this method was only able to clean water hyacinth as much as 20 ha per year. Meanwhile, Dinas PUSDATARU controlled water hyacinth population at Tuntang Canal by built klante, a series of pole to stem the spread of water hyacinth. Other efforts undertaken by government such as building check dam, river normalization and infiltration wells, didn't directly targeted water hyacinth blooming but effectively blocked sediment entering the lake.

Rawapening was dammed in 1936 by the Dutch East Indies Government with maximum water surface area 2,667 ha in rainy season (November-April) and 1,650 ha during drought season (May-October) [11,12]. The water surface area widens and narrows depending on the rainfall intensity in the water catchment area. Based on satellite imagery observation, it can be seen that the lake surface area has increased from 2012 to 2016, from 1.693 ha to 1.720 ha.

From the determination of water surface area using HRSI, it could be calculated the percentage of water hyacinth coverage. This study was not distinguishing water hyacinth with other aquatic plants because practically, the cleaning hadn't distinguished it anyway. Aquatic plants area in 2012 was 775,49 ha, while in 2016 its widens as much as 990,53 ha respectively. So, the coverage of aquatic plants was increased from 46 % to 58 % (Figure 1Figure 2).

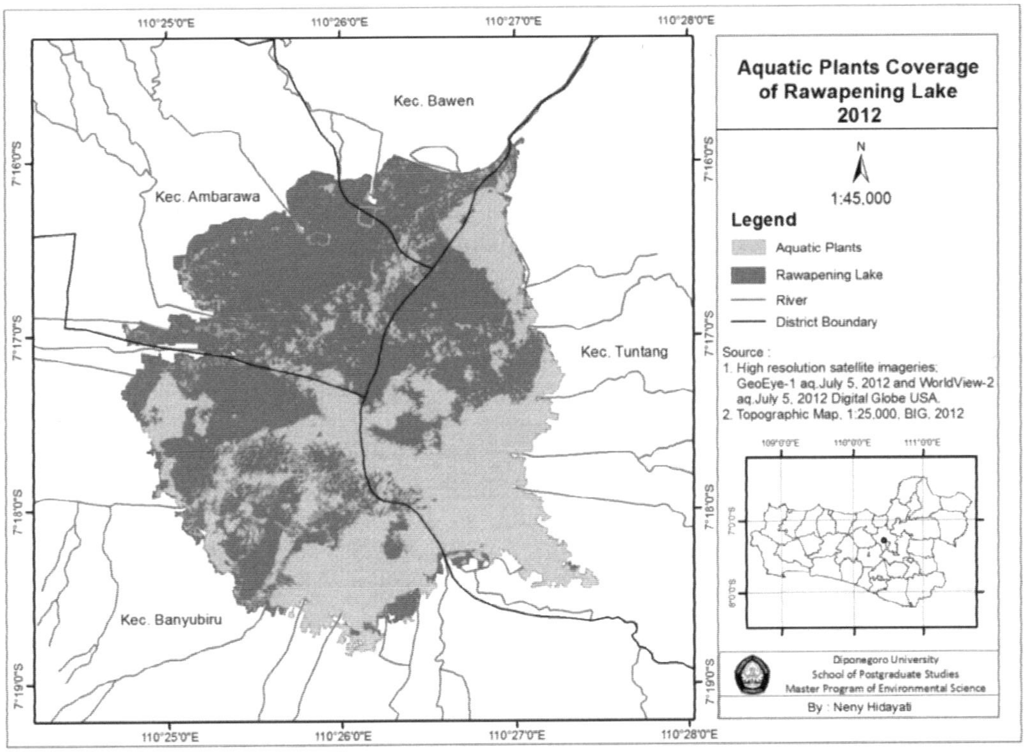

Figure 1. Aquatic plants coverage of Rawapening Lake 2012

In 2012, aquatic plants mostly distributed in Tuntang and half of Banyubiru district (Figure 1), while in 2016 the aquatic plants almost cover Banyubiru's part and invaded to Ambarawa district (Figure 2). The cleaning of these plants should conducted regularly since water hyacinth could be able to doubled within 12 days [13] or it may said that it took 52 days to form 1 m^2 colony. Water hyacinth is a floating plant that able to move following wind and water flow. During rainy season, it moves to peripheral, blocking waterways and make it difficult for farmers to cultivate their land. The existence of water hyacinth is not always harmful, it can be used as a sediment trap, bioaccumulator, fish shelter and protect shorelines from erosion [14].

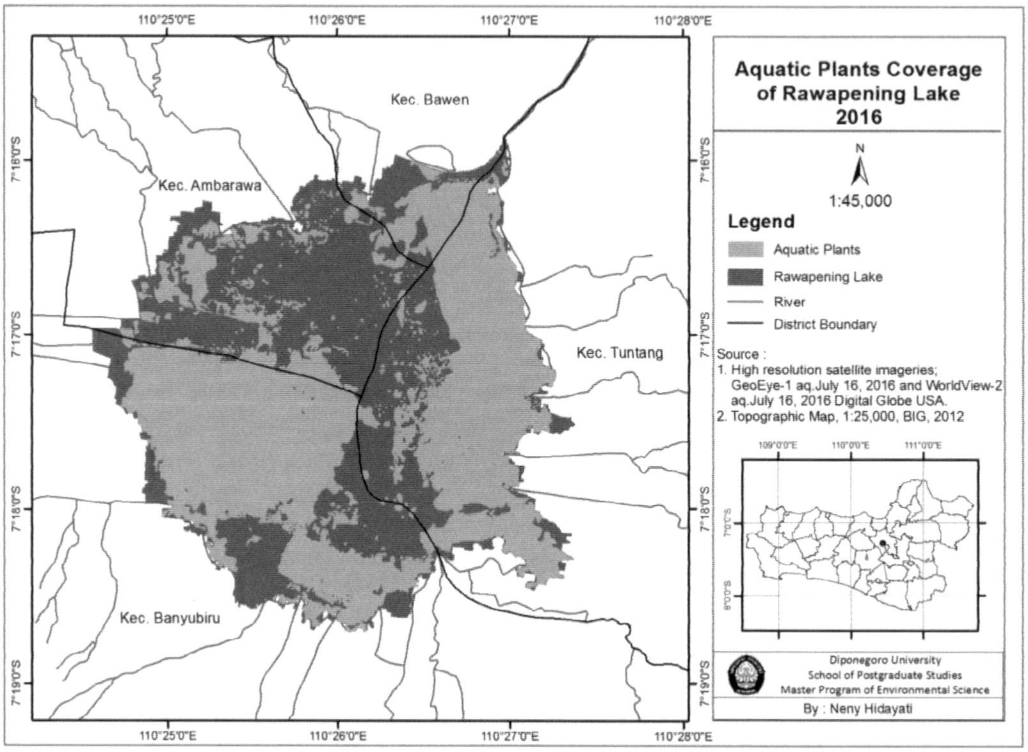

Figure 2. Aquatic plants coverage of Rawapening Lake 2016

Given the protracted water hyacinth problem, BBWSPJ feels the need to establish its own operating and control unit. In January 2017, it was set up Operasi & Pengendalian IV unit Rawapening (OP IV) which spesifically manages the lake. As the first step, OP IV equipped with more adequate equipment such as excavator long arm (1 unit), excavator short arm (1 unit), multipurpose dredger (1 unit), amphibious weed harvester (3 units) and dump truck (3 units). The capacity of these equipment is 1 ha/day. If it is operated everyday, it will take 3 years to remove all of water hyacinth, assuming no vegetation doubling. Mechanization may be effectively used in small area, but in wider lake, it consume more time, money and energy [15]. Mechanical method also threatening the ecology, affecting other species life and even ruining it [16]. In case of Rawapening, mechanization potentially caused social conflict with the owner of floating net cages and craftsmen.

The second activity of water hyacinth control program is reducing water hyacinth using biocontrol. Jawa Tengah provincial agency of marine and fisheries (Dinas Kelautan dan Perikanan/DKP) using grass carp fish *Ctenopharyngodon idella* (ikan koan). Grass carp targeted the stocking of 5.000.000 grass carp whose 100 g of weight every year. DKP was only able to provide 100.000 in 2012 and 2013, 105.000 in 2014, 148.300 in 2015 and none in 2016 due to the implementation of Law No.23/2014 on Regional Government. According to DKP, there is still unclearly explanation about the fisheries regional government authority at Rawapening Lake. Beside there are other priorities that need attention as well.

It could be said that controlling water hyacinth growth using grass carp is inefficient in Rawapening Lake because the amount is too small and fisherman still freely cought it. Nevertheless, 5.000.000 grass carp a year seems too much since Kerinci Lake in Jambi Sumatra whose success stories of biological method used only 2.000 grass carp per year. After 10 years using grass carp, the lake now clean from water hyacinth, but some of endemic fish species reported had been scarce to found [17]. Using grass carp to control water hyacinth growth is the most ecologically secure way, but it takes time and its need to maintain the number of predator and prey population. To overcome those problems, one of viable way is to combine several species such as grass carp and weevil (Neochetina spp) that have been proved to be more efficient and sustain [18].

The third method of water hyacinth control is using eco-friendly herbicide. This method is unconducted because the agency reluctant to do so. There is also unclear statement to whom the responsibility of this activity addressing to, because of lack coordination between the agencies. Cleaning up using herbicide had been done in 1964/1965. Despite its price and speed, cheaper and faster, herbicide no longer use because several damage effects, such as irritation and light poisoning. Even so, herbicides can still be an option to clean the water hyacinth in the middle of the lake that is difficult to reach by mechanical equipment.

Overall, after some of interviews with the stakeholders, the problems during program implementation are unsynergistic law and regulation, lack of coordination between stakeholder, low financial support, potential social conflict, and technical constraint in each method that described above.

4. Conclusion

The three indicators in water hyacinth control program which observed in this study were not achieved. The main indicator is Water hyacinth coverage area which not reduced and tend to increase during 2012 to 2016. We are suggesting a better coordination in order to avoid policies misinterpretation, as well as to clarify to the authority in each institution. We also suggesting the establishment of lake zonation plan and keep using all the three ways of cleaning water hyacinth with a maximum population remained at 20%.

References

[1] UNEP and UNDP 2015 *Indicators and Data Mapping to Measure Sustainable Development Goals (SDGs)* Targets (Jakarta) p 12
[2] Kementerian PPN/Bappenas 2014 *Rencana Pembangunan Jangka Menengah Nasional 2015-2019* (Jakarta) p 6:154
[3] Nasution Z, Sari Y D and Huda H M 2011 *J. Kebijak. Sos. Ekon. Kelaut. dan Perikan.* **1** 19–31
[4] Suryono T and Badjoeri M 2013 *Limnotek* **20** 169–77
[5] Soeprobowati T R and Suedy S W A 2010 *J. Sains dan Mat.* **18** 158–69
[6] Haryani G S 2013 *Limnol. LIPI* **2** 1–19
[7] Chislock M F, Doster E, Zitomer R A and Wilson A E 2013 Eutrophication: Causes, Consequences, and Controls in Aquatic Ecosystems *Nat. Educ. Knowl.* 4(4)10
[8] Cole G A and Weihe P E 2016 *Textbook of Limnology* 5th ed (Illinois: Waveland Press, Inc)
[9] Berka C, Schreier H a\nd Hall K 2001 *Water. Air. Soil Pollut.* **127** 389–401
[10] Soeprobowati T R 2015 *Procedia Environ. Sci.* **23** 368–74
[11] KLH 2011 *Profil 15 Danau Prioritas Nasional* (Jakarta:Kementerian Lingkungan Hidup)
[12] Sulastri, Henny C, and Handoko U 2016 *Oseanologi dan Limnol. di Indones.* **1** 23–38
[13] Lowe S, Browne M, Boudjelas S and De Poorter M 2000 *100 of The World's Worst Invasive Alien Species - A selection from the Global Invasive Species Database* (ISSG a specialist group of SSC of the IUCN)
[14] Strayer D L 2010 *Freshw. Biol.* **55** 152–74
[15] Villamagna A M and Murphy B R 2010 *Freshw. Biol.* **55** 282–98
[16] Ndimele P E, Kumolu-Johnson and Anetekhai M A 2011 *Res. J. Environ. Sci.* **5** 509–20
[17] Samuel, Suryati N K and Adiansyah V 2015 *Ind. Fish. Res. J.* **21** 9–18
[18] Ayyaru G, Mayalagu R, Jun S, Ajay P and Alagar V B 2011 *Chinese J. Oceanol. Limnol.* **29** 162–66

IOP Conf. Series: Earth and Environmental Science **142** (2018) 012017 doi:10.1088/1755-1315/142/1/012017

Wastewater treatment by nanofiltration membranes

R Mulyanti[1] and H Susanto[2]
[1] Master Program of Environmental Science, School of Postgraduate Studies Universitas Diponegoro, Jl. Imam Bardjo SH No 5, Pleburan, Semarang, 50241, Indonesia
[2] Departement of Chemical Engineering, Faculty of Engineering, Universitas Diponegoro, Jl. Prof Soedarto SH, Tembalang, Semarang, 50275, Indonesia

E-mail: heru.susanto@che.undip.ac.id

Abstract. Lower energy consumption compared to reverse osmosis (RO) and higher rejection compared to ultrafiltration make nanofiltration (NF) membrane get more and more attention for wastewater treatment. NF has become a promising technology not only for treating wastewater but also for reusing water from wastewater. This paper presents various application of NF for wastewater treatments. The factors affecting the performance of NF membranes including operating conditions, feed characteristics and membrane characteristics were discussed. In addition, fouling as a severe problem during NF application is also presented. Further, future prospects and challenges of NF for wastewater treatments are explained.

1. Introduction

Increased human activities in all sectors, i.e. industry, food, medical, fishery, etc., demands availability of adequate clean water. Low fresh water stock, only 3 % of total water on earth, requires sophisticated technology to process water and wastewater to produce safe output for the environment and even to be reused for various purposes.

The membrane technology which offers high selectivity, low energy requirement, cost advantage and environmental sustainability, can be combined with other separation process, without adding any chemical and it's easier in up-scaling [1]. Membrane technology can solve existing problems, especially in wastewater treatment.

The common membrane filtration technology is pressure driven membrane processes which are based on pore size, consisting of four processes, i.e. microfiltration, ultrafiltration, nanofiltration, and reverse osmosis.

2. Nanofiltration Membrane

The development of nanofiltration membrane application in all fields is very rapid in recent years. This membrane was already used in filtration process the end of 1980s [2] . The characteristics of NF which are 1- 5 nm pore size and 7- 30 operating pressure [3] are used to separate solutes with low molecular weight, e.g. lactose, glucose, salt, and it's effective in rejecting hardness, dye and heavy metal. NF as a membrane technology with pressure driven process is a very promising membrane application for future waste treatment, ending dependency to fresh water usage [4], [5] and ground water by wastewater reuse which can't be performed by conventional waste treatment. Relatively low operating pressure and higher permeate flux allow NF to replace reverse osmosis (RO) in various applications. Lower energy consumption compared to reverse osmosis (RO) and higher rejection compared to ultrafiltration (UF) make nanofiltration (NF) membrane get more and more attention for wastewater treatment.

3. Separation Mechanism of NF

Pollutant separation by NF membrane really depends on differences of particle sizes and charge effects (for ionic components) due to effects of steric and electrostatic interactions based on Donnan exclusion [2], [4], [6]. According to [3], components with high molecular weight are separated by sieving mechanism, unlike components with low molecular weight or ionic species which are separated by charge effect and solution diffusion mechanism.

4. NF Application

Nanofiltration is increasingly common in industry, pharmaceuticals, water purification, wastewater treatment, biotechnology, and brackish water desalination.

In industry, some uses of NF process are to separate colors in textile industry [7], [8], metal recovery, olive mill wastewater treatment [9], [10]. Integration of NF and RO effectively removed soluble monovalent and divalent ions in OME (olive mill Effluent) compared with conventional physicochemical process [11]. NF was also used in coke wastewater treatment [12], pulp and paper [13], oily wastewater treatment from oil and petroleum industries, acid sulphate removal from mine water [14].

There were also many applications on domestic sector, e.g. treatment of municipal wastewater [4], [15], Leachate [16], car wash wastewater [17], and restaurant effluent. NF-90 quite efficiently reduces BOD 5 and the conductivity was over 80 % compared with aerobic and anaerobic processes, chemical coagulation and electrocoagulation and electroflotation [18].

NF is also an alternative for separation in food processing, e.g. removal of phenolic compounds from pomegranate juice [19], wastewater treatment of instant tea powder [20], Combination of NF + RO with cascade operation in whey treatment gives optimal result for recovering protein and lactose [21]. NF is also used to concentrate coffee extract [22] and red wine [23]. In pharmaceuticals, NF process is selected in antibiotic separation process [24], [25].

Most commercial NF membranes have negative charge with varying Molecular Weight Cut Off (MWCO) e.g. 100-300 Da (ESNA1), 150 Da (TS80) , 150 - 300 Da (DL, HL, NFX) , 180 Da (UTC20), 200 Da (TS40, DK, TFC SR 100, SR3D, SPIRAPRO), 200-400 Da (NF270, NF200, NF 90) , 300 – 500 Da (NFW), 400 Da (TR60), 500 Da (XN45), 600- 800 Da (NTR7450, NFG), 2000 Da (CK) [2].

5. NF Performance

NF performance is represented by flux (J) calculation which is measured from the volume of flow which passes membrane (permeate volume) divided by the size of membrane surface (A) at particular time (t), such as in Equation (1)

$$J = \frac{V}{A.t} \tag{1}$$

and rejection (R), which is ratio of feed concentration reduced by permeate concentration (Cp) to feed concentration (Cf), can be calculated by Equation (2).

$$R = 1 - \frac{Cp}{Cf} x\ 100\% \tag{2}$$

NF performance in rejecting COD is approximately 70 % [26] to 99,4 % [18] and it's able to separate dyes by 90,6 – 95 % on average [27]. NF ability to remove salinity parameter varies from 75 % to 95 % [28] depending on the membrane in use. Meanwhile, the conductivity rejection level is 82,3 % [18].

Significant factors which determine membrane performance, especially NF, are operating condition, membrane characteristic and feed characteristic.

5. 1 Operating Condition
Membrane performance is strongly affected by operating conditions, including temperature, operating pressure and flow rate.

5.1.1 Temperature. Permeate flux increases along with temperature increase because when temperature increases, viscosity and level of concentration polarization will decline [29]. Removal of COD and Electrical Conductivity in oily wastewater declines when temperature and permeate flux increase [6]. A study on rubber waste treatment [30] shows that membrane permeability will decline above 24°C, but if flux permeate is maintained at 11-12 $Lm^{-2}h^{-1}$, the effect of temperature will be insignificant. However, if there is fouling, flux will decline despite increase of temperature [13]. Temperature change causes variation of diffusion coefficient and component absorption, and then influences flux.

5. 1.2 Operating Pressure. Increased pressure is essentially directly proportional with flux increase, but at certain point this doesn't apply due to fouling and concentration polarization phenomenon [31]. The study by [27] shows that increased operation pressure will increase dye rejection and transport rate of solvent.

5. 1.3 Flow Rate. Increased flow rate impacts increased permeate flux and mass transport, as well as declining possibility of concentration polarization [27].

5. 2 Membrane Characteristic
Membrane characteristics, such as charge, pore size, and hydrophilicity, are closely associated with membrane performance. Separation process by membrane must consider the charges of the membrane and solute. If membrane charge is different from solute charge, there will be gravitational force, so there's bigger chance of fouling. Fouling will eventually reduce flux and change membrane selectivity. Pore size of membrane is a determinant of membrane rejection level on uncharged contaminant. Meanwhile, hydrophilicity is related to very weak interaction to solute, but high to water, the opposite of hydrophobicity. The advantage of hydrophilic membrane compared by hydrophobic is it's not vulnerable to fouling, but hydrophobic materials tend to produce high flux [31].

5. 3 Feed Characteristic
Similar to membrane characteristic, feed characteristic significantly affects membrane performance, e.g. related to chemical structure, pH, charge, geometry and hydrophilicity of solute. Feed pH can influence membrane charge, even changing it.

"The occurrence of an isoelectric point means that at lower pH than the isoelectric point, the membrane is

positively charged and vice-versa by [32]. Isoelectric point may be different for each membrane type. For example, the isoelectric point on TFCS and NF 70 is 3 and 4, whereas TFCL is 5,2 [33]. while membrane FM NP010 is 4,2 [34]. pH change also influences hydrophilicity of membrane [35]. Furthermore, temperature of feed will increase mass transfer process in membrane [6].

6. Wastewater Reuse
Waste as a result of human activity is a challenge and has great potential to be reused for various needs. Waste is a renewable resource and can be a new water source alternative as sustainability point in minimizing environmental impact, and a solution for clean water scarcity. Application of wastewater reuse has spread beyond Middle East and United States as countries with the highest levels of wastewater reuse [36]. In Asian and European countries, wastewater reuse is increasing. Membrane

technology can produce high quality effluent, enabling reuse for various purposes, e.g. irrigation [4], tourism, public park, processed water [30], toilet, with certain criteria that are given in table 1.

Table 1. Application and wastewater reuse of NF.

Application	Type Of NF Membrane	Operating Condition	Membranes Performance	Wastewater Reuse
Rubber wastewater treatment [30]	Integrated process of NF270 and evaporation	NF : Pilot scale (continuous) Permeate flux 11L/m^{2}h Evaporator : pilot scale (batch), capacity of full size drum 5 m^3/d with the boiler blowdown.	COD removal was over 95 % and water recovery was 80-90 % if permeate flux maintain less than 12 L/m2h , energy costs decline until 55 %	Permeate could be reused as boiler feed water with reduction of total hardness concentrate.
Dairy industry [37]	Membrane Bioreactor (MBR) and NF	MBR : The flow rate was 0.80 L/h, and the permeate flux was 18.2 L/(h m^2). Cross-flow velocity of 7.8 m/s	MBR as secondary process was followed by NF process had rejected 99 % of COD, 93.1% of total solids	For boiler and cooling water
Coking wastewater treatment [38]	A combination of biological (full scale plant) processes with MBR and NF-RO integrated system (pilot scale)	Full scale : T : 35 – 38°C, average flow rate 200 m^3/h, pH and DO were maintained at 6.5-6.8 and 5-6 mg/L(in oxic reactor) Pilot scale : *MBR* :flow rate 10 m^3/h, 4 modules (flat-sheet submerged), DO & MLSS concentration were maintained 6 mg/L & 8500 mg/L *NF +RO* : feed pressure : 0.6 Mpa & 1,0 Mpa , pore size : 0.001 μm & 0.0001μm, filtration time : 2-3 months in 1 cycle	Removal of COD, hardness , conductivity, fluoride, TN, T-CN and chloride ion meets the standards for industrial reuse	Industrial reuse
Secondary effluent of textile wastewater treatment [28]	NF 90, NF 200, NF 270	The temperature and pressure of experiment were at 25°C and 6 – 22 bar, with resirculation flow were at 340 L/h	Best result was given by NF 90, which COD removal and salt retention were 99 % and 75 – 95 %	Textile industry reuse
Wastewater treatment of rendering plant [39]	Combination of SBR, sand filtration, UF (MW,CQ,GM), NF(NF279, NF90), RO (XLE)	laboratory scale (batch). Six UF / RO/NF cells connected in parallel. Operating pressure of MW, CQ, GM, were 2, 4, and 5 bar. Then NF270, NF90, XLE were at 10 bar	Best rejection was showed by combination of UF (MW) with NF90 and RO (XLE)	steam generation, washing factory and vehicle

7. Fouling Phenomena

Fouling is the main and inevitable problem in membrane application, including NF, beside it reduces membrane performance by flux reduction over time, as well as reducing cost efficiency [11]. Fouling can be defined as blockage on the surface and pores of membrane by soluble material or suspended material, such as colloid, organic component, inorganic component and biological component [2], [6].

Fouling can be solved physically, chemically and hydrodinamically. Fouling control by chemical cleansing will cause long term damage to membrane structure, reducing membrane performance [18].

8. Future prospects and challenges of NF

NF is the latest technology alternative for waste treatment which is very innovative, economical, and environmentally friendly. The integration of NF with other separation processes potentially yields optimal result in wastewater treatment, so the quality of the produced effluent can be reused. Beside saving resources, waste reuse has great prospects in long term water provision and it's more economical than desalination process. The future challenges and prospects of NF application is very great considering the broad range of selectivity of organic solute and salt, enabling further studies for various applications.

9. Conclusions

NF characteristics which are between UF and RO makes NF application increasingly favors across sectors due to low energy requirement and much better selectivity. Beside the size difference of the components to be separated, charge effect also influences separation process by NF. NF performance is also determined by membrane and feed characteristics, as well as operation condition.

Separation process by membrane technology, especially NF, has bright prospects in wastewater treatment and great potentials for further development. Integration with other processes potentially produces higher quality effluent.

References

[1] Mulder M 1996 *Basic Principles of Membrane Technology* vol 2 (Nederland: Kluwer Aca publisher)
[2] Mohammad A W, Teow Y H, Ang W L, Chung Y T, Oatley-Radcliffe D L and Hilal N 2015 *Desalination* **356** 226–54
[3] Shon H K, Phuntsho S, Chaudhary D S, Vigneswaran S and Cho J 2013 *Drink. Water Eng. Sci* **6** 47–53
[4] Bunani S, Yörükoğlu E, Sert G, Yüksel Ü, Yüksel M and Kabay N 2013 *Desalination* **315** 33–6
[5] Gómez M, Plaza F, Garralón G, Pérez J and Gómez M A 2007 *Desalination* **202** 369–76
[6] Rahimpour A, Rajaeian B, Hosienzadeh A, Madaeni S S and Ghoreishi F 2011 *Desalination* **265** 190–8
[7] Ellouze E, Tahri N and Ben Amar R 2012 *Desalination* **286** 16–23
[8] Ong Y K, Li F Y, Sun S P, Zhao B W, Liang C Z and Chung T S 2014 *Chem. Eng. Sci.* **114** 51–7
[9] Coskun T, Debik E and Demir N M 2010 *Desalination* **259** 65–70
[10] Paraskeva C A, Papadakis V G, Tsarouchi E, Kanellopoulou D G and Koutsoukos P G 2007 *Desalination* **213** 218–29
[11] Pulido J M O 2016 *J. Sci. Total Environ.* **563–564** 664–75
[12] Korzenowski C, Minhalma M, Bernardes A M, Ferreira J Z and De Pinho M N 2011 *Sep. Purif. Technol.* **76** 303–7
[13] Beril Gönder Z, Arayici S and Barlas H 2011 *Sep. Purif. Technol.* **76** 292–302
[14] Visser T J K, Modise S J, Krieg H M and Keizer K 2001 *Desalination* **140** 79–86
[15] Miralles-Cuevas S, Oller I, Agüera A, Llorca M, Sánchez Pérez J A and Malato S 2017 *J. Hazard. Mater.* **323** 442–51
[16] Linde K and Jönsson A S 1995 *Desalination* **103** 223–32
[17] Lau W J, Ismail A F and Firdaus S 2013 *Sep. Purif. Technol.* **104** 26–31
[18] Zulaikha S, Lau W J, Ismail A F and Jaafar J 2014 *J. Water Process Eng.* **2** 58–62
[19] Conidi C, Cassano A, Caiazzo F and Drioli E 2017 *J. Food Eng.* **195** 1–13

[20] Chen B, Xiong X, Yao Z, Yin N, Low Z and Zhong Z 2015 *Desalination* **355** 147–54
[21] Yorgun M S, Balcioglu I A and Saygin O 2008 *Desalination* **229** 204–16
[22] Pan B, Yan P, Zhu L and Li X 2013 *Desalination* **317** 127–31
[23] Banvolgyi S, Kiss I, Bekassy-Molnar E and Vatai G 2006 *Desalination* **198** 8–15
[24] Weng X D, Ji Y L, Ma R, Zhao F Y, An Q F and Gao C J 2016 *J. Memb. Sci.* **510** 122–30
[25] Koyuncu I, Arikan O A, Wiesner M R and Rice C 2008 *J. Memb. Sci.* **309** 94–101
[26] Miyaki H, Adachi S, Suda K and Kojima Y 2000 *Desalination* **131** 47–53
[27] Shaaban A M F, Hafez A I, Abdel-Fatah M A, Abdel-Monem N M and Mahmoud M H 2016 *Egypt. J. Pet.* **25** 79–90
[28] Gozálvez-Zafrilla J M, Sanz-Escribano D, Lora-García J and León Hidalgo M C 2008 *Desalination* **222** 272–9
[29] Agashichev S P 2009 *Desalination* **236** 252–8
[30] Xin G, Lopes M P, Crespo J G and Rusten B 2013 *Sep. Purif. Technol.* **119** 19–27
[31] Susanto H 2011 *Teknologi Membran* (Semarang: Badan Penerbit Universitas Diponegoro)
[32] Ahmad A L, Tan L S and Abd. Shukor S R 2008 *J. Hazard. Mater.* **154** 633–8
[33] Childress A E and Elimelech M 1996 *J. Memb. Sci.* **119** 253–68
[34] Childress A E and Elimelech M 2000 *Environ. Sci. Technol.* **34** 3710–16
[35] Mänttäri M, Pihlajamäki A and Nyström M 2006 *J. Memb. Sci.* **280** 311–20
[36] Lyu S, Chen W, Zhang W, Fan Y and Jiao W 2016 *J. Environ. Sci.* **39** 86–96
[37] Andrade L H, Mendes F D S, Espindola J C and Amaral M C S 2014 *Sep. Purif. Technol.* **126** 21–29
[38] Jin X, Li E, Lu S, Qiu Z and Sui Q 2013 *J. Environ. Sci.* **25** 1565–74
[39] Racar M, Dolar D, ˇSpehar A, Kosutic K, 2017 *Process Saf. Environ. Prot.* **105** 386–92

The competitiveness of medicinal plants in Central Java Indonesia

E W Riptanti, R A Qonita and R U Fajarningsih
Department of Agribusiness, the Faculty of Agriculture, Universitas Sebelas Maret, Surakarta, Indonesia.

E-mail: erlynawida@staff.uns.ac.id

Abstract. Indonesia hold a potential opportunity to be a pharmaceutical production country. In the other side, Indonesia facing agriculture product and market competition internationally and domestically. Improving competitiveness of specific products are aimed to hit the export market and also to be able to compete with import products in domestic market. Considering Indonesian market opportunity, therefore we perform this study aiming to examine the competitiveness of medicinal plants in Central Java area. The basic method applied in this study was descriptive and data sources are primary and secondary data. Data were analyzed using Revealed Comparative Advantage (RCA) and Trade Specialization Index (TSI) approaches. According to the study, we reveal that medicinal plant commodities in Central Java province have average score 8.52, indicating a strong comparative advantage in global markets, while the commodities appear to have average score of 0.97 at 0 to 1 range, representing a strong competitiveness. Those results are attributed to natural and environmental condition, as well as high demand of global market. This competitiveness will encourage traditional/modern medicinal industries in global market.

1. Introduction

Indonesia is one of potential pharmaceutical manufacturing countries. It is home of 30.000 out of 40.000 medicinal plants in the world and has been developed to be one of the most extensive places of natural beauty care regimen in the world [1]. Various types of medicinal plants have been produced as raw materials of both modern and traditional medicine (herb). The large demand of medicinal plants results in tremendous trade from local to international scales [2]. The development of medicinal plant production is filled with bright prospects due to some factors such f as flora, soil and climate conditions, the development of modern and traditional medicine industries, and the increasing of consumption and rising prices in medicinal commodities [3]. Out of 283 species of medicinal plants recommended by the Indonesia National Agency for Drug and Food Control, only 13 species were found have been intensively cultivated by the Indonesian people that are ginger (*jahe*), galangal (*lengkuas*), lesser galangal (*kencur*), turmeric (*kunyit*), bitter ginger (*lempuyang*), Javanese turmeric (*temulawak*), pink and blue ginger (*temu ireng*), Black Face General (*keji beling*), sweet grass (*dringo*), cardamom (*kapulaga*), fingerroot (*temukunci*), cheese fruit (*mengkudu*), and king of bitters (*sambiloto*). Table 1 showing production data of intensively-cultivated medicinal plants in Indonesia [4].

Table 1. The production of medicinal plants in Indonesia (2011-2015)

No	Commodity	Production Year (ton)					Growth
		2011	2012	2013	2014	2015	from 2014 to 2015 (%)
1	Ginger	94.743	114.538	155.286	226.115	313.064	38.45
2	Galangal	57.701	58.186	69.730	62.521	55.150	-11.79
3	Lesser galangal	34.017	·42.626	41.343	37.716	35.972	-4.62
4	Turmeric	84.803	96.979	120,726	112.088	113.101	0.90
5	Javanese turmeric	24.106	44.085	35.665	25.128	27.840	10.79
6	Cardamom	47.231.297	42.973.264	54.171.417	72.760.295	93.121.006	27.98
	Total	47.526.667	43.329.678	54.594.167	73.223.863	93.666.133	61.71

The data indicate that of six commodities, galangal is the only commodity which showed a decline production from 2011 to 2015. The production declining was resulted, probably, from a decrease in width of planting area, as well as in demand, and change in trend of demand of both domestic and foreign medicinal plant industry. Central Java is one of the largest centers of medicinal plant production which contributes to substantial proportion of medicinal plant production in Indonesia. Table 2 specifies the aforementioned fact.

Table 2. Medicinal plant production in Central Java (2015)

No	City/ Regency	Production (kg)					
		Ginger	Lesser galangal	Turmeric	Galangal	Javanese turmeric	Cardamom
1	Purworejo regency	2.508.370	350.229	400.160	116.987	3.024.783	768.540
2	Wonogiri regency	7.579.773	1.373.130	16.647.991	4.007.142	3.215.300	4.871
3	Karanganyar regency	2.214.017	122.570	1.089.981	1.141.087	787.790	53.000
4	Rembang regency	4.764.615	1.099.646	3.527.909	1.422.399	234.537	10
5	Semarang regency	8.163.670	75.549	2.777.461	457.896	534.349	2.332.981
6	Others	15.071.295	5.743.756	4.130.244	6.910.194	2.245.894	18.259.338
	Total	40.301.740	8.764.880	28.573.746	13.055.705	10.043.653	20.418.840

Source : Central Bureau of Statistics and General Directorate of Horticulture, 2016 [5].

Table 2 shows that the medicinal plant production in Central Java contributes varies between 15-50% of Indonesian total production. The main areas of medicinal plant production are Semarang regency, Semarang city, Boyolali, Sukoharjo, Karanganyar and Wonogiri. Meanwhile, the central areas of the development of medicinal plants in farmers' lands under the guidance of the government and the business are in Kudus, Kendal, Temanggung, Wonosobo, Magelang, Purworejo, Banyumas, Cilacap, and Sragen.

The medicinal plant products are intended not only to fulfill the needs of domestic industries, but also to be traded in international markets. Indonesia does not only export the products, but it also imports medicinal plant materials. Table 3 gives an overview of exports and imports of medicinal plants.

Table 3. Value (US$) and volume (kg) of Indonesia's medicinal plant exports-imports (2011-2015)

Export Commodity

Year	Cardamom/ HS 090830		Ginger / HS 091010		Turmeric/ HS 091030		Others / HS 091091	
	Value	Volume	Value	Volume	Value	Volume	Value	Volume
2011	17.213.756	7.150.538	1.209.189	1.175.524	4.502.349	2.672.115	1.534.115	245.282
2012	16.977.378	8.151.228	1.357.846	1.013.717	2.125.352	1.212.312	2.272.618	488.117
2013	10.729,475	6.737.734	14.909.488	22.471.685	2.100.919	1.946.541	1.622.003	731.877
2014	10.145.154	7.766.404	49.126.585	61.191.401	4.515.333	3.808.159	1.287.306	769/996
2015	7.785.335	6.248.064	18.230.204	25.935.366	10.499.058	8.670.791	2.468.081	1.186.806

Import Commodity

Year	Cardamom/ HS 090830		Ginger / HS 091010		Turmeric/ HS 091030		Others / HS 091091	
	Value	Volume	Value	Volume	Value	Volume	Value	Volume
2011	184.736	32.645	16.571.827	22.381.338	332.457	269.541	407.662	58.827
2012	17.134	29.355	16.703.921	22.885.392	390.248	119.373	255.124	60.718
2013	36.351	18.180	5.927.010	6.307.746	475.280	249.491	91.716	16.259
2014	85.067	22.695	2.464.877	2.763.949	323.984	245.413	235.096	73.166
2015	10.656	4.669	5.976.593	6.857.943	299.930	152.966	178.831	91.055

Source : United Nations Commodity Trade, 2017[6].

In the above table shows the value and volume of four export commodities are greater than those of import commodities. However, if we compare in each item's value, we will find that import value is higher than the export one. This indicating that the competitiveness of Indonesia's medicinal plants is higher compared to other countries. Not all intensively-cultivated medicinal plant commodities is traded in international markets. The export / import value of medicinal plant commodities products in Central Java province can be seen in the Table 4.

Table 4. Value (US $) and volume (kg) of medicinal plant exports of Java province during the periods of 2011-2015

Export Commodity

Year	Cardamom/ HS 090830		Ginger / HS 091010		Turmeric/ HS 091030		Others/ HS 091091	
	Value	Volume	Value	Volume	Value	Volume	Value	Volume
2011	4,454,631	1,561,126	173,787	65,152	276,312	177,565	563,696	208,975
2012	7,728,354	2,892,426	233,722	76,168	178,000	40,431	570,402	86,526
2013	3,676,129	2,358,899	1,347,201	1,994,415	95,773	50,000	497,782	180,075
2014	3,996,925	3,523,851	3,562,613	3,388,355	398,296	261,773	733,913	375,164
2015	4,386,925	3,458,805	2,353,129	2,992,882	433,461	397,211	1,004,484	206,853

Import commodity

Year	Cardamom/ HS 090830		Ginger / HS 091010		Turmeric/ HS 091030		Others / HS 091091	
	Value	Volume	Value	Volume	Value	Volume	Value	Volume
2011	0	0	28,202	33,957	0	0	92,311	249,490
2012	0	0	41,200	35,000	0	0	257,932	389,819
2013	28,350	14,175	0	0	0	0	62,475	168,852
2014	0	0	0	0	0	0	57,394	154,848
2015	0	0	0	0	0	0	28,990	78,350

Source : Department of Industry and Commerce of Central Java, 2016 [7].

Only three commodities including cardamom (in 2013), turmeric (only until 2012), and others were imported during the periods of 2011-2015. In the globalization era, Indonesia encounters product market and agricultural commodity competition. The competitiveness improvement is intended not only to hit the export market, but also able to compete with import products in domestic markets. The continuity of agricultural production, particularly production of *simplisia* (crude drugs), is essential to absorb employment, to reduce the number of poor population, to increase state revenue, as well as to improve the economic wellbeing of millions of farmers [8]. The purpose of this study is to examine the competitiveness of medical plant of *simplisia* for supporting traditional medicine industries in Central Java.

2. Methods

2.1. Basic Method of Research

Descriptive analytical method using time series during periods of 2011-2015 was employed in the present study.

2.2. Location Determination Method

Research location of Central Java was selected using purposive sampling under some considerations that are one of provinces which produces *simplisia* and its region has a potential to develop, and it contributes to the export of medicinal plant commodities.

2.3. Type and Source of Data

The data used in this research was primary and secondary data.

2.4. Analysis Method

2.4.1 Revealed Comparative Advantage (RCA) method RCA method belongs to one of methods used to find out export commodity competitiveness from a country or industry viewed from a comparative advantage [9]. The RCA method applied in the present research was not the common kind which was intended to compare commodity export of a country and that of other countries in the world, but rather the kind in the smaller scope which compared export medicinal plant commodity export of Central Java and that of medicinal plant commodity export of Indonesia. The value of RCA systematically used in the research was calculated using the following formula:

$$\mathrm{RCA} = \frac{X_{ij}/X_t}{W_{ij}/W_t} \quad \ldots\ldots\ldots\ldots\ldots\ldots\ldots(1)$$

Where, RCA represents competitiveness of medicinal plant commodity export of Central Java viewed from comparative advantage, Xij denotes value of medicinal plant commodity export of Central Java province, Xt is total value of export of Central Java, Wij is value of medicinal plant commodity export, and Wt is total value of export of Indonesia. The value of competitiveness resulted from the computation entails two possibilities: either the value of RCA>1 which indicates that medicinal plant commodity export of Central Java has strong competitiveness or the value of RCA<1, which reveals that medicinal plant commodity export of Central Java has weak competitiveness. The index of RCA shows the comparison between the current year value of RCA and the previous year value of RCA.

2.4.2 Analysis of Trade Specialization Index (TSI) Trade Specialization Index (TSI) was employed to compute competitive advantage of medicinal plant commodities of Central Java in Indonesia. The analysis of TSI is useful to depict Central Java's position of medicinal plant commodities, whether it tends to be an exporter or importer, as well as to put medicinal plant industry in a competitive position which enables to give an overview of capabilities obtained from performance of maximum advantage in the framework of additional value of technologies or government's policy instruments. The formula used is shown below:

$$\text{TSI} = \frac{N_x{}' - N_m{}'}{N_x{}' + N_m{}'} \quad \ldots\ldots\ldots\ldots\ldots\ldots\ldots(2)$$

Where, TSI denotes Trade Specialization Index, Nx' is value of Central Java's medicinal plant commodity export, and Nm' is value of Central Java's medicinal plant commodity import. A value of the index ranges between -1 and +1. If it has positive value ranging from 0 to 1, then medicinal plant commodities are considered to have strong competitiveness, or Central Java tends to be an exporter of medicinal plant commodities. on contrary, if it has a negative value ranging below 0 to -1, then medicinal plant commodities are said to be weak, or Central Java tends to be an importer [10].

In addition, TSI can be used to identify a competitive position of a commodity in trade, which is divided into several stages: (1) introduction, a stage in which forerunners of medicinal plant industry in other regions export new products and latecomers of medicinal plant industry in Central Java import their products , and a value of latecomer industry TSI ranges between -1.00 and -0.50, (2) import substitution, a stage in which a value of TSI rises to between -0.51 and 0,00, and Central Java's medicinal plant industry indicates low level of competitiveness due to its less high level of production to achieve its economies of scale; such industry exports products with inadequate quality, whereas in fact, the domestic production is less than the domestic demands (in other words, for such commodities, Central Java's medicinal plant industry imports more than it exports, (3) growth, a stage in which a value of TSI rises to between 0.01 and 0.80, Central Java's medicinal plant industry makes large-scale production and starts rapid scale up of exports in domestic markets, and the supply of the commodities is greater than the demand, and (4) maturity, a stage in which a value of TSI ranges between 0.81 and 1.00, medicinal plant products are in the stage of technology standardization, and Central Java is considered a net exporter.

3. Results and Discussion

3.1. Competitiveness of Medicinal Plant Commodity Export of Central Java viewed from Comparative Advantage

Revealed Comparative Advantage (RCA) in the present research is defined if export share of medicinal plant commodities in total commodity export of Central Java is greater than that of medicinal plant commodities in total commodity export of Indonesia. It is, therefore, expected that Central Java province possess comparative advantage in producing efficiency-based specialized medicinal plant commodities over other countries.

Table 5. The growth of RCA value of Central Java's medicinal plants during the periods of 2011-2015

Year	RCA value
2011	9.70
2012	16.31
2013	6.56
2014	4.17
2015	5.87
Total	42.61
Mean score	8.52

Table 5 indicates that mean score of the value of RCA for medicinal plant commodities of Central Java province during periods of 2011-2015 was above 1 (8.52), meaning that during the periods, medicinal plant commodities of Central Java has strong comparative advantage in international markets. The great value of comparative competitiveness of medicinal plant commodity export of Central Java depends very much on the change in exchange rate of Indonesian Rupiah against the U.S. dollars. This occurs since the existing RCA value is influenced by dollar value of exports. Either strengthening or weakening exchange rate of Indonesian Rupiah against the U.S. Dollar exerts a significant influence on competitiveness of medicinal plant commodities of Central Java province.

In addition, land resources availability is one of determinant factors of comparative analysis of the competitiveness seen from the availability of land resources used to produce medicinal plant commodities. Central Java province had width of planting area of medicinal plants of 6,530.39 ha with production of 131,890 tons in 2015 [7]. Table 2 presents regencies/ cities in Central Java province with the most abundant production of medicinal plants, and indicates that Wonogiri regency had the most abundant production in 2015. Table 4 provides an overview of exported production of medicinal plants. Medicinal plant products had been exported to more than 24 countries, but some exports had not been continuously done. Several countries which have continuously been export markets of medicinal plants include Bangladesh, the Netherlands, India, Japan, Germany, Malaysia, Pakistan, the Kingdom of Saudi Arabia, Singapore, Thailand, the United Arab Emirates, and the U.S.A. The biggest export markets are Malaysia, Singapore, Japan, and the Kingdom of Saudi Arabia.

The optimal production of medicinal plants as herbs raw material is determined by such factors as cultivation environment, cultivation techniques, as well as a means of production. The former includes such requirements for growth as climate, growing media, and elevation. Meanwhile, the middle and the latter involve seed varieties used, seed preparation and nursery, land cultivation, planting, care, and harvesting (the Standard Operating Procedure issued by Indonesian Spices and Medicinal Crops Research Institute (*Balai Penelitian Tanaman Rempah dan Obat*) (2004).

Products of herb plants sold by farmers in two categories: fresh products and crude drugs (chopped and dried medicinal plants). The former were produced by farmers in common, while the latter by farmers who also work as product collectors. Medicinal plants as material of herb under SNI No. 01-7084-2005 can be identified by the content of secondary metabolites (active compounds) with medicinal properties and by physical aspects/ properties of the material. Factors determining the content of secondary metabolites include agroclimatic environment, variety, growing media, as well as such physical environments during plant growth as precipitation, intensity of light, moisture, and preharvest-harvest-postharvest treatment. Physical properties of the material cover cleanliness, the growth of fungi, the presence of foreign material, age, as well as the condition of rhizome. In addition to rhizome age, such properties are influenced by harvest treatment and cleaning process. Less hygienic cleaning, or cleaning using dirty water, poor environmental condition and container lead to the growth of fungi and microbes on the material. Traditional medicine industry supplying herb plants requires of dryness levels, physical properties, appearance, color and cleanliness.

Table 6. The Production of Medicinal Plants according to Province in 2015

No	Province	The number of production (kg)
1	East Java	144.172.421
2	Central Java	100.739.724
3	West Java	86.369.331
4	South Sulawesi	47.841.950
5	Bengkulu	24.239.060
6	Others	94.337.970

Source : Central Bureau of Statistics, 2016 [11].

Efforts to develop medicinal plants bring about positive benefits for socio-economic aspect of local area in the form of employment absorption in order to reduce unemployment level since the medicinal plant cultivation is done by local farmers living in the surrounding neighborhood. In the aspect of cultivation, farmers do not find any significant difficulties/problems.

Marketing problems are the obstacles mostly faced by farmers in determining product selling price. Farmers occupy price-takers position and product collectors regulate the selling price. Problem dealing with price fluctuation, or the absence of price warranty, is another challenging obstacle. For improving farmers' bargaining power, the Agricultural Department has provided guidance for farming groups to join Association of Medicinal Plant Farmers (*APTO—Asosiasi Petani Tanaman Obat*) which establishes trading partnership program with merchants in Medicinal Plant Industries.

3.2. Competitiveness of Medicinal Plant Commodity Export of Central Java viewed from Competitive Advantage

Analysis for measuring competitive advantage of competitiveness of medical plant commodity export of Central Java was done using Trade Specialization Index (TSI) method. The competitive advantage is useful to depict Central Java's position of medicinal plant commodities, whether it tends to be an exporter or importer, as well as to put medicinal plant industry in a competitive position which enables to give an overview of capabilities obtained from performance of maximum advantage in the framework of additional value of technologies or government's policy instruments.

Table 7. The growth of TSI value of Central Java's medicinal plant commodities during the periods of 2011-2015

Year	TSI Value
2011	0.96
2012	0.93
2013	0.97
2014	0.99
2015	0.99
Total	4.84
Average	0.97

Table 7 demonstrates the average score of TSI of Central Java's medicinal plant commodities for the period of 2011 to 2015, which is 0.97, indicating that positive value of the index ranges between 0 and 1. Hence, medicinal plant commodities are considered to have strong competitiveness. In other words, Central Java tends to play a role as exporter of medicinal plant commodities. TSI can also be used to identify the position of competitiveness of a commodity in trade. The competitiveness of Central Java's medicinal plant commodities shows TSI average score ranging from 0.81 to 1.00. It is concluded that medicinal plant commodities of Central Java are in stage of maturity.

The usages of traditional medicines are often debatable among surrounding healthcare service providers. With the presence of the Minister of Health Regulation number 003/MENKES/PER/I/2010 on Traditional Medicine Scientific Validity, traditional medicines formulated from medicinal plants today can be consumed along with chemical medicines. As raw materials for traditional medicines in traditional medicine industry (TMI), medicinal plants have to compete with synthetic chemical materials/compounds and active compounds (bioactive materials) which are used by medicine manufactures or pharmaceutical industries. However, some of people's behaviors to consume traditional medicines and modern people's tendencies to seek for alternative medication with the principle of "going back to nature" are prospective opportunities for medicinal plants. The tendencies are attributable to the side effects of synthetic medicines and antibiotics, as well as opinion that natural materials are relatively safer than synthetic materials. According to Corinthian Infopharma Corpora-CIC (2000), the consumptions of traditional medicines appear to have increased 5.4% each year on the average. The opportunities are more widely opened along with the government's policies to develop Indonesian traditional medicines into phytopharmaceutical industry. The increase in medicinal plant global consumption and trade of 20.96% each year on the average is a chance for developing traditional medicine industry which directly increases the demand for medicinal plants to produce traditional medicines.

Central Java is the second biggest producer of medicinal plants in Indonesia [5]. The large-scale production can show that Central Java is able to meet the export demands of medicinal plant commodities with more competitive and relatively lower price compared to those of other countries (importers) in term of the largest export market share. This can be seen in Table 8.

Table 8. Price of exported medicinal plants in 2015

Country	Export Price				Average
	Cardamom/ HS 090830	Ginger / HS 091010	Turmeric/ HS 091030	Others/ HS 091091	
Indonesia	1.25	0.70	1.21	2.08	1.31
Malaysia	11.11	1.81	2.25	3.38	4.64
Singapore	7.28	0.66	1.76	4.55	3.56
Saudi Arabia	8.21	1.11	1.87	1.19	3.10
Japan	11.84	8.13	58.37	10.88	22.31

Source: United Nations Commodity Trade, 2017[6].

Export price of medicinal plants is lower for the exported products have not met the specification demanded by importers. This is because the contents of active materials/essential oils in Indonesian products are lower than those of products from other countries like Malaysia, Singapore, Saudi Arabia, and Japan.

On one side, several western countries open their markets to stimulate customs through trade contracts. On the other side, they strengthen supervision on imported medicines, decide or improve requirements of relevant technologies through various actions of medicine use safety and protection, including standards to examine the microorganisms content,antiseptics and standards of product quality, standard techniques and requirements of environmental protection to extract plants containing heavy metals, pesticide residue, and aflatoxin to improve acceptable threshold of technology trade obstacle and green trade barrier [12].

4. Conclusion

On the basis of the RCA analysis results for research data periods of 2011 to 2015, Central Java's commodities of medicinal plant of *simplisia* appear to have the average score of 8.52, indicating a strong comparative advantage in global markets. Meanwhile, referring to TSI analysis results, the commodities appear to have the average score of 0.97 at 0 to 1 range, representing a strong competitiveness. In other words, Central Java has more tendencies to be exporter. Based on the average score, the position of the competitiveness of Central Java's medicinal plants ranges from 0.81 to 1.00, and hence, the commodities are at stage of maturity.

Acknowledgements

We would like to express our sincere gratitude to Universitas Sebelas Maret for funding this research entitled "Analysis of Competitiveness of Central Java's *Simplisia* with Medicinal Properties for Supporting Traditional Medicine Industries". Our sincere appreciation also extended to all parties who have provided information to complete this article.

References

[1] Ministry of Trade of the Republic of Indonesia 2009 *Indonesian herbal: The tradisional therapy* (Republic of Indonesia : Trade Research and Development Agency Ministry of Trade)

[2] Dagmar L 2004 Medicinal and aromatic plants: Trade, production, and management of botanical resources. *Proc. XXVI IHC – Future for Medicinal and Aromatic Plants Eds. : 177-197.* Can. Int. Dev. Agency (CIDA)

[3] Zuhud E A, Ekarelawan M and Riswan S 1994 Hutan tropika Indonesia sebagai sumber keanaekaragaman plasma nutfah tumbuhan obat. *Jurnal Pelestarian Pemanfaatan Keanekaragaman Tumbuhan Obat Hutan Tropika Indonesia*. (Bogor :The study program of Forest Conservation, The Faculty of Forestry of IPB and Lembaga Alam Tropika Indonesia)

[4] Badan Pusat Statistik 2016 *Indonesia dalam angka 2016* (Jakarta : Badan Pusat Statistik)

[5] Badan Pusat Statistik 2016 *Indikator utama pertanian Jawa Tengah 2015* (Jawa Tengah : Badan Pusat Statistik)

IOP Conf. Series: Earth and Environmental Science **142** (2018) 012018 doi:10.1088/1755-1315/142/1/012018

[6] United Nations Comodity Trade 2017 *Ekspor impor tanaman obat 2011-2015 (*Statistic Division)

[7] Dinas Perindustrian dan Perdagangan 2016 *Statistik perdagangan ekspor impor Jawa Tengah 2011-2015* (Jawa Tengah : Disperindag)

[8] Suradisastra K, Suherman, and Dariah A 2014 *Inovasi kreatif untuk membangun daya saing komoditas pertanian, dalam memperkuat daya saing produk pertanian* (Jakarta : Badan Penelitian dan Pengembangan Pertanian. Kementerian Pertanian)

[9] Balassa B A 1965 *Trade liberalization and revealed comparative advantage*. Manchester School of Economic Studies **32** 99-122

[10] Tambunan T 2004 *Globalisasi dan perdagangan internasional* (Bogor : Ghalia Indonesia)

[11] Badan Pusat Statistik 2016 *Statistik tanaman biofarmaka* (Jakarta : Badan Pusat Statistik)

[12] Yao L 2003 *An analysis on Chinese and foreign market of traditional Chinese medicine* (Beijing: China Medical Science and Technology Press)

The growth response of *Artemisia annua* L. to organic fertilizer type in lowland

L Prasetyo[1], Y Widyastuti[2], A Yunus[3] and Samanhudi[3]
[1] Student of Agrotechnology Study Program, Faculty of Agriculture, Universitas Sebelas Maret Surakarta, Indonesia
[2] Research Centre for Medicinal Plant and Traditional Drug, Tawangmangu, Indonesia
[3] Lecturer of Study of Agrotechnology, Faculty of Agriculture, Universitas Sebelas Maret Surakarta, Indonesia

E-mail: pras.latif@gmail.com

Abstract. *Artemisia annua* L. is a medicinal plant known in long period of time. *Artemisia annua* has a drug content therein, the compound is artemisinin, these compounds are useful as anti-malarial compounds. Growth of *Artemisia annua* L. in normal conditions is on the plateau. Planting can be done in lowland, but there is a risk that must be faced. This study was conducted to evaluate the response of the growth of the plant *Artemisia annua*.conducted in lowland. This research was conducted at the Laboratory Jumantono of the Faculty of Agriculture, University of Sebelas Maret Surakarta. Data were analyzed using analysis of variance and if there is a significant difference continued with Duncan Multiple Range Test (DMRT) level of 5%. The results showed that the application of goat manure has a positive effect on plant height by 201.9 cm, the number of branches by 57, 30.67 ml root volume and root length of 25 cm, and weight 12.4 grams interest.

1. Introduction

Malaria can be prevented by keeping the environment clean so that mosquitoes do not breed. Someone who has been suffer with malaria should immediately get a response. Drugs used to treat malaria may be obtained from various sources and one of them is from the plant *Artemisia annua*. *Artemisia annua* L. is a medicinal plant known in a fairly long period of time. This plant comes from China and has been spread to many countries such as America, Europe, and South America. Artemisia annua has a drug content therein, artemisinin compounds, these compounds are useful as antimalarial compounds [1].

Cultivation of *Artemisia annua* L. can be done in generative and vegetative. Growth of *Artemisia annua* L. in normal conditions is on the plateau. Planting can also be done at lower altitudes, but there will be a risk that must be faced. Lowlands is the location with a very different situation from the highlands, ranging from temperature, humidity, and also the intensity of the light received. Indonesia also started planting this plant, but there are still some obstacles. It is a constraint is the content of artemisinin remains low. Another problem is the growing point of the plant, Artemisia is a plant that grows in the highlands. Altitude can reach 1,000 m, while in Indonesia mostly been growing vegetables at that height [2].

Planting in the lowlands require special treatment to speed up the production process of the plant Artemesia annua L., one of them with organic fertilizers that serve to support the adequacy of nutrients on plant growth. Organic fertilizer is considered safe for the environment and does not give adverse effects to the environment. Organic fertilizers commonly used in agriculture are manure.

2. Methods

2.1. Location and Time Research
This research was conducted from July 2015 to December 2015. This research was located in Jumantono Laboratory research at the Faculty of Agriculture of March University of Surakarta

2.2. Tools and Materials
The tools used in this research are plastic pots, hoe, and ruler. The materials used are *Artemisia annua* and manure.

2.3. Research Design and Data Analysis
This research used a complete randomized design (CRD) with one levels of treatment and three replications. Data were Analyzed using analysis of variance and if there is a significant difference continued with Duncan Multiple test (Duncan Multiple Range Test) level of 5%.

3.Results and Discussion

3.1. General Conditions Research Sites
Location of planting is situated in the Jumantono Laboratory research at the Faculty of Agriculture of March University of Surakarta in the village Sukosari, District Jumantono, Karanganyar. The exact location of the location is at 7° 37' 42.1" N and 110° 57' 6.9" S. This research location has a height of 196 m above sea level. Laboratory Jumantono has some area with various plants.

3.2. Variable of Plant Growth
Table 1. Treatment of manure application on several variables observation.

Manure	Plant Height (cm)	Number of branch	Roots Volume (ml)	Roots Length (cm)	Flowers Weight(g)
Goat	201.9	57.08	30.67	25	12.47
Chickens	187	53	29.75	24.4	11.2
Cow	147.3	42.5	25.33	22.1	8.07

3.2.1. Plant height Plant height is one of the indicators or parameters used to determine the effect of a given treatment. Plant height can indicate how much nutrients are met from the growing media used. In Table 1 it can be seen that the combination treatment showed the highest plant height was treated by goat manure with an average height of 201.9 cm. While the lowest plant height is shown by the treatment of cow manure with an average height of 147.3 cm. Previous study [3] revealed that the giving of NPK proven to accelerate the growth of plant height. Plants in the vegetative phase requires a high N fertilizer. Nutrient deficiencies can cause a variety of disorders of the plant. Some of them are plant growth may decline, occur yellowish color of the leaves, the process of the formation of chlorophyll disturbed [4].

3.2.2 Number of Branch Number of branches on plants to determine how many leaves and flowers as well as indicators in observing plant growth. The more the number of branches on a plant, the interest earned will also be more. In Table 1, the average number of best branches present in goat manure treatment with the average number of branches 57.08. Treatment using cow manure show the lowest branch with 42.5. In the other study [5], revealed that the use of manure can increase plant height and number of branches of the plant. This is because the content of Nitrogen, Phosphorus and Potassium that contained in manure. The addition of manure can provide nutrients is quite high, thus affecting the growth of plants and the number of branch plants. Fertilizer given since the plants are still small where the plant is actively growing, making the number of branches increased [6].

3.2.3 Roots volume Root on plant become a very important part of the growth. The process of absorption of water and nutrients also be an important basis for growth is done by root. Plants can stand tall and live in various soil conditions caused by roots that can sustain a good crop in the field or in pots. In table 1 shows that the root volume is highest in the treatment of goat manure. As for treatment with the use of cow manure showed the smallest volume. Manure on crops can provide a positive influence is the increasing volume of the roots of plants. That is because the provision of P fertilizer to the plants, of P into nutrients essential for reproductive role in the development section [7].

3.2.4. Length of the root Length of the plant roots become one of the important indicators to determine growth. Conditions normal root can help the absorption of nutrients and water from the soil, so that plants grow better. Long roots can absorb more nutrients and water when compared with short rooted plants. In Table 1 indicate that goat manure showed the average length of the roots of the best that is 25 cm. whereas treatment with the shortest root is with cow manure by 22.1 cm. The study [8] revealed that the organic fertilizer can help the growth of the roots and make the plant roots become longer. This can provide benefits for the plant because the roots have a wider range to take nutrients. It is also in accordance with the conclusions from [9], which revealed that the addition of goat and chicken manure can increase the growth.

3.2.5. Flowers weight Flowers weight on the plant became one of the important indicators in the study the response of *Artemisia annua* in the lowlands. Plants that generate a lot of interest indicates that the plant can grow well. *Artemisia annua* is a plant that can be propagated with flowers. In Table 1 goat manure treatment showed the greatest result with 12.47 g. The treatment of cow manure has only average with 8.07 g. Treatment of manure give real effect and gives good results on the weight of the *Artemisia annua* flowers. This is according to the previous study [10] that the use of an assortment of manure showed significantly different results to the flowers.

4. Conclusions

The treatment of manure increased the growth of the observed variables in plant height, number of branches, root volume, and flowers weight. Goat manure gives the best results compared to other fertilizers.

Acknowledgement

This research was supported by the Ministry of Research Technology and Higher Education and Dr. Ir. Yuli Widyastuti as field from Research Centre for Medicinal Plant and Traditional Drug, Tawangmangu, Indonesia. Artemisia annua seeds in this research were obtained from B2P2TOOT, Tawangmangu, Karanganyar.

References

[1] Bhakuni R S, Jain D C, Sharma R P, Kumar S 2001 Secondary metabolites of *Artemisia annua* and their biological activity *Current science* **80**(01) 35-48

[2] Lestari E G, Purnamaningsih R, Gratitude M, Yunita R 2010 Diversity somaklonal to repair the plant artemisia (Artemisia annua L.) through in vitro culture *Journal agro.biogen* **6**(01) 26-32

[3] Napitulu D, Winarto L 2010 Effect of N and K fertilizer application on the growth and production of red onion *Horti Journal* **20**(1) 27-35

IOP Conf. Series: Earth and Environmental Science **142** (2018) 012019 doi:10.1088/1755-1315/142/1/012019

[4] Ruhnayat A 2007 Determination of basic needs nutrients NPK plant growth vanilla (Vanillaplanifolia Andrews) *Littro Bulletin* **18**(1) 45-59

[5] Baskoro D, Purwoko B S 2011 Influence of Plant Propagation Material and Organic Fertilizer Types Against Binahong Plant Growth (*Anredera cordifolia* (Ten.) Steenis) *J. of Horti Indonesia* **2**(1) 6-13.

[6] Susanti H, Aziz S A, Jasmine M 2007 Production of biomass and bioactive ingredient ginseng *(Talinumtriangulare* (Jacq.) willd) of various origin seedlings and doses of chicken manure *Agronomy Bulletin* **36**(1) 48-55.

[7] Noor A 2003 Influence of natural phosphates and phosphate solvent bacteria combination with manure on available P and soybean growth in Ultisol *Agronomy Bulletin* **31**(3) 100-106

[8] Rizqiani N F, Ambarwati E, Yuwono N W 2007 The effect of dose and frequency of administration of liquid organic fertilizer on growth and yield of beans *(PhaseolusvulgarisL.)* lowlands *J. of Soil and Environ. Sci.* **7**(1) 43-53

[9] Agus M 2014 Growth, leaf biomass production, and secretory structures bitter (Andrographispanicula Ness.) in the shade treatment and organic fertilizer *Essay. Agricultural Institute of Bogor* Bogor Asrob 2010

[10] Susilowati Y E, Tujiyanta 2008 Use of Various Organic Fertilizers on results Several varieties of strawberries *Agron. J.* **30**(2) 126-139

IOP Conf. Series: Earth and Environmental Science **142** (2018) 012020 doi:10.1088/1755-1315/142/1/012020

Polyploid response of *Artemisia annua* L. to colchicine treatment

A Yunus[1], Parjanto[1], Samanhudi[1], M P Hikam[2] and Y Widyastuti[3]
[1] Lecturer of Study Program of Agrotechnology, Faculty of Agriculture, Universitas Sebelas Maret Surakarta, Indonesia
[2] Student of Study Program of Agrotechnology, Faculty of Agriculture, Universitas Sebelas Maret Surakarta, Indonesia
[3] Research Center for Medicinal Plant and Traditional Drug, Tawangmangu, Indonesia

E-mail: yunus.uns7@yahoo.com

Abstract. Artemisia (*Artemisia annua*) is a a medicinal herb originated from Asia, its contains Artemisinin for malaria (caused by *Plasmodium falciparum*) treatment. . Artemisinin content in *A. annua* are relatively low, ranging from 0.01% -0.5%. In order to increase the Artemisinin content, polyploid induction could be one effort to be done. For that, this experiment aims to examine the effect of colchicine on morphological characteristics and the induction of polyploidization in Artemisia plants. Polyploid induction on *Artemisia annua* L. seeds was performed by soaking the Artemisia seeds in colchicine (0%, 0,05%, 0,1% and 0,2%; concentration based) for 2 hours. The experimental design was Completely Randomized Design, one factor, 4 colchicine treatments and in each treatment 7 replicate. The results showed that polyploid occur in plants treated with 0.05% colchicine concentration and its morphological characteristic are 89.4 cm height, 30 branches, 15.9 CCI chlorophyll content , 0.78 cm stem diameter , and chromosome number 2n = 27. In the stomata density of polyploid plants (treated by 0.05% colchicine) was 130 number/mm^2 with stomata diameter of 22.8 μm.

1. Introduction

Artemisia (*Artemisia annua*) is one of Asia's most well-known herb used in Chinese traditional medicine. *A. annua* L is known as *Qinghao* in China, and is also known as *sweet wormwood*, *sweet annie*, *sweet sagewort* and *annual wormwood*. which contains Artemisinin compounds. The World Health Organization (WHO) recommends Artemisinin-based Combination Therapy (ACT) for the treatment of malaria from *Plasmodium falciparum* due to Plasmodium is resistant to quinidine or quinidine. Artemisinin is effective in treating parasitic diseases, some viral infections and various neoplasms [1]. The artemisinin compound is derived from *Artemisia annua* L extract [2]. One of the obstacles to develop artemisia plants in Indonesia is because this plant is a sub-tropical plant and its rare to cultivate [3]. The other problem is [4] the genotype available on this plant is a 2n = 18 chromosome with resulting very low content of artemisinin, ranging from 0.01% -0.5%. With this low content, it is not meet to the artemisinin on the global market. Therefore, a study aiming to obtain polyploid genotypes of Artemisia for higher artemisinin content purpose (up to 0.5%) need to be done. Colchicine is one of chemical commonly to induce polyploid in plants with high success rates [5]. In addition to the number of different chromosomes, polyploid plants are characterized by different morphology compared to diploid plants.

2. Methods

The materials used in this study were artemisia seeds, colchicine, soil, manure, nail polish, immersion oil, and other chemicals. The chemicals needed for the preparation of chromosome preparate include aquadest, 45% glacial acetic acid, HCl 1 N, and 2% aceto-orcein. The nursery process was conducted from November 2016 - May 2017 in Tawangmangu, Karanganyar (-7° 39.880' S and 111° 07.881' E; 1062 m above sea level; 19°C - 23°C), while planting and observation were in Screen house of Agriculture Faculty, Sebelas Maret University (-7° 33.702 'S and 110° 51.501' E;53 m above sea level;26°C - 30°C).

Seeds at 3 weeks old after the nursery were followed with soaking in colchicine concentrations (treatments; each 7 replicates) 0%, 0.05%, 0.1% and 0.2% for 2 hours and then seeds were planted in polybag. Plants at 4 weeks old after planting were transferred to Screen house of Agriculture Faculty, Sebelas Maret University. Stomatal data collection was done by collecting the sample at 10.00 am from the leaves that has been 11 weeks old. The copy of stomata leaf were then observed under microscope at 400x and 1000x magnifications. Chromosomes were calculated using squash method.

3. Results and Discussion

Colchicine treatment may cause side effect on treated plants [6]. One of indication to identify polyploid in plants is observing the morphological aspects which is height. The plant height data were shown in Table 1.

Table 1. Artemisia height from various colchicine concentration treatment

Colchicine (%)	Lowest (cm)	Highest (cm)	Range (cm)	Average (cm)
0	63	102	63 – 102	85.93
0.05	30	89.4	30 – 89.4	66.73
0.10	52	118	52 – 118	77.53
0.20	26	97	26 – 97	62.47

Plants treated with 0.2% colchicine have the lowest plant height of all treatments which is 26 cm. This is suggested due to colonic mutation factor from polyploid which resulting in slow growth rate. According to the previous study [7], the general effect of polyploid in plants is slower growth rate compared to diploid. Colchicine treatment to plants will result to different effect on morphology of each plant, including the number of branches. The calculation results of the number of branches were shown in Table 2.

Table 2. Number of branches of Artemisia from various concentration of colchicine

Colchicine (%)	Lowest	Highest	Range	Average
0	24	42	24 – 42	36.57 b
0.05	16	40	16 – 40	24.71 a
0.1	17	52	17 – 52	32.43 ab
0.2	11	40	11 – 40	26.57 a

Description : Number followed by same character shows no significant different based on DMRT 5%

The lowest number of branches of the various treatments is shown by plants treated with.2% colchicine was 11 branches. According to previous study [8], leaves characteristics such as leave area, size of stomata, and chlorophyll content in polyploid plants were significantly higher than diploid plants. Polyploid plants usually have a higher performance, thicker leaves, more green leaf color, larger number of branches, larger lengths and leaf widths, and larger flowers [9]. The results of leaf measurements can be seen in Table 3 and Table 4.

Table 3. Artemisia leaves length from various colchicine treatment

Colchicine (%)	Lowest (cm)	Highest (cm)	Range (cm)	Average (cm)
0	3.67	6.17	3.67 – 6.17	5.55
0.05	4	9.17	4 – 9.17	6.73
0.1	4.82	6.33	4.82 – 6.33	5.62
0.2	6	8.17	6 – 8.17	6.6

Table 4. Artemisia leaves widht from various colchicine treatment

Colchicine (%)	Lowest (cm)	Highest (cm)	Range (cm)	Average (cm)
0	2	6.3	2 – 6.3	3.38
0.05	1.5	4.83	1.5 – 4.83	3.14
0.1	3	3.84	3 – 3.84	3.57
0.2	2	4.67	2 – 4.67	3.62

The highest leaf length of all treatments was 9.17 cm at 0.05% colchicine concentration treatment. As seen from the highest leaf lengths value, it showed that the treated plants had a longer size compared with the highest leaf length of the control treatment. The effect of ploidation on each plant will be different according to the gene reaction and physiological reaction of the plant. As shown from the highest leaf width size of all treatments, shows that the treated plants with colchicine have a lower leaf width compared with the highest leaf width in control treatment. The result of [10] study show that the length and width of polyploid *Artemisia annua* L. leaf by colchicine treatment has smaller size compared with diploid plants. The leaves appearance of potentially polyploid of rubbery plants is greener, more rounded, thicker, and wavy than the control plants [11]. The thickness of plant leaves was done at the end of the observation (16 weeks). The results of the plant leaf thickness measurement are presented in Table 5.

Table 5. Artemisia leaves thickness from various colchicine treatment

Colchicine (%)	Lowest (cm)	Highest (cm)	Range (cm)	Average (cm)
0	0.04	0.07	0.04 – 0.07	0.06
0.05	0.04	0.06	0.04 – 0.06	0.1
0.1	0.04	0.09	0.04 – 0.09	0.06
0.2	0.04	0.06	0.04 – 0.06	0.05

The lowest leaf thickness in all treatments given to Artemisia plant was 0.04 cm. In the control treatment, the highest leaf thickness was 0.07 cm. In plants treated with a concentration of 0.05% and 0.2% had the highest leaf thickness with the same value of 0.06 cm, while the treatment of 0.1% colchicine concentration yielded the highest artemisia leaf thickness of 0.09 cm. This is not in accordance with research [12] stating that the higher the concentration of colchicine, the thicker leaves produced on the Siam Simadu crop. The result of stem diameter measurement shown in Table 6.

Table 6. Artemisia stem diameter from various colchicine treatment

Colchicine (%)	Lowest (cm)	Highest (cm)	Range (cm)	Average (cm)
0	0.38	0.54	0.38 – 0.54	0.47
0.05	0.25	0.79	0.25 – 0.79	0.49
0.1	0.3	0.69	0.3 – 0.69	0.48
0.2	0.32	0.54	0.32 – 0.54	0.53

The treatment of 0.05% colchicine concentration produce the lowest stem diameter of 0.25 cm and the highest stem diameter of 0.79 cm. The highest stem diameter of all treatments was 0.79 cm in the treatment of 0.5% colchicine concentration. Based on research results [13], watercolors plant treated with colchicine showed larger stem diameter than control treatment. Chlorophyll is a green coloring pigment in plants, algae, and photosynthetic bacteria. These pigments play a role in the photosynthesis of plants by absorbing and converting light energy into chemical energy. According to [14], polyploid tissue is suspected to contain more chlorophyll so the leaf color looks greener. Calculation of chlorophyll content is done using chlorophyll meter, as shown by Table 7.

Table 7. Artemisia chlorophyll content from various colchicine treatment

Colchicine (%)	Lowest (CCI)	Highest (CCI)	Range (CCI)	Average (CCI)
0	8.8	11,8	8.8 – 11.8	10.3 a
0.05	13.1	24,7	13.1 – 24.7	18.21 b
0.1	8.5	21,6	8.5 – 21.6	14.66 ab
0.2	9.3	19,7	9.3 – 19.7	15.31 b

Description : Number followed by same character shows no significant different based on DMRT 5%

The highest chlorophyll content of all treatments was found in plants treated with 0.05% colchicine concentration with a value of 24.7 CCI. As shown from the chlorophyll content, it concluded that plants with the highest levels of chlorophyll are indicated as polyploid plants. When seeds are treated with colchicine, mitosis in embryonic cells is followed by proplastid cleavage, although the chromosomes that have been doubled may fail to split on the anaphase due to damage to microtubule formation of spindle by colchicine compounds to produce plants with higher chlorophyll content [15]. Stomata are pores formed by a pair of special cells (guard cells), which are found on the surface of plant leaves and can be openned and closed to control the gas exchange between the plant and its environment. According to [16], stomata analysis is a functional and economical method of determining the level of plant ploidies. The result of stomata density measurement of Artemisia can be seen in table 8.

Table 8. Average density of Artemisia stomata

Colchicine (%)	Unindicated Plant (number/mm^2)	Indicated Plant (number/mm^2)
0	210	-
0.05	270	130
0.1	250	190
0.2	190	160

Source: Observation result

Plants that are not indicated as polyploid have greater stomatal density. Plants induced by colchicine-treated polyploids have lower stomatal density averages with an average stomatal density of 130 number/mm^2 at 0.05% colchicine concentration treatment. At a concentration treatment of 0.1%, the density of stomata was 190 number/mm^2, while in the treatment of 0.2% colchicine concentration yielded stomata density 160 number/mm^2. This may occur due to the smaller diameter of the stomatal diameter of the diploid plant when compared to the stomata size by the treatment of some polyploid-induced concentration of colchicine. Here is a picture of leaves stomata of various colchicine concentration treatment at 1000x magnification.

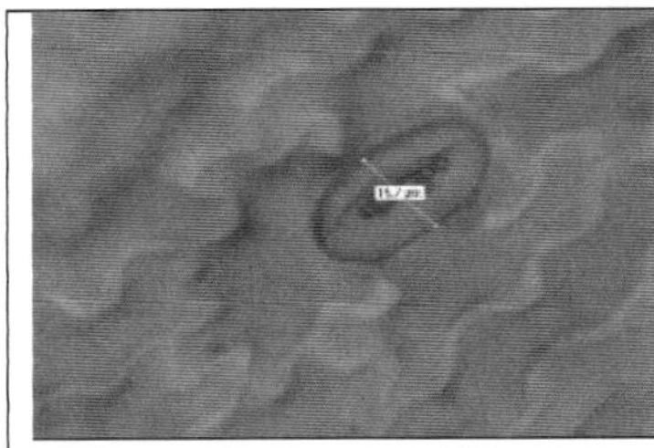

Figure 1. Stomata Diameter Without Colchicine Treatment (15.7 µm)

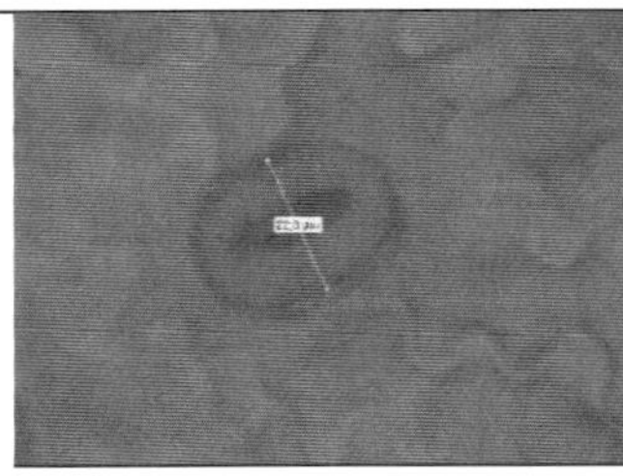

Figure 2. Stomata Diameter from Polyploid-Indicated Plant of 0.05% Colchicine Treatment (22.8 µm)

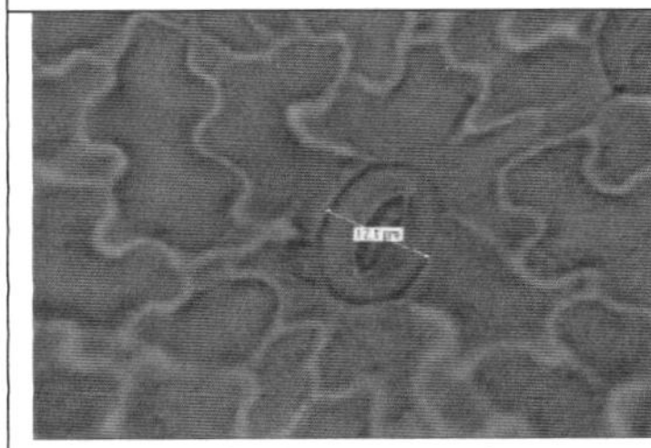

Figure 3. Stomata Diameter from Polyploid-Indicated Plant of 0.1% Colchicine Treatment (17.1 (µm)

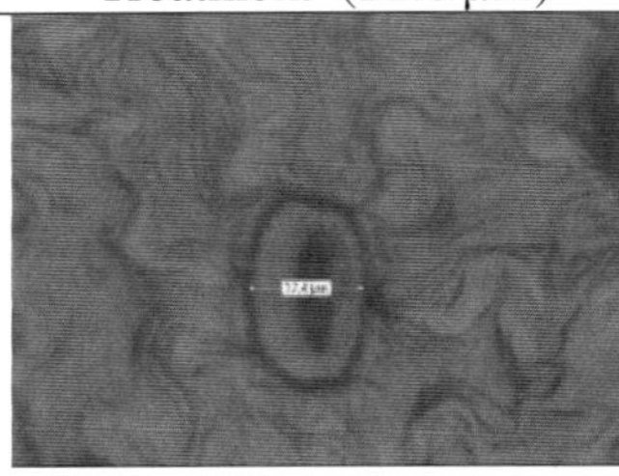

Figure 4. Stomata Diameter from Polyploid-Indicated Plant of 0.2% Colchicine Treatment (17.1 (µm)

Based on Figure 1, it was shown that in the control plants, the size of stomatal leaf diameter was 15.7 µm. Plants indicated as polyploid have larger stomata size than diploid plants. This is in accordance with the opinion of [11] which states that the higher the level of ploidy chromosome, the bigger stomata leaves will be. The density of stomata of polyploid plants is lower than diploid plants because of the size of stomata and epidermal cells are larger in polyploid plants [17]. According to [16], chromosom number on mitotic root cells were performed to confirm the extent of plant ploidies in new cells formed after colchicine treatment. Colchicine induction is a mechanism often used to induce mutations, resulting in changes in shape, size and number of chromosomes (reduction or addition) [18]. Chromosome calculation was done by squash method, as shown on the images below.

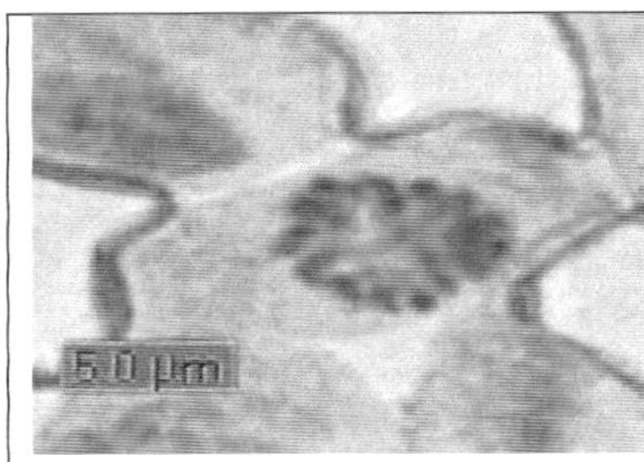

Figure 5. *A. annua* Chromosome Without Colchicine Treatment (2n =18)

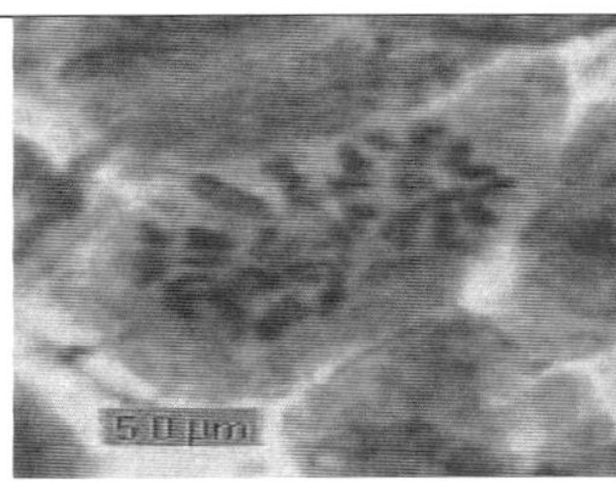

Figure 6. *A. annua* Chromosome from 0,05% Colchicine Treatment (2n = 27)

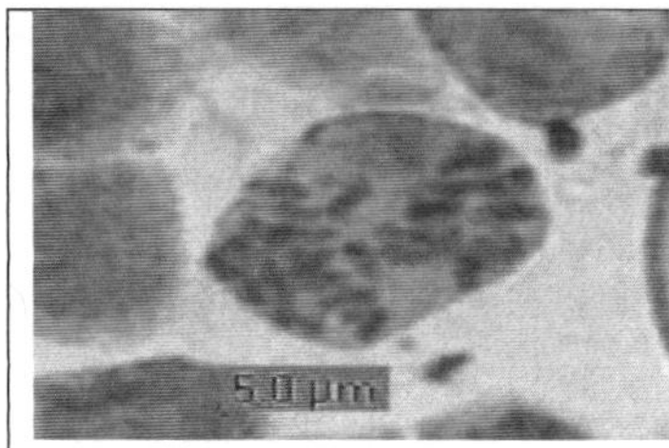

Figure 7. *A. annua* Chromosome from 0,05% Colchicine Treatment (2n = 24)

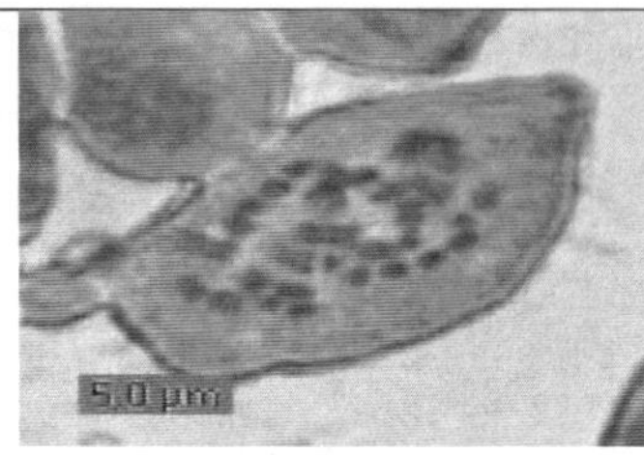

Figure 8. *A. annua* Chromosome from 0,2% Colchicine Treatment (2n = 32)

According to [3], the number of chromosomes of the Artemisia genus is x = 8 or x = 9. Based on research results [19], before polyploid was induced, *A. annua* was tested cytologically to confirm the amount of diploid chromosomes (2n = 18). In Figure 5, the number of chromosomes of diploid plants (without treatment) was 2n = 18. Figure 6 shows the number of chromosomes of *A. annua* due to treatment of colchicine concentration of 0.05% was 2n = 27. Figure 7 shows the number of chromosome of the same concentration (0.05%) was 2n = 24. Figure 8 shows that the 0.2% concentration of colchicine treatment to Artemisia seed yields the number of plant chromosomes of 2n = 32.

4. Conclusions

The polyploid were shown in plants treated by 0.05% colchicine concentration with plant characteristics having a height of 89.4 cm, number of branches of 30, chlorophyll content of 15.9 CCI, stem diameter of 0.78 cm, and number of chromosomes of 2n = 27. The density of stomata in polyploid indicated plants treated by 0.05% colchicine concentration has a density of 130 number/mm^2 with stomata diameter of 22.8 μm.

Acknowledgement

This research was supported by the Ministry of Research Technology and Higher Education and Dr. Ir. Yuli Widyastuti as field from Research Center for Medicinal Plant and Traditional Drug, Tawangmangu, Indonesia. *Artemisia annua* seeds in this research were obtained from B2P2TOOT, Tawangmangu, Karanganyar.

References

[1] Weathers P J, Arsenault P R, Covello P S et al. 2011 Artemisinin production in *Artemisia annua*: studies in plantaand results of a novel delivery method for treating malaria and other neglected diseases *Phytochem Rev* **10(2)** 173–183 DOI: 10.1007/s11101-010-9166-0

[2] [WHO] World Health Organization 2006 Monograph on good agricultural and collection practices (GACP) for *Artemisia annua* L. (Geneva (CHE): WHO Press)

[3] Ermayanti T M, Hafiizh E A, Rahmawati S, et al. 2009 Transformasi *Artemisia cina* dan *Artemisia annua* dengan *Agrobacterium rhizogenes J. of Applied and Industrial Biotechnology in Tropical Region* **2**(2) 1-5 ISSN: 1979-9748

[4] Lestari E G, Purnamaningsih R, Syukur M et al. 2010 Keragaman somaklonal untuk perbaikan tanaman artemisia (*Artemisia annua* L.) melalui kultur in vitro *J. AgroBiogen* **6**(1) 26-32

[5] Rantau D E, Hafiizh E A, Rahman W et al. 2014 Analisis ukuran dan kerapatan stomata pada *Artemisia annua* L hasil perlakuan kolkisin dalam Sigit, Sihono, Prayitno et al. (eds) *Prosiding Seminar Nasional ke 53 temu-ilmiah jaringan kerjasama kimia Indonesia* Yogyakarta, 13 November 2014 Jaringan kerjasama kimia Indonesia yayasan media kimia utama P45-52 ISSN: 0854-4778

[6] Zeng S H, Chen C W, Hong L et al. 2006 In vitro induction, regeneration and analysis of

IOP Conf. Series: Earth and Environmental Science **142** (2018) 012020 doi:10.1088/1755-1315/142/1/012020

autotetraploids derived from protoplasts and callus treated with colchicine in *Citrus* sp. *Plant Cell, Tiss Organ Culture* **87** 85–93

[7] Sinaga E J, Bayu E S, Hasyim H 2014 Pengaruh konsentrasi kolkhisin terhadap pertumbuhan dan produksi kacang hijau (*Vigna radiata* L.) *J. Online Agroekoteknologi* **2**(3) 1238-1244 ISSN: 2337-6597

[8] Herawati M M, Pudjihartati E, Pramono S 2015 Obtaining *Artemisia cina* polyploidy through plant growth regulator treatment in shoot culture *Argivita* **37**(2) 178-184 ISSN: 0126-0537.

[9] Amiri S, Kazemitabaar S K, Ranjbar G, dan Azadbakht M 2010 The Effect of trifluralin and colchicine treatments on morphological characteristics of Jimsonweed (*Datura stramonium* L.) *Trakia J. Sci.* **8**(4) 47–61 ISSN: 1313-3551

[10] Banyai W, Sangthong R, Karaket N, et al. 2010 Overproduction of artemisinin in tetraploid *Artemisia annua* L *J. Plant Biotechnology* **27** 427-433

[11] Sukamto L A, Ahmad F, Wawo A H 2010 Pengaruh oryzalin terhadap tingkat ploidi tanaman garut (*Maranta arundinacea* L.) *Bul. Littro* **21**(2) 93-102

[12] Yulianti F, Purwito A, Husni A, Dinarti D 2015 Induksi tetraploid tunas pucuk jeruk siam simadu (*Citrus nobilis* Lour) menggunakan kolkisin secara *In vitro J. Agron. Indonesia* **43**(1) 66-71

[13] Wiendra N M S, Pharmawati M, Astiti N P A 2011 Pemberian kolkisin dengan lama perendaman berbeda pada induksi poliploidi tanaman pacar air (*Impatiens balsamina* L.) *J. Biologi* **15**(1) 9-14 ISSN: 1410 5292

[14] Rahayu E M D, Sukma D, Syukur M et al. 2015 Induksi poliploidi *Phalaenopsis amabilis* (L.) Blume dan *Phalaenopsis amboinensis* J. J. Smith dengan kolkisin dalam kultur in vitro. *J. Agron. Indonesia* **43**(3) 219-226

[15] Syaifudin A, Ratnasari E, Isnawati 2013 Pengaruh pemberian berbagai konsentrasi kolkhisin terhadap pertumbuhan dan produksi tanaman cabai (*Capsicum annum)* varietas Lado F1 *LenteraBio* **2**(2) 167-171 ISSN: 2252-3979

[16] Miguel T P, Leonhardt K W 2011 In vitro polyploid induction of orchids using oryzalin *Scientia Horticulturae* **130** 314–319. DOI: 10.1016/j.scienta.2011.07.002

[17] Gantait S, Mandal N, Bhattacharyya S et al. 2011 Induction and identification of tetraploids using in vitro colchicine treatment of *Gerbera jamesonii* Bolus cv. Sciella *Plant Cell, Tissue and Organ Culture* **106** 485-493 DOI: 10.1007/s11240-011-9947

[18] Suminah, Sutarno, Setyawan A D 2002 Induksi poliploidi bawang merah (*Alium ascalonicum* L.) dengan pemberian kolkisin *Biodiversitas* **3**(1) 174-180 ISSN: 1412-033X DOI: 10.13057/biodiv/d030102

[19] Gonzales D J, Weathers P J 2003 Tetraploid *Artemisia annua* hairy roots produce more artemisinin than diploids *Plant Cell Rep* **21**: 809-813 DOI 10.1007/s00299-003-0587-8

Artemisia annua respon to various types of organic fertilizer and dose in lowland

A Yunus[1], Samanhudi[1], N Brahmanto[2] and Y Widyastuti[3]

[1] Lecturer of Study Program of Agrotechnology, Faculty of Agriculture, Universitas Sebelas Maret Surakarta, Indonesia

[2] Student of Faculty of Agriculture, Universitas Sebelas Maret (UNS) Surakarta, Indonesia

[3] Research Center for Medicinal Plant and Traditional Drug, Tawangmangu, Indonesia

E-mail: yunus.uns7@yahoo.com

Abstract. *Artemisia annua* belongs to asteraceae genus which has many benefits in the medical field. Artemisia contains artemisinin which is used to cure malaria disease. The obstacle of artemisia development in Indonesia is low artemisinin content and the fact that artemisia only able to grow well in the highland area. For that this experiment aimed to increase the artemisinin content through enhancing artemisia biomass in the lowland using the application of organic fertilizer. Experiment was conducted in GreenhouseLab, Faculty of Agriculture, Sebelas Maret University, Surakarta from October 2015 to January 2016. Two factor of treatment and three replications was performed during experiment. The first factor is the organic fertilizer type and the second is the application dose. Result showed that Rabbit manure at 40% application dose give best influence on the plant height (172,62 cm), number of branches (68,3 branch), flowering time (102,67 day after planted), fresh weight (56,47 g) and dry weight (43,15 g), moreover Rabbit manure at 80% dose give the best influence on the root length (27,33 cm).

1. Introduction

Artemisia (*Artemisia annua* L.) is asteraceae plant having distinctive aroma or unique taste and benefits in the medical field. Artemisia is a medicinal plant originated from China and generally cultivated in Asia, America, and Europe [1]. The main problem of artemisia development in Indonesia are low content of artemisinin and the fact that artemisia only grow well on highlands with altitude more than 1,000 m [2].

Many factors are influencing the artemisinin content and one of them is the biomass. Artemisia usually grown in the long day field. The flowering stage of artemisia will be faster when artemisia planted in the tropical field. The faster of flowering stage will affected the biomass and the biomass will decreased. In the tropical field, artemisia will be able to produce high content of artemisinin if planted in highlands or using artemisia variety with longer flowering stage [3].

According to [4], organic and inorganic fertilizers have a positive effect on the artemisias growth. Application of fertilizer positively effect to the vegetative growth and biomass. Moreover, biomass can increase the level of artemisinin content. In this regard, organic fertilizers are more preferable due to have sustain nutritional benefit.

2. Methods

2.1. Location and Time Research
This experiment was conducted at , Faculty of Agriculture, Universitas Sebelas Maret, Surakarta from October 2015 to January 2016. Analysis of the yield was conducted in Ecology Plant Production Physiology and Biotechnology Laboratory, Faculty of Agriculture, Universitas Sebelas Maret, Surakarta in February 2016.

2.2. Experimental Design and Data Analysis
This research used two factors of treatment and three replications. The first factor is the type of fertilizer which consists of rabbit manure (P1), goat manure (P2), cow manure mixed with husk (P3) and without fertilizer treatment (P0). The second factor is the percentage of dose which consists of 20% dose (D1), 40% dose (D2), 60% dose (D3), 80% dose (D4), and 100% dose of fertilizer (D5). Growth data resulted from the experiment were analyzed using ANOVA and P value was set at 0.05.

3. Results and Discussion

3.1. Plant height
Interaction between fertilizer type and the percentage of dose have no significant effect (P> 0.05) on the height. The average height is shown on Figure 1. The height is varies between 172.67 cm to 96 cm. The treatment resulted highest average of height is the rabbit manure with the 20% of dose (172,62 cm), while lowest average of height is the goat manure with 100% of dose (96 cm).

Figure 1 showed the fertilizer application with certain dose give positive effect to artemisia height. However, if organic fertilizeris applied in the excessive way, it will cause obstructed growth of artemisia. According to [5] and [6], excessive application of organic fertilizer affected to increment of soil acidity. For that, plants cannot utilize N, P, K and other nutrients needed. Moreover, in the acid soils, plants have high probability to be poisoned by Al. Thus right amount of organic fertilizer application must be done to obtain optimum results.

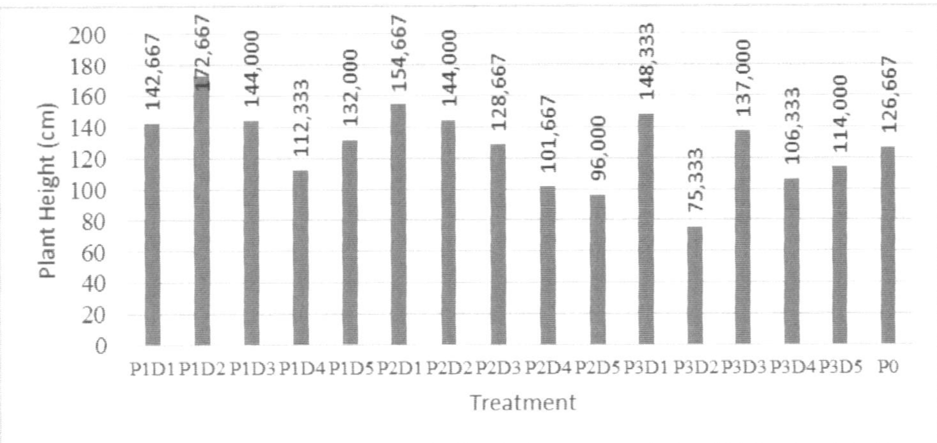

Figure 1. Effect the fertilizer type and percentage of dose to artemisia height

3.2. Number of Branch
Interaction between fertilizer type and the percentage of dose have no significant effect (P> 0.05) on the artemisia branches. The average branches that can be seen in Figure 2 varies between 68.3 and 42.3. The treatment showed highest number of branches of plants is the rabbit manure with 40% percentage of dose (68.3), while lowest number of branches is the goat manure with 100% percentage

of dose (42.3). Increasing the doses can stimulate the lateral meristem activity and nutrient uptake, especially for N, because N is important for vegetative growth and form a new branch [7].

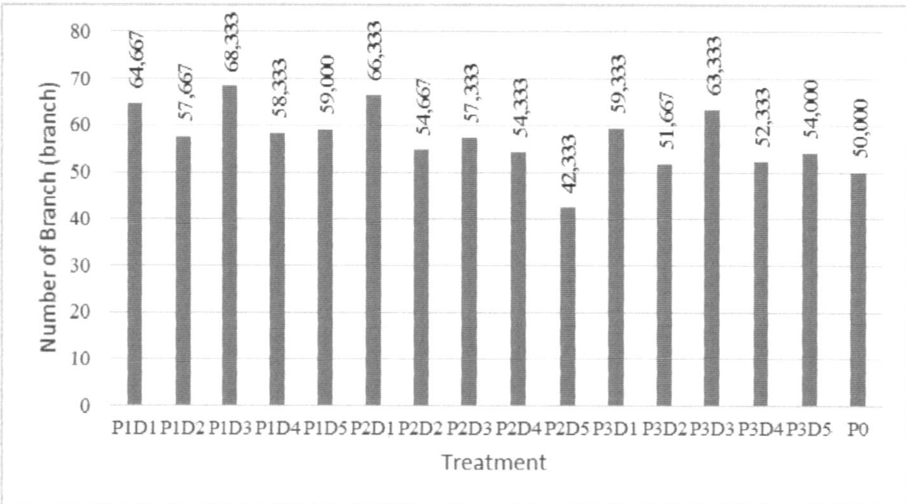

Figure 2. Effect the fertilizer type and percentage of dose to number of branch

Nitrogen is very important for the chlorophyll formation. Chlorophyll in sufficient quantities can increase the leaves ability to absorb sunlight, so the photosynthesis process will run smoothly. The product of the photosynthesis will be revamped through the respiration process and generate the energy that will be needed by the cell to perform activities such a cell division and enlargement. Phosphorus that contained in rabbit manure has a functions to develop the meristem tissue. Potassium is an activator for various essential enzymes, photosynthetic and respiratory reactions, and protein synthesis enzymes [8].

3.3. Root Lenght

Interaction between fertilizer type and the percentage of dose have no significant effect ($P > 0.05$) on the root length. The average length of the roots shown in Figure 3 varies between 27.33 and 13.33 cm. The treatment showed highest average of the root length is rabbit manure with 80% dose (27,33 cm), while lowest average of is in the control treatment (13.33 cm).

According to [9] application of organic fertilizers can increase the pH, affect the aggregates, and increase the porosity of soils. Soils that has a good porosity will have sufficient water availability. Rabbit manure can be used to improve soil physical properties. Crumbly soil can be easily penetrated by roots. P in the fertilizer also plays an important role for the growth and root branching. These factors will affect the absorption of roots for water or for nutrients in the soil [10].

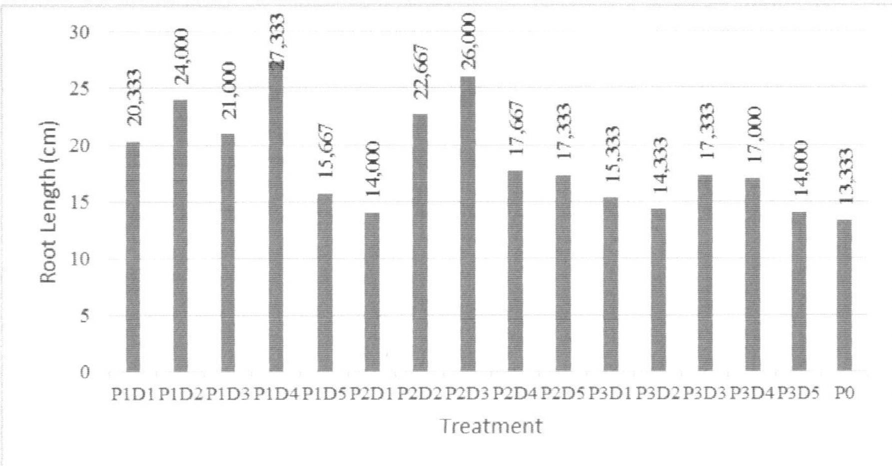

Figure 3. Effect the fertilizer type and percentage of dose to root length

3.4. Flowering Day

Interaction between fertilizer type and the percentage of dose have no significant effect (P> 0.05) on the flowering day. The average flowering day varies between 86.33 to 102.67 day after planted. The treatment that has highest average flowering day is the rabbit manure with 40% dose (102.67 day) after planted. The treatment that has the lowest average flowering day is the rabbit manure with 100% dose (86.33 day) after planted.

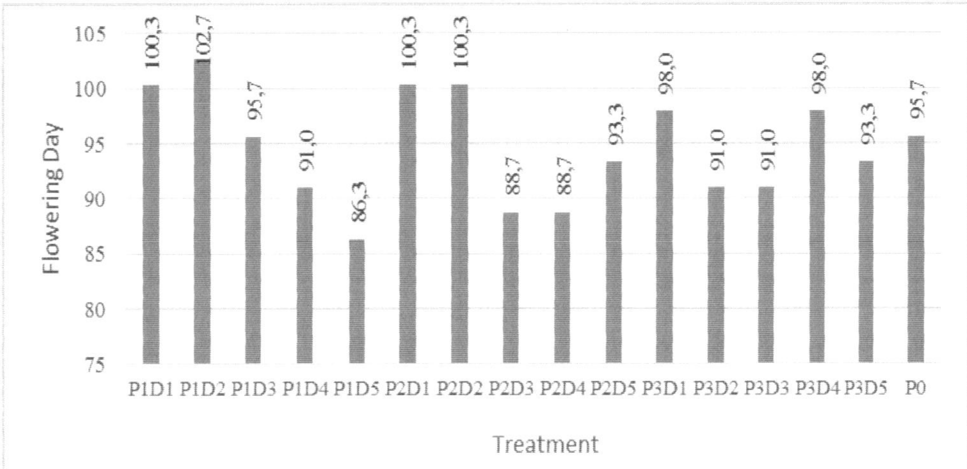

Figure 4. Effect the fertilizer type and percentage of dose to flowering day

Lack of the N will inhibit the vegetative growth and the flowering period will delay. The application of N fertilizer can accelerate the appearance of the first flowers when it compared with the control treatment. The application of N fertilizer in certain doses can also slowing the appearance of the first flowers. Increasing the dose of N fertilizer can stimulate the vegetative growth and multiply the number of flowers. The vegetatif growth can delay appearance of the first flower. If the appearence of the first flower is delay, the plants will maximize the vegetative growth so that the height, number of branches, the number of flowers, and the biomass of the plant will increase [11].

3.5. Fresh Weight

Interaction between fertilizer type and the percentage of dose have no significant effect (P> 0.05) on the fresh weight. The average weight of fresh plant varies between 7.57 to 56.47 g. The treatment that has the highest average fresh weight is the treatment of rabbit manure with 40% dose (56.47 g). The treatment that has the lowest average fresh weight is the treatment of cow manure and husk with 60% dose (7.57 g).

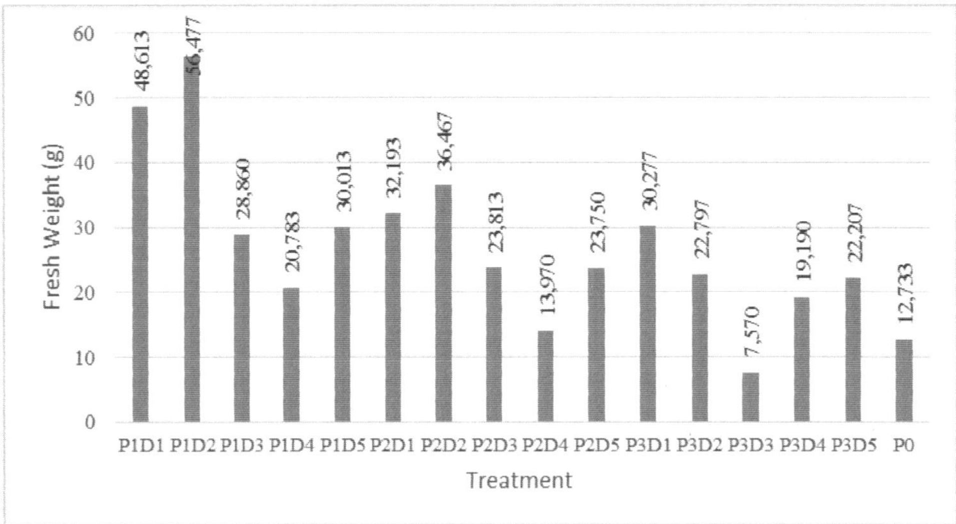

Figure 5. Effect the fertilizer type and percentage of dose to fresh weight

According to [12] the addition of organic fertilizer will increase the process of decomposition in the soil and the availability of nutrients. Organic fertilizers can also be used to improve soil physical properties. Treatment with rabbit manure have the best average fresh weight. Rabbit manure have sufficient nutrient and can be easily absorbed by plants.

3.6. Dry Weight

Interaction between fertilizer type and the percentage of dose have no significant effect (P> 0.05) on the dry weight. The average weight of the dry plant varies between 14.01 and 43.15 g. The treatment that has the highest average dry weight is the rabbit manure with 40% dose (43.15 g). The treatment that has the lowest average of dry weight is the cow manure and husk with 60% dose (14.01 g). All the fertilizers that applied in this experiment able toincrease the dry weight. According to [13], the provision of organic fertilizer can stimulate vegetative growth and increase dry weight of plants. The treatment showed best yield of dry weight has the best nutrient intake.

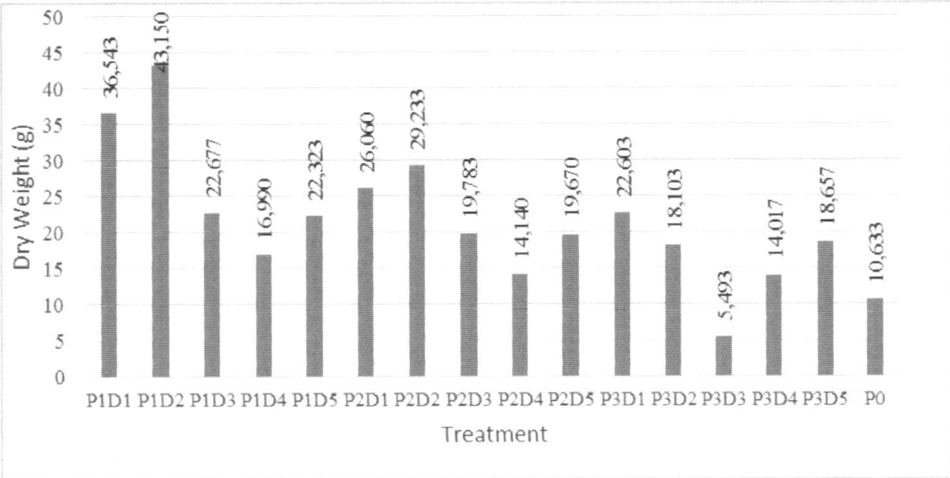

Figure 6. Effect the fertilizer type and percentage of dose to dry weight

4. Conclusion

Rabbit manure with 40% of dose give best influence on the plant height (172,62 cm), number of branches (68,3 branch), flowering time (102,67 day after planted), fresh weight (56,47 g) and dry weight (43,15 g). Rabbit manure with 80% of dose give the best influence on the root length (27,33 cm).

Acknowledgement

This research was supported by the Ministry of Research Technology and Higher Education and Dr. Ir. Yuli Widyastuti as field from Research Center for Medicinal Plant and Traditional Drug, Tawangmangu, Indonesia. Artemisia annua seeds in this research were obtained from B2P2TOOT, Tawangmangu, Karanganyar.

References

[1] Ozguven M, Sener B, Orhan I et al. 2008. *Ind Corps Prod.* **27** 60-64
[2] Purnamaningsih R, Lestari E G, Syukur M, and Yunita R 2010 *J. Ilm. Apl. Iso. Rad.* **6**(2) 139-146
[3] Ferreira J F S, Laughlin J C, Delabays, and Magalhaes P M 2005 *Plant Gen. Reso.* **3**(2) 206-229
[4] Yeboah S, Akromah R, and Quansah C 2012 *Afri. J. Agr. Res.* **7**(2) 177-182.
[5] Efendi 2010. *J. Floratek* **5** 65 -73
[6] Patti P S, Kaya E and Silahooy C 2013 *J Agrologia* **2**(1) 51-58
[7] Sukarman 2012 *J. Littri.* **18**(2) 81-87
[8] Sitompul H F, Simanungkalit S, and Mawarni L 2014 *J. On Agroekotek.* **2**(3) 1064-1071
[9] Masulili A 2010 *J. Agric. Sci.* **2** (1) 39-47
[10] Kabir R, Yeasmin S, Mominul A K and Sarkar A R 2013 *Int. J. Bio-Sci Bio-Tech.* **5**(3) 51-60
[11] Aminifard M H, Aroiee H, Nemati H, Azizi M and Khayyat M 2012 *J. Pln.t Nutri.* **35** 235–242
[12] Tiamiyu R A, Ahmed H G, and Muhammad A S 2012 *Nig. J. Bas. Apl. Sci.* **20**(2) 213-216
[13] Olaniyi J O and Ojetayo A E 2011 *J. Anim. Plnt. Sci.* **12**(2) 1573-1582

IOP Conf. Series: Earth and Environmental Science **142** (2018) 012022 doi:10.1088/1755-1315/142/1/012022

Shading and vermicompost effect on growth and flavonoid content of Tapak Liman (*Elephantopus scaber* L.)

R Y A Dawiyah[1], A Yunus[2], Samanhudi[2], Y Widiyastuti[3] and Widodo[3]
[1]Undergraduate Student of Agrotechnology Major, Agriculture Faculty Universitas Sebelas Maret, Surakarta, Indonesia
[2]Lecturer Staff of Agrotechnology Major, Agriculture Faaculty of Universitas Sebelas Maret, Surakarta, Indonesia
[3]Research center for Medicinal Plant and Traditional Drug, Tawangmangu, Indonesia

E-mail:yunus.uns7@yahoo.com

Abstract. Tapak Liman (*Elephantopus scaber L*) is one of Indonesian medicinal plants which is well known as weed. In Thailand, Tapak Limanthis plant is use for traditional medicine due to its flavonoids contains. Flavonoid is compound with red, yellow, purple and blue pigments, used for cancer, aphrodiasiac and anti-radical treatments. One obstacle of Tapak liman cultivation is the effort to increase its flavonoids compound. There is a bridge between flavonoids compound with growth and yield of Tapak Liman. For that, this research aims to find out the effect of shade intensity combined with vermicompost dosage on Tapak Liman growth and yield. This research was conducted in Mei to August 2016 at Medicinal Plantation of BPTO, Tanjungsari Village, Tegal Gede, Karanganyar. Complete Randomized Design compiled with split plot and two factors: shade intensity (0%, 50%, 75%) and vermicompost dosage per plant (0 g, 250 g, 500 g, 750 g) used as the experimental design. The variables observed are leaves number,leaves length, canopy diameter, fresh weight, dry weight, root length, chlorophyl analysis and flavonoid identification. Data were analyzed using ANOVA, any significant treatments followed with Duncan's Multiple Range Test (DMRT) at $\alpha = 10\%$. Result showed that 75% shade intesity and 750 g of vermicompost has gave highest yield of leaf and total simplicia of Tapak Liman. Shade intensity of 50% with 250 g of and 500 g/plant of vermicompost dosage showed highest flavonoid rendement (Rf 0,5) with highly contrasting spot colors.

1. Introduction

Tapak Liman (*Elephantopus scaber*) is a plant containing several compounds such as dihydro deoxyelephantopus, lupeol, stigmasterol and flavonoids [1] which can be used as medicine for various diseases. Types of diseases that can be treated with Tapak Liman include various inflammations such as inflammation of the tonsils, influenza, sore throat, inflammation of the eyes, acute inflammation of the kidney and inflammation of the uterus and vaginal discharge [2].

The study shows that flavonoids are affected by different light intensity [3]. A similar trend to increase flavonoid content by light intensity is also seen in some medicinal plants, as shown in [4] research. Many flavonoid compounds contained in the leaves, so the simplicia leaves of Tapak Liman often used as a medicinal. To increase the quantity and quality of simplicia leaf of Tapak Liman, it require intensive cultivation.

One effort to increase the simplicia of the Tapak Liman is by applying vermicompost to the soil. Vermicompost is a high quality fertilizer due to its various nutrients content which needed by plants. Vermicompost contains a lot of N in available form that is used for the formation of plant organs especially leaves, but also N influential in flavonoid formation process [5]. Thus, the application of shade treatment and vermicompost expected to give effect to the flavonoids content in Tapak Liman.

2. Methods

The research was conducted in BPTO Medicinal Plant Garden, Tanjungsari Village, Tegal Gede, Karanganyar. The field experiment was conducted from June to August 2016. The geographical location of the research site was -7^0 36' 687" S and 110^0 57' 766" E. The altitude of the site was approximately 215 meters above sea level. Daily temperature ranging from 27^{0}C - 30^{0}C. The materials used in this research were the seeds of Tapak Liman (72 seedlings), planting medium in the form of soil, vermicompost fertilizer, polybag, bamboo, 50% and 75% paranet, treatment board, ethanol 0.1N, Tapak Liman leaf extract and aquades.

This research was performed using CRD experimental design which arranged in split plot. Treatment consists of shade as main plot and vermicompost dose as sub plot. Shade treatments consists of 0% (without shade), 50% and 75%. The dose of vermicompost (per polybag) consists of 0 g, 250 g, 500 g and 750 g. Each treatment was repeated 3 times and each experimental unit consist of 2 sample plants. The data obtained were analyzed using ANOVA, any significant treatments were followed with Duncan Test at $\alpha = 10\%$.

3. Results and Discussion

3.1. Tapak Liman leaves simplicia

3.1.1. Leaves weight The result of shade and vermicompost treatment level on leaf weight of Tapak Liman is shown in Figure 1 and 2.

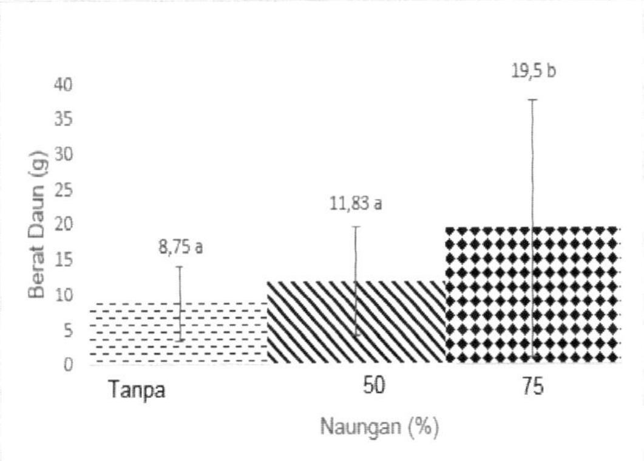

Figure 1. Effect of shade on Tapak Liman leaves weight

Plants with a shade treatment of 75% gave significant difference results compared to plants without shade and 50% shade. The 75% shade treatment has the highest average leaves weight value of 19.5 grams/plant, whereas without shade and 50% shade treatment produce 10,29 gram/plant leaf weight. [6]. Shade serves to reduce the radiation received leaves and reduce water loss so dehydration can be avoided. This causes the weight of the leaves to increase in the shaded condition, because the leaves store more water.

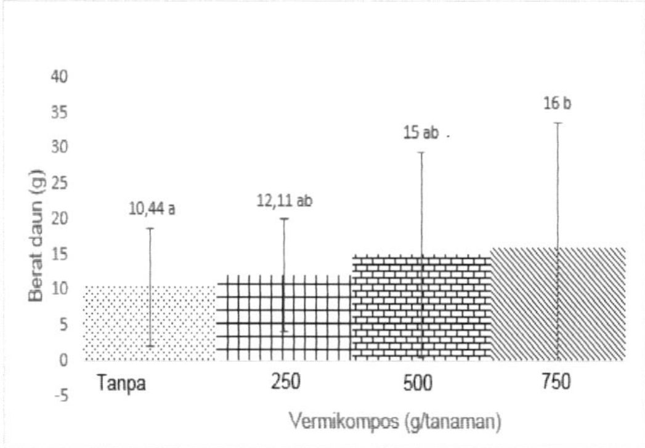

Figure 2. Effect of vermicompost on Tapak Liman leaves weight

Plants with 750 g of vermicompost treatment shows the highest average leaves weight of 16 g/plant, while the dose of 250 g/plant and 500 g / plant treatment resulted in average leaves weight of 13.55 g/plant. The lowest average leaves weight produced by plants without vermicompost which is 10.44 g/plant [7]. The formation of leaves by plants is strongly influenced by the availability of nutrients N and P on planting medium. In the treatment without vermicompost the plants undergo nutrient deficiency, because the planting media lacksof nutrients.

3.1.2. Number of leaves The effect of shade intensity and vermicompost treatment on leaves number shown in Figure 3 and 4.

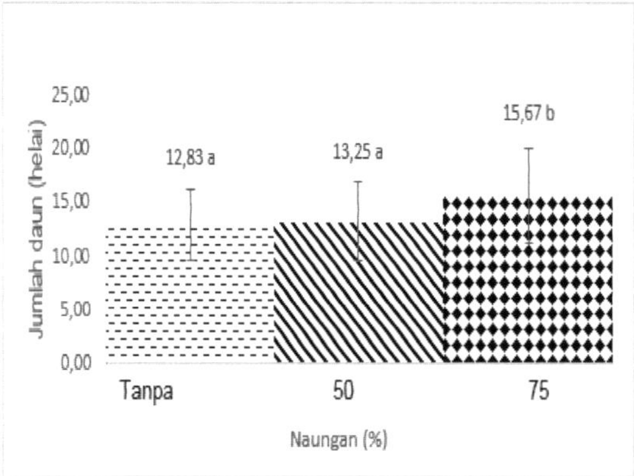

Figure 3. Effect of shade on the number of leaves of Tapak Liman

Plants with 75% shading treatment gave the highest average yield of leaves of 15.67 pieces, while the plants with shade treatment of 50% and without shade resulted in average leaves number of 13.04 pieces.

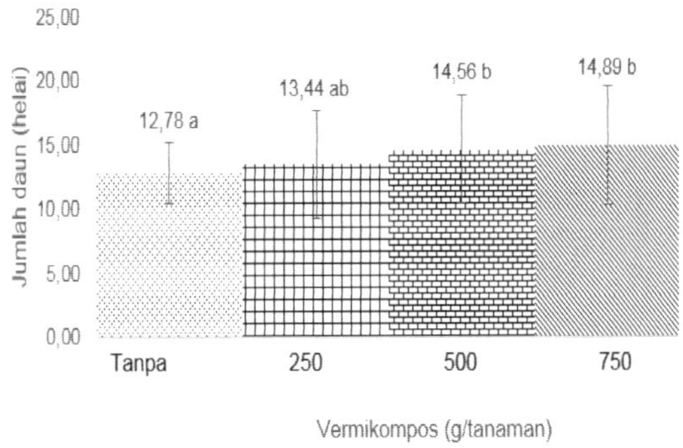

Figure 4. Effect of vermicompost on the number of leaves of Tapak Liman

Plants without vermicompost produce the lowest average of 12,78 sheets, while the highest average leaves number of 14.72 pieces was produced by the treatment of vermicompost with a dose of 500 g / plant and 750 g / plant. Plants with a dose of vermicompost 250 g / plant yields a total of 13.44 pieces of leaves. Tapak Liman on the treatment without the application of vermicompost fertilizer experienced obstacles in the formation of plant leaves. [8] The availability of N in sufficient quantities will facilitate plant metabolism and will eventually affect the growth of organs such as stems, leaves, roots for the better. Leaves number variables have significantly different correlation (r = 0,342) with root length. That was, the greater length of the root, the more the number of leaves produced.

3.1.3. Leaf length and canopy diameter The result of analysis of variation of shade intensity and vermicompost dose to leaf length and canopy diameter showed no significant difference to Tapak Liman leaf length. Treatment of shade and vermicompost on leaf length variables also did not cause interaction. The influence of shade on the leaf length and vermicompost of Tapak Liman can be seen that the average length of leaf Tapak Liman was 13.31 cm. The average canopy diameter produced by the shade and vermicompost treatment was 25 cm.

3.1.4. Chlorophyl content The result of shade treatment effect on chlorophyll content is shown in Figure 5.

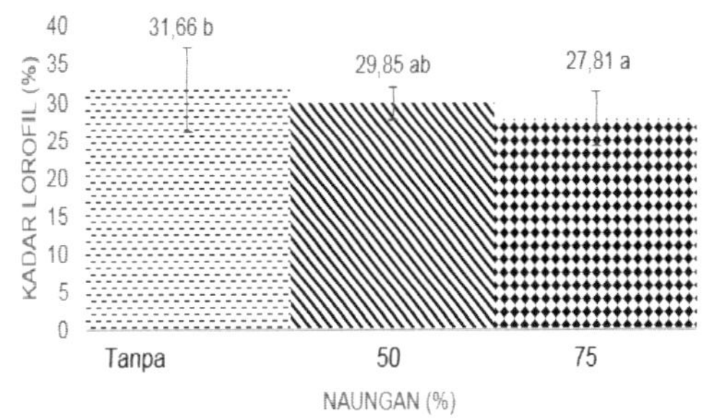

Figure 5. Effect of shade on the chlorophyll content of Tapak Liman

Plants without shade showed highest average yield of the chlorophyll content while the 75% shade treatment resulted in the lowest average chlorophyll content of 27.81%. 50% shade treatment resulted in average chlorophyll content of 29.85%. The higher shade level, , the less chlorophyll produced. This happens because, in a state of light with low intensity, protein synthesis for chlorophyll formation decreases and more leads to the occurrence of leaf senescene [9].

3.2. Tapak Liman root simplicia

3.2.1. Root weight and root length The result of analysis of variance (anova) of shade and vermicompost effect showed no significant difference to the root weight and root lenght of the Tapak Liman. It is known that the average root weight of the Tapak Liman produced by the shading treatment and vermicomost treatment is 5.79 grams. The result of shading treatment and vermicompost treatment yielded an average root length of 35.80 cm.

3.3. Total simplicia of Tapak Liman

3.3.1. Fresh weight The result of analysis of variance (anova) on the effect of shade and dose of vermicompost to fresh weight of plant showed significant difference (Figure 6)

.

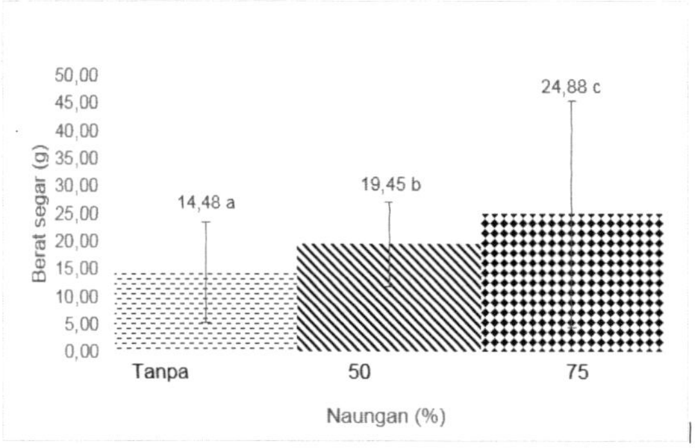

Figure 6. Effect of shade on the fresh weight of Tapak Liman

Plants with 75% shading treatment showed the highest average fresh weight yield of 24.88 g/plant. While without shade treatment showed the lowest average fresh weight of 14.48 g/plant. The shade treatment of 50% shows the average yield of fresh weight of 19.45 g/plant.

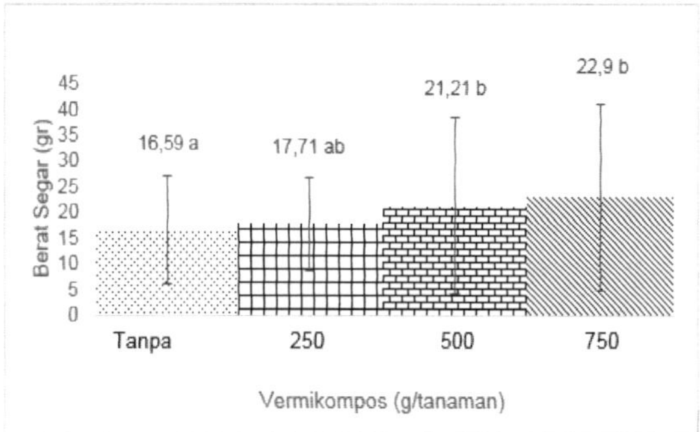

Figure 7. Effect of vermicompost on the fresh weight of Tapak Liman

The highest average fresh weight produced by plants with treatment of 500 g and 750 g vermicompost is 22.9 g/plant, while the lowest average fresh weight produced by plants without vermicompost of 16.59 g/plant (Figure 7). Plants with a vermicompost treatment of 250 g/plant yield average fresh weight of 17.71 g/plant. The fresh weight of the plant is influenced by the increase in the length and volume of the plant caused by expansion of plant cells [10]. This is what causes the increase in fresh weight of the plant.

3.3.2. Dry weight The results of variance analysis showed that only shade treatment had a significant effect on the dry weight of Tapak Liman (Figure 8).

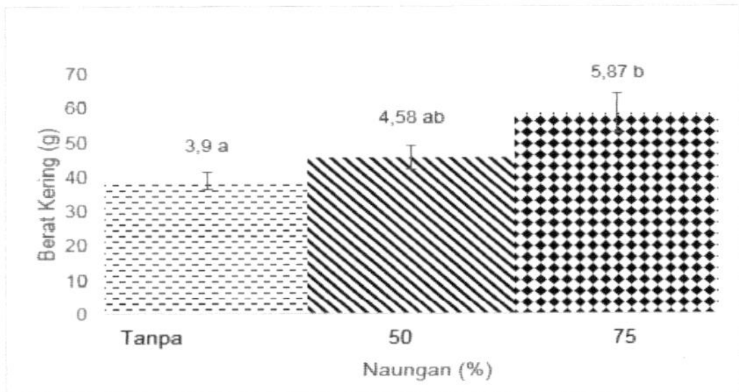

Figure 8. Effect of shade on the fresh weight of Tapak Liman

The highest average dry weight was produced by plants with 75% shade treatment of 5.87 g/plant, while the lowest average dry weight was shown in no shade treatment which is 3.9 g/plant. Plants with a shade of 50% produce an average dry weight of 4.58 g / plant. [11] says that light with high intensity can lead to growth inhibition associated with the balance of photosynthesis and respiration. Light with high intensity, increase temperature, so that respiration increased and the lower dry weight produced. The lower light intensity results in a decrease in photosynthesis rate so that the dry weight is lower.

3.4. Flavonoid analysis

3.4.1. Extract content It can be seen the highest extract level of Tapak Liman is produced by 50% shade treatment with 500 g vermicompost of 13,33% (Figure 9). The highest extract content was then shown by 50% shade treatmen with 750 gram vermicompost of 10%. From the 12 trreatment combination, 10 treatment combinations were on average yielding an extract level of 6.67%. The content of this extract shows the secondary metabolite content present in plant. The greater level of extract, the more secondary metabolite content in the plant, whereas the fewer extract content, the lower secondary metabolite content in the plant will be. These active compounds will later be identified as flavonoid compounds.

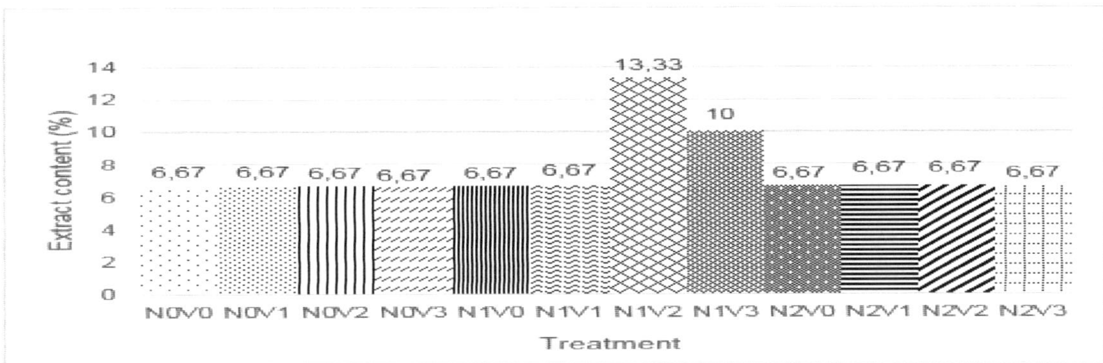

Figure 9. Effect of shade intensity and vermicompost dose treatment on Tapak Liman extract concentration of active compound.

3.5. Flavonoid of Tapak Liman

Flavonoids are accumulated or secreted in the vacuole, as in the roots, but are more often produced in flowers and leaves. These compounds are **usually** formed due to stressful environmental conditions and to protect themselves against pathogens and pests. Flavonoid of Tapak Liman analyzed using KLT test and the results shown in Figure 10.

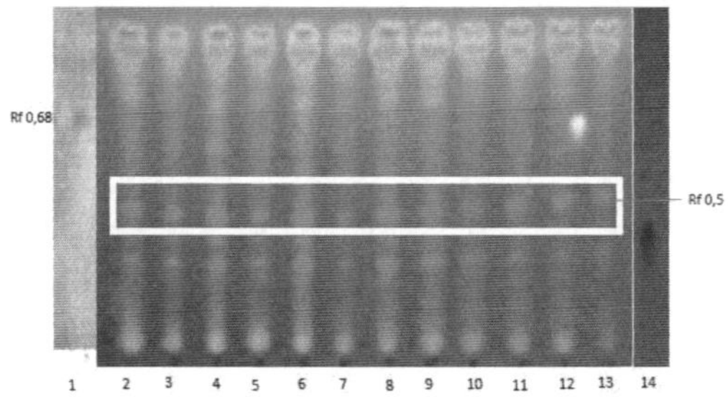

Figure 10. Test result of KLT of flavonoid content

Table 1. Result of flavonoid test of Tapak Liman

Code	Treatment	Spot Diameter	Color	Flavonoid	Harkat
1	Standard	4	3	7	Average
2	Without shade intensity and without vermicompost	3	3	6	Average
3	Without shade intensity and 250 g vermicompost	2	3	5	Low
4	Without shade intensity and 500 g vermicompost	3	4	7	Average
5	Without shade intensity and 750 g vermicompost	3	2	5	Low
6	Shade intensity 50% and without vermicompost	3	4	7	Average
7	Shade intensity 50% and 250 g vermicompost	2	1	3	Very low
8	Shade intensity 50% and 500 g vermicompost	3	4	7	Average
9	Shade intensity 50% and 750 g vermicompost	1	2	3	Very low
10	Shade intensity 75% and without vermicompost	2	1	3	Very low
11	Shade intensity 75% and 250 g vermicompost	1	2	3	Very low
12	Shade intensity 75% and 500 g vermicompost	2	2	5	Low
13	Shade intensity 75% and 750 g vermicompost	2	2	4	Low

Information:

Spot diameter score:	Color score:	Flavonoid score :
1: diameter 0 – 24%	1: nuclear	2 – 3 : Very low
2: diameter 25 – 54%	2: not contrast	4 – 5 : Low
3: diameter 55 – 74%	3: contrast	6 – 7 : Average
4: diameter 75 – 100%	4: very contrast	8 : High

The result of TLC analysis of flavonoids content (Table 1) shown that there are other flavonoid group compounds at Rf 0.5. The compound of flavonoids successfully isolated in the range of Rf 0.45

which is quercetin (Figure 10 No. 14) to Rf 0.68 (Figure 10 Number 1) which is isodeoxyelephantopin [12]. So the spot that appears on Rf 0.5 can be expressed as a class of flavonoids. To know the name of the compound is needed further quantitative analysis. The spots generated from the TLC test were performed on 12 treatment combinations, most of them in the same Rf. This indicates that the flavonoid group found in the 12 treatments were the same. These spots can be measured by scores so that the data obtained into semi-quantitative to find out how much spot generated by each treatment to know whether the amount exceeds the standard value or not. The table of flavonoid scales of Tapak Liman shown as follows.

Without shade treatment with 500 g of vermicompost produce a spot of 66.6% with very bright color brightness in a flavonoid score of 7 (medium) equal to standard (isodeoxy elephantopin), whereas wtihout shade treatment with 750 g of vermicompost produced a spot of equal magnitude of 66.6% but the pale color brightness belongs to a flavonoid score of 5 (low). Differences in the types of flavonoids according to [13] occurs because of the type missing / not appearing due to the possibility of shade. This is evident from the type of flavonoids that decrease in number along with increasing shade levels.

Such spots contained in a 50% shade treatment with 250 g vermicompost and a 75% shade without vermicompost resulted in an almost invisible spot or with very pale color brightness resulting in flavonoid scoring of 3 (very low). Flavonoids have a correlation with dry weight (r = 0.723). This shows that there is a relationship between dry weight and flavonoids. The process of photosynthesis produces a primary metabolite in the form of photosynthate, then some of the photosynthate processed for secondary metabolites, one of the materials for the formation of flavonoid formation.

4. Conclusion

Level of shade of 75% yield the highest result on leaves simplicia variable and total simplicia of Tapak Liman with leaves weight of 19,5 gram, number of leaves of 24,85 piece, and fresh weight of 24,85 gram. The dosage of vermicompost of 750 grams/plant produce the highest yield on the simplicia varieties of leaves and the total simplicia of Tapak Liman with 22.9 grams leaves weight. The shade treatment of 50% with 250 g/plant vermicompost dosage yields the highest extract content of 13%. Without shade treatment and 50% shade with 500 g/plant vermicompost produced the greatest spot of 66.6% with very bright red brightness defined as flavonoids at Rf 0.5 with flavonoid score 7 defined as moderate.

Acknowledgement
This research was supported by the Medicinal Plant and Traditional Drug, Tawangmangu, Indonesia.

References

[1] Ho W Y, Ky H, Yeap S K, Rahim R A, Omar A R. *J Med. Plant Res* **3**(13)
[2] Setyari W and Sudjarwo SA 2008 *J Penelitian Medikal Eksakta* **7**(1) 16-22
[3] Ghasemzadeh A, Jafaar H Z E, Rahmat A 2010 *Molecules* **15**: 6231-43
[4] Chen H G , Yu Y G and Zeng O X 2002 *J Food Sci* **23** 69-71
[5] Dewick P M 2009 Medicinal natural products 3rd edition (United Kingdom: A John Wiley and Sons, Publication)
[6] Edmond J B, Senn F S A and Halfacre R G 1979 Fundamentals of horticulture (New Delhi, Tata Mc Graw-Hill Pub Co) 560
[7] Parashurama T R, Vasanthakumari M M and Shivanna M B 2013 *Int. J. Pharma Bio. Sci.* **4**(2) 871-883 ISSN 0957-6299
[8] Samavatipour P, Salehi R and Abdossi V 2015 *Int. Res. J. Appl. Basic Sci.* **9**(8) 1345-1347 ISSN 2251-838X
[9] Aziez A F, Indradewa D, Yudhono P and Hanudin E 2014 *J Agrineca* **14**(2) 144-127 ISSN:0854-2813

[10] Morais H, Medri M E, Marur J C, Caramori P H, Ribeiro A M and Gomes J C 2012 *Braz. Arch. Biol. Technol.* **47**(6) 863-871 ISSN 1516-8913
[11] Hermanto N, Purnomo D and Dewi WS 2014 El vivo 2(2) 58-66 ISSN: 2339-1901
[12] Kementrian Kesehatan Indonesia 2008 Farmakope herbal indonesia 1st edition Kementrian Kesehatan Republik Indonesia
[13] Kurniawati A, Darusman L K and Rachmawaty R Y 2005 *Buletin Agronomi* **33**(3) 62-67

IOP Conf. Series: Earth and Environmental Science **142** (2018) 012023 doi:10.1088/1755-1315/142/1/012023

A review of phytoremediation technology: heavy metals uptake by plants

A Sumiahadi[1,2] and R Acar[3]
[1]PhD Student of Dept. of Field Crops, Selcuk University, Konya, Turkey
[2]Dept. of Agrotechnology, Muhammadiyah University of Jakarta, Indonesia
[3]Dept. of Field Crops, Selcuk University, Konya, Turkey

Email: ade.sumiahadi@gmail.com

Abstract. Heavy metal is one of the serious environmental pollutions for now days as impact of industrial development in several countries. Heavy metals give toxic effects on human health and cause several serious diseases. Several techniques have been using for removing heavy metal contaminants from the environmental but these techniques have limitations such as high cost, long time, logistical problems and mechanical complexity. Phytoremediation can be used as an alternative solution for heavy metal remediation process because of its advantages as a cost-effective, efficient, environment- and eco-friendly technology based on the use of metal-accumulating plants. According to previous studies, several plants have a high potential as heavy metals bioaccumulator and can be used for phytoremediation process of heavy metals.

1. Introduction

Environmental pollution has became a serious public health concern because it becomes a major source of health risk and causes several serious deseases throughout the world [1]. One of the serious environmental pollutions is heavy metals. Although the health effects of heavy metals have been known for a long time, exposure to heavy metals continues and is even increasing in some areas. The effects of heavy metals on human health can even lead to death [2]. The different heavy metal gives different toxic effects on human health as showed by table 1. Environmental pollution by heavy metals have increased as a influence of industrial development and it was shown that many heavy metals in high level was found in industrial areas [3,4,5].

Heavy metals become a primary concern than other environmental pollutions because heavy metals can't be destroyed by degradation. The remediation process of contaminated soils, groundwater, and surface water by heavy metals needs some methods to remove the metals from contaminated areas [6]. Several methods have being used for removing the pollutants from the contaminated environments. Soils that are contaminated with heavy metals can be treated by acid leaching, soil washing, physical or mechanical separation of the contaminant, electro-chemical treatment, electrokinetics, chemical treatment, thermal or pyrometalurgical separation and biochemical processes [7,8,9,10]. Remediation techniques can be used for removing heavy metals from contaminated ground water are extraction and treatment by activated carbon adsorption, microbes use, air stripping [11], chemical, biological, biochemical and biosorptive treatment technologies [9].

The use of some of these remediation techniques requires a high cost [10,12,13], a long time [11], logistical problems [14] and technical complexity [15]. Therefore alternative sollution is needed for heavy metals removing from the environment. Bioremediation is an innovative and promising technology available for removal of heavy metals and recovery of the heavy metals in polluted water and lands [16].

Table 1. Toxic effects of some heavy metals on human health [16]

Heavy Metal	EPA[a] Regulatory Limit (ppm)	Toxic Effects
Ag	0.10	Exposure may cause skin and other body tissues to turn gray or blue-gray, breathing problems, lung and throat irritation and stomach pain.
As	0.01	Affects essential cellular processes such asoxidative phosphorylation and ATP synthesis
Ba	2.00	Cause cardiac arrhythmias, respiratory failure, gastrointestinal dysfunction, muscle twitching and elevated blood pressure
Cd	5.00	Carcinogenic, mutagenic, endocrine disruptor, lung damage and fragile bones, affects calcium regulation in biological systems
Cr	0.10	Hair loss
Cu	1.30	Brain and kidney damage, elevated levels result in liver cirrhosis and chronic anemia, stomach and intestine irritation
Hg	2.00	Autoimmune diseases, depression, drowsiness, fatigue, hair loss, insomnia, loss of memory, restlessness, disturbance of vision, tremors, temper outbursts, brain damage, lung and kidney failure
Ni	0.20	Allergic skin diseases such as itching, cancer of the lungs, nose, sinuses, throat through continuous inhalation, immunotoxic, neurotoxic, genotoxic, affects fertility, hair loss
Pb	15.00	Excess exposure in children causes impaired development, reduced intelligence, short-term memory loss, disabilities in learning and coordination problems, risk of cardiovascular disease
Se	50.00	Dietary exposure of around 300 μg day^{-1} affects endocrine function, impairment of natural killer cells activity, hepatotoxicity and gastrointestinal disturbaces
Zn	0.50	Dizziness, fatigue etc.

[a]EPA : United State Environmental Protection Agency

Phytoremediation is one of bioremediation techniques can be used as an alternative solution for heavy metal remediation proscess. The phytoremediation of metals is a cost-effective, efficient, environment- and eco-friendly 'green' technology based on the use of metal-accumulating plants to remove toxic metals, including radionuclides as well as organic pollutants from contaminated soils and water [12,15]. The objectives of this review is to give the general information about phytoremediation and use of plants for phytoremediation processes of heavy metals from the environment.

2. Source of heavy metals in the environment

Elements with metallic properties and an atomic number >20 is the conventionally definition of heavy metals. Naturally, metals are normal components in soils. However, in high levels, metals can be toxic for plants, animal and microbes [17]. The most common and important heavy metals as contaminant in the environment are As, Sr, Cs, U [12], Cd, Cr, Cu, Hg, Pb and Zn [12,17]. Some of these metals are micronutrients necessary for plant growth and development, such as Zn, Cu, Mn, Ni, and Co, while others have unknown biological function, such as Cd, Pb, and Hg [18].

Heavy metals in the environment come from natural and anthropogenic (human intervention) sources. Minerals weathering, erosion and volcanic activity are the most significant natural sources while for anthropogenic sources are mining, smelting, electroplating, use of pesticides and fertilizers as well as biosolids in agriculture, sludge dumping, industrial discharge, atmospheric deposition, etc. [15,16]. The anthropogenic sources of several heavy metals in the environment presented in table 2.

Table 2. Anthropogenic sources of several heavy metals in the environment [15]

Heavy Metals	Sources
As	Pesticides and wood preservatives
Cd	Paints and pigments, plastic stabilizers, electroplating of cadmium containing plastics, phosphate fertilizer
Cr	Tanneries, steel industries, fly ash
Cu	Pesticides, fertilizers
Hg	Release from Au-Ag mining and coal combustion, medical waste
Ni	Idustrial effluents, kitchen appliances, surgical instruments, steel alloys, automobile batteries
Pb	Aerial emission from combustion of lead petrol, battery manufacture, herbicides and insecticides

2.1 Phytoremediation and its mechanisms

The term of phytoremediation is relatively new, started from 1991. The term "*phytoremediation*" consists of the Greek prefix *phyto* which is means 'plant' and the Latin root *remedium* which is means 'to correct or remove evil'. Basic information for phytoremediation comes from a variety of research areas including constructed wetlands, oil spills, and agricultural plant accumulation of heavy metals. The term has been used widely since its inception, with a variety of specific meanings [19].

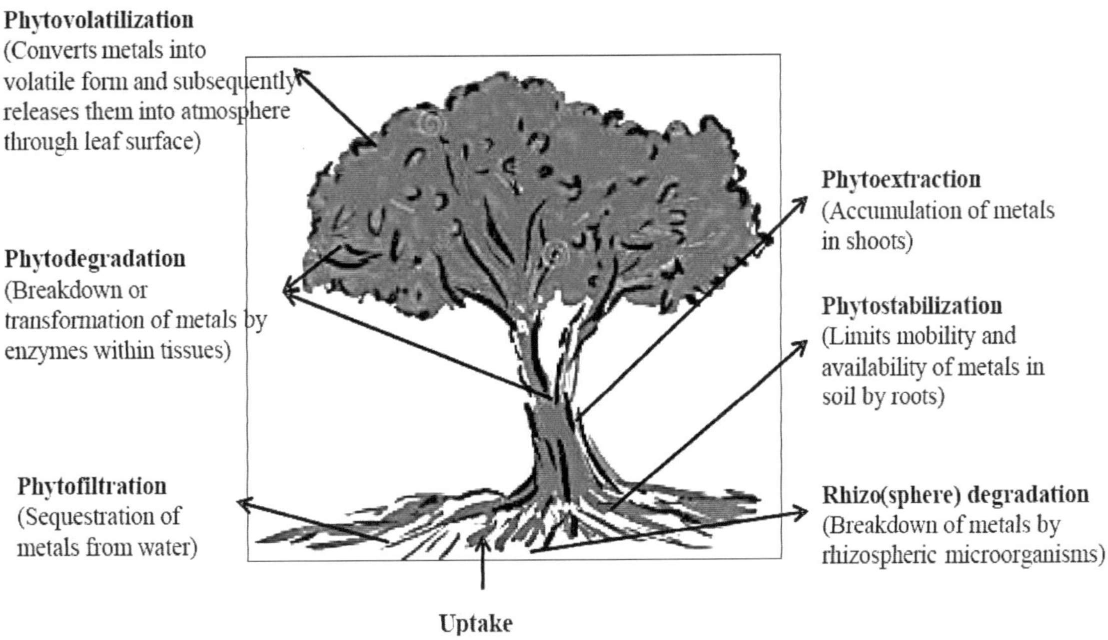

Figure 1. Various mechanisms involved in the phytoremediation of heavy metals [16]

Many definitions of phytoremediation have been given by researchers. According to those definitions, [10] have concluded and made the general definition of phytoremediation as an emerging technology using selected plants to clean up the contaminated environment from hazardous contaminant to improve the environment quality. [19] has said that phytoremediation has been receiving attention lately as an innovative, cost-effective alternative to the more established treatment methods used at hazardous waste sites. [15] have called phytoremediation as 'green technology' because of its advantages as a cost-effective, efficient, environment- and eco-friendly technology.

There are numerous plant mechanisms for remediating heavy metal contaminants from the environment. As it functions to remediate contaminant from soils and water, at least there are six mechanisms of plants on phytoremediation process include phytoextraction, phytofiltration, phytostabilization, phytodegradation, phytovolatilation, rhizodegradation [15,16,19]. The summary of the phytoremediation mechanisms is shown in figure 1.

2.2 Plants heavy metals uptake and responses

Several previous studies have described the potential of plant as heavy metals bioaccumulator from contaminated soil and water. The studies have indicated the use of plants through phytoremediation technology is an alternative solution to treat heavy metal contaminated areas and can be used to remediate the environment. Table 3 summarizes the list of several plants reported for heavy metals remediation. Each different plant has also different responses to different heavy metals exposure. Several plants are sensitive while other plants have a high tolerance to several heavy metals. As a consequence of plant-metal interaction, several plants accumulate heavy metals from soil and have growth and development decreases. However, some plants have a high tolerance and can keep the growth and development as well under heavy metals stress.

Different responses of plants to heavy metals exposure depend on its level of heavy metals tolerance. For examples, Chives plants (*Allium schoenoprasum*) got wilting, yellowing and growth inhibition on Ni, Co and Cd at 0.25 mM consentrations [20]. On chickpea plants (*Cicer auratinum*), Pb and Cr inhibited the seed germination and decreased the dry wight of plants with increase in metal consentrations and time intervals [21]. Cd stress with 20 μM consentration did not significantly affect root dry weight, shoot height, shoot dry weight, leaf number and total chlorophyll concentration (a and b) of pea plant cv. Kelvedon Wonder except root length compared with the plants grown without Cd treatments [22]. The dry weight of maize plant (*Zea mays*) extremely decreased on Zn-amanded soil with increase in Zn doses. At 270 mg kg^{-1} dose of Zn, shoot and root dry matter production of maize was 468% and 250% lower than control, approximately. The presence of Zn also changed chlorophyll a flourescence and antioxidant system parameters [23].

The most important factor affecting the rate of metal removal in phytoremediation is plant selection to be used as accumulator. [17] has described some considerations for the selection of remediating plants:

- The plant biomass, the metal removal rate depends on the plant biomass harvested and metal consentration in harvested biomass.
- Ecosystem protection, native species are preferred to exotic plants, which can be invasive and endanger the harmony of the ecosystem. To avoid propagation of weedy species, crops are in general preferred, although some crops may be too palatable and pose a risk to grazing animals.
- Physical characteristics of soil contamination, for the remediation of surface-contaminated soils, shallowrooted species would be appropriate to use, whereas deep-rooted plants would be the choice for more profound contamination.

Table 3. Several crop plants used to heavy metals phytoremediation studies.

Plants	Contaminated areas	Heavy Metals	References
Allium schoenoprasum L. (Chive)	Soil	Ni, Co, Cd	[20]
Brassica juncea L. (Indian mustard)	Soil and water	Cd, Cu, Zn, Pb	[24,25,26,27,28]
Brassica napus L. (canola)	Soil	Cd, Cu, Zn, Pb	[26,29,30]
Cajanus Cajan (L.) Milsp. (pigeon pea)	Soil	As, Cd	[31]
Cicer aeritinum L. (chickpea)	Soil	Cd, Pb, Cr, Cu	[21,32,33]
Cucumis sativus L. (cucumber)	Water	Pb	[25]
Eichhornia crassipes L. (water hyacinth)	Water	As, Cr, Zn, Cs, Co	[34,35,36]
Jatropha curcas L. (purging nut, physic nut)	Soil	Fe, Al, Cu, Mn, Cr, As, Zn, Hg	[37,38,39]
Lantana camara L. (lantana)	Soil	Pb	[40]
Lens culinaris Medic. (lentil)	Soil	Pb	[41]
Lepidium sativum L. (cress)	Soil	As, Cd, Fe, Pb, Hg	[42,43]
Lactuca sativa L. (lettuce)	Soil	Cu, Fe, Mn, Zn, Ni, Cd, Pb, Co, As	[42,44,45,46]
Medicago sativa L. (alfalfa)	Soil	Cd	[47]
Oryza sativa L. (rice)	Soil	Cu, Cd	[48]
Pistia stratiotes L. (water lettuce)	Water	Cr, Cd, As	[49,50,51]
Pisum sativum L. (pea)	Soil	Pb, Cu, Zn, Fe, Cd, Ni, As, Cr	[31,52,53,54,55]
Rapanus sativus L. (radish)	Soil	As, Cd, Fe, Pb, Cu	[42,56]
Spinacia oleracea L. (spinach)	Soil	Cd, Cu, Fe, Ni, Pb, Zn, Cr	[57,58,59,60,61]
Solanum nigrum L. (black nightshade)	Soil	Cd	[62,63,64]
Sorghum bicolor L. (sorghum)	Soil	Cd, Cu, Zn, Fe	[65]
Zea mays L. (corn)	Soil	Cd, Pb, Zn, Cu	[23,61,66]

3. Conclusions

Phytoremediation is a remediation technology to clean up the contaminants from environment by using green plants. Phytoremediation can be an alternative solution as a green technology to treat heavy metal contaminated areas. According to previous studies, several plants have a high potential as heavy metals bioaccumulator and can be used for phytoremediation process of heavy metals.

References

[1] Briggs D 2003 Environmental pollution and the global burden disease *British Medical Bulletin* **68** 1-24
[2] Jarup L 2003 Hazards of heavy metal contamination *British Medical Bulletin* **68** 167-82
[3] Suvaryan Y, Sargsyan V and Sargsyan A 2011. The problem of heavy metal pollution in The Republic of Armenia: overview and strategies of balancing socioeconomic and ecological development *Environmental Heavy Metal Pollution and Effects on Child Mental*

Developmnet: Risk and Prevention Strategies ed L I Simeonov, M V Kochubovski and B G Simeonova (Springer Science and Business Media) pp 309-15

[4] Adesuyi A A, Hjoku K L and Akinola M O 2015 Assassment of heavy metals pollution in soil and vegetation around selected industries in Lagos State, Nigeria *J. Geosci. Env. Prot.* **3** 11-19

[5] Jiao X, Teng Y, Zhan Y, Wu J and Lin X 2015 Soil heavy metal pollution and risk assessment in Shenyang industrial district, Northeast China *Plos One* **10(5)** 1-9

[6] Henry J R 2000 *An Overview of the Phytoremediation of Lead and Mercury* (Washington : US Environmental Protection Agency)

[7] Cunningham S D. Berti W R and Huang J W 1995 Phytoremediation of contaminated soils *TIBTECH* **13** 393-97

[8] Mulligan C N, Yong R N and Gibbs B F 2001 Remediation technologies for metal-contaminated soils and groundwater: an evaluation *Engineering Geology* **60** 193-207

[9] Hashim M A, Mukhopadhyay S, Sahu J N and Sengupta B 2011 Remediation technologies for heavy metal contaminated groundwater *Journal of Environmental Management* **92** 2355-88

[10] Tangahu B V, Abdullah S R S, Basri H, Idris M, Anuar N and Mukhlisin M 2011 A review on heavy metals (As, Pb, and Hg) uptake by plants through phytoremediation. *International Journal of Chemical Enineering* 939161(2011)31

[11] Susarla S, Medina V F and McCutheon S C 2002 Phytoremediation: an ecological solution to organic chemical contamination *Ecological Engineering* **18** 647-58

[12] Raskin I, Smith R D and Salt D E 1997 Phytoremediation of metals: using plants to remove pollutants from the environment *Current Opinion in Biotechnology* **8** 221-228

[13] Padmavathiamma P K and Li L Y 2007 Phytoremediation technology: hyperaccumulation metals in plants *Water Air Soil Pollut.* **184** 105-26

[14] Vangronsveld J, Herzig R, Weyens N, Boulet J, Andriaensen K, Ruttens A, Thewys T, Vassilev A, Meers E, Nehevajova E, van der Lelie D and Mench M 2009 Phytoremediation of contaminated soils and groundwater: lesson from the fiel. *Environ. Sci. Pollut. Res.* 859(2009)30

[15] Ali H, Khan E and Sajad M A 2013 Phytoremediation of heavy metals – concepts and applications *Chemosphere* **91** 869-81

[16] Dixit R, Wasiulah, Malaviya D, Pandiyan K, Singh U B, Sahu A, Shukla R, Singh B P, Rai J P, Sharma P K, Lade H and Paul D 2015 Bioremediation of heavy metals from soil and aquatic environment: An overview of principles and criteria of fundamental processes *Sustainability* **7** 2189-212

[17] Lasat M M 2000 Phytoextraction of metals from contaminated soil: a review of plant/soil/metal interaction and assessment of pertinent agronomic issues *J. Hazardous Substance Res.* **2** 1-25

[18] Gaur A and Adholeya A 2004 Prosfects of arbusclar mycorrhizal fungi in phytoremediation of heavy metal contaminated soils *Current Science* **86**(4) 528-34

[19] EPA (U.S. Environmental Protection Agency) 2000 *Introduction to Phytoremediation* (Ohio: State Environmental Protection Agency) pp 104

[20] Goland-Goldhirsh A 2006 Plant tolerance to heavy metals, a risk for food toxicity or a means for food fortification with essential metals: the *Allium schoenoprasum* model *Soil and Water Pollution Monitoring, Protection and Remediation* ed I Twardowska, H E Allen and M M Haggblom (Amsterdam: Spinger) pp 479-86

[21] Dasgupta S, Satvat P S and Mahinrakar A B 2011 Ability of *Cicer arientinum* (L.) for bioremoval of lead and chromium from soil *IJTES* **2**(3) 338-41

[22] Rahman M F, Islam M, Begum M C, Kabir A H and Alam M F 2016 Genentic variation in cadmium tolerance is related to transport and antioxidant activities in field peas (*Pisum sativum* L.) *Arch. Agron. Soil Scie.* **63**(4) 578-85

[23] Tiecher T, Ceretta C A, Ferreira P A A, Lourenzi C A, Tiecher T, Girotto E, Nicoloso F T, Soriani H H, De Conti L, Mimmo T, Cesco S and Brunetto G 2016 The potencial of *Zea mays* L. in remediating copper and zinc contaminated soils for grapevine production *Geoderma* **262** 52-61

[24] Belimov AA, Hontzeas N, Safranova V I, Demchinskaya S V, Piluzza G, Bullitta S and Glick B R 2005 Cadmium-tolerant plant growth-promoting bacteria associated with the roots of Indian mustard (*Brassica juncea* L. Czern.) *Soil Biol. Biochem.* **37** 241-50

[25] Takeda R, Sato Y, Yoshimura R, Komemushi S and Sawabe A 2006 Accumulation of heavy metals by cucumber and *Brassica juncea* under different cultivation conditions *Proc. Ann. Int. Conf. on Soil Sediments Water Energy (Massachusetts)* **11** (California: The Barkeley Electronic Press) pp 293-99

[26] Turan M and Esringu A 2007 Phytoremediation based on canola (*Brassica napus* L.) and Indian mustard (*Brassica juncea* L.) planted on spiked soil by aliquot amount of Cd, Cu, Pb, and Zn *Plant Soil Environ.* **53(1)** 7-15

[27] Singh A and Fulekar M H 2012 Phytoremediation of heavy metals by *Brassica juncea* in aquatic and terrestrial environment *The Plant Family Brassicaceae: Contribution Towards Phytoremediation* ed N A Anjum, I Ahmad, M E Pereira, A C Duarte, S Umar and N A Khan (Amsterdam: Springer Science+Business Media) pp 153-69

[28] Sharma H 2016 Phytoremediation of lead using *Brasica juncea* and *Vetiveria zizanoides Int. J. Life Sci. Res.* **4**(1) 91-96

[29] Sheng X F and Xia J J 2006 Improvement of rape (*Brassica napus*) plnat growth and cadmium uptake by cadmium-resistant bacteria *Chemosphere* **64** 1036-42

[30] Dell'Amico E, Cavalva L and Andreoni V 2008 Improvement of *Brassica napus* growth under cadmium stress by cadmium-resistant rhizobacteria *Soil Biol. Biochem.* **40** 74-84

[31] Garg N, Singla P and Bhandari P 2014 Metal uptake, oxidative metabolism, and mycorrhization in pegeon pea and pea under arsenic and cadmium stress *Turk. J. Agric. For.* **39** 234-50

[32] Wani P A, Khan M S and Zaidi A 2007 Impact of heavy metal toxicity on plant growth, symbiosis, seed yield and nitrogen and metal uptake in chickpea *Australian J. Exp. Agric.* **47** 712-20

[33] Kambhampati M S and Vu V T 2013 EDTA enhanced phytoremediation of copper contaminated soils using chickpea (*Cicer aeritinum* L.) *Bull. Environ. Contam. Toxicol.* **91** 310-13

[34] Alvarado S, Guedez M, Lue-Meru M P, Nelson G, Alvaro A, Jesus A C and Gyula Z 2008 Arsenic removal from water by bioremediation with the aquatic plants water hyacinth (*Eichhornia crassipes*) and lesser duckweed (*Lemna minor*) *Biosource Tech.* **99** 8436-40

[35] Mishra V K and Tripathi B D 2009 Accumulation of chromium and zinc from aqueous solutions using water hyacinth (*Eichornia crassipes*) *J. Hazardous Materials* **164** 1059-63

[36] Saleh H M 2012 Water hyacinth for phytoremediation of radioactive waste simulate contaminated with cesium and cobalt radionuclides *Nuclear Engineering and Design* **242** 425-32

[37] Jamil S, Abhilash P C, Singh N and Sharma P N 2009 *Jatropha curcas*: a potential crop for coal fly ash *J. Hazards Materials* **172** 269-75

[38] Yadav S K, Juwarkar A S, Kumar P, Thawale P R, Singh S K and Chakrabarti T 2009 Bioaccumulation and phyto-translocation of arsenic, chromium, and zinc by *Jatropha curcas* L.: impact of diary sludge and biofertilizer *Biosource Tech.* **100** 4616-22
[39] Marrugo-Negrete J, Durango-Hernandez J, Pinedo-Hernandez J, Olivero-Verbel J and Diez S 2015 Phytoremediation of mercury-contaminated soils by *Jatropha curcas Chemosphere* **127** 58-63
[40] Alaribe F O and Agamuthu P 2015 Assessment of phytoremediation potencials of *Lantana camara* in Pb impacted soil with organic wasted additives *Ecological Engineering* **83** 513-20
[41] Wani P A and Khan M S 2012 Bioremediation of lead by a plant growth promoting *Rhizobium* species RL9 *Bacteriology J.* **2**(4) 66-78
[42] Gunduz S, Uygur F N and Kahramanoglu I 2012 Heavy metal phytoremediation potencials of *Lepidium* sativum L., *Lactuca sativa* L., *Spinacia oleracea* L. and *Raphanus sativus* L. *Herald J. Agric. Food Sci. Res.* **1**(1) 1-5
[43] Smolinska B and Szczodrowska A 2016 Antioxidative response of *Lepidium sativum* L. during assisted phytoremediation of Hg contaminated soil *New Biotechnology* NBT896(2016)10
[44] Achakzai A K K, Bazai Z A and Kayani S A 2011 Accumulation of heavy metals by lettuce (*Lactuca sativa* L.) irrigated with different levels of wastewater of Quetta City *Pak. J. Bot.* **43**(6) 2953-60
[45] Rashid A, Mahmood T, Mehmood F, Khalid A, Saba B, Batool A and Riaz A 2014 Phytoaccumulation, competitive adsorption and evaluation of chelators-metal interaction in letuce plant *Environ. Eng. Management J.* **13**(10) 2683-92
[46] Quainoo A K, Konadu A and Kumi M 2015 The potential of shea nut shells in phytoremediation of heavy metals in contaminated soil using lettuce (*Lactuca sativa*) as a test crop *J. Bioremed. Biodeg.* **6**(1) 1-7
[47] Ghnaya T, Mnassri M, Ghabriche R, Wali M, Poschenriender C, Lutts S and Abdelly C 2015 Nodulation by *Sinorhizobium meliloti* originated from a mining soil alleviates Cd toxicity and increases Cd-phytoextraction in *Medicago sativa* L. *Frontiers in Plant Science* **6** 1-10
[48] Li P, Wang X, Zhang T, Zhou D and He Y 2008 Effect of several amendments on rice growth and uptake of copper and cadmium from a contaminated soil *J. Environ. Sci.* **20** 449-55
[49] Akter S, Afrin R, Mia M Y and Hossen M Z 2014 Phytoremediation of chromium (Cr) from tannery effluent by using water lettuce (*Pistia stratiotes*) *ASA University Review* **8**(2) 149-56
[50] Das S, Goswami S and Talukdar A D 2014 A study on cadmium phytoremediation potential of water lettuce, *Pistia statiotes* L. *Bull. Environ. Contam.* Toxicol. **92** 169-74
[51] Farnese F S, Oliveira J A, Lima F S, Leao G A, Gusman G S and Silva L C 2014 Evaluation of the potential of *Pistia stratiotes* L. (water lettuce) for bioindication and phytoremediation of aquatic environments contaminated with arsenic *Braz. J. Biol.* **74**(3) 103-12
[52] Malecka A, Piechalak A and Morkunas I 2008 Accumulation of lead in root cells of *Pisum sativum Acta Physiol Plant* **30** 629-37
[53] Wani P A, Khan M S and Zaidi A 2008 Effects of heavy metal toxicity on growth, symbiosis, seed yield and metal uptake in pea grown in metl amanded soil *Bull. Environ. Contam. Toxicol.* **81** 152-58
[54] Hegedusova A, Jakabova S, Vargova A, Hegeus O and Pernyeszi T J 2009 Use of phytoremediation techniques for elimination of lead from polluted soils *Nova Biotechnologica* **9**(2) 125-32
[55] Sharma S, Sharma P and Mehrotra P 2010 Bioaccumulation of heavy metals in *Pisum sativum* L. gowing in fly ash amandd soil *J. American Sci.* **6(6)** 43-50

[56] Hatano K, Kanazawa K, Tomura K, Yamatsu T, Tsunoda K and Kubota K 2016 Molases melanoidin promotes copper uptake for radish sprouts: the potential for an accelerator of phytoextraction *Environ. Sci. Pollut. Res.* **23** 17656-63

[57] Patel M and Subramanian R B 2006 Effect of a chelating agent on lead uptake by *Spinacia olearea Poll. Res.* **25**(1) 77-79

[58] Salaskar D, Shrivastava M and Kale S P 2011 Bioremediation potential od spinach (*Spinacia oleracea* L.) for decontamination of cadmium in soil *Current Sci.* **101**(10) 1359-63

[59] Pathak C, Chopra A K and Zivastava S 2013 Accumulation of heavy metals in *Spinacia oleracea* irrigated with paper mill effluent and sewage *Environ. Monit. Assess.* **185** 7343-52

[60] Jahanbakhshi S, Rezaei M R and Sayyari-Zahan M H 2014 Optimization of phytoremediation in Cd-contaminated soil by using Taguchi method in *Spinacia oleracea Proceedings of the International Academy of Ecology and Environmental Sciences* vol 4 ed W Zhang (Hongkong: International Academy of Ecology and Evironmental Sciences) pp 185-93

[61] Abhilash M R, Srikantaswamy S, Shiva Kumar D, Jagadish K and Shruthi L 2016 Phytoremediation of heavy metal industrial contaminated soil by *Spinacia oleracea* L. and *Zea mays* L. *Int. J. Applied Sci.* **4**(1) 192-99

[62] Wei S, Zhou Q and Koval P V 2006 Flowering stage characteristic of cadmium hyperaccumulator *Solanum nigrum* L. and their significance to phytoremediation *Sci. Total Environ.* **369** 441-46

[63] Wei S, Li Y, Zhou Q, Srivastava M, Chiu S, Zhan J, Wu Z and Sun T 2010 Effect of fertilizer amandments on phytoremediation of Cd-contaminated soil by newly discovered hyperaccumulator *Solanum nigrum* L. *J. Hazardous Materials* **176** 269-73

[64] Ji P, Sun T, Song Y, Ackland M L and Liu Y 2011 Strategies for enhanching the phytoremediation of cadmium-contaminated agricultural soils by *Solanum nigrum* L. *Environ. Poll.* **159** 762-68

[65] Pinto A P, Mota A M, de Varennes A and Pinto F C 2004 Influence of organic matter on uptake of cadmium, zinc, copper and iron by shorgum plant *Science of the Total Environ.* **326** 239-47

[66] Mojiri A 2011 The potencial of corn (*Zea mays*) for phytoremediation of soil contaminated with cadmium and lead *J. Biol. Environ. Sci.* **5**(13) 17-22

IOP Conf. Series: Earth and Environmental Science **142** (2018) 012024 doi:10.1088/1755-1315/142/1/012024

The effect of protected sardine fish oil as feed supplement on ruminal fermentation

A Pramono[1], D T Widayati[2] and E Handayanta[1]
[1]Department of Animal Science, Faculty of Agriculture, Universitas Sebelas Maret, Surakarta, Indonesia
[2]Faculty of Animal Science, Universitas Gadjah Mada, Yogyakarta, Indonesia

E-mail: ahmad_pram@staff.uns.ac.id

Abstract. The research aims to evaluate the influence of protected sardine fish oil as feed supplement on ruminal fermentation (pH rumen fluid, ammonia concentration and volatile fatty acids production in the rumen). Protected feed supplement was produced from sardine fish oil and soybean meal, through two protection methods, they were saponification and microencapsulation. The experiment consists of two treatments i.e. P0: basal diet (control) and P1: basal diet + 3 % protected feed supplement. Each treatment was repeated 10 times. The kinetics observation of the pH rumen fluid, ammonia concentration and volatile fatty acids production were performed at incubation times 0, 2, 4 and 6 hours respectively. Data were analyzed using independent samples t-test. Results in cow with protected feed supplement showed that kinetics of pH rumen fluid: 7.23; 7.13; 6.90 and 6.76 respectively; ruminal ammonia concentration: 26.70; 31.06; 19.75 and 15.52 respectively; and volatile fatty acids production: 22.75; 26.08; 29.19 and 25.79 respectively. The results could be concluded that the effect of supplementation of protected sardine fish oil have an optimal of pH rumen fluid, ammonia concentration, and volatile fatty acids production so it did not interfere the ruminal fermentation in the rumen.

1. Introduction

Energy and protein are nutrients as needed for growth, milk production and reproduction in dairy cows [1]. Increasing the energy intake for ruminant livestock can be conducted by increasing the energy density of feed by using fat (oil). One of potential energy source that widely available and cheap is sardine fish oil. Sardine fish oil contains high unsaturated fatty acid Unsaturated fatty acid supplementation is proven to be able to increase energy efficiency through increasing energy density and sustain the efficiency improvement of milk protein synthesis through increasing the flow of nitrogen (N) non-ammonia into the duodenum [2].

Protection treatment is needed to gain the real benefit from the presence of energy source supplementation or protein in feed. In the unsaturated fatty acid supplementation, protection is required to prevent unsaturated fatty acids from ruminal microbes biohydrogenation. Protection is also useful to eliminate the negative effects of high concentration of unsaturated fatty acid supplementation, and reduction in fiber degradability [1]. Based on the above statement this study aim to evaluate the effect of the protected sardine fish oil as feed supplement on ruminal fermentation (pH rumen fluid, ammonia concentration and volatile fatty acids production in the rumen).

2. Material and Methods

2.1 Sample preparation

This research was used two cows with fistula. The requirement of dry matter intake for each cow was calculated 3% of their weight. The balance of forage and concentrate is 70:30 (control). The diet containing for 12% crude protein and 60% total digestible nutrients. Feed materials used to construct the concentrate consist of concentrate, bran pollard, urea and mineral. King grass was used as forage in this research. Feed giving of protected fat as a supplement feed was 3% of dry matter intake. Comparison for forage : concentrate : feed supplements was 70 : 27 : 3. Feed is given twice daily at 08.00 and 15.00 WIB, unless protected fat is given once at 08.00 am, while drinking water for livestock is given ad libitum. The rumen fluid intake for pH, NH3 and VFA was performed using an aspirator to represent the entire rumen portion. Sampling was done in 4 times that is divided to before feeding (at 8.00 AM) and after feeding, that time is: 8.00 AM, 10.00 AM, 12.00 AM, and 2.00 PM, so the total of 4 samples to be analyzed. Each treatment was repeated for 10 times observation.

2.2 Measurement of pH rumen, NH_3 and VFA

pH of rumen was measured immediately after sampling of the rumen using pH meters. Ammonia nitrogen (NH_3) content was calculated using the Conway method. Finally, VFA was determined using gas chromatography method.

2.3 Data analysis

Data obtained in the current study were analyzed by independent Student's t-test using SPSS version 22.

3. Results and Discussion

3.1 pH kinetic of rumen fluid

The comparison of *pH kinetic* of rumen fluid between treatment (control vs protected) is shown in Figure 1. There were no significant different between treatment.

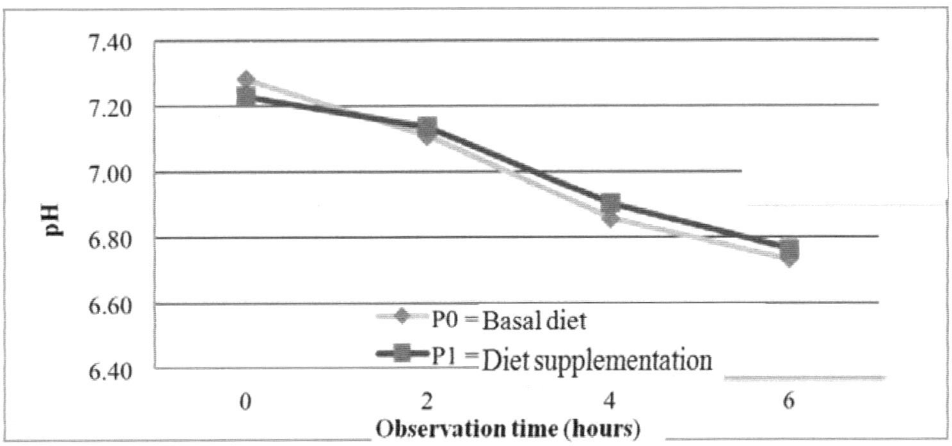

Figure 1. pH of rumen

Changes in pH occurring in the rumen are affected by the type of feed given [3]. The results indicate that the provision of protected sardine fish oil supplementation (P1) does not give a negative effect on the rumen microbial environment, because the pH value of rumen fluid 6.76 to 7.23 in this study is still within the normal condition. Kinetic pH of rumen fluid, after feeding tends to decrease gradually, because of fermentation in the rumen. The crude fiber will be degraded by cellulotic bacteria in the rumen resulting VFA. The decrease of pH in the rumen is due to VFA is an acidic compound [4].

3.2 Amonia (NH_3)

The concentration of NH_3 in rumen fluid of basal diet and protected sardine fish oil supplementation (P1) can be seen in Figure 2.

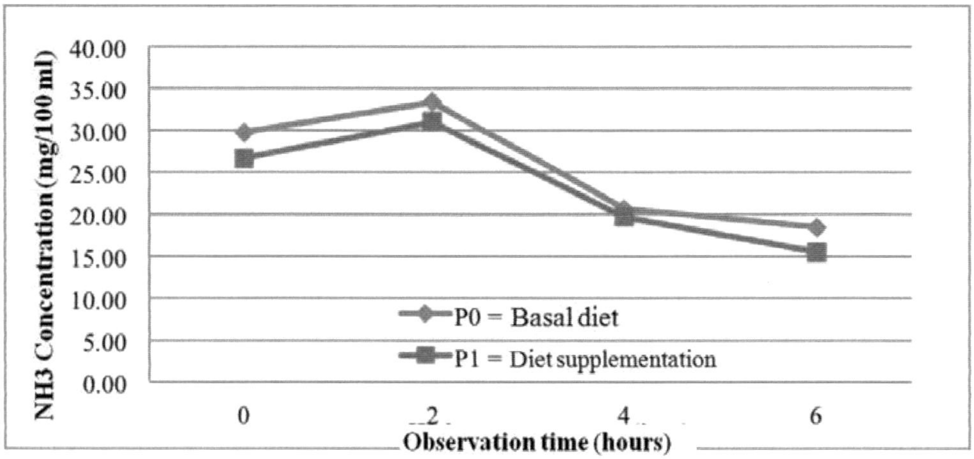

Figure 2. Concentration of NH_3

NH_3 concentration of rumen fluid in P1 is lower than P0. This is presumably because soybean meal protein as microencapsulation agent of Sardine fish oil is also protected then it is not degraded completely lead to the less concentration of NH_3. Feeding of protected proteins will decreasing NH_3 rumen concentrations [5]. Moreover, protecting soybean meal protein with formaldehyde is also shown to decrease protein degradation [6].

3.3 Volatile fatty acid (VFA)

Kinetics of total VFA concentration of cow rumen fluid at 0, 2, 4, and 6 hours after feeding can be seen in Figure 3.

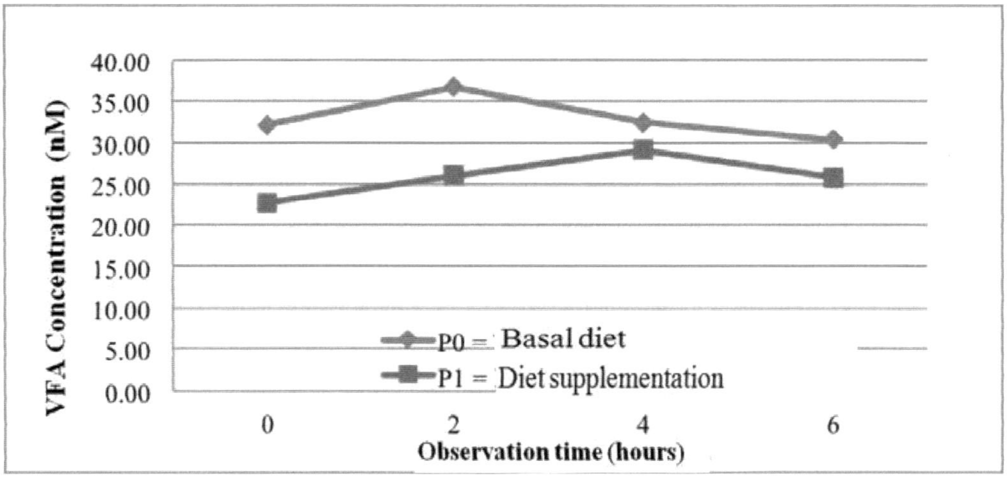

Figure 3. VFA Concentration

VFA concentration of rumen fluid in P1 shown to have lower total VFA concentration than P0, although there was no significant difference (P>0.05). This indicated that protected of sardine fish oil supplementation able to decrease rumen microbial activity in producing VFA and NH3 in the rumen fluid. The protected feed treatment may reduce/limit the ability of rumen microbial cell surface to contact with feed [7].

4. Conclusion

It is concluded that effect of supplementation of protected sardine fish oil has an optimal of pH rumen fluid, ammonia concentration, and volatile fatty acids production, therefore did not interfere the ruminal fermentation in the rumen.

References

[1] Pramono A, Handayanta E, Widayati D T, Putro P P and Kustono 2017 Dietary protected feed supplement to increase milk production and quality of dairy cows *Proc. International Conference On Food Science and Engineering* Surakarta

[2] Johnson K A, Kincald R L, Westberg H H, Gaskins C T, Lamb B K and Conrath J D 2002 *J. Dairy Sci.* **85** 1509–15

[3] Owens F N and A L Goest 1988 Ruminant fermentatition. In : D.C. Church (Ed), The Ruminant animal. Printice Hall, Englewood Cliifs, New Yersey. Pp: 145 - 171

[4] Kerley M S, Fahey G S, Berger L L and Merchen N R 1987 Effect of treating wheat straw with pH regulated solution alkaline hydrogen peroxideon nutrient digestion by sheep *J. Dairy. Sci.* **70** 2078 - 2084

[5] Rodriguez L A, Stallings C C, Herbein J H, and Mc Gilliard M L 1997 Effect of degradability of dietary protein and fat on ruminal, blood, and milk components of Jersey and Holstein cows *J. Dairy. Sci.* **80** 353 - 363

[6] Utomo R 2010 Modifikasi metode penetapan kecernaan in vitro bahan kering atau bahan organik *Sintesis* **5** 1 - 11

[7] Schaefer D M 2000 *Potential For Altering Quality Of Muscle And Milk From Ruminants*: Antioxidants In Muscle Wiley Interscience. New York. pp: 293 – 312

IOP Conf. Series: Earth and Environmental Science **142** (2018) 012025 doi:10.1088/1755-1315/142/1/012025

Nutrient digestibility of vegetables waste flour on male quail (*Coturnix coturnix japonica*)

A Pramono[1], M S Primadhani[1], W Swastike[1] and J Sutrisno[2]
[1] Department of Animal Science, Faculty of Agriculture, Universitas Sebelas Maret, Surakarta, Indonesia
[2] Department of Agribusiness, Faculty of Agriculture, Universitas Sebelas Maret, Surakarta, Indonesia

E-mail: ahmad_pram@staff.uns.ac.id

Abstract. The aim of this research is to determine the nutrient digestibility of vegetables waste flour on of male quail. Four hundred male quails were divided into four groups with five replications. The experiment is Completely Randomized Design and the data were analyzed by analyses of variants. The experimental diets were P0 = basal diet, P1 = 97% basal diet + 3% vegetables waste flour, P2 = 94% basal diet + 6% vegetables waste flour, and P3 = 91% basal diet + 9% vegetables waste flour. The observed variables were the digestibility of dry matter, crude protein and extract ether. Result showed that of the addition of vegetable waste flour in the diet had no effect on crude protein digestibility ($P>0.05$), however shown significant effect on dry matter ($P <0.01$) and extract ether ($P <0.01$) digestibility.

1. Introduction

Male quail is a type of poultry that has potential to be developed for protein source. Feed is one of important factor in quail management [1]. Feed cost is one of the highest costs in poultry breeding, ie 70 - 80% of all components of production costs incurred [2]. The scarcity of feed ingredients and the high price of feed caused the problem on farmers. The efforts to minimize feed cost is by using feed materials based on agricultural waste as alternative feed [3]. Vegetables waste is cheap that can be obtained from the market for example cabbage, mustard green and chicory [4]. Physically the vegetables waste is decayed because contains high of water, but chemically contains vitamins and minerals which required by quail [5].

Vegetables waste has a low crude protein content of 1-15%. This vegetable waste will be more valuable if used as a feed through processing. The through processing of vegetable waste flour is an alternative effort to save the nutrient content and can be used for a long time [6]. The digestibility can be used to see the benefits of vegetable waste flour as feed materials. Measurement of digestibility value of vegetable waste flour as feed material in diet can be done directly, because quail has fast growth [7]. Therefore, the aim of this study is to evaluated the digestibility of nutrient that is provided by the addition of vegetables waste flour in the diet at different level on male quail diet.

2. Methods

2.1. Sample preparation

Four hundred of 7 days male quail were employed in this study with the initial body weight is 14.72 + 1.51 gram (cv = 10.23%). Diet treatments were given when quail at 8 - 42 days consisting of basal diet and vegetables waste flour with calculated composition. The experiment consists of four treatments i.e.: P0 = basal diet 100%, P1 = basal diet 97% + 3% vegetables waste flour, P2 = basal diet 94% + 6% vegetables waste flour and P3 = basal diet 91% + vegetables waste flour 9%. Each treatment was repeated 5 times and each replication consisted of 20 quails. The nutrient content of feed ingredients for the diet can be seen in Tables 1 and 2.

Table 1. Nutrient content of the diet

Feedstuff	Metabolic Energy	Crude protein	Extract ether	Crude fiber	Ash	Ca	P available
	Kcal/kg	%					
Yellow corn	3958.33[3)]	6.89[2)]	2.83[2)]	0.94[2)]	1.55[2)]	0.02[1)]	0.08[1)]
Dedak halus	3607.00[3)]	10.02[2)]	10.57[2)]	7.21[2)]	6.85[2)]	0.07[1)]	0.21[1)]
Bungkil kedelai	2992.10[3)]	37.65[2)]	1.20[2)]	7.83[2)]	8.08[2)]	0.29[1)]	0.27[1)]
Tepung Ikan	2258.32[3)]	30.90[2)]	2.78[2)]	11.63[2)]	19.91[2)]	5.11[1)]	2.88[1)]
Limestone	-	-	-	-	-	38[1)]	-
Premikx	-	-	-	-	-	-	-
NaCl	-	-	-	-	-	-	-
Mustard green waste	3879.49[3)]	2.27[2)]	0.36[2)]	0.36[2)]	1.45[2)]	-	-
Cabbage waste	3876.37[3)]	1.30[2)]	-	0.34[2)]	1.09[2)]	-	-
Chicory waste	3862.23[3)]	1.09[2)]	-	0.32[2)]	1.48[2)]	-	-

Source: [1)] NRC (1994) [8]
[2)] Analysis Laboratory of Center for Food and Nutrition Studies of Gadjah Mada University, Yogyakarta (2017)
[3)] Sibbald *et al.* (1980). [9]
EM = 3951 + (54.4 x EE) – (88.7 x CF) – (40.8 x Ash)

Table 2. Feedstuff composition and nutrient content of treatment diet

Feedstuff	Proportion (%)			
	P0	P1	P2	P3
Basal diet	100	97	94	91
Vegetables waste flour	0	3	6	9
Jagung kuning	46.5	43.5	40.5	35.5
Dedak halus	17	17	17	19
Bungkil kedelai	25	25	25	25
Tepung Ikan	10	10	10	10
Limestone	1	1	1	1
Premiks	0.25	0.25	0.25	0.25
NaCl	0.25	0.25	0.25	0.25
Flour of Mustard Green Waste	0	0.3	0.6	0.9
Flour of Cabage Waste	0	0.9	1.8	2.7
Flour of Mustard White Waste	0	1.8	3.6	5.4
Total	100	100	100	100
Nutrient content[*)]				
Metabolyc Energy (Kkal/kg)	3427.67	3425.14	3422.61	3413.05
Crude protein (%)	17.41	17.26	17.11	17.02
Extract ether (%)	3.69	3.61	3.53	3.61
Crude fiber (%)	4.78	4.77	4.75	4.86
Ash (%)	5.90	5.89	5.89	5.99
Calsium (%)	0.98	0.98	0.98	0.98
P available (%)	0.43	043	042	0.42

*) Calculated based on the diet nutrient content in Table 2 and percentage of Table 3

2.2 Data analysis
Data obtained in current study were analyzed using ANOVA.

3. Result and Discussion
The results of analysis of dry matter, fat, protein and ash digestibility were presented in Table 3.

Table 3. Digested nutrients of male quail by adding vegetable waste flour in diet to 9% concentrations

Parameter	Treatment				
	P0	P1	P2	P3	P value
Dry Matter Digestibility	11.76	11.64	11.55	11.48	<0.05
Extract Ether Digestibility	0.50	0.50	0.51	0.51	<0.05
Crude Protein Digestibility	1.43	1.46	1.49	1.48	0.08
Ash Digestibility	0.09	0.11	0.12	0.11	0.97

3.1 Dry matter digestibility
The result of variance analysis showed that there is an effect by giving of vegetables waste flour in diet to dry matter digestibility (Table 3). The value of dry matter digestibility P0 was 11.76, while dry matter values P1, P2, and P3 respectively were 11.64; 11.55; and 11.48. The results showed that the dry matter value of digested basal diet was greater than basal diet supplemented with vegetables waste flour. The higher concentration of vegetables waste flour which is supplemented in basal diet can be reduce the value of dry matter digestible. Dry matter digestibility in broiler chickens increases as the digestibility of organic matter increases [10].

3.2 Extract ether digestibility
Supplementing of vegetables waste flour in the diet to a concentration level up to 9% showed significant different results on extract ether digestibility. The extract ether digestibility values P0, P1, P2, and P3 are respectively 0.50; 0.50; 0.51; and 0.51. The extract ether digestibility is influenced by fat composition, animal age, species and diet composition [11]. Treatments diet have extract ether content of between 3.97-5.99% so the quail can still increase the extract ether digestibility.

3.3 Crude protein digestibility
Supplementation of vegetables waste flour up to 9% concentration in basal diet had no effect on crude protein (Table 3). Protein digestibility of this research is relatively similar due to the protein content in the same diet [12]. Crude protein digestibility is not significant may be due to the protein content in the diet meet with the requirement therefore supplementing of vegetables waste flour in the quail diet did not increase the activity of phytase to hydrolyze protein with phytate.

4. Conclusion
Supplementation of vegetables waste flour in the diet up to 9% concentration has no effect on crude protein digestibility ($P>0.05$), however the supplementation give significant effect on dry matter ($P < 0.01$) and extract ether ($P < 0.01$) digestibility of male quail.

References
[1] Mahfudz L D, Ratnawati Y, Suprijatna E and Sarengat W 2009 *Performance of male quail carcass caused by Distribution of Alcoholic Beverage Distillation in Ransum. Proc.* Nat. Seminar on Livestock Rising. Faculty of Animal Husbandry. Diponegoro University pp 589 – 595.
[2] Sari D T I, Sudjarwo E and Prayogi H 2014 Effect of addition fresh earthworm *(Lumbricusrubellus)* in Feeding on Egg Weight, Haugh Unit (HU) and Thickness of Mojosari Ducker Shell *J. Tropical Livestock* **15** 23-30

[3] Kurniawan D, Widodo E and Natsir M H 2014 Effects of Tomato Flour as a Feed Material on the appearance of quail production. (Malang: Brawijaya University) *J. Animal Science* pp 1 – 7
[4] Hadiwiyoto S 1983 *Handling and Waste Utilization* (Jakarta: Indayu Foundation)
[5] Abun R D and Deny S 2015 Effect of Organic Vegetable Waste Treatment on Digestibility Value in Ayam Kampung Super JJ-101 (Bandung: Padjajaran University) *J. Animal Science* pp 81 – 86
[6] Susangka I, Haetami K and Andriani Y 2015 Evaluation of Nutritional Value of Vegetable Waste Product Different Processing Way and Its Influence on Tilapia Growth (Bandung: Padjajaran University)
[7] Hamid S A 2013 Protein and Energy Digestibility in Village Chicken Rations with Several Levels of Tofu tofu Differentiated with Yeast Tape (Makassar: Hasanuddin University)
[8] National Research Council 1994 Nutrient Requirements of Poultry Eighth Revised Edition (Washington DC: National Academy of Sciences)
[9] Sibbald I R 1980 Metabolic plus endogenous energy ad nitrogen losses of adult cockerels: the correction used in bioassay for true metabolizable energy *Poultry Science* **60** 805-811
[10] Abun D Rusmana and Indriani N P 2003 The determination of the digestibility of the ration contains the garut bulb dregs (*Maranta arundinacea* Linn.) In broiler chickens by the method of cutting *J. Bionatura* **3** 227–238
[11] Leeson S 1993 Recent advances in fat utilization by poultry. In: Recent Advance in Animal Nutrition in Australia, Farrell, D J (Ed.) (Australia: University of New England) pp: 170–181
[12] Widodo A R, Setiawan H, Sudiyono, Sudibyo dan R Indreswari 2013 Nutrient digestibility and quill performances (*Coturnix coturnix japonica*) males dregs tofu fermented in rations *J. Tropical Animal Husbandry* **2** 51-57

Study of antidiarrheal and hematology profile of laboratory rat fed with yogurt containing local probiotic and purple sweet potato extract

A I N Tari, C B Handayani and S Hartati
Agriculture faculty, Universitas Veteran Bangun Nusantara Sukoharjo, Indonesia
Jl. Letjen S. Humardani No.1 Jombor Sukoharjo 57521 Phone (0271) 593156

Email : intanniken@gmail.com

Abstract. The aim of this study was to evaluate the effectiveness of local probiotic in yogurt with purple sweet potato extract supplementation on the hematological parameters of albino rats (*Spraque dawley*). The study was conducted using a Completely Randomized Design with 30 rats divided into 6 groups. In group K-, rats were fed with distilled water from day 1 to 21. In group YTP, Rats were fed with yogurt without probiotics from day 1 to 21. YDP group was rats were fed with probiotic yogurt from day 1 to 21. In group YTP+E, rats were fed with yogurt without probiotic from day 1 to 7, interspersed with exposure to enteropathogenic *Escherichia coli* (EPEC) on day 8 to 14. In group YDP+E, rats were fed with probiotic yogurt from day 1 to 7, interspersed by EPEC on day 8 to 24. In group K +, rats were fed with water from day 1 to 7, then fed with EPEC on day 8 to 14, after which water was given back on day 15 to 21. The result showed that probiotic yogurt treatment with supplement of purple sweet potato extract had a significant effect (P<0,05) on feces water content, number of erythrocyte, leucocyte, and hemoglobin. The treatment of YDP had water content in feces 48.422% and the number of erythrocyte, leucocytes, and hemoglobin were 8.578 $10^6/\mu l$, 14.152 $10^6/\mu l$ and 13.98 g / dL respectively.

1. Introduction

Probiotic is a biological microorganism which still survives at the time of reaching the digestive tract, its provides the health benefit to the body through microbiotic balance [1]. According to previous study [2], probiotics are beneficial to increase physical immunity,. and also reported to be able to overcome diarrhea caused by *E.coli*, either enterotoxigenic (ETEC) [3] or enterohemorrhagic (EHEC) [4]. Such effect emerges if a number of biological bacteria reaching the digestive tract are more than 10^6CFU/g or 10^6CFU/ml [3]. The required minimum amount of biological cell of probiotics in yogurt is 10^7CFU/g [5], which is expected to be able to anticipate the decrease in the amount of cell during passing the extreme environment in the digestive tract [6]. The ability of probiotics to protect intestine is assumed through several mechanisms, such as the ability to compete with nutrition in such a way that it inhibits the growth of enteral bacteria in the human digestive tract. Probiotics is also able to compete for the adhesive side of the surface of intestinal cell against the pathogenic bacteria causing the infection of digestive tract in such a way that it can inhibit the adhesion or exclude the pathogenic bacteria or displace the adhesive side of pathogenic bacteria which has previously been adhesive to the intestine [7,8].

Nowadays, experts are intensely promoting the concept of food as medicine. Such concept is a way to optimize functional food in order to overcome diseases. One of the functional food types from own is sweet potato. Sweet potato contains such antioxidants as phenolic acid, anthocyanin, and tocopheryl, which can prevent several diseases [9]. The improvement in the savor of sweet potato in order to maximize its utilization as alternative diet, such as converted into yogurt.

Some local lactic acid bacteria have been isolated such as *Lactobacillus plantarumMut* 7 isolated from dried cassava fermentation [10],*Lactobacillus plantarum* Dad 13 isolated from buffalo fermented milk [11], and *Lactobacillus acidophilus* SNP-2 and *Lactobacillus acidophilus* SNP-2 which are isolated from the baby's stools consuming breast milk [12].

Probiotic bacteria that given orally has the ability to affect to body metabolic system, including the status of hematology or blood function [13]. There is no information about local probiotic on hematology profile in suppressing the growth of EPEC ATCC 35218. Therefore, this study aims to evaluate the effectiveness of local probiotic in yogurt with purple sweet potato extract supplementation on the hematological parameters of albino rats (*Spraque dawley*).

2. Method

2.1. Materials and Equipment

Materials used in this research were purple sweet potato (*Ipomoea batatas* L) and yogurt starter culture: *Streptococcus thermophillus* FNCC 0040, *Lactobacillus bulgaricus* FNCC 0041 and indigenous probiotic lactic acid bacteria : *Lactobacillus plantarum* Dad13, obtained from Food and Nutrition laboratory on Inter-university UGM Yogyakarta. MRS media (de Mann Rogossa Sharp) Bacteriological Agar/Broth (Oxoid), for Lactic Acid Bacteria culture, Nutrient Broth medium (Merck) for EPEC ATCC 35218 culture. Other materials were skim milk, alcohol 70 %, spiritus, and distilled water obtained from Biology, Chemical, and Microbiology Laboratory of Agriculture Department of Veteran Bangun Nusantara University. Equipment used in this research were glassware (test tube, Beaker glass, Erlenmeyer, and Petridish), autoclave (All America), incubator (Inko), Oven (Binder), and entkas.

2.2.The making of yoghurt with the supplement of sweet potato extract

Fresh milk, skimmed milk (5% b/v), and the sweet potato extract (10% v/v) were pasteurized at the temperature of 72°C for 15 minutes followed with cooling down to the temperature of 40-45°C. Next, it was inoculated with *Streptococcus thermophillus* and *Lactobacillus bulgaricus* along with the indigenus probiotic bacteria with the ratio of 1 : 1 : 0.5 aseptically at the temperature of 43-45°C, as much as 5 % (v/v), and then it was shaken up for becoming homogeneous and incubated at the temperature of 40°C for 17 hours in such a way that yoghurt with the supplement of sweet potato extract is yielded.

2.3. Administration of the Experimental Animal

This research was conducted by using 20 male rats of *Sprague dawley* strain, 5 – 6 weeks old, at 100 – 120 g body weight. The cage used was an individual cage with the size of 17.5 x 23.5 x 17.5 cm without base. The room temperature was adjusted into 23-24°C [14]. The ration was administered as much as 15 g for each rat for each day at 06.00 – 07.00 AM (GMT+7).

The water was provided ad libithum. The rest of the ration was collected on a daily basis in order to be weighed and found out for the consumption of ration by each rat for each day. Weighing the body of the rats was conducted every 3 days. The composition of the basal ration was set up on the basis of the American Institute of Nutrition (AIN)-93 in 1 kg consists of 620.7 g of cornstarch, 140 g of casein, 100 g of sucrose, 40 g of oil, 50 g of fiber, 35 g of mineral, 10 g of vitamin, 1.8 g of L-cysteine, 2.5 g of choline, and 0.008 g of tert-Buthyhydroquinone (TBHQ). The experiment flow provided in Figure 1.

IOP Conf. Series: Earth and Environmental Science **142** (2018) 012026 doi:10.1088/1755-1315/142/1/012026

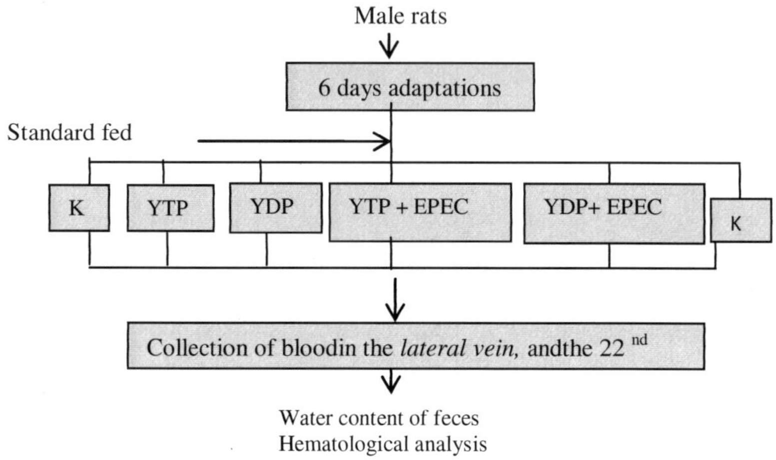

Figure 1. Research Flow chart

Note 1

- Control	Rats were fed with distilled water from 1st -21st day
YTP	Rats were fed with yogurt without probiotics from 1st -21st day
YDP	Rats were fed with probiotic yogurt from 1st- 21st day
YTP +EPEC	rats were fed yogurt without probiotic from 1st- 21st day, interspersed by EPEC on 8th– 14th day
YDP +EPEC	= Rats were fed with probiotic yogurt from 1st- 21st day, interspersed EPEC on 8th-14th day
+ Control	Rats were fed with water from 1st- 7th day, then fed EPEC on 8th- 14th day, after which water was given back on 15th- 21st day

Note 2

- Probiotics is administered orally as much as 1 mL/day from the 1st day - the 2st day with the population of LAB 10^9 CFU/ml by using feeding hose
- Infection by EPEC ATCC 35218 is administered orally as much 1 mL/day from the eighth day to the fourteenth day with the population of EPEC 10^7 CFU/mL
- The termination of rat was conducted each time against six rats on the twenty second day

2.4. Research Design and Statistical Analysis

This research used the Completely Randomized Design with six treatments, namely K-, YTP, YDP, YTP+E, YDP+E and K+ and 5 repetitions, each consisted of 1 rat. The data were analyzed using ANOVA, any significant treatment followed with DMRT (Duncan's Multiple Range Test) [15]

2.5. Observation Parameters

2.5.1. Fecal water content The water content of animal feces was observed on the 14th day, using the thermogravimetric method [16]

2.5.2. Hematology analysis Hematology analysis included number of erythrocytes, leucocytes, and hemoglobin levels wereperformed. Hematologic analysis was performed according to [13] methods. Blood samples were collected in the tube containing EDTA. The analysis was performed using Hemavet HV950FS multispecies hematology analyzer.

3. Result and Discussion

3.1. Water content of feces

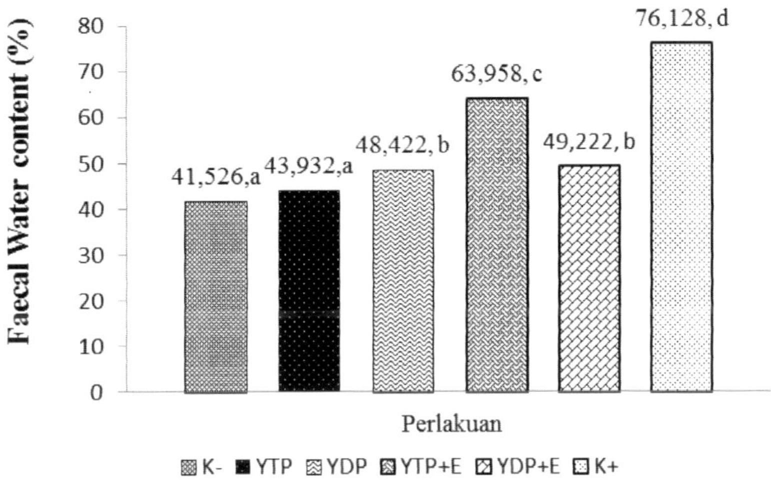

Figure 2. Histograms of fecal water content (%) of experimental animals in various treatments of probiotics yogurt with supplement of purple sweet potato extract and after an exposure to enteropathogenic Escherichia coli (EPEC) ATCC35218

The water content of rat's feces is presented in Figure 2. Figure 2 showed that the treatment of probiotic yogurt with supplementary of purple sweet potato extract and after exposure of enteropathogenic *Escherichia* was significant (P<0,05) to the water content of rat's feces. Rats group treated with YTP + E (yogurt without probiotics interspersed and EPEC) and K + (positive control) had high level of fecal water and were classified under diarrhea (above 60%). [17] reported that the normal stool water content of rats was below 60%.The condition of diarrhea in rats was characterized by soft feces and water content above 60%, whereas very severe diarrhea in rats was characterized by liquid feces and water content above 80%

The group of YDP + E rats may have no diarrhea occurrence, although the water content was slightly higher than that of the K-rats, YTP, and YDP. This was possible because of the probiotics in purple sweet potato extract yogurt was able to protect against EPEC bacteria ATCC 35218. According [7] mechanism of probiotic bacteria protection against pathogenic bacteria, among others through attachment competition on bonding and nutrient side, modulate the immune system and secretion of an anti microbial compound

3.2. Blood Test (Hematology)

Blood is an important component because it functions to circulate blood that goes into body as well as produced by body from metabolism process. Hematology value (blood profile) is useful to assess health condition and as baseline or control in research.

3. 2.1. Erythrocytes

The number of normal erythrocytes ranges from 6.6-9.0 x 10^6 / μL [18]. Groups of rats in K + and YTP + E had lower erythrocytes (P<0.05) than the rats in K-group, YTP, YDP (Figure 3).EPEC treatment caused cell membrane damage and disrupted the cell wall permeability properties, resulting in leakage of cells and loss of some metabolites, and ended with the reducing of erythrocytes [13]. The probiotic (YDP) group of rats had the highest erythrocyte (8,578 x 10^6 / μL), this was because probiotic bacteria had bacteriostatic and bactericidal properties by protecting the plasma membrane.

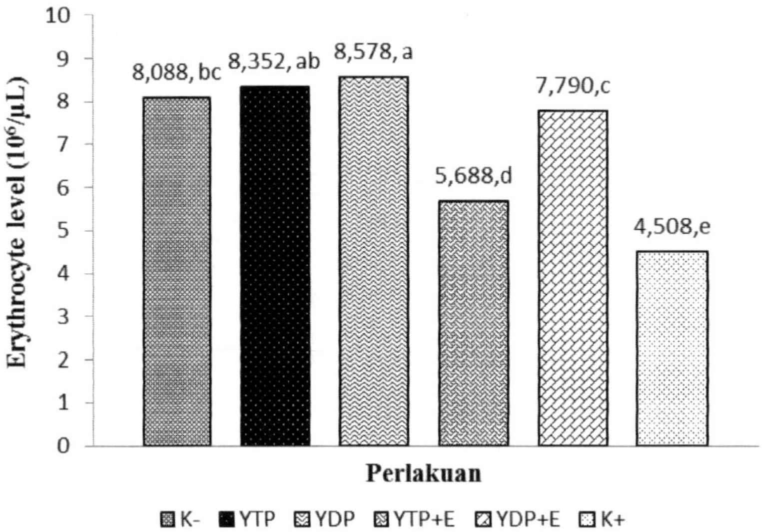

Figure 3. Histogram of erythrocyte level 10^6 / μL) of experimental animals on various treatments of probiotic yogurt with purple sweet potato extract supplementation and after exposure to enteropathogenic *Escherichia coli* (EPEC) ATCC35218

3.2.2. Leukocytes

Groups K + of rats had significant lower leukocytes ($P < 0.05$) than in the K-group, YTP, YDP, YTP + E and YDP + E (Figure 4). The high number of leukocytes in the YTP and YDP rats when compared to K +, was possible due to the probiotic ability of the immunomodulator, which can boost the immune system [18]. Rats group with YTP treatment could increase the number of probiotics because the presence of prebiotic in the form of purple sweet potato could increase the number of probiotics. [19] reports that probiotic bacteria can attach to the intestinal surface to protect a host from pathogenic bacterial colonization by different mechanisms.

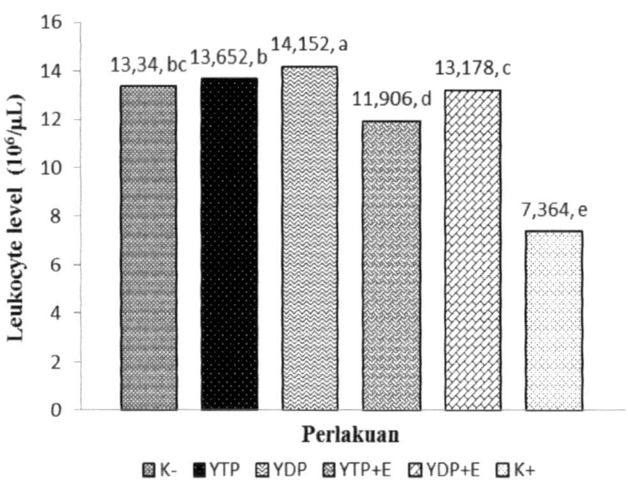

Figure 4. Histogram of leukocyte level 10^6 / μL) of experimental animals on various treatments of probiotic yogurt with purple sweet potato extract supplementation and after exposure to enteropathogenic *Escherichia coli* (EPEC) ATCC35218

The decrease of the number of leukocytes in the group of rats treated with YTP + E and YDP + E compared to K-was possible injury to the intestinal mucosa due to EOEC infection so that leukocytes were used for body defense. This condition caused the number of leukocytes in the blood circulation to be reduced.

3.2.3. Haemoglobin

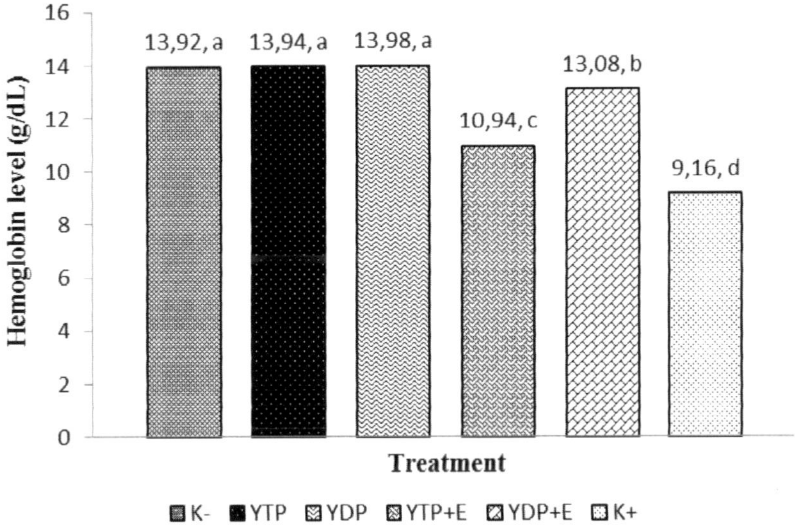

Figure 5. Histogram of hemoglobin level (g/dL) of experimental animals on various treatments of probiotic yogurt with purple sweet potato extract supplementation and after exposure to enteropathogenic *Escherichia coli* (EPEC) ATCC35218

Hemoglobin (Hb) is an erythrocyte pigment consisting of a conjugated complex protein containing iron. Normal hemoglobin level is 12.0-17,5 g/dL Figure 5 showed that groups of rats treated with K + had lower hemoglobin ($P < 0.05$) than the K-group, YTP, YDP, YTP + E or YDP + E.

According to [13], pathogenic bacteria such as EPEC can damage the permeability of cell membranes and cause cell wall damage, resulting in the release of hemoglobin from cells and causing decreased hemoglobin levels.Group YTP + E rats and YDP + E rats had higher hemoglobin levels than the group K rats. In the group of YTP rats, a prebiotic of purple sweet potato could increase the number of probiotics, Whereas in the group of YDP rats there were probiotics Probiotics in both groups of rats had the ability to produce antimicrobials which was capable of inhibiting EPEC growth, thereby reducing intestinal epithelial damage.

4. Conclussions

Probiotic yogurt treatment with supplement of purple sweet potato extract had a significant effect ($P < 0.05$) on feces water content, the number of erythrocyte, leucocyte, and hemoglobin. The treatment of YDP rats had water content of feces 48.422%, while number of erythrocyte, leucocytes, and hemoglobin were 8.578 $10^6/\mu l$, 14.152 $10^6/\mu l$ and 13.98 g / dL.

Acknowledgement
The gratitude of the researchers go to the Directorate of Research and Community Service of the Ministry of Research, Technology, and Higher Education for having granted fund for this research through Applied Product Research.Grant of The year 2017.

IOP Conf. Series: Earth and Environmental Science **142** (2018) 012026 doi:10.1088/1755-1315/142/1/012026

References

[1] FAO 2002 Guidelines for the evaluation of probiotics in food *Report of Joint FAO/WHO Working Group on drafting Guidelines for the evaluation of probiotics in food* London Ontario Canada

[2] Parvez S, Malik K A, Ah-Kang S and Kim H Y 2006 Probiotics and their fermented food products are beneficial for health Review article *J. Appl. Microbio.* **100** 1171-1185

[3] Oyetayo V O 2004 Performance of ratsorograstically dosed with faecal strain of *Lactobacillus acidophilus* and chalenged with *Escherichia coli.Afr.J.Biotecnol* **3** *409-4011.*

[4] Madellin-Pena M J dan M W Griffith 2009 Effect of molecules secreted by *Lactobacillus acidophilus* strainLa-5 on *Escherichia coli* O157:H7Colonization *Applied Environmental Microbiology* **75** 1165-1172

[5] Codex standard 243 2003 http :// www.codexalimentarius.com/codex stan 243-2003[7 Agustus 2016]

[6] Shah P N 2000 Probiotic bacteria : selective enumeration and survival in dairy food *Jurnal Dairy Science* **83** 894-907

[7] Collado M C, L S Surono, J Meruluoto and Salminen 2007 Potential probiotic characteristics of *Lactobacillus* and *Enterococcus* strains isolated from traditional dadihfermented milk against pathogen intestinal colonization *Journal of Food Protection* **70** 700-705

[8] Lee Y, K Puong, A C Ouwehand, S Salminen 2003 Displacement of bacterial pathogens from mucus and Caco-2 cell surface by lactobacilli **52** 925-930

[9] Woolfe J 1993 *Sweet Potato: An Untapped Food Resource* Cambridge: Cambridge University Press

[10] Rahayu E S, Indrati R, Utami T, Harmayani E, Nur M 1996 *Bahan Pangan Hasil Fermentasi* Food Nutrition Culture Collection PAU Pangan dan Gizi UGM Yogyakarta

[11] Ngatirah, Harmayani E, Rahayu E S, Utami T 2000 *Seleksi bakteri asam laktat sebagai agensia probiotik yang berpotensi menurunkan kolesterol* Paper dalam prosiding Seminar Nasional Industri Pangan Surabaya

[12] Purwandani S N and Rahayu, E.S 2003 Isolasi dan seleksi *Lactobacillus* yang berpotensi sebagai agensia probiotik. *Agritech* **23** 67-74

[13] Aboderin F I, Oyetayo V O 2006 Haematological studies of rats fed different doses of probiotic, *Lactobacillus plantarum*, isolated from Fermenting Corn Slurry. Pakistan *J. Of Nutrition* **5** (2) 102-105

[14] Muchtadi, D 1993 *Teknik Evaluasi Nilai Gizi Protein* Program Pascasarjana Institut Pertanian Bogor

[15] Steel R G D and Torrie J H 1995 *Principles and procedures of statistic : a biometrical approach 2nd edition* New York : McGraw Hill Book Co

[16] AOAC 2005 *Official Methods of Analysis* Washington DC

[17] Spehlmann M E, S M Dann, P Hruz, E Hanson, D F Mc-Cole and L Eckmann 2009 CXCR2-dependen Mucosal Neutrophil Influx Protects Agains Collitis-associated Diarrhea caused by An Attackhing/Efficacing Lesion-forming Bacterial Pathoge. *J. Immunilogy* 183: 3333-3343

[18] Campell T W 2004 Mammalian hematology : Laboratory animals and miscellaneous species In Thrall MA (ed) Veterinary Hemaology and Clinical Chemistry Lippincott Williams and Wilkins

[19] Tannock GW 1999 Probiotics : a critical review. Nortfolk, England : Horizon Scientific Press

[20] Walker WA 2008 Role of nutrients and bacterial colonization in the development of intestinal host defense *J. Ped Gastroenterol. Nutr* **30** 22000

IOP Conf. Series: Earth and Environmental Science **142** (2018) 012027 doi:10.1088/1755-1315/142/1/012027

The effect of CMC and arabic gum stabilizer combination on the characteristics of soursop velva (*Annona muricata L*.)

N H R Parnanto, B Yudhistira, S R Pertiwi and A Pangestika
Food Science and Technology Department, Agriculture Faculty, Universitas Sebelas Maret, Surakarta, Indonesia

E-mail: barayudhistira@staff.uns.ac.id

Abstract. The aims of this study were to determine the effect of the combination of CMC and Arabic Gum stabilizer toward the soursop velva characteristics and to determine the best stabilizer combination of soursop velva. This study was performed using Completely Randomized Design (CRD) with one factor: combination of CMC and Arabic Gum stabilizer using two sample replications and the analysis was repeated three times. The result showed that the use of the combination of CMC and Arabic Gum stabilizer gave a significant effect on the overrun value, melting power, total dissolved solids, moisture content, dietary fiber, taste, texture and overalls. Moreover, there were no significant effect on color and flavor of the soursop velva. The experiment showed that soursop velva F4 (3:1) was the best formula with overrun value 9.93%, the melting power was 22 minutes 52 seconds, the total dissolved solids 19,10°*Brix*, the moisture content 71.508%, dietary fiber 3.301% and it has sensory values of color, taste, flavor, texture, *overall* at 3.66, 3.267, 3.33, 4.06, 3.10 respectively.

1. Introduction

Soursop (*Annona muricata L*) is tropical fruit that can grow in lowland with dry climates. The large number of national soursop production, makes this fruit can be easily found and available in the abundant amounts. The ripe soursop fruit is delicious, sweet, or sweet sour. Nevertheless, public interest to processed soursop is still low. Food diversification is needed so that the availability of abundant soursop fruit can be utilized. One form of processing that can be applied is producing soursop velva products.Velva is one type of desserts made from fruits or vegetables and frozen with ice cream maker. It has low fat content and high fiber. The fat contained in velva is very low because it does not use fatty ingredients, such as milk, in the production process. The main difference between fruit velva and ice cream is the consistency of the fruit used. Fruit velva uses fruit pure as the main ingredient, while ice cream uses fruit juice as an additional ingredient.

One important component in the production of velva is the stabilizer. Stabilizers were used to prevent the formation of rough ice crystals, form a soft texture, and provide better endurance [1]. The problem faced in making the fruit velva is a rough texture and quick melt. To produce velva with smooth texture required stabilizer ingredient with the type and concentration in accordance with the character of the fruit. According to the previous study [2] the use of two or more different stabilizers combination on ice cream gives better results. Therefore, in this study a combination of stabilizer ingredients CMC and Arabic Gum was applied. According to [3], the use of CMC as stabilizer does not need short aging time. The addition of Arabic Gum aims to prevent the formation of larger ice crystals by binding some of the water. In this study, the researchers tried to use a combination of CMC and Arabic Gum stabilizers which aimed to produce the best velva with better overrun, better melting power and texture. According to [4], generally the choice of the stabilizer type is based primarily on its effect on the overrun and texture. By the combination of stabilizers used in this study, it is expected that production of velva product can be more optimal.

2. Methods

2.1. Equipment
Blender (National), mixer (Miyoko), refrigerator (Sharp), ice cream maker (Gelatio II), freezer (Modena), digital scales (Cis analytical), stopwatch, measuring cup (Pyrex), Erlenmeyer, refractometer (RHH-92ATC), destilator, water bath, oven, desiccator

2.2. Ingredient
Javanese varieties of ripe soursop ripe condition obtained from Pasar Legi Surakarta, sugar, CMC and Arabic Gum stabilizer.

2.3. Preliminary Research
Preliminary research was conducted to determine the combination of *puree* (fruit : water) i.e. 1:1, 1:2, 1:3 and sugar concentration 15%, 20%, 25%. The result showed that the selected *puree* was 1:1 and the sugar concentration 20%. From the observations and opinions of 30 panelists the average of panelist preferences is puree 1:1 with 20% sugar concentration because has a good texture and it is not too sour taste. On a 1:2 puree ratio it looks too fluid. In Comparison of pure 2:1 looks very thick and too sour taste that causes panelists do not like.

2.4. Formulation and Production
Soursop was peeled and the seeds were removed. Then, it was blanched in 5 minutes and shattered with a blender. After that, CMC and Arabic Gum were added and mixed. The dough was stored in refrigerator temperature for 8-12 hours then done quick freezing with ice cream maker. After that, the velva was stored in freezer temperature [5].

2.5 Analysis of Physical, Chemical, and Sensory Characteristic of Sousop Velva
Physical characteristics observed were overrun [6], melting power [7]. The observed chemical properties were total soluble solids [8], moisture content [8], dietary fiber [9]. While sensory analysis included color, taste, flavor, texture, overall.

3. Result and Discusion

3.1. Physical characteristic of soursop velva

The highest overrun value was produced by the product with formula F2 using combination of CMC and Arabic Gum stabilizer 1:3, that was equal to 13.84% (Figure 1). While the lowest overrun value produced by product F5 with CMC and Arabic Gum stabilizer 1:0 that was equal to 9.15%. Arabic Gum has the characteristic of forming a solution with a low viscosity so that the velva with a combination of stabilizers with greater Arabic Gum ratio increases the overrun as the air can penetrate the dough surface and the product can expand [3]. CMC has the characteristic of forming high viscosity, so the product can expand which causes a low measured overrun. The overrun value of carrot velva reference with the same stabilizer CMC and Arabic Gum that was equal to 11.05% - 19.79% [10], was not much different from the velva soursop overrun.

3.2 Melting Power
The highest melting power value was produced by the product with the formula F4 (Figure 2) using a combination of stabilizer CMC and Arabic Gum 3:1, i.e. 22 minutes 52 seconds. While the lowest melting power value was produced by F1 product with stabilizer CMC and Arabic Gum 0:1, 19 minutes 17 seconds. CMC could form a high viscosity, so the velva with a combination of stabilizers

using a larger CMC ratio could produce small ice crystals so they did not melt quickly. The value of the carrot velva reference was 20.25-27.00 minutes [10], it was not much different from the soursop velva.

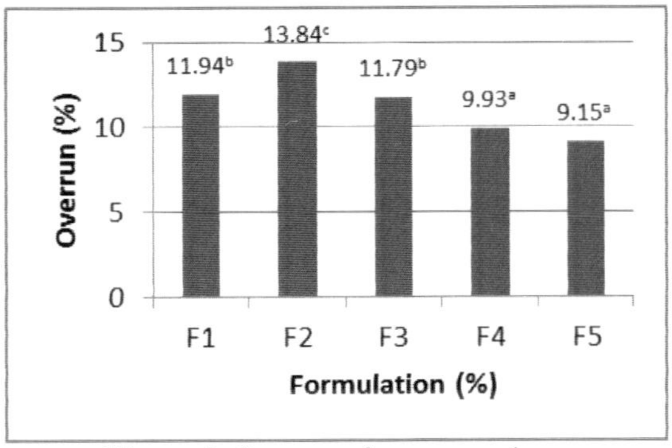

Figure 1. Overrun of soursop velva

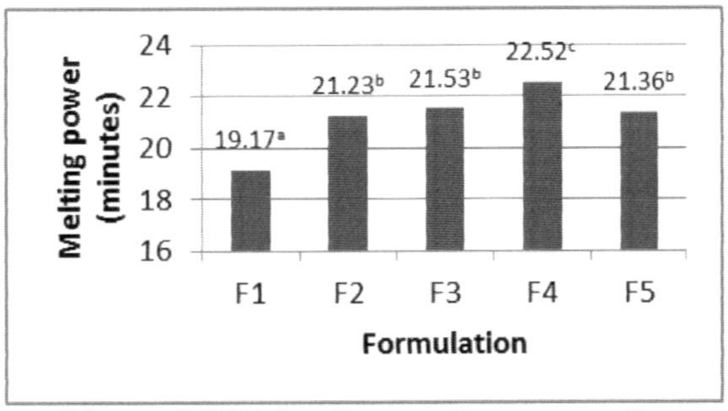

Figure 2. Melting Power of soursop velva

3.3. Chemical characteristic of soursop velva

3.3.1. Total Dissolved Solids The highest value of total dissolved solids was produced by the product with formulation F4 using a combination of CMC and Arabic Gum stabilizer 3:1 at 19.10°Brix. While the total value of the lowest dissolved solids was produced by F1 product with CMC and Arabic Gum stabilizer 0:1 which was equal to 16.80°Brix. The total value of dissolved solids of carrot velva reference was 18.75-19.25°Brix [10] and on parang machete velva 15.1-18.9°Brix [11] was not much different from the total solid value of soursop velva. The role of the stabilizer in the increased soluble solids of the soursop velva because CMC contains linear and water-soluble polysaccharida [12]. While Arabic Gum are simple sugars or descendants, can gradually hydrolyzed into smaller parts [13]. So that the use of a combination of stabilizers of CMC and Gum Arab 3: 1 is optimal in the increase of dissolved solids of soursop velva.

3.3.2. Moisture Content The highest value of moisture content was produced by the product with formula F1 using the ingredients of CMC and Arabic Gum 0:1 i.e., 75.103%. The lowest moisture value was produced by F5 product with stabilizer of CMC and Arabic Gum 1:0 i.e., 70.249%. According to [14] CMC has the ability to absorb water greater than other stabilizers so that the moisture content is measured lower. The value of moisture content of carrot velva reference [10] that was 74.12 – 75.25% not much different from soursop velva.

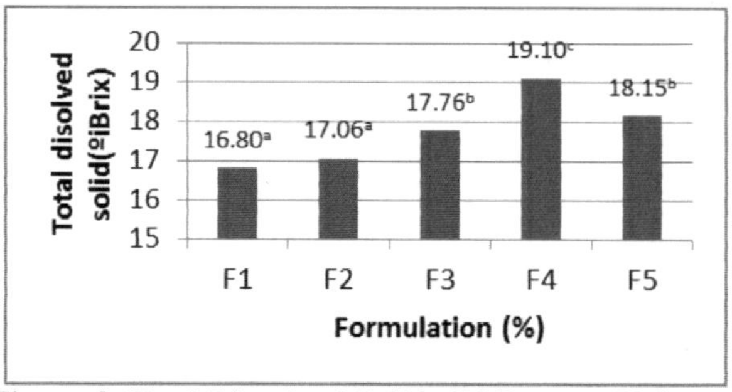

Figure 3. Total Dissolved Solids Of Soursop Velva

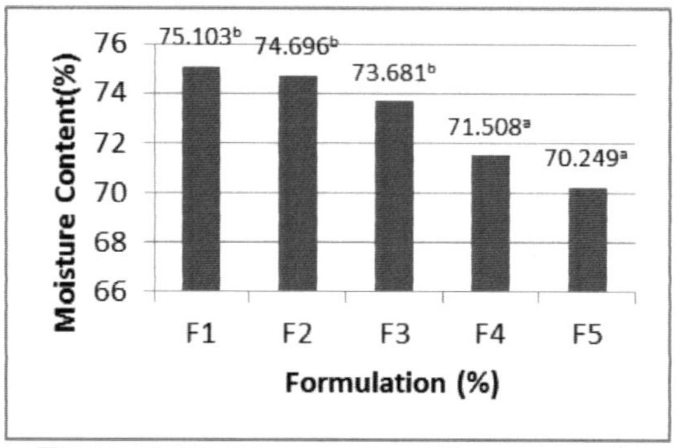

Figure 4. Moisture content of soursop velva

3.3.3. Dietary fiber The highest value of dietary fiber (Figure 5) was produced by product with formula F4 using combination of stabilizer of CMC and Arabic Gum 3:1 that was equal to 3.301% .While the lowest fiber value was produced by F1 product with stabilizer CMC and Arabic Gum 0:1 that was 1.329%. CMC is a polymer fiber food chain consisting of cellulose molecule units having fiber and Arabic Gum is a group of soluble fiber [15] so as to increase the food fiber of soursop velva.

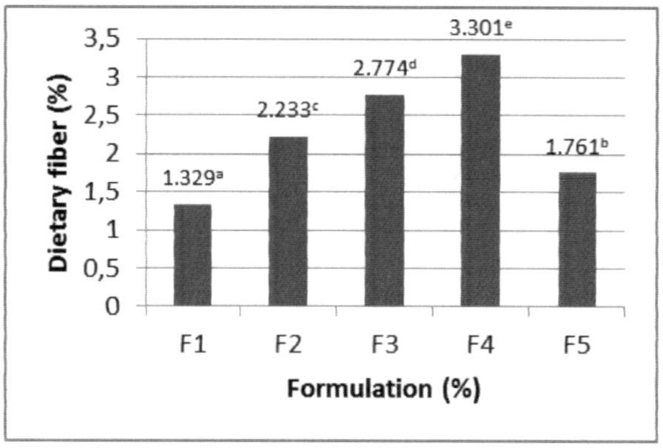

Figure 5. Dietary Fiber Of Soursop Velva

3.4. Sensory Characteristic of Soursop Velva

Table 1. Sensory characteristic of soursop velva

	Characteristic				
	Colour	Taste	Flavor	Texture	Overall
F1	3.70	3.43^{ab}	3.36	2.50^{a}	3.00^{a}
F2	3.80	3.83^{c}	3.56	2.56^{a}	4.10^{b}
F3	3.76	3.56^{b}	3.53	3.06^{a}	3.30^{a}
F4	3.66	3.26^{ab}	3.33	4.06^{b}	3.10^{a}
F5	3.63	3.10^{a}	3.30	2.83^{a}	3.00^{a}

Table 1 shows the stabilizers CMC and Arabic Gum effect on taste, texture and overall parameters, but it does not show any significant effect on the color and flavor of the soursop velva. Stabilizer is a hydrocoloid that does not have a volatile compound so that it will not affect the color and flavor of a product but it will affect the taste significantly [16]. According to [17], CMC has a threshold of taste so that at certain concentrations will experience a decrease in sweet taste and increase the sour taste that make panelists do not like.

3.5. Determination of the best velva concentration

Determination of the best formula of soursop velva with stabilizers of CMC and Arabic Gum was determined from the analysis of physical, chemical and sensory characteristics. It used compensatory method [18]. The F4 velva product using CMC and Arabic Gum stabilizer 3:1 was the best velva formulation. It had an overrun value of 9.93%, the power of melting reached 22 minutes 52 seconds, the total dissolved solids 19.10°Brix, moisture content of 71.508%, 3.301% dietary fiber, and had sensory value of color, flavor, texture, overalls respectively i.e., 3.66, 3.26, 3.33, 4.06, 3.10.

4. Conclusion

The use of CMC and Arabic Gum stabilizer combination gave a significant effect on the overrun, melting power, total dissolved solids, moisture content, dietary fiber, taste, texture and overalls; and there was no significant effect on color and flavor of the soursop velva. Velva with CMC and Arabic Gum stabilizers (3:1) was the best concentration of soursop velva.

References

[1] Bahramparvar M and Ve Tehrani M M 2011 Application and functions of stabilizers in ice cream *Food Reviews International* **27** 389-407
[2] Considine D M and Considine G 1982 *Food and Food Production Encyclopedia* (New York: Van Nonstran Reinhold Company)
[3] Arbuckle, W S 1986 *Ice Cream* AVI Publishing Company, Inc. Westport, Connecticut
[4] Soomer, H H 1947 *The Teory and Practise of Ice Cream Making*. Publised by Author Madison Wincosin
[5] Tressler D K and Woodroof J G 1976 *Food Product Formulary Vol* **3** *Food Vegetable and Nut Product.* The AVI Publishing Company Wesport Connecticut
[6] Marshall R T and Arbuckle W S 1996 *Ice Cream* New York: Chapman and Hill
[7] Bodyfelt F W, Tobias J and Trout G 1998 *The Sensory Evaluation of Dairy Product.* AVI Publishing company Westport Connecticut
[8] AOAC 1995 *Official Method of Analysis of The Association of Official Analytical Chemist* (Washington, D.C : Benyamin Franklin Stasion)
[9] Asp N G, Johanson C G, Halmer H, and Siljestrom M 1983 Rapyd enzimatic assay of insoluble and soluble diatary fiber *J. Agric. Food.Chem* **31** 476-482
[10] Rini K A, Ishartani D and Basito 2012 Pengaruh kombinasi bahan penstabil cmc dan gum arab terhadap mutu velva wortel (*Daucus carota* L.) varietas selo dan varietas Tawangmangu *Jurnal Teknosains Pangan* **1** 86-94 ISSN: 2302-0733

[11] Aisyah I 2003 *Penggunaan Kombinasi Bahan Penstabil pada Pembuatan Velva Labu Parang (Cucurbita moschata).* [Skripsi] Jurusan Gizi Masyarakat dan Sumberdaya Keluarga, Fakultas Pertanian, Institut Pertanian Bogor
[12] Furia T E 1975 *Handbook of Food Additive* (New York: AP Press)
[13] Rahmawati S 2014 *Karakteristik Fisiko-kimia dan Sensori Velva Pepaya (Acrica papaya L.) dengan Pemanis Madu.* [Skripsi] Fakultas Pertanian Universitas Sebelas Maret Surakarta
[14] Kusbiantoro B, Herawati H, Ahza A B 2015 *Pengaruh Jenis dan Konsentrasi Bahan Penstabil Terhadap Mutu Produk Velva Labu Jepang* Jurusan Teknologi Pangan dan Gizi Institut Pertanian Bogor
[15] Winarno F G 2002 *Kimia Pangan dan Gizi* (Jakarta : PT. Gramedia Pustaka Utama)
[16] Glicksman M 1969 *Gum Technology in The Food Industries* (New York: Academic Press)
[17] Ganz A Z 1977 *Cellulose Hydrocolloids.* Dalam H.D. Graham (eds,). Food Colloids. The AVI Publissing Company Westport Connectitude
[18] Sullivan and William G 2015 *Engineering Economy* (United State of America: Person Higher Education, Inc)

Characteristic study of chitosan addition in Tilapia (*Oreochromis niloticus*) bone based gelatin film

W Atmaka, B Yudhistira and M I S Putro
Food Science and Technology Program of Faculty of Agriculture, Universitas Sebelas Maret, Indonesia

E-mail: barayudhistira@staff.uns.ac.id

Abstract. Tilapia (*Oreochromis niloticus*) is one of popular fish species in Indonesia. The high number in tilapia's production and export of tilapia resulting in the increase of bone fish waste. an attempt to decrease the amount of the aforementioned waste, the fish bones were turned into gelatine. The gelatine produced from this waste can be put to good use by turning it into edible film due to its high water resistance and low tensile strength value. However, in order to make a proper film, both the water resistance and the tensile strength value needs another appropriate additional biopolymer. In this case, the appropriate biopolymer needed both to form the film and to repair its characteristics is chitosan. The purpose of this research is to find out the effect of the chitosan addition on the tilapia bone based gelatine film. The research used several mixtures of gelatine (G) and chitosan (C) with the following ratio: G100:C0 (GC1), G75:C25 (GC2), G50:C50 (GC3), G25:C75 (GC4), G0:C100 (GC5). ANOVA results ($P<0.05$) shows that the addition of chitosan on the gelatine film affected its characteristics in thickness, solubility, tensile strength, elongation at break, Fourier Transform Infrared (FTIR), color a, and color b but no significant effects on the vapor permeability and color L. The best result is shown on GC2 with thickness 0.119 mm; solubility 74.95%; tensile strength 2.635 Mpa; elongation at break 68.26%; water vapor permeability 5.897 g/h m^2 and FTIR. The parameters in GC2 shows good compatibility between the two biopolymers.

1. Introduction

Fishes are are one of many popular foodstuffss consumed by the community and also as export commodity. One of many popular fish species are tilapia (*Oreochromis niloticus*). According to the Data, Statistics and Information Center of Indonesia, in 2011 the number of tilapia production reached 328,473 tons and in 2012 it reached 338,659 tons. The higher the number of the exported tilapia, the higher the amount of its bones wasted. The bone of tilapia is a very good source of protein containing high amount of collagen that can be potential to be turned into gelatin [1]. The skins and the bones of fishes cover 10% to 20% of their total weight [2]. The percentage of gelatin yield on fish bones are generally estimated for around 12%. Therefore, it can be estimated that the expected amount of gelatin produced from 6,703 tons of fish bones are around 804.6 tons [3].

Gelatin is a protein derived from collagens which was extracted from bones [4]. Collagen is the main structural component of connective tissues, it has white color and on fishes, it can be found on their skin, bone and cartilage [5]. Gelatins can be turned into films and coatings with a high optical quality due to its superior gas barrier and mechanical properties on weak moistures which make it possible to be used as a bio-packaging material [6].

Chitosan is a polysaccharide obtained from chitin deacetylation which is generally produced from crustacean skin waste. Chitosan kills bacteria [7], inhibits the growth of spoilage bacteria [8], is not harmful to health [9] and disrupting the activity of the outer membrane of Gram-negative bacteria [10]. Chitosan also functions as antimicrobial agent against the spore germination of *Botrytis cinerea, Fusarium oxysporum* [11], *Penicillium sp* [12] and *Pseudomonas sp* [13].

The purpose of this research is to decrease the amount of tilapia bone waste by turning it into gelatin-based edible film. However, the gelatin itself contains weak vapor permeability and tensile strength resulting on its less-optimal potential as a packaging. Thus, a composition of the gelatin with another renewable film-forming biopolymer is necessary [14]. Chitosan is known to be able to improve the characteristics of the gelatin film. However, its addition to the gelatin as a packaging innovation still lacks of further information. Chitosan and gelatin offers a great potential in completing food nutritive value and are shown to be biodegradable and biocompatible [15]. It is thus, the characterization of chitosan-added tilapia bone gelatin edible film is required in order to observe the characteristics of the edible films.

2. Methods

The materials used in this study are the bones of tilapia which are collected from fresh tilapia fillet residuals in MAJU 57 fishing spot of Janti, Polanharjo, Klaten and chitosans. The chemicals used were sorbitols, aquadest, NaCl, and 0.85% liquid glacial acetic acids. The instruments used in producing tilapia bone gelatin are a knife, a basin, a pan, beaker glasses, a waterbath, whatman papers, a cabinet dryer, a pHmeter, a scale, and a mixer. The instrumentss used in producing the edible film are an analytical scale, graduated cylinders, hot plates, a magnetic stirrer, a cabinet dryer, graduated pipettes, propipettes, edible film trays, and a thermometer. The testing instruments used in this research are weighing bottles, an oven, a desiccator, a 0.01 mm *Krisbow* micrometer, a *Lloyd Instrument* TA plus texture analyzer, WVTR cups, silica gels, a hygrometer, a *Shimadzu* IR Prestige21 FTIR Spetrophotometer, and a colorimeter.

The film making procedures are as follow: first, tilapia bone gelatin. The bones of the tilapia were cleansed from both dirt and flesh residuals, then they were degreased in 80°C for 30 minutes. The bones were then re-cleaned to make sure that neither flesh residuals nor any outer high-fat containing surface were left. The clean bones were cut into 3-5 cm pieces and were demineralized in a 4% HCl (v/v) with the ratio of 1:4 (b/v) for 48 hours. The 4% HCl was replaced every 24 hours. After 48 hours of demineralization, *ossein* was obtained. The *ossein* was cleansed on flowing water to neutralize its pH (6-7). The *ossein* of which pH had been lowered was put into a beaker glass, aquadest was added with the ratio of 1:3 (b/v). The extraction process was executed in the water bath with the temperature of 85°C for 6 hours. The extract obtained from the process was later filtered using whatman paper and then dried using oven with the temperature of 50°C for 24 hours. Finally, the size of the extract was minimized into powder using mixer, producing powder of gelatin. Second, the gelatin solution. The process of gelatin solution production was started by putting 4 grams of gelatin into a beaker glass placed on a hot plate. Next, 100 ml of aquadest was also put into the glass slowly. Then the mixture was stirred using magnetic stirrer (4% gelatin). Finally, 1.2 ml of sorbitol was dissolved into the solution and heated with the temperature of 40°C for 30 minutes. Third, chitosan solution. In producing chitosan solution, the first step was started by putting 3 grams of chitosan into a beaker glass placed on a hot plate followed by 100 ml of 0.85% acetic acid poured into the glass slowly. Then, the mixture was stirred using magnetic stirrer (chitosan 3%). Finally, 1 ml of sorbitol was dissolved into the solution and was heated with the temperature of 40°C for 30 minutes. Fourth, gelatin-chitosan solution. The process of gelatin-chitosan production was started by composing both gelatin solution and chitosan solution on different beaker glasses placed on hot plates with the following ratios: 100%:0%, 25%:75%, 50%:50%, 75%:25%, and 0%:100%. Next, each mixture was stirred using the magnetic stirrer. Finally, each solution was heated within the temperature of 40°C for 30 minutes. The produced solutions later were quenched to the room temperature and after a while, they were dried inside a cabinet dryer within the temperature of 60°C for 24 hours.

The method of the analyses was ran through several tests, namely: thickness test via 0.01 mm *Krisbow* micrometer [16], solubility test [17], tensile strength and elongation at break test via *Lloyd Instrument* TA plus texture analyzer [18], vapor permeability test [18], functional groups test via *Shimadzu* IR Prestige21 FTIR Spetrophotometer [19], and color test via colorimeter. This research applied randomized complete block design with one factor, namely comparison between addition of gelatin and chitosan solution. The various ratios of gelatin (G) and chitosan (C) mixtures were as follows: G100:C0 (GC1), G75:C25 (GC2), G50:C50 (GC3), G25:C75 (GC4), G0:C100 (GC5). The sampling of each mixture was repeated three times and was analyzed two times. The collected data were later analyzed using one-way ANOVA with 5% significance level and were analyzed further using Duncan's Multiple Range Test (DMRT) with 5% significance level when significant difference was observed.

3. Result and discussion

Table 1. Chitosan-added Tilapia Bone Gelatin Edible Film Characteristics Text Result

Edible Film	Edible Film Characteristics Test				
	Thickness (mm)	Solubility (%)	Tensile Strength (Mpa)	Elongation at Break (%)	Vapor Permeability (g/h m^2)
GC1	0.173^b±0.060	89.52^d±0.016	0.583^a±0.224	254.40^e±7.762	5.987^a±0.739
GC2	0.119^a±0.016	74.95^c±0.032	2.635^b±0.209	68.26^d±6.459	5.897^a±0.708
GC3	0.117^a±0.021	62.51^b±0.075	4.064^c±1.073	35.85^c±4.384	5.966^a±0.958
GC4	0.116^a±0.027	49.31^a±0.033	4.536^c±0.472	21.65^b±1.417	6.865^a±1.866
GC5	0.115^a±0.015	47.40^a±0.043	4.580^c±0.339	13.16^a±2.037	7.204^a±1.958

$^{a, b, c}$ *different subscripts showed significant difference at P<0.05*

3.1. Thickness

The thickness of film determines its capability in preventing any substance exchanges from its surroundings and vice versa [20], affects its biological characteristics and its shelf life [21]. The result of this research is in accordance with the findings of a research conducted [14] in which shows that addition of chitosan on gelatin-based film would significantly lower its thickness. The higher the amount of solid in the solution the thicker the film produced [22].

The result of this research shows that the gelatin-based film tends to be thicker than the chitosan-based film because it holds more solids than the chitosan-based film does. Due to its good physical properties of gel, gelatins are used in food productions in contributing and thicker food textures [23]. The finding of this research shows that the best thickness was resulted in the mixture of GC2 (75% Gelatin 25% Chitosan) with the thickness of 0.119 mm, and overall thickness around 0.173-0.115 mm, which it is stated by JIS (Japanesse Industrial Standard) that the maximum thickness of plastic films food packaging is 0.25 mm. Therefore, the thickness of the gelatin-based film in this research already fulfilled the standard of plastic film food packaging [24].

3.2. Solubility

The solubility of film determines its biodegradability as a packaging. The solubility level of a film depends on the product's characteristics [25]. Based on the research of the composition of cuttlefish gelatin edible film and chitosan [14], gelatin-based film shows high level solubility of (90.68%), while chitosan-based film only shows lower solubility level (around 50.35%). The solubility of the samples in this research ranged from 89.523% to 47.403%. The best solubility was resulted from the mixture of GC2 (75% Gelatin 25% Chitosan) with the solubility of 74.954%. This value of solubility was close to the value of solubility in the previous research.

There are several factors affecting the solubility of an edible film: hydrophilic and hydrophobic properties, pH, polarity [26] particle size, and molecular structure [27]. This research found that the more chitosan added to the edible film, the more its solubility decreased. Polysaccharide film has a strong barrier to O_2 and CO_2 due to its tight molecular structure of hydrogen bonding which makes it less soluble [16]. The addition of chitosan will raise water resistance properties which shows interaction between polymers disrupting polypeptide chain inside the gelatin's matrix. The interaction between the two biopolymers is resulted from electrostatic activity and low solubility is resulted from hydrogen bonding [14].

3.3. Tensile Strength

Film requires an adequate tensile strength and elongation at break in so it can hold external weight, preserving its integrity and may guard the product when applied as packaging [28]. In the previous research, the edible film composed of fish gelatin showed tensile strength at 2.17±0.97 Mpa [29]. This finding was closely related to this research's in which in the composition of GC2 (75% Gelatin 25% Chitosan), tensile strength was shown at 2.63±0.20 Mpa. As for chitosan was shown at 8.41 Mpa [15] which is far different from this research's finding in the composition of GC5 (100% Chitosan) where tensile strength was shown at 4.58±0.33 Mpa.

According to JIS, the minimum value of tensile strength which can be classified as food packaging film is 0.39 Mpa [24]. The tensile strength of the samples on this research ranged from 0.58 Mpa to 4.580 Mpa, meaning that they already fulfilled the minimum standard of food packaging film. The addition of chitosan on film causes increment on its tensile strength, while gelatin-only-based film does not. Tensile strength and elongation at break are usually related to micro structure of the film network and intermolecular interaction [14].

Compositions of protein with different ratios of polysaccharide may improve tensile strength. In the composition of gelatin-chitosan film, the addition of chitosan improves significant tensile strength compared to the gelatin alone [29]. The level of chain elongation and sequence of amino acid residue may affect the mechanical properties of protein-based film [30]. This condition results on the strengthening of the interaction between gelatin and chitosan due to hydrogen bonding and complex polyanion forming between the gelatin and the chitosan [31]. Chitosan with higher molecular weight or deacetylation degree may contribute more reactive groups in one chain to interact with the gelatin through the hydrogen bonding which results on film strengthening [32].

3.4. Elongation at Break

Elongation at break is a parameter which is inversely proportional to tensile strength. The lower the value of tensile strength, the higher the value of elongation [24]. In the previous research, edible film composed of fish gelatin showed elongation at 82.64±20.11% [29] which is far different from the result of this research in which elongation was shown at 254.40%. As for chitosan the elongation at break was shown at 19.55% [15] which is close to the result of this research in the composition of GC4 (25 % Gelatin 75% Chitosan) in which elongation at break was shown at 21.65%. The elongation at break of the samples in this research ranged from 254.49% to 21.65%. The finding of this research shows that the best elongation was resulted in the mixture of GC2 (75% Gelatin 25% Chitosan) with the elongation at break of 68.260%.

According to JIS, the minimum value of elongation at break which can be classified as food packaging film is 70% [24]. Therefore, only the mixture of GC1 managed to fulfill the standard. However, the mixture of GC2 was also close the standard value with the elongation at break of 68.260%. Gelatin-based film is more flexible yet fragile compared to chitosan-based film. Thus, the composition of gelatin and chitosan will strengthen gelatin-based film and strengthen chitosan-based film [29]. The elasticity of film depends on its material which can be crystals or amorphous solids. Gelatin is a more amorphous material compared to chitosan. The elasticity of film composed of gelatin and chitosan is improved due to the lowering of its crystallinity that makes the composite film more amorphous compared to the chitosan film. The lowering of the composite film's crystallinity is mainly caused by damaged hydrogen bonding on the molecule of chitosan. Groups of amino on the hydrogen bonding of chitosan and groups of carboxyl on gelatin formed complex gelatin-chitosan polyelectrolytes. This form shatters the hydrogen bonding of amino groups and hydroxyl groups on the chitosan and results on the forming of amorphous structure of complex polyelectrolyte, causing improvement of value of elongation at break [33].

3.5. Water Vapor Permeability

Water vapor permeability consists of dissolution process and active diffusion in which vapor dissolves on one of either side of the film and then diffuses on the other side of the film. The resistance to water vapor transmission rate is determined by the thickness of the film, temperature, and water vapor partial gradient pressure [16]. The result of this research shows higher water vapor permeability than that of the result of the research conducted [24], in which is stated that according to JIS, the maximum value of water vapor transmission rate which can be classified as food packaging film is 7 g/m^2/day. As for commercial LDPE (Low-density polyethylene) has the value of water vapor permeability of 9.14 g/m s and HDPE (High-density polyethylene) has the value of water vapor permeability of 2.31 g/m s [23]. Generally, protein and polysaccharide-based film has a high value of water vapor transmission rate [16]. High value of water vapor transmission rate affects the packaged product shelf life; the higher the product's water vapor transmission rate, the less durable the product will be.

Gelatin-based edible film has lower water vapor permeability than chitosan-added edible film, as stated [34], the more polysaccharides added to a film, the weaker its barrier toward water vapor permeability and oxygen will be. It occurs due to the gelatin-based film thickness which is thicker than chitosan-based film [24]. In addition, the low water vapor permeability of a fish gelatin-based film is due to its high hydrophobicity of which the amount of its proline and hydroxyproline decreases by [35]. Chitosan-based film is also polysaccharide which tends to be more useful as sacrificing agent than as barrier because it absorbs water and thus gives a temporary protection to the product [23]. The use of chitosan to block the water vapor transmission has an optimal value which in this research was shown in the mixture of GC2 where chitosan functioned as sacrificing agent and the composition of gelatin was higher than chitosan. In the other mixtures, the water vapor permeability was shown growing due to the composition of chitosan which temporary blocked the water vapor transmission and the low composition of gelatin which resulted on higher water vapor permeability.

3.6. FTIR Test

Fourier Transform Infrared (FTIR) is one of the methods to detect compound molecular structure through the functional group identification of its composition.The chemical structure, molecular shape and functional group of the tested sample function as the basic shape of a spectrum which later was obtained from the result of the analysis [36]. Theoretically, hydrogen bonding or other chemical polymer interaction causes shift in peak position [37]. FTIR is useful in completing the characterization of composite film microstructure due to its function as an evaluation instrument to the film inter-component interaction.

Gelatin has similar structure which generally owned by protein, namely carbon, hydrogen, hydroxyl group, (OH), carbonyl group (C=O), and amine group (NH). The result of the analysis on the mixture of GC1 using via FTIR showed a stretching vibration of OH group with the absorption peak of 3424.76 cm^{-1}. The presence of OH group was strengthened by absorption in the region of amide II (1538.30-1320.33 cm^{-1}) which exhibited bending vibration of OH group. The absorption peak in the region of 2921.32-2852.84 cm^{-1} showed stretching vibration of CH group which was bonded to NH group in the amide region. Stretching vibration of C=O group (amide I) was shown on the absorption peak in the region of 1632.81 cm^{-1}. The absorption peak region of 1271.14 cm^{-1} showed amide III (CN stretching and NH bending) which was the absorption region of gelatin functional group.

In previous research, gelatin powder showed wavelength of amide A group around 2930 cm^{-1}. Absorption peak of 1565-1644 cm^{-1} in the region of amide I, absorption peak of 1560-1335 cm^{-1} in the region of amide II, and absorption peak of 670-1240 cm^{-1} in the region of amide III was shown in the absorption region of gelatin functional group [38]. In the other hand, gelatin powder made of sharpnose stingray *(himantura gerrardi)* skin soaked in 4% HCl showed absorption peak of 1647 cm^{-1} which exhibited stretching of C=O [39].

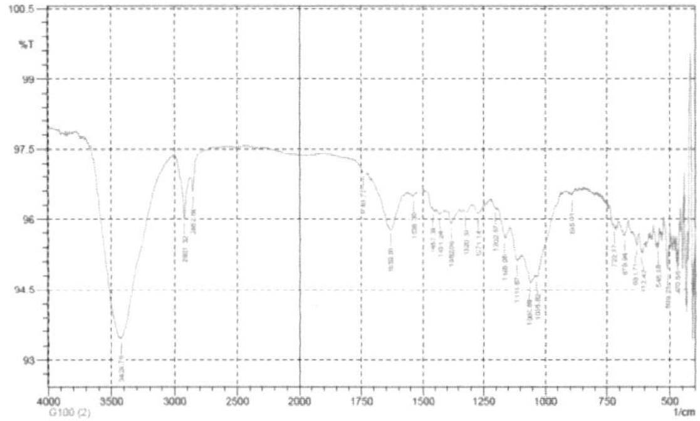

Figure 1. Infrared Spectrum of Gelatin-based Edible Film (GC1)

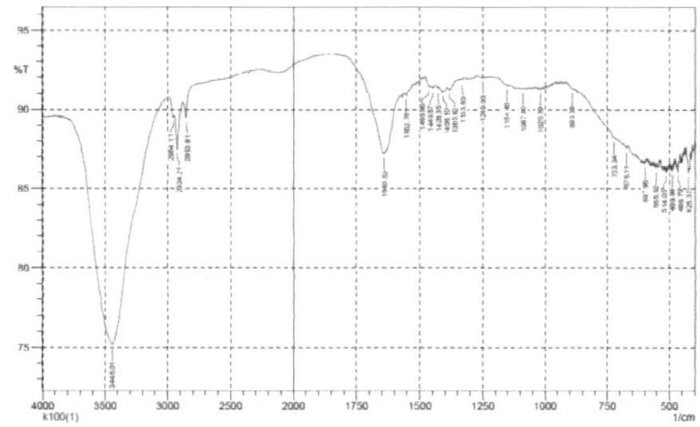

Figure 2. Infrared Spectrum of Chitosan-based Edible Film (GC5)

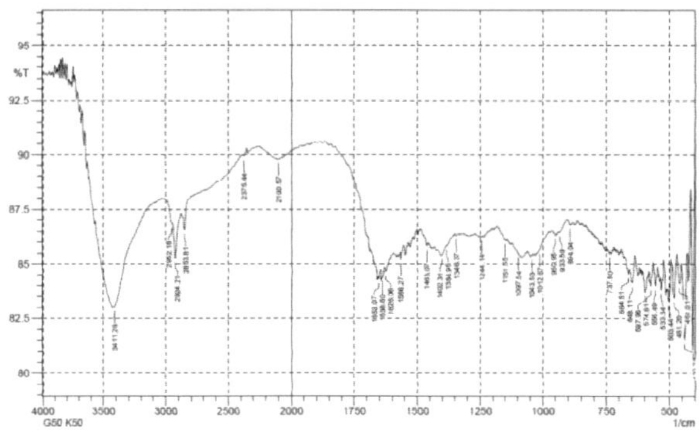

Figure 3. Infrared Spectrum of Gelatin-Chitosan-based Edible Film (GC3)

Figure 2 represents the result of the FTIR test on the chitosan-based gelatin film mixture of GC5. The absorption peak of this film was shown in the region of 3445.01 cm^{-1}, exhibiting stretching vibration of OH and NH group. The presence of OH group was strengthened by absorption in the region of amide II (1552.76-1333.83 cm^{-1}) which exhibited bending vibration of OH group. The absorption peak in the region of 1640.53 cm^{-1} exhibited stretching vibration of C=O group (amide I) and the absorption peak in the region of 2954.11-2853.81 cm^{-1} exhibited stretching vibration of CH group. In previous research, chitosan exhibits amide I bonding on the absorption of 1637 cm^{-1}, indicating the presence of acetyl group (C-N), and the region of 3500-3400 cm^{-1} is the absorption band to the stretching vibration of NH and OH group of chitosan compound [40].

Figure 3 represents the result of the FTIR test on the gelatin-chitosan-based film mixture of GC3. The absorption peak of this film was shown in the region of 3411.26 cm^{-1}, exhibiting stretching vibration of OH and NH group. The absorption peak in the region of 1645.35 cm^{-1} exhibited stretching vibration of C=O group (amide I) and the absorption peak in the region of 2954.21-2853.81 cm^{-1} exhibited stretching vibration of CH group.

In the region of 1653.07-1463.07 cm^{-1} (the absorption peak of amide I and amide II), compared to gelatin-based and chitosan-based edible films, a change of the gelatin's polypeptide chain structure occurred [14], which was due to the addition of chitosan. Gelatin-chitosan-based edible film experienced intensity increment in the region of 1700-1500 cm^{-1} [41]. This statement was in accordance with the finding of this research in which in the region of 1653.07-1463.07 cm^{-1}, an increment of vibration intensity emerged in the samples of GC1 and GC5. Such pattern occurred due to the electrostatic interaction between amino group and carbonyl group in the formation of polyelectrolyte complex.

3.7. Color

Table 2. Chitosan-added Tilapia Bone Gelatin Edible Film Color Test Result

Edible Film	Color		
	L	a	B
GC1	$45.91^{a} \pm 2.222$	$2.38^{b} \pm 0.499$	$-0.80^{a} \pm 0.317$
GC2	$48.13^{a} \pm 3.763$	$1.74^{ab} \pm 0.272$	$3.77^{b} \pm 1.062$
GC3	$46.62^{a} \pm 5.062$	$1.9^{ab} \pm 0.530$	$3.03^{b} \pm 0.674$
GC4	$47.55^{a} \pm 5.704$	$1.78^{ab} \pm 0.577$	$4.82^{b} \pm 1.817$
GC5	$44.69^{a} \pm 3.798$	$1.24^{a} \pm 0.745$	$11.38^{c} \pm 4.864$

The color of the edible film is an important factor in its use as packaging when covering the surface of the product. The color of the edible film is expected not to disrupt the original color of the packaged product in order to make it attractive to the consumer. The data of collected colors in the research is shown in **Table 2.** The result of this research is in accordance with the previous research which shows that chitosan film has lower values of L (lightness) dan a (redness) than gelatin film and composite film do, but has higher value of b (yellowness) which shows that the color of chitosan tends to be yellow [14].

Table 3. Edible Film Effectiveness Test Result

Edible Film	Effectiveness Value
GC1	0.736
GC2	0.527
GC3	0.487
GC4	0.324
GC5	0.226

4. Conclusion

Chitosan addition on the tilapia bone-based gelatin edible film affected its characteristics in thickness, solubility, tensile strength, elongation at break, Fourier Transform Infrared (FTIR), Color A, and Color B but no effect on the vapor ability and Color L. The best result for gelatin-chitosan film is shown on GC2 with thickness 0.119 mm; solubility 74.95%; tensile strength 2.635 Mpa; elongation at break 68.26%; water vapor permeability 5.897 g/h m^2 and FTIR shows good compatibility between the two biopolymers.

References

[1] Maryani, Surti T and Ibrahim R 2010 Aplikasi Gelatin Tulang Ikan Nila Merah (*Oreochromis niloticus*) terhadap Mutu Permen Jelly *J. Saintek Perikanan* **6** (1) pp 62-68

[2] Surono N D, Budiyanto D, Widarto, Ratnawati, Aji U S, Suyuni A M and Sugiran 1994 *Penerapan Paket Teknologi Pengolahan Gelatin dari Ikan Cucut* Jakarta: BPPMHP

[3] Junianto, Kiki H and Ine M 2006 Produksi Gelatin dari Tulang Ikan dan Pemanfaatannya Sebagai Bahan Dasar Pembuatan Cangkang Kapsul. *Penelitian Hibah Bersaing IV Tahun I Fakultas Perikanan dan Ilmu Kelautan Universitas Padjajaran*

[4] Rahayu F and Nurul H F 2015 *Pengaruh* Waktu Ekstraksi Terhadap Rendemen Gelatin dari

Tulang Ikan Nila Merah *Seminar Nasional Sains dan Teknologi Fakultas Teknik Universitas Muhammadiyah Jakarta* ISSN: 2407-1846 e-ISSN: 2460-8416

[5] Bennion M 1980 *The Science of Food* New York: John Willey and Sons

[6] Gennadios A, McHugh T H Weller C L and Krochta J M. 1994 Edible Coatings and Films Based on Proteins *Lancaster, P, A, Technomi.* pp 201-278

[7] Hui L, Yumin D, Xiaohui W, and Liping S 2004 Chitosan Kills Bacteria through Cell Membrane Damage *Int. J. of Food Microbiology* **95** pp 147– 155

[8] Morhsed A, Bashir A, Khan M H and Alam M K 2011 Antibacterial Activity of Shrimp Chitosan Against some Local Food Spoilagebacteria and Food Borne Pathogens *Bangladesh J. Microbial* **28** (1) pp 45-47

[9] Henriette M C Azeredo, de Britto D and Assis O B G 2010 Chitosan Edible Films and Coating – Review. *Embrapa Tropical Agroindustry Fortaleza CE Brazil* ISBN 978-1-61728-831-9

[10] Helander E L, Nurmiaho L, Ahvenainen R, Rhoades J and Roller S 2001 Chitosan Disrupts The Barrier Properties of The Outer Membrane of Gram-Negative Bacteria *Int. J. of Food Microbiology.* **71** pp 235–244

[11] Susilowati, Endang, Maryani, M Masykuri and Arista N D.2011 Sintesis Kitosan Hidrolisat dari Limbah Udang Putih (Penaeus merguinensis) Secara Enzimatis Menggunakan Papain dan Uji Aktivitas Antibakteri Terhadap Klebsiella pneumonia. *Prosiding Seminar Nasional Kimia dan Pendidikan Kimia III Program Studi Pendidikan Kimia Jurusan PMIPA FKIP UNS*

[12] Pratama T 2013 *Pengaruh Penambahan Edible Coat Kitosan Sebagai Anti Jamur pada Tepung Kentang.* Skripsi Program Studi Kimia Fakultas Pendidikan Matematika dan Ilmu Pengetahuan Alam Universitas Pendidikan Indonesia Bandung

[13] Simpson B K, N Gagne, I N A Ashie and E Noroozi 1997 Utilization of Chitosan for Preservation of Raw Shrimp (Pandalus borealis) *Food Biotech* **11** (1) 25-44

[14] Jridi, Mourad, Sawssan H, Hanen B A, Imen L, Aicha M, Maher K, Nabil S and Moncef N 2014 Physical, Structural, Antioxidant and Antimicrobial Properties of Gelatin-Chitosan Composite Edible Film *Int. J. of Biological Macromolecules.* pp 1-34

[15] Pereda A, Ponce G A, Marcovich, E N, Ruseckaite A R and Martucci F J 2011 Chitosan-Gelatin Composites and Bi- Layer Films with Potential Antimicrobial Activity *Food Hydrocolloids.* **25** pp 1372- 81

[16] Mc Hugh T H and Krochta J M 1994 *Permeability Properties of Edible Film.* In: Krochta J M, Baldwin E A, and Nisperos-Carriedo M O (Eds). *Edible Coating and Film to Improve Food Quality* Technomis Publishing Co. Inc. Lancester. Bacel

[17] Ghanbarzadeh, Babak, Hadi A and Ali A E 2011 Improving The Barrier and Mechanical Properties of Corn Starch-Based Edible Films: Effect of Citric Acid and Carboxymethyl Cellulose *Industrial Crops and Products* **33** pp 229-235

[18] Gontard N, Guilbert S and Cuq J L 1993 Water and Glycerol as Plasticizer Effect Mechanical and Water Barrier Properties of an Edible Wheat Gluten Film *J. Food Science.* **58** (1) pp 206-211

[19] Zuraida A, Yusliza Y, Anuar H and Mohd K M R 2012 The Effect of Water and Citric Acid on Sago Starch Bio-Plastics *Int. Food Research J.* **19** (2) pp 715-719

[20] Guiterrez, Tony J, Maria S T, Elevina P and Lucia F 2015 Structural and Mechanical Properties of Edible Films Made from Native and Modified Cush-Cush Yam and Cassava Starch *Food Hydrocolloids.* **45**

[21] Skurtys O, Acevedo C, Pedreschi F, Enrione J, Osorio F and Aguilera J M 2010 *Food Hydrocolloid Edible Films and Coatings.* Departmen of Food Science and Technology Universidad de Santiago de Chile

[22] Handayani A 2010 *Pembuatan dan Karakterisasi Film Biodegradable dari Kitosan/PLA (Polu Lactid Acid) dengan Pemlastis Polietilen Glikol (PEG)* Skripsi Fakultas Teknologi Pertanian Institut Pertanian Bogor

[23] E Milda and C Kerry 2009 *Edible Films and Coatings for Food Applications* Springer
[24] Nurindra, Azka P, Moch A A and Sudarno 2015 Karakterisasi Edible Film dari Pati Propagul Mangrove Lindur (Bruguiera gymnorrhiza) dengan Penambahan Carboxymethyl Cellulose (CMC) sebagai Pemlastis *J. Ilmiah Perikanan dan Kelautan.* **7** (2) pp 125-132
[25] Nurjannah W 2004 *Isolasi dan Karakterisasi Alginat dari Rumput Laut Sargassum sp. Untuk Pembuatan Biodegradable Film Komposit Alginat Tapioka* Skripsi Fakultas Teknologi Pertanian Universitas Gadjah Mada Yogyakarta
[26] Widyastuti, Eny S, Abdul M and Ria A P 2008 Pengaruh Penambahan Mentega dan Perlakuan pH terhadap Karakterisasi Kimia Edible Film Gluten. *J. Ilmu dan Teknologi Hasil Ternak.* **3** (1) pp 24-34
[27] Suwardi, Yuniarto, Lukman A and Fahimah M 2011 Pengaruh Variasi Larutan Asam pada Isolasi Gelatin Kulit Ikan Patin (Pangasius hypothalmus) terhadap Sifat-Sifat Kimia dan Fisik. *Paper Seminar Jurusan Kimia Fakultas Matematika dan Ilmu Pengetahuan Alam Institut Teknologi Sepuluh Nopember Surabaya*
[28] Rao M S, Khanatt S R, Chawla S P and Sharma A 2010 Chitosan and Guar Gum Composite Films: Preparation, Physical, Mechanical, and Antimicobial Properties *Carbohydrate Polymers.* **82** pp 1243-47
[29] Hosseini, Seyed F, Masoud R, Mojgan Z and Farhid F G 2013 Preparation and Functional Properties of Fish Gelatin-Chitosan Blend Edible Films. *Food Chemistry.* **136** pp 1490-95
[30] Krochta J M and DeMulder-Johnston C 1977 Edible and Biodegradable Polymer Films; Challenges and Opportunities. *Food Technology.* **51** (2) pp 61-74
[31] Sionkowska A, Wisniewski M, Skopinska J, Kennedy C J and Wess T J 2004 Molecular Interactions in Collagen and Chitosan Blends. *Biomaterials.* **25** (5) pp 795-801
[32] Liu, Zunying, Xiaojun G, Yuan L, Shiyuan D, Yuanhui Z and Mingyong Z 2012 Effects of Chitosan Molecular Weight and Degree of Deacetylation on The Properties of Gelatine-Based Films *Food Hydrocolloids* **26** pp 311-317
[33] Cheng M, Deng J, Yang F, Gong Y, Zhao N and Zhang X 2003 Study on Physical Properties and Nerve Cell Affinity of Composite Films from Chitosan and Gelatin Solutions.*Biomaterials* **24** pp 2871-80
[34] Elsabee, Maher Z and Entsar S A 2013 *Chitosan Based Edible Film and Coatins: A Review.* Material Science and Engineering
[35] Chiou B S, Avena-Bustillos R J, Bechtel P J, Jafri H, Narayan R, Imam S H 2008 Cold Water Fish Gelatin Films: Effects of Cross-Linking on Thermal, Mechanical, Barrier, and Biodegradable Properties *European Polymer J.* **44** pp 3748-53
[36] Sari M 2011 *Identifikasi Protein Menggunakan Fourier Transform Infrared (FT-IR).* Skripsi Program S-1 Ekstensi Teknik Kimia Departemen Teknik Kimia Universitas Indonesia Depok
[37] Wanchoo R K and Sharma P K 2003 Viscometric Study on The Compatibility of Some Water-Soluble Polymer-Polymer Mixtures *Eur. Polym. J.* **39** pp 1481-1490
[38] Karlina I R and Lukman A. 2010. Ekstrak Gelatin dari Tulang Rawan Ikan Pari (Himantura gerarrdi) pada Variasi Larutan Asam untuk Perendaman *Program Studi Kimia MIPA Institut Teknologi Sepuluh Nopember Surabaya*
[39] Martianingsih N and Lukman A 2009 Analisis Sifat Kimia, Fisik, dan Termal Gelatin dari Ekstraksi Kulit Ikan Pari (Himantura gerarrdi) melalui Varisasi Jenis Larutan Asam. *Program Studi Kimia MIPA Institut Teknologi Sepuluh Nopember. Surabaya*
[40] Lima C G A, de Oliveira R S, Figueiro S D, Wehmann C F, Goes J C and Sombra A S B 2006 DC Conductivity and Dielectric Permittivity of Collagene Chitosan Film *Materials Chemistry and Physics* **99** pp 284-288

Local durian (*Durio zibethinus* murr.) exploration for potentially superior tree as parents in Ngrambe District, Ngawi

E Yuniastuti[1], A Anggita[2], Nandariyah[1] and Sukaya[1]
[1]Study Program of Agrotechnology, Faculty of Agriculture, Universitas Sebelas Maret, Surakarta, Indonesia
[2]Undergraduate Student of Faculty of Agriculture, Universitas Sebelas Maret, Surakarta, Indonesia

E-mail:yuniastutisibuea@staff.uns.ac.id

Abstract. The characteristics durian based on specific area gives a wide diversity of phenotype. This research objective was to build an inventory of the local durian of Ngrambe as well as to obtain potentially superior local durian as prospective parent trees. The research was conducted in Ngrambe sub-district, on October 2015 until April 2016 using the explorative descriptive method. The determination of sample point used the non-probability method of snowball sampling type. Primary data include the morphology of plant characters, trunks, leaves, flower, fruits and seeds and their superiority. The data of the research were analyzed using SIMQUAL (Similarity for Qualitative) function based on the DICE coefficient on NTSYS v.2.02. The data cluster and dendrogram analyses were determined by Unweighted Pair-Group Arithmetic Average (UPGMA) method. The result of DICE coefficient analyses of 58 local durian accession based on the phenotypic character of vegetative organs ranged from 0.84-1.0. The phenotypic character of the vegetative and generative organ from 3 local durian accession superior potential ranged from 0.7 to 0.8. In conclusion, the accession of local durian which were Miyem and Rusmiyati have advantage and potential as prospective parent trees.

1. Introduction

The diversity of the Indonesian germplasms is a great asset for plant breeders to carry out evaluations of naturally superior characters and then to combine them into new, better genetic products [1]. Indonesia has various types of tropical fruits, such as durian. Durian has a wide variety of phenotypes exhibited by plant height, first branch height, stem circumference, leaf length, fruit length, fruit circumference, fruit weight, fruit flavor, the thickness of fruit pulp and seed weight.

According to Setiono [2], the potential development of durian as one of the plantation commodities is very large. The nature of the region-specific durian plants provides different production variations. In addition, the pollinating nature of open flower crops produces variation in production and form [3]. The people's favorite durian according to Ihsan [4] that has a medium size, sweet taste, soft texture, thick fruit pulp and small seeds.

The number of durian species in Indonesia provides a great opportunity to produce superior durian cultivars. Ngrambe sub district was one of durian centers in Ngawi Regency. Topography in Ngrambe district meets as a requirement for growing durian to grow well. Based on Horticultural Commodity

Information [5], durian production in 2012, 17.83% durian in Indonesia was from East Java Province. Ngawi District ranks tenth in East Java Province in 2011 with 2.184 tons of production.
The prospect of developing durian was very good but it has not been taken care adequately. Therefore, it was necessary to conduct a research on the exploration of potentially superior local durian that aims to build an inventory of local durian crops grown in Ngrambe Subdistrict as a resource of germplasm and to obtain local durian crops Ngrambe which has the potential to excel as prospective parent trees.

2. Methods

2.1. Location and Time Research

This research was held out at the village of Giriharjo, Wakah, Babadan, Setono, and Sambirejo, Ngrambe subdistrict, Ngawi district. The altitude of this village was 400-500 meters above the sea level. The topographic conditions were flat to hilly. The observations were carried out at the plant breeding laboratory, Faculty of Agriculture, Sebelas Maret University from October 2015 to April 2016.

2.2. Materials

The material and tools were 58 accessions of local durian Ngrambe that was more than 50 years based on the results of the research location. The materials were stationary, ruler, digital camera, crop descriptor for durian (*Durio zibethinus* Murr.) that was published by Biodiversity International [6], official PVT news (Registration of Plant Varieties) local durian, analytical scales, caliper, and hand refractometer.

2.3. Research Design and Data Analysis

The quantities and qualities character data identified by the descriptive analysis. The kinship relationship was observed by scoring the presence or absence morphological character in an observational variable [6]. The results were translated into binary data if there was morphological character get score 1, and if none get score 0. The analysis was done by cluster analysis [7] used NTSYS 2.02 program with Unweighted Pair-Group Method Arithmetic Average (UPGMA) SIMQUAL (for qualitative similarity). Similarity matrix used DICE coefficient [8].

$$S = \frac{2n_{ab}}{n_a + n_b}$$

Description:
S = coefficient of equation
nab = two individuals are compared
na = the same number of characters between individuals (1 and 1)
a = number of character in individual a only
b = number of character in individual b only

3. Results and Discussion

3.1. General Explanation about Durian

The research activities had begun with that exploration, which searches, collects and identify local durian germplasm that was older than 50 years old. Simultaneously, durian exploration activities were characterized to identify the characters possessed by an accession. Based on the exploration results at five locations obtained 58 accessions of local durian.

3.2. The Diversity of Durian

The identification of durian studied consisted of the accession of local durian from exploration on the vegetative part which were trunks and leaves and the generative parts of the flower, fruit, and seed. Local durian accession criteria that were observed were fruitful, over 50 years old and located in the administrative area of Ngrambe sub-district. This was according to Prastowo [9] that the plants from the seeds to be used as parent trees must have been produced at least five seasons, to determine the stability of the nature that was carried.

a b c

Figure 1. The morphology of diameter (a and b) and canopy of the durian

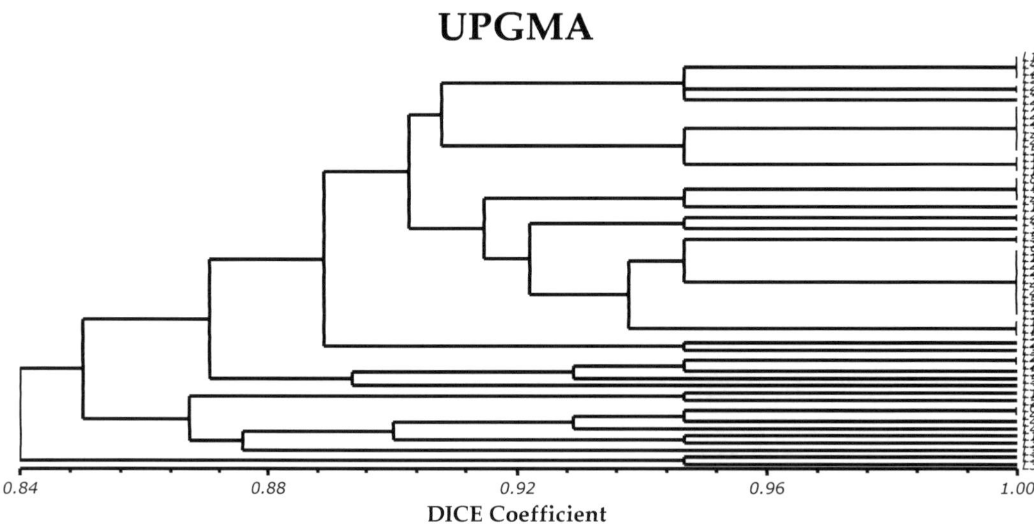

Figure 2. Dendogram based on UPGMA analysis of 58 local durian Ngrambe based on phenotypic characters of vegetative organs morphology processed with NTSYS program version 2.02

The result of durian accession grouping was based on phenotypic character of plant vegetative morphology which includes 1) plant character, i.e age, plant height, crown shape, growth direction, branching, 2) trunks, i.e cross-sectional shape, trunk texture, trunk color, 3) leaves, i.e leaf shape, leaf tip, leaf margin, leaf blade, leaf surface, leaf length, leaf width, the color over the leaf surface, and the bottom surface of leaf color on 58 durian accessions indicate wide diversity.

The morphology of the durian height was 20 m for the shortest and reached 50 m for the highest. The durian was more than 100 years old, tree height reached 50 m [9]. The morphology of the canopy form of the durian that found was (1) triangle (Pyramidal), (2) spherical and (3) irregular. The morphology of diameter of the durian plant ranges from 40 cm for the smallest and 180 cm for the largest (fig 1).

3.3 Identification of Vegetatif Organ

Accessing the local durian Ngrambe, does not yet have a variety name. The name of the local durian variant, usually given by the local community based on morphological variations, the name of the owner or the area where the durian grows [11]. The result of durian accession grouping was based on the phenotypic character of plant vegetative organ morphology, which includes 1) plant character, ie age, plant height, crown shape, growth direction, branching, 2) trunks, in cross-sectional shape, rod texture, trunk color, trunk, 3) leaves, ie leaf shape, leaf tip, leaf margin, leaf blade, leaf surface, leaf length, leaf width, upper surface of leaf color, and lower surface of leaf color on 58 accessions durian show that at the coefficient of similarity limit 0. 92 was divided into twelve groups (fig.2).

Group A consist of seven accessions, including accession (local durian) L1, L4, L25, L52, L2, L47 and L26. Group B consist of ten accessions, including L7, L20, L27, L39, L41, L49, L50, L11, L17 and L18. Group C consist of five accessions, including L8, L13, L38, L42 and L15. Group D consist of eighteen accessions, including L28, L45, L34, L5, L33, L9, L19, L23, L24, L31, L43, L46, L53, L54, L55, L10, L16 and L58. Group E consist of two accessions, including L29 and L30. Group F consist of four accessions, including L21, L22, L56 and L57. Group G consist of one accession, L37. Group H consist of two accessions, including L12 and L51. Group I consist of four accessions, including L6, L40, L44 and L48. Group J consist of two accessions, including L14 and L32. Group K consist of one accession, namely L3. Group L consist of two accessions, including L3 and L35.

Dendrogram of 58 durian accession based on vegetative organ morphology, also showed eleven clusters that showed the similarity of coefficient reaches 1.00. Cluster 1 consist of four accessions namely accession durian L1, L4, L25, and L52. Cluster 2 consist of two accessions namely accession durian L2 and L47. Cluster 3 consist of seven accessions that were durian accessions L7, L20, L27, L39, L41, L49, and L50. Cluster 4 consist of three accessions namely accession durian L11, L17, and L18. Cluster 5 consist of four accessions namely accession durian L8, L13, L38, and L42. Cluster 6 consist of two accessions namely accession durian L28 and L45. Cluster 7 consist of two accessions namely accession durian L5 and L33. Cluster 8 consist of ten accessions of accession durian L9, L19, L23, L24, L31, L43, L46, L53, L54, and L55. Cluster 9 consist of three accessions namely accession durian L10, L16, and L58. Cluster 10 consist of two accessions namely accession durian L21 and L22. Cluster 11 consist of two accessions namely accession durian L6 and L40. Each cluster shows that the accession of durian in one cluster has the same morphological character.

Dendrogram based on morphological of vegetative organs showed the existence of the lowest characteristic similarity of 0.84. This means that there was a mismatch based on morphological characters of 16%. The value indicates that there was still variation between durian accession, this can be due to genetic factors and environmental conditions.

Grouping occurs because accession has a morphological characteristic similarity due to growth in the same or adjacent location and the proliferation of the durian from seed or vegetative originating from the same parent. While the differences between individuals durian can be caused by the cross-pollination naturally in durian with the help of bats, birds, squirrels and some types of beetles [14].

3.4 Identification of Generatif Organ

Based on the exploration of local durian accession, from 58 accessions that were identified the morphologically of durian fruit and seeds had only three accessions, namely L5 (durian Miyem), L8 (durian Amin) and L10 (durian Rusmiyati). The accession of local durian that were identified by fruit was an accession that has potentially superior characteristics according to interviews with the owners and local residents. In addition, some durian trees that have superior potential experience failure in flowering, pollination and even have molted to death, so cut by the owner. Observation data of durian fruit shape was presented in table 1.

Table 1. Observation Data on 3 Local Duian Fruit Accessions

Fruit character	L5 (D. Miyem)	L8 (D Amin)	L10 (D. Rusmiyati)
Fruit clustering habit	*Combination*	*Combination*	*Combination*
Fruit Shape	*Oval*	*Elliptic*	*Oblong*
Fruit weight	*2.6 kg* *Heavy* (2.6 – 4.0 kg)	*1.25 kg* *Light* (0.9 – 1.5 kg)	*2.1 kg* *Medium* (1.6 – 2.5 kg)
Fruit rind colour.	*Green*	*Green*	*Green*
Fruit rind colour (at maturity)	*Greenish Yellow*	*Greenish Yellow*	*Yellowish Green*
Fruit rind thickness	*Medium*	*Thin*	*Medium*
Fruit spine shape	*Hooked*	*Convex*	*Convex*
Flesh colour	*Creamy White*	*Creamy White*	*Lemon Yellow*
Flesh taste	*Sweet*	*Slightly Sweet*	*Bitter Sweet*
Flesh creaminess	*Intermediate*	*Intermediate*	*Intermediate*
Flesh aroma	*Moderate*	*Mild*	*Moderate*
Aril thickness	*1,2 cm* *Medium* (0.6 – 2.0 cm)	*0,5 cm* *Thin* (< 0.5 cm)	*1,5 cm* *Medium* (0.6 – 2.0 cm)
Fruit Length	31 cm	21 cm	34 cm
Fruit diameter	56 cm	45.5 cm	51 cm
Number of carpels per fruit	*Five*	*Five*	*Five*
Seed shape	*Oblong*	*Spheroid*	*Oblong*
Seed coat colour	*Yellow Brown*	*Yellow Brown*	*Yellow Brown*

According Sundari [11], durian have no striking difference in the flower character, shape, and color of seeds. The texture and flesh colour on the local durian has several variations. Some are white and some are yellow. Moisture content is one of the most important physicochemical characteristics, as it will influence the flavor, texture, appearance and shelf life of fruit. Durian fruits had moisture content within the range of 58.80-79.35% [12] which is in accordance with those reported by Charoensiri [13].

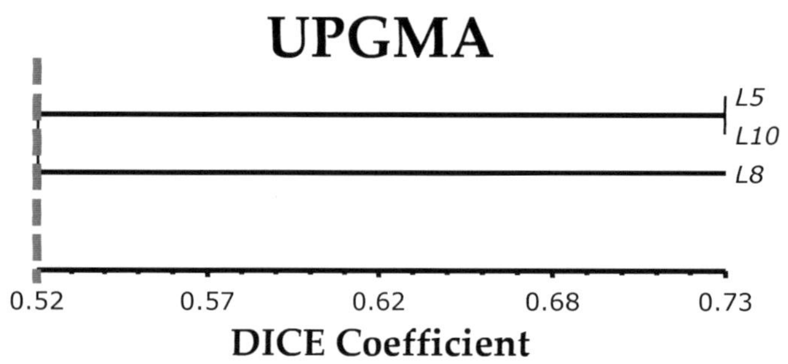

Figure 3. Dendogram based on UPGMA analysis of 3 local durian accession based on phenotypic characters of generative organs morphology processed with NTSYS program version 2.02

The results of classification of 3 local durian accession which have superior potential based on the morphological character of the generative organ were presented in figure 3.Based on the dendrogram (fig. 3), it was found that at a coefficient of 0.52 the level of similarity of the lowest properties. This means there was a 48% mismatch. The coefficient was divided into two groups, namely group A consist of accession L (5) durian Miyem and L (10) durian Rusmiyati, and group B consist of accession L (8) durian Amin. Accession durian Miyem and Rusmiyati, is the most similar accession with the coefficient of similarity 0.73. The value indicates that there were still variation between durian accession, this can be due to genetic factors and environmental conditions. Grouping the sample into a single cluster, indicating a high degree of similarity between generative characters shown through the coefficient of similarity [15].

The variables that cause differences between groups A and B were variations in morphology of fruit shape, fruit weight, fruit rind colour (at maturity), fruit rind thickness, fruit spine shape, flesh colour, flesh taste,flesh aroma, rail thickness, fruit length, fruit diameter and seed shape. According to Sri Lestari [16], the weight of fruit was positively correlated with fruit length, fruit weight, and fruit diameter. The fruit with heavyweight has a long fruit shape, large diameter, and thick fruit flesh. While the fruit with a lightweight has a short fruit shape, small diameter, and little pulp.

3.5 Identification of Potentially Superior Durian

The accession of local durian of exploration in Ngrambe subdistrict, Ngawi indicates that the potential of local durian development has the potential to be superior as the parent tree was large enough. Adjid [17] and Sunarjono [18] provide criteria for excellence include, (a) on the short age has been reproducing, (b) high productivity, (c) flavor and aroma of the fruit to satisfy the tastes of the consumer, (d) fruit flesh was thick, the texture was soft and it has the small size of seeds, (e) the shape and color relatively uniform and (f) it has resistant to disease.

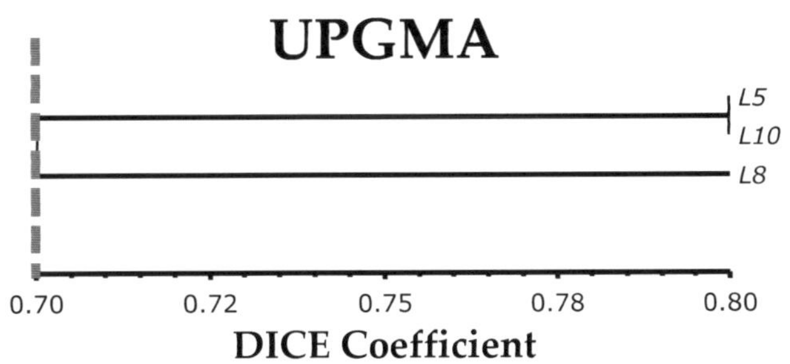

Figure 4. Dendogram based on UPGMA analysis of 3 local durian accession based on phenotypic characters of vegetative and generative organs morphology processed with NTSYS program version 2.02

Based on dendrogram (fig.4) obtained that the coefficient of similarity 0.70 divided into two groups. Group A consist of accession L5 (durian Miyem) and L10 (durian Rusmiyati) and Group B consist of accession L8 (durian Amin), Identification of morphology of vegetative and generative organ showed the similarity of the highest character based on the value of coefficient of similarity 0,80 that was seen at accession L5 (Durian Miyem) and L10 (durian Rusmiyati). The degree of diversity of durian accession based on the morphological characteristics of fruits and seeds was high, presumably, this was due to the influence of plant genetic factors as well as environmental factors.

The variables that were differentiating between groups A and B were the variation in the morphology of crown shape, branching, leaf width, fruit shape, fruit weight, fruit rind colour (at maturity), fruit rind thickness, fruit spine shape, flesh colour, flesh taste, flesh aroma, rail thickness, fruit length, fruit diameter and seed shape.

Selection of 58 durian accessions that have potentially superior character based on fruit length, fruit diameter, fruit weight, fruit spine shape, rail thickness, flesh colour, flesh taste and sugar content, flesh creaminess, flesh aroma, fruit rind thickness, and seed shape obtained 2 accessions Durian, namely accession durian Miyem (L5) and durian Rusmiyati (L10). Durian Amin (L8) was not included in the potentially superior because plants were being attacked by plant-disturbing organism caused fruit identification result not matched with the result of interview with the owner.

The recognition of local varieties of regions that were superior and preferred consumers need to be done by proposing these varieties through efforts to release the regional varieties as a superior variety approved by the Minister of Agriculture [19].

The first durian accession (L5) had located in Munggur Hamlet, Giriharjo Village at ± 500 m above sea level. Durian superiority was 31 cm in length, 56 cm fruit diameter, 2.6 kg in fruit weight, with shaped spines. Aril thickness was 1.2 cm, creamy white color with medium in flesh creaminess, medium aroma, 1.5 cm in fruit rind thickness, sweet taste with sugar content was 28.8 Brix°, longitudinal round shape, brownish yellow (Figure 5).

The second durian accession was L10 that located in Gunung Hamlet, Wakah Village at ± 500 m above sea level. The superiority of durian fruit has 41 cm in fruit length, 51 cm fruit diameter, fruit weight was 2.1 kg (medium), convex spikes (Convex). Aril thickness was 1.5 cm (medium), medium yellow (Lemon yellow) with medium texture (Intermediate), medium aroma, 1.5 cm (medium) fruit rind thickness, bittersweet taste with sugar content > 30 Brix°, Yellow brown longitudinal (Oblong) shape (Yellow-brown) (fig. 5).

Durian ripening is characterized by increasing sugar content in fruits [12]. Color and appearance are critical factors of food for both its esthetic value and for quality judgment. Dark product is usually less attractive for the consumers [20].

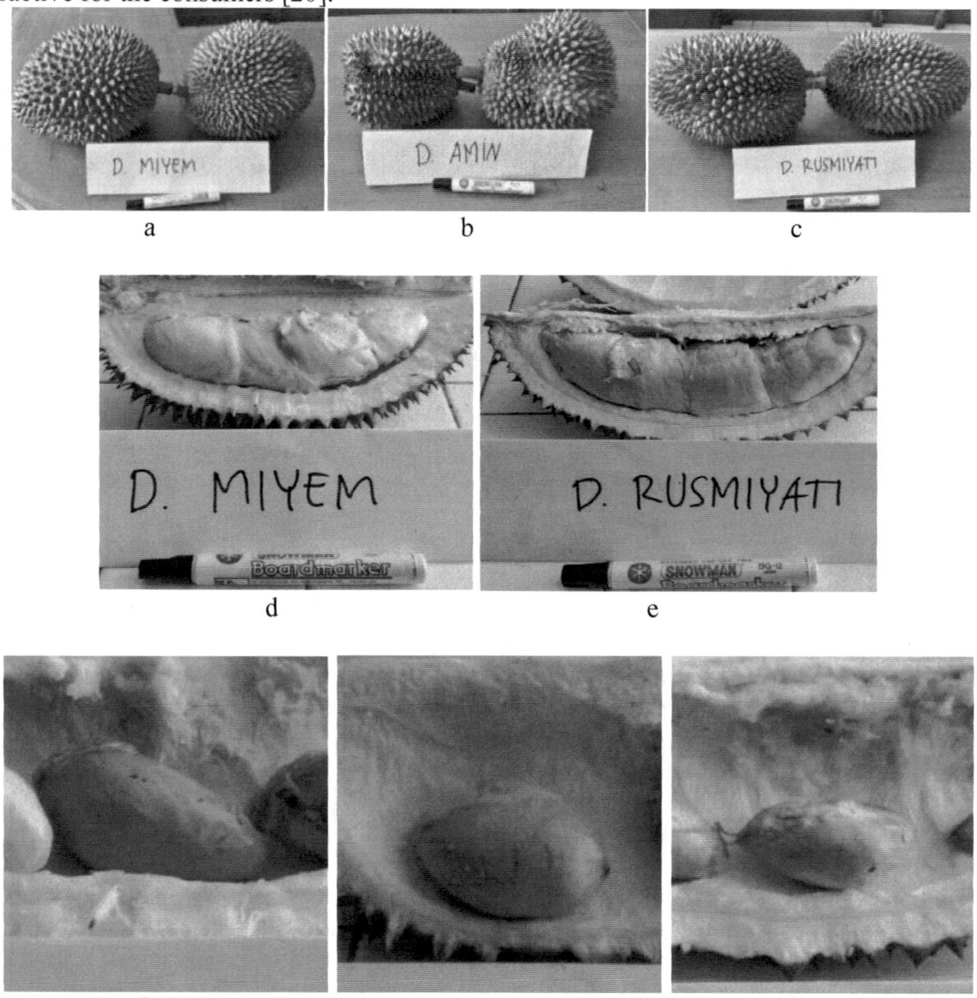

Figure 5. The fruit shape of durian miyem was oval (a), durian amin was elliptic (b), durian rusmiyati was round longitudinal (c), the pulp color of durian miyem was creamy white (d), durian rusmiyati lemon yellow (e) seed coat colour of durian miyem (f), amin (h) and rusmiyati (h) are yellow brown.

Wiryanta [10] added that durian could be classified into 3, that was quality I, quality II and quality III. Durian shape was grouped into four kinds, which was very good, good, less good, and not good. Durian fruit that was very nice and good shape generally has five rooms (lokulus) full of flesh and seeds. While the fruit was less good and not good, the number of five or less room and not all the room contains seeds and fruit pulp. Durian Miyem and Rusmiyati can be included in grade I and the quality of fruit form was very good.

The overall results, indicating that there are variability found among potentially superior local durian accessions. This research, is the first step of inventory and identification of local durian, before a local varieties are proposed to be regional and national superior varieties [19].

4. Conclusions

Based on the result, it can be concluded that 58 accessions of local durian at research sites that were more than 50 years old with similarity coefficients ranged from 0.84 to 1.0 based on the phenotypic character of the vegetative organ. The result of coefficient analysis of phenotypic resemblance of the vegetative and generative organ from 3 local durian accessions potentially superior ranged from 0.7 to 0.8. Accession durian Miyem and durian Rusmiyati had the advantage and potential as a prospective parent plant of durian.

Acknowledgement

Authors would like to acknowledge Penelitian Unggulan of Sebelas Maret University year 2012-2016 for the financial support.

References

[1] Saleh M, Mawardi, Khairullah I. 2013. Diversity of durian phenotype of local cultivars in Karang Intan Kalimantan Selatan. Agricultural Research Institute of Swamp Land, Agroscientiae 20(1): 22-25. [Indonesian]

[2] Setiono. 2012. Economic boosts develop durian cultivation. Government of Ngawi Regency. Disrict Goverment Ngawi URL: http://www.ngawikab.go.id/ home/2012/12/dongkrakperekonomian-kembangkan-budidaya-durian/ Diakses 31 Agustus 2015. [Indonesian]

[3] Sudjijo. 2009. Influence of rootstock and stem size on growth durian Monthong, Hepe, and DCK-01. J. Hort. 19(1):89-94. [Indonesian]

[4] Ihsan, F, Sukarmin, Engkos, K. 2012. Durian crossword technique for assembling new improved varieties. Agricultural Engineering Bulletin. 17(1):14-17. [Indonesian]

[5] Information on Horticultural Commodities. 2013. Durian. Jakarta: Data and Information System Center of the Ministry of Agriculture. 01/07/1 hal 1-4. [Indonesian]

[6] Bioversity. 2007. Descriptors for Durian (*Durio zibethinus* Murr.) International Plant Genetic Resources Institute (IPGRI). Rome, Italy: International Bioversity International.

[7] Supranto, J. 2004. Multivariate Analysis of Meaning and Interpretation. Publisher: Rineka Cipta, Jakarta. [Indonesian]

[8] Rohlf, F. J. 1998. NTSYSpc Version 2.0. Exeter Software, New York.

[9] Prastowo NH, James MR, Gerhard ESM, Erry N, Joel MT et al. 2006. Nursery Techniques and Vegetative Propagation of Fruit Crops. world agroforestry centre (ICRAF) dan Winrock International. Bogor, Indonesia. p.100 ISBN 979 - 3198 -28 – 1. [Indonesian]

[10] Wiryanta BTW. 2002. Planting Durian. Jakarta (ID): AgroMedia Pustaka. [Indonesian]

[11] Sundari, Estri Laras Arumingtyas, Luchman Hakim, Rodiyati Azrianingsih. 2015. Exploration and Morphological Character Identification of Local Durian (*Durio zibethinus* Murr.) from Tidore Island, North Maluku. Proceeding of 6th ICGRC (International Conference on Global Resource Conservation).

[12] Belgis, M., Wijaya, C.H., Apriyantono, A., Kusbiantoro, B. and Yuliana, N.D. 2016. Physicochemical Differences and Sensory Profiling of Six Lai (*Durio kutejensis*)and Four

Durian (*Durio zibethinus*) Cultivars Indigenous Indonesia. International Food Research Journal 23(4): 1466-1473.

[13] Charoensiri, R., Kongkachuichai, R., Suknicom, S. andSungpuag, P. 2009. Beta-carotene, lycopene andAlpha-tocopherol Contents of Selected Thai Fruits. J.Food Chemistry 113: 202–207

[14] Yumoto T. 2000. Bird-Pollination of Three Durio Species (*Bombacaceae*) in A Tropical Rainforest in Sarawak, Malaysia. American Journal of Botany 87 (8) : 1181-1188.

[15] Yulita, KS. 2013. Molecular Identification of parent Trees of Some Varieties Durian from Jepara using Random Amplified Polymorphic DNA. J. Hort. 23(2): 99-106. [Indonesian]

[16] Sri Lestari, Fitmawati dan Ninik Nihayatul Wahibah. 2011. Diversity of Durian (*Durio zibethinus* Murr.) in Bengkalis Island basedon morphological characters. Kebun Raya Bulletin 14(2): 29-44. [Indonesian]

[17] Adjid DA. 1994. Guide of horticultural cultivation (Durian). Project for Increasing Food Crop Production. Bengkulu Provincial Food Crops Agency. Bengkulu. [Indonesian]

[18] Sunarjono H. 1995. Various Problems Durian and Solution. Penebar Swadaya, Jakarta. [Indonesian]

[19] Nandariyah. 2009. Breeder Role in Partcipatory Plant Breeding by Using of Genetic Resource of Local Varieties. J. Agrosains 11(1): 28-35. [Indonesian]

[20] Norjana, I. and Aziah, N. A. A. 2011. Quality of Durian(*Durio zibethinus* Murr) Juice after Pectinase EnzymeTreatment. International Food Research Journal 18:1117-112

IOP Conf. Series: Earth and Environmental Science **142** (2018) 012030 doi:10.1088/1755-1315/142/1/012030

Weed diversity identification on growth phases of twelve maize varieties

M Rahayu[1,3], P Yudono[2], D Indradewa[2] and E Hanudin[2]
[1]Faculty of Agriculture, Universitas Sebelas Maret, Surakarta, Indonesia
[2]Faculty of Agriculture, Gadjah Mada University, Yogyakarta, Indonesia
[3]Post Graduate Program of Gadjah Mada University, Yogyakarta, Indonesia

Email: rahayu_imuj@yahoo.com

Abstract. Weed is one of the factors that disturb maize cultivation. Each maize variety has specific characteristics which will lead to weeds diversity in particular farm. This study aimed to identify the effect of maize variety on weed diversity on maize's growth phases. The research was conducted from March to July 2016 in Banguntapan, Bantul, Yogyakarta, Indonesia. The field research was arranged by using Randomised Complete Block Design with three replications. The treatments were maize varieties, consist of 12 levels: 'Bisi 18', 'Bisi 2', 'Pertiwi 3', 'NK 22', 'NK 33', 'DK 959', 'P23', 'DK 771', 'DK 95', 'Bima 19 Uri', 'Bisma', and 'Sukmaraga'.The results showed that at four weeks after planting (WAP) the most of weeds were growing on varieties 'Bisi 18', 'NK 33', 'P 23', and 'DK 95', there were 14 species. *Bulbostylispuberula* and *Cyperusrotundus* were growing on all maize varieties with the highest SDR on 'Sukmaraga' and 'DK 959'. At six WAP, weeds that grew, most consists of four species of sedges, six species of grasses and 24 species of broadleaf. While at the age of eight WAP, the weeds that grow at least on 'NK 33' that were 13 species and the most on 'P23' that were 21 species. *Dactyloctenium aeghypthium* having highest SDR in all maize varieties except Bisi 18. The highest SDR value on Bisi 18 was *Boerhaavia erecta* about 16.23%.

1. Introduction

Weed is a plant that grows where it is not desired or out of places [1-2-3-4-5]; thus needs to be controlled. Weed is an important factor in the management of all land and water resources, but their effective impact is the greatest on agriculture. Weed compete with crop for nutrient, soil moisture and sunlight. The intensity of weed competition depends on type of weed species, severity of weed infestation, duration of weed infestation, competing ability of crop and climate conditions which affect weed and crop growth [3].

In cultivated crops, weeds can cause reduction in crop yield. The magnitude of reduction in crop yield caused by the presence of weeds varies greatly. On some plants, the presence of weeds in farm can reduce in crop yield by 15 – 100 % and weeds can also reduce product quality [6]. Reduction in corn yield by 40 – 60% depending on the intensity and type of weed [7]. It is reported that maximum yield loss by 24 – 34% if barnyard grass (*Echinochloa crus-galli* L.) appears at an early stage, and about 9% if barnyard grass appeared after V4 stage growth [8]. Maize seeds yield loss caused by weed in land farmer by 15 – 30% or 0.75 – 1.5 tons per hectare [9].

The types of weed which grow in land are very diverse. They depend on type of soil, environment, and the kinds of plants cultivated. The dominant weed which grow on soil clays, among others

Amaranthus retroflexus, Chenopodium album, Xanthium strumarium, sorghum halepense, Portulaca oleracea [10]. Some weeds that grow on red dragon fruit cultivation with sandy soils among other *Amaranthus lividus, Hedyotis corymbosa, Eleusine indica, Aneilema nudiflorum, Ludwigia hyssopifolia* and *Portulaca oleracea*. While on unsandy soil, the kinds of weed among other *Eleusine indica, Cyperus compressus, Eragrostis unioloides, Digitaria ciliaris, Echinochloa colonum, Bulbostylis puberula* and *Hedyotis corymbosa* [11].

Each plant variety has specific characteristics which will influence its environmental conditions and affecting the growth of other organisms around them, including weeds. Each variety will exhibit a diverse morphological characteristic such as plant height, leaflenght, leaf width, leaf shape, leaf angle, leaf area, rooting, etc. In addition, each variety also have physiological characteristics such as net assimilation rate, crop growth rate, relative growth rate, and specific leaf weight. It is reported that six cultivars, (KSC 302, KSC 647, KSC 500, KSC 700, and KSC 704) can produce various net assimilation rate [12], crop growth rate, relative growth rate, specific leaf weight and the parameters also influence plant ability to compete against weeds. The competitive ability will affect the resistance of corn to weeds existence. Crop types and cultivars differed in their competitive abilities, measured as the ability to suppress weed and maintain grain yield under weed competition. The selection of cultivar can have an impact on the ability of the crop to achieve high grain yield and suppress weeds [13].

The composition of weed at a farm also depends on the age of plant. In general, each plant growth phase or age will influence the diversity of weeds that grow in the farm. The time of weeds occurrence was also determining the magnitude of competition. Weed disturbance until 14th will decrease corn leaf area index through decrease in leaf area development and accelerate of bottom leaves senescence [14]. On the conditions where weeds exist in a prolonged period, the producedbiomass of *Phaseolus vulgaris* is the lowest, while the highest biomass produced in conditions which were free of weeds. The increased presence of weeds in farm also decreases crop growth rate and net assimilation rate [15]. This research aims are to identify the influence of maize varieties on weed diversity on some growth phases.

2. Material and Methods

The research was conducted from March to July 2016 in Banguntapan, Bantul, Yogyakarta and Management and Crop Production Laboratory, Agriculture Faculty, Gadjah Mada University, Yogyakarta. The field research arranged using Randomised Complete Block Design with three replications. The treatments were 12 maize varieties, consist of 'Bisi 18', 'Bisi 2', 'Pertiwi 3', 'NK 22', 'NK 33', 'DK 959', 'P23', 'DK 771', 'DK 95', 'Bima 19 Uri', 'Bisma', and 'Sukmaraga'.

Vegetation analysis was conducted before the research was carried out, using quadratic method with size 50 cm x 50 cm. Each block was taken 5 sample plots. The observed parameters include type of weed, amount of weed, and dry weight of weed. Weed vegetation analysis by computing the absolute density of a weed type, absolute frequency of a weed type, absolute dominance of a weed type, and summed dominance ratio

$$\textbf{SDR a weed species} = \frac{\textbf{Relative density} + \textbf{Relative frequency} + \textbf{Relative dominance}}{\textbf{3}}$$

To observe the weed uniformity community, then the coefficients of weed community is calculated using the following equation:

$$\mathbf{C} = (\frac{\mathbf{2W}}{\mathbf{a+b)}}\mathbf{x\ 100\%}$$

Where:
C = coefficient of community (%)
W = the amount of the lowest SDR from each pair of species of weeds from the two communities are compared
a = the amount of SDR from all weed specieson first community
b = the amount of SDR from all weed species on second community

If the value of C ≥ 75%, then the two weeds communities are no different or weed community real uniform. The land can be used for research [2].

During maize growth, weed data recorded at 4 week after planting (WAP), 6 WAP and 8 WAP. The Observed of weed using quadratic method with the size sample plots 70 cm x 20 cm. Weed data analysis by computing the absolute density of a weed species, absolute frequency of a weed species, and absolute dominance of a weed species and summed dominance ratio (SDR).

3. Result and Discussion

3.1. Vegetation Analysis

Vegetation analysis showed that there were 20 different species of weed, consists of 2 species of sedges, 4 species of grasses, and 14 species of broadleaf. Based on the Summed Dominance Ratio (SDR), three weeds being dominant on the field were *Panicum distachyum* (33.43 %), *Cyperus rotundus* (14.96%) and *Murdania blumei* (6.09%) (Table 1).

Table 1. Summed Dominance Ratio of weed on vegetation analysis

No.	Species of weed	Type of weed	SDR (%)			
			I	II	III	Average
1	*Cyperus rotundus*	S	13.22	17.08	14.59	14.96
2	*Bulbostylis puberula*	S	2.31	6.61	3.14	4.02
3	*Eragrostis tenela*	G	0.87	0	0	0.29
4	*Ischaemum thimorense*	G	4.54	5.4	1.48	3.81
5	*Panicum distachyum*	G	32.32	34.14	33.83	33.43
6	*Dactyloctenium aegyptium*	G	0	1.93	1.87	1.27
7	*Aeschinomene americana*	B	0	1.1	0.88	0.66
8	*Boerhaavia erecta*	B	2.91	3.12	3.67	3.23
9	*Commelinano diflora*	B	4.27	0	0.84	1.7
10	*Croton hirtus*	B	1.74	1.08	2.12	1.65
11	*Euphorbia hirta*	B	2.88	2.84	6.19	3.97
12	*Euphorbia hysopifolia*	B	6.4	2.09	1.03	3.17
13	*Lindernia ciliata*	B	4.28	1.28	1.98	2.51
14	*Lindernia crustacea*	B	3.16	4.04	5.03	4.08
15	*Murdania blumei*	B	4.1	8.17	5.99	6.09
16	*Ocimum sanctum*	B	0.75	1.37	0	0.71
17	*Oldenlandiadicotoma*	B	1.58	1.21	3.55	2.11
18	*Philanthus niruri*	B	0.78	0	2.51	1.1
19	*Richardias cabra*	B	7.77	4.65	4.37	5.6
20	*Tridaxprocumbens*	B	6.11	3.87	6.93	5.64

Remarks: S: Sedges; G: Grass; B: Broadleaf

Table 2. Weed communities coefficient (%) between blocks

Comparison between blocks	Weed community coefficient (%)	Average of weed community coefficient (%)
I : II	80.34	
I : III	80.82	81.93
II : III	84.63	

Based on the SDR (table 1) can be calculated coefficients of weed communities. Weed community coefficients (C) on comparison blocks I, II, and III produce C value greater than 75%, about 81.93% (Table 2). The value of weed community coefficient indicates the land is having uniform weeds community. If the value of C ≥ 75%, it means the two weeds communities are no different or significantly uniform. The land can be used for research [2].

3.2. Weed Diversity at 4 Weeks after Planting

Weed diversity at 4 weeks after planting in each variety was quite varied. Weeds that grow on all varieties of maize consists of sedges, grasses and broadleaf. *Cyperus rotundus* and *Bulbostylis puberula* grew between all maize varieties with varies value of SDR. The weeds growing between varieties of 'Bisi 2' were about 10 weeds species, while 'Bima 19 Uri' were about 8 species. Between varieties of 'DK 959' and 'Bisma' were as much as 13 species. Between varieties of "Pertiwi 3', 'NK 22', and 'Sukmaraga' were 12 species. The weeds growing between varieties of 'Bisi 18', 'NK 33', 'P 23', and 'DK 95' were 14 species (Table 3).

On paddy fields, a type of weed that grows on corn field were vary, depends on the age of the plant. On four weeks old corn plant, weeds that grow were 8 families which consists of 16 species, among other *Amaranthus spinosus*, Ageratum *conoyzoides, Cyperusiria, Euphorbia hirta, Phyllanthus urinaria, and Portulaca oleraceae.* At the age corn of 6 weeks, weed that grow as many as 10 families comprising 23 species. Weeds grow at age 8 weeks as many as 14 families consist of 28 species and at the age of 10 weeks, weeds that grow as many as 20 families comprising 36 species [16].

Based on value of SDR, the highest SDR were on 'Bisi 18', 'Bisi 2', 'P 23', 'Bisma' and 'Sukmaraga' were *Bulbostylis puberula* by 18.03%, 18,14%, 18,59%, 27.61% and 24.50%. The highest SDR on varieties of Pertiwi 3, 'NK 22', 'NK, 33', 'DK 959', 'DK 771' and Bima 19 Uri were *Cyperus rotundus* by 26.82%, 21.09% 20.57, %,, 30.74% and 24.35%. Whereas the highest SDR on 'DK 95' varieties was *Boerhaavia erecta* by 18.51%. Each plant variety has specific characteristics, such as leaves shape, leaf length, leaf angle, rooting, etc. Fujiyoshi et al. [17] reported that pumpkin cultivation among corn can reduce total weed biomass. The dominant weed is *Amaranthus retroflexus* and *Convolvulus arvensis*. Shade of the pumpkin was the main mechanism in weed suppression.

3.3. Weed Diversity at 6 Weeks after Planting

At the age of maize was 6 WAP, there are some changes in the composition of the weed that grows. This weed type that grows at 6 WAP more than those on 4 WAP. *Cyperus rotundus* and *Bulbostylis puberula* were growing in all maize varieties with various value of SDR. In addition, on 'NK 33' varieties also grow *Cyperusrotundus* and *Cyperusiria* (Table 4).

Weeds that grew between maizes of varieties of 'NK 33' about 27 species, 'Bisi 18' as many as 26 species, and 'NK 22" as many as 25 species. Weeds that grew on varieties of 'DK 959', 'DK 771' and Bisma were as much as 22 species. Weeds growing between maizes of varieties of 'Pertiwi 3', 'DK 95', 'Bisi 2', 'P 23', 'Bima 19 Uri and 'Sukmaraga' were as many as 14, 19, 20, 20, 21 and 23 species.

Table 3. Summed Dominance Ratio of weeds at 4 week after planting

No	Species of weed	Summed Dominance Ratio (%)											
		V1	V2	V3	V4	V5	V6	V7	V8	V9	V10	V11	V12
	Sedges												
1	*Bulbostyli spuberula*	18.03	18.14	20.20	17.21	13.86	5.82	18.59	15.11	7.19	9.85	27.61	24.50
2	*Cyperus rotundus*	10.15	15.66	26.82	21.09	20.57	30.74	17.23	6.54	22.94	24.35	2.32	16.29
	Grasses												
3	*Dactyloctenium aeghypthium*	14.74	16.61	4.47	4.26	7.80	4.33	18.37	11.00	9.31	22.59	9.63	14.79
4	*Eleusine indica*	0.00	3.53	4.39	2.77	1.89	4.91	1.74	3.72	3.93	0.00	2.27	0.00
5	*Ischaemumtimorense*	0.00	0.00	0.00	3.00	0.00	0.00	0.00	0.00	0.00	0.00	0.00	0.00
6	*Panicum distacyum*	6.48	13.69	5.05	8.17	9.64	9.29	2.60	3.95	6.23	6.35	0.00	3.95
	Broadleaf												
7	*Aeschynomeneamericana*	0.00	0.00	0.00	0.00	0.00	2.06	0.00	0.00	1.46	0.00	2.35	0.00
8	*Amaranthusspinosus*	0.00	0.00	0.00	0.00	0.00	0.00	0.00	0.00	0.00	0.00	0.00	3.17
9	*Boerhaavia erecta*	7.88	11.55	9.66	23.22	7.71	12.58	1.84	18.51	10.29	17.67	13.24	15.72
10	*Cleome rubidospermae*	9.33	8.14	8.37	2.28	9.66	8.73	0.00	2.45	9.58	0.00	13.37	0.00
11	*Centrosemapubescens*	0.00	0.00	0.00	0.00	0.00	3.79	3.26	0.00	0.00	0.00	0.00	5.15
12	*Croton hirtus*	4.80	0.00	0.00	5.64	1.85	0.00	5.94	2.83	4.01	4.21	5.40	0.00
13	*Euphorbia heterophylla*	2.38	0.00	0.00	5.57	0.00	0.00	3.21	12.22	6.71	0.00	2.86	0.00
14	*Gynandropsisgynandra*	5.92	3.29	2.53	0.00	0.00	2.35	5.61	5.38	1.60	0.00	3.18	0.00
15	*Ipoemoeacongesta*	2.20	0.00	0.00	0.00	5.97	0.00	0.00	3.96	0.00	0.00	3.36	4.90
16	*Linderniaciliata*	0.00	0.00	0.00	0.00	0.00	2.07	0.00	0.00	0.00	0.00	0.00	0.00
17	*Melochiapiramidata*	5.20	5.16	2.70	0.00	2.36	7.02	3.70	9.55	1.94	0.00	2.60	0.00
18	*Murdania*	1.79	0.00	0.00	3.37	3.42	0.00	4.90	0.00	0.00	0.00	0.00	0.00
19	*Ocimum sanctum*	0.00	0.00	4.30	0.00	0.00	0.00	0.00	0.00	0.00	0.00	0.00	0.00
20	*Portulcaoleracea*	2.51	0.00	0.00	0.00	0.00	0.00	0.00	0.00	0.00	0.00	0.00	0.00
21	*Pylanthusniruri*	0.00	0.00	0.00	0.00	4.37	0.00	7.65	2.26	1.62	0.00	0.00	2.39
22	*Pysalisangulata*	0.00	0.00	0.00	0.00	0.00	0.00	0.00	0.00	3.44	0.00	0.00	0.00
23	*Richardiascabra*	8.60	4.24	5.03	3.41	7.65	6.30	5.36	2.52	9.73	11.64	11.82	5.67
24	*Spenocleazeylanica*	0.00	0.00	6.48	0.00	3.26	0.00	0.00	0.00	0.00	3.34	0.00	3.45

Remark: V1: Bisi 18, V2: Bisi 2, V3: Pertiwi 3, V4: NK 22, V5: NK 33, V6: DK 959, V7: P 23, V8: DK95, V9: DK 771, V10: Bima 19 Uri, V11: Bisma, V12: Sukmaraga

Table 4. Summed Dominance Ratio weeds at 6 weeks after planting

No	Species of weed	Summed Dominance Ratio (%)											
		V1	V2	V3	V4	V5	V6	V7	V8	V9	V10	V11	V12
	Sedges												
1	*Bulbostylispuberula*	14.09	8.05	11.05	18.15	23.69	11.90	9.99	9.39	17.35	8.94	18.44	4.70
2	*Cyperusrotundus*	5.87	1.55	19.94	5.25	2.14	7.95	4.89	11.90	9.24	7.21	9.80	7.28
3	*Cyperuscompresus*	0.00	0.00	0.00	0.00	1.23	0.00	0.00	0.00	0.00	0.00	0.00	0.00
4	*Cyperusiria*	0.00	0.00	0.00	0.00	1.32	0.00	0.00	0.00	0.00	0.00	0.00	0.00
	Grasses												
5	*Cynodondactylon*	5.47	1.18	0.00	1.29	1.58	1.77	0.00	0.00	1.05	1.97	0.00	2.35
6	*Dactylocteniumaeghypthium*	7.17	18.34	8.41	5.19	3.42	4.16	6.59	0.00	12.05	3.05	7.14	10.33
7	*Eleusineindica*	10.05	7.69	2.51	5.31	3.96	6.63	5.65	7.72	3.79	7.64	1.94	14.31
8	*Eragrotistenela*	0.00	0.00	0.00	0.00	0.77	0.00	0.00	0.00	2.33	0.00	0.00	0.00
9	*Ischaemumtimorense*	1.48	0.00	0.00	3.10	1.35	3.39	3.67	2.62	0.00	2.40	0.00	0.00
10	*Panicumdistacyum*	1.23	4.68	4.67	1.52	4.01	15.56	1.72	7.91	6.48	7.38	1.38	2.95
	Broadleaf												
11	*Aeschynomeneamerikana*	1.03	1.15	3.66	1.12	1.10	0.97	0.00	0.00	1.00	4.37	0.00	1.81
12	*Acallyphaindica*	1.85	1.64	0.00	2.65	0.00	0.00	0.00	0.00	0.00	2.40	1.56	2.43
13	*Althernantherabrasiliensis*	0.93	3.96	0.00	2.52	1.53	2.74	1.66	1.34	2.30	0.00	3.73	3.30
14	*Amaranthusspinosus*	0.00	0.00	0.00	0.00	0.00	0.00	0.00	0.00	0.00	0.00	1.55	1.19
15	*Boerhaaviaerecta*	5.12	7.30	5.80	8.48	16.60	7.60	10.00	9.97	3.98	4.16	5.43	10.33
16	*Centrosemapubescens*	1.07	0.00	0.00	0.00	0.00	0.00	0.00	0.00	0.00	0.00	1.26	0.00
17	*Cleome rubidospermae*	7.48	9.50	13.91	5.14	4.15	4.10	6.05	5.34	4.23	6.99	4.25	5.11
18	*Commelinadiffusa*	2.91	0.00	1.88	1.22	5.14	0.97	5.39	2.07	1.48	0.00	3.44	1.39
19	*Commelinabenghalensis*	2.75	0.00	0.00	2.07	0.00	0.98	1.69	5.41	3.50	1.97	2.93	1.33
20	*Croton hirtus*	1.06	1.27	0.00	0.00	1.26	2.50	4.18	6.19	1.27	2.19	5.25	2.91
21	*Euphorbia heterophylla*	7.20	6.12	14.36	4.89	5.00	6.95	16.38	5.74	5.99	6.54	19.19	4.13
22	*Euphorbia hirta*	2.30	3.32	2.48	7.18	3.18	2.95	3.99	2.78	3.53	2.40	1.50	3.03
23	*Ipoemoeatriloba*	0.00	0.00	3.89	1.56	3.39	2.46	0.00	0.00	0.00	0.00	0.00	0.00
24	*Linderniaciliata*	0.00	0.00	0.00	1.67	0.00	0.00	0.00	0.00	0.00	0.00	1.11	1.40
25	*Melochiapiramidata*	14.09	8.05	11.05	18.15	23.69	11.90	9.99	9.39	17.35	8.94	18.44	4.70
26	*Murdania*	5.87	1.55	19.94	5.25	2.14	7.95	4.89	11.90	9.24	7.21	9.80	7.28
27	*Ocimum sanctum*	0.00	0.00	0.00	0.00	1.23	0.00	0.00	0.00	0.00	0.00	0.00	0.00
28	*Oldenlandriadicotoma*	0.00	0.00	0.00	0.00	1.32	0.00	0.00	0.00	0.00	0.00	0.00	0.00
29	*Portulcaoleracea*												
30	*Pylanthusniruri*	5.47	1.18	0.00	1.29	1.58	1.77	0.00	0.00	1.05	1.97	0.00	2.35
31	*Pysalisangulata*	7.17	18.34	8.41	5.19	3.42	4.16	6.59	0.00	12.05	3.05	7.14	10.33
32	*Richardiascabra*	10.05	7.69	2.51	5.31	3.96	6.63	5.65	7.72	3.79	7.64	1.94	14.31
33	*Sidarhombifolia*	0.00	0.00	0.00	0.00	0.77	0.00	0.00	0.00	2.33	0.00	0.00	0.00
34	*Tridaxprocumbens*	1.48	0.00	0.00	3.10	1.35	3.39	3.67	2.62	0.00	2.40	0.00	0.00

Remarks: V1: Bisi 18, V2: Bisi 2, V3: Pertiwi 3, V4: NK 22, V5: NK 33, V6: DK 959, V7: P 23, V8: DK95, V9: DK 771, V10: Bima 19 Uri, V11: Bisma, V12: Sukmaraga

Table 5. Summed Dominance Ratio weeds at 8 week after planting

No.	Species of weed	Summed Dominance Ratio (%)											
		V1	**V2**	**V3**	**V4**	**V5**	**V6**	**V7**	**V8**	**V9**	**V10**	**V11**	**V12**
	Sedges												
1	*Bulbostylispuberula*	8.36	8.67	6.72	9.83	8.52	9.79	5.37	5.91	6.67	9.21	11.00	14.57
2	*Cyperusrotundus*	5.01	6.38	2.43	5.68	0.00	7.16	0.00	6.20	2.46	1.93	8.71	12.70
	Grasses												
3	*Cynodondactylon*	1.80	0.00	0.00	0.00	0.00	0.00	0.00	1.60	1.78	4.07	0.00	0.00
4	*Dactylocteniumaeghypthium*	15.00	27.87	36.35	25.61	46.50	36.18	19.79	20.53	33.57	41.06	27.66	13.82
5	*Eleusineindica*	0.00	0.35	1.29	0.00	0.00	0.00	0.00	0.00	1.78	0.00	2.76	1.66
6	*Eragrotistenela*	1.85	0.00	0.00	1.51	4.44	1.58	0.00	1.66	0.00	0.00	1.84	0.00
7	*Ischaemumtimorense*	1.67	4.50	0.00	0.00	0.00	0.00	0.00	0.00	0.00	0.00	0.00	12.64
8	*Panicumdistacyum*	0.00	0.00	2.67	6.87	0.00	2.61	3.85	8.68	9.14	6.98	6.66	6.92
	Broadleaf												
9	*Aeschynomeneamerikana*	0.00	0.00	3.17	1.49	0.00	0.00	1.38	0.00	1.55	0.00	0.00	1.41
10	*Acallyphaindica*	0.00	0.00	0.00	0.00	0.00	0.00	2.71	0.00	1.57	0.00	0.00	2.29
11	*Althernantherabrasiliensis*	0.00	0.35	1.53	1.31	0.00	2.23	1.49	1.78	1.67	0.00	0.00	0.00
12	*Amaranthusspinosus*	0.00	1.32	0.00	0.00	0.00	0.00	0.00	0.00	0.00	0.00	0.00	3.07
13	*Boerharaviaerecta*	16.23	16.03	9.71	15.52	0.00	10.90	4.29	7.90	1.89	4.85	5.28	8.65
14	*Centrosemapubescens*	0.00	0.00	1.65	0.00	0.00	1.87	0.00	0.00	1.63	0.00	0.00	0.00
15	*Cleome rubidospermae*	10.62	6.24	6.10	5.59	7.92	7.78	2.61	6.06	7.40	2.34	3.33	4.71
16	*Commelinadiffusa*	3.04	1.61	1.55	0.00	0.00	2.10	0.00	3.63	0.00	2.81	2.93	0.00
17	*Commelinabenghalensis*	0.00	0.70	1.60	1.64	2.47	2.00	3.99	1.55	2.02	1.95	2.44	1.53
18	*Croton hirtus*	8.93	5.51	2.14	3.85	2.91	0.00	2.51	4.24	4.66	2.14	6.37	1.69
19	*Euphorbia heterophylla*	6.43	3.04	4.21	7.11	7.22	0.00	17.90	17.91	4.39	9.95	5.40	1.70
20	*Euphorbia hirta*	1.53	1.34	0.00	3.14	4.62	3.28	5.46	1.99	1.76	1.92	0.00	4.24
21	*Ipoemoeatriloba*	0.00	3.08	1.80	0.00	0.00	2.04	0.00	0.00	0.00	0.00	0.00	0.00
22	*Linderniaciliata*	0.00	1.33	1.53	1.27	2.00	0.00	4.59	0.00	0.00	1.87	1.84	1.99
23	*Melochiapiramidata*	0.00	1.42	2.06	4.42	4.89	0.00	2.66	1.40	1.95	2.79	0.00	3.07
24	*Murdania*	5.21	1.32	4.26	1.26	1.82	2.96	3.51	3.11	3.30	0.00	0.00	1.35
25	*Ocimum sanctum*	8.36	8.67	6.72	9.83	8.52	9.79	5.37	5.91	6.67	9.21	11.00	14.57
26	*Oldenlandriadicotoma*	5.01	6.38	2.43	5.68	0.00	7.16	0.00	6.20	2.46	1.93	8.71	12.70
27	*Pylanthusniruri*												
28	*Pysalisangulata*	1.80	0.00	0.00	0.00	0.00	0.00	0.00	1.60	1.78	4.07	0.00	0.00
29	*Richardiascabra*	15.00	27.87	36.35	25.61	46.50	36.18	19.79	20.53	33.57	41.06	27.66	13.82
30	*Vernomiacinera*	0.00	0.35	1.29	0.00	0.00	0.00	0.00	0.00	1.78	0.00	2.76	1.66

Remarks: V1: Bisi 18, V2: Bisi 2, V3: Pertiwi 3, V4: NK 22, V5: NK 33, V6: DK 959, V7: P 23, V8: DK95, V9: DK 771, V10: Bima 19 Uri, V11: Bisma, V12: Sukmaraga

Grasses weeds that grows on all maize varieties are *Dactyloctenium aeghypthium, Eleusine indica, Panicumdistacyum*, while broadleaf weeds growing on all maize varieties is *Boerhaavia erecta,*

Cleome rubidospermae, Euphorbia heterophylla, Euphorbia hirta and *Pysalisangulata*. The highest of SDR on varieties of 'Bisi 18', 'NK 22', 'NK 33', 'DK 771', 'Bima 19 Uri' and 'Bisma' was *Bulbostylis puberula* by 14.09%, 18.15%, 23.69%, 17.35%, 8.94% and 18.44%. On varieties of 'Pertiwi 3' and DK 95', the highest SDR were *Cyperus rotundus* about 19.94% and 11.90%. The highest of SDR on 'Bisi2 was Dactyloctenium *aeghypthium* (18.34%), 'DK 959' was *Panicum distacyum* (15.5%), 'P23' was *Euphorbia heterophylla* (16.38%) and type of weed in 'Sukmaraga' were *Eleusine indica* (14.31%).

3.4. Weed Diversity at 8 Weeks After Planting

The weed diversity growing in field was changes at 8 WAP. There were 2 species of sedges, consist of *Bulbostylis puberula* and *Cyperus rotundus*. *Bulbostylis puberula* grows on all maize varieties with the highest value of SDR in 'Sukmaraga' (Table 5).

Based on Table 5, there were several species of weeds that appear on all maize varieties, there are *Bulbostylispuberula, Dactyloctenium aeghypthium, Cleome rubidospermae, Croton hirtus, Euphorbia heterophylla,* and *Richardiascabra. Dactyloctenium aeghypthium* having the highest of SDR in all maize varieties, except on 'Bisi 18'. The highest SDR on 'Bisi 18' was *Boerhaaviaerecta* with SDR value about 16.23%. Types of weeds growing at least on 'NK 33' (13 species) while most on 'P 23' (21 species). Weeds that grow in farm is very diverse. [18] reported some weeds that grow on maize field among other *Borreriaalata, Ageratum conyzoides L., Digitariasetigera, Porophyllum ruderale R, Phyllanthus niruri Cupheacarthagenensis, L., Axonopuscompressus, Cyperuskyllingia,* and *Echinocloacololonum* L.

4. Conclusions

This experiment indicated that on the vegetation analysis, weeds growing on farm consist of 2 species of sedges, 4 species of grasses and 14 species of broadleaf with coefficient of weed communities value about 81,93%. At 4 weeks after planting the most weed growing on varieties 'Bisi 18', 'NK 33', 'P 23', and 'DK 95', there were 14 species. *Bulbostylispuberula* and *Cyperusrotundus* were growing on all maize varieties with the highest SDR on 'Sukmaraga' and 'DK 959'. At 6 WAP, weeds that grow most consists of 4 species of sedges, 6 species of grasses and 24 species of broadleaf. At the age of 8 WAP, the weeds growing at least on 'NK 33' that were 13 species and the most on 'P23' that was 21 species. *Dactyloctenium aeghypthium* having highest SDR in all varieties, except Bisi 18. The highest SDR value on Bisi 18 was Boerhaavia *erecta* about 16.23%.

References

[1] Klingman G C and Ashton F M 1975 *Weed Science: principles and practices.* A-Wiley Intercsience publication.431 p
[2] Tjitrosoedirdjo S, Utomoand I H and Wiroatmodjo J 1984 *Pengelolaan Gulma di Perkebunan* Kerjasama Biotrop Bogor Jakarta Gramedia 210p
[3] Rao V S 2000 *Principles of Weed Science* Science Publishers USA 555p
[4] Naylor R E L and P J Lutman 2002 What is a weed? *In*: *Weed Management Handbook.* (Eds: Naylor, R.E.L.) Ninth Edition. Blackwell Publishing Company USA 423 p
[5] Anderson W P 2007 *Weed Science Principles and Applications.*Third Edition. Waveland Press. Long Grove Illinois 388 p
[6] Saeed M, Khaliq A, Cheema Z A and Ranjha A M 2010 *Effect of Nitrogen Levels and Weed-Crop Competition Duration on Yield and Yield Components of Maize J. Agric Res.* **48(4)**:471-481
[7] Sunitha N and Kalyani D L 2012 *Weed Management in Maize (Zea mays L.)* A Review Agricultural Research Communication Centre.Agri.Reviews **33(1)** 70-77
[8] Travlos I S, Economou G, and Kanatas P J 2011 Corn and Barnyardgrass Competition as Influenced by Relative Time of Weed Emergence and Corn Hybrid *Agronomy Journal* **103** Issue 1, 2011

[9] Pasaribu A 2015 *Roundup ReadyCornTechnology to Support Indonesia GovernmentProgram in Self-Sufficiency of Corn* Presentation materialof National Seminar HIGI Bandung 17 November 2015

[10] Mohammadi G R, Mozafari S, Ghobadi M E and Najaphy A 2012 *Weed Suppressive Ability andProductivity of Berseem clover (trifoliumalexandrianum L.) Interseeded at Different Rates and Times in Corn (Zea mays L.)Field.*Philipp Agric Scientist.**95** (4) 452-357

[11] Perdana E O, Chairul, and Syam Z 2013 *AnalisisVegetasiGulmaPadaTanamanBuah Naga Merah (Hylocereuspolyrhizus, L.) di KecamatanBatangAnai, Kabupaten Padang Pariaman, Sumatera Barat* J. Bio. UA. **2**(4) – Desember 2013: 242-248

[12] Mohammadi G R 2007 *Growth parameter enhancing the competitive ability of corn (Zea mays L.) against weeds* Weed Biol and Manag **7** 232-236

[13] Mason H, Navabi A, Frick B, Donovan J O and D Spaner 2007. *Cultivar and Seeding Rate Effects on The Competitive Ability of Spring Cereals Grown Under Organic Production in Northern Canada.* Agron J. **99** 1199 – 1207 (2007).

[14] Hall M R, Swanton C J and Anderson G W 1992 *The Critical of Weed Control in Grain Corn (Zea mays).* Weed Science **40** 441 - 447

[15] Ghamari H and Ahmad vand G 2013 *Growth Analysis of Dry Bean (Phaseolus vulgaris L.) in Different Weed Interferent Situation* Notulae Scientia Biologicae **5**(3) 394-399

[16] Suryaningsih M J and Darmadi A A K 2011 Inventarisasi Gulma Pada Tanaman Jagung (Zea mays L.) di Lahan Sawah Kelurahan Padang Galak, Denpasar Timur, Kodya Denpasar, Provinsi Bali *Jurnal Simbiosis.* **1**(1): 1 – 8 Jurusan Biologi MIPA Universitas Udayana

[17] Fujiyoshi P, Gliessman S R and Langenheim J H 2007 *Factor in The Suppression of Weeds by Squash Interplanted in Corn* Weed Biology and Management. **7** 105 – 114

[18] Solfiyeni, Chairul and Muharrami R 2013 *Analisis Vegetasi Gulma pada Pertanaman Jagung (Zea mays* L.) *di Lahan Kering dan Lahan Sawah di Kabupaten Pasaman* Proceedings of Semirata FMIPA Universitas Lampung351 – 356

Effectiveness and efficiency of potassium fertilizer application to increase the production and quality of rice in entisols

S Hartati[1], Suryono[1] and D Purnomo[2]
[1]Soil Science Department, Agriculture Faculty, Universitas Sebelas Maret, Surakarta, Indonesia
[2]Agrotechnology Department, Agriculture Faculty, Universitas Sebelas Maret, Surakarta, Indonesia

Email : tatikyth@gmail.com

Abstract. The study aimed to determine the appropriate dose of K-type fertilizer on the production and quality of IR-64 rice varieties in Entisols. The study was conducted on Entisols, Nglarang, Basin, Kebonarum, Klaten, Central Java, Indonesia. Randomized Complete Block Design (RCBD) was used with 2 factors and three replicates. Factor I: types of K fertilizer: KCl, ZK, and KNO_3. Factor II: K fertilizer dose consists of four levels: 0, 50, 100 and 150 kg K_2O ha^{-1}. The results showed that the type of K fertilizer KCl, ZK, and KNO_3 gave similar effect on growth and yield of rice variety IR 64 on Entisols. Dose of K fertilizer increased the growth and yield of rice but an increased dose provides improved meaningful results. K fertilizer with a dose of 50 kg ha-1 K_2O is the most efficient, with the agronomic efficiency for dry grain highest yield of 13.6 kg / kg K_2O achieved by ZK fertilizer.

1. Introduction

Rice is the staple food most of Indonesians, therefore, play a major role in the economic life and national politics. Indonesia's rice demand is around 32.947 million tons per year based on consumption of 139 kg per capita per year for a population to 237 million. If rice production is not increasing in 2030 there will be a shortage of rice by 26 million tonnes equivalent of 12 million hectares [1]. Constraints in increasing rice production including: weather anomalies that have the potential to climate change impact on water shortage [2], the rate of conversion of agricultural land [3][4], and a decrease in productivity intensification of land due to the changes in irrigation systems [5].

Potassium (K) contributes to [6][7] the quality and plant resistance to water stress because of lower transpiration so that water consumption is low [6]. K role in the process of transpiration or water transport within the plant body is as a regulator of cell osmotic potential [8][9]. Another opinion states that K plays a multifunctional: as activating enzymes, stimulating and transport of assimilates, a stabilizing anion or cation in the transport of water through stomatal movement. Likewise, some studies indicate that the addition of K fertilizer increases the number of branches cause on the crop yields [10]. Regosol soil generally contains enough elements of P and K which are still fresh and not ready to be absorbed by plants, but often lack of Nitrogen (N) [11]. On the other hand, Entisols is soil with low organic matter content and generally responsive to nitrogen fertilization [12]. The content of organic matter in soil Entisols is low and the land is vulnerable to erosion.

2. Methods

Research was conducted with experimental methods in wetland Regosol soil types, from April to November 2016. The experiment consists of two factors, types of fertilizers K (KCl, ZK, and KNO_3) and doses (0, 50, 100, and 150 kg K_2O ha^{-1}), in order to obtain 12 combined factorial treatments. The experimental design was a randomized complete block design (RCBD), any combination is repeated three times so that there are 36 experimental unit. Trial begins with processing the soil using hoes to the deep layers of silt (20 cm). This plot area is divided into three blocks (replications), the distance between blocks 50 cm. Each block is divided into 12 experimental unit (plot) measuring 2 x 2.5 m^2 each spaced 50 cm.

The planting of rice variety IR-64, carried out after the processing of land with a spacing of 25 x 25 cm and 5 seedlings per plot. The seedlings were obtained from the nursery were 18 days old. The day before planting dispersed Urea was given as basal fertilizer with a dose of 400 kg ha^{-1} and SP 36 fertilizer 300 kg ha^{-1}. Potassium fertilizer is given once, Urea and SP-36 is given twice at the rest at the age of six weeks with appropriate doses of the treatment. Watering was done with land flooded as high as about 5 cm to 3-week-old plants, after it was drained, irrigated every 3 days. Weeding was done at the age of 2 weeks after planting manually. Pest control (dominant pest snails) was using pesticides, while for the disease although the infection does not occur pesticide spraying was conducted as a preventative measure. Harvesting was done after all the plants in one plot dries (visually brown), about 90 days old plants.

Data was collected with direct observations on plant height, counting the number of productive tiller, tiller number nonproductive and weighing of grain harvested dry weight and the weight of milled rice, as well as the weight of 1000 seeds. Laboratory analysis of the soil before planting consisted of: texture methods pipette, heavy volume gravimetric method, specific gravity (BJ) gravimetric method, organic matter content method of Walkley and Black, acidity (pH) method elektrometri (soil:H_2O = 2:5), the cation exchange capacity (CEC) method of saturation of ammonium acetate, N (Kjeldahl method), K (method of saturation of ammonium acetate), available P (method Bray 1), Ca (method of saturation of ammonium acetate), and Mg (method of saturation of ammonium acetate). Analysis of the soil afterharvest wereincluding pH, total N content, available P and K available. Data were analyzed using analysis of variance with 0.05 significance level (F-test). If significantly different treatment werefound, Duncan multiple range test (DMRT) were done to confirm the difference between pairwise treatments. Relationships between variables were analyzed with Pearson's correlation.

3. Results and Discussion

3.1 Initial soil conditions

The results of the analysis of baseline soils have physical and chemical properties as follows: BV 1.12 g cm-3, BJ 2,17 g cm^{-3}, 48.89% sand, dust 33.80% and clay 15.31% that included loam, pH 6.85 (rather acid), CEC cmol 20,36 kg^{-1} (moderate), total N 0,13% (low), available P 5.3 ppm (low), K available cmol 0.15 cmol kg^{-1} (low), Ca 9.55 cmol kg^{-1} (moderate) and 0.89 Mg cmol kg^{-1} (low). Based on the chemical properties of the soil the content of macro nutrients needed crops including lower categories so it needs the input of fertilizer.

3.2 Last soil conditions

Diversity test results showed that the treatment had no effect on the final soil pH, soil total N content of the final, and the available soil P end with an average value of 7,38 in a row; 0,23% and 9,04 ppm. The test results showed that a wide diversity of K fertilizer had no effect on K provided soil and K fertilizers significantly affect K available and there was no interaction between the two. The results of the final analysis of available K presented by **Table 1**.

The test results showed that a wide diversity of K fertilizer, potassium fertilizer dosage and the interaction between them show the influences that are not significant. The results of initial soil analysis showed a pH of 6.85. The provision of a wide range of K fertilizer on various doses of K fertilizer had lower soil pH value. A decrease in pH was due to the flooding.

Table 1. The effect of Fertilizer K Dose to CEC and K available

Dose of K Fertilizer	CEC (cmol.kg^{-1})	K available (cmol.kg^{-1})
0 kg K_2O/ha	19,17^{a}	0,13^{a}
50 kg K_2O/ha	20,84^{b}	0,19^{b}
100 kg K_2O/ha	22,27^{b}	0,21^{bc}
150 kg K_2O/ha	23,31^{b}	0,27^{c}

^{a,b,c}Different superscripts indicated significant differences (P<0.05)Analysis of the cation exchange capacity (CEC) of the soil and K available soil after harvest showed that types of fertilizers K influence was not evident, as well as their interaction effect was not significant. K fertilizer application significantly affected the CEC and K available in soil. The K fertilizer application was significantly different compared with K fertilizer to the soil CEC and K available land. With increasing doses of fertilizers applied improve both the CEC and K available in the soil. K fertilizer application dose of 50 kg $K_2O.ha^{-1}$, 100 kg $K_2O.ha^{-1}$ and 150 kg kg $K_2O.ha^{-1}$ provides enhanced soil CEC but statistically not significant, the higher the dose given higher the soil CEC. K available on soil indicated K available after harvest, after transportation of K nutrients by plants. K available is determined by the amount of fertilizer given, transport of nutrients by plants and soil properties. Similarly, the higher the dose of K fertilizer is given higher the K available in soil. Good fertilizer such as KCl, KNO_3 ZK when given into the soil will undergo decomposition into the K^+ which may be available and absorbed by plants. Correlation test results showed that a significant correlation was obtained between the cation exchange capacity with plant height (r = 0.422), a thousand seeds (r = 0.344) and K available land late (r = 0.417). However, from various kinds and doses of K fertilizer given on Regosol soil K availability is still relatively moderate. Likewise with the cation exchange capacity is still relatively moderate.

3.3 Plant growth

Results of analysis of variance showed that the treatment effect on plant height. The types of fertilizer K and dose significantly affected plant height, while their interaction effect was not. Based on Figure 1, on average fertilizer KCl give the best results (92.7 cm high) and not significant when compared with KNO_3 fertilizer (90.7 cm), while ZK fertilizer low (88.7 cm).

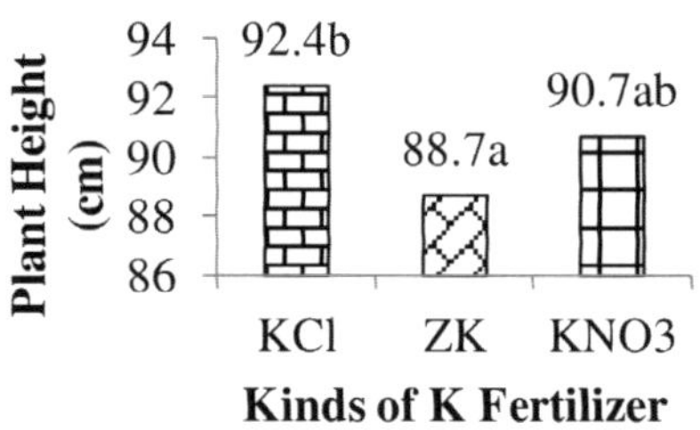

Figure 1. Effect of kinds of K fertilizer to plant height

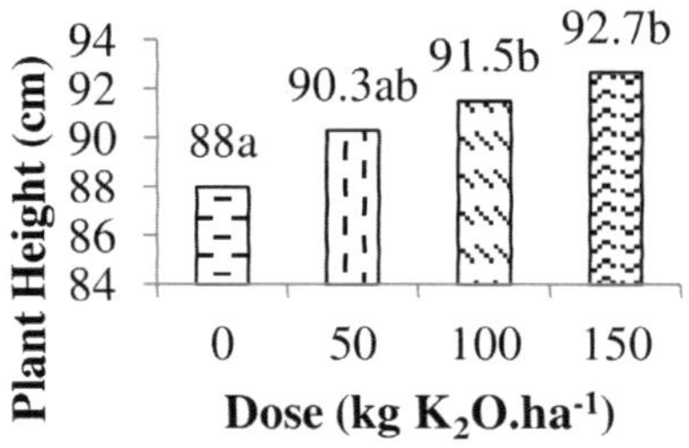

Figure 2. Effect of doses of K fertilizer to plant height

Description: Numbers followed by the same letter are not significant with DMRT 5%

Figure 2 shows that with the increase of K fertilizer dose will increase plant height at all dose levels 50 kg $K_2O.ha^{-1}$, 100 kg $K_2O.ha^{-1}$ kg and 150 kg $K_2O.ha^{-1}$. However, the height increase among dose levels were not significant. It can be understood that with the increase of K fertilizer given increase the availability of high-K so that the plant increases.

The plant grows high in relation to the apical dominance due to the interaction between auxin and light [8]. The response of plants to light for a clump of rice depends on the density of which is reflected in the number of tillers. Auxin as stimulants of endogenous growth (enlargement and elongation of cells) become active (since the lower light intensity), so the plants become taller.

3.4 Yields component

3.4.1 *The number of vegetative tillers* A wide diversity of K fertilizer effect no significant effect on the number of tillers per cluster during the vegetative phase. K fertilizer doses significantly affect the number of vegetative tillers per cluster. Figure 3 shows that with the addition of K fertilizer 50 kg $K_2O.ha^{-1}$ increase the total number of tillers from an average of 23 stem per cluster to an average of 24.6 stem per cluster but this increase was not statistically significant. If the dose of K fertilizer increased to 100 kg $K_2O.ha^{-1}$ total number of tillers per cluster may significantly increase but if the dose of K fertilizer increased again to 150 kg $K_2O.ha^{-1}$ had lower total tiller number although not significant.

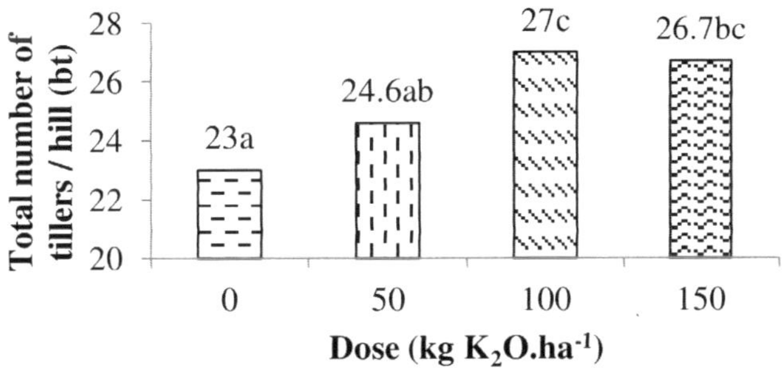

Figure 3. Effect kinds of K fertilizer on the number of vegetative tillers maximum total current
Description: Numbers followed by the same letter are not significant with DMRT 5%

3.4.2 *The number of productive tillers* The number of productive tillers can be seen in Figure 4 and Figure 5. The test results show that a wide diversity and K fertilizers significantly affect the number of productive tillers while the interaction between the two influential not evident.

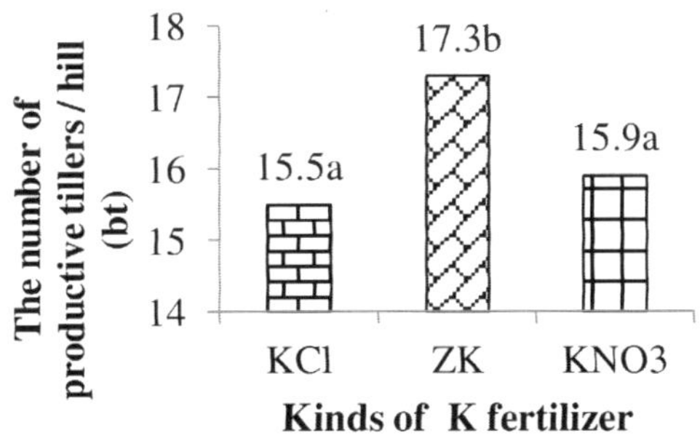

Figure 4. Effect of kinds of K fertilizer on the number of productive tillers per hill

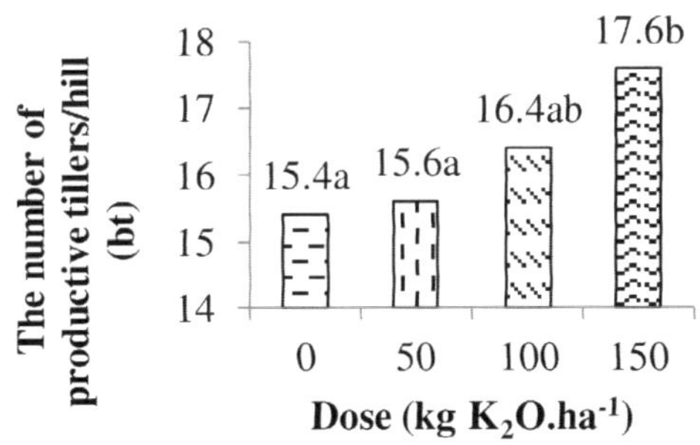

Figure 5. Effect of doses of K fertilizeron the number of productive tillers per hill

Description: Numbers followed by the same letter are not significant with DMRT 5%

In Figure 4 shows that the average number of productive tillers highest achieved at ZK fertilizer application and significantly different when compared with KCl and KNO3 fertilizer. The highest number of productive tillers reached at ZK fertilizer. This is because the sulfate content in the soil is low, then the follow-up of ZK fertilizer can increase the levels of S in the soil. Sulfur or sulfur required in the production of chlorophyll. S is required to produce amino acids (cysteine, methionine and cystin) in plants [13]. In Figure 5 shows that with increasing doses of K fertilizer will increase the number of productive tillers significantly in a dose of 150 kg $K_2O.ha^{-1}$. The number of productive tillers as one determinant of productivity of rice is supported by the condition of the reproductive organs (panicle length, number of grains per panicle, length and number of grains per panicle.

3.4.3 Dry grain harvest and dry milled grain The test results show that a wide diversity and K fertilizers significantly affect the weight of dry grain harvest. Results of dry grain harvest is presented in Figure 6 and 7. Based on the figure 6 that the types of fertilizers ZK weight of dry grain harvest produce the highest with an average of 88.104 $tons.ha^{-1}$ and the lowest KNO_3 fertilizer with the results weight of dry grain harvest average 7.466 $tons.ha^{-1}$. In figure 7 it can be seen that K fertilizer dose of 100 kg $K_2O.ha^{-1}$ produced the highest yield that is 8.127 $tonnes.ha^{-1}$ but when the dose is increased to 150 kg $K_2O.ha^{-1}$ there was no significant decrease of the yield. This can be explained due to sulfur in the soil are low so that the provision of fertilizer ZK (K_2SO_4) can increase the sulphate in the soil. Sulfate is absorbed by plants can increase chlorophyll of plants as well as nitrogen. Sulfate plays a role in the preparation of proteins, especially cystine, cysteine and methionine. Similarly, an increase in the dose of K fertilizer will increase available K in the soil.

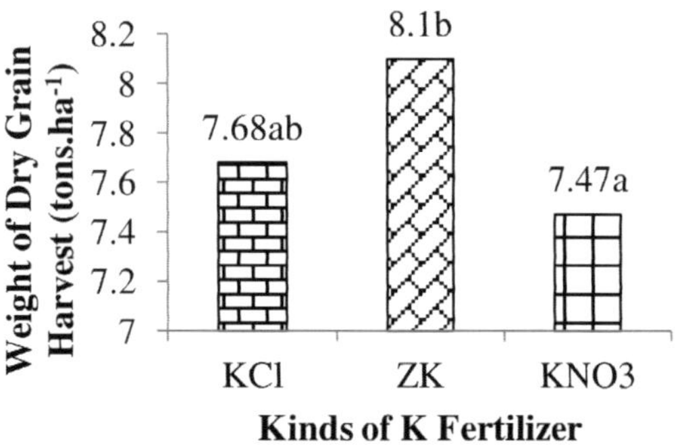

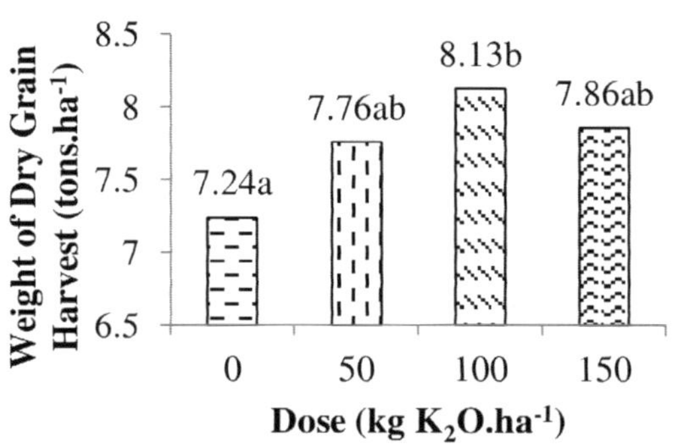

Figure 6. Effect of kinds of fertilizer K to weight of dry grain harvest

Figure 7. Effect of doses of K fertilizer to weight of dry grain harvest

Description: Numbers followed by the same letter are not significant with DMRT 5%

Production of rice in the form of grain and then into rice after going through several stages of processing. Role sorts K fertilizer to the weight of dry grain harvest and dry milled were not significantly different, while the dose of fertilizer real differences occur only in plants cultivated and nurtured (not controls). K fertilizer increased the weight of dry grain harvest and dry milled. It shows that the character in the form of plant growth is high and the number of seedlings that affect the production (weight of dry grain harvest and dry milled).

The presence of K-available enough in the soil, guaranteeing vigor of the plant. According [14], when potassium is reasonably available in the plant then the plant more resistant to pathogens and stimulate root growth. Better root growth will create more nutrient absorption so that it can be used in the process of metabolism, especially protein synthesis from amino acids and ammonium ions. The result of this synthesis can affect the growth and production of plants. Potassium special effect in the absorption of nutrients, regulation of breathing, transpiration, enzyme and affect the translocation fotosintat. Potassium plays a role in the translocation photosynthates as potassium arrange transportation system, consequently photosynthates can be distributed accordingly so it does not happen at the photosynthesis site where the photosynthates were piling [15].

Functions include potassium, helps maintain osmotic potential and extraction of water, hence the plants are fairly K only a slight loss of water and have a positive influence on stomata closure. Potassium may also serve to balance the charges of anions and affect uptake and transport of the anion. Potassium can help reduce outbreaks of certain diseases and fall, because potassium will increase the hardness of the plant. This is because K affect the lignin content of tissues sklerenkim [16].

Based on the figure 8 that the highest weight of dry milled grain at ZK fertilizer with an average of 4,408 tons.ha^{-1}. With increasing doses of potassium fertilizer up to a dose of 100 kg K_2O.ha^{-1} was obtained a yield increase of dry milled grain weight, but when the dose is increased again to 150 kg K_2O.ha^{-1} of dry milled grain yield is decreasing as is the case of dry milled grain.

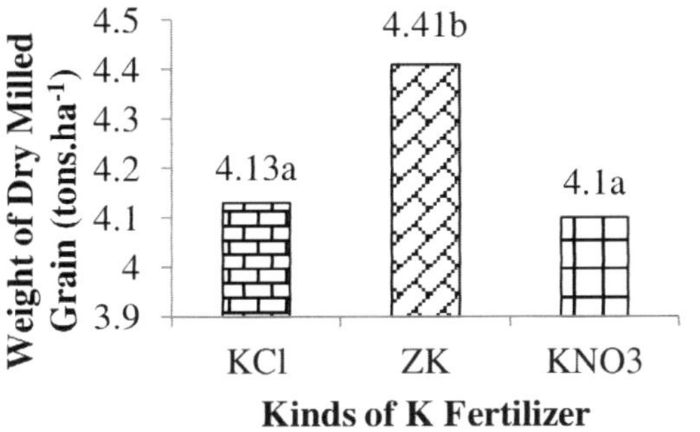

Figure 8. Effect of kinds of fertilizer K to dry milled grain

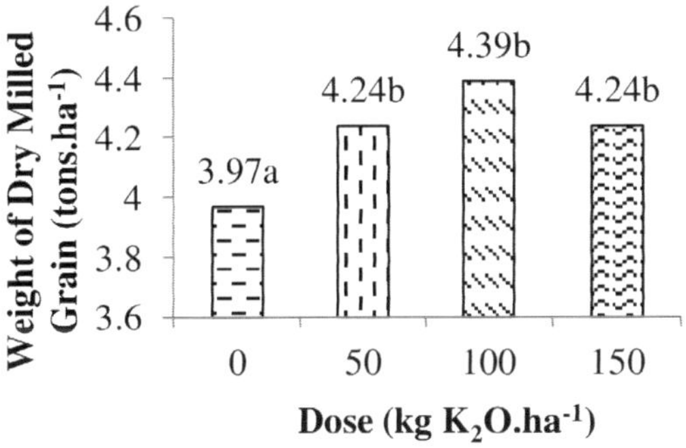

Figure 9. Effect of doses of K fertilizerto dry milled grain

Description: Numbers followed by the same letter are not significant with DMRT 5%

3.4.4 The weight of 1000 seeds The test results show that a wide diversity and K fertilizers significantly affect the weight of 1000 seeds. Weight of 1000 seeds results presented in Figure 10 and 11.Weight of 1000 seeds is one indicator of the quality of the crop. Calculation of the weight of 1000 seeds is necessary because it is one of the characteristics listed in the description of a plant variety. According [17], the average weight of 1000 seeds tend to become permanent traits of each plant variety. Use of the weight of 1000 seeds is to find the average weight that can cause constant grain size in some varieties due to the large number of sample, it is able to cover the variations in each individual plant.

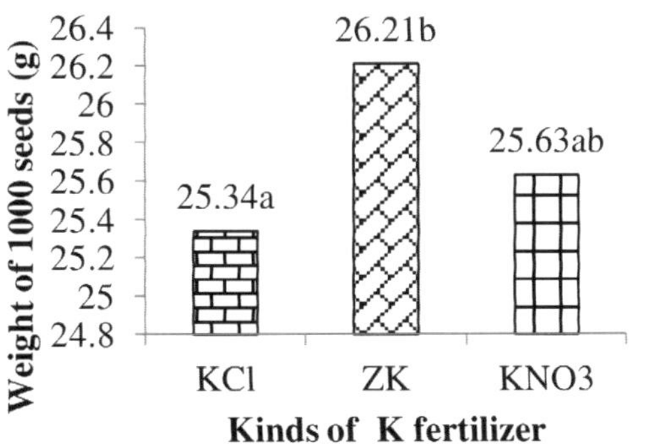

Figure 10. Effect of kinds of K fertilizer to the weight of 1000 seeds

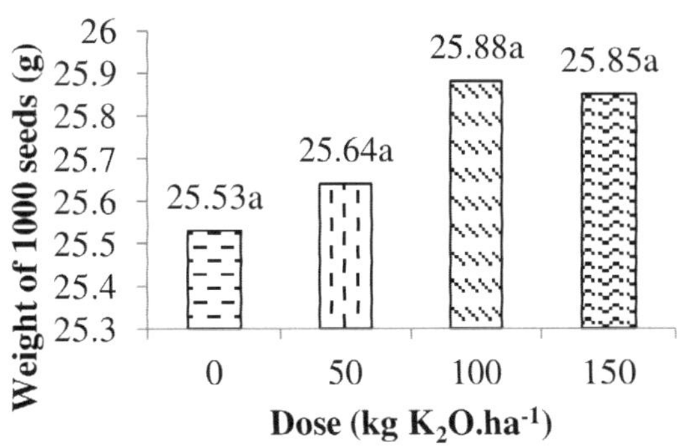

Figure 11. Effect of doses of K fertilizer to the weight of 1000 seeds

Description: Numbers followed by the same letter are not significant with DMRT 5%

One application of heavy use of 1000 seeds is to determine the optimal seed number in a plot of land, knowing the large seeds or weight means signifies the seed when harvested in a state that is really ripe for good seed to be planted or used as seed is a seed which actually matured. In addition, people often choose large seed as a seed for a reason: usually crop of good races, have seeds that are bulkier and heavier because of large and heavy seeds is a trait that is inherited, large seeds and the weight is usually sufficiently ripe and the seeds are large and heavy has a lot of food reserves [18].

3.5 Agronomic efficiency

Agronomic efficiency is calculated based on the formula: (kg per hectare results P1 - P0 kg per hectare results) divided kg per hectare of nutrients given = kg / kg nutrient [19]. Agronomic efficiency for weight dry grain harvest is presented in Figure 12. Of the various fertilizers K is given, the higher the dose given to lower the agronomic efficiency of dry grain harvest. Highest agronomic efficiency was achieved by treatment with a dose of fertilizer ZK 50 kg $K_2O.ha^{-1}$ ie 13.60 $kg.kg^{-1}$. According [19], that the agronomic value ranges between 10-30, the rate of> 25 indicates that the plant has been getting good farming management system.

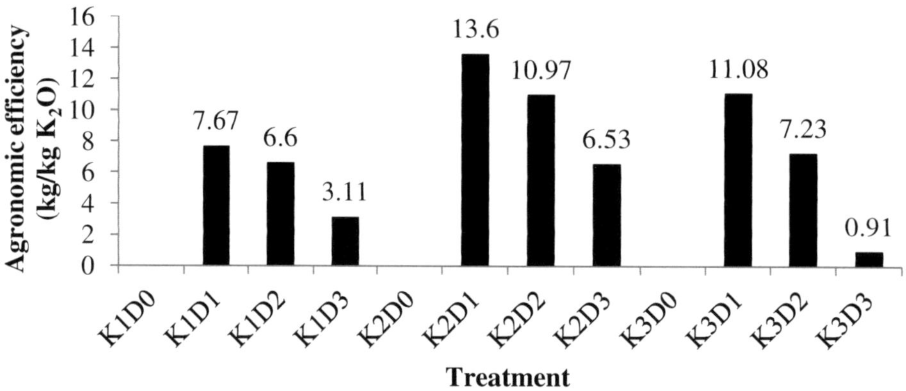

Figure 12. Effect of treatment to the agronomic efficiency
Description: Numbers unfollowed by the letter are not significant with DMRT 5%

4. Conclusion

Various K fertilizer in the form of KCl, ZK, and KNO_3 almost have the same effect on the growth and yield of rice variety IR 64 on Entisols. Dose of K fertilizer increased the growth and yield of rice but an increased dose provides improved meaningful results. Fertilizers K at a dose of 50 kg ha^{-1} K_2O relatively efficient, with the agronomic efficiency for dry grain highest yield of 13,6 kg / kg K_2O achieved by ZK fertilizer.

Reference

[1] Suswono 2012 Kebijakan Kementerian Pertanian Dalam Penyediaan Lahan dan teknologi Untuk Terwujudnya Kedaulatan Pangan Prosiding *Simposium Bersama PERAGI-PERHORTI-PERIPI-HIGI IPB Bogor* 7-13

[2] Budiastuti S 2010 Fenomena Perubahan Iklim Dan Kontinyuitas Produksi Pertanian: Suatu Tinjauan Pemberdayaan Sumberdaya Lahan Ekosains *Jurnal Ilmiah Lingkungan Hidup* **2**(1) 5-13

[3] Hairiah K 2004 *Pertanian Organik Suatu Harapan Atau Tantangan* Makalah Seminar Jurusan Tanah FakPertanian Universitas Brawijaya Malang

[4] Suprayogo D, Widianto, Purnomosidi P, Widodo R H, Rusiana F, ZAini Z, Khasanah N, Kusuma Z 2004 Degradasi Sifat Fisik Tanah Sebagai Akibat Alih Guna Lahan Hutan Menjadi Sistem Kopi Monokultur: Kajian Perubahan Makro Porositas Tanah *Agrivita* **26** 60-68

[5] Purnomo D, Yunus A, Budiastuti S 2011 Budidaya Padi Berwawasan Lingkungan Dengan Metode System Rice Intensification dan Penggunaan Pupuk Organik Cair *Ekosains Jurnal Ilmiah Lingkungan Hidup* **3**(1):25-31

[6] Chaize T 2010 Potash wwwdani2989com diakses 5 Mei 2015
[7] Miah M M A, Saha P K, Islam A, M N Hasan, Nosov V 2006 Efficiency of potash fertilizer application in a rice-rice and a rice-wheat cropping system in Bangladesh Paper presented at the IPI-PAU International Symposium on Balanced Fertilization for Sustaining Crop Productivity Ludhiana India 22-25 November
[8] Taiz L and E Zieger 2010Plant Physiology Fifth ed Sinauer Associates Inc
[9] Grzebisz W, Gransee A, Szczepaniak W, Diatta J 2012 The Effects of Potassium Fertilization on Water-Use Efficiency in Crop Plants *J Plant Nutr Soil Sci* **176**(3):355-374
[10] Yassen A A, Habib A M, Sahar, Aghloul M Z, Khaled S M 2010 Effect of different sources of potassium fertilizers on growth yield, and chemical composition of *Calendula Officinalis J. of American Sci* **6**(12):1044-1048
[11] Munir M 1996 *Tanah-tanah Utama di Indonesia* PT Dunia Pustaka Jaya Jakarta
[12] Foth HD 1998 *Dasar-dasar Ilmu Tanah* Gadjah Mada University Press Yogyakarta
[13] Marschner H 1995 Mineral Nutrition of Higher Plants (2nd Edition) Academic Press London
[14] Sarief S 1983 Ilmu Tanah Umum Bagian Ilmu Tanah Universitas Padjadjaran Bandung
[15] Poerwowidodo 1992 Telaah Kesuburan Tanah Bandung Angkasa
[16] Gardner F P, R B Pearce, R L Mitchell 1991 *Fisiologi Tanaman Budidaya* Terjemahan Herawati dan Subiyanto UI Press Jakarta
[17] Mugnisyah W 1990 *Pengantar Produksi Benih* Rajawali Press, Jakarta
[18] Anonim 2010 Berat 1000 Butir Biji http://daunmudhablogspotcom/2010/02/-berat-1000-butir-bijihtml Diakses tanggal 20 April 2015
[19] Dobermann A 2007 Nutrient Use Efficiency-Measurement University if Nebrasja-Lincoln, USA

IOP Conf. Series: Earth and Environmental Science **142** (2018) 012032 doi:10.1088/1755-1315/142/1/012032

Application of Azolla and intermittent irrigation to improve the productivity and nutrient contents of local black rice variety

Sulandjari[1,2] and M B Yunindanova[1]
[1]Department of Agrotechnology, Faculty of Agriculture, Universitas Sebelas Maret, Surakarta, Indonesia
[2]Biotechnology and Biodiversity Research Center, Universitas Sebelas Maret, Surakarta, Indonesia

E-mail: sulandjari@staff.uns.ac.id

Abstract. Black rice is a local rice variety that contains a high level of anthocyanin pigment. Anthocyanin has been reported to be very effective in reducing cholesterol levels as well as cancer cell invasion. One of the main problems in rice cultivation is lack of water. System of Rice Intensification (SRI) has shown to be able to increase rice productivity by increasing the number of tillers. This system is known as a water-efficient cultivation. Other rice cultivation barrier is related to the use of nitrogen fertilizer. One of replacement of nitrogen fertilizer is by adding azolla. The objective of this research was identifying growth and yield of organic black rice with intermittent irrigation and application of azolla. The plant material used was black rice Cempo variety from Sleman, Yogyakarta. This experiment utilized 4 dosages of azolla as the first treatment: 100 gm^{-2}, 200 gm^{-2} and 400 gm^{-2}. The second treatment was water supply consisted of continuous flooded 2 cm; flooded 2 cm every 3 days; flooded 2 cm every 6 days. The results depicted that the application of azolla was able to increase the growth of black rice. Azolla of 200 gm^{-2} and 400 gm^{-2} and intermittent 3 days to 6 days generated higher dry grain, anthocyanin and antioxidant. Azolla 200 gm^{-2} with intermittent irrigation 3 days could be a good combination to improve plant growth, yield and properties of local black rice.

1. Introduction

Nitrogen is essential for improving plant growth and yield. On the other hand, excessive use of inorganic fertilizers will cause adverse effects to soil fertility. Therefore, currently the growers are interested to find alternative of inorganic fertilizer. Providing biofertilizer and organic fertilizer is a priority to solve this problem. The use of azolla as a biofertlizer has been successful in some countries as well as in Indonesia [1]. Azolla could be a substitution for urea and beneficial for maintaining plant growth [2]. Azolla is widely applied because of its ability to supply nitrogen with fixation capability of nitrogen reaching 1.2 kg ha^{-1} daily [3].

Rice is the major cereal that consumed as the staple food by more than half the world's population. Various colors of rice depend on the pigment in the epidermis or aleurone. At this time, the black rice is widely consumed as a functional food. A functional food is food that naturally or through a specific process comprising one or more compounds which may have physiological functions that are beneficial to health. Black rice in Indonesia is widely available in a variety of areas with different local names. Black rice in Java known as: Cempo rice in Sleman; Gadog rice in West Java, Melik rice in

Bantul, and Wulung rice in Solo. Black rice has a number of advantageous nutrients compared to white rice and red rice. It contained of high protein, vitamins and minerals which were varies among cultivars and planting sites [4]. Anthocyanin pigment has been reported to be very effective in reducing cholesterol levels in the human body [5], treatment of allergic reactions, and also inhibits cancer cell invasion [6].

Problems in rice cultivation are lack of water and nitrogen excessive fertilizer. SRI (System of Rice Intensification) is known as water-efficient cultivation system. This system applied young seedlings that are transplanted in wider planting distance and intermittent irrigation in vegetative phase [7]. This system is able to increase rice productivity through increasing number of tillers.

In order to support black rice as a functional food crop, it is recommended to cultivate this crop by organic method using organic fertilizers. One replacement for nitrogen fertilizer is azolla. Soil organic matter, soil pH and soil nutrients (N, P, K, Ca, Mg and Na) improved with the rate of azolla [8]. Azolla could boost seedling growth. The objective of this research was to increase the productivity of organic black rice [9]. Specific objectives were to identify the growth and yield, anthocyanin, antioxidant and protein contents as the response of different water supply system and azzola application.

2. Material and methods

The experiment was conducted in the net house located in Ngemplak village, Wedomartani, Yogjakarta. The plant material used was black rice Cempo variety from Sleman, Yogyakarta. Sources of plant nutrient are cow manure, Bio Natural^{TM} phosphates (6.4 g/plant) and Z-ka (6.4 g/plant). Azolla was utilized to replace urea fertilizer. Rice were planted in plastic pot with 40 cm in diameter. This experiment employed 2 treatments. The first treatment was dose of azolla comprised three levels (100 gm^{-2}, 200 gm^{-2} and 400 gm^{-2}). The second treatment was water supply consisted of flooded 2 cm tall every day (continuous flooded); flooded 2 cm every 3 days; flooded 2 cm every 6 days. On the flowering and grain filling period, plants were supplied by 2 cm high of water for all treatments. Irrigation was stopped in 2 weeks before harvesting. The experiment employed Randomize Complete Block Design with three replications.

In order to investigate the soil fertility, soil pH, texture, soil chemical properties were analysed in this experiment. Soil chemical properties comprised Nitrogen (N) total, available P, exchangeable K, organic material, and cation exchange capacity. N was analysed by Kjehdahl method. To determine the differences in vegetative growth, plant height was measured. Generative and plant quality parameters covered dry grain yield, anthocyanin, protein and antioxidant.

3. Result and discussion

Azolla and inttermittent performed significant effects on plant height, dry grain yield, anthocyanin, protein and antioxidant. Interaction occurred between azolla and intermittent to plant height, protein and antioxidant.

Interaction between azolla 200 gm^{-2} and 400 gm^{-2} with 3 days intermittent irrigation produced higher plant (Figure 1.). This was due to higher nitrogen availability. Besides, the low fertility of the soil (Table 1.) contributed to the plant response. It meant that in unfertile soil, addition of azolla will be beneficial to rice growth. Flooded soil will effect to mineral conditions in the soil. In flooded soil, there will be some changes in soil chemistry. Some of those changes were the decline in the value of oxidation reduction potential (redox) of some nutrients. A decrease in these value would be toxic to plants [10].

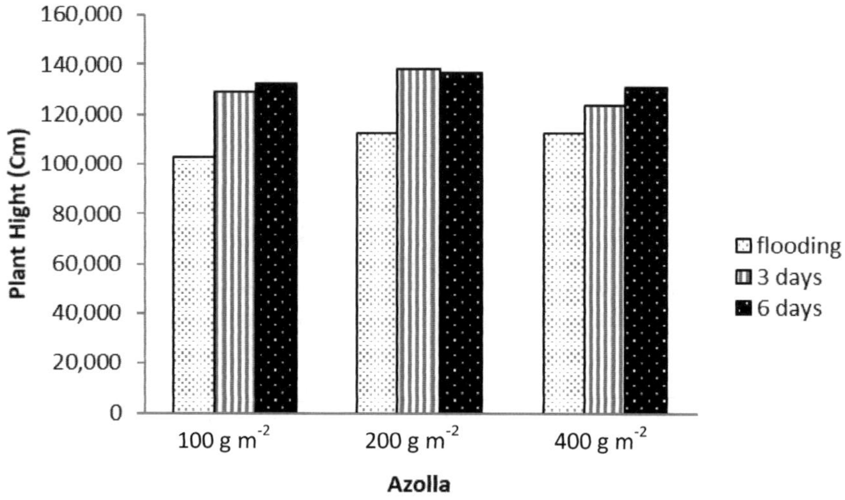

Figure 1. Interaction between azolla doses and intermittent water supply to plant height

Table 1. Soil analysis result

No	Parameters	
1	Type	Regosol
2	Chemical properties	
	N total (%)	0.36
	Available P (ppm)	11.70
	Exchangeable K (me/100 g)	0.24
	Ca (me/100 g)	11.20
	Organic Material (%)	3.30
	Cation Exchange Capacity (%)	28.15
3	Texture	
	Silt (%)	29.46
	Clay (%)	22.55
	Sand (%)	47.99
	pH H_2O	5.64

Table 2. Grain yield and anthocyanin content due to azolla doses

Azolla dose	Dry grain yield (g)	Anthocyanin (ppm)
100	73.77 b	6033 b
200	86.32 a	7187 a
400	82.34 ab	6625 a

Note: The same letter in the same column showed no significant difference

Table 3. Grain yield and anthocyanin content due to intermittent irrigation treatment

Intermittent	Dry grain yield (g)	Anthocyanin(ppm)
Continuous flooded	76.60 b	6459 b
3 day	84.60 a	6539 b
6 day	81.23 ab	6846 a

Note: The same letter in the same column showed no significant difference

The higher dry yield and anthocyanin compound were generated by application azolla 200 gm^{-2} and 400 gm^{-2}. Intermittent 3 days significantly boosted the rice yield but it declined the production of anthocyanin. On the contrary, intermittent every 6 days yielded higher anthocyanin. Intermittent every 6 days meant longer dry period experienced by plants so rice faced water stress period and produced higher anthocyanin as the response. This result was in line with Daneshmand [11] who conveyed that anthocyanin increased under drought. Intermittent irrigation is a concept of water saving through the regulation of water conditions in the field. At intermittent irrigation, the land is arranged in flooded and dried conditions alternately according to land conditions and growth phases.

Paddy fields flooded after the cultivation process will have consequences for the physicochemical changes in the soil. In flooded soil, the oxygen levels declined drastically so that anaerobic microorganisms became active. As a result, organic matter would decompose slower and less perfect. In addition, Flooding had a real impact on soil microbial characteristics [12]. Conversely, intermittent irrigation produced a positive response to aerobic bacteria in the soil [13]. Intermittent irrigation created higher root oxidizing activity rate, higher root length density, greater plant biomass, as well as it improved plant yield [14]. Moreover, intermittent drainage contributes to the alleviation of CH_4 emission and thus may decline net global warming potential in paddy cultivation [15].

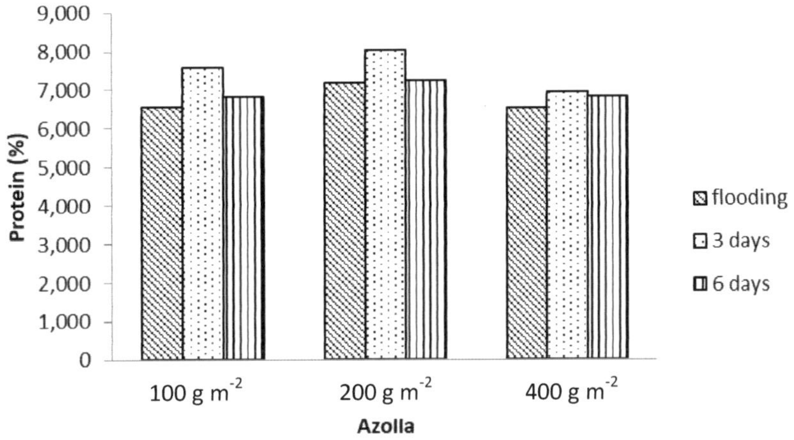

Figure 2. Protein content as the effect of azolla dan water supply

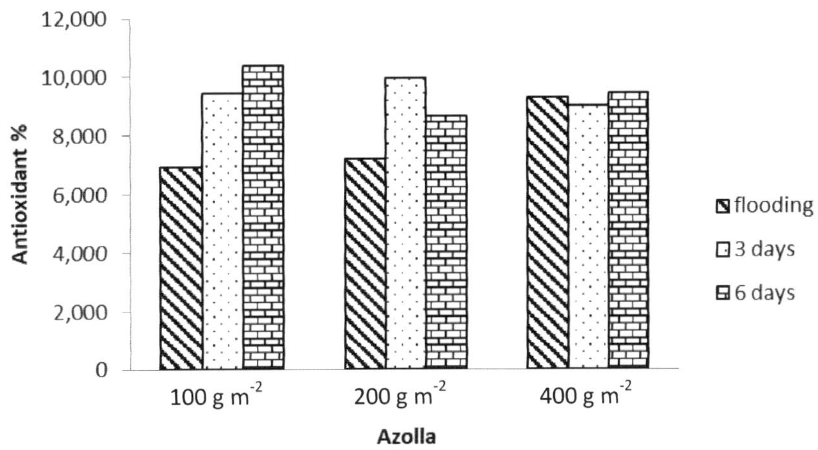

Figure 3. Antioxidant as the effect of azolla and water supply

Intermittent 3 to 6 days and azolla100 to 200 gm^{-2} can increase the percentage of protein and antioxidant. The greatest protein content exhibited by utilization of azolla 200 combined with 3 days intermittent water supply of 8.027%. This value was higher than those stated by Juliano [16] that protein content of rice was in the range of 7%. This record correlates with the vegetative growth. The protein content of milled rice was strongly influenced by degree of post-harvest processing and the soil conditions. Rice grown on soil rich in N elements will tend to have high levels of protein.

Treatment of 100 azolla combined with 6 days intermittent irrigation created the highest antioxidant. Lower nutrient and longer drying period lead to stress condition resulting more antioxidant. This revealed that antioxidant activity was consistent with levels of anthocyanin. The dark purple colour of black rice is due to the high anthocyanin content located in the pericarp layers. Anthocyanin contributed to the antioxidant activity [17]. Anthocyanin is a polyphenol compound that acts as an antioxidant [18] because it is very reactive towards reactive oxygen species (ROS) or reactive nitrogen species (RNS) [19,20]. The phenolic components on the anthocyanin pigment contributed to the antioxidant capacity of black rice [21].

4. Conclusion

Azolla and intermittent escalated the yield grain, anthocyanin, protein and antioxidant. Azolla 200 gm^{-2} combined with intermittent irrigation 3 days could be a good combination to improve plant growth, yield and properties of local black rice.

References

[1] Cheng W, Sakai H, Matsushima M, Yagi K and Hasegawa T 2010 Response of the floating aquatic fern Azolla filiculoides to elevated CO_2, temperature, and phosphorus levels Hydrobiologia **656** 5–14

[2] Jumadi O, Hiola SF, Hala Y, Norton J, and Inubushi K 2014 J. Soil Sci. Plant Nutr. **60** 722-730

[3] Talley SN, Talley BJ, and Rains DW 1977 Nitrogen fixation by Azolla in rice fields. In *Genetic Engineering for Nitrogen Fixation,* Ed. Hollaender A (New York and London: Plenum Press) p 259–281

[4] Suzuki M, Kimura T, Yamagishi K, Shinmoto H, and Yamaki K 2004 Comparison of mineral contents in 8 cultivars of pigmented brown rice *Nippon Shokuhin Kagaku Kogaku Kaishi* **51** 424–427

[5] Lee JC, Kim JD, Hsieh FH, and Eun JB 2008 Production of black rice cake using ground black rice and medium-grain brown rice *Int. J. Food Sci. Technol.* **43** 1078–1082

[6] Choi SP, Kang MY, Koh HJ, Nam SH, and Friedman M 2007 Antiallergic activities of pigmented rice bran extracts in cell assays *J. Food Sci.* **72** 719-726

[7] Hasan M and Sato S 2007 Water saving for paddy cultivation under the system of rice intensification (SRI) in eastern Indonesia *J. Tanah dan Lingkungan* **9** 57-63

[8] Awodun MA 2008 Effect of azolla (Azolla species) on physiochemical properties of the soil *WJAS* **4** 157-160

[9] Bindhu KB 2013 Effect of azolla extract on growth performance of Pisum sativum *Int. Res. J. Biol. Sci.* **2** 88-90

[10] Regazzoni O, Sugito Y, and Suryanto A 2013 Intermittent Irrigation on Rice (*Oryza sativa* l.) Inpari-13 varieties in a System of Rice Intensification (SRI) *J. Produksi Tanaman* **1**

[11] Daneshmand F, Arvin MJ, and Kalantari KM 2010 Physiological responses to NaCl stress in three wild species of potato in vitro *Acta. Physiol. Plant.* **32** 91–101

[12] Suzuki C, Kunito T, Aono T, Liu CT, and Oyaizu H 2005 Microbial indices of soil fertility *J. Appl. Microbiol.* **98** 1062–1074

[13] Unger IM, Kennedy AC, Muzika RM 2009 Flooding effects on soil microbial communities *Appl. Soil. Ecol.* **42** 1–8

[14] Mishra A 2013 Intermittern irrigation enhances morphological and physiological effeiciency of rice plants *The Journal of National Agricultural and Food Centre* **58** 121-130

[15] Haque MdM, Biswas JC, Kim SY, and Kim PJ 2017 Intermitternt drainage in paddy soil: ecosystem carbon budget and global warming potential Paddy Water Environ. **15** 403-411
[16] Juliano BO 1972 A simplified assay for milled rice amylose measurement *J. Cereal Sci.* **16** 334-336
[17] Takashi I, Bing X, Yoichi Y, Masaharu N, and Tetsuya K 2001 Antioxidant activity of anthocyanin extract from purple black rice *J. Med. Food* **4** 211–218
[18] Miguel MG 2011 Anthocyanins: Antioxidant and/or anti-inflammatory activities *J. Appl. Pharm. Sci.* **1** 7-15
[19] Meltem T, Seref T, Ufuk D, and Mehmet O 2013 Effects of various pressing programs and yields on the antioxidant activity, antimicrobial activity, phenolic content and colour of pomegranate juices *Food Chem.* **138** 1810–1818
[20] Rice-Evans AC, Miller NJ, and Paganga G 1996 Structure-antioxidant activity relationships of flavonoids and phenolic acids *Free Radic. Biol. Med.* **20** 933–956
[21] Pitija K, Nakornriab M, Sriseadka T, Vanavichit A, and Wongpornchai S 2013 Anthocyanin content and antioxidant capacity in bran extracts of some Thai black rice varieties *Int. J. Food. Sci. Technol.* **48** 300–308

Stomata character and chlorophyll content of tomato in response to Zn application under drought condition

A T Sakya[1], E Sulistyaningsih[2], D Indradewa[2] and B H Purwanto[2]
[1]Agriculture Faculty, Universitas Sebelas Maret, Surakarta, Indonesia
[2]Agriculture Faculty, Universitas Gadjah Mada, Yogyakarta, Indonesia

E-mail:amaliatetrani@staff.uns.ac.id

Abstract. This experiment was performed in order to evaluate the effects of Zn application under drought condition on tomato, especially its chlorophyll content and stomata character. This experiment was arranged in factorial using randomized complete block design with three replications. The treatment consisted of the Zn application method, namely: soil and foliar, the Zn dosage, namely: 0, 40 and 60 mg $ZnSO_4$ kg^{-1} soil and two cultivars of tomato, namely: 'Tyrana' F_1 and 'Permata' F_1. The stress condition was induced by watering every 12 days of 3 weeks after transplanting until harvesting. The results showed that the soil with a Zn application under drought conditions increased the aperture stomata, chlorophyll b and chlorophyll a/b ratio. The response of stomata character, chlorophyll a and total chlorophyll in both cultivars was similar.

1. Introduction

Water shortages in tomato plants will affect a number of physiological and biochemical processes, including the ability of plants to obtain water and nutrients, which directly or indirectly affect the growth of plants and crops. Several studies of drought stress in tomato plants show that drought stress decreases yield, yield components and dry matter production significantly both insensitive and tolerant cultivars [1,2]. Drought stress on tomato growth fell by about 60% [3].

Various cultivation technologies are done to reduce the impact of drought stress; one of them is by using micronutrient Zink (Zn). Zinc possibly reduces the impact of drought because Zn is involved in membrane integrity and protects cells from damage caused by reactive oxygen species [4-6]. Other literature suggests that Zn involvement in stomata opens possibilities due to Zn as part of the carbonate anhydrase enzyme which is necessary to maintain the HCO_3^-content in guard cells [7]. In addition, Zn is one of the micronutrients involved in carbohydrate metabolism. Zn deficiency can lead to a reduction of net photosynthesis by 50-70%, depending on the plant species and the severity of the deficiency. Decrease in the efficiency of photosynthesis can be caused, among others, by the decrease in enzyme activity, chlorophyll conditions, stomata conditions that affect the availability of CO_2 in plants as well as environments such as temperature and water. Decreased efficiency of photosynthesis may be due, among others, to a decrease in enzyme activity of carbonic anhydrase and ribulose enzyme 1.5 bisphosphate carboxylase (RuBPC) [8].

However, there is no research report on the effect of Zn on tomato plants on drought condition. Therefore, this article will examine the application of Zn on tomato under drought conditions, especially its stomata character and chlorophyll content.

2. Methods

2.1. Place and Time
Tomato was grown in Research Field and observation of stomata and chlorophyll analysis was done in Plant Science Laboratory of Faculty of Agriculture, UGM Yogyakarta, Indonesia. Starting from March to August 2014.

2.2. Research design
The study was designed using 3 factors, namely: Zn dosages of 0, 40 and 60 mg kg $^{-1}$ soil given in the form of $ZnSO_4 7 H_2O$, cultivars, namely: 'Tyrana' F_1 and 'Permata' F_1 and the third factor is application method, namely: soil and foliar. The study was conducted in factorial based on a randomized complete block design. Drought conditions are made by watering every 12 days. Besides, control plants were planted that watered every 2 days. $ZnSO_4$ application through soil was done by broadcasting around the plants when the plants were 1 week after planting. Foliar application was done by spraying ZnSO4 solution with concentration of 1%. Fertilization in the form of manure and NPK was given as basal fertilizer.

Stomatal and chlorophyll measurements were taken when the plants were 6 weeks after transplanting. Stomata observation was done by smearing the leaves with nail polish after dry lift the layer by using insulation and then attached to the preparation. The measurement of chlorophyll content was done by measuring the chlorophyll content taken from the leaf which has been perfectly cleared on the 9-week-old plant after planting. The measurement of chlorophyll content was performed as described by Islam [10], one gram of leaf sample was cut into pieces and crushed in mortar and then added by 20 ml of 80% acetone. The solution was put aside for a while, then filtered with Whatman filter paper number 42. The filtrate was fed into cuvette until the boundary line and then its absorbance was measured by spectrophotometer at λ 645 and 663 nm. Calculation of chlorophyll content is determined by the formula:

Chlorophyll a = (12.7 x A663 - 2,69 x A645) x (20 ml / 1000 x 1 g)
Chlorophyll b = (22.9 x A645 - 4,68 x A663) x (20 ml / 1000 x 1 g)
Total chlorophyll content = (20.2 x A645 + 8.02 x A663) x (20 ml / 1000 x 1 g)

The data obtained were analyzed by Anova and continued with Duncan test. The analysis was done using SPSS.

3. Result and discussion

3.1. Stomata character
Stomatal observations include stomatal density and stomatal opening width. Stomatal density and stomatal opening width showed no interaction of 3 factors, however, there is interaction between the method and the dosage as well as the dosage and the cultivar.

Table 1 illustrates the difference in stomatal opening widths due to $ZnSO_4$ dosage applications depending on the application method. The width of stomatal openings under drought conditions is narrower than the width of stomatal openings in plants with sufficient water. The decrease in stomatal opening width under drought conditions is due to the reduced moisture content in the plant which will affect the turgor of guard cell that controls the opening and closing of the stomata. Stomatal closure is an early response to drought-induced plants [10] The application of $ZnSO_4$ through the soil under drought conditions widened the opening of stomata, while the application through the leaves did not affect the width of the stomatal opening. As $ZnSO_4$ increases, Zn in the crop will increase. This is also due to the application of $ZnSO_4$ which increase the surface area and root length [11] so that it will increase water uptake and increase the cell turgor. Some researchers have reported that Zn application influences root length on barley [12], as well as Malik [13] stated that 200 ppm Zn increase root length of Amaranthus. Increased Zn in plants is thought to be able to reduce damage to stomata, because one of the functions of Zn is to maintain the integrity of the membrane [4]. However, too high Zn concentrations will also damage stomata morphology and narrow stomata openings [14]

Table 1. Stomatal aperture (μm) of tomato plants at watering intervals 2 and 12 days with soil and foliar $ZnSO_4$ application

Method application	Watering interval (day) + $ZnSO_4$ (mg Zn kg^{-1} soil)				Mean
	2 (normal)	12 (drought)			
	- $ZnSO_4$	0	40	60	
Soil	23.82 a-b	9.51 d	12.66 c	12.36 c	14.59
Foliar	23.87 a	10.22 d	6.98 e-f	6.11 e-f	11.79
Mean	23.85	9.89	9.82	9.23	(+)

Description: CV = 7.40%. Values in the same row or column followed by the same letter are not different on the Duncan 5% test. Sign (+): there is an interaction.

Under normal watering (every 2 days) without $ZnSO_4$ application, 'Tyrana' F_1 has a significant greater stomatal opening width than the 'Permata' F1. The application of $ZnSO_4$ under drought conditions increased stomatal opening width in both 'Permata' F_1 and 'Tyrana' F_1, but the stomatal opening width of the $ZnSO_4$ application showed no significant difference with stomatal opening width without the $ZnSO_4$ application. Both cultivars showed no significant difference in stomatal openings (Table 2).

Table 2. The stomatal aperture and stomatal density of two tomato plants at the watering interval of 2 and 12 days with the application of $ZnSO_4$.

Cultivar	Watering interval (day) + $ZnSO_4$ (mg Zn kg^{-1} soil)				Mean
	2 (normal)	12 (drought)			
	- $ZnSO_4$	0	40	60	
Aperture stomata (μm)					
Permata F_1	22.45 b	9.62 c-e	10.22 c-e	10.59 c-d	13.22 a
Tyrana F_1	25.24 a	10.11 c-e	11.42 c	10.88 c-d	14.41 b
Mean	23.85	9.86	10.82	10.74	(+)
Density stomata (per μm^2)					
Permata F_1	31.5 b	24.5 d-e	22.0 d-f	24.8 d-e	25.71
Tyrana F_1	47.8 a	28.3 b-c	29.0 b-c	23.8 d-e	32.25
Mean	39.67	26.42	25.50	24.33	(+)

Description: CV aperture stomata = 7.40%. CV stomatal density = 18.18%. Values on the same row followed by the same letter are not different on the Duncan 5% test. (+): There is interaction.

The stomata density response of the two cultivars to the application of $ZnSO_4$ under drought conditions is different. In the 'Permata' F_1, stomata density on application of 40 mg Zn kg^{-1} soil is less than stomatal density in 60 mg Zn kg^{-1} and without $ZnSO_4$ application, although it shows no significant difference in density. Different responses are shown by the 'Tyrana' F_1, in the application of 40 mg Zn kg^{-1} soils have a large density, although it is not significantly different from the density in plants without the application of $ZnSO_4$. The application of 60 mg Zn kg^{-1} soil on 'Tyrana' F_1 decreased stomatal density significantly compared with no application or with application of 40 mg Zn kg^{-1} (Table 2). The difference in response may be due to the genetic Zn efficiency from the two cultivars are different.

3.2. Chlorophyll Content

Chlorophyll measurements include chlorophyll a, chlorophyll b, total chlorophyll and chlorophyll ratio. There is interaction between dosage and $ZnSO_4$ application method to chlorophyll b and chlorophyll ratio, while chlorophyll a is influenced by Zn dosage and total chlorophyll method is only influenced by the Zn application method.

Chlorophyll a content under drought conditions either with $ZnSO_4$ application or not is lower than the chlorophyll content in plants with normal watering (Table 3). The content of chlorophyll a in $ZnSO_4$ applications with a dosage of 40 and 60 mg kg^{-1} under drought conditions was not significantly different from that without the application. The content of chlorophyll a on the foliar application was less than chlorophyll a in the soil application of $ZnSO_4$. Similarly, the content of chlorophyll a in both cultivars was not significantly different.

Table 3. Chlorophyll a and total chlorophyll content of two tomato cultivars at watering intervals 2 and 12 days with $ZnSO_4$ foliar and soil application

Treatment		Chlorophyll a (mg g^{-1} fresh weight)	Chlorophyll total (mg g^{-1} fresh weight)
Watering interval (day) + $ZnSO_4$ (mg Zn kg^{-1} soil)			
2 (normal)	0	0.258 a	0.468 k
12 (drought)	0	0.195 b	0.433 k
	40	0.193 b	0.438 k
	60	0.195 b	0.455 k
Application method $ZnSO_4$			
Soil		0.232 a	0.489 k
Foliar		0.186 b	0.397 l
Cultivar			
Permata F_1		0.209 a	0.449 k
Tyrana F_1		0.211 a	0.448 k
Interaction		(-)	(-)
CV (%)		19.39	18.28

Description: Values on the same column followed by the same letter are not different on the Duncan 5% test. (-): There is no interaction.

Table 4 shows the different responses of chlorophyll content b to the increase in $ZnSO_4$ influenced by the $ZnSO_4$ application method. In soil applications, chlorophyll b content in drought condition plants was more than chlorophyll b in plants with sufficient water. The application of $ZnSO_4$ on drought conditions decreased the chlorophyll b content compared to without $ZnSO_4$ application and the chlorophyll content b in the application with $ZnSO_4$ is not significantly different from the chlorophyll content of b in the plant by watering enough water. In the foliar application through leaves, the chlorophyll b content in drought plants without the application of $ZnSO_4$ is less although is not significantly different, whereas in plants with $ZnSO_4$ applications had more chlorophyll content than chlorophyll in plants with normal watering.

There is also interaction between dosage and method to chlorophyll ratio (Table 4). The chlorophyll ratio under drought conditions is generally lower than in normal watering. The application of $ZnSO_4$ through the soil increased the chlorophyll ratio compared with no application of $ZnSO_4$ under drought conditions, even the chlorophyll ratio in a dosage application with 60 mg Zn kg^{-1} soil equaled the chlorophyll ratio at normal watering without the application of $ZnSO_4$. Different things are seen in the application of $ZnSO_4$ through the leaves, which decreases the chlorophyll ratio than without the application of $ZnSO_4$ in both drought and normal watering conditions. Wang also reported that Zn did not affect chlorophyll b and total chlorophyll contents under drought-stressed leaves, instead it caused an increase of chlorophyll b and total chlorophyll contents in well-watered leaves. Chlorophyll a contents in leaves did not change and chlorophyll a/b ratio was increased by Zn treatment [15].

Although the method and dosage affect the chlorophyll ratio, the total chlorophyll is only affected by the $ZnSO_4$ application method. Table 4 shows the total chlorophyll content of the $ZnSO_4$ application through the leaves was less than in the soil. It is thought that foliar application resulted in Zn toxicity affected biosynthesis and chlorophyll degradation resulting in a decrease in chlorophyll content. Excess Zn in the plant results in a decrease in chlorophyll was also reported by Sagardoy [16] in sugar beet grown hydroponically. Table 4 also shows that both cultivars had the same total chlorophyll content, and the total chlorophyll content in both drought conditions without $ZnSO_4$ application and with $ZnSO_4$ application is no different from the total chlorophyll content at normal watering and the total chlorophyll content shows only difference in the $ZnSO_4$ application method.

Table 4. Chlorophyll b content and ratio chlorophyll of tomato plant at watering interval 2 and 12 days with soil and foliar $ZnSO_4$ application

Application method	Watering interval (day) + $ZnSO_4$ (mg Zn kg^{-1} soil)				Mean
	2 (normal)	12 (drought)			
	- $ZnSO_4$	0	40	60	
	Chlorophyll b content (mg g^{-1} fresh weight)				
Soil	0.193 b-e	0.308 a	0.254 a-b	0.272 b-d	0.507
Foliar	0.227 a	0.170 b-f	0.237 b-c	0.247 a-b	0.469
Mean	0.210	0.239	0.246	0.260	(+)
	Chlorophyll a/b ratio				
Soil	1.481 k	0.675 m-o	0.819 m	0.905 k-l	0.985
Foliar	1.104 k-l	1.106 k-l	0.751 m-n	0.591 n-p	0.942
Mean	1.292	0.891	0.785	0.749	(+)

Description: CV chlorophyll b content = 20,31 %; CV chlorophyll ratio = 20,46 %. Values in the same row or column followed by the same letter are not different on the Duncan 5% test

4. Conclusion

The stomata opening width under drought conditions increases by increasing $ZnSO_4$ application under drought condition, but still smaller than stomata opening width under no drought stress condition. Chlorophyll a and chlorophyll total content are higher in $ZnSO_4$ soil application than foliar application under drought stress. $ZnSO_4$ application method on drought conditions gives a different pattern on chlorophyll b and the chlorophyll ratio along with increasing of $ZnSO_4$ dosage.

References

[1] Castrillo M, Fernandez D, Calcagno A, Trujillo I, and Guenni I 2001 *Photosynthetica* **39** 221-6
[2] Rahman S M L, Mackay W, Quebedeaux B., Nawata E, Sakuratani T and Uddin A S M M 2002. *Subtropical Plant Science* **54** 16-22
[3] Sakya A T, Sulistyaningsih E, Indradewa D, Purwanto B H 2013 *Proc. of International Conference on Sustainable Agriculture and Environment*. June 26th-29th Solo Indonesia 2nd Series ed P Setiyono et al (Indonesia: Sebelas Maret University Press) p 162
[4] Cakmak I 2000 *New Phytologist* **146** 185–205
[5] Fageria N K 2009 *The Use of Nutrients in Crop Plants* (Boca Raton London: CRC Press) p 448
[6] Marschner H 1995 *Mineral nutrition of Higher Plants*. 2nd Edition. (London, U K: Academic Press) p 889
[7] Sharma P, Tripathi A and Bisht S S 1995 *Plant Physiology* **107** 751–6
[8] Alloway B J 2008 *Zinc in Soils and Crop Nutrition* (Brussels: IZA and IFA) p 137
[9] Islam M S, Hasanuzzaman M, Rokonuzzaman M, Nahar K 2009 *Inter. J of Plant Production* **3(1)** 51-61
[10] Medrano H, Escalona J M, Bota J, Gulias J and Flexas J 2002 *Annals of Botany* **89(7)** 895-905
[11] Sakya A T, Sulistyaningsih E, Indradewa D, Purwanto B H 2016, *STJSSA* **13(2)** 74-80
[12] Genc Y, Huang C Y and Langridge P 2007, *J Exp Botany* **58(11)** 2775-84

[13] Malik N J, Chamon A S, Mondol M N, Elahi S F, Faiz S M A 2011. *J Bangladesh Association of Young Researchers* **1(1)** 79-91
[14] Sagardoy R, Vázquez S, Florez-Sarasa I D, Albacete A, Ribas-Carbó M, Flexas J, Abadíaand J and Morales F 2010 *New Phytologist* **187(1)** 145-158
[15] Wang H, Liu R and Jin J Y 2009 *Biologia Plantarum* **53(1)** 191-4
[16] Sagardoy R, Morales F, Lopez-Millian A F, Abadia A and Abadia J 2009. *Plant Biology* **11** 339-350

IOP Conf. Series: Earth and Environmental Science **142** (2018) 012034 doi:10.1088/1755-1315/142/1/012034

The potential of turmeric (*Curcuma xanthorrhiza*) in agroforestry system based on silk tree (*Albizia chinensis*)

D Purnomo[1], M S Budiastuti[1], A T Sakya[1] and M I Cholid[2]
[1] Study Program of Agrotechnology, Faculty of Agriculture, Universitas Sebelas Maret, Jl. Ir. Sutami 36A, Surakarta, Central of Java, Indonesia 57126
[2]Study Program of Agrotechnology, Faculty of Agriculture, Universitas Sebelas Maret, Jl. Ir. Sutami 36A, Surakarta, Central of Java, Indonesia 57126

Email: djpuruns@gmail.com

Abstract. Turmeric (*Curcuma xanthorrhiza* Roxb.) is a traditional medicinal plant. In Indonesia, it is generally cultivated in village home gardens. Famers conducted very simpple cultivation method of turmeric, without specific maintenance and below varies tree. The experiment was conducted by cultivating turmeric below silk trees as in agroforetry system. The experiment was arranged split plot design, the first factor was three level of irradiation (turmeric monoculture/full irradiation, turmeric below silktree with pruning canopy, and turmeric below silk tree no pruning). The second factor was fertilizer NPK 15-15-15 with three levels of doses (100, 150, and 200 kg ha^{-1}). Cultivating turmeric in agroforestry system based on silk tree which were one year old and not yet needed pruning, application of NPK 15-15-15 fertilizer 100 kg ha^{-1} was enough. The rhizome yield of turmeric 3 months age reaches 139 g per plant (fresh weight). Litter fall from a silk tree one year old in one year is 30 kg per tree per year.

1. Introduction

Traditionally the yard in rural Indonesia, turmeric (*Curcuma xanthorrhiza* Roxb.) is grown simply to produce rhizomes commonly used as a spice vegetable and traditional medicine. Turmeric contains protein, fat, and carbohydrates, respectively 6.3, 5.1, and 69.4%, as well as minerals and water of 3.5 and 13.1% [1]. The use of the ginger rhizome as a traditional medicine other than in Indonesia, also in China and India [2]. Until now the compounds on the stems and rhizome of tumeric has been identified as many as 235 kinds, most of them are phenolics and terpenoids. Curcuminoid or curcumin (diarylheptanoid) (especially accumulates in the rhizome) and essential oils (essential oils) are major bioactive ingredients. The essential oils of leaves and flowers are generally dominated by monoterpenes while the root and rhizomes are dominant sesquiterpenes [3]. Positive correlations occur between fenolate contents and total antioxidant capacity in the turmeric species *A. chinensis* [4]. In Indonesia the original *C. colon* from Bogor (West Java) contains curcumin 2315 - 3059 mg g^{-100} [2].

The content of these compounds make turmeric useful to enhance the immune system, as an anti: inflammatory, oxidant, carcinogenic, mutagenic, coagulants, fertility, diabetic, bacterial, fungal, protozoa, viruses, fibrotic, venom, ulcers, and hypotensive and hipokholesteremik [1]. Turmeric then become raw material of beverages industry, cosmetics, and natural dyes [5, 6]. The need of turmeric for industry is quite big [7]. Turmeric including the top five among the 31 species of medicinal plants for industry, medicine and herbs, even the two highest ranks as a base material for 'jamu gendong' (carryied herbal medicine).

The areas of turmeric development in Indonesia are 13 provinces in North Sumatera, Java, Bali, Kalimantan and South Sulawesi Island. The increase of national turmeric production rate in 2010 to 2014 was 11% per year, and in 2015, the production of tumeric was 27,840 tons from the harvest area of 14,803 ha (1.88 tons ha^{-1}) [8], 2019 production is expected to reach 47,795 tons. The productivity is very low when compared with the turmeric in Bangladesh that can reach 16.77 tons ha^{-1} [9]. The increase of turmeric productivity in Indonesia is very potential because in a production center, turmeric reaches 17,3 ton ha^{-1} with curcumin 0,5-1,7%, it is possible to be increased to 25 ton ha^{-1} with curcumin level of 2-4% [10].

The cause of productivity of turmeric in Indonesia is low: turmeric is not at all a priority crop [11] so that the level of cultivation is still simple (not intensive). Most farmers grow tumeric under trees such as mango, 'rambutan', jackfruit, 'durian', and bananas in the yard. In relation to the needs of Indonesian special plant extracts as a drug ingredient, especially turmeric and temulawak [12], the potential for production to meet the needs should be attempted. Increased tumeric production through intensification is potentials to be done, but an extensive increase in conventional farmland cause in competition with staple plant. One of the opportunities of planting location is in forest land or under yard or stands in agroforestry system. Silk tree (*Paraserianthes falcataria* L. or *Albizia chinensis*) is currently a potentially cultivable tree, both personal (farmer) and institutional such as forestry or plantation.

Cultivation of turmeric in silk tree based agroforestry system has the opportunity to practice environmentally and healthy cultivation (Good and Health Agriculture Practices). Silk tree includes legumes, thus the soil is healthy as it provides natural N nutrients. The silk tree grows relatively quickly, the short harvest time, the small leaf canopy thereby passing the light to the underlying plants, the leaves easily fall out [13] containing high N elements (low C / N) so easily decompose then maintain the physical properties of the soil The condition is very suitable for medicinal plants such as tumeric. [7] stated that based on technical efficiency and high criterion economics, the turmeric cultivation can be done by low energy cultivation (Low External Input Sustainable Agriculture) using NPK fertilization of 200 kg ha^{-1} (urea) and 100 kg ha^{-1} (SP-36 and KCl).

This study attempted to reveal: the light escaped from the canopy of silk tree within 3x3 m, the increase of light escaped from the canopy of the tree when pruned by a third of the lower canopy, the growth of tumeric under the conditions of light, and the contribution of tree litter (especially leaf) to nutrient needs for the turmeric, especially nitrogen. The results of this study are expected to contribute to the development of agricultural science and organic technology cultivation of medicinal plants in the silk tree based agroforestry system.

2. Material and Methods

The research was a field trial on individual silk tree forest land in Bakalan Village, Karanganyar, Central Java (geographical position 07° 41 '42.1 "LS and 110° 58' 46.7" BT and 400 m elevation above sea level), from November 2016 to March 2017. Turmeric is planted between silk trees (10 months aged in agroforestry system, plant spacing of 3 x 3 m), plant spacing tumiric is 50 x 50 cm, the tip of tumeric is 75 cm from silk tree, the area of each plot Unit experiments 150 x 150 cm. The experimental design was split plot based Randomized Complete Block Design (between block 3 m).

The first factor is the intensity of light (full intensity, the light under the canopy of silk tress is not pruned, and 1/3 of the lower part head is prunned). While the second factor is NPK compound fertilizer 15-15-15 (dose 100, 150, and 200 kg ha^{-1}), so there were 9 treatment combinations, each repeated 3 times resulting in 27 experimental units (13.5 x 40.5 m in size). The data of the observation were analyzed using variance (F0,05 test), if different was followed by Duncan test 0,05. Total N content (Kjedhal method), P (Olsen method) and K available (NH4O Ac 1 N pH7 extraction method), organic material (Walkley & Black method), pH (Walkley & Black type pH meter) and C-Organic (Walkley & Black method), as well as C/N (Walkley & Black method) were analyzed before planting.

The tillage was done two weeks before planting using hoe as deep as layers (20 cm) then added the basic fertilizer (manure) 20 ton ha^{-1}. The amount of litter silk trees observed within 3 months about polyphenol content, lignin, and selullose, as well as quantity.

Planting material was the rhizome of tumiric cultivation in Central Research and Development of Medicinal Plants and Traditional Medicines (B2P2TOOT Tawangmangu). The rhizomes are first planted in polybag with a mixture of soil and manure (3:1, weight/weight). Small rhizomes (aged one day) is planted in the field by burying it into the planting hole, the rhizome is covered with soil, manually. Maintenance includes: irrigation using a bucket every 1 week based on soil conditions until the soil looks wet under the field capacity. Weed management was done manually every time weeds grow, various weeds include: Putri Malu (*Mimosa pudica*), Teki-tekian (*Cyperus rotundus*), and Bandotan (*Ageratum conyzoides*). Fertilizer was given according to treatment in three stages that is when the plants are 1 week, 1 month and 2 months, each 1/3 dose. Fertilizer was immersed into the hole around the plant (4 holes were made).

The growth observation consisted of the plant height (measured from the base of the stem to the highest leaf shoot), the number of leaves, the number of shoots, and the leaf area (Gravimetri method). Leaf N analysis using Kjeldahl method on third leaf from above, as well as chlorophyll analysis using destruction and spectrophotometer method, as a solvent is 95% diethyl ether (colorimetric method). Harvesting is done when the plants were 12 weeks (3 months) by removing and taking the rhizomes, rhizomes were then washed and dried (weighed to produce fresh rhizome weight). The rhizomes were then subsequently cut into and put in the oven at 80 °C for 24 hours resulting in dry weights of the rhizomes. The observation of silk tress litter using a rectangular plot of 1x1 m was placed under a tree canopy, for 3 months (November-January), then weighed.

3. Results and discussion

3.1. Silk Tree, Micro Climate, and Soil

High silk tree at the age of 1 year 2 months about 4.5 m, canopy width ± 1.5 m, so tree with a distance of 3x3 m is still have a free shade space. The light escapes from the canopy of 27%, 8440 lux from 31200 lux (light in the open space), pruning tree canopy of the lower third of the canopy increases the light passes to 47% (14550 lux). The intensity of light is related to temperature and humidity. Temperatures under the canopy were not pruned during the morning, day and afternoon on average by 30.1°C, and increased to 32.5°C on the pruned canopies. Silk tree produced litter (leaves and twigs) 6.1 kg per tree, for 3 months (rainy season), so it was assumed to produce litter of about 30 kg per tree each year (in dry season litter is assumed to increase by one third).

The litter content of N-total, C-organic, and organic matter were 3.73, 33.92 and 57.67%, respectively, so that the potency of litter as source of N, C-organic, and organic matter was 113,76, 1034.56, and 1759.93 g per tree per year (1,264.6, 11,495.11, and 19,543.7 g per ha). Contribution of the year-old silk tree to soil fertility improvement was not yet apparent, soil analysis at the beginning of the experiments of N, P, and K content were 0.15% (low), 14.39 ppm (medium) and 0.16% (low). Some of the possible reasons why it happens are: the initial growth of silk tree was still fresh, nutrients are used by weed, or eroded.

3.2. Growth of Turmeric

3.2.1. *Plant height, number of leaves, and number of tillers* There was an interaction between light intensity and fertilizer dose at plant height responses of turmeric at 12 weeks (3 months). At full light, plant height of turmeric was highest in NPK 15-15-15 fertilization with doses of 100 kg ha^{-1}, however at lower light intensity (silk tree pruned 1/3 of the lower canopy) has a higher plant and the highest plant was in application 150 kg ha^{-1} fertilizer. Turmeric under the canopy of silk tree was not pruned, the plant was higher (reach 145.3 cm) also on the crop fertilized 150 kg ha^{-1}. Plants contain auxin, a plant regulators, on the tip (root and shoot) promote the enlargement and elongation of the cell.

Auksin is more active when light intensity is lower [14,15] making plants taller. At high light, high temperatures become high. When light is high, soil temperature becomes high resulting in increased activity of soil microorganisms. Such activity requires nutrient especially N so competition between plants with soil microorganism occurs. The competition is reduced when the soil temperature is low and the light areas are lower, so the plants are higher at higher fertilizer doses.

The number of leaves ranged from 9.2 to 10.57 strands, was not significantly different between light variation and fertilization. Plant height is related to the number of leaves as the leaves grow from the node stems. The internodes of the stem become wider if the plant is etiolated so that the number of nodes is lower. The number of leaves is almost the same both in higher plants (due to low light intensity) or lower plants, means the plants not yet experiencing etiolation so there has been no growth disorder.

Turmeiric planted from tubers that have more than one bud so that form clump. The bud grows first as the parent plant while the other shoots as saplings. Leaves growing on the parent plant and also from the the saplings. The number of saplings on all three light treatment was three, meaning that the disorders growth because of low light intensity has not occurred

3.2.2. Leaf area index, biomass and tuber weight Leaves as light harvesters play the main role of photosynthetic organ. One of the leaf characters reflecting the potential of light absorption is the area of the leaf expressed in the leaf area index (LAI: the area of a leaf canopy of a plant per unit ground area where the plant grows). Turmeric under silk tree stands having higher LAI (1.09-1.36 than 0.99) indicates that the leaf was wider when the plant receives a lower light intensity. Plants are grown with wider distances, LAI is also lower [16]. Leaf area index of about one means intercepting light about 50% [17], it happens because the turmeric is cultivated under silk tree stands.

The process of photosynthesis by the leaves produces photosyntate, most of which accumulates in the vegetative part expressed in plant biomass. The biomass is then divided into: as deposits, metabolism, and translocation to the sink [18] The turmeric biomass at the age of 3 months reaches 59.19 (under full light) and about 70 g per plant under silk tree stands. This means that the turmeric was subjected to photosynthesis when the light is full. Turmeric is a C3 plant, when full light can decrease in photosynthesis rate due to photorespiration process.

Photosyntate in biomass then undergoes remobilization to the bottom, in this case the rhizomes, then used by humans (harvest). Quantitatively the harvest in the form of rhizomes is expressed in the rhizomes weight. The weight of the fresh rhizome (at harvest, aged of three months) about 139 g (under stand) and 129 g (full light). Low rhizome weight due to young harvest, turmeric is generally harvested at the age of 12 months and the rhizome reaches 380 g per plant [10]. The yield of rhizomes under stands higher than under full light indicates that photorespiration does occur means that turmeric plant is an adaptive plant to low light intensity.

4. Conclusions

The following conclusions were obtained from the research results:

1. Turmeric is cultivated in agroforestry system based on one year-old silk tree grow well, biomass and yield higher than full light (139 compared to 129 g of wet rhizome per plant).
2. Silk tree age of year as base agroforestry system does not need pruning.
3. Pruning a third of the lower canopy can be done as nutrient substitution especially N.
4. Silk tree produces litter as much as 30 kg per tree per year.

References

[1] Chattopadhyay I, Biswas K, Bandyopadhyay U and Banerjee R K 2004 Turmeric and Curcumin: Biological Actions and Medicinal Applications *Current Sci.* **87** (1): 44-53.

[2] Hayakawa H, Minaniya Y, Ito K, Yamamoto Y, and Fukuda T 2011 Different of Curcumin Content in *Curcuma Longa* L. (Zingiberaceae) Caused by Hybridization with Other *Curcuma* Species *American J. of Plant Sci.* **2**: 111-119.

[3] Li S, W Yuan, G Deng, P Wang, P Yang and B B Aggarwal 2011 Chemical Composition and Product Quality Control of Turmeric (*Curcuma longa* L.) *Pharm. Crops.* **2**: 28-54.
[4] Sharmin T, F Islam, M A Kaisar, Md A Al-Mansur, Md A A Sikder and M A Rashid 2014 Chemical and Biological Investigations of *Albizia chinensis* (Osbeck.) Merr. *J. of Physical Sci.* **25**(2): 29–38.
[5] Nurjanah N, S Yuliani dan A B Sembiring 1994 Temulawak (*Curcuma xanthorrhiza*). Review Hasil-Hasil Penelitian. *Balai Penelitian Tanaman Rempah dan Obat.* **10** (2): 43-57.
[6] Ferry Y, Bambang E T dan Enny R 2009 Pengaruh Intensitas Cahaya Dan Umur Panen Terhadap Pertumbuhan, Produksi, dan Kualitas Hasil Temulawak Diantara Tanaman Kelapa. Balai Penelitian Tanaman Rempah Dan Aneka Tanaman Industri. *Bul. Littro* **20** (2): 131 - 140
[7] Pribadi E R dan Rahardjo M 2007 Kajian ekonomi budidaya organik dan konvensional pada 3 nomor harapan temulawak (*Curcuma xanthorrhiza* Roxb.). *Buletin Penelitian Tanaman Rempah dan Obat.* **18**: 73-85
[8] Badan Pusat Statistika Indonesia (BPS) 2017 Produksi dan Luas Panen Temulawak Tahun 2016. http://www.bps.go.id/. Diakses 1 April 2017.
[9] Islam F, Karim M R, Shahjahan M, Hoque M O, Alam M R, and Hossain M A 2002 Study on the Effect of Plant Spacing on the Production of Turmieric at Farmer Field. *Asian J. Plant Sci.* **1**(6):616-617.
[10] Nihayati E 2016 *Peningkatan Produksi dan Kadar Kurkumin Temulawak* (Malang: UB Press)
[11] Kemala S, Sudiarto, Pribadi E R, Yuhono J T, Yusron M, Mauludi L, Rahardjo M, Y Ferry, Waskito B, dan Nurhayati H 2003 Studi Serapan Pasokan dan Pemanfaatan Tanaman Obat di Indonesia. Laporan Teknis Penelitian. *Balai Penelitian Tanaman Rempah dan Obat Bogor*. Bogor. pp. 143-241.
[12] Gosta D R 2016 Permintaan Ekstrak Kunyit dan Temulawak Paling Besar. *Bisnis Indonesia.* http://industri.bisnis.com/. Diakses 1 April 2017.
[13] Priyadi A 2015 Teknik Budidaya Sengon atau Albasia menggunakan Pupuk Organik NASA (Natural Nusantara). *Agro Bisnis. Langkah bisnis dot.com.* 8 hal. [14] Taiz L and Eduardo Z 1998 *Plant Physiology* 2nd ed Sinauer Associates Inc. Pub. USA.
[15] Taiz L, Eduardo Z, Ian Max M, and Angus M 2015 *Plant Physiology and Development* 6th ed. Sinauer Associates Inc. Pub. USA.
[16] Mattera L A, Romero L A, Cuatrin A L, Cornaglia P S, and Grimoldi A A. 2013. Yield Component, Light Interception, and Radiation Used Efficiency of Lucerne (*Medicago sativa* L.) in Respon to Row Spacing. *Europ. J. Agron.* **45**:87-95
[17] Sinclair T R and Gardner F P 1998 Invironment Limit to Plant Production in Principles of Ecology in Plant Production. CAB Int. 63-78
[18] Purnomo D, Sakya A T, dan Rahayu M 2010 *Fisiologi Tumbuhan, Dasar Ilmu Pertanian* (Surakarta: UNS Press)

IOP Conf. Series: Earth and Environmental Science **142** (2018) 012035 doi:10.1088/1755-1315/142/1/012035

Influence of spacing and depth of planting to growth and yield of arrowroot (*Marantha arundinacea*)

M Qodliyati, Supriyono and S Nyoto
Department of Agrotechnology, Faculty of Agriculture, Universitas Sebelas Maret, Surakarta, Indonesia

E-mail: supriyono_uns@yahoo.com

Abstract. This study was conducted to determine the optimum spacing and depth of planting to the growth and yield of arrowroot. This research was conducted at the Experimental Field of Agriculture Faculty, Sebelas Maret University on Jumantono, Karanganyar. This research was conducted using Randomized Completely Block Design (RCBD) with two treatment factors of plant spacing and depth of planting. Plant spacing consists of 3 levels, including J1 (30x30 cm), J2 (30x40 cm) and J3 (30x50 cm). Depth of planting consists of 2 levels which are K1 (10 cm) and K2 (20 cm). Data were analyzed by DMRT (Duncan's Multiple Range Test) at 5% significance level. The results showed that spacing of 30x50 cm have significantly higher plant height, tuber (common names of rhizome) length, and tuber weight per plant. The depth of 20 cm gives a higher yield on the number of tubers per plant and tuber weight per plot variables. Both treatments have no significant interaction on growth and yield.

1. Introduction

The Indonesian population is increasing, causing food demand to increase; meanwhile the availability of food is not enough to meet those needs. One way that can be done to overcome the shortage of food in Indonesia is by increasing the productivity of food crops. One of the crops that can be used as alternative food is arrowroot plant. Arrowroot plants are currenly mostly cultivated for its starch content [1][2]. Fresh arrowroot contains 69-72% water, 1.0-2.2% protein, 0.1% fat, 19.4-21.7% starch, 0.6-1.3% fiber, and organic ashes 1.3 -1.4%. Arrowroot starch can be used for wheat substitution up to 50-100% [3]. Therefore, arrowroot starch has a potential to decrease the flour imports that have red 4.10 million tons per year with a value of 3.40 trillion rupiah [4].

Arrowroot is a plant that grows by using rhizomes and usually planted at the beginning of the rainy season. This plant is often found in the countryside and planted in the yard [5]. Arrowroot has a rhizome under the ground that grows long and from the rhizome will grow new shoots [6]. Effort on arrowroot plant development need to be improved because arrowroot plant in Indonesia is only cultivated as an intercrop under the annual crops in the yard [7]. The majority of Indonesia's population still has not planted arrowroot plants as the main crops on a field and the cultivation management of the crops is still relatively low [8]. Increasing the amount of production can be done by continuously improving the cultivation [9]. One point to note in the cultivation of arrowroot is the spacing and the depth of planting. Arrowroot planting by farmers is done without considering the spacing and proper depth of planting so the planting form irregularly.

Plant spacing and depth of planting adjusment is important in the cultivation of arrowroot, because with proper planting spacing and proper depth of planting can increase the plant growth and development, also increase the yield of arrowroot [8]. This study was conducted to determine the spacing and the proper depth of planting for the growth and yield of arrowroot planted by 6 months age transplanting vegetative seed.

2. Methods

The research was conducted at Experimental field of Agriculture Faculty, Sebelas Maret University Surakarta in Jumantono, Karanganyar on March 2016 until September 2016. The materials used in this research are vegetative seeds of 6 months old arrowroot and organic fertilizer. The tools used in this study include hoes, plow hoe, crowbar, trowel, rulers, scissors, hole stick, watering can, plastic rope, bucket, scales, oven, calipers, and treatment boardsign.

This research uses a Randomized Complete Block Design (RCBD) factorial consisting of two treatment factors of plant spacing and depth of planting. The plant spacing factor consist of 3 treatment levels of plant spacing 30x30 cm (J1), 30x40 cm (J2), and 30x50 cm (J3), while the depth of planting consist of 2 treatment levels of 10 cm (K1) depth and 20 cm depth (K2), so there are 6 treatment combinations obtained. Treatment combination was repeated four times, resulting of 24 experimental units. The arrowroot plant is palnted in the treatment plots measuring 1.8 m x 3.2 m.

The variables observed were plant height, number of leaves, number of tillers, number of productive tillers, fresh weight, dry weight, number of tubers (commonly names of rhizome produce)per plant, tuber length, tuber diameter, tuber weight per plant, and tuber weight per plot. Data were analyzed using DMRT (Duncan's Multiple Range Test) with 5% level.

3. Results and Discussion

3.1General Condition of Research

This research was conducted at 170 meters altitude above sea level and located at 7° 37' 48.3" S and 110° 56' 51.2" E. The rainfall rate on March 2016 to October 2016 at 1517 mm with an average rainfall of 7.09 mm/month. The air temperature of the location is ranged 27-28°C with an average air humidity of 80-86%. The experimental field has a type of Alvisol soil. The Alvisol soil is usually reddish-brown to dark red with a soil structure from grain to pole and soil texture from sandy clay to clay. The content of organic matter on alvisol soils is generally low [10].

3.2 Plant Spacing and Depth of Planting

Result of research on plant spacing and depth of planting on the growth and yield of arrowroot (*Marantha arundinacea*) shown in Table 1 and Table 2. Plant spacing has significant effect on plant height, number of leaf, fresh weight, dry weight, length of arrowroot tuber weight per plant and tuber weight per plot. The depth of planting significantly affected the number of tubers per plant and tuber weight per plot. Both treatments no significant interaction on growth and yield.

Table 1. Influence spacing to growth and yield of arrowroot (*Marantha arundinacea*)

Variabel	Plant Spacing (cm)		
	30x30	30x40	30x50
Plant Height (cm)	50.00[a]	52.50[a]	62.63[b]
Number of Leaf (sheet)	12.63[a]	14.63[ab]	16.75[b]
Number of Tillers (tiller)	2.38	2.38	2.63
Number of Productive Tillers (tiller)	2.25	2.63	3.75
Fresh Weight (g)	67.50[a]	127.13[ab]	210.38[b]
Dry Weight (g)	13.25[a]	30.00[b]	41.75[b]
Numbet Of Tuber Per Plant (tuber)	5.38[a]	6.50	7.38
Tuber Length (cm)	16.63[a]	17.50[a]	20.25[b]
Tuber Diameter(cm)	1.88[a]	2.00	2.25
Tuber Weight Per Plant (g)	85.00[a]	121.00[a]	191.25[b]
Tuber Weight Per Plot (kg)	3.75[a]	4.63[b]	5.13 b

Description: Value in rows followed by the same letters are not significantly different in DMRT of 0.05 levels

Table 2. Influence depth of planting to growth and yield of arrowroot (*Marantha arundinacea*)

Variable	Depth of Planting	
	10 cm	20 cm
Plant Height (cm)	54.08	42.00
Number of Leaf (sheet)	14.33	11.25
Number of Tillers (tiller)	1.75	2.58
Number of Productive Tillers (tiller)	2.42	3.33
Fresh Weight (g)	101.33	168.67
Dry Weight (g)	21.83	34.83
Numbet of Tuber Per Plant (tuber)	5.33[a]	7.50[b]
Tuber Length (cm)	17.75	18.50
Tuber Diameter (cm)	2.08	2.00
Tuber Weight Per Plant (gr)	119.83	145.00
Tuber Weight Per Plot (kg)	3.06[a]	4.92[b]

Description: Value in rows followed by the same letters are not significantly different in ANOVA of 0.05 levels

3.3 Plant Height
Treatment of J1 and J2 was significantly different from treatment J3. The highest average of plant height is found in treatment J3 which is 62.63 cm. The wider planting space will increase the availability of nutrients and water for individual plants so that the growth of the plant increases [11]. Plant spacing adjustment which are too close will cause the availability of nutrients and water to be limited, causing the growth of plants to be inhibited [12][13].

3.4 The Number of Leaf
Treatment J1 was significantly different from treatment J3. The highest number of leaves is found in the treatment of J3 which is 16.75 and the least amount of leaf is found in treatment J1 which is 12.63. Plants with a wide spacing will receive more light than plants with a narrow spacing. Plants that receive more light tend to have more leaves than the less one [14]. This is because the more light received, the photosynthesis will goes well [15].

3.5 The Number of Tillers
Spacing and depth of planting did not significantly affect the number of tillers. This can be due to the ability of the plant to produce the seedlings, the age of the planting material and the unsuitable environmental conditions. Age of planting materials that are almost uniform cause the formation of tiller not much different. In the dry season, nutrients and water are scarce than the rainy season [16]. These conditions cause the plant is not able to produce the maximum tillers on both treatments. The occurance of shoots is also influenced by the temperature of the soil. Temperatures that are too high due to high amount of sunlight would cause the occurance time of buds to be longer, because those environmental conditions are not suitable for shoot growth [17].

3.6 The Number of Productive Tillers
Spacing and depth of planting also did not significantly affect the number of productive tillers of arrowroot. This is because the number of tillers that have formed are almost in the same number, resulting in the number of productive tillers that did not show a significant difference. The formation of productive tillers is closely related to the number of maximum tillers, the more the number of the maximum tillers, the more number of productive tillers tends to be [18].

3.7 Fresh Weight

Treatment J1 was significantly different from treatment J3. Treatment of J3 has the most fresh weight at 210.38 gram and J1 treatment has fresh weight of 67.50 gram. The wider use of plant spacing will result in higher fresh plant weight compared to the use of narrow spacing [19]. Tenous spacing provides an opportunity for plants to absorb more water so as to increase fresh weight per plant [20].

3.8 Dry Weight

Treatment J1 was significantly different from treatment of J2 and J3. The lowest result obtained from treatment J1 that is equal to 13.25 gram. Dry weight is a net result of photosynthesis activities thoroughly. Sunlight is a major factor in the photosynthesis process, the more sunlight it receives, the photosynthesis process can run well, resulting in more photosynthate made. Larger photosynthetic production will form larger plant organs which those affecting the increase of dry weight of plants [21].

3.9 The Number of Tuber per Plant

K2 treatment shows more response in the number of tuber per plant than K1 treatment. The treatment of K2 yielded the number of tuber per plant as much as 7.50 tubers, while treatment of K1 have 5.33 tubers. The depth of planting is closely related to the growing season. Planting in the dry season should be planted deeply in order that the roots can rper the deepest layer of soil. The deeper depth planting in potato plants produces more tubers than the shallow planting [22]. Competition causes a decrease in tuber yield [23].

3.10 Tuber Length

Treatment of J1 and J2 was significantly different from treatment J3. The longest tuber length found in treatment J3 of 20.25 cm. Wide spacing produces long tubers. This is because at a wide spacing there will be more empty space and no competition in the use of nutrients, water and light that affect photosynthesis. Tubers formed from the accumulation of photosynthate as form of long tubers ([23];[24];[25])

3.11 Tubers Diameter

Plant spacing and depth of planting have no significant effect on diameter of arrowroot. This can be influenced by environmental factors. Environmental factors here derived from water availability factor, nutrients, and soil conditions. In the dry season the availability of water and nutrients tends to be less when compared to the rainy season [26].The condition of the tubers produced has a size that is relatively uniform. Tuber yield and quality are influenced by texture and soil structure [27]. Alvisol soils at 0-20 cm deep have the same texture of clay. This soil conditions cause plants grown with a depth of 10 cm and 20 cm has a relatively same tuber diameter.

3.12 Tuber Weight per Plant

Treatment of J1 and J2 was significantly different from treatment J3. Treatment of J3 produces tuber weight per plant of 191.25 gram. Treatment J1 yields the smallest tuber weight of 85.00 grams. The tuber fresh weight per plant at a wide spacing is greater than the tuber fresh weight at a narrow plant spacing. This is because the nutrients, water, and sunlight can be received optimally and the competition between plants can be minimized, thereby enlarge the yield of plants in the form of arrowroot tubers.

3.13 Tuber Weight Per Plot

Plant spacing and depth of planting significantly affect the weight of arrowroot tubers per plot. Treatment J1 was significantly different from treatment of J2 and J3. The lowest tuber weight obtained from treatment J1 that is 3.75 kg. This is inconsistent with [28] statement, which says that the narrow spacing treatment resulting higher tuber yield per plot because at a narrow spacing the plant has a larger number of individual plants. This disaccordance can occur because the results of the study indicate a narrow spacing, although the number of plant populations are many but the tubers produced have the number and size is smaller than the wide spacing treatment. In addition, a narrow plant spacing will only increase tuber yield per plot to a certain spacing limit. The narrower spacing of the optimum spacing limit will decrease tuber yield per plot.

The K1 treatment was significantly different from the treatment of K2. The most weight of tubers obtained from treatment K2 is 4.92 kg. Depth of planting is a factor that can affect tuber yield [29. In the contrast, the depth of planting has significant effect on potato tuber weight. Potato planting with a depth of 15 cm in the dry season produces the largest tuber weight than the depth of 5 cm and 10 cm. This is because the plant is able to use water in the deepest layer of soil.

4. Conclusion

The plant spacing of 30x50 cm shows higher growth and yield response on plant height, length of arrowroot, and tuber weight per plant variable compared to planting distance treatment of 30x30 cm and 30x40 cm. The 20 cm planting depth brought the higher yield on variables of tubers per plant and tuber weight per plot compared to 10 cm plant depth treatment. Plant spacing and depth of planting have no interaction in all of observated variables.

References

[1] Handler J S 1971 The history of arrowroot and the origin of peasantries in the british West Indies *J. Carib*. Hist. **2**: 46-93.

[2] Harmayani E, Kumalasari I D, Marsono Y 2011 Effect of arrowroot (Maranta arundinacea L.) diet on the selected bacterial population and chemical properties of caecal digesta of sparague dawley rats *International Research Journal of Microbiology (IRJM)* **2**(8): 278-284. ISSN: 2141-5463.

[3] Djaafar T F, Sarjiman, Arlyna B P 2010 Pengembangan budidaya tanaman garut dan teknologi pengolahannya untuk mendukung ketahanan pangan. *J Litbang Pertanian* **29**(1): 25-33.

[4] Gusmaini, Sudiarto, Nurhayati H 2003 Pengaruh macam bahan tanam terhadap pertumbuhan produksi umbi-umbian dan pati garut. *J Ilmiah Pertanian* IX **1**(2): 13-21.

[5] Maspary 2010 Budidaya tanaman garut secara intensif. http://www.gerbang pertanian.com/2010/06/budidayatanaman-garut-secara-insentif.html.

[6] Stephens J M. 1994. *Arrowroot-Maranta arundinacea L.*. Ifas Extention. University of Florida.

[7] Capina M V, Capina V L L 2017 *Arrowroot (Maranta arundinacea): starch extraction, processing, and by-products utilization* 4th International Conference on Civil, Environment and Waste Management (CEWM-17) Manila (Philippines).

[8] Yudianto A A, Fajriani S, Aini N 2015 Pengaruh jarak tanam dan pembumbunan terhadap pertumbuhan dan hasil tanaman garut (*Marantha arundinaceae* L.*). J Produkti Tanaman* **3**(3): 172-181.

[9] Lestari G W, Solichatun, Sugiyarto 2008 Pertumbuhan, kandungan klorofil, dan laju respirasi tanaman garut (*Maranta arundinacea* L.) setelah pemberian asam giberelat (GA_3*). J Bioteknologi* **5**(1): 1-9.

[10] Wijanarko A, Sudaryono, Sutarno. 2007. Karakteristik sifat kimia dan fisika tanah alfisol di Jawa Timur dan Jawa Tengah. *J IPTEK Tanaman Pangan* **2**(2): 214-226.

[11] Mawazin, Suhaendi H. 2008. Pengaruh jarak tanam terhadap pertumbuhan diameter *Shorea parvifolia* Dyer *J Penelitian Hutan dan Konservasi Alam* **5**(4):381-388.

[12] Nurlaili 2010 Respon pertumbuhan tanaman jagung (*Zea mays* L.) dan gulma terhadap berbagai jarak tanam. *J Agronobis* **2**(4): 19-29 ISSN 1979-8245X.

[13] Sohel M A T, Siddique M A B, Asaduzzaman M *et al.* . 2009. Varietal performance of transplant rice under different hill densities. *J Agril Res* **34**(1): 33-39.

[14] Milthorpe F L, Mourby M T 1979 *An introduction to crop physiology*. 2ndEd. London: Canbridge UNW Press.

[15] Polnaya F, Patty J E 2012 Kajian pertumbuhan dan produksi varietas jagung lokal dan kacang hijau dalam sistem tumpang sari. *J Agrologia* **1**(1): 42-50.

[16] Suhariyono G, Menry Y 2005 *Analisis karakteristik unsur-unsur dalam tanah di berbagai lokasi dengan menggunakan XRF*. Prosiding PPI-PDIPTN 2005. Puslitbang Teknologi Maju-Batan Jogjakarta 12 Juli 2005. ISSN 0216-3128.

[17] Fanos T, Belew D, Nebiyu A 2015 Effect of planting depth and time of earthing-up on potato (*Solanum* tuberosum L.) yield and yield components at jimma university college of

agriculture and veterinary medicine, south west ethiopia. *International Journal Of African and Asian Studies***15**: 61-78 ISSN 2409-6938.

[18] Misran 2014 Efisiensi penggunaan jumlah bibit terhadap pertumbuhan dan produksi padi sawah *J Penelitian Pertanian Terapan* **14**(1):39-43 ISSN 1410-5020.

[19] Abadi I J, Sebayang H T, Widaryanto E. 2013. Pengaruh jarak tanam dan teknik pengendalian gulma pada pertumbuhan dan hasil ubi jalar (*Ipomoea batatas* L.). *J Produksi Tanaman* **1**(2): 8-16 ISSN 2338-3976.

[20] Darma W A, Susila A D, Dinarti D 2013 Pertumbuhan dan hasil bawang merah asal umbi varietas tuk tuk pada ukuran dan jarak tanam yang berbeda. *J Agrivor* **8**(2): 1-7 ISSN 1979-5777

[21] Sawitri R 2013 Pengaruh jarak tanam ganda dan pangkas pucuk terhadap pertumbuhan dan hasil tanaman kedelai (*Glycine max*, L. Merr.) varietas wilis.http://jurnal.utm.ac.id/index.php/jp/artice/view/230.

[22] Pangaribuan D H 2000 *Pengaruh berbagai ukuran umbi bibit dan kedalaman tanam terhadap pertumbuhan dan hasil produksi kentang dari umbi bibit benih botani.* Prosiding seminar nasional III pengembangan wilayah lahan kering. Bandar Lampung, 3-4 Oktober 2000. P230-236.

[23] Arab H R, Afshari H, Daliri M S, Laei G, Toudar S R 2011 The effect of planting date, depth and density on yield and yield components of potato in Shahrood (Iran). *Journal of Research in Agricultural Science* **7**(2): 141-149.

[24] Mangani R, Mazarura U, Tuarira A M, Shayanowako A 2015 Growth, yield, and responses to plant spacing in irish potato: a review. *African Journal of Agricultural Research* **10**(7): 727-730. ISSN 1991-637X.

[25] Sasvita W, Hanum C, Purba E 2013 Pertumbuhan dan hasil tiga klon ubi jalar pada jarak tanam yang berbeda.*J Online Agroteknologi* **2**(1): 462-473 ISSN: 2337-6597.

[26] Sutiyono 2007 Pengaruh jarak tanam terhadap pertumbuhan bambu tutul (*Bambusa maculata* Widjaja) umur 3 tahun. *Info Hutan* **4**(5): 477-486.

[27] Putra A A G 2010 Pengaruh jarak tanam dan dosis pupuk kandang ayam terhadap pertumbuhan dan hasil bawang merah (*Allium ascalonicum* L.) di lahan kering beriklim basah *J Ganec Swara* **4**(1): 22-24.

[28] Nugraheni T 2013 Respon tanaman bawang merah (*Allium* ascalonicum L.) varietas tuk tuk terhadap pengaturan jarak tanam dan konsentrasi pupuk organik cair nasa. *J Ziraa'ah* **36**(1): 60-65. ISSN: 1412-1468.

[29] Khaoula A, Sayed C, Bouthaina D, Kamel B M. 2012. Influence of mechanical planting depth on the agronomic behavior of a potato crop conduced on two soil types. *The American Journal of Plant Science and Bioteknology* **7**(1): 75-78.

IOP Conf. Series: Earth and Environmental Science **142** (2018) 012036 doi:10.1088/1755-1315/142/1/012036

The development of the food barn in achieving food security of poor household in Wonogiri Regency

Suminah
Communication and Extention Agriculture Department, Faculty of Agriculture, Universitas Sebelas Maret, Surakarta, Indonesia

Email: sum_anan@yahoo.com

Abstract. Food barn has function as food stock which managed by households both individually and collectively. It has important role in food security. This study aims to determine the factors that influence the development of food barns, and formulate alternative strategies in the development of food barns in Wonogiri regency. The samples of this research were determined by random sampling. The data collection technique was done by interview, field observation and focus group discussion (FGD). The data analysis was done quantitatively with multiple regression, and qualitative information. The results showed that the age of respondents included in the category of middle age, their level of education was mostly from elementary school, most of their occupation were farmers, they had narrow knowledge of land management, and their income was around 1-2 million rupiah/ month. In fact, the perception of food barns, institutional regulation of food barns, and the Economic institutions of food barn have a positive significant influence toward food barn development, filling the food barn and institutional capacity of food barn groups in realizing the food security of poor household in Wonogiri Regency. In food barn development, there are positive and clear perceptions of the food barn, institutional regulations. It indicates that significant as well as the economic institution of food barn are needed so that the food security can be realized.

1. Introduction

Food is a basic necessity and a strategic commodity in human life to maintain its healthy and productive survival. However, the reality shows that there are still many poor people who cannot meet their needs for food that they are experiencing shortage of consumption and food insecurity. The percentage of poor people in Wonogiri Regency is 12.19% [1]. The poor are people who have an average monthly per capita expenditure below the poverty line. Wonogiri poverty line is 245.129 rupiah per capita per month. These poor people (grassroot groups) are at high risk and vulnerable to food insecurity level so that in such situation, a food barn is needed. The food barn is an institution formed by rural and/or urban communities aimed at the development of provision of food reserves with a delay sale, storage, distribution, processing and trading system managed in groups [2]. The purpose of empowering the food barn is to increase the supply of food so that food security can be realized. The food barn is as a food reserve which is controlled by households individually or collectively. The food barn has an important role in: (1) anticipating the lack of food supply in the famine season and (2) anticipating the threat of crop failure due to natural disasters such as pest and disease attacks, climate anomalies and floods.

In addition, the problem of food security in the future becomes increasingly complex, both as the impact of the global food crisis, global economic crisis and the impact of global warming, so the food barn as a buffer institution of food storage becomes strategic to be developed in every area. indicates that the food barn can not be relied upon as an institution that is able to absorb marketable plus at harvest time. Moreover, it is expected to be the stability of food storage and help to secure grain prices from the fall. In Central Java, the whole capacity of food barns absorbs only 0.92 percent marketable plus [3]. Similarly, the role of Bulog is also not fully capable in the formation of national food security.

Actually, food barn has been developed in some areas in Central Java, including in Wonogiri Regency but its development still faces many problems, including; (1) lack of support from the local government in the form of allocation of activities, (3) lack of roles of village government in developing the food barns for food security of poor households and (4) lack of commitment of policy makers in food barns development. Therefore, the potential of this food barn needs to be revitalized through a systematic, integrated and sustainable empowerment process involving all related elements.

In order to increase family food security, some efforts are made through strengthening the food storage of society in the form of institutional food barns. Thus, the institution of food barns will be a supporting element of food security that needs to be developed in order to be able to give more significant contribution to realize the welfare of society. Related to the condition above, the food barn should be developed through empowerment process in a systematic, integrated and sustainable by involving all related elements. In the development of the food barn, firstly the community needs to know what factors affect the development of food barn, so that it can formulate alternative strategy in the development of food barn in realizing the food security of poor households in Wonogiri Regency.

2. Method

This research uses quantitative method with survey technique. The location of research is determined purposively in Wonogiri Regency because it is a district with high poverty rate and there is a food barn. The population in this study are the members of the poor food barn in Wonogiri regency.

The samples were taken by random sampling. The qualitative information which generated from the preliminary survey, it shows that in Wonogiri district, there are currently 281 food barn groups which spread over 112 villages, from 294 villages in 25 sub-districts. Then from 25 districts in Wonogiri Regency, five rural districts were taken randomly with the food barn activities. From each sub-district, five groups of food barn were taken with the most active member. Furthermore, from each group of food barn, 6 random samples were collected including poor households so that the total number of respondents from 5 districts which interviewed were as many as 150 people. The technique of collecting data was done by interview, field observation and focus group discussion (FGD). The data analysis technique was done by using multiple regression.

3. Results and Discussion

3.1 Socio-Economic Characteristics of Poor Household

Poor households are a group of people who live in some or all of the buildings and generally eat together from one kitchen and those who with poor condition. Socio-economic characteristics of respondents of households include some data about age, education level, occupation, land area and income of poor households. The characteristics of respondents will influence the availability, access, and consumption of family food which will ultimately impact on household food security of household. Respondents in this study were members of food barn group who were in Wonogiri District. The samples were 150 husbands and wives. The characteristics of respondents can be seen in Table 1.

Table 1. The distribution of Poor Household Respondents' Characteristics in Wonogiri District

Variables	Categories	Persentase (%)
The Age of Father:		
20 – 40 th	Young adults	12.3
41 – 60 th	Mature adults	56.5
60 th <	Old	28.1
Mother's age:		
20 – 40 th	Young adults	31.5
41 – 60 th	Mature adults	63.7
60 th <	Old	4.7
Last education of father	No school	14.0
	Did not pass elementary school	13.0
	Graduated from elementary school	45.0
	Graduated from SLTP	18,0
	Graduated from SLTA	20.0
	University/ college	1.0
Last education of mother	No school	19.0
	Did not pass elementary school	2.0
	Graduated from elementary school	45.8
	Graduated from SLTP	19.0
	Graduated from SLTA	11.1
	University/ college	0.7
The main job of father	Labors	32.4
	Farmers	57.4
	Traders	8.8
	PNS/ civil servant	1.4
The main job of mother	No work	24.8
	Labors	21.5
	Farmers	40.9
	Traders	12.0
	PNS/ civil servant	0.67
Land area (ha):		
<0,25	Very narrow	62.7
0,25 – 0.50	Narrow	32.7
0,51 – 1,00	Medium	3.3
1,00 <	Large	1.3
Family income /month :		
< 1.000	Low	36.6
1.000-2.000	Medium	45.7
2.000 <	High	17.7

From the table 1, it can be explained that the age of father and mother are included in the medium category. The age of father and mother can be said in productive groups, so that it can work optimally and can enough income to meet the need of household food. The education will affect on one's knowledge and insight. Most of the education level of them is still low, that is primary school or elementary school. The low level of their education is because of limited cost. This condition can be understood because the income is also low. In addition they are also difficult to access transportation, and it is common for them to help their parents to work to support their life. The work of poor fathers is mostly work as farmers. In this case, there are still mothers who do not have a job or do not work. Meanwhile, having job or working will have an impact on their family income so that it influences the availability of household food. The land area of poor household will affect the total of agricultural production which will ultimately affect the food consumption of family, and the land production will contribute to the availability of family food. The average size of poor households' ownership is very narrow. It is less than 0.25 ha where 96 are from 153 people most of them owned it because of heritance. The income of poor households is mostly in the medium category around 1 - 2 million per month, this condition will affect the purchasing power of food.

3.2 Normality Test

One kinds of tests which can be used to test the normality of data is to see the value of skewness ratio and the ratio of kurtosis. Data is said to be normally distributed if it is in the range -2.58 and +2.58. Based on the normality test by analysing at the value of the skewness ratio and the kurtosis ratio, the results are obtained as shown in this following table:

Table 2. The Normality Test of Research Variables

Variables	**Skewness**			**Kurtosis**		
	Statistics	**Std.Error**	**Ratio**	**Statistics**	**Std.Error**	**Ratio**
X1	0.241	0.219	0.589	0.312	0.422	0.881
X2	0.518	0.209	2.478	0.505	0.422	1.031
X3	0.293	0.209	1.402	0.331	0.422	0.790
Y	0.246	0.209	1.177	0.913	0.422	2.107

Notes: From the normality test above, the results in Table 2 show that the value is in the range -2.58 and +2.58 so that the data can be interpreted as in normal distribution.

3.2 Multicollinearity Test

The analysis of the presence of multicollinearity symptoms in this study is done through the measurement of Varian Inflation Factor (VIF) on the regression results. To show the presence of multicolinearity is tolerance <0.50 or it is equal to VIF value <10 [4]. The test results of tolerance and VIF values can be seen in Table 3 below:

Table 3. The Results of Multicollinearity Test of Research Variables

Variables	Collinearity Statistics	
	Tolerance	VIF
Perception on the benefits of food barns	0.217	2.873
Institutional regulation of food barn	0.204	4.172
Economic institutions of food barn	0.273	3.576

3.3 The Development of Food Barns in Realizing the Poor Household Food Security

The development of food barn in this study refers to the criteria of the development stage of food barn which explained by [5] it includes the procurement of food for the filling process of barns and the development of the capacity of the food barn itself. The variables suspected to influence the development of the food barn include: the perception of group members toward the benefits of food barns, institutional regulation of food barns, and various economics institution of food barn itself. The result of the development analysis of food barn in realizing the food security of poor household is presented in Table 4.

Table 4. The Results of Factor Analysis which Influence the Development of Food Barns in Realizing the Food Security of Poor Household

Variables	Coefficient Regression	t-count	p
Constant	11.009	3.599	0.000
Perceptions on the benefits of food barn	0.297	0.259	0.038**)
Institutional regulation of food barns	0.201	0.301	0.019**)
Economic institutions of food barns	0.227	0.222	0.044**)
F Count	17.986		0.000***)
Adjusted R Square	0.568		

Notes: ***) significant at $\alpha = 0.01$; **) significant at $\alpha = 0.05$

3.4 The Influence of Perception of Group Member of Food Barn Concerning to the Benefit of Barrowing Food toward the Development of Food Barns

From the analysis result, it indicates that the perception of member of food barn group toward the benefit of food barn, it has significant effect to the development of food barn, it happens because of food procurement dimension to fill the food barn in realizing food security of poor household. Perception is a process in which individuals organize and interpret their sense impressions to give meaning to their environment [6]. [7] who argued that the perception of a person is an active process that plays a role, not only the stimulus that relate to it but also the individual as a whole with his experiences, motivation and relevant attitude in giving response to the stimulus.

In this study, the perception of poor households on the development of food barns shows that all respondents (100%) stated that food barns can be useful in providing, storing and marketing the crops. A good perception of the benefits of the food barn of the food procurement dimension for filling the food barn is indicated by the awareness of the members to be willing to keep or save some crops in the food barn, for example, Beji Village with 455 Poor Households of 800 households. In addition, besides the to the village barn, there is also a RT, / RW barn. Based on the observations in the institutional field of the RT food barn, although the management is simple but it is very useful for its members. The food barn in RT 01 RW 05 Pandan hamlet, Beji, it currently has 3.5 tons of grain reserves from 42 members. The grain which are as many as 3.5 are obtained from the member's deposit when harvest time, the amount is not determined but the average member of grain according to the leader of RT is approximately 10 beruk. Before the famine is not shared, the grains are shared, but if there are members who need to borrow, for example a new person who has a job, a bayenan or a member who really needs to consume them. Some members who want to borrow 10 beruk, they are obliged to return 11 beruk and multiples. Almost all the activities of the barn are in the form of savings and loans, either in the form of grain, corn, rice or cashed. Food barns which run by members can be useful to overcome difficulties when famine time, crop failure due to pests, diseases or due to natural disasters. [8] research which states that food barns can act as food reserves, especially in rural areas. In addition, the food barn is not only effective in serving the needs of its members' food in crisis condition but also serving the financial needs of its members from the management of barns. [9], which explains that the food barn group has function to serves to overcome food insecurity at the time of famine and assist the members in the provision of capital.

The perception of the benefits of food barns toward its development in the dimensions of group capacity in this study was good, there were 89 percent of members who stated that group capacity always increase. This increase is beginning to show when there is capital aid for filling the food barn from the government. Associated with the name of the group since the establishment of the food barn group, the group which already existing with its name, and it already has the chairman, and the manager, the address is also clear.

This study shows that not all food barns have buildings built through the Special Allocation Fund for Agriculture field. Based on the ownership of warehouse (warehouse building), there are 4% of food barns which already have warehouse, namely that is food barn village, and the rest do not have warehouse. During this time, the barn is in the home of people who are trusted by their members such as in the house of the group leader, head of RT/RW, or at a member's home based on agreement. In term of the land use for buildings, some members still own the land, and most of the members of the food barn are poor. The groups are also actively conduct regular group meetings which can be proven by the attendance list. In the meeting routine, not all groups of food barn do activity in the meeting, because their meetings were just for arisan, and savings and loan.

3.5 The Influence of Institutional Regulation on Food Barns toward the Development of Food Barns
The results of the study in Wonogiri district show that regulation or rule is often becomes problem in an institution as occurs in food barns. There are still many food barns whose regulations are set by the administrators only without approval of members. In addition, there are also no written rules and sanctions for both the board and members as set in the AD/ART.

The institutional regulation of food barns in this study, the majority 79% food barns do not have good regulation. There are still merely barns with large loan interest, using collateral for grain lending, excessive payback periods, and unsuitable returns on loan. There are some specified loan interests depending on the local barn policy. The interest is set from 1% to 10% of the loan, for example, borrowing 10 kg of grain, she or he must return a number of 15 kg of grain. Lending money in some food barns must use some securities such as BPKB as grain loan guarantee. The payback period is also determined from 3 months to 1 year with monthly installments accompanied by interest on the loan. Some cases also mention that the return is not appropriate with the loan, such as borrowing grain, and then he or she returns using money.

The existing management in the food barn runs simply. According to some respondents, the management is held by local RT leaders or local institutional figures who are believed manage the food barns. It actually does not become a problem, but it will be a problem when there is no cohesiveness in the barn management. Some activities which are undertaken by food barns, such as savings and loan institutions are only savings and loan activities without other activities such as counseling extension. The frequency of meetings is rare so that the members' togetherness fades. This condition is one of the causes of the absence of food barns.

3.6 The Influence of the Economy Institution of Food Barn toward Food Barn Development.
The results of the analysis in this study indicate that the economic institutions of food barn significantly influence the development of food barns. This condition means that the food barn whose economic function is running well then the development of the food barn will also run well. The indicator of economic institution of food granary in this research is seen from the existence of clear bookkeeping, monthly dues regularly, procurement of special money for saving book, and marketing of crops.

The conditions on the field show only 8 percent of food barns which already have a good and clear bookkeeping, the rest of them, their bookkeeping is very simple. Acting as the manager or administrator of food barn, the RT assisted by his family who only has a simple notebook containing who deposit, who borrows without writing when they deposit, when they borrow, and when they return, they just trust each other.

Mostly the condition of the food barn which still exist, especially the food barns that are in a simple classification such as the barn RT / RW caused by the need which arise from the community itself. Seeing the institutional condition of the simple barn, they are now in need of physical warehouse, because the institute has been running. Relating to the role of the simple barn, it is very helpful in supporting the food security, especially for poor households. In addition, the number of grains / assets needs to be increased, both from the amount of the deposit and external capital assistance. Bookkeeping related to the amount of grain which is stored or taken in the time of storage and taking, especially for the food barn of the RT which also does not exist, in their notes, it is just written who entered. Taking no provisions, and any time it can be taken. The village barn which was founded in

IOP Conf. Series: Earth and Environmental Science **142** (2018) 012036 doi:10.1088/1755-1315/142/1/012036

2010/2011 under the government, the bookkeeping is good because it is still new. The capital should be given to a food barn that runs well. In term of the new food barn, it seems that its paradigm is behind it, because the board is more focus on seeking profit rather than social.

Routine monthly dues which based on the basis for simple food barn, such as RT or RW barns do not exist. The food barn facilitated by the government, also has no monthly dues, but when its member enter the barn, they have to pay for stocks, either in the form of money or unhulled grain, for example, the village barn which was built in the village of Pijiharjo sub-district, Manyaran, it was given the name of Source of Food which was harvested by Mr. Warsa. The village barn is going very well. Since 2011, the condition is similar to the food barn in Beji village which has an initial capital from the government of 20 million, until June 2012, the existing grains if they are changed into the money, the value is as much as 46 million. The difference is obtained from the members as many as 26 million in the form of grain deposits, each of 25 kg of grain from the number of members are as many as 255 people. Then, the activity of buying and selling grain, and savings and lending grain runs well. There is good bookkeeping from the board. To socialize the food barns to the community through the procurement of cheap markets and basic necessities of daily needs in the food barn itself.

The procurement of a special saving book for money where in all food barns which are as samples in this study, they do not have a food barn that holds a special book for note the money, but they can save money in the form of grains. In term of marketing results, simple barns like the RT and RW barns do not do so, but the village barns that get help from their government, they do the marketing of the results of their members.

4. Conclusion

Based on research result of perception of member of food barn group on the benefit of food barn, institutional regulation of food barn, and institution of food barn have significant influence to the development of food barn itself because those three variables have a significant influence in the development of food barn in realizing the food security of poor households. We should pay attention to these three things in order to maximize its development.

The condition of the food barn in the study are various, there is the regulation and the economic institution which run well and some of them are not good, for that reason, it need different way of management, for example the regulation of food granary and economic institutions are good but they do not have buildings, the government should provide assistance to build buildings, so that the development of food barns can be maximum. Almost all barns in this study are not merely as food stocks of their members during famine condition, but the existence of food barns can serve as social media. This condition is potential to be developed through the process of empowering systematically, intact, integrated and sustainable by involving all related elements.

4.1 Alternative Strategies In Food Lumbung Development

Based on the results of strategy studies in the development of food barns can be formulated as follows: The results of the study it is necessary to have the same regulation for all members of the barn related to: (1) grain storage regulation which determined its quantity, for example 50 kg per harvest per member; (2) giving balance interest, for example every 50 kg of 1% interest; (3) the repayment period is based on the total amount of loan, for example 50 kg for a period of 2 months of return, 100 kg for a period of 4 months. The election of the board is done by mutual agreement: (1) the election of the board must be attended at least 50% + 1 member; (2) there are representatives of members who are as administrators. Provision of adequate facilities for barns: (1) providing facilities in the form of granting funds for the manufacture of barns which are suitable for use as storage; (2) supporting facilities such as grinding machines, so that the members have no difficulty in grinding the grain. Development of food barn activities is needed: (1) socialization of barns, there are still many people who do not understand about the barn; (2) general counseling on agriculture, especially in the production and post-harvest of grains; (3) providing information on how to apply for funding to the government.

IOP Conf. Series: Earth and Environmental Science **142** (2018) 012036 doi:10.1088/1755-1315/142/1/012036

In term of bookkeeping related to the number of grains that are stored or taken need to be accompanied by the time of storage and retrieval, especially for the barn of RT. Monthly dues on a regular basis need to start, because it can serve as cash or capital for the barn if at any time it is required by its members. The procurement of a special money-saving book should also have begun to be considered by the managers of barns. Saving book has function to facilitate the members in the activities of storage and borrowing money which is non grain. The marketing of the members' harvests is done by cooperating with other parties, for example with a separate sales place in the market, so that members have no difficulty in marketing or distributing their crops.

References

[1] [BPS] Badan Pusat Statistik 2014 Jawa tengah dalam angka 2014 Badan Pusat Statistik Propinsi Jawa Tengah
[2] Anonymous 2007 Pedoman umum pemberdayaan lumbung pangan Jakarta: Badan Ketahanan Pangan Departemen Pertanian RI
[3] Khudori 2006 Urgensi lumbung pangan Harian Republika
[4] Hair J F, Black W C, Babin B J, Anderson R E, Tatham R L 1998 Multivariate data nalysis Sixth Edition New Jersey Prentice-Hall International
[5] Rusdiyanto 2017 http://www.academia.edu/29061028/Pengembangan_Lumbung_Pangan
[6] Gibson J L, Ivancevich, J M, Donnelly J H 1997 Organizations: Behavior, Structure, Processes Chicago: IRWIN
[7] Suryabrata S 1993 Psikologi Kepribadian Jakarta: CV Rajawali
[8] Soemarmo 2015 Model pengembangan LPMD lumbung pangan masyarakat desa FP Universitas Brawijaya Malang http:// marno.lecture.ub.ac.id
[9] Prasmatiwi F E, Rosanti N, Listiana I 2013 Kajian cadangan pangan rumah tangga petani padi di Provinsi Lampung

IOP Conf. Series: Earth and Environmental Science **142** (2018) 012037 doi:10.1088/1755-1315/142/1/012037

The yield and quality of black rice varieties in different altitude

E Purwanto, W Hidayati and Nandariyah
Study program of Agrotechnology, Agriculture faculty, Universitas Sebelas Maret, Surakarta, Indonesia

Email: edipur_uns@yahoo.com

Abstract. This study aims to determine the optimal environmental conditions and corresponding black rice varieties in order to produce high production and optimum quality rice. The study using nested design, first was location (2 levels: the highlands and lowland), second was varieties (3 levels: Cempo Ireng, IPB, and Gagak) the study was conducted from November 2015 until May 2016 in Karanglo and Gutanon village, Karanganyar. Anthocyanin analysis conducted in laboratory of Nutrition and Food, Faculty of Agriculture, Sebelas Maret University. Data were analyzed using analysis of variance and significant difference continued with DMRT (Duncan Multiple Range Test) level of 5%. Results showed that cultivation in highlands and use of diverse varieties showed different quantity of rice that can be seen on panicle length, and weight of grain crops. Improved quality of results showed same things, anthocyanin content and iron increased on black rice which cultivated in highlands.

1. Introduction

One of several types of rice consumed is black rice which contains anthocyanin in its pericarp [1]. This day black rice is known as a functional food and widely consumed as a staple food that is not only as a source of carbohydrates, but also source of fiber which for good health. Generally, rice was grown in lowland about 0-600 meters above sea level. The quality of ricecan be influenced by the environment for growth. The differences in elevation will lead different climate character elements such as temperature, light, and moisture. The lowlands have a relatively high temperature with high light intensity, while the highlands have a relatively low temperature with low light intensity as well. Temperature and light intensity affect the synthesis of primary metabolites and secondary metabolites in plants.

Black rice cultivation in Indonesia has not been so prevalent. Only few farmers were willing to cultivate black rice, because of the long age of cropping and use local seed production less preferred. This resulted in increased declivous productivity of rice nationally. In 1981-1990 the production of rice increased by 4.08% per year, from 1991 to 2000 years of 1.31% per year, while in 2001-2007 by 0.58%. The declivous of productivity becomes a problem to fulfill community needs for rice, so it needs a cultivation technique that can improve productivity. This research aims to study the the quantity and quality of yield of black rice from various variety grown in the highlands and lowlands to.

2. Methods

The study was conducted in November 2015 to May 2016. Research about altitude implemented at two locations with different elevation. The first location in Gutanon village, Jaten, Karanganyar with an altitude of 141 meters above the sea level in position 07° 35 '48.8' 'South Latitude and 110° 55' 23.0 " BT. The second location is in the village Karanglo, Tawangmangu with elevation 665.0 meters above sea level right at the position 07° 39 '43.3' 'South Latitude and 111° 05' 11.6 " BT. Iron content was analyzed in the Laboratory of Chemistry and Soil Fertility, Faculty of Agriculture UNS and total anthocyanin analyzed in the Laboratory of Nutrition and Food, Faculty of Agriculture, Universitas Sebelas Maret, in May 2016.

This study used a nested design with two factors. The first factor was the elevation: lowland (Gutanton village, subdistrict Jaten) and Highlands (village Karanglo, subdistrict Tawangmangu). The second factor was the varieties: Cempo Ireng, IPB, and Gagak. Variable observations include flowering time, variable quantity and quality of yield. Variables of quantity include the number of total tillers, number of productive tiller, panicle length, number of grain per panicles, grain weight per plant, crop productivity. While variables of quality include total anthocyanin content and iron content. Data were analyzed using analysis of variance (ANOVA) on 95% confidence level. If there was a significant effect on the results of the test continued using DMR (Duncan Multiple Range) on 95% confidence level.

3. Results and Discussion

Rice can grow well in the temperature range 20-35°C. Based on the research that has been done, the researcher not able to show the complete data of plant samples as expected, because the Cempo Ireng varieties sample plants which were grown in lowland were lost due to technical errors. *General conditions research*

The location were in the lowland (Gutanon village, Jaten sub district) with elevation 141 meters above sea level and second location on highland (Karanglo village, Tawangmangu sub district) with elevation 665 meters above sea level.

Table 1. Elements of climate in Gutanon village, Jaten sub district and Karanglo village, Tawangmangu sub district

Elements of climate	Lowland	Highland
Elevation (mdpl)	141	565
Geographic	'07° 35' 48,8'' SL-110° 55' 23,0'' EL	)7° 39' 43,3'' SL-111° 05' 11,6'' EL
Daily temperature (^{o}C)	29 -34	26 -31
Soil temperature ^{o}C)	25- 29	24 – 27
Humidity (%)	50	55 %
Humidity around plant (%)	67	79 %
Light Intensity (lx)	832-930 x 100	1891, 2508, 2683 x 10

The differences of elevation affect micro-climatic conditions of both locations. Daily temperature in Karanglo and Gutanon village respectively between 29 °C-34 °C and 26 °C-31 °C. The temperature value is inversely to the humidity, which if the temperature high, the humidity will be low, and vice versa. Observations in humidity at high and low elevation show those such things. The humidity were 60-65 % and 55-50 % respectively. The light intensity of both is different, which is 261.5 x 103 lux for lowland and 70.82 x 103 lux on highland. The difference of light intensity in this observation is quite large about 190.68x103 lux. Elevations effect on temperature and light intensity. Temperature and light intensity will be smaller following the high point of growth [2]. Effect of elevations on plant growth tends indirectly, meaning that the difference elevation will affect the environment by temperature, humidity, oxygen in the air, and soil conditions.

3.1. Observation of flowering time

The transition from the vegetative phase to the generative phase is shown by the flowering time of plant. The development of the plant at this stage begins with the initiation of flower. Flowering (anthesis) on rice begins when the tip of stamens on the panicle has looked out of the grains and fertilization process begin [3].

Table 2. Differences of flowering time in black rice

Location	Flowering time (Days after planting)		
	Cempo Ireng	IPB	Gagak
Highland	116	83	85
Lowland	110	80	83

Flowering in Cempo Ireng variety was 5 days faster than those in the highlands, 3 days on IPB variety, and 2 days on Gagak variety (Table 2). This shows that Cempo Ireng variety has vegetative and flowering time period longer than IPB and Gagak. [4] states that flower initiation depends on genotype interaction and specific environmental conditions such as low temperature period, the length of the day or photoperiodism. Rice plants flowering time affected by environmental conditions during pollination and flowering. High temperatures can accelerate the rate of initiation time devoted with the cultivation of lowland showing flowering time faster than the highland.

3.2. Effect of treatment to quantities of product

3.2.1. The number of total tillers Based on the observations, the tillers at three different varieties used have different number of the average. Number of tillers in highland locations tend to be lower than the lowlands.

Table 3. Average of Total Tillers in Paddy at Different Altitude

Location	Total Tillers (Days after planting)		
	Cempo Ireng	IPB	Gagak
Highland	14.95±3.27a	16.65±1.81c	17.1±2.81b
Lowland		19.7±3.66 c	15.95±3.30 b
Mean	14.95 a	18.17 c	16.53 b

Addition: The figure is followed by different letters on the same line, significantly different (P < 0.05).

Results of analysis of variance showed that the location does not affect the number of tillers, while varieties those are nested within location show significant difference in the number of tillers. Cempo Ireng have the lowest Number of tillers, followed by IPB and Gagak (Figure 1). The highest number of tillers obtained Gagak varieties which cultivated in the highlands with a mean of 19.7 and the lowest number of tillers obtained Cempo Ireng by the number of tillers 14.95. The number of tillers are influenced by genotype factors in plant and also can be influenced by cultivation technique factors such as the wide spacing can increase the production of rice tillers because it can minimize the occurrence of light and nutrient competition. [5] explains that the ability of the rice plant to produce tillers are genetically different. However, the number of tillers can be increased because it depends on environmental factors and cultivation technique.

3.2.2. Length of Panicle Overall, rice flower called panicles. Panicles consists of 8-10 nodes that produce primary branches then produces a secondary branch. From the panicle nodes generally it will appear only one primary branch, but in certain condition the nodes can produce 2-3 primary branches [6].

Table 4. Average of length of panicle at different elevation

Location	Length of Panicle		
	Cempo Ireng	IPB	Gagak
Highland	22.55 a±2.09	24.40 b±1.30	23.67b±1.76
Lowland		22.68 b±1.05	22.93b±1.00
Mean	22.55 a	23.54 b	23.30 b

Addition: The figure is followed by different letters on the same line, significantly different ($P < 0.05$).

Results of analysis of variance showed no significant difference occurred in the treatment location, but there was a significant difference in varieties that are nested within the location. Figure 2 shows that the highest panicle length obtained IPB varieties (highlands) with an average 24.4 cm, the lowest one was Cempo Ireng (lowland) with panicle length 22.55 cm. As explained by [7] panicle length character controlled by a genetic trait, so the difference in the character of each variety is a genetic trait. Panicle length is one indicator of the high productivity of rice. The longer panicles, the more number of grain that increases the weight of grain per crops and can increase productivity of rice.

3.2.3. Grain weight per crop Crop production in addition influenced by genetic factors may also be influenced by environmental factors and cultivation. Environmental factors such as temperature, humidity, light intensity and others will strongly support the plants for optimal production. Crop production can also be improved by enhancing cultivation technique such as spacing, selection of appropriate seedlings, irrigation scheduled and others.

Table 5. Average weight of Paddy crop in Different Elevation

Location	Grain weight per crop		
	Cempo Ireng	IPB	Gagak
Highland	16.81±8.93 a	23.94±5.50 b	23.08±7.75 ab
Lowland		17.07±5.58 b	17.19±4.80 ab
Mean	16.81 a	20.50 b	20.14 ab

Addition: The figure is followed by different letters on the same line, significantly different ($P < 0.05$).

Results of analysis of variance showed significant difference in location and varieties those are nested within the location treatment. Grain weight per plant (highland) derived IPB varieties with highest average about 23.94 g, 23.07 g Gagak, and the lowest one Cempo Ireng 16.81 g. Grain weight per plant affected by the amount of grain per panicle and panicle length. It is influenced by products of photosynthesis which produced during the ripening grain period [8]. Environmental factors one of them is temperature effect on the formation of grain, which at the optimal temperature (± 30 °C) flowering and fruiting can occur well. If the temperature is high, then the stigma that there is liquid sugar for the attachment of pollen can be dried up, pollen cannot stick on it, so there is no fertilization and formed empty grain. Each plant has a capability production themselves, also have different response to fertilizer or environment.

3.2.4. Crop Production Crops have their own agronomic characteristic in production. Crop production can be affected by the nature of crops as well as the treatment given to the empire itself. Spacing, the addition of certain fertilizers, or proper watering can support the productivity of rice.

Table 6. Average Crops Production at Different Elevation

Location	Production		
	Cempo Ireng	IPB	Gagak
Highland	2.30±0.14 a	4.44±0.15 b	4.40±0.05 b
Lowland		3.24±1.06 b	3.17±0.34 b
Mean	2.30 a	3.84 b	3.78 b

Addition: The figure is followed by different letters on the same line, significantly different ($P < 0.05$).

Results of analysis of variance showed a significant difference did not occur at treatment location, but significant difference occur on varieties those are nested within locations treatment. Although the analysis of variance treatment of different locations does not occur significantly, Table 6 above shows that highland rice production was higher than the lowlands. Table 6 shows that rice production in upland is higher than the lowlands. This can be due to on highland harvesting is done manually by removing the grain by hitting them, while in the lowlands harvesting using a thresher machine. Yield loss at harvest rice can be influenced by several factors, such as harvesting age, water content, and tools, and how to harvest. Besides overripe age with low water levels cause grain fall easily at harvest. [9] reported the results of postharvest loss is still quite high, more than 20%. The critical point of yield loss occurs at the stage of harvesting and threshing. With the loss rate is still quite high, and on the stages of harvesting loss is still around 9%, and at what stage threshing still more than 4%.

3.3. Treatment Effect on the Quality Produce

3.3.1. Anthocyanin Anthocyanin is secondary metabolites of the flavonoid formed due to environmental stress. The results are consistent with the statement [10] stated that light affects the setting of production of primary metabolites such as enzymes, carbohydrates, lipids and amino acids and secondary metabolites such as anthocyanins, flavonols and carotenoids [10]. [11] stated, the darker color of rice, the higher anthocyanin content and antioxidant. It makes high potential as a functional food.

Table 7. Average levels of anthocyanin in different elevation

Location	Anthocyanin (ppm)		
	Cempo Ireng	IPB	Gagak
Highland	521.9±8.27 c	113.99±6.30 a	131.31±7.47 b
Lowland		40.82±3.91 a	79.75±4.52 b
Mean	521.9 c	77.40 a	105.53 b

Addition: The figure is followed by different letters on the same line, significantly different ($P < 0.05$).

Results of analysis of variance showed that there was a significant difference in treatment location and varieties nested within the location. The highest total anthocyanin content owned by rice varieties Cempo Ireng, then Gagak, and the lowest IPB grown in the highlands with an average of 521.97 ppm, 131.31 ppm and 113.99 ppm (Table 7). Anthocyanin content was lower in Gagak and IPB varieties grown in the lowlands are 79.76 ppm and 40.82 ppm. The results are consistent with [12] in their research on the effect of temperature on anthocyanin, explained that at low temperatures anthocyanins can accumulate maximum compared with the high temperature treatment. The optimal temperature to accumulate anthocyanin is about 25-28 °C, whereas at 35 ° C anthocyanin in lower production.

[13] reported lower total anthocyanin content is a trait that is influenced by fewer genes (the qualitative nature). It is [14] appropriate opinion that the qualitative nature is controlled by one or two genes, are little affected the environment so that the election / selection can be done visually.

3.3.2. Iron Iron content in rice is at the aleurone pericarp and the rest are in the endosperm of the rice [15]. Aleurone layer thickness can affect the iron content in it.

Table 8. Mean of iron content on different elevation

Location	Anthocyanin (ppm)		
	Cempo Ireng	IPB	Gagak
Highland	62,83±14,26 a	59.06±11,63 a	75,35±30,05 a
Lowland		57,51±11,16 a	39,51±30,34 a
Mean	62,83 a	58,28 a	57,43 a

Addition: The figure is followed by different letters on the same line, significantly different (P < 0.05).

Result of analysis of variance showed the significant difference in location's treatment in iron content. However, the treatment is nested within location there is no significant difference in iron content. Varieties with low iron content obtained Gagak varieties on cultivation in the lowlands (39.51 ppm), and the highest in Gagak varieties on cultivating in highland (75.34 ppm). The differences of iron concentration is affected by plant genetic trait. Black rice has a high iron content compared to white rice or red. A high iron content can also be influenced by factors such as post-harvest handling milling degree also affect the iron content of rice, which is caused by the loss of aleurone layer during the milling and the iron minerals that dissolve during the washing process [13].

4. Conclusions

The conclusions from the study are, rice cultivation in highlands able to increase rice production which correlated with parameters number of total tillers, panicle length, and the number of grain per crops. The best on quantity (highest yield) were IPB and Gagak varieties, while the best variety on quality was Cempo Ireng. Difference elements of climate on the highland may increase the anthocyanin content and iron on black rice than in lowland

Reference

[1] Yoshimura Y, Zaima N, Moriyama T, Kawamura Y 2012 Different Localization Patterns of Anthocyanin Species in the Pericarp of Black Rice Revealed by Imaging Mass Spectrometry PLoS *ONE 7(2): e31285 doi:101371/journalpone0031285*

[2] Putri IR 2009 Pengaruh Intensitas Cahaya Matahari Terhadap Pertumbuhan Jenis *Shorea* Parvifolia dan *Shorea Leprosula* dalam Teknik TPTI Intensif (Studi Kasus Di Areal IUPHHK-HA PT Sarpatim, Kalimatan Tengah) *Skripsi* Departemen Silvikultur Fakultas Kehutanan Institut Pertanian Bogor

[3] Makarim AK dan Suhartatik E 2009 Morfologi dan Fisiologi Tanaman Padi Balai Besar Penelitian Tanaman Padi Hal 296-330 http://www.litbangpertanian.go.id/special/padi/bbpadi_2009_itkp_11pdf

[4] Lang A 1965 *Physiology of flower initiation Differentiation and development* Encyclopedia Plant Physiology XV/1 (Berlin: Springer) pp 1380-1536

[5] Purwanto 2009 Pertumbuhan dan Hasil Empat Varietas Padi (Oryza sativa L) Pada Sistem Pertanian Organik, Semiorganik dan Pertanian Konvensional *Thesis Agronomi* Universitas Gadjah Mada Yogyakarata

[6] Mungara E, Indradewa D, dan Rogomulyo R 2013 Analisis Pertumbuhan dan Hasil Padi Sawah (Oryza sativa L) pada Sistem Pertanian Konvensional, Transisi Organik, dan Organik *Vegetalika* **2** 1-12

[7] Tahir M, Wadan D, and Zada A 2002 Genetic variability of different plant yield characters in rice Sarhad J. *Agriculture* **18**

[8] Fathelrahman S A, Alsadig A I and Dagash Y I 2015 Genetic variability in rice genotypes (*Oryza* Sativa L) in yield and yield component under semi-arid zone (Sudan) *Journal of Forest Products & Industries* **4** 21-32
[9] Nugraha S et al 2009 Metode Menekan Kehilangan Hasil Padi Balai Besar Litbang Pascapanen Pertanian http://wwwlitbangpertaniangoid/berita/one/930/file/MetodeMenekanKehilanganHaspdf
[10] Siregar L A M, Keng C L, dan Lim B P 2010 Pengaruh Kasein Hidrolisat dan Intensitas Cahaya Terhadap Produksi Biomassa dan Alkaloid Canthinone di dalam Kultur Suspensi Sel Pasak Bumi (*Eurycoma longifolia* Jack) *MAKARA SAINS* **14** 15-21
[11] Anggraini T, Novelina, Limber U and Amelia R 2015 Antioxidant activities of some red, black and white rice cultivar from West Sumatra, Indonesia *Pakistan Journal of Nutrition* **14** 112-117
[12] Zhong J J and Yoshida T 1993 Effect of Temperature on Cell Growth and Anthocyanin Production in Suspension Cultures of *Perilla frutescens*
[13] Kristamtini, Taryono, Basunanda P dan Murti RH 2014 Keragaman Genetik dan Korelasi Parameter Warna Beras dan Kandungan Antosianin Total Sebelas Kultivar Padi Beras Hitam Lokal *Jurnal Ilmu Pertanian* **17** 90 - 10
[14] Mangoendidjojo, 2003 Dasar-Dasar Pemuliaan Tanaman Kanisius, Yogyakarta
[15] Indrasari S D 2015 Kandungan Besi Varietas Padi Balai Besar Penelitian Tanaman Padi http://pustakalitbangpertaniangoid/publikasi/wr286066pdf
[16] Kementrian Pertanian 2015 Rencana Strategis Kementrian Pertanian Tahun 2015-2019 http://wwwpertaniangoid/file/RENSTRA_2015-2019pdf

IOP Conf. Series: Earth and Environmental Science **142** (2018) 012038 doi:10.1088/1755-1315/142/1/012038

Effect of manure and plants spacing on yield and flavonoid content of *Elephantopus scaber* L.

D Riyana[1], Y Widiyastuti[2], H Widodo[2], E Purwanto[3] and Samanhudi[3]
[1]Study Program of Agrotechnology, Faculty of Agriculture, Universitas Sebelas Maret, Surakarta, Indonesia
[2]Research Center for Medicinal Plant and Traditional Drug, Tawangmangu, Indonesia
[3]Study Program of Agrotechnology, Faculty of Agriculture, Universitas Sebelas Maret, Surakarta, Indonesia

E-mail: yunus.uns7@yahoo.com

Abstract. This experiment is aimed to observe the growth and flavonoid contain of Tapak Liman (Elephantopus scaber L.) with different manure types and plants spacing treatment. This experiment is conducted at Tegal Gede Village, Karanganyar District on June until August 2016. The analysis of secondary metabolism was done in B2P2TOOT, Tawangamangu. This experiment is conducted with Randomized Complete Block Design (RCBD) with two treatment factors, those are manure and plants spacing. Animal manure treatment had 3 levels, those are without manure, cow manure with 20 ton/ha dose, and chicken manure with 20 ton/ha dose. Plants spacing treatment had 3 phrase, those are 20 cm x 20 cm; 30 x 30 cm; 40 cm x 40 cm. The result of this experiment shows that chicken manure with 20 ton/ha dosage increase the development of leaves' lengthiness, header's diameter, plant's fresh weight, and plant's dry weight. Plants spacing 40 cm x 40 cm increase for the development of leaves' lengthiness, header's diameter, plant's wet weight, and plant's dry weight. The combination between chicken manure with 20 ton/ha dose and plants spacing 40 cm x 40cm treatments show the highest amount of tapak liman extract and alleged having the biggest amount of flavonoid substance.

1. Introduction

Organic fertilizer is a fertilizer composed of living matter such as the remains of plants, animals and humans. The use of organic fertilizers should be encouraged to restore soil conditions as well as increase land productivity and human health. It is also beneficial to increase agricultural production. The use of organic fertilizer is very important as a buffer of physical, chemical, and biological properties so that soil can improve fertilizer efficiency and land productivity [1]. In addition to nutrient factors, plant spacing also plays an important role in increasing agricultural production. Plant spacing determines the population of plants in a certain area [2], so a good arrangement can reduce the competition of water, light and nutrients.

Tapak Liman *(Elephantopus scaber* L.) including in the family of Asteraceae which was often regarded as a weed in cultivated plants. Even so, the elephant's foot contains chemicals that have pharmacological effects. The tapak liman contains chemicals which one of them is flavonoid [3]. Flavonoids are lactagogous substances in which they can increase the hormone prolactin and oxytocin as a trigger exclusive breastfeeding [4]. The efficacy of tapak liman is undoubtedly primarily for the development of herbal medicine, but its excessive use will lead to scarcity if not balanced with cultivation. The good and right tapak liman procedure of cultivation such as the use of organic fertilizer and plant spacing arrangement was expected to increase the production both quality and quantity.

2. Methods

The research was conducted in June to August 2016 in experimental garden of Balai Besar Penelitian dan Pengembangan Tanaman Obat dan Obat Tradisional (B2P2TOOT) of Tegal Gede Village, Karanganyar Regency. Flavonoid content analysis was performed at the B2P2TOOT Phytochemical Laboratory. Materials in this study are including of tapak liman seeds, soil media, manure (chicken and cow), ethanol, methanol, hexan, and ethyl acetate. The tools used in this study including of hoes, gauges, ovens, knives or scissors, analytical scales, kilogram scales, rulers, cameras, plastic tubs/plastic trays, stationery, grinders, thin layer chromatography (TLC), chamber, porcelain , 50 ml beaker, waterbath, sonicator, dropper, tweezers and filter paper.

This research use Randomized Complete Block Design (RCBD)factorial pattern with two factors (manure and plant spacing). Different types of manure as the first factor consists of 3 levels: control/no manure, cow manure, and chicken manure. While the second factor of plant spacing consists of 3 levels of plant spacing (20 cm x 20 cm, 30 cm x 30 cm, and 40 cm x 40 cm). From these two factors, there were 9 treatment combinations and each treatment combination was repeated 3 times. Observation variables such as leaf length, crown diameter, leaves number, root length, fresh stover weight, dry stover weight, total extract content, flavonoid content analysis. Data analysis was used variance analysis if the real effect continued with DMRT (Duncan Multiple Range Test) test at the level of 10%.

3. Results and Discussion

3.1 General condition of Research

This research was conducted at B2P2TOOT Medicinal Plant Garden, Tanjungsari, Tegal Gede Village, Karanganyar. The geographical location of the research site is 7°36.687‘ S and 110° 57.766’ E. The height study site is about 215 meters above sea level. The temperature at the study site ranged from 27°C to 30°C. The soil type in this area is red latosol land, textured clay and with the medium nutrients content.

3.2 Leaves Number

The result of variance analysis was showed that manure has no significant effect on the tapak liman leaves number (Table 1). The average number of tapak liman leaves without manure is 32.03, with cow manure is 32.42 and with chicken manure are 36.61.

Table 1. Influence of manure on the leaves number

Manure	Leaves number
Without manure	32.03 ± 1.32
Cow	32.42 ± 2.05
Chicken	36.61 ± 2.11

The result of variance analysis on plant spacing treatment did not give significant effect to the leaves number (Table 2). The average of leaves number on 20 cm x 20 cm of plant spacing treatment is 32.25, at 30 cm x 30 cm of plant spacing is 34, and at 30 cm x 40 cm plant spacing is 36.80. The closer the spacing between plants will be the lower rate of photosynthesis, because there is competition in obtaining water, nutrients, and sunlight needed for the photosynthesis process [5].

Table 2. The effect of plant spacing on the leaves number

Plants spacing (cm)	Leaves number
20 x 20	32.25 ± 2.46
30 x 30	34.00 ± 2.10
40 x 40	36.80 ± 1.98

3.3 Leaves length

Leaves length observation was done by measuring the longest leaves of each sample. The result of variance analysis was showed that manure significantly affected leaves length (Figure 1). The average leaves length on the without manure treatment is 11.8 cm, cow manure is 12,26 cm, and chicken manure is 13.86 cm. It was stated that chicken manure has more nutrients and is easily decomposed so that roots can absorb more nutrients ions to be used in photosynthesis [6].

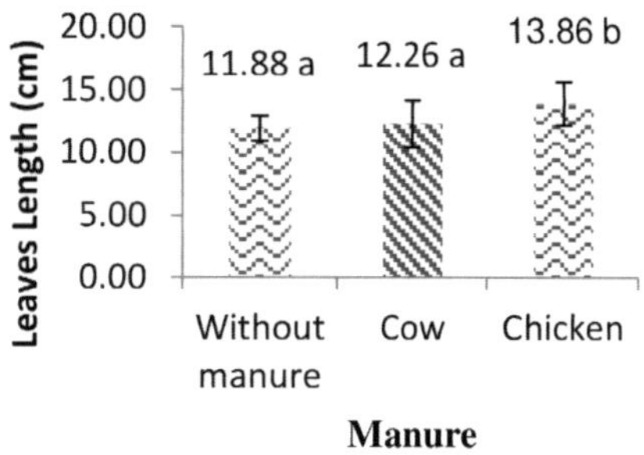

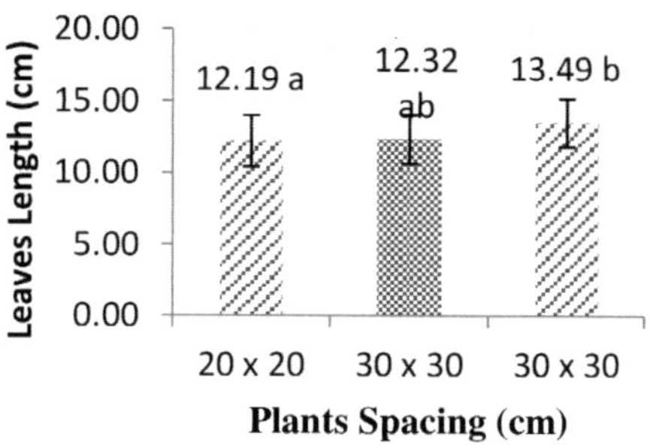

Numbers were followed by the same letter does not differat the real level of DMRT test of 10%

Figure 1. Effect of manure on leaves length

Figure 2. Effect of plant spacing on leaf length

Treatment of plant spacing had significant effect on the leaves length of the tapak liman (Figure 2). Average of leaves length at plant spacing of 20 cm x 20 cm is 12.19 cm, 30 cm x 30 cm plant spacing is 12.32 cm, and 40cm x 40cm plant spacing is 13.49 cm. This is an accordance to the research [7] which says the more spaced of plants more solar energy captured by the plants used for photosynthesis.

3.4 Crown Diameter

The result of variance analysis showed that manure significantly affect the crown diameter (Figure 3). The average of crown diameter on the treatment without manure is 23.92 cm is not significantly different with cow manure of 24.14 cm, but the real difference in chicken manure of 27.65 cm. This is because chicken manure contains a higher N element. [8] stated nitrogen is necessary to support vegetative growth of plants [8]. In a research also stated that the provision of chicken manure can increase growth such as plant height and leaf area [9].

The result of variance analysis was showed that the plant spacing treatment had significant effect on crown diameter (Figure 4). At plant spacing of 20 cm x 20 cm has a crown diameter average of 23.72 cm, plant spacing of 30 cm x 30 cm is 24.74 cm and plant spacing of 40 cm x 40 cm is 27.27 cm. According to [10] the more planting spacing will reduce competition in using nutrients and sunlight.

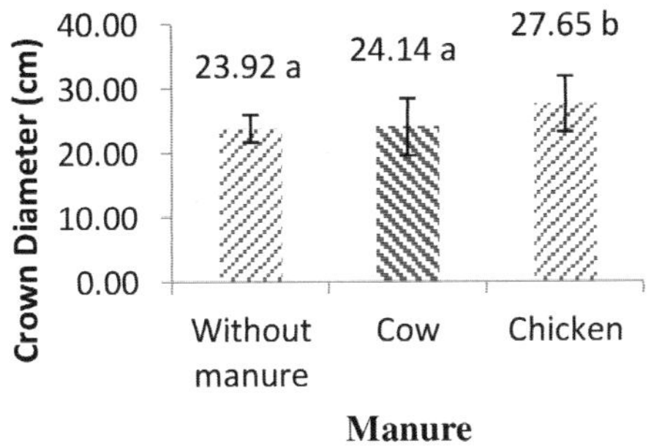

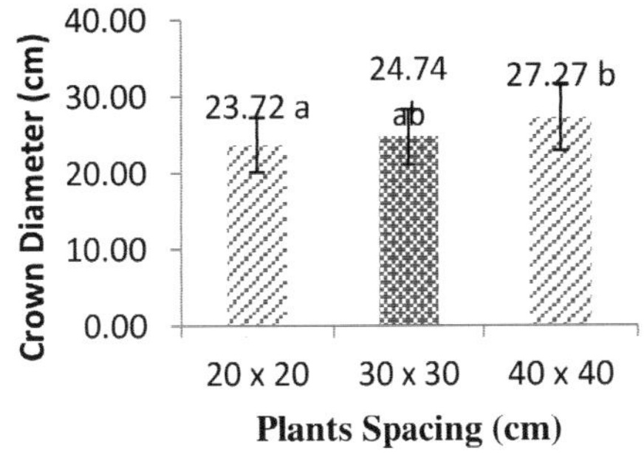

Numbers were followed by the same letter does not differat the real level of DMRT test of 10%

Figure 3. The effect of manure on crown diameter

Figure 4. The effect of plant spacing on crown diameter

3.5 Root Length

The result of variance analysis was showed that manure has no significant effect on root length (Table 3). The roots length average on the treatment without manure is 26.69 cm, cow manure is 28.54 cm and chicken manure is 27.09 cm.

Table 3. The effect of manure to root length (cm)

Manure	Root Length (cm)
Without manure	26.69 a ± 2.83
Cow	28.54 a ± 3.13
Chicken	27.09 a ± 1.74

Description: The numbers followed by the same letters show no significantdifference in the Duncan test level of 10%

Plant spacing treatment did not have a significant effect on root length (Table 4). The average of root length of tapak liman at plant spacing of 20 cm x 20 cm is 26.62 cm, plant spacing of 30 cm x 30 cm is 27.08 cm and plant spacing of 40 cm x 40 cm that is 28.62 cm. [11] was argued that root growth was influenced by many things, one of which is the soil properties. In addition, soil water availability also affected root growth.

Table 4. The effect of plant spacing on root length (cm)

Plants spacing	Root Length (cm)
20 x 20	26.62 a ± 2.02
30 x 30	27.08 a ± 2.74
40 x 40	28,62 a ± 2.98

Description: The numbers followed by the same letters show no significantdifference in the Duncan test level of 10%

3.6 Plants Wet Weight

The result of variance analysis showed that manure significantly affected the plants wet weight (Figure 5). Plants wet weight on the without manure treatment is 59.47 g, cow manure is 69.78 g, and chicken manure is 98.69 g.According by opinion of [12], the fresh weight of the plant is influenced by the increase in the length and volume of plants caused by the increase of plant cells. N elements contained in chicken fertilizer can stimulate the formation of auksin which serves to soften the cell wall so as to increase the ability of plants to absorb water. This is what causes the increase in fresh weight of the plant.

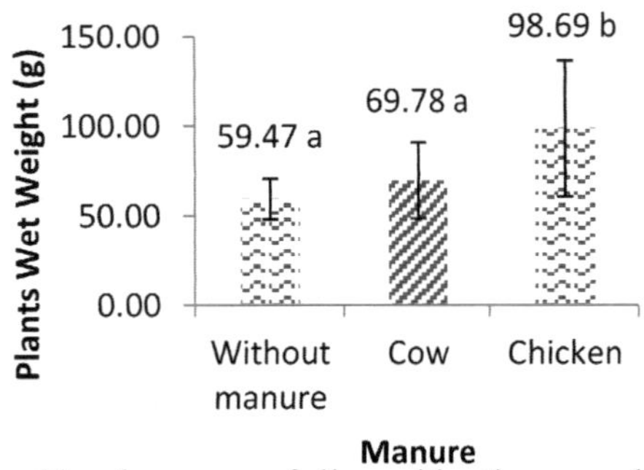

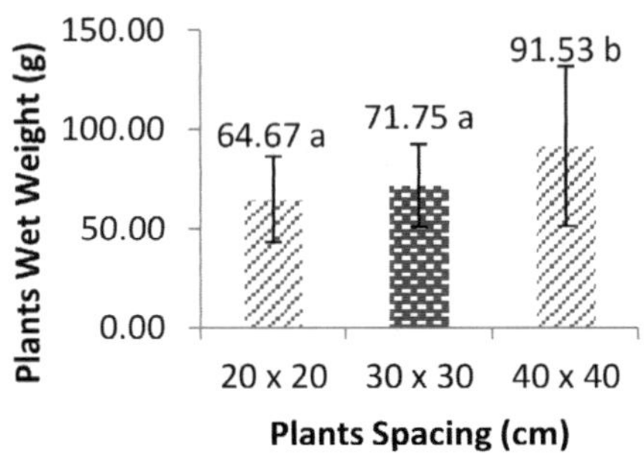

Numbers were followed by the same letter does not differat the real level of DMRT test of 10%

Figure 5. The effect of manure on plants wet weight

Figure 6. The effect of plant spacing on plants wet weight

The result of variance analysis showed that plant spacing had significant effect on fresh stover weight (Figure 6). Plant spacing of 40 cm x 40 cm have a significant effect on the fresh stover weight, in line with the research results of [13], the wider plant spacing, the greater the fresh forage production. In the treatment of plant spacing 40 cm x 40 cm was resulted a highest fresh stover weight of 64.67 g, significantly different with the plant spacing of 30 cm x 30 cm of 71.75 g, and also the plants spacing of 20 cm x 20 cm of 91.53g.

3.7 Plants Dry Weight

The result of variance analysis was showed that the application of manure significantly affected the dry stover weight (Figure 7). The treatment of chicken manure resulted in average dry stover weight higher than cow manure and without manure which amounted to 17.14 g. In cow manure only produce average of dry stover weight of 13.02 g, while without manure is 11.76 g. This is in accordance with the opinion [14] that the high N elements in chicken manure can spur the growth of plant biomass. [15] also said that N elements in chicken manure cause formation of flavonoid.

The plant spacing treatment was showed a significant effect on the dry stover weight (Figure 8). The average of dry stover weight at 20cm x 20 cm plant spacing is 12.21 g and plant spacing of 30 cm x 30 cm is 13.63 g. Different with plant spacing of 40 cm x 40 cm is to 16.07 g. According to [16] it is caused by the large percentage of light that can be forwarded by plants at a wider range so that plants can take advantage of growth and light for its development.

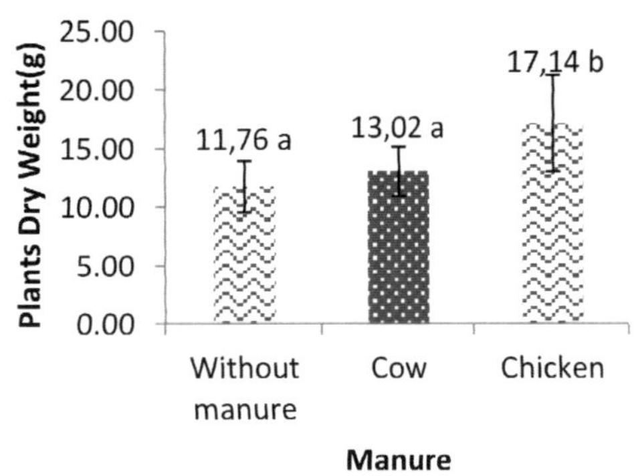

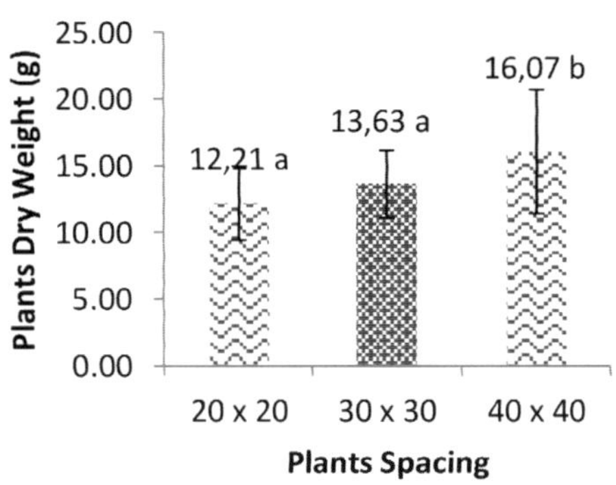

Numbers were followed by the same letter does not differat the real level of DMRT test of 10%

Figure 7. The effect of manure on plants dry weight

Figure 8. The effect of plant spacing on plants dry weight

3.8 Flavonoid Analysis

The first step before doing flavonoid analysis is calculated the percentage of total extract content of the tapak liman. The highest percentage of total extract content (Figure 9) resulted in the manure treatment at the plant spacing of 40 cm x 40 cm, that is 22%, while the lowest extract level was produced at the treatment without the manure at the plant spacing of 20 cm x 20 cm, cow manure at plant spacing of 30 cm x 30 cm and chicken manure at plant spacing of 20 cm x 20 cm is 16%. The percentage of extract content was indicated the presence of secondarymetabolite content within the plant.

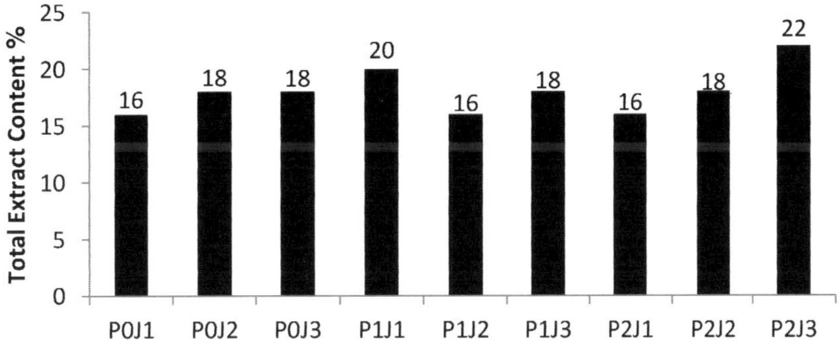

Figure 9. The influence of manure and plant spacing to the total extract content

Analysis of flavonoid compounds was used the method of Pharmacopoeia Herbal Indonesia Edition I Year of 2008 with TLC. TLC plates were used made of silica gel with size of 8 cm x 9 cm. The TLC plates of silica gel were activated by oven at 100° C for 1 hour. The viscous extract of the extraction was then dissolved with hexane as much as ± 5 ml. Hexane works to dissolve fat and fatty acids. It is then bottled along the plate using a micro pipette at a distance of 1 cm from the bottom line and each treatment is 1 cm apart. Then eluted to give the best separation result on TLC, i.e. n-Hexane-atil acetate-methanol by comparison of 5: 5: 1. After the formation of propagation then TLC plat aerated and examined under UV light at a wavelength of 366 nm (Figure 10).

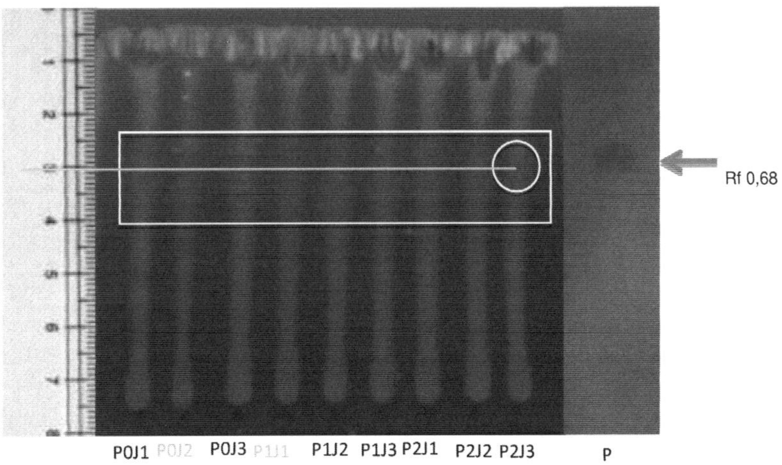

Figure 10. The result of TLC test of secondary metabolite compound of tapak liman

The brighter the spot color was indicated the higher the concentration of certain compounds. So in the flavonoid compounds, the brighter the spot color the higher the level of flavonoid compound. When plants containing flavonoids will produce orange, pink, and red spot colors. From Figure 10, the spot color of chicken manure treatment and plant spacing of 40cm x 40cm is at Rf 0,68 which is closest to comparative compounds and has a red-orange color indicated as a flavonoid compound in the tapak liman.

4. Conclusion

Based on the research results, can be concluded that, chicken manure has a good yield potential for the growth of tapak liman seen from leaves length component reaches to 13,86 cm, the highest crown diameter of 27,65 Cm, fresh stover weight of 98,69 g and dry stover weight of 17,14 g. Plant spacing treatment of 40 cm x 40 cm can increase the leaves length of tapak liman of 13.49 cm, crown diameter of 27,27 cm, fresh stover weight of 91,53 g, and the dry stover weight of 16,07g. The combination of chicken manure and plant spacing of 40 cm x 40 cm can increase the total extract content by 22% and indicated flavonoid compound

Based on this research, it is necessary to do research by using other types of fertilizers in order to determine which the best fertilizer in improving the growth and production of tapak liman. Need to do the analysis until the quantitative stage to known how much flavonoid levels in the tapak liman. Another secondary metabolite analysis was required to know what compounds are contained in the tapak liman.

Acknowledgement

This research was supported by the Medicinal Plant and Traditional Drug, Tawangmangu, Indonesia.

References

[1] Afif T, D Kastono, Prapto Y 2014 Pengaruh macam pupuk kandang terhadap pertumbuhan dan hasil tiga kultivar kacang hijau (*Vigna radiata* L Wilczek) di lahan pasir pantai bugel, Kulon Progo *J. Vegetalika* 3(3): 78-88
[2] Amara DG, SM Mourad 2013 Influense of organic manure on the vegetative growth and tuber production of potato (*solanum tuberosum* L varspunta) in a Sahara desert region *J. Agri. CropSci* 5(22): 2724-2731
[3] Arwani A, T Harwati, S Hardiatmi 2013 Pengaruh jumlah benih per lubang terhadap pertumbuhan dan hasil jagung manis *(Zea Mays Saccharata Sturt) J. Inovasi Pertanian* 12(2): 27-40
[4] Aziz Ha, Arman 2013 Respons jarak tanam dan dosis pupuk organik granul yang berbeda terhadap pertumbuhan dan produksi tanaman jagung manis *J. Agrisistem* 9(1): 16-23
[5] Budiarti V, Alifia N 2012 Pemanfaatan zat laktagogum pada daun tapak liman (*elephantopus scaber* l) Sebagai alternatif memperlancar hormon oksitosin dalam pengeluaran ASI eksklusif *J. BIMABI* 1(3): 12-18
[6] Desyanto E, Herman BS 2014 Pengaruh jarak tanam terhadap pertumbuhan hijauan dan hasil buah jagung (*zea mays* L) Pada varietas bisi dan pioneer di lahan marginal *J. Agro.UPY* 5(2): 51-66
[7] Dewick PM 2009 Medicinal natural products 3rd edition United Kingdom: A John Wiley and Sons, Publication
[8] Handayani IA, Benbasyar E Dea DU 2016 Perbandingan kadar flavonoid ekstrak buah mahkota dewa (*Phaleria macrocarpa* [Scheff] Boerl) secara remaserasi dan perkolasi J Ilmiah Ibnu Sina 1(1): 79-87
[9] Ho WY, Ky H, Yeap SK, Rahim RA, Omar AR 2009 Traditional practice, bioactivities, and commercialization potential of *Elephantopus scaber* Linn *J. Medicinal Plants Research* **3** (13): 1212-1221

[10] Marsiwi T, Setyastuti P, Djoko P 2015 Pengaruh jarak tanam dan takaran pupuk npk terhadap pertumbuhan dan hasil benih kacang hijau (*Vigna Radiata* L Wilczek) *J. Vegetalika* 4(2): 124-132

[11] Mayadewi NYA 2007 Pengaruh jenis pupuk kandang dan jarak tanam terhadap pertumbuhan gulma dan hasil jagung manis *J. Agritrop* 26(4): 1153-159

[12] Morais H, Medri ME, Marur JC, Caramori PH, Ribeiro AM, Gomes JC 2004 Modifications on leaf anatomy of coffea arabica caused by shade of pigeonpea (*Cajanus cajan*) International *Journal of Brazilian Archives of Biology and Technology* 47(6): 863-871

[13] Pangli M 2014 Pengaruh jarak tanaman terhadap pertumbuhan dan hasil kedelai (*glycine max* l merril) J AgroPet 11 (1): 1-9

[14] Plaster EJ 2003 Soil science and mangement Delmar Learning Inc 4th ed United States 384p

[15] Pujiasmanto B, Pratignya S, Toeranto et al 2009 Pengaruh macam dan dosis pupuk organik terhadap pertumbuhan dan hasil tanaman sambiloto (*Andrographis Paniculata* Ness) *J Ilmu Tanah dan Agroklimatologi* 6 (2): 81-90

[16] Supartha INY, Gedhe W, GM Adnyana 2012 Aplikasi jenis pupuk organik pada tanaman padi sistem pertanian organik *E-jurnal Agroteknologi Tropika* 1(2): 98-106

IOP Conf. Series: Earth and Environmental Science **142** (2018) 012039 doi:10.1088/1755-1315/142/1/012039

Analysis of rice purchase decision on rice consumer in Bandung city

K Kusno, A Imannurdin, N Syamsiyah and E Djuwendah
Agriculture Economic and Social Science Department, Faculty of Agriculture, Padjadjaran University, Jalan Raya Bandung Sumedang KM 21, Jatinangor, 45363, Indonesia

E-mail: kuswarini.kusno@unpad.ac.id

Abstract. This study was conducted at three kinds of purchase location which were traditional market, rice kiosk, and supermarket in Bandung City, with survey data of 108 respondents which were selected by systematic random sampling. The aim of this study is to (1) identify consumer characteristics, (2) identify which atribute is considered by consumer in buying rice, and (3) analyze the relationship between purchase decision and income class. Data were analyzed by descriptive analysis and Chi Square test. The results showed most consumers in the traditional market were middle-educated and lower middle-income, at the rice kiosk, the consumer were generally middle-educated and middle-income, and in the supermarkets, the majority were high-educated and upper middle-income consumers. "Kepulenan" be the first priority of most consumers, but for the lower-middle class, the main priority was price. Thus, in case of scarcity and rice price increase, the government should immediately arrange market operations which targeting to lower-middle class consumers. There was a significant relationship between (1) the quality of rice consumed, (2) the frequency of rice purchase per month, and (3) attitudes toward rice price increase; each with the income class. Although the price of rice increase, consumers of middle and upper-middle were remain loyal to the quality of rice they consumed. This indicates rice market in Bandung city is an ideal market for premium rice so that traders and producers are expected to maintain the quality of rice, such as keep using superior seeds and applying good cultivation based on Good Agricultural Practice (GAP) rules.

1. Introduction

Rice is one of food commodity that has an important role from the side of producers, consumers, government, society and also from the environment in general [1]. Ninety persen of Indonesians consume rice as staple food [2].

There are 3 types of retail (purchase location) of rice that most chosen by consumers that is traditional market (55%), rice kiosk (25%), and supermarket (13%) [3]. This is largely determined by forces of demand and supply which has changed dynamically. Rice is consumed by consumers, both individual consumers, households and business service consumers.

The wide choice of rice product, whether based on the type of rice, packaging, price, taste, and other things as well as the differences and effects of the cultural environment, social class, purchasing power, motivation, and lifestyle shape different consumer behaviors. In addition, there has been a fundamental change in consumer preferences of agricultural products [4]. Changes in consumer preferences from just buying commodity to buying product leads to increased demand on the quality of rice which makes rice not only viewed as a staple food but also as a product with its various attributes. This requires producers and market actors in the rice trading system to provide rice that matches the consumers' wishes, in accordance with the intended market segment. The shifting rice demand on premium quality and high consumer's Willingness to Pay for better quality indicated a great opportunity for increased income of farmers if the market signal was understood by rice farmers [5].

The pattern of rice consumption is influenced by the level of income [6]. Increasing one's income will not increase the quantity of rice but more on improving the quality of rice consumed. Therefore, the higher the level of one's welfare, the proportion of spending on rice tends to be smaller, and vice versa.

Bandung city has a diverse community structure covering ethnicity, work background, income, culture and economy level. The diversity will certainly affect the people in the city in making decisions about the consumption of a product, including the consumption of rice. The most important stage in the purchase decision process is purchase stage where at this stage consumers take decisions about when to buy rice, where to buy, and how to buy.

The problems identified are how is the characteristics of rice consumers at traditional market, rice kiosk, and supermarkets? Is there any relationship between purchase decision and income? The aim of the study is to (1) determine consumer characteristics in the three types of purchase locations, (2) determine the rice attributes that consumers consider in buying rice and (3) test the relationship between purchase decision and income class.

2. Methods

This study was carried out in February to August 2016 at three types of purchase locations which were traditional market, rice kiosk, and supermarket in Bandung City, with survey data of 108 respondents which selected by systematic random sampling. The research design used was quantitative with survey technique. Variable of consumer characteristics were education level of the household head and the total income level of the family. Rice atributes were: 'kepulenan', rice price, type of rice (quality of rice), physical properties, and other. Variable of purchase decision were: a) frequency of purchases per month, b) way of purchase, c) type of rice, d) attitudes toward rice price increase, e) rice price per kg, f) attitude toward unavailability of rice usually consumed.

The data collected consists of primary and secondary data. Primary data was obtained through interviews of respondents by questionnaires; while secondary data were obtained from the Central Bureau of Statistics, scientific papers, and other sources relevant to the study. The selection of respondents can be explained as follows: Of 30 districts in Bandung, there are only 10 districts that have the three types of purchase locations. Then 3 districts were chosen randomly. In each selected district, the number of each type of purchase location was recorded. From each type of purchase location in each district, randomly selected 1 traditional market location, 1 rice kiosk and 1 supermarket location, so there were 9 purchase locations. 108th of respondents proportionally spread to each district and type of purchase location, so that 36 respondents were taken from each district and 12 respondents from each type of location. At every selected purchase location, samples were chosen by Systematic Random Sampling, in which the first sample had to be randomly selected, and the next sample was selected after a certain time interval. The first sample taken is considered random since the total number of consumers who come to a retail is never known, so that each consumer has the same opportunity to be selected as a sample. The second sample of respondent to 108th was chosen every half an hour. This half an hour determination is the approximate time needed to interview a respondent.

Data were analyzed descriptively by cross-tabulation and inferentially by using Chi Square test. There is a grouping of education and income level. The grouping of education level was based on Constitution Number 20 Year 2003 regarding National Education System, which is low educated class (elementary education), middle educated class (junior-senior high school), and high educated class (graduate-post graduate). The income level grouping was based on World Bank classification which are lower middle class (less than IDR 2.6 million), middle class (2.6 - IDR 6 million) [7] and upper middle class (more than IDR 6 million). We then cross-tabulated consumer characteristics variables with the type of purchase location, and the rice attributes with income class. To test whether there is a relationship between each of the purchase decision variables and the income class, we used Chi Square test which apply the following formula [8]:

$$\chi^2 = \sum_{i=1}^{r}\sum_{j=1}^{k}\frac{(O_{ij} - E_{ij})^2}{E_{ij}} \qquad (1)$$

χ^2: Chi Square
O_{ij}: observed frequency
E_{ij}: expected frequency

The hypothesis were H_0: there is no relationship between the two variables versus H_1: there is a relationship between the two variables. At level of significance (α) = 5% with a certain degree of freedom; if the calculated χ^2 is $\leq$ the table value, accept H_0, while if the calculated χ^2 is > the table value, then reject H_0.

3. Results and Discussion

The selected district were Bandung Kulon, Buah Batu and Cidadap. Selected traditional markets were Pasar Cijerah, Pasar Gordon, and Pasar Gandok; and selected rice kiosks were Kios Beras, Jalan Nilem, and Zalfaa; and selected supermarkets were Borma Cijerah, Griya, and Griya Setiabudi.

3.1 Characteristics of Rice Consumers

One's level of education will affect one's thinking and perception of something [9]. Table 1 shows the level of consumer head education class at each purchase location.

Table 1. Consumer Characteristics Based on Education Level at Each Purchase Location

Education Level	Purchase Location					
	TM		RK		SUP	
	f_i	%	f_i	%	f_i	%
Low	9	25	4	11	1	3
Middle	19	53	22	61	7	19
High	8	22	10	28	28	78
Total	36	100	36	100	36	100

Note: IC1=Lower-middle; IC2=Middle; IC3=Upper-middle

It can be seen that consumer in traditional markets and rice kiosks were dominated by middle-educated consumers, while in supermarket, the majority of consumers were highly educated. Consumer characteristics according to income level are presented in Table 2. It shows that the majority of traditional market consumers were lower middle income, at ricem kiosk, it was dominated by middle-income consumers and in supermarket, the majority were upper-middle consumers. This condition is consistent with consumers' purchasing power where the average price of rice in traditional markets and rice kiosks is relatively lower than that happened in supermarkets.

Table 2. Consumer Characteristics Based on Income Class at Each Purchase Location

Income Class	Purchase Location TM		RK		SUP	
	f_i	%	f_i	%	f_i	%
IC1	17	47	12	33	2	6
IC2	14	39	18	50	10	27
IC3	5	14	6	17	24	67
Total	36	100	36	100	36	100
AVG (IDR/Kg)	10,556		10,969		13,628	

Note: TM=Traditional Market; RK=Rice Kiosk; SUP=Supermarket. AVG=Average Rice Price

3.2 The Rice Attributes that Consumer Consider in Purchasing Rice

Based on the income class, there is a difference in the consumer's consideration of the price of rice. Lower middle income class was more considering the price of rice than *kepulenan*. This findings is interpreted from Table 3.

General product purchasing could be distinguished from fully planned purchasing, semi-planned, and unplanned purchasing. Since rice is a staple food with a high level of necessity, it is more appropriate if rice is classified in the planned purchase process, and every consumer will surely set aside some of his or her income for buying rice. This is consistent with the data presented in Table 4, where 65% of consumers plan to purchase rice with certain cultivar, brands, or price; while 32% of consumers made plans in half; they

Table 3. *The Rice Attribute Considered By Consumers Based On Income Class*

Rice Attributes	Income Class IC1		IC2		IC3	
	f_i	%	f_i	%	f_i	%
'Kepulenan'	12	39	17	40	13	37
Rice price	15	48	9	21	3	9
Type of rice	1	3	5	12	11	31
Physical properties	0	0	6	14	5	14
Other	3	10	5	13	3	9
Total	31	100	42	100	35	100

Had already known that they are going to buy rice but had not yet decided on the cultivar or brand they will buy until they got complete information from the seller or from the store display.

Table 4. *Frequency Distribution of Relationship Between Way of Purchase and Income Class*

Way of Purchase	Income Class Lower-middle	Middle	Upper-middle	Total
Planned	20	30	20	70 (= 65%)
Semi-planned	11	10	14	35 (= 32%)
Unplanned	0	2	1	3 (= (3%)
Total	31	42	35	108 (= 100%)

3.3 The Relationship Between Purchase Decision and Income Class

The relationship between the way of purchase and the income class was non significant. This is because at the level of significance (α) = 5%, the calculated value of the χ^2 test (= 3.73) is smaller than the table value (= 9.488) at the degrees of freedom = 4. Thus, there was not enough evidence to reject the null hypothesis.

In general, the rice consumed by consumers was rice that tasty and/or aromatic. Field observation found that rice is practically distinguishable based on price and physical rice itself. Average Medium-quality rice price was less than IDR 11,000/kg, while premium quality was more than IDR 11,000/kg. Precisely the average price of rice consumed by consumers was based on income class can be seen in Table 5. The table also shows that the number of consumers who buy premium rice was 4 times larger than those who buy medium rice. The higher the income class, the higher the average rice price.

Table 5. Frequency Distribution of Relationship Between Rice Type and Income Class

Type/Cultivar		Income Class			**Total**
		Lower-middle	Middle	Upper-middle	
Premium	Pandan Wangi	3	11	6	20
	Cianjur	6	5	8	19
	Setra Ramos	2	8	7	17
	Kurmo	7	5	4	16
	Setrawangi	0	5	4	9
	Ciparay	0	2	2	4
	Rojolele	0	0	3	3
	Sub-total				88 (= 81.5%)
Medium	Sumedang	3	2	0	5
	IR 64	3	1	1	5
	Subang	2	0	0	2
	Other	5	3	0	8
	Sub-total				20 (= 18.5%)
Total		31	42	35	108 (= 100%)
Average Rice Price (IDR/Kg)		10,848.39	11,645.24	13,220	11,926.9

This indicates that the higher the income class, the higher the purchasing power of consumers on rice. This result is in line with the χ^2 test indicating that there was a significant relationship between the quality of rice (i.e. premium or medium) consumed and income class because the calculated value of χ^2 test (= 18.85) at degree of freedom = 2 is greater than the table value (= 5.99).

Table 6. Frequency Distribution of Relationship Between Frequency of Rice Purchase and Income Class

Frequency of Rice Purchase	Income Class			Total
	Lower-middle	Middle	Upper-middle	
1time	8	23	18	49 (= 45%)
2 times	5	4	3	12 (= 11%)
3 times	2	2	7	11 (= 10%)
4 times	10	7	6	23 (= 21%)
> 4 times	6	6	1	13 (= 12%)
Total	31	42	35	108 (= 100%)

The frequency of rice purchase by the consumers can be seen in Table 6. It is clear that the majority of consumers (45%) bought rice once per month, followed by the purchase of 4 times per month as much as 21% of consumers. It was obtained that the calculated value of χ^2 test (= 16.124) is greater than the table value (= 15.507) at degree of freedom =8. The conclusion was to reject the null hypothesis or in other words, there was a significant relationship between the frequency of consumers buying rice and income class.

Table 7 describes that if the price of rice increased, 80% of consumers will continue to consume the same rice (which they usually buy). The result explained that the increase or decrease in rice price will have relatively a little effect on the changes in rice demand [10]. This is because people will not increase or decrease their consumption of rice significantly.

Table 7. Frequency Distribution of Relationship Between Consumer Attitudes toward Rice Price Increase and Income Class

Attitude toward Rice Price Increase	Income Class			Total
	Lower-middle	Middle	Upper-middle	
Buy cheaper rice	14	5	0	19 (= 18%)
Not affected	17	35	33	85 (= 80%)
Buy the same rice but reduce the amount	0	2	2	4 (= 3%)
Total	31	42	35	108 (= 100%)

The calculated value of χ^2 test (= 25.471) is larger than the table value (= 9.488) at degree of freedom = 4, so it can be that there is a significant relationship between attitudes on rice price increase and consumer income class.

The consumer's attitude if the desired rice is not available at the purchase location is presented in Table 8. Most consumers (70%) bought other types of rice in the same place if the rice usually consumed is not available at the purchase location. This is because

Table 8. Frequency Distribution of Relationship Between Consumer Attitudes toward Unavailability of Rice Usually Consumed and Income Class

Attitudes toward Unavailability of Rice	Income Class			Total
	Lower-middle	Middle	Upper-middle	
Buy other type of rice at the same place	23	33	20	76 (= 70%)
Buy the same rice elsewhere	7	8	12	27 (= 25%)
Delaying rice purchase	1	1	3	5 (= 5%)
Total	31	42	35	108 (= 100%)

Consumers have long term patronage to the seller. The distance and convenience of shopping also makes consumers reluctant to buy rice elsewhere. The calculated value of χ^2 (= 4.976) is smaller than the table value (= 9.488) at degree of freedom = 4, so there is no significant relationship between attitude toward unavailability of rice consumed and income class.

4. Conclusion

Rice consumers in Bandung city who buy rice to traditional markets were generally middle-educated and lower-middle income; those who buy rice to rice kiosks were generally middle-educated and middle-income; whereas consumers who buy rice to supermarkets were generally highly educated and upper middle income. "Kepulenan" be the first priority of most consumers, but for the lower-middle class, the main priority was price. There is a significant relationship between (1) the quality of rice consumed, (2) the frequency of rice purchase per month, and (3) attitudes toward rice price increase;

each with the income class. Although the price of rice increases, consumers of middle and upper-middle were remain loyal to the quality of rice they consumed. This indicates rice market in Bandung city is an ideal market for premium rice so that traders and producers are expected to maintain the quality of rice, such as keep using superior seeds and applying good cultivation based on Good Agricultural Practice (GAP) rules.

References

[1] Syaukat Y 2016 *Proposal Penelitian Beras* (Bogor: Perhimpunan Ekonomi Pertanian Indonesia)
[2] Center for Data and Agricultural Information 2015 *Outlook Komoditas Pertanian Subsektor Tanaman Pangan* (Jakarta: PADI)
[3] Natawidjaja, Ronnie. S dan Sulistyowati, L 2010 Analisis rantai pasok dan rantai nilai beras di Jawa Barat *J Agrikultura* **21** 2 pp 128-136
[4] Tambunan M 2010 *Membangun Industrialisasi yang Tangguh BerbasisAgroindustri dalam Menghadapi Perdagangan Bebas dalam Refleksi Agribisnis : 65 Tahun Profesor Bungaran Saragih* (Bogor: IPB Press)
[5] Natawidjaja RS, Sulistiowaty L, Kusno K, Aryani D, dan Saefudin BR 2016 *Analisis Preferensi, Kepuasan, dan Willingness to Pay Konsumen Beras di Kota Bandung*. (Bandung: PERHEPI Komisariat Bandung)
[6] Selamet R 2003 *Analisis Proses Keputusan Konsumen dalam Pembelian Beras dan Strategi Pemasaran Beras* (Bogor: Program Studi Agribisnis IPB)
[7] Widiatmanti H 2015 *Penghasilan Kelas Menengah Naik = Potensi Pajak? Badan Pendidikan dan Pelatihan Keuangan* (Jakarta: Kementerian Keuangan RI)
[8] Siegel S 2009 *Nonparametric Statistics for The Behavioral Sciences* (Singapore: McGraw-Hill Book Company) p. 171
[9] Kotler dan Keller 2007 *Manajemen Pemasaran edisi kedua belas Jilid 1 dan 2* (PT Indeks)
[10] Sari, Nina T 2007 *Analisis Dampak Kenaikan Harga Beras Terhadap Pola Konsumsi Beras Rumah Tangga di Cipinang, Jakarta Timur*. (Bogor: Skripsi. Program Studi Manajemen Agribisnis. Fakultas Pertanian. Institut Pertanian Bogor)

IOP Conf. Series: Earth and Environmental Science **142** (2018) 012040 doi:10.1088/1755-1315/142/1/012040

First record of entomopathogenic fungi on autumn leaf Caterpillar (*Doleschallia bisaltide*)

A K Dayanti[1], Sholahuddin[2], A Yunus[2] and D Subositi[3]
[1]Undergraduate Student of Agrotechnology Major, Agriculture Faculty Universitas Sebelas Maret, Surakarta, Indonesia
[2]Lecturer Staff of Agrotechnology Major, Agriculture Faaculty of Universitas Sebelas Maret, Surakarta, Indonesia
[3]Research center for Medicinal Plant and Traditional Drug, Tawangmangu, Indonesia

E-mail:yunus.uns7@yahoo.com

Abstract. Caricature plant is one of the medicinal plants in Indonesia to cure hemorrhoids, menstruation, and others. The cultivation constraints of caricature plant is autumn leaf caterpillars (*Doleschallia bisaltide*). Utilization of synthetic insecticides is not allowed to avoid bioaccumulation of chemical residues. Entomopathogenic fungi is an alternative way to control *D. bisaltide*. The objective of the research was to obtain isolates of entomopathogenic fungi of *D. bisaltide*. The research conducted by two steps, which were exsploration of infecfted D. bisaltide. The second step was identification of the fungi. Exploration results of16 pupae of *D. Bisaltide* were infected by fungi. Identification done by classify the mcroscopic and microscopic fungi isolate characteristic. One from five fungal isolates were entomopathogenic fungi from Verticillium genera.

1. Introduction

Caricature plant (Graptophyllum pictum L.) is one of the medicinal plants in Indonesia. The leaves are efficacious to cure diseases such as hemorrhoids, urinating, menstruation and others [8]. One of the constraints of caricature plant cultivation is a pest attack. Doleschallia bisaltide (autumn leaf caterpillar) is a main pest [10].

Pest controlled can be done with the application of synthetic insecticides. But, the use of synthetic pesticides has many negative impacts on ecology, especially the chemical residues. Therefore, applied synthetic pesticides is not permitted on medicinal plants to avoid bioaccumulation of chemical residues in the human body when it consumed. Biological control is one of the alternative pest control with environmentally friendly. One commonly used in biological control is to use entomopathogenic fungi [9].

Exploration is the first step in the implementation of biological control techniques. The first step in developing entomopathogenic fungi as the biological control agent that needs to be done is the collection of isolates from the host, the isolate characterization obtained and the pathogenicity test (Koch Posatultes) [5]. There are no research on entomopathogenic fungi of *D. bisaltide* before. Therefore, the preliminary research is needed to determine the species of potential entomopathogenic fungi to control *D. bisaltide*.

IOP Conf. Series: Earth and Environmental Science **142** (2018) 012040 doi:10.1088/1755-1315/142/1/012040

2. Methods

The research was conducted at B2P2TOOT Field (Central Research and Development of Medicinal and Traditional Medicinal Plants) located in Tawangmangu, Karangpandan and Karanganyar, Central Java, and also in Laboratory of Pest and Plant Pest Control. The study was conducted for 6 months, from November 2016 to May 2017. The research design was adapted from Rosmini and Lasmini [13]. The research was conducted in two phases, the first is exploration of *D. bisaltide* that infected by entomopathogenic fungi and the second stage is the identification.

The first phase using survey method to get infected *D. bisaltide* by entomopathogenic fungi in B2P2TOOT Karangpandan field with purposive sampling. The second stage is the isolation and identification of entomopathogenic fungi that infected *D. bisaltide*. Fungi identifcation by macroscopic and microscopic identification based on Humber [6], Ando [1] and Barnett & Hunter [3],[4]. The observed data were analyzed descriptively.

3. Results and Discussion

3.1. General condition of research location

The experiment was conducted in B2P2TOOT Field in Toh Kuning Village, Karangpandan Sub-district, Karanganyar Regency, Central Java Province. The average temperature is 25.3^0C and an average humidity is 88.4% during the research.

3.2. Exploration and identification entomopathogen fungi

Exploration done by collecting D. bisaltide infected with fungi in the caricature plant on B2P2TOOT Karangpandan and Tawangmangu field. There are 16 pupae of *D. bisaltide* found during exploration. The similarity symptoms of infection is appears white hyphae that enveloped the surface of the pupae, and when the pupae dissected, it will appear white hyphae in the pupae body. Entomopathogenic fungi will penetrate into the host's body through the cuticle as an initial step of infection [15].

Infected *D. bisaltide* dissected using sterile tweezers and a scalpel. The inner hyphae of the pupa is drawn with an ose needle and isolated on a PDA medium. Isolation was performed several times until some fungi isolates grew. The isolation of the fungus was done to separate the entomopathogenic fungi from the infected host, to obtain a pure culture isolate [14]. The result of isolation, get 5 isolates, namely isolate 1, isolate 2, isolate 3, isolate 4 and isolate 5. Isolates of fungi subcultured into other PDA media. Subcultures were performed for purification and propagation of isolates.

Isolates observed until it full in petridish. Observation of isolates was done macroscopically and microscopically. Macroscopic observations include patterns of fungal colony spreading, colony shapes and colony color from either the upper surface or the bottom. Pure fungi grown in media and then identified in macroscopic morphology and microscopic morphology.

Table 1. Data of macroscopic morphology identification of fungi isolate

No.	Isolate Fungi	Color		Pattern of Colonies Distribution	Time Meets Petri
		Surface	Basic		
1.	Isolate 1	White Yellow	White Yellow	The Colony Spreads Flat	> 30 Days
2.	Isolate 2	Dark Green Whites	Brown Chocolate	The Colony Spreads Flat	9 Days
3.	Isolate 3	White	White Tanned	Unfavorable Colony Edge	9 Days
4.	Isolate 4	White	White	The Colony Spreads Flat	7 days
5.	Isolate 5	White	White	The Colony Spreads Flat	9 Days

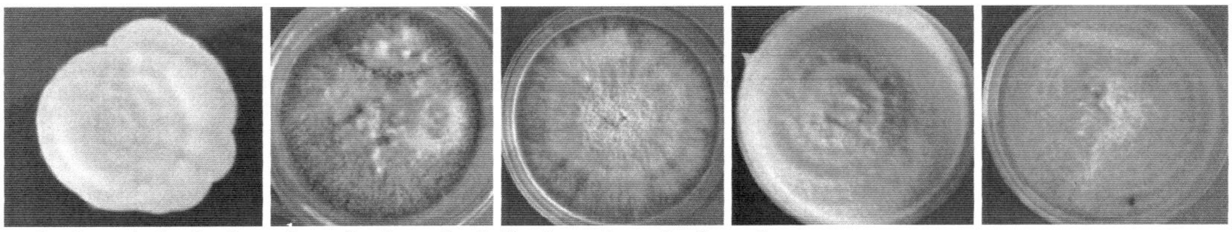

a. Isolate 1, b. Isolate 2, c. Isolate 3, d. Isolate 4, e. Isolate 5

Figure 1. Pure isolate culture fungus on PDA media

Microscopic identification based on conidia is an important component of identification. When the sexual structure associated with the conidial, then can be a valid taxonomic classification [3],[4]. Conidial morphology is required in the identification of imperfect fungi. Conidia shape in isolates 1 is oblong, fusiform-elliptical is isolates 2, isolates 3 is cylindrical, isolates 4 is catenate catenulate and isolates 5 is elliptic fusiform. The number of cells in conidia between 5 isolates also had diversity, but most isolates (isolates 2, 4 and 5) have more than one septa (Pragmospore). Isolate one does not have septa (Amerospore), whereas isolates 3 only have one septum (Didymospore). The size of the conidia on each isolate was different, but all isolates ranged from 7.6 to 16.4 μm.

Table 2. Data of microscopic morphology by conidia identification of fungi isolate

No.	Isolate Fungi	Konidia			
		Form	Number of Cells in Konidia	Konidia Distribution Pattern	Size (μm)
1.	Isolate 1	Oblong	Amerospore	Clustered on conidiofor	± 4 -12
2.	Isolate 2	Fusiform-elliptical	Phragmospor	Clustered on Conidiofor	± 18 -28
3.	Isolate 3	Cylindrical	Didymospore	Clustered on Conidiofor	± 5 - 10
4.	Isolate 4	Catenate catenulate	Phragmospor	Clustered on Conidiofor	± 6 - 13
5.	Isolate 5	Elliptic-fusiform	Phragmospor	Clustered on Conidiofor	± 5 - 19

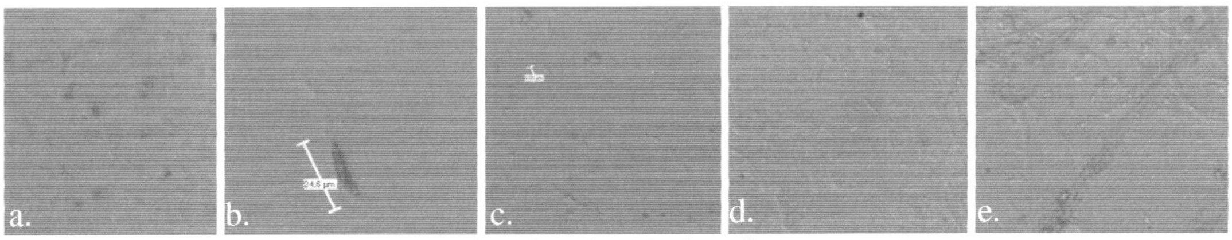

a. Isolate 1, b. Isolate 2, c. Isolate 3, d. Isolate 4, e. Isolate 5

Figure 2. Microscopic observation (Konidia)

The macroscopic and microscopic morphological diversity of fungal isolates indicates that there is a classification difference between one isolate and the other, whether the species, genus or family are different. Identification was carried out to the genus level. Identification of entomopathogenic fungi in rhizosphere in vegetable crops was carried out to the genus level by observing the characteristics of the fungus macroscopically and microscopically [16]. The morphology identification can not describe phylogeny to the species level, therefore molecular identification techniques are also performed [18].

Table 3. Data of macroscopic and microscopic morphological identification of entomopathogen fungi isolate

Isolate fungi	Macroscopic Color		Hypha	Konidia		Genus Fungus
	Surface	Basic		Form	Spread	
Isolate 1 (Isolate)	White Yellow	White Yellow	Branch	Oblong	Clustered on conidiofor	Verticillium*
Isolate 2 (2.1)	Dark Green Whites	Brown Chocolate	Branch	Fusiform-elliptical	Clustered on conidiofor	Cercosporella
Isolate 3 (B2.1)	White	White Tanned	No	Cylindrical	Clustered on conidiofor	Monilochaetes
Isolate 4 (1.1A)	White	White	No	Catenate catenulate	Clustered on conidiofor	Thielaviopsis
Isolate 5 (A1 White)	White	White	Branch	Elliptic-fusiform	Clustered on conidiofor	Fusarium

*Entomopathogenic fungi

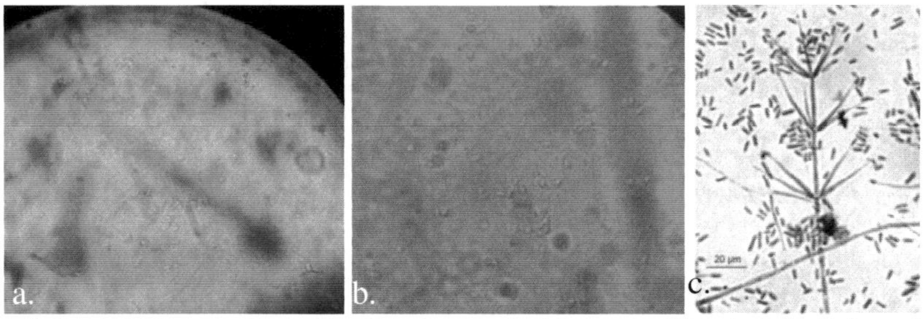

a. Infected Cocoon from Exploration with 40x magnification, b. Infected Cocoon from Application with 100x magnification, c. Verticillium fungus spores [7]

Figure 3. Microscopic Observation of Konidia Fungus Isolate

The results of identification based on macroscopic and microscopic morphology indicate that there is one isolate (isolate one) that has macroscopic color features on the surface and bottom of a yellowish white petri dish that turns into a pinkish brownish gray. Verticillium macroscopic morphology, the colony's color is pale yellowish white and becomes pink brown, red, green or gray. Microscopic observation, hyphae sectional, oblong-shaped and spreading conidia on conidiophores clustered so identified genus Verticillium [7]. The microscopic features of this fungus are slim conidiofor and circularly branched. Single or clustered Konidia [2].

There is one isolate which has morphology of surface color and base color is white. This is consistent with research Trizelia and Winarto [17], the results of the morphological characterization of fungi on PDA 14 days of incubation, colonies of Fusarium color is white. Isolates 5 white can grow full of petridish within 9 days, this is according to research Nuraida and Hashim [12], the rate of growth in colony diameter Fusarium exploration results shows that the growth of the colony faster than fungi Aschersonia spp., Paecilomyces, Metarhizium and Beauveria. At 7 days incubation, the diameter of the colony has reached an average of more than 6 cm. Microscopic hyphae sectional, shape Elliptic-fusiform conidia and conidiophores conidia clustered on, so identified include the genus Fusarium. Macroconidia generally crescent-shaped sectional three and measuring 30-40 x 4.5-5.5 µm and microconida cells round-shaped or oval that is characteristic of Fusarium [11].

4. Conclusion and Suggestion

There are five fungi isolate from isolation, one of those isolates was from Verticillium genera. Symptoms of an infected *D. bisaltide* is pupa will become darker brown, pupa body hardened and stiff, and the nascent white dots (fungal hyphae that grow). Suggestion that can be given by authors about this research are Verticillium fungi pathogenicity test against *D. bisaltide* with several treatment conidia, stadia *D. bisaltide* and location of the research should be conducted.

Acknowledgment

This research is supported by Balai Besar Penelitian dan Pengembangan Tanaman Obat dan Obat Tradisional Tawangmangu, Indonesia.

References

[1] Ando K. 1998. Workshop identifikasi fungi imperfekti dareah tropis 19-21 November 1998 in Yogyakarta.
[2] Arif A, Muin M, Kuswinanti T et al. 2008. *J Perennial,* 5 (1) 15-22.
[3] Barnett HL and Hunter BB. 1972. *Illustrated genera of imperfect fungi third edition.* America (US).
[4] Barnett HL and Hunter BB. 1972. *Illustrated genera of imperfect fungi fourth edition.* America (US).
[5] Hamdani, Trizelia and Yaherwandi. 2010. *J Manggaro,* 11 (2) 71 -76.
[6] Humber RA. 1998. *ARS Plant protection Research Unit.* America (US).
[7] Kidd S, Halliday C, Alexiou H et al. 2016. *Australian and New Zealand Mycoses Interest Group..* Australia (AUS).
[8] Lestari P, Khumaida N, Sartiami D, et al. 2015. *SABRAO J of Breeding and Genetics,* 47 (2) 172-184.
[9] Lovett B, J. Raymond dan Leger ST. 2017. *American Society for Microbiology Press.* America (US)
[10] Mardiningsih TL, Sartiami D, Sukmana C. 2008. Pros. Seminar Nasional Tumbuhan Obat Indonesia XXXV. Hal. 59-65. Serpong, Indonesia. 13-14 November 2008.
[11] Ngittu YS, Mantiri FR, Tallei TE, et. al. 2014. *J Ilmiah Farmasi,* 3 (3) 156 – 161.
[12] Nuraida dan Hasyim A. 2009. *J Hortikultura,* 19 (4) 419-432.
[13] Rosmini dan Lasmini SA. 2010. *J Agroland,* 17 (3) 205 – 212.
[14] Sanjaya Y, Ocampo VR dan Caoili BL. 2013. *J Agrivita,* 35 (1) 64 – 73.
[15] Shang Y, feng P dan Wang C. 2015. *J PloS Pathog,* 11 (8) 1-6.
[16] Trizelia, Armon N dan Jailani H. 2015. *Prosiding Seminar Nasional Masyarakat Biodiversiti Indonesia,* 1 (5) 998-1004.
[17] Trizelia dan Winarto. 2016. *Prosiding Seminar Nasional Masyarakat biodiversity Indonesia,* 2 (2) 277 – 281.
[18] Yusuf EV, Nuryani W dan Hanudin. 2016. *J Hortikultura,* 26 (2) : 2017-2222.

IOP Conf. Series: Earth and Environmental Science **142** (2018) 012041 doi:10.1088/1755-1315/142/1/012041

The utilization of microbes as a fermentation agent to reduce saponin in Trembesi leaves (*Sammanea saman*)

A K Sariri, A M W Mulyono and A I N Tari
Livestock Production Program, Agriculture Faculty, Veteran Bangun Nusantara University, Sukoharjo, Indonesia

E-mail: ak_sariri@ymail.com

Abstract. This objective of this research was to observe the utilization of microbes as a fermentation agent of trembesi leaves that can increase the quality of trembesi leaves as ruminants feed. Before fermentation, trembesi leaves were divided into three treatments. They were control = non-agentic in fermentation, D-An = the addition of *Aspergillus niger* as fermentation agent, and D-Lp = the addition of *Lactobacillus plantarum* as fermentation agent. Each treatment experienced five repetitions. The experimental design used a randomized direct pattern group design. The analysis included proximate analysis consisting of water content, crude protein content, crude fiber content, lipid content, mineral content (ash) and saponin content after fermentation. It could be concluded that the utilization of *Aspergillus niger and Lactobacillus plantarum* in fermentation could decrease saponin content and could increase the nutrient content of trembesi leaves by increasing crude protein content otherwise by decreasing crude fiber content of trembesi leaves.

1. Introduction

Ruminant livestock has a stomach with four plural compartments and is constantly in a process of rumination. Sustainability of rumination process depends on the presence of feed material which has a high content of crude fiber. Feed ingredients with high crude fiber content are found in many forage crops. Cells within the plants contain cell walls which typical nutrient composer is cellulose. Cellulose is a polymer of D-glucose by binding ß-1 and 4 glycoside [1].

Indonesia is a tropical country. A special feature on this climate is that there are two seasons namely rainy season and dry season in areas, beside that the tropical climate has high temperatures and humidity. This condition is actually less favorable for livestock especially ruminants cattle because the availability of forage is not guaranteed. During the rainy season, there are abundant mount of forage but during the dry season is very lacking, even in certain areas there is absolutely no forage stock.

In drought conditions, for example Wonogiri area, there will be massive sales of existing stock. This is due to the absence of staple forage ruminants, so to maintain the existing livestock, the farmers will provide improvised forage that is usually obtained from annual crops.

Trembesi is a perennial plant that is often referred to "ever green" from the family of Mimosoideae with their leaves, seeds and bark contain saponins. Moreover, Trembesi leaves and seeds contain polyphenols [2]. Saponins exist in all parts of the plant, for example in leaves, stems, roots, flowers and seeds while the amount varies according to time of cutting.

Saponins are glycosides that after hydrolyzed will produce sugar (glycan) and sapogenin (aglycone). Surface-active compounds of saponin has the quality as soap and is detected based on the ability to form foam on agitation and has a bitter taste that has the effect of decreasing the surface tension so can damage cell membranes and activate cell enzymes and also damage the cell proteins.

Saponins can affect the body's biological processes and nutrient metabolism by inhibiting enzymes such as productivity of chymotrypsin enzymes work that inhibit productivity and growth of livestock. The main biological effects of saponins are able to cause haemolysis of red blood cells due to interaction of saponins with membrane (proteins, phospholipids and cholesterol) of erythrocytes. Haemolysis is the release of hemoglobin into the blood plasma due to destruction of erythrocytes.

Feed containing more than 0.20% saponin would be bad for growth, feed intake and feed efficiency. Saponins in alfalfa can cause bloating in ruminants because saponins are the active agents in producing a foamy surface of soap. Low levels of alfalfa meal usage decrease the amount of average growth in poultry, the main effect of the saponin content is the palatability and feed intake compared to the effects of metabolism. The use of low strain increases the level of alfalfa saponins into food for growth ruminants without performance degradation.

Fermentation is the fission of carbohydrates into alcohol, lactic acid, butyric acid and carbonic acid, and heat release. Protein reforms into ammonia, amino acids, amides, acetic acid, butyric acid and water. In fermentation, there is a removal of anti nutritional substances that are toxic such as glucoside [3]. Furthermore, fermentation of cassava leaves by *Aspergillus niger* increases protein digestibility and decreases crude fiber value [4].

2. Methods

The research was conducted in Laboratory of Biology, Chemistry and Microbiology of Agriculture Veteran Bangun Nusantara Sukoharjo University. Proximate analysis was done at Laboratory of Agricultural Technology of Sebelas Maret University and saponin analysis was done in Biology Laboratory of Pharmacy Faculty Gadjah Mada University. The study was conducted by following these procedures:

2.1. *Preparation of fermentation media*

The trembesi leaves were separated from tertiary stems (*dipritil*) then being collected and weighed. Each experimental unit required 0.5 kg of trembesi leaves.

2.2. *Preparation of Aspergillus niger inoculum*

Creating media to grow *Aspergillus niger* which is potatoes dextrose broth (PDB). In this study, 150 ml of GDP were poured into 10 ml of each reaction tube for 4 treatments of An-1 and 15 ml in 4 test tubes of An-2 treatment. Then being sterilized. Growing the *Aspergillus niger* in GDP media for each treatment. Incubating for 5 days at room temperature aerobically.

2.3. *Implementation of fermentation*

The collected trembesi leaves were then divided into treatments:

Control : fermentation of trembesi leaves without microbial addition as an agent
D-An : addition of *Aspergillus niger* inoculum in fermentation of trembesi leaves
D-Lp : addition of *Lactobacillus plantarum* in fermentation of trembesi leaves

Each treatment was repeated 5 times then stirred evenly and put into polyethylene plastic, pressed compressed and then pressed with siller. After three days of dismantling, we measured the content of fermented trembesi leaves nutrients through proximate analysis and saponin analysis.

In this study the variables observed were: nutrient content including water content, crude protein, crude fat, ash and crude carbohydrate through proximate analysis and saponin content of trembesi leaves [5].

The data obtained were analyzed by using a complete randomized design (RAL) on unidirectional pattern. If the treatment factor showed significant effect ($P < 0.05$) Duncan's Multiple Range Test (DMRT) at 5% level were performed [6].

3. Results and Discussion

3.1. *The nutrient content of fermented trembesi leaves*

To know the difference of quality of fermented trembesi leaves with the addition of different microbes such as *Aspergillus niger* and *Lactobacillus plantarum*, it is needed to do the proximate analysis. The result of proximate analysis can be seen in Table 1.

Table 1. The nutrient content in fermented trembesi leaves

Treatment	Nutrient Content (%)				
	Water	Ash	CL	CP	CF
Kontrol	8.78	3.87^{a}	5.85^{c}	10.80	70.67^{c}
D-An	8.79	4.15^{b}	5.26^{a}	20.74^{c}	60.72^{b}
D-Lp	8.57	4.40^{c}	5.44^{b}	9.54^{b}	60.31^{a}

Different superscripts on the same column showed significant differences (P <0.01)

From Table 1 it was seen that with the addition of water content, the addition of *Aspergillus niger* and *Lactobacillus plantarum* provided a significant effect on the controls in water content, lipid, crude protein, crude fiber and ash. This showed that with the addition of microbes both *Aspergillus niger* and *Lactobacillus plantarum* in the fermentation of trembesi leaves could streamline the fermentation process. Fermentation was an effort to utilize microbes to increase the added value of a substance or substance with food or feed products [7].

The addition of *Aspergillus niger* and *Lactobacillus plantarum* in fermentation may increase the water content in the trembesi leaves although this increase is not significantly different when compared to the controls. Increased water content is due to fermentation produced molecular transformation so that the composition of nutrients contained in the leaves of trembesi experienced changes. This could be due to the activity of *Aspergillus niger* would produce acid namely citric acid. the acid state would inhibit the activity of microorganisms in the fission of carbohydrates and proteins which products are water vapor.

Table 1. showed that the ash content of the fermented leaves was relatively low but the addition of *Aspergillus niger* and *Lactobacillus plantarum* gave a marked difference compared to the subsequent controls. *Aspergillus niger* addition gives a marked difference compared to the addition of *Lactobacillus plantarum*. The ash content in the proximate analysis indicated the mineral content of the material. [8] stated that all human body tissues, animals and plants contained inorganic substances called minerals. Meanwhile, according to [9], proteins in plant leaves were existed in the cytoplasm of about half to one-third of the total protein, chloroplasts of which one-third to one-half of the total and the remainder of the nucleus were often called nucleoproteins.

Crude fiber content was found only in plant products. The presence of crude fiber in plants was due to the presence of lignin content that surrounds cellulose and hemicellulose in plant cell walls. The older the age of the plant the higher the crude fiber content of the plant [10]. Fermentation means changing the structure of feed ingredients into more easily digested, reducing allergen, anti-nutritive or indigestible components, to adding important anti-pathogenic, antioxidant and anti-carcinogenic metabolites [11]. Cellulose and hemicellulose are carbohydrates, when forage is fermented, microbes multiply rapidly and ferment carbohydrates into organic acids [12]. The existence of *Aspergillus niger* and *Lactobacillus plantarum* in the fermentation process turned out to significantly lower the content of crude fiber of trembesi leaves. With the decline in the content of crude fiber will be more easily digested forage. The addition of *Lactobacillus plantarum* in the fermentation of trembesi leaves was more able to reduce the coarse fiber content compared with the addition of *Aspergillus niger*. This suggested that bacteria more rapidly multiplied to have a higher ability in break the lignin bonds in cell walls. By breaking the lignin bond it would decrease the lignin content in the cell wall which would further decrease the crude fiber content. The reduction of the content of crude fiber, the forage would be more easily digested.

IOP Conf. Series: Earth and Environmental Science **142** (2018) 012041 doi:10.1088/1755-1315/142/1/012041

3.2. Saponin content of fermented trembesi leaves

The purpose of this study was to determine the effectiveness of microbes as a fermentation agent in trembesi leaves which could improve the quality of trembesi leaves as ruminants feed. To see the quality of a material as animal feed, it was necessary to see the nutrient and anti-nutrient content of the material. Trembesi leaves contain saponin anti-nutrient substances [13]. The content of saponin of trembesi leaves which has been fermented by the addition of *Aspergillus niger* and *Lactobacillus plantarum* can be seen in Table 2.

Table 2. Saponin content in fermented trembesi leaves

Treatment	Saponin Content (%)
Kontrol	1.468[a]
D-An	0.920[b]
D-Lp	0.888[b]

Different superscripts on the same row and column showvery significants differences (P <0.01)

Table 2. showed that the addition of *Aspergillus niger* and *Lactobacillus plantarum* in the fermentation of the trembesi leaves gave a significant effect on the decrease of saponin content in trembesi leaves compared with the control. Although it had a significant effect but the decrease in saponin content in this study has not yet reached the tolerable limit of saponin content in feed. [14] states that feeds containing more than 0.20% saponin will adversely affect growth, feed consumption and feed efficiency.

The addition of *Aspergillus niger* to the fermentation of trembesi leaves was not significantly different with the addition of *Lactobacillus plantarum* in reducing the saponin content of trembesi leaves. This was in accordance with the opinions of [8] stating that fermentation is the utilization of microbes to increase the added value of a substance or material. And with the fermentation occurs elimination of anti-nutritional substances that are toxic [4].

4. Conclusion

The utilization of *Aspergillus niger* and Lactobacillus plantarum in fermentation could reduce saponin content in trembesi leaves and could increase the trembesi leaves' nutrient content by increasing the crude protein content otherwise by decreasing the crude fiber content.

References

[1] Carlile MJ, SC Watkinson and GW Gooday 2001 *The Fungi* 2nd (London-California: Academy Press)
[2] Lade HS, MP Chitanand, G Gyananath and TA Kadam 2006 Studies on some properties of bacteriocins produced by Lactobacillus species isolated from agro-based waste *The internet J. microbial* **2** pp 1937-8289
[3] Suliantari dan WP Rahayu 1990 *Teknologi Fermentasi Biji-bijian dan Umbi-umbian* 9(Bogor: PAU-IPB)
[4] Balitnak 1994 *Pemanfaatan Limbah Pertanian dan Limbah Pengolahan Tapioka/Sagu sebagai Pakan Ternak* (Warta Penelitian dan Pengembangan Pertanian)
[5] Soejono M 2004 *Petunjuk Laboratorium Analisis dan Evaluasi Pakan* (Yogyakarta: UGM)
[6] Ali-Mursyid WM 2011 *Buku Ajar : Rancangan Percobaan* (Yogyakarta: Kepel Press)
[7] Rachman A 1989 *Pengantar Teknologi Fermentasi* (Bogor: PAU-IPB)
[8] Prawirokusumo S 1993 *Ilmu Gizi Komparative* (Yogyakarta: BPFE)
[9] Dwidjoseputro D 1990 *Pengantar Fisiologi Tumbuhan* (Jakarta: PT Gramedia Pustaka Utama)
[10] Parakkasi A 1995 *Ilmu Nutrisi dan Makanan Ternak Ruminan* (Jakarta: Universitas Indonesia)
[11] Brouk M and R Belyea 1993 Chewing activity and digestive responses of cows few alfafa forages *J. Dairy Sci* **76** pp 175:-182

[12] Bosch MW, SCW Lammers-Wienhoven, GA Bangma, H Boer, and PWM van Adrichem 1992 Influence of stage of maturity of grass silages on digestion processesin dairy cows 2 Rumen contens, passage rates, distribution of rumen and faecal particles and mastication activity *Livest ProdSci* **32** pp 265-281
[13] Sariri AK 2012 *Fermentasi dengan Menggunakan Berbagai Jenis Mikrobia untukMenurunkan Kandungan Saponin Daun Trembesi* (*Albizia saman*) (Sukoharjo: Laporan Penelitian Kompetitif Universitas Veteran Bangun Nusantara)
[14] Francis, George; Zohar Kerem, Harinder P S Makkar and Klaus Becker 2002 The biological action of saponins in animal systems: a review *British J. Nutrition* **88** pp 587–605

Development of small and medium business (SMES) of mango dodol processing to increase the added value (a case study in Ujungjaya Village, Indramayu District, West Java)

L Sulistyowati, P Pardian, N Syamsyiah and Y Deliana
Department of Agricultural Economics and Social Science, Faculty of Agriculture, Padjadjaran University, Indonesia

E-mail : lies.sulistyowati@unpad.ac.id

Abstract. In the national economic development in Indonesia, Small and Medium Enterprises (SMEs) become a priority to be developed, because SMEs can be the backbone of the populist economic system to reduce the problem of poverty. In addition, the development of SMEs is able to expand the economic base and can contribute to the increase of added value, in addition it would also serve to open employment opportunities in rural areas. Indramayu is one of the three mango production centers in West Java that face the problem that there are about 20% of the mangoes that is not worth selling. This opportunity is utilized by women who are members of KUB (Joint Business Group) to be processed into mango dodol at household scale. But this effort has not been widespread, only pioneered by a small portion of women. This study aims toobserve the driving force of women to participate in the processing of mango dodol, and whether the mango processing business to become mango dodol is profitable, also how much added value obtained. This study uses case study method with interview for data collection, participant observation and documentation study. While the data analysis technique using Hayami Value-added Method and descriptive analysis. The results revealed that the factors that affect the women's participation in the processing of dodol is to increase family income, take advantage of spare time and take advantage of rejected mangoes. The added value obtained in mango dodol processing is Rp.50.600,00 per kilogram of input, with a value-added ratio of 52.8%. For the development of SMEs mangoes Training and socialization are needed for the good dodol processing and hygienic according to SOP (Standard Operational Procedure) from the relevant institutions, innovation in packaging, pioneering business partnerships with stores in the city of Indramayu and surrounding areas, and support financing from banks with an affordable interest rate.

1. Introduction

In the national economic development in Indonesia, Micro-Small and Medium Enterprises (SMEs) become a priority to be developed, because SMEs can be the backbone of the populist economic system to reduce poverty problems [16]. In addition, the development of SMEs able to expand the economic base and can contribute to increase added value, in addition to opening employment in rural areas. Indramayu is one of the 3 mango production centers in West Java. Data from the Department of Agriculture shows that the number of mango plants in Indramayu is the largest in West Java, which is about 1,391,926 trees, which consists of various types of mango. The most dominant varieties are cengkir (50%), then *harummanis* (18%), and *gedong –gincu* (11%) and others (18%). (Figure 1)

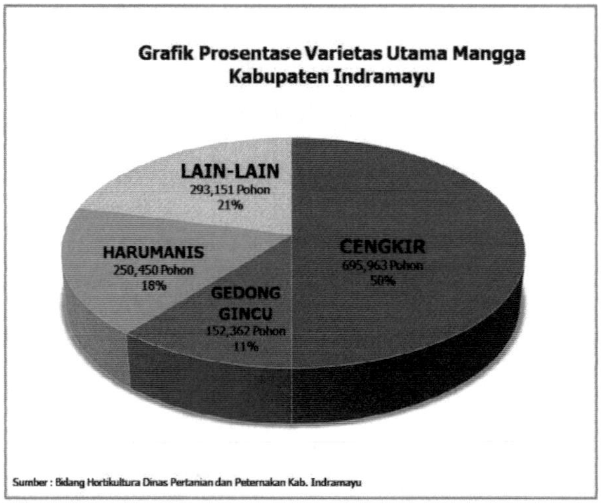

Figure 1. Number and Percentage of Mango Crops based on Varieties.

The high mango production in Indramayu turns out to be the problem of about 20% of reject mangoes that are not worth selling, or if sold at very low prices. This opportunity is utilized by womens who are members of Joint Business Group (KUB) to be processed into mango dodol at household scale. With so many mango production, mango processing business become dodol, manisan, syrup, puree and others, will not lack of raw materials.

According to one KTNA Mango from Situbondo, with mango processing, the price is 7 times higher. Not only the price is high, the shelf life becomes longer, can last up to 6 months [2]. But the reality in the field, not all women are interested to process mango, although in Ujung Jaya, mango production is abundant. The mangoes dodol processing business has not expanded. Newly pioneered by a small part of women who are members of KUB Barokah.

This study aims to: 1). Know the driving factors of women interested in processing mango into dodol and 2). Analyze whether the mango processing business to be mango dodol is profitable, and how much added value it gained.

2. Research Methods

This study uses Case Study method, while data collection method through interview, participant observation and documentation study. According to [11] case study is a descriptive research that aims to give a detailed picture of the background, nature, character that is then made general things.

Collection technique is done by:

1. Observation is a complex process, a process composed of various biological and psychological processes. Two of the most important are the processes of observation and memory [15].
2. Interview, namely the process of obtaining information by means of direct question and answer with the respondent, where the data obtained in the form of things needed to complete the results of observation.

While the data analysis technique using Hayami Value-added Method [6] and descriptive analysis. Quantitative data were used to know the added value of mango dodol agroindustry, analyzed using Hayami method as follows:

Table 1. Procedures of the Added Value Calculation of Hayami Method

No	Variable	Value
	I. *Output*, Input and price	
1	*Output* (Kg)	1
2	Input (Kg)	2
3	Labor (HOK)	3
4	Conversion Factor	4 = ½
5	Labor coefficient	5 = 3/2
6	*Output Price* (Rp/Kg)	6
7	Direct labor wage (Rp/HOK)	7
	II. Revenue and Profit	
8	Raw material price (Rp/Kg)	8
9	Another input contribution (Rp/Kg)	9
10	*Output Value* (Rp/Kg)	10 = 4 x 6
11	a. Value added (Rp/Kg)	11a = 10 – 9 – 8
	b. Value added ratio (%)	11b = 11a/10 x 100%
12	a. Direct labor income(Rp/Kg)	12a = 5 x 7
	b. Labor contribution (%)	12b = 12a / 11a x 100%
13	a. Profit (Rp/Kg)	13a = 11a – 12a
	b. Percentage of profit (%)	13b = 13a/11a x 100%
	III. Reply Services to Owners of Factors of Production	
14	Margin (Rp/Kg)	14 = 10 – 8
	a. Direct labor income	14a = 12a /14 x 100%
	b. Another input contribution (%)	14b = 9/14 x 100%
	c. The Benefits of the Company Owner (%)	14c = 13a/14 x 100%

Source : [6]

3. Results and Discussion

3.1. Potency of mango in Indramayu Regency

Ujung Jaya is one of the villages in Widasari Sub-district of Indramayu Regency located east of Indramayu Regency which has an area of ± 406, and has a population of 3,376 people, with a population of 1739 women and a population of 1,637 men [9]. Villages that have a residential area of 35 ha / this has access to good and affordable by limited by two major roads, namely pantura and the old road, so passed by public transportation between cities, such as buses and transport in the city, such as Kopayu, so accessibility very good.

The majority of Ujung Jaya villagers work as farmers and farm laborers, with a total population of 654 workers and 312 people. This is because the village has a large agricultural land, which is 344ha /, with the number of heads of households who have rice field is 202 Head of Family. In addition to farming, many people also become traders, craftsmen, private employees, domestic workers, as well as TKI (Tenaga Kerja Indonesia).

Ujung Jaya village is also known as Indramayu mango center. In addition to the garden area, many mango trees adorn the homes of their citizens. Therefore, during the harvest season, food production in Ujung Jaya will be very abundant. Ujung Jaya Mango has been widely marketed to various surrounding areas. Not only the abundant production, Indramayu original mango was known to have a distinctive sweet taste. Meanwhile, when viewed from the volume of production, the production of mango cengkir most compared to other mangoes[7]. (Figure 2).

IOP Conf. Series: Earth and Environmental Science **142** (2018) 012042 doi:10.1088/1755-1315/142/1/012042

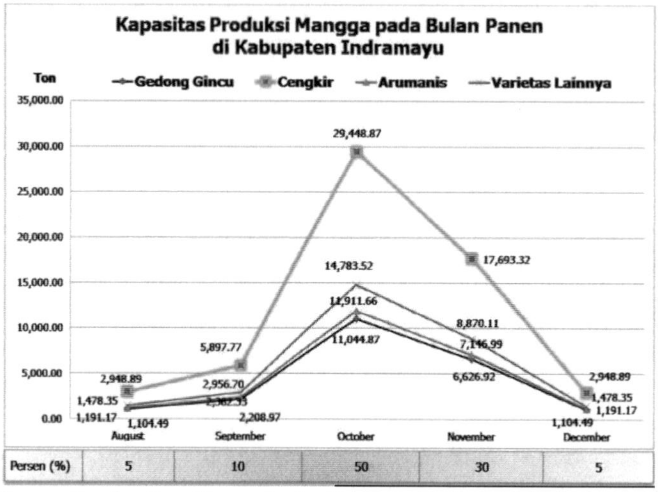

Figure 2. Mango Production Capacity in Harvest Period in Indramayu District

In addition to selling in the form of fresh fruit, Indramayu mango is often also processed into processed food products such as dodol, syrup, juice, jam and rujak mango packaging. Most processed products made from mango are produced by home industry. The business location of mango processing industry is spread all over Indramayu. One of mango processed mangoes from Ujungjaya village, Widasari sub-district, which started to produce processed mango Indramayu since 2008 [7]. Joint Group (KUB) Barokah is one of the centers in Ujung Jaya village in producing processed mango, such as dodol manga. In producing dodol mango branded 'Dodol Mangga Syauqi', KUB Barokah uses the main ingredient of original mango. To obtain the results of sweet mango dodol, mango used comes from a mixture of three types of ripe mango, namely sweet fragrant mango, mango gedong gincu, and mango cengkir. Processing mango into dodol is included in the agro-industrial sub-system, which is likely to increase the income of mango farmers.

The definition of Agroindustry according to the [3] is an activity that processes primary agricultural commodities into processed products either intermediate-product or end-product. This is in line with [1] and [5] which states that agro-industry is the business of processing agricultural raw materials into various products that consumers need.

Further, [12] and [13] suggests that there are five main reasons why agro-industries are important to become locomotives for national economic growth, namely: (a) Processing industries capable of transforming comparative advantage into competitive advantage, which in turn will strengthen the competitiveness of Indonesian agribusiness products; (b) The product has added value and a large market share so that progress can affect the growth of the national economy as a whole; (c) Has a great linkage both upstream and downstream, so as to attract the progress of other sectors; (d) Have a local raw material base (comparative advantage) that can be upgraded to ensure sustainability; (e) Has a slab to transform the national economic structure from agriculture to industry with agro-industry as its driving force.

Strategies in the development of agroindustry should be adjusted to the characteristics and problems of the agroindustry concerned. In general, according to [14] the problems faced in the development of agro-industry is:

1. The nature of agricultural products that are easily damaged, bulky so that required technology of packaging and transportation that can overcome the problem.
2. Most agricultural products are seasonal and heavily influenced by climatic conditions so that the aspect of continuity of agro-industry production is not guaranteed.
3. The quality of agricultural products and agro-industry produced is generally still low so that there is difficulty in market competition both domestically and in the international market. Most small industries (SMEs) use simple technology.

3.2. Processing of mango dodol

In producing dodol mango branded 'Dodol Mangga Syauqi', using the main ingredient of original mango. To obtain the result of sweet mango dodol, mango used comes from a mixture of three types of mango mature. That is sweet sweet mango, mango gedong gincu, and mango cengkir. Once peeled and cleaned, the mango is then sliced in a container. The sliced mango was then blended to become a mango porridge. Then the mango porridge is divided into two parts. First dough, added sticky rice flour, rice flour, coconut milk and salt. Second donation added red dangula sugar. Then it is heated, and the dough is mixed. Keep heated until the slices of mango become porridge and water content decreases, so the calis and leaving mango juice. Then pour into the pan, cooled overnight. After that can be cut, then packed for ready to market (Figure 3). "This mango dodol can hold up to a month without using preservatives. I only use mango and sugar only "said craftsman dodol mango.

If averaged, craftsmen in the tip of this Jaya can spend about 50 pounds of mango per day to make dodol. Production of mango dodol it is marketed to the area of Indramayu, Cirebon and brass. In fact, through the online sales system, he can also market their production to the outside of Java Island. With the help of Diskopindag Indramayu, Craftsman in KUB Barokah has completed its production with halal certificate from MUI as well as certificate of Household Food Industry (PIRT) from Dinas Kesehatan. This is to convince their customers that the mango dodol they produce is completely halal and healthy. Most of the dodol processing crafters are women, aged between 35-55 years. The reason for the workforce is to help the husband to increase the daily needs of the family, because the husband's work has not been able to meet the daily needs (50%) and use the mango reject (20%). In addition, there is a small proportion which states that working to fill the spare time (10%), helping his brother's business (10%) and to socialize with friends (5%) and others (5%).

Figure 3. Flow Chart of Mango Dodol Processing

While the reason work in SMEs processing dodol because the location of SMEs close to home, comfortable, strict or loose regulations, not many sanctions, permits if there is family needs, working hours are not too early, the owner of SMEs friendly, a good work environment. Related to the Added Value, [6] mentioned that, the added value is the difference between the value of the commodities that are treated at a certain stage with the value of the production cost used during the production process takes place. There are two factors that influence the added value of technical factors and market factors. Engineering factors that influence the product capacity, the amount of raw materials, and labor. Market factors include output prices, wage labor, fees and salaries.

3.3. Analysis of added value of mango dodol

Added value may be defined as the incremental value to a commodity as it undergoes processing in the production stream.[6]. [13] defined added value as the "difference between output value and the input costs".

Hayami value-added analysis method is one of the analysis that is often used and has advantages such as:

1. More appropriate for agricultural products. Can know the amount of added value, output value and productivity.
2. It is known to the extent of remuneration to the owners of factors of production.
3. Hayami's value added principle can be applied also to other subsystems outside processing, for example for marketing activities.

Value added analysis on Hayami method also has weaknesses:

1. The average approach is not appropriate if applied to a business unit that produces many products from one type of raw material.
2. Can't explain by products.
3. It is difficult to determine which comparator can be used to conclude whether the remuneration of the factors of production is appropriate.

From Table 2. It can be concluded that mango dodol processing business is very profitable, that is Rp.44.405 per kilogram, or 87.75% of output value per kilogram. While the added value obtained from the processing of mango dodol is Rp. 50,600 per kilogram of output, or 52.80% of the output value.

Therefore, the mango dodol processing business is potential to continue to be socialized and developed, in an effort to increase the income of mango farmers and the communities involved in mango agro-industry. However, efforts to develop SMEs mangoes are not running smoothly, because they face various constraints, both internal and external, as faced by this Barokah KUB.

Although the production of mango dodol from KUB Barokah is favored by consumers, and demand is high, it is still experiencing difficulties. It mainly concerns raw material of mango that is not continuous. In the mango harvest season is over, often have difficulty obtaining raw materials of mango. In addition, also often difficult to obtain labor. Often, the workers choose to work in the fields during the growing season and rice harvest, rather than making mango dodol.

Related to the development of SMEs, [8] explains that the classical problem of SMEs in Indonesia in developing its business can be divided into two, namely funding and marketing. The government has tried to provide a program for such a solution, but in reality it has not been effective yet, due to lack of rationality in implementation and tend not to see SMEs as actors who can progressively participate in the competition.

Table 2. Value Added Analysis of Mango Dodol Processing Based on Hayami Method.

No	Variable	Value
	I. ***Output*, Input and price**	
1	*Output* (Kg)	12,5
2	Input (Kg)	8
3	Labor (HOK)	9,5
4	Conversion Factor	1,56
5	Labor coefficient	1,18
6	*Output Price* (Rp/Kg)	61.430
7	Direct labor wage (Rp/HOK)	5.250
	II. **Revenue and Profit**	
8	Raw material price (Rp/Kg)	16.000
9	Another input contribution (Rp/Kg)	29.230
10	*Output Value* (Rp/Kg)	95.830
11	a. Value added (Rp/Kg)	50.600
	b. Value added ratio (%)	52,8
12	a. Direct labor income(Rp/Kg)	6.195
	b. Labor contribution (%)	12,24
13	a.Profit (Rp/Kg)	44.405
	b.Percentage of profit (%)	87,75
	III. **Reply Services to Owners of Factors of Production**	
14	Margin (Rp/Kg)	79.830
	a. Direct labor income (%)	7,76
	b. Another input contribution (%)	36,61
	c.The Benefits of the Company Owner (%)	55,62

Furthermore [1] also identified the problem of SMEs namely limited access to global markets and low capacity to increase value added (Value added). Therefore, for the development of SMEs of mango dodol processing, it is necessary to train and socialize the correct and hygenic processing of dodol from the related Office, innovation in packaging, pioneering business partnership with shop in the city of Indramayu, and support financing from banks with an affordable interest rate.

4. Conclusions

The results revealed that the factors that affect the participation of women in SMEs mango dodol processing is to increase family income, take advantage of spare time and take advantage of reject mango. The added value obtained in mango dodol processing is Rp.50.600,00 per kilogram of input, with a value added ratio of 52.8%. While the profit earned by Rp.44.405 per kilogram, or 87.75% of the output value per kilogram.

Acknowledgment

The author would like to thank the Rector of the Padjadjaran University, which has provided funding through "***Competency Research of Unpad Lecturers Scheme (RKDU)*- 2017**, so that this research can be carried out.

References

[1] Abonyi G 2005 *Transformation of Global Production, Trade and Investment: Global Value Chain and International Production Net Work* In Expert Group Meeting on SME's Participation in Global and Regional Supply Chains (Bangkok: UNESCAP)

IOP Conf. Series: Earth and Environmental Science **142** (2018) 012042 doi:10.1088/1755-1315/142/1/012042

[2] Anitasari et al 2014 *Supply Chain Management of Arumanis Mango Products in Situbondo Regency*

[3] Directorate General of Agro and Chemical Industry Ministry of Industry 2009 *Road Map Industyi Fruit Processing* Ministry of Industry Jakarta

[4] Firmansyah 2010 *Strategy Development of Agroindustry Tape Assorted Rasa, Cases Study in PD Sumber Rahayu, Sumedang* (Jatinangor: Faculty of Agriculture Padjadjaran University)

[5] Gumbira Sa'id and A Harizt 2001 *Agribusiness Management (*Jakarta: PT Ghalia Indonesia)

[6] Hayami Yujiro 1987 *The Value Added Method* Through <http://community.gunadarma.ac.id/user/blogs/view/yudhistira/id_7389/title_%22-input-proses-output-usaha-business%22/> [21/03/17]

[7] Histifarina D and Deliana P A 2009 Review of Implementation of Dry Mango Sweet Processing Technology in Indramayu Regency *J. Agr. Tech. Assessment and Development* **12** 91-98

[8] Hill H 2001 *Small and Medium Enterprices in Indonesia: Old Policy Chalenges for a New Administration* Asian Survey **41** 248-270

[9] KKNM Unpad 2014 *Ujung JayaVillage -Widasari Indramayu District* Available at http://kknm.unpad.ac.id/ujungjaya/desa-ujungjaya-2014/display-desa/ Accessed February 5, 2017

[10] Muhammad R 2010 *Analysis of Value Added Cassava to Tapioca Flour* (Jatinangor: Padjadjaran University)

[11] Nazir M 2005 *Research Methods* (Jakarta: PT Ghalia Indonesia)

[12] Soekartawi 2000 *Introduction to Agro-industry* (Jakarta: PT Raja Grafindo Persada)

[13] Soekartawi 2005 *Agroindustry in Socio-Economic Perspective* (Jakarta: PT Raja Grafindo Persada)

[14] Suprapto 2005 Characteristics, Implementation, and Development Agroindustry of Agricultural Products Indonesia http://research.mercubuana.ac.id/proceeding / implementing and carryinggan_agroindustrial.pdf. [accessed April 20, 2017]

[15] Sugiyono 2008 *Understanding Qualitative Research* (Bandung: CV Alfabeta)

[16] News of KUKM 2016 *Strengthening SMEs for Economic Growth Qualified* News KUKM **5** (Jakarta :Directorate of SME and Cooperative Development)

IOP Conf. Series: Earth and Environmental Science **142** (2018) 012043 doi:10.1088/1755-1315/142/1/012043

Response of *Artemisia annua* L. to shade and manure fertilizer application in lowland altitude

H H Permana[1], Y Widyastuti[2], Samanhudi[3] and A Yunus[3]

[1]Student of Study Program of Agrotechnology, Faculty of Agriculture, Universitas Sebelas Maret Surakarta, Indonesia

[2]Research Center for Medicinal Plant and Traditional Drug , Tawangmangu, Indonesia

[3]Lecturer of Study Program of Agrotechnology, Faculty of Agriculture, Universitas Sebelas Maret Surakarta, Indonesia

Email : yunus.uns7@yahoo.com

Abstract. Artemisia is a plant producing artemisinin substance which is the main compound in the treatment of malaria. Artemisia comes from China, usually grows wild in native habitats in the plains with an altitude of 1,000-1,500 meters above the sea level. Artemisia development efforts in Indonesia hampered by limited land with the required altitude due to their competition with vegetable crops. Based on this reason, this research is conducted to observe the growth of artemisia planted in lowland with the help of shade and manure. This study aims to determine the level of shade and best manure on the growth of Artemisia. Research conducted at the Laboratory of the Faculty of Agriculture UNS Jumantono using nested design with two factors, shade as main factor and manure fertilizer as sub factor. The data analysis used F test with confidence level of 5%, if significant, then continued with DMRT (Duncan Multiple Range Test). The results showed the treatment of shade gave no difference in growth within 50% shade, 75% shade as well as without shade treatment. Goat manure fertilizer gave the highest result and able to increase plant height, number of branches, flower weight and root volume.

1. Introduction

A. annua L. is a medicinal plant that has been used since a long time ago in China as an antimalarial medicine. Artemisia containing terpenoids complex compounds, called artemisinin which known to be able to eradicate Plasmodium falciparum as a major cause of malaria and treat breast cancer[1]. Artemisinin is a compound that is effective for the types of malaria that resistant to quinine and chloroquine and cerebral malaria caused by Plasmodium falciparum. In Artemisia, artemisinin is synthesized in the roots and accumulated in the leaves. The artemisinin content in flower is quite high and can be compared with artemisinin contained in the leaves [2]. Artemisinin production is influenced by climate, soil conditions, plant age and genetic variations [3].

Treatment for malaria usually use quinine that has been used for a long time by traditional society. Along with the development of malaria, resistency of malaria occuring to various existing drugs. Resistency happens to drugs such as chloroquine and sulfadoxine-pyrimethamine (SP) so that medical efforts are made with treatment of Artemisinin based Combination Therapy [4]. Artemisia development as the only producer of artemisinin have a major problem, that is the existing genotype provided relatively low artemisinin content between 0.01% -0.5% with the time of planting about 6 months plus the time of extraction to produce finished materials between 2-5 months [5]. Other problems are in the tropic area, this plant grows in the highlands at an altitude of over 1,000 meters, while in Indonesia the area with that altitude usually used for the cultivation of vegetables and fruit, so that there is competition of land with vegetable crops in the cultivation of these crops.

2. Methods

2.1. Location and Time Research
The study was conducted in the village of Sukosari, District Jumantono, Karanganyar, Surakarta from July to *November* 2015

2.2. Tools and Materials
The tools used are 30cm x 20cm size planting pot (36 units), paranet 50%, 75% paranet, bamboo poles, hoes, tape measure, ruler, measuring cups and analytical scale. The material used in the research are the seeds of *A. annua L.*, Jumantono Alfisol soil (growing media), goat manure fertilizer, chicken manure fertilizer and cow manure fertilizer.

2.3. Research Design and Data Analysis
This research used a nested design with manure fertilizer nested in shade. The main factor was the level of shade and sub factors that nested on the main factors was manure fertilizer. The shade consisted of 3 levels: without shade (N0), 50% shade (N1) and a shade 75% (N2). The manure fertilizer consisted of 4 levels which were without fertilizers (P0), goat manure (P1), chicken manure (P2) and cow manure (P3) with a dose of 200 gr on each pot. In total 12 combinations of treatments were obtained. Every combination treatment was repeated three times, then the total number of samples obtained was 36 units. The variables observed in this study were plant height, branch number, date of flowering, flower weight, root volume and root length. The data were tested using the F test with significance level of 5%. If significant then continued by Duncan Multiple Range Test (DMRT).

3. Results and Discussion

3.1. Plant height
Treatment of shade does not provide a remarkable difference on plant height, while the treatment of manure that nested in the shade effect provided significant difference on plant height. Based on the data table (Table 1), it can be seen that the combination of goat manure and without shade gave the highest plant height with an average of 178.67 cm and increase in plant height by 22% when compared to treatment without fertilizer. Goat manure fertilizer also gave an average of the highest plants with significant results at the level of shade 50% and 75%. At the level of 50% shade highest yield was 177.33 cm with an increase in plant height by 35% of the treatment without fertilizer and at the level of 75% shade, the plant height was 181.33 cm and a high increase of 34% compared to without fertilizer treatment. Manure fertilizer on the cultivation of *A. annua* L. aims to fulfill essential nutrient needed. With sufficient soil nutrients then productivity of biomass will increase. Goat manure containing 1.19% N, 0.92% P and 1.58% K_2O. From the various elements, nutrient elements nitrogen (N) is the most influential element on plant growth, especially vegetative growth [6]. Manure has a nitrogen content that serves to form amino acids that are the result of protein degradation and function to form growth hormone [7]. Cell division that affects the growth of plant height depends on the supply of carbohydrates resulting from photosynthesis process involving chlorophyll and nitrogen elements play a role in chlorophyll formation [8].

3.2. Number of Branch

The shade treatment does not give significant difference to the number of branches, whereas the treatment of manure fertilizer nested in the shade gave a very significant difference to the number of branches, presented in table 1. Based on the table it can be seen that in the treatment without shade, the largest number of branches produced in the treatment of chicken manure fertilizer with branch number of 53.67 and able to increase the number of branches as much as 26.7% compared to without fertilizer. Goat manure fertilizer produces the highest number of branches in the shade treatment of 50% with the number of branches as much as 55.67 and increasing the number of branches by 37.5%. Goat manure fertilizer also give the largest number of branch in the shade 75% with the number of branches 52 and 33% increase in the number of branches than the treatment without fertilizer. Increased vegetative growth includes plant height, number of leaves and number of branches is strongly influenced by the presence of nitrogen (N), Phospor (P) and Potassium (K) [9]. Increased macro nutrients, especially nitrogen, will provide an increase in vegetative growth. Nitrogen serves as a builder of chlorophyll, proteins, fats and enzymes in cells that stimulate the growth of stems, leaves and branches. Chicken manure produces the largest number of branches on each shade level. Chicken manure has the most nitrogen content than other types of manure [10]. At 50% shade level and 75% goat manure also gives a significant effect with the highest value both on the variable height of the plant and also the variable number of branches. This is because goat fertilizer has more nutrient variant and optimal amount compared to chicken or cow manure [11], so it will give the best effect on the number of branches.

3.3 Days of Flowering

Treatment of shade does not give a significant difference to the days of flowering, as well as treatment of manure fertilizer that is nested in the shade does not provide a real difference to the life of flowering. From the results obtained, the analysis variance calculation showed that none of the treatments give significant result on days of flowering. However, when seen in Table 1, the average age of flowering in the shade of 75% was slightly higher than the other shade level even though no significant difference in the analysis of variance calculation. At the days of flowering table shows the average days of flowering ranges on day 70 until day 95. These results are consistent with the results of Firdaus [12] which showed that *A. annua* L. grown in the lowlands will flowering at the age of 2-3 months. *A. annua* L. is a plant that in its natural habitat, these plants will flowering at day 190 to day 200 [5]. Artemisia itself is a short-day plant that will became flowering when the sun shines shorter than the critical exposure period. With their natural habitat located in a sub-tropical climate that has long photoperiodism > 13 hours/day, this duration of radiation can cause the time the emergence of flower in *A. annua* L. will be longer, so that harvesting will also be longer. Therefore, when the planting is done in areas with radiation are short (<13 hours/day) this plant will quickly form a flower [13].

Table 1. Table of average plant height, number of branches and day of blooming of *A. annua* L. on treatment of shade and manure fertilizer

Treatment		Plant Height (cm)	Number of Branch	Day of flowering (days)
Shade (N)	Manure Fertilizer (P)			
Without Shade (N0)	Without Fertilizer (P0)	145.00±1.70[a]	42.33±6.40[a]	79.33±5.80[a]
	Goat (P1)	178.67±7.80[abc]	50.67±4.00[abc]	85.33±16.60[a]
	Chicken (P2)	169.67±15.50[abc]	53.67±8.50[abc]	73.67±2.00[a]
	Cow (P3)	160.00±5.20[ab]	43.00±6.60[ab]	86.00±14.00[a]
Shade 50% (N1)	Without Fertilizer (P0)	131.67±24.08[ab]	40.00±12.50[a]	90.00±12.10[a]
	Goat (P1)	177.33±6.60[c]	55.67±3.20[bc]	82.00±14.00[a]
	Chicken (P2)	157.00±6.50[abc]	45.67±6.70[abc]	86.67±16.10[a]
	Cow (P3)	128.33±27.20[a]	42.00±5.20[ab]	73.00±1.70[a]
Shade 75% (N2)	Without Fertilizer (P0)	135.57±25.10[ab]	39.00±2.60[ab]	90.00±15.60[a]
	Goat (P1)	181.33±9.60[c]	52.00±5.50[bc]	85.33±12.90[a]
	Chicken (P2)	164.67±19.60[bc]	49.00±4.90[abc]	94.00±19.70[a]
	Cow (P3)	112.00±10.40[a]	35.00±2.00[a]	95.33±18.60[a]

Description:The numbers followed by different letters in each shade levels and variable show different significant values ($P<0.05$) on duncan multiple range test

3.4. Flower Weight

Flowers on *A. annua* L. have components such as antioxidants, anti-microbial, anti-fungal and artemisinin content. These chemical components are affected by harvest age, fertilizer and soil pH, geographical location, plant clones and extraction methods [14]. According to [15] the concentration of artemisinin reaches its peak when flowering plants bloom entirely. The shade treatment does not give any significant difference to the weight of the flower, while the treatment of manure fertilizer nested in shade gave a very significant difference to flowering age, presented in table 2. Based on the table it can be seen that at the highest shade level the highest observation of flower weight is the treatment of goat manure with the weight of 14.18 gram and increase the average weight of flower by 35% compared to the treatment without fertilizer. At 50% shade, the highest yield weight on goat manure treatment with the weight of 9.06 gram and 50% increase compared to the treatment without fertilizer. Goat manure also has the highest yield at 75% shade level with weight 10.40 grams and raising the flower weight by 62% compared to the treatment without fertilizer. Treatment of manure fertilizer gives a very significant effect for the flower weight. This shows that plants need nutrients to form generative organs (flowers). It is also supported by Ahmad et al. [16] which states that the application of nitrogen (N) and phosphor (P) elements contained in the fertilizer gives positive results on flower weight and flower blooming. Nitrogen element is the most needed element in plant growth. In general, nitrogen absorbed by plants is used to form chlorophyll that makes the plants green and amino acids that play a role in cell division and plant development. Phospor is very important when the plant enters the generative phase. The presence of a phosphorus element is used in the transfer of energy for all plant metabolism, so in addition to stimulate the plant growth, phosphorus is useful to spur the formation of flowers and fruit maturation, so the percentage of flower formation and fruit ripening will increase [17].

3.5. Root Volume

Root is a part of Artemisia for synthesized artemisinin. [18] revealed in artemisia's root hair are contained artemisinin could reach up to 0.4%. Shade treatment does not give a real difference to the volume of roots, while the treatment of manure fertilizer that is nested within the shades provides very real difference to the volume of roots, presented in (Table 2). Based on the table it can be seen that the manure fertilizer can increase root volume significantly, especially goat manure. Goat manure fertilizer got the highest volume results at every level of shade. At without shade level, goat manure fertilizer application is able to produce root volume amounted to 34.67 ml and improve root volume by an average of 67% than treatment without fertilizer. In the shade level of 50%, the goat manure fertilizer treatment had the highest root volume results amounted to 30.67 ml and an average increase root volume by 43% compared to the treatment without fertilizer. In the shade level of 75%, the goat manure fertilizer treatment had the highest root volume amounted to 37.33 ml and an average increase root volume by 85% compared to treatment without fertilizer. Manure contains nutrients and organic matter which is good for the soil. The organic material is this which is the source of nitrogen, phosphorus and sulfur for plant growth [19]. The addition of nitrogen does not only affect the plant biomass but also on root growth [20]. Nutrients are not the only factor that determines the root volume. Volume roots other than influenced by the level of root distribution and the availability of nutrients, are also affected by the water content in the soil. Organic materials other than as a primary source of nutrients establishment also serves as a binder of water in the soil. Manure fertilizer input into the soil, will improving soil organic matter content so that the water stored in the soil also increased. The availability of sufficient water in the soil will affect the optimal nutrient absorption by the roots. The roots will respond when the plants are in condition of water shortage by reducing transpiration rate conserve water [21]. The longer transpiration occurs without water supply from the soil will lead to reduce sized of plant roots. Position of roots spread out and supported by sufficient soil nutrient availability will increase the volume of roots.

3.6. Root Length

Treatment of shade does not give a real difference to the length of the roots, as well as treatment of manure that is nested within the shades provides very real difference to the length of the roots. Based on table 2, the treatment of manure fertilizer can be seen that the highest yield observations of root length is on chicken manure fertilizer. Chicken manure fertilizer give root length results highest at all levels of shade treatment although based on the statistical calculation influence of manure that is nested in the shade does not provide a real different on the length of the root as well as the treatment level of shade, so that the variable length of the root is not followed by DMRT, Excellence manure is able to improve soil structure and restore the soil ecosystem that increases the activity of microorganisms in the soil. According to [22], the increased of microorganisms activity in the soil will lead to increased availability of nutrient elements from the activity of bacterial binding nitrogen [23] and also from the organic material was decomposed to available nutrient elements to plants by the activity soil microorganisms, this affects root development which then allows increased absorption of nutrient elements. In line with the [24] which States, land that had a lot of activity of the organism has a porous soil that much that can facilitate the growth of the roots penetrate the ground through soil pores that can absorb water and nutrient elements are dissolved.

Table 2. Table of average flower weight, root volume and root length of *A. annua* L. on treatment of shade and manure fertilizer.

Treatment		Flower weight (gram)	Root volume (ml)	Root length (cm)
Shade (N)	Manure Fertilizer (P)			
Without Shade (N0)	Without Fertilizer (P0)	10.45±5.1^{ab}	20.00±9.1^{a}	22,80±0,8^{a}
	Goat (P1)	14.18±1.8^{bc}	34.67±4.1^{c}	25,43±5,1^{a}
	Chicken (P2)	11.09±2.6^{abc}	29.33±5.1^{abc}	26,40±1,0^{a}
	Cow (P3)	8.18±0.8^{a}	21.67±6.6^{ab}	24,03±0,9^{a}
Shade 50% (N1)	Without Fertilizer (P0)	6.03±1.2^{ab}	21.00±6.2^{ab}	21,20±3,4^{a}
	Goat (P1)	9.06±0.6^{abc}	30.67±4.0^{abc}	21,03±3,3^{a}
	Chicken (P2)	8.43±0.3^{abc}	29.67±4.7^{abc}	26,00±3,1^{a}
	Cow (P3)	5.72±1.6^{a}	20.33±1.5^{a}	23,40±4,7^{a}
Shade 75% (N2)	Without Fertilizer (P0)	6.42±1.8^{a}	20.33±3.2^{a}	21,60±4,6^{a}
	Goat (P1)	10.40±0.6^{abc}	37.33±5.5^{c}	25,50±1,6^{a}
	Chicken (P2)	10.03±1.2^{abc}	31.67±4.6^{abc}	26,63±2,1^{a}
	Cow (P3)	7.32±0.5^{ab}	22.67±2.5^{ab}	20,10±4,9^{a}

Description:The numbers followed by different letters in each shade levels and variable show different significant values on duncan multiple range test

4. Conclusions

The conclusion of the study are the various shade as main factor showed that there were no difference in growth within 50% shade, 75% shade as well as without shade treatment. Treatment of goat manure fertilizer as a sub factor nested in 3 levels of shades overall gave the highest results in the variable plant height, number of branches, flower weight and root volume.

Acknowledgement

This research was supported by the Ministry of Research Technology and Higher Education and Dr. Ir. Yuli Widyastuti as field from Research Center for Medicinal Plant and Traditional Drug, Tawangmangu, Indonesia. Artemisia annua seeds in this research were obtained from B2P2TOOT, Tawangmangu, Karanganyar.

References

[1] Ferreira JFS 2004 *Artemisia annua* L.: The Hope Against Malaria and Cancer. Medicinal and Aromatic Plants: Production, Business & Applications Proceedings of the Jan 15-17/2004 meeting. Mountain State University, Beckley, WV.

[2] Kardinan A 2008 Artemisia (*Artemisia annua*) Tanaman Anti Malaria *Warta Penelitian dan Pengembangan Tanaman Industri Vol.* **14** No. 2. Badan Penelitian dan Penelitian dan Pengembangan Pertanian.

[3] Klayman DL 1985 Qinghaosu (Artemisinin): An Antimalarial Drug From China. *Science* Vol **228**.

[4] Bjorkman A, Bhattarai A 2005 Public Health Impact of Drug Resistant *Plasmodium falsiparum* Malaria. *Acta Tropica* **94** (2005) 163-169.

[5] WHO (World Health Organization) 2006 WHO Monograph On Good Agricultural and Collection Practices (GACP) For *Artemisia annua* L.

[6] Hikmah 2008 Pemberian Beberapa Bahan Organik Pada Budidaya Tumpang Sari Tanaman Brokoli (Brassica oleracea) dan Petai (*Brassica pekinensis*) Serta Pengaruhnya Terhadap Pertumbuhan dan Serapan Cu dan Zn. Skripsi. Institut Pertanian Bogor. Bogor.

[7] Dewi WW 2016 Respon Dosis Pupuk Kandang Kambing Terhadap Pertumbuhan Dan Hasil Tanaman Mentimun (*Cucumis Sativus* L.) Varietas Hibrida. *Journal Viabel Pertanian* **10**(2) 11-29.

[8] Dapoigny L, Fleury A, Robin P 1997 Relation between relative growth rate and nitrogen content during growth of lettuce (*Lactuca sativa* L.) effect of radiation and temperature *Agronomie* **17**(1):13-41.

[9] Hertos M 2015 Pengaruh Pemberian Pupuk Kandang Kotoran Ayam Dan Pupuk NPK Mutiara Yaramila Terhadap Pertumbuhan Dan Hasil Tanaman Terung (*Solanum melongena* L.) Pada Tanah Berpasir. *Anterior Jurnal*, Volume **14** Nomor 2, Juni 2015, Hal 147-153.

[10] Lingga P 1991 Jenis dan Kandungan Hara pada Beberapa Kotoran Ternak. Pusat Pelatihan Pertanian dan Pedesaan Swadaya (P4S) ANTANAN Bogor.

[11] Nurshanti DF 2009 Pengaruh Pemberian Pupuk Organik Terhadap Pertumbuhan dan Hasil Tanaman Sawi Caisim (*Brassica Juncea* L.) *Jurnal Agronobis*, Vol. **1**, No. 1, Maret 2009.

[12] Firdaus R, Muhamad S, Endang GL 2009 Evaluasi Daya Hasil Artemisia (*Artemisia annua* L.) Hasil Mutasi Di Dataran Tinggi, Dataran Sedang dan Dataran Rendah Makalah Seminar Departemen Agronomi Dan Hortikultura Fakultas Pertanian Institut Pertanian Bogor.

[13] Gusmaini dan Hera N 2007 Potensi Pengembangan Budidaya *Artemisia annua* L. di Indonesia. *Jurnal Perspektif* Vol. **6** No. 2 Desember 2007 Hal 57 – 67.

[14] Bilia AR, Francesca S, Cristiana S, Maria CB and Rosa D 2014 Essential Oil Of *Artemisia annua* L.: An Extraordinary Component with Numerous Antimicrobial Properties. Evidence-Based Complementary and Alternative Medicine Volume **2014**.

[15] Ferreira JFS, JC Laughlin, N Delabays and PM Magalhaes 2005 Cultivation and Genetics Of *Artemisia annua* L. for Increased Production of the Antimalarial Artemisinin. *Plant Genetic Resources* **3**(2): 206 – 229.
[16] Ahmad I, Asif M, Amjad A, Ahmad S 2010 Fertilization Enhances Growth, Yield and Xanthophyll Contents of Marigold. *Research Article Turk J Agric For* **35** (2011) 641 - 648.
[17] Rina D 2015 Manfaat Unsur N, P dan K Bagi Tanaman. Balai Litbang Pertanian Kalimantan Timur. Kementrian Pertanian Republik Indonesia.
[18] Patra N, Srivastava AK and Sharma S 2013 Study Of Various Factors for Enhancement Of Artemisinin in *Artemisia annua* Hairy Roots. *International Journal Of Chemical Engineering and Applications*, Vol. **4**, No. 3, June 2013.
[19] Mujiyati dan Supriyadi 2009 Pengaruh Pupuk Kandang dan NPK terhadap Populasi Bakteri Azotobacter dan Azospirillum dalam Tanah pada Budidaya Cabai (*Capsicum annum*).
[20] Wang ZR, Rui YK, Shen JB and Zhang FS 2008 Effects Of Fertilizer On Root Growth in *Zea mays* L. Seedlings. *Spanish Journal Of Agricultural Research* 2008 **6**(4), 677 - 682.
[21] Nio SA dan Torey P 2013 Karakter Morfologi Akar Sebagai Indikator Kekurangan Air Pada Tanaman. *Jurnal Bioslogos* Vol. **3** No. 1. Februari 2013.
[22] Dyan MSP 2006 Pengaruh Jenis Media terhadap Pertumbuhan Begonia imperialis dan Begonia 'Bethlehem Star'. *Biodiversitas* Volume **7**, Nomor 2, Hal: 168-170.
[23] Coyne MS and Mikkelsen R 2015 Soil Microorganism Contribute to Plant Nutrition and Root Health. *Better Crops Journal* Vol. **99**. No.1.
[24] Kroon HD, Hendriks M, Van Ruijven J, Ravenek J, Padilla FM, Jongejans E, Visser EJW and Mommer L 2012 Root Responses to Nutrients and Soil Biota: Drivers of Species Coexistence and Ecosystem Productivity. *Journal of Ecology* Vol **100** : 6-15.

Yields and content of ursolic acid in pearl grass (*Hedyotis corymbosa*) when utilizing cow manure fertilizer in different shades

L D Jayanti[1], A Yunus[2], B Pujiasmanto[2] and Y Widyastuti[3]
[1]Study Program of Agrotechnology, Faculty of Agriculture, Universitas Sebelas Maret, Surakarta, Indonesia
[2]Study Program of Agrotechnology, Faculty of Agriculture, Universitas Sebelas Maret, Surakarta, Indonesia
[3]Research center for Medicinal Plant and Traditional Drug, Tawangmangu, Indonesia

E-mail: yunus.uns7@yahoo.com

Abstract. This study aims to examine the intensity of shade and proper dosage of fertilizer to maximize the content of ursolic acid and pearl grass yield. The field-run study was carried out at the Medicinal Plant Garden of the Research and Development of Medicinal Plants and Traditional Medicines (B_2P_2TOOT) and the laboratory research was conducted at the B_2P_2TOOT Phytochemistry Laboratory. The experiment design was Randomized Complete Block Design with Split Plot pattern with the intensity (three levels) as main plot and fertilizer dosage (four levels) as sub plot. The data obtained were analyzed using the variance analysis, if there were significant differences tested further by using Duncan's Multiple Range Test (DMRT) with 95% confidence level. The results showed that shade treatment gave significant effects on plant height, number of branches, root length, fresh weight and dry weight. The treatment of a fertilizer dosage 15 tons/ha gave the best results on fresh weight and dry weight. The combination of N_0K_0 treatment (without shade, without cow manure) resulted in the highest quality of ursolic acid as it featured a light blue color when detected under UV_{366} light.

1. Introduction

Modern medicine recently is costly, as an alternative, many members of the community back to use trusted traditional medicine. Plants have been widely used by the community as a source of food and medicines for their generation [1]. One of the herbs that has been used as an herbal medicine and widely consumed by the community is pearl grass [1].

The pearl grass (*Hedyotis corymbosa*) is an herbaceous plant that almost all of the communities recognize it as weed or pest plants. Viewed in terms of the benefits this grass contains substances that has potential as an anti-cancer. Ursolate acid compounds which are contained by pearl grass are proved to prevent the development of cancer cell division to a more vicious stage [2]. [3] states that whole pearl grass plants can be used as medicines. The efforts in improving the yield and quality of the pearl grasses are expected to take an important role to maintain the availability and maintain the quality of the processed post-harvest. One effort that can be done is by applying cow manure as organic fertilizer and providing shade above the plantation.

According to [4] organic farming is considered able to produce products that are quite safe and nutritious, so it can increase the positive effects in its use and once can increase the competitiveness of the agribusiness products. The use of the shade is expected to be able to know the increase of biomass of pearl grass harvest.

2. Methods

The research was conducted in July-November 2016 in Tanjungsari Village, Tegal Gede, Karanganyar. The analysis of secondary metabolite content was conducted at Indonesian Center for Research and Development of Traditional Medicinal and Medicinal Plants of Tawangmangu (B2P2TOOT). The tools used during planting are bucket, mortar, hoe, silver black mulch, gauge, knife or scissor, ruler, pegs, knapsack sprayer and hammer. The tools used to analyze secondary metabolites are TLC (Thin Layer Chromatography), pipettes, petri dishes, ovens, waterbaths, funnels, bottles, filter paper, sonicators and Chamber UV_{366} and UV_{256}. The materials used for planting include pearl grass seed (*Hedyotis corymbosa*), moss soil media, sand, cow manure, paranet, bamboo and nail. Materials used for the analysis of sine metabolites are methanol (extraction agent), silica gel, ethyl acetate and formic acid.

This study uses Randomized Complete Block Design (RCBD) split plot pattern with 2 factors, they are cow manure and shade in three replicates. The shade that becomes the main plot in this study consists of 3 levels, they are control or shade 0% (N_0), 50% shade (N_1) and 75% shade (N_2). Sub plot of this research is manure consisting of 4 levels which were control or without cow manure (K_0), cow manure 10 tons/ha (K1), cow manure 15 tons/ha (K_2) and cow manure 20 tons/ha (K_3). The data obtained were analyzed using variance analysis (Anova) and if there were significant differences followed by DMRT with 95% confidence level. The observed variables are plant height, number of branches, root length, fresh weight of plant, dry weight of plants and secondary metabolite content of pearl grass.

3. Results and Discussion

3.1. Growth

3.1.1. Plant heigh Plant height is one variable that is often used to observe the growth process because the process of measuring plant height can be done in a simple way and does not disturb the plant. [5] and [6] argue that plant height is one of the most frequently observed plant sizes and is used as a parameter to measure the size of the effect of the applied treatments.

Table 1. The effect of shading on plant height

Shade	Plant Height (cm)
0% (N_0)	28.80b
50% (N_1)	23.85a
75% (N_2)	24.04a

Description: The numbers followed by the same letter show no significant difference ($P<0.05$).

Table 1 shows the average height of the highest pearl grass obtained in the treatment without shade (N0) which is 28.80 cm. The average height of the lowest pearl grass obtained at the shade treatment of 50% (N1) is 23.85 cm. the provision of shading can reduce the acceptability of light by plants because plants receive the light that comes as is [6].

Based on the test results of the further given treatment, the dosage of cow manure has no significant effect on the height of pearl grass. It is suspected because the dosage given is still low and has not been able to give the real effect to the height of the plant. The given fertilizer is organic fertilizer which has advantages and disadvantages as explained by [7], cow manure is an organic fertilizer that has less nutrient content than inorganic fertilizer or artificial fertilizer.

The growth of pearl grass at the age of 0-7 weeks after planting (WAP) is presented in Figure 1.

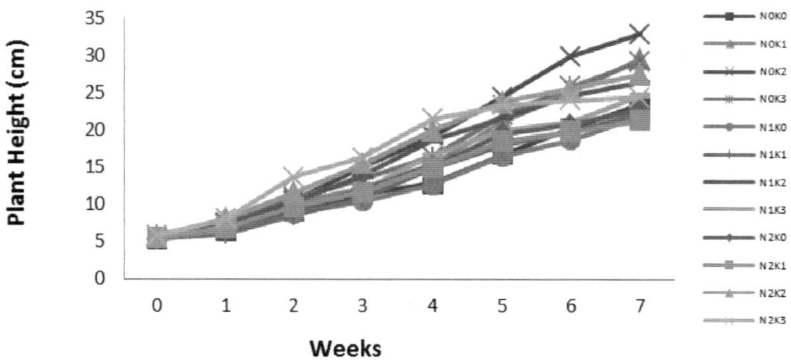

Figure 1. The graph of pearl grass growth in 8 weeks.

Figure 1. shows an increase of shade intensity from shade treatment without shade, 50% shade and 75% shade affect the average of pearl grass height per week. From this study it can be obtained the highest average pearl grass at the combination of N_0K_2 treatment (without shade and manure 15 tons/ha). Organic fertilizer has functions as a soil aggregate feeder and as an important source of soil and plant nutrients [5]. Basically pearl grass is a weed that can live in damp places and enough of light. Pearl grass does not demand on too much treatment, it can be done by simply watering in an adequate way, maintaining moisture and fertilizing the basic fertilizer and placed in a sufficient place for the sun [1].

3.1.2. Number of Branches In environmental conditions with the sufficient light, photosynthesis process will be going well and produce high energy[8]. The higher the energy formed, the higher the number of plant organ size such as the number of branches and leaf area.

Table 2. The effect of shading on the number of branches

Shade	Number of Branches
0% (N_0)	13.3b
50% (N_1)	5.4a
75% (N_2)	6.5a

Description: The numbers followed by the same letter show no significant difference (P<0.05).

Table 2 shows the treatment without shade (N_0) yielding the highest average number of branches(13.3). The average number of the lowest branches was obtained in 50% shade treatment (N_1) of 5.4. [19] state that un-shaded plants will get high light intensity so the photosynthesis process will increase because light is a source of energy for photosynthesis process then this photosynthesis results will then be used to respond the shoot growth.

Based on the result of the analysis of variety effects of cow manure toward the number of branches, there is no significant difference was found. The treatment of cow manure at various levels does not have any effects on the observed variables, it might be due to the nutrient elements in the fertilizer are not enough for the formation of the new cells. It was explained that when plants get enough nutrients to actively form new cells, causing enlargement of plants toward the diameter of the stem and the formation of primary branches [10].

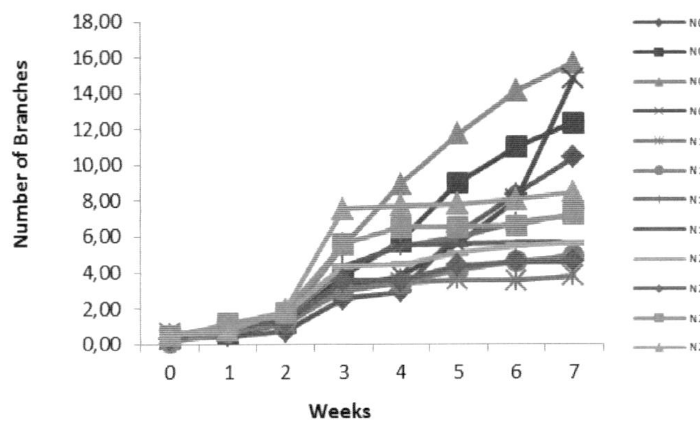

Figure 2. The graph of pearl grass increment in 8 weeks.

The treatment without shade with the provision of fertilizer dosage 15 tons/ha (N_0K_2) gives the highest number of branches with 15 branches. Plants increase body size in overall is the result of increasing the size of plant parts (organs); it is the effect of the increase of cell tissue produced by the increase of cell size [6].

The observed branches are the primary branches that can grow from various directions and cover the main stem. The branches on a plant form a certain angle with the main stem that cause the direction of growing branches are different [11]. In a shaded plant condition, plants also maximize their light capture by modifying the position of the leaves and the other organs to avoid a shade called avoiding response to shade [12].

3.1.3. Root Length Roots are part of the plant that has an important role in the growth of a plant.[13] explain that root is a part of the plant that has an important role in the adaptation of plants to water shortages because the roots are able to absorb water by maximizing the root system. In contrast with other grasses that have a fiber root system, [2] states that the root of the pearl grass is a tread root with a yarn-shaped root.

Table 3. The effect of shade level on the root length

Shade	Root Length (cm)
0% (N_0)	12.2b
50% (N_1)	9.5a
75% (N_2)	9.4a

Description: The numbers followed by the same letter show no significant difference ($P<0.05$).

Table 3 shows the longest root length average obtained at the non-shaded treatment (N_0) that is 12.2 cm long. The average of the lowest root length was obtained at the treatment of 75% (N_2) shade which is 9.4 cm long. A shaded condition or a low light condition causes the auxin hormone works [14]. [8] explain that the auxin hormone will only stimulate the growth of new roots and does not cause a long increase on plant roots. In the [16] study the length and dryness data between roots and canopy in a shade treatment has a lower mean length and dry weight compared with controls. The result of the analysis of the variety of cow manure influence on the length of the pearl grass roots there is no significant difference in each treatment level. The possibility that can be predicted is the lack of fertilization dosage so the roots growth of pearl grasses does not make a significant difference.

3.1.4. Fresh Weight of Plant The fresh weight of the plant is a measure of the result of photosynthetic deposits in plants in the form of heavy plants that still contain water content which is not constant. Explains that plant weight can also show the result of plant metabolism activity and the wet weight is also influenced by the tissue water content.

Table 4. The effect of shade level on the fresh weight of plant

Shade	Fresh Weight (g)
0% (N_0)	8.92b
50% (N_1)	2.05a
75% (N_2)	2.14a

Description: The numbers followed by the same letter show no significant difference (P<0.05).

Table 4 shows the highest average fresh weight obtained in the treatment without shade (N_0), it is 8.92 grams. The lowest average fresh weight was obtained at 50% shade treatment (N_1). The results are directly proportional to the average shade effect on plant height contained in Table 1. Fresh weight of pegagan in shade treatment gives less result compared with unregistered pegagan [17] while shaded arrowroots plants produce fewer end biomasses than un-shaded arrowroots plants due to less light availability to form assimilates [18].

The fertilizer used in this research was cow manure. Cow manure is one of organic fertilizers that is often used by farmers in general. The addition of organic materials into the soil such as fertilizer would have stronger influence towards the improvement of soil properties and not in particular to increase the nutrients in the soil [19].

Table 5. Effect of different dosages of cow manure on fresh plant weight.

The Dosage of Fertilizer	Fresh Weight (g)
0 (K_0)	2.11a
10 tons/ha (K_1)	4.53ab
15 tons/ha (K_2)	5.86b
20 tons/ha (K_3)	4.97b

Description: The numbers followed by the same letter show no significant difference (P<0.05).

Table 5 shows the highest average fresh weight of pearl grass obtained in the treatment of manure 15 tons/ha it is 5.86 grams (K_2). The lowest average fresh weight of pearl grass was obtained in the treatment without fertilizer (K_0). This proves that the addition of cow manure or organic fertilizer can improve the yield of pearl grass. The study of [20] stated that organic fertilizers can give beneficial effects because it has the ability to improve soil quality.

3.1.5. Dry Weight of Plant Dry weight of plants is a measure used to know the results of photosynthesis of a plant with a constant water content. The dry matter of a plant is a asymilat reflection which is produced from the process of photosynthesis of the plant [8]. If the dried material produced by the plant is low then the asymilat produced by the plant is also low [20].

Table 6. The effect of shade level on the dry weight of plants

Shade	Dry Weight (g)
0% (N_0)	4.40b
50% (N_1)	1.46a
75% (N_2)	1.64a

Description: The numbers followed by the same letter show no significant difference (P<0.05).

Table 6 shows the highest average dry weight obtained at the treatment without shade (N_0) of 4.40 grams. The lowest average dry weight was obtained at the 50% (N_1) treatment of 1.46 grams. The pearl grass grows well and produces high dry weight at full light intensity or treatment without shade (N_0). [21] who state that the dry weight of plants in the shade treatment will decrease because the intensity of light is reduced then followed by the rate of photosynthesis that will decrease as well.

The growth of a plant can decrease due to less optimum conditions, hence the input is needed to support the growth of the plant [22]. One of them is by doing fertilization. Fertilization is the most important factor in growth and development until the result of production [23]. He also states that fertilizing using organic fertilizer is as important as fertilizing by using inorganic fertilizer for plant growth process.

Table 7. The effect of different dosages of cow manure on the dry weight of plant

The Dosage of Fertilizer	Dry Weight (g)
0 (K_0)	1.77a
10 tons/ha (K_1)	2.34a
15 tons/ha (K_2)	3.37b
15 tons/ha (K_3)	2.52ab

Description: The numbers followed by the same letter show no significant difference (P<0.05).

Table 7 shows the highest average dry weight obtained in the treatment of fertilizer dosage of 15 tons/ha (K_2). The lowest average dry weight was obtained in the treatment without fertilizer (K_0). Chemical fertilizers and organic fertilizers are equally important for plant growth, but it is recommended to use more organic fertilizer than chemical fertilizers [24]. In the study of [27] proves that the biomass of beans by using organic fertilizer is higher than the bean plant biomass by using urea fertilizer. Table 7. shows the increasing dry weight of no fertilizer (K_0), the fertilizer dosage of 10 tons/ha (K_1) to the fertilizer dosage of 15 tons/ha and decreased at fertilizer dosage 20 tons/ha (K_3). In accordance with Table 7, it can be concluded that the dosage of fertilizer that can maximize the dry weight of pearl grass in this study is the dosage of fertilizer 15 tons/ha (K_2).

3.2. Secondary Metabolite

3.2.1. Secondary Metabolite Content Analysis Plants produce secondary metabolite compound (secondary metabolite) [25] that can be used to cure chronic and lethal disease [26]. In the utility as a traditional medicine, pearl grass is formed into a dosage. It shows that pearl grass has a great potential to be developed into useful herbs and in more advanced production.

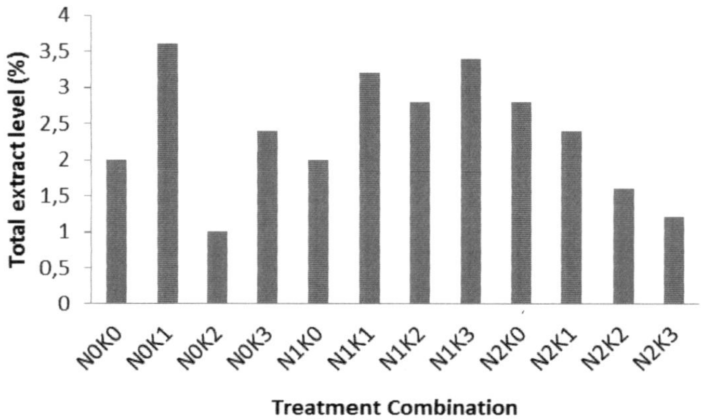

Figure 3. Total extract level of pearl grass.

Based on the figure 3, it can be seen that the highest total extract level is obtained from the combination of treatment without shade and the dosage of manure 10 tons/ha (N_0K_1). The lowest extract level is obtained from the combination of treatment without shade and 15 tons/ha (N_0K_2) fertilizer dosage. This result is in contrast to the growth of pearl grass. The growth of the plant height and the number of pearl grass branches in the combination of treatment without shade and cattle manure 15 tons/ha (N_0K_2) yielded the highest result. Primary metabolite in plants is used by plants for growth [27]; whereas secondary metabolite is produced by plants do not directly have a role for the plant growth. The higher the total extracts level of pearl grass, the higher the secondary metabolite content contained in it.

3.2.2. *Ursolate Acid Content Analysis* One of the metabolites which is contained in pearl grass is *ursolic acid*. Pearl grass is a good source of *oleanolate acid* (OA) and *ursolic acid* (UA) and both compounds have many beneficial effects for human health [28]. This compound is an active anti-cancer compound that has long been used in the field of health. In Chinese health science, pearl grass has efficacy in the healing of fever, anti-inflammatory, diuretic, ulcers and also activate the blood circulation [28]. That statement is also supported by [29] which explain that *ursolic acid* has the potential to be developed as an anticancer compound due to its ability to inhibit several stages in carcinogenesis and supported by its widespread presence in many plants.

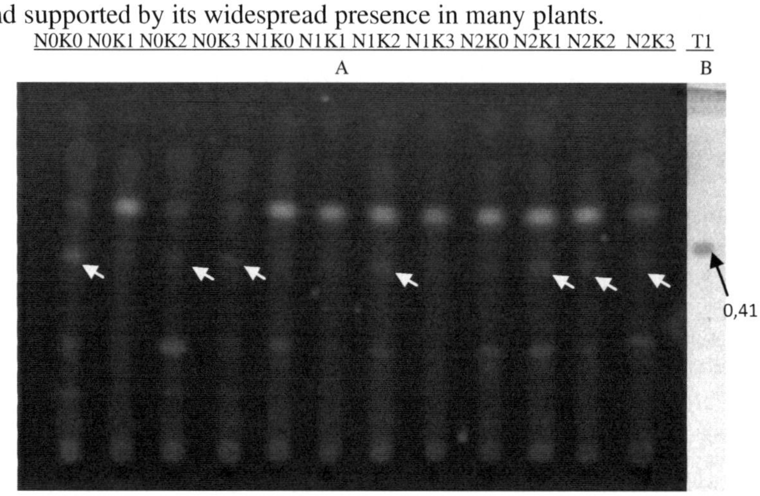

Figure 4. The results of thin layer chromatography analysis with UV_{366} (A) standard light detection of ursolic acid standard (B).

Ursolic acid is a chemical compound of triterpenoid pentacyclic group which is found in plants. The triterpenoid group belongs to terpenoid groups which are synthesized through the mevalonate pathway. Here are the results of Thin Layer Chromatography that has been done at the B_2P_2TOOT Phytochemistry Laboratory.

The combination of treatment without shade and without fertilization (N_0K_0) produces a stronger color than other treatments. This can be expected because of the condition of pearl grass that experienced stress both light stress and nutrient stress so it produces side results to defend itself.

Various types of plants are able to produce secondary metabolite compounds in certain condition and secondary metabolite serves as a defense mechanism of plants, both from biotic and abiotic stress. On the other hand the combination of shade treatment of 50% and 20 tons / ha manure application (N_1K_3) produces a weak and almost faded violet color. The treatment produces the highest total extract content in the secondary metabolite content analysis. This proves that the pearl grass does not only contain *ursolic acid* and in this combination there is another stronger compound and it is not discussed in this research. The differences in the environmental conditions where the plants grow can also cause the differences of the amount of secondary metabolites contained in plants.

4. Conclusions

Based on the results of the study it can be concluded that, a treatment without shade (N_0) is expected to provide the highest ursolic acid quality in ursolic acid quality analysis by providing a bright blue spot color. A fertilization with dosage of cow manure 15 tons/ha (K_2) can produce the highest fresh weight with 5.86 grams and the highest dry weight 3.37 grams. A treatment without shade (N_0) and without cow manure (K_0) can give best quality of pearl grass appropriate with ursolic acid quality analysis.

Acknowledgement
This research was supported by the Medicinal Plant and Traditional Drug, Tawangmangu, Indonesia.

References

[1] Soemardji A S, Anisa I N, Damayanti N A 2015 Study on rumput mutiara (hedyotis corymbosa) herbs as medicine. J. Medicine and Health 1(2) 187-199
[2] Sirait N 2014 Potensi rumput mutiara (hedyotis corymbosa) sebagai antikanker Warta Penelitian dan Pengembangan Tanaman Industri 20(3) 11-13
[3] Endrini S 2011 Antioxidant activity and anticarcinogenic properties of "rumput mutiara" (hedyotis corymbosa (L.) Lam.) and "pohpohan" (pilea trinervia (Roxb.) Wight) J Med. Plant Res. 5(16) 3715-3718
[4] Roidah I Y 2013 Manfaat penggunaan pupuk organic untuk kesuburan tanah. J Universitas Tulungagung Bonoworo 1(1) 31-42
[5] Sonbai J H H, Prajitno D and Syukur A 2013 Pertumbuhan dan hasil jagung pada berbagai pemberian pupuk nitrogen di lahan kering regosol Ilmu Pertanian 16(1) 77-89
[6] Sitompul S M and Guritno B 1995 Analisis pertumbuhan tanaman Yogyakarta UGM Press
[7] Sompotan S 2013 Hasil tanaman sawi (brasicca junea L.) terhadap pemupukan organik dan anorganik Geosains 2(1) 14-17
[8] Suminarti N E 2016 Analisis pertumbuhan tanaman dan hasil ubi jalar (ipomoea batatas (L.) Lam.). Prosiding Seminar Nasional Pembangunan Pertanian
[9] Fauzi R, Meiriani and Barus A 2016 Pengaruh persentase naungan terhadap pertumbuhan bibit Mucuna bracteata D.C. asal setek dengan konsentrasi IAA yang berbeda J Agroteknologi 4(3) 2114-2126
[10] Zamriyetti and Rambe S 2006 Kajian pertumbuhan dan produksi varietas jagung lokal dan kacang hijau dalam sistem tumpangsari J Penelitian Bidang Ilmu Pertanian 4(2) 70-7
[11] Tjitrosoepomo G 1990 Morfologi tumbuhan Yogyakarta UGM Press

[12] Merelo P, Paredes E B, Heisler M G et al 2017 The shady side of leaf development: the role of the revoluta/kanadi1 module in leaf patterning and auxin-mediated growth promotion J. Curr. Opin. Plant Biol. 35 111-116
[13] Ai N S, Torey P 2013 Karakter morfologi akar sebagai indikator kekurangan air pada tanaman. J Bioslogos 3(1) 31-39
[14] Buntoro B H, Rogomulyo R and Trisnowati S 2014 Pengaruh tekanan pupuk kandang dan intensitas cahaya terhadap pertumbuhan dan hasil temu putih (curcuma zedoaria L.) J Vegetalika 3(4) 29-39
[15] Husein E, Saraswati R and Hastuti RD 2008 Rizobakteri pemacu tumbuh tanaman. balittanah.litbang.pertanian.go.id. [5 Mei 2017]
[16] Khoiri M 2010 Pengaruh naungan terhadap pertumbuhan dan laju fotosintesis tanaman cabe merah (capsicum annuum L.) sebagai salah satu sumber belajar biologi. Download.portalgaruda.org. [30 April 2017]
[17] Tripatmasari M, Wasonowati C and Alianti VR 2010 Pemanfaatan naungan dan pupuk kotoran sapi terhadap pertumbuhan dan kandungan tripenoid pegagan (centella asiatica L.) Agrovivor 3(2) 137-145
[18] Begna S H, Dwyner L M, Cloutier D et al 2002 Decoupling of light intensity effects on the growth and development of C3 and C4 weed species trough sucrose supplementation J Exp. Bot. 53(376) 1935-1940
[19] Janmohammadi M, Mahmoudi Z S, Ahadnehzad A et al 2014 Influence of chemical and organic fertilizer on growth, yield and essential oil of dragonhead (dracocephalum moldavica L.) plant. Acta Agric. Slov. 103(1) 73-81
[20] Ogbonna P E and Nweze N J 2012 Evaluation of growth and yield responses of cocoyam (colocasia esculenta) cultivars to rates of NPK 15:15:15 fertilizer. Afr. J Agric. Res. 7 6553-6561
[21] Gardner F P, Pearce R B, Mitchell R L 2005 Fisiologi tanaman budidaya Jakarta UI Press
[22] Musyarofah N, Susanto S, Aziz A A et al 2007 Respon tanaman pegagan (centella asiatica L. Urban) terhadap pemberian pupuk alami di bawah naungan Buletin Agronomi 35(3) 217-224
[23] Şener S 2016 Effects of several cultivars', mulch and fertilizer aplications on plant growth and development criteria and plant's nutrition elements uptake in organic strawberry plantation in nevşehir city Asian Economic and Social Society 6(11) 221-228
[24] Rafaat J G, Hussein R A, Ahmad M M 2015 Effect of fertilizer application on vegetative growth characters of broad bean (vicia faba L.). International J of Plant, Animal and Environtmental Sciences 6(1) 103-109
[25] Lakitan B 2015 Dasar-dasar fisologi tumbuhan Jakarta Rajawali Press
[26] Kahsyap D, Tuli H S and Sharma A K 2016 Ursolic acid (UA): A metabolite with promising therapeutic potential J Life Sci 216 201-213
[27] Yang Y C and Wei M C 2016 A combined procedure of ultrasound-assisted and supercriticalcarbon dioxide for extraction and quantitation oleanolic and ursolicacids from Hedyotis corymbosa Ind Crops Prod. 79 7-17
[28] Liang Z, Jiang Z, Fong D W et al 2009 Determination of oleanolic acid and ursolic acid in Oldenlandia difussa and its substitute using high performance liquid chromatography. J. Food Drug Anal. 17(2) 69-77
[29] Haryanti S and Katno 2011 Aktivitas sitotoksik Ocimum sanctum L pada sel kanker kolon WiDr. Simposium Nasional XV PERHPBA Surakarta

IOP Conf. Series: Earth and Environmental Science **142** (2018) 012045 doi:10.1088/1755-1315/142/1/012045

The potential of arbuscular mycorrhizal fungi application on aggregrate stability in alfisol soil

J Syamsiyah, A Herawati and Mujiyo
Soil Science Department, Agriculture Faculty, Universitas Sebelas Maret, Surakarta, Indonesia

E-mail: ninukts@staff.uns.ac.id

Abstract. The aim of this study was to determine the soil aggregate stability and its relationship with another variable in alfisol. The research used completely randomized design with four treatments: two sterilization levels (no sterilization and with sterilization) and two levels of mycorrhizal inoculation (no mycorrhizal and with mycorrhizal). Mycorrhizal (5 grams/pot) was inoculated before planting rice seeds. The soil aggregate stability was measured by wet-sieving and turbidimetric measurements. The results showed that soil aggregate stability was higher in mycorrhizal inoculated than non-mycorrhizal inoculated treatment, by 5% in sterilization soil and 3.2% in non-sterilization soil. The correlation analysis indicated that soil aggregate stability has a tight relationship with spore population, total glomalin, available glomalin, dry weight, tiller number of plant, and soil organic C. Inoculation of mycorrhizal contributed to stabilize soil aggregates in alfisol

1. Introduction

Soil structure is crucial to crop management and sustainable agricultural ecosystems [1] and is usually expressed by the degree of aggregate stability [2]. Poor arrangement of aggregates will induce serious problems, such as soil compaction, which can easily occur in poorly drained clay soils and detrimental to plant cultivation due to reduce the permeability to air, water, and roots [3], and causes nutrient stress and poor emergence of seeds. Appropriate measures for the alleviation of soil compaction were established through agriculture practices. Studies concerning the ecological role of arbuscular mycorrhizal fungi (AMF) put the emphasis on the benefits to host plants. For instance, AMF are widely known to enhance plant growth and improve nutrient uptake [4], and protect the host plant from drought, temperature, and toxic metal stress [5]. Recently, some studies have focused on the effects of AMF on soil structure [6]. The AMF have been found to provide great benefits in improving soil structure by the combined function of hyphae [7], which can physically entangle soil particles over an extended region and release an immune-reactive glycoprotein called glomalin, which is highly correlated with soil aggregate stability [8].

The AMF are the form of mutualism symbiotic association between plant roots and fungal mycelium [9], with more than 70% of plant families [10]. The other studies showed that the growth of plants with mycorrhizal inoculation was higher than without mycorrhizal inoculation due to more provided nutrients to the host plant [11], increase the resistance of drought [12] and protect the plants from fungal pathogens [11, 13, 14], and nematodes. In addition, mycorrhizal can improve the soil structure by forming stable soil aggregates through the external hyphae tissue that it produces. Application of AMF is one of the ways to increase the crops production in upland that is cheap, easy and environmental friendly. Inoculation of mycorrhizal is expected to assist in rehabilitating critical land. It can be regarded as bio-fertilization, both for food crops, plantations, forestry and greening crops.

Soil organisms are known to play a crucial role in soil ecological processes such as organic matter turnover, nutrient cycling and engineering of the soil physical properties. Therefore, they are essential to soil fertility and nutrient uptake by plants [15, 16]. The soil physical habitat is widely assumed to be of prime importance in determining and regulating biological activities [17]. As a result, interactions between soil physics and the biological and chemical processes are key determinants of ecosystem health [18], but are still largely to be deciphered.

Plant roots, microorganisms (bacterial and fungal mycelium) and soil fauna are considered to be major cause of soil aggregation and porosity [2, 19, 20, 21]. The AMF influence soil physical processes with soil particles entanglement and glue substance secretion such as glomalin, but the respective effects of roots, but there is still controversy due to the symbiotic relationship between plants and AMF [22, 23, 24].

Glomalin is one of the deposits C in the soil, because it has a long-life time in the soil and contains soil organic C as the main binding material that will stabilize the soil [21], thus, soil aggregate stability increases. The working principle of mycorrhizal is to infect the root system of host plants and produce hyphae tissue intensively so it will be able to increase the capacity of nutrients absorption. Glomalin is one of the deposits C in the soil, because it has a long-life time in the soil and contains soil organic C as the main binding material that will stabilize the soil [21], so that soil aggregate stability increases. The working principle of mycorrhizal is to infect the root system of host plants and produce hyphae tissue intensively leads to increase the capacity of nutrients absorption.

The AMF can contribute to soil aggregate stability directly by a physical effect of a network around soil particles, and indirectly by the hyphal exudation of an iron-containing, heat stable glycoprotein named glomalin as an aggregate binding agent [25, 26]. Glomalin has been operationally defined as glomalin-related soil protein (GRSP) by extraction and detection conditions from soil, and it is detected in large amounts in diverse ecosystems [27]. The sticky GRSP acts as biological glue, helping to bind soil tiny particles into small aggregates of different sizes [28]. Well-aggregated soil is stable enough to resist wind and water erosion, and has better air and water infiltration rates favorable for plant and microbial growth [29]. The present study was conducted to determine the soil aggregate stability and its relationship with another variable in alfisol due to soil sterilization and inoculation of mycorrhizal.

2. Methods

The alfisol was sampled from the top layer (0-20 cm) at the Jumantono experimental site of the Universitas Sebelas Maret. The experiment was arranged in pot experiment used completely randomized design with four treatments: two sterilization levels, no sterilization (T0) and with sterilization (T1) and two levels of mycorrhizal inoculation, no mycorrhizal (M0) and with mycorrhizal (M1). Each treatment had six replicates, which made a total of 24 pots. Soil sterilization used autoclave, **7** kilograms of alfisol passed through 5 mm sieve was sterilized using autoclave at 0.11 MPa and 121 °C for 2 h. The AMF were provided by Soil Biology Laboratorium of Agriculture Faculty, Gadjah Mada University. Mycorrhizal at 5 grams pot^{-1} was inoculated before planting rice seeds. The base fertilizer was used 100 kg Nitrogen Ha^{-1}, 40 kg P_2O_5 ha^{-1} and 50 kg K_2O ha^{-1}.

Portions of soil samples were dried by spreading them out on paper in ambient air and at room temperature for two weeks, and then were homogenized, grounded in an agate mortar and sieved to 0.2 mm and 0.5 mm. for. Soil pH was determined in a suspension with deionized water (soil/water, 1:2.5) using digital pH meter electrode. Soil organic matter (SOM) was measured by the wet combustion method using a mixture of potassium dichromate and sulfuric acid under heating. Total nitrogen (TN) concentration was determined according to the semi-micro Kjeldhal method. Available phosphorus concentration was measured colorimetrical. Exchangeable cations were extracted with ammonium acetate at pH 7.8 and measured by Atomic Absorption Spectrophotometry (Shimadzu

AA). The blank reagent and standard reference soil were assayed for quality assurance and quality control.

The soil parameter that measured was soil aggregate stability, C organic, total glomalin, available glomalin, spore population. Total glomalin related-soil protein was analyzed with 8 ml 50 mM sodium citrate (pH = 8), and then mixed using autoclave for 90 minutes. The supernatant was determined as a glomalin related-soil protein with standard bovine serum albumin [9]. The plant growth parameters included plant height, tiller number, and dry weight. Spearman correlation coefficients between soil aggregate stability and other variables were calculated using Proc-Corr procedure in program SAS version 9.13 portable. Correlation test was used to determine the relationship between variables.

3. Results and Discussion

3.1. Soil chemical and physical properties

Soil chemical and physical property of sample are presented in Table 1. Soil pH was acid (5.93), total C organic concentration was low (0.95%), total Nitrogen was low (0.11%), C/N ratio was low (8.74), CEC was moderate (20.00 $cmol^{+}kg^{-1}$), available phosphate was low (6.90 ppm), exchangeable potassium and calcium were low (0.31 and 3.24 $cmol^{+}kg^{-1}$), exchangeable magnesium and sodium were moderate (2.95 and 0.77 $cmol^{+}kg^{-1}$), base saturation was moderate (36.55%), and the texture was clay. So, based on the chemical and physical properties, the soil sample that used in this research have a low fertility.

Table 1. The soil properties of samples

Soil Properties	Value	Level*
pH H_2O (1:2,5)	5.93	Acid
C organic (%)	0.95	Low
Total Nitrogen (%)	0.11	Low
C/N Ratio	8.74	Low
P_2O_5 Bray (ppm)	6.90	Low
Exchangeable K (cmol(+) kg^{-1})	0.31	Low
Exchangeable Ca (cmol(+) kg^{-1})	3.24	Low
Exchangeable Mg (cmol(+) kg^{-1})	2.95	Moderate
Exchangeable Na (cmol(+) kg^{-1})	0.77	Moderate
CEC (cmol(+) kg^{-1})	20.00	Moderate
Exchangeable Al (cmol (+) kg^{-1})	2.50	
Exchangeable H-(cmol (+) kg^{-1})	0.70	
Base saturation (%)	36.55	Moderate
Soil Texture: Sand (%)	27.49	Clay
Silt (%)	16.96	
Clay (%)	55.56	

3.2. Effects of sterilization soil and mycorrhizal inoculation on soil parameters and aggregate stability

Soil sterilization and inoculation of mycorrhizal had a high correlation with soil aggregate stability from the highest to the lowest; C organic, total glomalin, available glomalin, and spore population (Table 2). Inoculation of mycorrhizal contribute to heat stable glycoprotein named glomalin as an aggregate binding agent. Some previous studies indicated that SOC affected the growth of AMF [30, 31] and was positively correlated with hyphal length density [32], which is sufficient to positively affect soil aggregation in the absence of other soil biota [7]. The effects of SOC on AMF were induced both by compounds released during the decomposition process and by secondary metabolites produced by microorganisms during organic matter decomposition [33]. Conversely, fungal inoculation enhanced the decomposition of organic matter in soil [34] and carbon could be lost from mycorrhizal roots into soil and external hyphae [35], which might be the main reasons for the increase in SOC when inoculating with AMF.

Table 2. Spearman correlation coefficients between soil aggregate stability to sterilization soil and mycorrhizal inoculation on several soil parameters

	Spearman Correlation Coefficients
Total glomalin	0.1856
Available glomalin	0.304
C-organic	0.111
Spore population	0.9853

3.2.1. Effects of sterilization soil and mycorrhizal inoculation on soil aggregate stability and total glomalin A significant positive correlation between total glomalin and soil aggregate stability was observed in this study (Figure 1). Mycorrhizal increased the aggregates stability. This suggested that mycorrhizal contributed to the improvement of soil aggregates stability partly via the function of glomalin, which bound soil primary particles into larger aggregates.

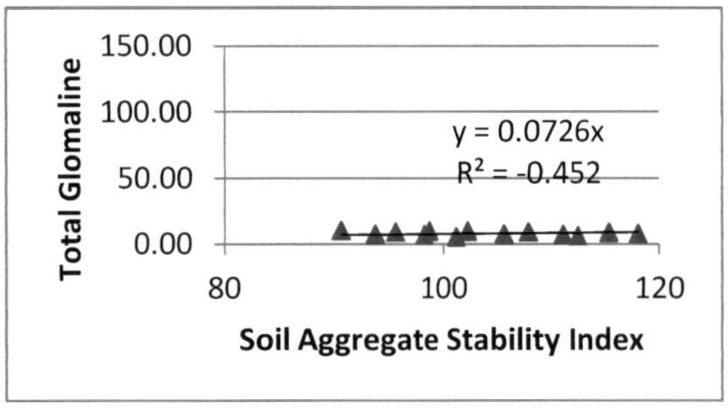

Figure 1. Correlation between soil aggregate stability with total glomalin.

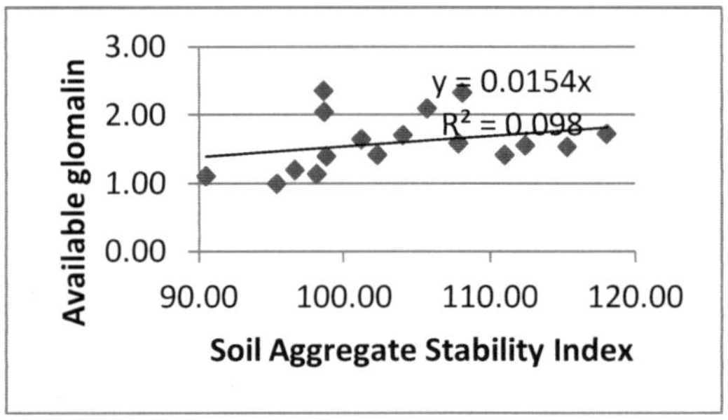

Figure 2. Correlation between soil aggregate stability with available glomalin

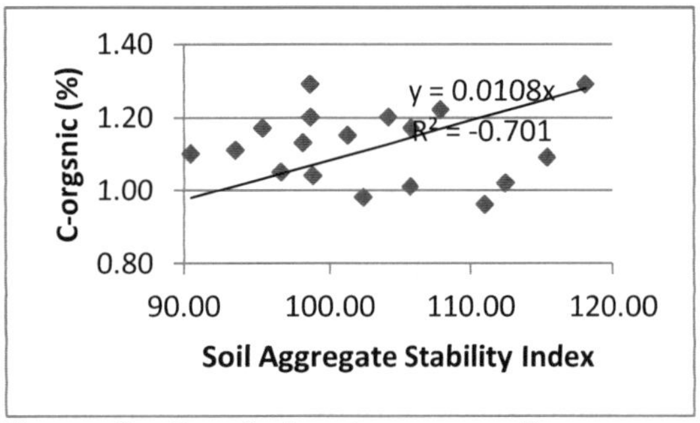

Figure 3. Correlation between soil aggregate stability with C-organic

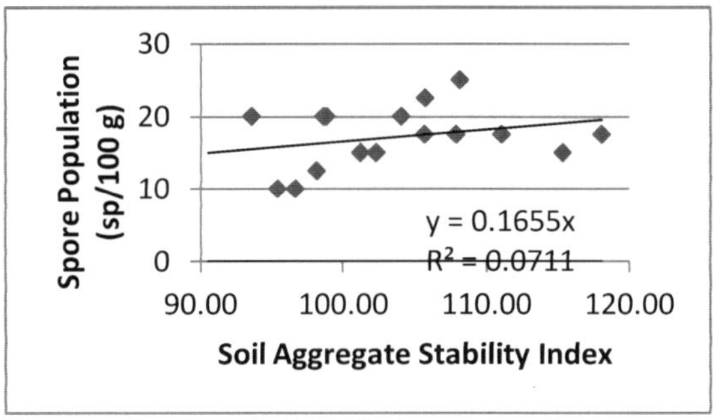

Figure 4. Correlation between soil aggregate stability with spore population

3.2.2. Effects of sterilization soil and mycorrhizal inoculation on soil aggregate stability with available glomaline and C-organic A significant positive correlation between soil aggregate stability with available glomalin and C organic were observed in this study (Figure 2 and 3). The significant positive relationships between SOC and >1-mm aggregates in a five-year tillage experiment, and a significant positive correlation between C-organic was found. Mycorrhizal reduced the proportion of

<0.25-mm aggregates while there was an increase in the proportion of >1-mm aggregates [36]. These results suggested that mycorrhizal contributed to the improvement of soil aggregate stability via the function of soil organic carbon, which bound soil primary particles into larger aggregates.

3.2.3. *Effects of sterilization soil and mycorrhizal inoculation on soil aggregate stability with spore population.*

Correlation test showed that there was a positive correlation between spore population and soil aggregate stability because of sterilization soil and inoculation of mycorrhizal (Figure. 4). Inoculation of mycorrhizal increased spore population up to 21%. Inoculation of mycorrhizal increase spore population up to 6% in unsterilized soil and up to 43% at sterilized soil (without inoculation of mycorrhizal). Partly, inoculation of mycorrhizal significant increased soil aggregates stability. This result showed that soil microorganism did not influence the forming of spore in soil.

3.3. Effects of sterilization soil and mycorrhizal inoculation on plant growth parameters

Soil sterilization and inoculation of mycorrhizal increased plant height, dry weight and tiller number of rice. A positive correlation between soil aggregate stability with plant growth showed at Table 3. The highest correlation, from the high to the weak are tiller number, dry weight, and plant height of rice.

Table 3. Spearman correlation coefficients between soil aggregate stability with several plant growth parameters

	Spearman Correlation Coefficient
Plant height	0.5455
Dry Weight	0.4931
Tiller number	0.3978

There were positive correlation between sterilization soil and inoculation of mycorrhizal and plant growth parameters. Sterilization soil and inoculation of mycorrhizal increase plant height, dry weight, and tiller number. Tiller number incerased up to 38% than unsterilized soil without inoculation of mycorrhizal. Unsterilized soil, inoculation of mycorrhizal improved tiller number up to 33 % than unsterilized soil without inoculation of mycorrhizal (Table 3).

4. Conclusion

There was a correlation between soil aggregate stability with the other variables. Soil sterilization and inoculation of mycorrhizal increased total glomalin, available glomalin, C organic, spore population, plant height, tiller number, dry weight of rice. Inoculation of mycorrhizal increased total glomalin up to 16 %, available glomalin 20%; C organic 11.44%; spore population up to 43% in sterilized soil. Inoculation of mycorrhizal increased plant height up to 9% at sterilized soil and 4% at unsterilized soil, tiller number up to 21% at sterilized soil and 16% at unsterilized soil, dry weight of rice up to 16% at sterilized soil and 34% at unsterilized soil.

References

[1] Rillig M C, Wright S F, Eviner V T 2002 *Plant Soil* **238** 325–33
[2] Amézketa E 1999 *J. Sustain. Agric.* **14** 83–151
[3] Batey T 2009 *Soil Use Manage* **25** 335–345.
[4] Hu J L, Lin X G, Wang J H, Dai J, Cui X C, Chen R R, Zhang J B 2009 *Soil Biol. Biochem.* **41** 2460–2465
[5] Zhang X H, Lin A J, Gao Y L, Reid R J, Wong M H, Zhu Y G 2009 *Soil Biol. Biochem.* **41** 930–935
[6] Rillig, M C 2004 *EcolLett.* **7** 740–754

[7] Rillig M C, Mardatina N F, Leifheita E F, Antunes P M 2010 *Soil Biol. Biochem.* **427** 1189–1191.
[8] Wright S F, Upadhyaya A 1998 *Plant Soil* **198** 97–107
[9] Schuessler A, Schwarzott D and Walker C 2001 *Mycol. Res.* **105** 1413-1421
[10] Newman E I and Reddel P 1987 *New Phytol.* **106** 745-751
[11] Smith SE and Read J 2008 *Mycorrhizal Symbiosis* (New York: Academic Press)
[12] Auge R M, Duan X G, Ebel R C and Stodola A J W 1994 *Planta* **193** 74-82
[13] Matsubara Y, Ohba N, Fukui H 2001 *J. The Japan. Soc. Hortic. Sci.* **70** 202-206
[14] Thygesen K, Larsen J, Bodker L 2004 *Eur. J.Plant Pathol.* **110** 411-419
[15] Bradford M A, Jones T H, Bardgett R D, Black H I J, Boag B, Bonkowski M, Cook R, Eggers T, Gange A C, Grayston S J, Kandeler E, Mc Caig A E, Newington J E, Prosser J I, Setala H, Staddon P L, Tordoff G M, Tscherko D, Lawton JH 2002 *Science* **298** 615–618
[16] Wardle D A, Bardgett R D, Klironomos J N, Setala H, van der Putten W H, Wall D H 2004 *Science* **304** 1629–1633
[17] Young I M, Crawford J W 2004 *Science* **304** 1634–1637
[18] Feeney D S, Crawford J W, Daniell T, Hallett P D, Nunan N, Ritz K, Rivers M, Young I M 2006 *Microb. Ecol.* **52** 151–158
[19] Jastrow J D, Miller R M 1991 *Agric. Ecosyst. Environ.* **34** 279–303
[20] Six J, Bossuyt H, Degryze S, Denef K 2004 *Soil Tillage Res.* **79** 7–31
[21] Tisdall J M, Oades J M 1982 *J. Soil. Sci.* **33** 141–163
[22] Hallett P, Feeney D, Bengough A, Rillig M, Scrimgeour C, Young I 2009 *Plant Soil* **314** 183–196
[23] Jastrow J D, Miller R M, Lussenhop 1998 *J. Soil Biol. Biochem.* **30** 905–916
[24] Thomas R S, Franson R L, Bethlenfalvay G 1993 *J. Soil Sci. Soc. Am J.* **57** 77–81
[25] Gonzalez-Chavez M C, Carrillo-Gonzalez R, Wright S F, Nichols K A 2004 *Environ. Pollut.* **130** 317-323
[26] Rillig, M C 2004 *Can. J. Soil Sci.* **84** 355-363
[27] Rillig, M C. Mummey D L 2006 *New Phytol.* **171** 41-53
[28] Bronick C J 2005 *Lal. R. Geoderma* **124** 3-2
[29] Wu Q S, Xia R X, Zou Y N 2008 *Eur. J. Soil Biol.* **44** 122-128
[30] Quilliam R S, Hodge A, Jones D L 2010 *Appl. Soil Ecol.* **46** 247–250
[31] Olsson O, Olsson P A, Hammer E C 2014 *Mycorrhiza* **24** 443–541
[32] Wang P, Zhang J J, Xia R X, Shu B, Wang M Y, Wu Q S, Dong T 2011 *Spanish J. Agric. Res.* **9** 1307–1318
[33] Gryndler M, Hršelová H, Cajthaml T, Havránková M, Řezáčová V, Gryndlerová, H, Larsen J 2009 *Mycorrhiza* **19** 255–266
[34] Ritz K, Young I M 2004 *Mycologist* **18** 52–59
[35] Jakobsen I, Rosendahl L 1990 *New Phytol.* **115** 77–83
[36] Liang A Z, McLaughlin N B, Zhang X P, Shen Y, Shi X H, Fan R Q 2011 *Acta Agricult. Scand. Sect. B Soil Plant Sci.* **616** 535–542

IOP Conf. Series: Earth and Environmental Science **142** (2018) 012046 doi:10.1088/1755-1315/142/1/012046

Effect of utilization of tomato extract and foliar fertilizer as media on shoots multiplication of banana cv Ambon in vitro

L A Bidhari[1], E Purwanto[2] and A Yunus[2]
[1] Departement of Agronomy, Graduate School of Universitas Sebelas Maret, Surakarta 57126, Central Java, Indonesia
[2] Departement of Agrotechnology, Faculty of Agriculture, Universitas Sebelas Maret, Surakarta, 57126, Central Java, Indonesia

E-mail: yunus.uns7@yahoo.com

Abstract. The good quality banana seeds are still difficult to obtain. There are two ways to provide seeds, namely conventional and tissue culture (*in vitro*). Tomato extract contains natural ZPT or phytohormone which can be utilized in modification of banana tissue culture media. The aim of this study was to determine the influence of media types and tomato extracts in various concentrations for multiplication of banana cv. Ambon *in vitro*. The study was conducted from October - December 2016 at the Tissue Culture Laboratory of Horticulture Seed Center, Salaman, Magelang. The experimental design used was completely randomized design with two treatment factors. The firs factor was media type with the addition of foliar fertilizer, the second factor was modification of tomato extract with 4 levels. The results showed that the different of the treated media treatment did not affect the emerge of leaf and leaf length, the number of roots and root length. The emerge of the leaves of all treatments occurred at 6 days after planting with the highest average length was obtained in MS treatment with a combination of tomato extract 50 ml/l (10.3 cm). The use of MS medium with a combination of tomato extract 50 ml/l generated the average root number 15.5 with a root lengths 7.5 cm. Substitution of MS medium with tomato extract and foliar fertilizer did not show better results compared to the use of MS media in the multiplication of banana shoots in tissue culture.

1. Introduction

Bananas production in Indonesia is quite high, and tends to increase from year to year, but so far, the potential has not been optimally utilized. In Indonesia, the largest consumption of bananas is for the domestic market. The technology of banana cultivation has widely investigated, with the most basic and very decisive result in the provision of quality seeds [1]. The enhancement of banana production in Indonesian requires expansion of planting. Banana plantations need high quality seeds. There are two ways to provide seeds, namely conventional and tissue culture (*in vitro*). Conventional propagation through suckers, tubers and hemispheres takes a long time, the seeds produced are few, not uniform and their health is not guaranteed. While tissue culture techniques (*in vitro*) can produce healthy and large uniform banana seedlings in a relatively short period of time and are not climate dependent, thus, the availability of seeds is guaranteed [2].

Substitution of media can be done by using tomato extract containing natural ZPT or phytohormone and addition of compound fertilizer, specially leaf fertilizer as modification of tissue culture media. The use of organic material that is cheaper and easier to obtain is expected to replace the micro nutrient supply that is in vitro media [3]. The aims of this research were to know the influence of media type and tomato extract in various concentrations on the multiplication of bananas cv. ambon *in vitro*, and to obtain alternative regeneration media suitable for multiplication of ambon banana plant *in vitro*.

2. Methods

The study was conducted from October - December 2016 at the Tissue Culture Laboratory of Holticulture Seed Center Salaman, Magelang. The explant material used is plantlet of banana cv. Ambon plant from Horticulture Seed Center, Magelang. The basic ingredients media MS is a stock solution of macro nutrients composed of KNO_3, NH_4NO_3, $CaCl_2$, $MgSO_4$ and KH_2PO_4, micro-nutrients are composed of $MnSO_4$, $ZnSO_4$, H_3BO_3, KI, Na_2MoO_4, $CoCl_2$ and $CuSO_4$, vitamin solution composed of *Myo-inositol*, *Thiamine* HCl, *Pyridoxine* HCl, dan *Nicotinic acid*; as well as *a buffer* solution composed of $FeSO_4$ and Na_2 EDTA.Other ingredients used are for media hardener, sugar as carbon source, BAP and IAA as growth regulator and NaOH and HCl as pH controller (5,8-6,2). The research used ripe tomatoes, fresh and red, as well as foliar fertilizer *Growmore* brands. Sterilization of media use *an autoclave* for 60 minutes at a pressure of 17.5 psi and a temperature of 121 ^{0}C followed by a *drying* process for 15 minutes.

The experimental design used was completely randomized design (CRD) with two treatment factors which was media type with the addition of foliar fertilizer and modification of tomato extract with 4 levels with an addition of coconut water. The first factor was media types which was consist of P_0 = MS, P_1 = ½ MS + ½ foliar fertilizer, and P_2 = foliar fertilizer. The second factor was tomato extract concentrations with an addition of coconut water 150 ml/l that were T_1 = tomato extract 50 g/l, T_2 = tomato extract 100 g/l, T_3 = tomato extract 150 g/l, and T_4 = tomato extract 200 g/l. Based on these two treatment factors, 12 treatment combinations were repeated three times, thus there were 36 experimental units. The data were analyzed using analysis of variance by F test level of 5%. If there is a significant difference, the analysis is continued with a DMRT test of 5% level.

3. Results and Discussion

3.1. The Emerge of Leaf and Leaf Length

The increase in leaf length caused by the acceleration of cell division and the process of differentiation. Cell division requires energy that can be obtained from auxin or cytokines with other nutrients. The length and width of the leaf are closely related to the direction of cleavage, enlargement, quantity, and cell distribution. The more leaf area, the number of stomata grows. Stomata play a major role in the absorption of nutrient and substances needed in the process of plant metabolism [4]. Based on the results of data analysis obtained from this study are presented in Table 1. There was no interaction as well as no effect of treatments on the emerge of leaf and leaf length (Table 1). In all media and treatments, the leaf appeared at 6 days after planting (DAP).

[5] showed that the addition of adenda increased the leaf length, root number and wet weight of plant. Organic fertilizers *i.e.* coconut water, banana pulp, carrot extract, tomatoes, and pineapple added in ½ MS medium supplied essential nutrients, vitamins, carbohydrates and growth regulators required for growth of *Phalaenopsis orchid* seedling *in vitro*. Table 1 showed that in MS medium generated a higher leaf length compared with other media treatments. The highest yield was obtained in MS medium treatment with the addition of 50 ml/l tomato extract with 11.5 cm leaf length. [6] reported that adenda organic materials can be added in the culture medium to enrich the basic media.

Table 1. The emerge of leaves and leaf length on media treatment and concentration of tomato extract in multiplication of banana shoots with tissue culture

Treatment Media	Tomato Extract (mL/L)				Averages
	50	100	150	200	
	Day of Emerge Leaves (DAP)				
MS	6	6	6	6	6.0
1/2 MS + 1/2 FF	6	6	6	6	6.0
Foliar Fertilizer	6	6	6	6	6.0
Averages	6.0	6.0	6.0	6.0	6.0
	Leaf Length (cm)				
MS	11.5	8.7	10.3	10.8	10.3
1/2 MS + 1/2 FF	10.9	7	8.3	9.3	8.9
Foliar Fertilizer	11	7.8	8.3	7	8.5
Averages	11.1	7.8	8.9	9.0	9.2

3.2. Number of Roots and Roots Length

According to [7] the roots are needed for nutrient absorption from the media. The more the number of roots the more extensive the absorption of nutrients by the roots. The results of [8] showed that the addition of tomato extract into ½ MS medium increased the number of roots of *Phalaenopsis orchid*. The number of roots and roots length analysis in multiplication of banana cv. ambon are presented in Table 2. There was no effect of various media and addition of tomato extract at various concentrations as well as their interaction on the number of roots and root length. The roots yielded on MS medium treatment with the addition of tomato extract with various concentrations tended to be higher than the number of roots obtained from the use of ½ MS + ½ FF or full foliar fertilizer media.

Table 2. Number of roots and root lengths in media treatment and concentration of tomato extract in the multiplication of banana shoots with tissue culture

Treatment Media	Tomato Extract (mL/L)				Averages
	50	100	150	200	
	Number of Roots				
MS	18	10	18	16	15.5
1/2 MS + 1/2 FF	15	12	15	13	13.8
Foliar Fertilizer	9	0	0	0	2.3
Average	14	7.3	11	9.7	10.5
	Roots Length (cm)				
MS	9.1	3.3	9.1	8.6	7.5
1/2 MS + 1/2 FF	8.3	4.6	4	9.1	6.5
Foliar Fertilizer	0.8	0	0	0	0.2
Average	6.1	2.6	4.4	5.9	4.8

Roots length affects the explant growth because long-rooted plants have better ability to absorb water and nutrients than short-rooted plants [9]. [10] argue that organic additives include one of the basic compositions of culture media compounds other than minerals, carbon sources, amino acids and growth regulators. The addition of foliar fertilizer can be done as an alternative media in multiplication of banana in tissue culture. Compound Fertilizer Growmore (20:20:20), Hortigro (19:19:19) and Kristalon (18:18:18) have potential as media substitutes for MS because they have macro and micro nutrients that useful for plant growth and development [11].

4. Conclusion
Substitution of MS media with tomato extract and foliar fertilizer did not show better results compared with the use of MS media in the multiplication of banana cv. ambon shoots in tissue culture.

References
[1] Rugayah, Hapsoro D, Ulumudin A, and Feria W M 2012 Study of vegetative propagation technique of banana cv. ambon kuning with tube cleavage (Corm) *J. Agrotropika* **17(2)** 58
[2] Maslukhah U 2008 *Banana extract as MS media supplement in culture media of banana cv. raja bulu* (*Musa paradisiaca* L. AAB Group) *in vitro* (Bachelor Thesis, Study Program of Hortikultura, Department of Agronomy and Horticulture, Faculty of Agriculture, Institut Pertanian Bogor)
[3] Rianti, Suhermiatin T, Ermawati N 2017 Optimazation of the growth cattleya plantlet through strength combination of murashige-skoog and organic substrances *J. Agriprima* **1** 59
[4] Widyastoety D 2014 Effect of auxin and cytokinin on the growth of mokara orchid plantlets *J. Hort.* **24** 230
[5] Maera Z 2015 Growth response planlet orchid phalaenopsis hybrids against giving two types of leaf fertilizer and benziladenin during acclimatization *J. Enviagro* **7** (2)
[6] Yusnita 2004 *Tissue culture how to expand plants efficiently* (Jakarta: Agromedia Pustaka)
[7] Yuniastuti E, Praswanto, Harminingsih I 2010 the effect of bap concentration of anthurium's (anthurium *andraeanum* linden) shoot multiplication on some nutrient mediums by in vitro *Carakatani* **XXV** (1)
[8] Jayanti A 2011 *effect of tomato and tripton extract on enlargement seedling orchid phalaenopsis hybrids in vitro* (Lampung: Thesis Universitas Lampung)
[9] Andini N 2016 The use of coconut water and banana extract on the multiplication of temulawak shoots (*Curcuma xanthorriza* Roxb.) on in vitro (Surakarta: Universitas Sebelas Maret)
[10] Hartmann H T, Kester D E, Davies F T and Geneve R L 2002 *Plant propagation principles and practice* 7th Ed (New Jersey: Prentice Hall)
[11] Husnul K, Kendarini N, and Soetopo L 2016 The effect of compound fertilizer on chrysanthemum growth (*Dendranthema grandiflora* Tzvelev) on in vitro *J. Prod. Tan.* **4** 352

IOP Conf. Series: Earth and Environmental Science **142** (2018) 012047 doi:10.1088/1755-1315/142/1/012047

Analysis of the effectiveness of fertilizer subsidy policy and its effect on rice production in Karanganyar Regency

A T Mulyadiana, S Marwanti and W Rahayu
Department of Agribusiness, Universitas Sebelas Maret, Surakarta, Indonesia

E-mail: amaliatiara11@gmail.com

Abstract. The research aims to know the factors which affecting rice production, and to know the effectiveness of fertilizer subsidy policy on rice production in Karanganyar Regency. The fertilizer subsidy policy was based on four indicators of fertilizer subsidy namely exact price, exact place, exact time, and exact quantity. Data was analyzed using descriptive quantitative and qualitative and multiple linear regression. The result of research showed that fertilizer subsidy policy in Karanganyar Regency evaluated from four indicators was not effective because the distribution of fertilizer subsidy to farmers still experience some mistakes. The result of regression analysis showed that production factors such as land area, use of urea fertilizer, use of NPK fertilizer, and effectiveness of fertilizer subsidy policy had positive correlation and significant influence on rice production, while labor utilization and use of seeds factors had no significant effect on rice production in Karanganyar Regency. This means that if the fertilizer subsidy policy is more effective, rice production is also increased.

1. Introduction

Subsidized fertilizer according to Regulation of the President of Republic of Indonesia Number 77 Year 2005 is fertilizer where the procurement and distribution gets subsidy from the government for the needs of farmers implemented on the basis of government programs in the agricultural sector. This regulation also determines the type of subsidized fertilizer consisting of inorganic fertilizers (urea, SP36, ZA, NPK) and organic fertilizers. According to [1], the main objective of fertilizer subsidy is that the price of fertilizer at the farm level can remain affordable by farmers, so that it can support productivity and food security program.

The effectiveness level of the fertilizer subsidy policy is measured based on six indicators called the 6 (six) principles appropriately. Based on the Regulation of the Minister of Trade No. 15 / M-DAG / PER / 4/2013, these principles are the principle of procurement and distribution of subsidized fertilizer covering the exact type, quantity, price, place, time and quality. The indicators used in this study focused on the precise four consisted of price, place, time, and amount. These four indicators are selected because it's can be quantified and then interpreted.

[2] explains that effectiveness (results) is the relationship between output with goals or targets to be achieved. The notion of effectiveness is basically related to the achievement of the objectives or targets of the policy. Operational policy is said to be effective if the process of the activity reaches the final goals and objectives of the policy. According to [3], effectiveness is an agreed target for joint efforts. Based on the understanding of effectiveness according to the two experts mentioned above, it

can be summarized that the effectiveness is a measure of achievement targets that show the realized output that has been achieved from the output that should be achieved.

2. Methods

2.1 Basic Method of Research

The basic method used in this research was descriptive method. This method focused on solving the problems that exist in the present and the actual. The collected data was compiled, described and then analyzed [4]. The research used survey technique by taking samples from one population and using questionnaire as the main data collection tool [5].

2.2 Method of Determination of Research Location

Determination of research location was done purposively based on certain considerations according to the purpose of research [5]. This research was conducted in Lemahbang Village, Jumapolo Subdistrict and Kaliboto Village, Mojogedang Subdistrict; both in Karanganyar Regency. Determination of the location was done with consideration of the district that has the largest rice field area in Karanganyar Regency.

2.3 Types and Data Collection Methods

The data collected in this research consisted of primary and secondary data. The primary data was obtained from interviews with respondents (rice farmers). The secondary data was a complementary data scope of general condition of Karanganyar Regency which was obtained from BPS Karanganyar Regency and other supporting data from related institutions.

2.4 Data Analysis Method

2.4.1 Effectiveness Analysis of Fertilizer Subsidy Policy The effectiveness of the fertilizer subsidy policy was calculated by the effectiveness formula in **Table 1**.

Table 1. Formulation of Effectiveness of Urea and NPK Fertilizer Distribution with the Four Precise Indicator

Exact Price	Exact Place
$\text{Exact Price} = \frac{n_h}{N} \times 100\%$	$\text{Exact Place} = \frac{n_t}{N} \times 100\%$
Note : n_h = Number of sample farmers who obtained fertilizer according to HET (people) N = Total number of sample farmers (people)	Note : n_t = Number of sample farmers who buy fertilizer at authorized retailers (people) N = Total number of sample farmers (people)
Timely	**Exact Quantity**
$\text{Timely} = \frac{n_w}{N} \times 100\%$	$\text{Exact quantity} = \frac{n_j}{N} \times 100\%$
Note : n_w = Number of sample farmers who think fertilizer is available when needed (people) N = Total number of sample farmers (people)	Note : n_j = Number of sample farmers who do the fertilization according to the recommended dose (people) N = Total number of sample farmers (people)

Source: Arisandi [6]

The criteria of effectiveness of fertilizer subsidy policy are appropriately presented in Table 2.

Table 2. Criteria Assessment Effectiveness

Percentage Effectiveness Interval(k)	Criteria
$k \leq 40\%$	Very ineffective
$40\% \leq k \leq 60\%$	Ineffective
$60\% \leq k \leq 80\%$	Quite effective
$80\% \leq k \leq 90\%$	Effective
$90\% \leq k \leq 100\%$	Very effective

Source: Arisandi [6]

2.4.2 Analysis of Factors Affecting Rice Production To examine other factors affecting rice production in Karanganyar Regency, multiple linear regression model was applied:

$$Y = a + b_1X_1 + b_2X_2 + b_3X_3 + b_4X_4 + b_5X_5 + b_6X_6 + e$$

Note :
Y = Rice production (kilograms)
X_1 = Land area (hectare)
X_2 = Use of urea fertilizer (kilograms)
X_3 = Use of NPK fertilizer (kilograms)
X_4 = Use of labor (HKP)
X_5 = Use of seed (kilograms)
X_6 = Effectiveness of fertilizer subsidy policy (%)

2.4.3 Analysis of Effectiveness of Fertilizer Subsidy Policy on Rice Production The effectiveness of fertilizer subsidy policy on rice production in Karanganyar Regency can be seen in t-test of effectiveness variable of fertilizer subsidy of linear regression equation.

3. Results and Discussion

3.1 Effectiveness Analysis of Fertilizer Subsidy Policy on Rice Production

Urea fertilizer had the highest retail price set by the government of IDR 1,800 / kg. But in fact the price of urea fertilizer received by respondents on average amounted to IDR 1,880 / kg so there was a difference of IDR 80 / kg from the actual price. The price of NPK fertilizer at the highest retail price set by the government was IDR 2,300 / kg. But in fact, the average price of NPK fertilizer to be paid by respondents was IDR 2,430 / kg so there was a difference of IDR 130 / kg with the highest retail price of NPK fertilizer. This price difference occured because retailers wanted to take a greater return for their transportation costs to be incurred retailers.

Table 3. Average Purchase Price of Subsidized Fertilizer by Farmers

Component	Urea	NPK
Average purchasing price (IDR /kg)	1,880	2,430
The highest retail price (IDR /kg)	1,800	2,300
Difference (IDR /kg)	80	130

Source : Primary Data Analysis, 2017

Percentage of each indicator of the effectiveness of fertilizer subsidies is presented in Table 4. About 43.75% of the farmers obtained subsidized fertilizer at the highest retail price. This indicated that the distribution of urea fertilizer and subsidized NPK was not effective, because it was included in the interval ($40\% \leq k \leq 60\%$).

Table 4. Percentage of Effectiveness Rate of Fertilizer Subsidy Policy

Effectiveness Level Indicators	Exact (%)	Not Exact (%)	Total (%)
Price	43.75	56.25	100
Place	65	35	100
Time	42.5	57.5	100
Quantity	20	80	100
Average	42.80	57.20	100

Source: Primary Data Analysis, 2017

Based on the exact place indicators, 65% of farmers bought fertilizer in the village, while 35% outside the village. From the percentage of accuracy of the place, it could be quite effective because the number of farmers who bought in the village are includd in the intervals of 60% $\leq$ k $\leq$ 80%. Based the exact time indicator, 57.5% of farmers who stated that "fertilizer was not always available when needed" usually had to make purchases outside the village to obtain the necessary fertilizer during rice production. This showed that when viewed from the indicators on time, fertilizer subsidy was ineffective because only 42.5% of farmers experienced that fertilizer was always available when needed. This percentage was included in the interval 40% $\leq$k $\leq$ 60% (ineffective). Moreover, only about 20% of farmers used urea and NPK fertilizer with the exact quantity as recommended, while 80% not as recommended. The use of fertilizer by farmers is usually done in accordance with the experience of farming and soil conditions, and farmers do not pay attention to the recommended dose of fertilizers. The accuracy of the amount was only 20% which mean very ineffective, because included in interval k $\leq$ 40%.

Thus, overall based on the four indicators of the effectiveness of fertilizer subsidy policy in Karanganyar Regency was ineffective. This was indicated by the average rate of accuracy of fertilizer subsidy of 42.80%, where this percentage was included in the interval 40% $\leq$k $\leq$ 60%.

3.2 Factors Affecting of Rice Production

The statistical test results for the F test are shown in Table 5.

Table 5. F test results Factors Affecting Rice Production in Karanganyar Regency

Model	$F_{arithmetic}$	F_{table}	Sig.
Regression	87.007 *	2.490	0.000
Residual			
Total			

Source: Primary Data Analysis, 2017

Based on Table 5 the factors comprise of land area, use of urea fertilizer, use of NPK fertilizer, use of labor, use of seeds, and effectiveness of fertilizer subsidy policy simultaneously influenced rice production. Thus, further analysis can be done. The t test results are shown in Table 6.

The variable of land area, use of urea fertilizer, use of NPK fertilizer, and effectiveness of the fertilizer subsidy policy partially influenced the rice production due to the t arithmetic value $\geq$ t table (Table 6). While the variable use of labor and the use of seeds partially did not affect the production of rice because the value of t arithmetic $\leq$ t table.

The value of the regression coefficient showed how much the independent variable influences the dependent variable. If there is an increase of independent variable (Xi) of 1 unit then the production of rice increases by β unit. If there is an increase of land area of 1 hectare then the production of rice increased by 2072.751 kg, while other factors are considered constant or constant. Increasing the use of urea fertilizer by 1 kg increased rice production by 858.035 kg, while another factor is considered constant. Meanwhile, if there is an increase in the use of NPK fertilizer of 1 kg then the production of rice increased by 662.05 kg, while other factors are considered constant or constant.

Table 6. Coefficient Analysis Factors Affecting Rice Production in Karanganyar Regency

Variabel	Regression Coefficient	t arithmetic	t tabel
Constant	3847,500	37.751	
Land area	2072.751 ***	20.002	2.032
Use of urea fertilizer	858.035 ***	8.313	2.032
Use of NPK fertilizer	662.057 ***	6.414	2.032
Use of labor	-9.418 ns	-0.091	2.032
Use of seeds	-83.204 ns	-0.806	2.032
Fertilizer subsidy policy effectiveness	289.444 ***	2.804	2.032

Source: Primary Data Analysis, 2017

Model equation of factors influencing rice production in Karanganyar Regency was as follows:

$$Y = 3847.500 + 2072.751\,X_1 + 858.035\,X_2 + 662.057\,X_3 - 9.418\,X_4 - 83.204\,X_5 + 289.444\,X_6$$

The value of R^2 square or coefficient of determination of 0.941 or 94.1%. This showed that variant of rice production variables could be explained by the variable of land area, the use of urea fertilizer, the use of NPK fertilizer, the use of labor, the use of seed, and the effectiveness of fertilizer subsidy policy of 94.1%. The remaining 4.9% was influenced by other variables outside the model such as the use of pesticides, the use of fertilizers other than urea and NPK, the type of irrigation used, and others.

3.3 The Effectiveness Effect of Fertilizer Subsidy Policy on Rice Production

Effect of effectiveness of fertilizer subsidy policy on paddy production in Karanganyar Regency can be seen from t test result. T test was used to determine the effect of independent variables on rice production partially. The value of t arithmetic (2.804) > t table (2.032) then H_0 was rejected, meaning the hypothesis $\beta_6 = 0$ was unacceptable. This showed that the variable effectiveness of fertilizer subsidy policy partially influences the production of rice in Karanganyar Regency. The regression coefficient value of effectiveness of fertilizer subsidy policy was 289.444. The positive coefficient indicated a direct relationship between the effectiveness of fertilizer subsidy policy and rice production in Karanganyar Regency. This means that if there is an increase in the effectiveness of fertilizer subsidy policy of 1% then the production of rice increased by 289.444 kg, while other factors are considered constant or constant.

The effectiveness of fertilizer subsidy policy was closely related to the use of fertilizer by farmers. According to indicator of price and quantity, if farmers get fertilizer according to the highest retail price then farmers can use fertilizer according to the recommended dose without replacing or reducing the quantity of fertilizer used for each land. While seen from the exact indicators of the place and time, if the fertilizer was always available when needed and can be obtained in the village, it will facilitate farmers in getting fertilizer. Thus, the use of fertilizer can be optimized and can increase rice production.

This study was in accordance with [7] where rice production in Malaysia cannot be solved without fertilizer subsidies and farmers cannot afford to buy their own fertilizer. If the fertilizer compensation is complete, it must be given to encourage farmers to buy their own. In order to improve yields and meet increasing demand for rice. This study simulated the impact of NPK fertilizer subsidies on rice production. Simulation results with the decline of incidents died of 4.502 tons per hectare to 3.081 tons per hectare.

The other result was consistent with [8] research in Sri Lanka where the price of subsidized fertilizer influences the use of fertilizer by rice farmers. This study also discussed the fertilizer subsidy program "Kethata Aruna" to increase the use of fertilizer per unit area in previous rice farmers. This program can also control the use of fertilizer, where previously rice farmers use more urea fertilizer from TSP and MOP fertilizer.

Accordingly, [9] research in Nandi North Disctrict, Kenya observed the relationship between corn production with fertilizer subsidy, the amount of fertilizer and the use of certified seed. The utilization of fertilizer also had a positive relationship with the land area owned by farmers. Another result of this study was inadequate transportation costs, input prices and inadequate input supplies were the main challenges being experienced by farmers in Nandi Utara District, Kenya. Therefore, it can be concluded that fertilizer subsidy affected to the production of farming.

4. Conclusion

The policy of fertilizer subsidy in Karanganyar Regency was evaluated from four indicators consisted of exact price, exact place, exact time, and exact quantity. This policy was not effective because the distribution of fertilizer subsidy to farmers still experience some inaccuracies. The production factors of land area, the use of urea fertilizer, the use of NPK fertilizer, and the effectiveness of fertilizer subsidy policy had a positive influence on rice production, while the factor use of labor and the use of seed had no effect on rice production in Karanganyar Regency. The effectiveness of fertilizer subsidy policy had a positive effect on rice production in Karanganyar Regency. This means that the more effective the fertilizer subsidy policy is, the more rice production increases.

The government should improve the mechanism of distribution of fertilizer subsidies because fertilizer subsidy policy was categorized as not effectively. The fertilizer subsidy policy still needs to be implemented and paid more attention in the distribution because based on the research result that the effectiveness of the fertilizer subsidy policy had a positive effect on rice production. In addition, farmers should also pay more attention to other factors such as land area, use of labor, and the use of seeds to provide an increase to rice production.

References

[1] Handoko R and Patriadi P 2005 *Evaluasi Kebijakan Subsidi Non BBM. Kajian Ekonomi dan Keuangan* **9** 4

[2] Mahsun M 2006 *Pengukuran Kinerja Sektor Publik* (Yogyakarta: Penerbit BPFE)

[3] Gibson 2002 *Organisasi Perilaku-Struktur-Proses Terjemahan Edisi V* (Jakarta: Penerbit Erlangga)

[4] Surakhmad W 2004 *Pengantar Penelitian Ilmiah Dasar Metode Teknik* (Bandung: Tarsito)

[5] Effendi S dan Tukiran 2012 *Metode Penelitian Survei* (Jakarta: PT. Pustaka LP3ES)

[6] Arisandi NWW, I-Made S, dan I-Ketut R 2016 *Efektivitas Distribusi Subsidi Pupuk Organisk dan Dampaknya Terhadap Pendapatan Usahatani Padi Sawah* di Subak Sungsang, Desa Tibubiu, Kabupaten Tabanan. *E-Jurnal Agribisnis dan Agrowisata* **5** 1

[7] Ramli N N, Mad N S, Zainalabidin M, Alias R 2012 The impact of fertilizer subsidy on malaysia paddy/rice industry using a system dinamics approach *Intr J. Soc. Sci. Human* **2** 3

[8] Herath H M K V, Gunawardena E R N, and Wickramasinghe W M A D B 2013 The Impact of "Kethata Aruna" Fertilizer Subsidy Programme on Fertilizer Use and Paddy Production in Sri Lanka *Trop. Agric. Res.* **25** 14-26

[9] Bunde A O, Kibet K, Daphen O O, Mugo S W and Chomboi K C 2014 Impact of fertilizer input subsidy on maize production in Nandi North District, Kenya *Int. J. Sci Basic Appl. Res.* **15** 1

Performance of Arrowroot (*Marantha arundinacea*) in various light intensities

M B Oktafani[1], Supriyono[2], MTh S Budiastuti[2] and D Purnomo[2]
[1]Bachelor of Study Program of Agrotechnology, Faculty of Agriculture, Universitas Sebelas Maret, Jl. Ir. Sutami 36A, Surakarta, Central of Java, Indonesia 57126
[2]Lecturer Staff on Study Program of Agrotechnology, Faculty of Agriculture, Universitas Sebelas Maret, Jl. Ir. Sutami 36A, Surakarta, Central of Java, Indonesia 57126

E-Mail: supriyono_uns@yahoo.com

Abstract. Arrowroot (*Marantha arundinacea*) is one of the potential food crops to support food security programs. Light intensity is one of the important factors for plant growth. Arrowroot cultivation technology still need further development. Traditionally, arrowroot grows wild under canopy without intentisification of cultivating which have low productivity. The purpose of research was to investigate the suitable light intensity for arrowroot. The experiment was conducted at Jumantono as Experimental Field of Faculty of Agricultural, University of Sebelas Maret Surakarta located in Karanganyar, from March to September 2016. The experiment used a complete randomized block design (CRBD) of light intensity level there are 27400 lux (full sun light), 18900 lux (shaded 31%), 13500 lux (shaded 51%) and 7400 lux (shaded 72%). Each treatment was replicated six times so there were 24 experimental units. The results showed that arrowroot is a low light adaptive plant. Arrowroot under the light intensity 7400 lux (27% full light), the number of leaves and tillers is not significantly different than under full light, although the plant is higher. The highest tuber diameter and length were 1.91 and 25.06 cm, respectively, and tuber weight reached 607.5-651.67 g per plant.

1. Introduction

Food security requires the attention of the Indonesian government in relation to the rate of function change on agricultural land especially for rice field. Increased population requires an increase in crop production. One of the efforts to overcome food demand is cultivation of alternative plant on suboptimum land or others such as agroforestry system [1], hydroponics [2], or dry land (gripped water) [3]. Arrowroot is one source of potential food in support of food security programs because it has high carbohydrate content [4]. Carbohydrate content of arrowroot bulbs ranged from 19.4% to 21.7%. Cultivation of arrowroot has not been intensive, mostly left to grow wildly without maintenance in the yard which has low productivity. Intensive development of arrowroot cultivation from the beginning of soil processing until harvest will optimize the growth and yield. Arrowroot is able to grow on low soil fertility and relatively low pest disease [5]. Light intensity is one of the important factors for plant growth. The light intensity requirements of each plant vary according to the plant character. Arrowroott is a C_3 plant that has a certain upper limit of light intensity [6]. Light arrangement (low) using shade to avoids exposure to direct sunlight for optimum level for plant. The optimum conditions make growth and yield of arrowroot obtain optimum yield too [7]. The research investigated the optimum irradiation for arrowroot and intensive cultivation technology.

2. Methods

The experiment was conducted from March to September 2016 at Experimental Field of Faculty of Agriculture, University of Sebelas Maret Surakarta Indonesia (7° 37' 47.1" S and 110° 96' 35" E), with altitude of 187 m above sea level. The experimental design was a complete randomized block design (CRBD) with four levels of light intensity consist of 27.400 lux (full light), 18,900 lux (31% shaded), 13,500 lux (51% shaded), and 7.400 lux (72% shaded). Each treatment replicated six times so there were 24 experimental units. Planting material was arrowroot seedling 6 months age. Polybag 60×60 cm size filled with 25-30 kg mixture of heavy alvisol soil and manure 3:1 (manure 30 ton ha^{-1}) as planting medium. The maintenance of crop included: irrigation (using watering can every two days until a level below field capacity if no rainfall), weed grazing (manually), pest control (taken manually or removed), fertilization (N, P, K fertilizer with dose 150, 75, and 75 kg ha^{-1}). The plants were harvested at 12 months of age (> 50% of the leaves have dried out), the tubers were cut from the plants then air dried. The observed variables included growth components such as plant height (measured from base of the stem to the ends of the plant using a ruler), the number of leaves (counting all the leaves), and the number of tillers (crops grown from the tubers after the parent plant). Production components include tuber weight (weighing tubers after drying by wind), number of tubers (counting the tubers formed), tuber length (measured from tip to base of tuber using ruler), tuber diameter (measured by sliding). Observational data were analyzed using ANOVA 0.05, if among the treatment differ significantly, further analysis by DMRT level of 0.05.

3. Results and discussion

3.1 Growth components

3.1.1 Plant height Plant height as indicator of arrowroot growth was not affected by light intensity. Plant height obtained 43 - 69.16 cm under high light intensity (27400 lux) and low (13500 lux). Other research showed that the high of arrowroot seedlings in various manure between 39.60 until 98.60 cm [8]. Significant effect of treatment only between 43 and 79.33 cm (in middle and low light intensity) (Table 1). This indicates that the arrowroot is adaptive to low light intensity. However, physiologically it appears that at the lower light intensity, the plant grow higher [9]. Correlation coeffient between light intensity and height of plant was very high (correlation coefficient or r: -0.85). Arrowroot height under coconut trees was 113.8 cm [10]. Plants contain auxin as growth regulators, which antagonistic activity with light [11]. In low light intensity auxin activity is high so that the plant becomes longer or taller. Whether low light intensity result arrowroot etiolation will be discussed through the number of leaves.

3.1.2 Number of leaves The number of arrowroot leaves (between 16.16 – 42.33) was not affected by light intensity (Table 1). The number of leaves was calculated from several stems of tillers. When the leaves grow on one stem that experiencing etiolation, the number of leaves decreases because the low light intensity extending the internodes so decrease the number of nodes in which branches or leaves grow. This means that arrowroot has not been etiolated. The reasoning of this explanation is stronger when associated with the number of tillers.

3.1.3 Number of tillers The arrowroot tiller grows from the shoots of the tubers, grows not simultaneously, the earliest growing shoots produce the parent plant, while the later produce the tillers. The number of tillers gained varied from 4.67 to 10.67, higher than that previous observation, which is only 5-6 [12]. The number of tillers highly correlated (r: 0.91) with the number of leaves reinforcing the assumption that the number of leaves originated from the parent plant and tiller, ensuring that the etiolation was not occured.

3.1.4 Root canopy ratio Environmental conditions affect the growth of plants. The light intensity from high to low did not affect root and canopy growth (Table 1). This means that the arrowroot has not been etiolated too although it grows under very low light. The plants which have an etiolated growth were elongated but low plant weight, while roots grow normally. Etiolation results in a lower root canopy ratio.

Table 1. Arrowroot growth on various light intensity

LI (x 1000 lux)	PH (cm)	NL	NT	R/C
274	48.16^{ab}	34.83	10.67	0.12
189	43^{a}	16.16	4.67	0.13
135	69.16^{ab}	26.50	7.83	0.08
74	79.33^{b}	42.33	9.83	0.11

Description: Number in a column followed by different letter showed significant difference in Duncan test 0.05; LI, light intensity; PH, plant height.

Regression analysis show that light intensity determines the height of plant with determination coeffient (R^2) 0.73, anddetermines to number of leafs and tillers with R^2 0.96 and 0.83, respectively (Figure 1 left).

3.2 Yield Component

The arrowroot yielded tubers which are roots modification. Roots as a reinforcing organ ensure the plant to grows upright which also used to absorbs nutrients and water, when the mature stage reached, it swell and turns as a food storage or sink. The weight of arrowroot is the yield quantity obtained at harvest. The weight of the arrowroot tuber is highly dependent on the number of tillers and tuber size determined by the length and diameter. The determination coefficient (R^2) of the number of tillers to the weight of tuber, number leafs and tuber were 0.81, 0.83, and 0.55, respectively. There is show that the role of number of tillers to the weight of tuber and number of leafs more dominant than to the number of tuber (Figure 1 right and Table 2).

Table 2. Arrowroot tuber yield on various light intensity

LI	TW (g)	NT	TL (cm)	TD (cm)
274	705.0	20.0	24.08	1.76
189	410.83	12.50	20.83	1.72
135	651.67	17.16	22.65	1.88
74	607.50	14.50	25.06	1.91

Description: LI, light intensity (x100 lux); TW, tuber weight; NT, number of tuber; TL, tuber length; TD, tuber diameter.

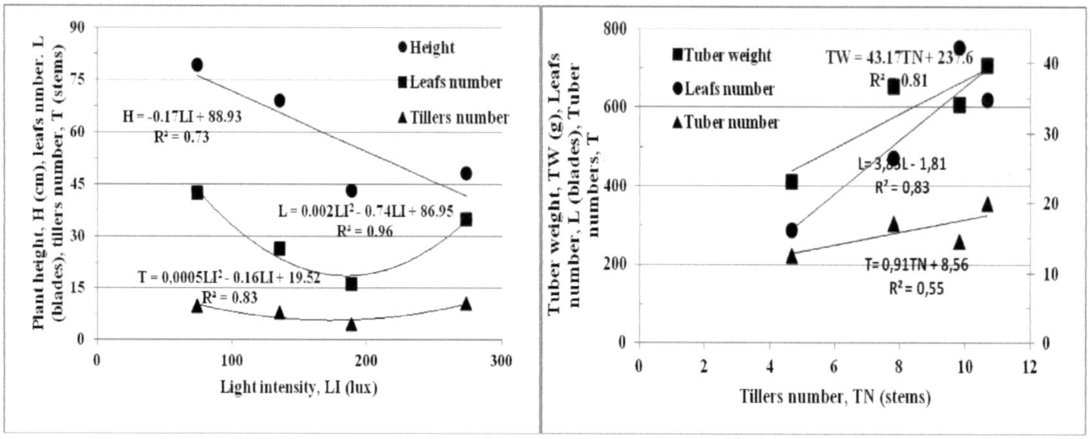

Figure 1. Arrowroot response to light intensity in form of plant height, number of leaves, and number of tiller (left curve). Tiller number of tillers ditermine to to number of tubers and greatly determine number of leaf and tuber weght (right curve).

4. Conclusions

Arrowroot is a low light adaptive plant. Arrowroot under the light intensity 7400 lux (27% full light), the number of leaves and tillers were not different than under full light, although the plant is higher. The highest tuber diameter and length were 1.91 and 25.06 cm, respectively, and tuber weight reached 607.5 to 651.67 g per plant.

References

[1] Budiastuti Mth S, dan Purnomo D 2012 *Agroforestri, Bentuk Pengelolaan Lahan Berwawasan Lingkungan* (Surakarta: UNS Press)

[2] Purnomo D, Dwi H, Trijono D S 2016 Budidaya Cabai Rawit Sistem Hidroponik Sunstrat Dengan Variasi Media Mutrisi *Caraka Tani* **31** 129-136

[3] Ratri A D Y S, Pujiasmanto B, Yunus A 2015 *Caraka Tani* **30** 1-6

[4] Sukamto A, Fajarudin A dan Albertus H 2010 Pengaruh oryzalin terhadap tingkat ploidi garut (Marantha arundinacea*) J Bul Litro* **21** 93-102

[5] Sukarsa E 2011 Tanaman umbi garut *http://wwwbbpp-lembanginfo/indexphp/arsip/artikel/artikel-pertanian/500-tanaman-umbi-garut*

[6] Taiz L, Eduardo Z, Ian Max M, Angus M 2015 *Plant Physiology and Development* 6th ed (Sinauer)s

[7] Nurkhasanah N, Wicaksono K P, Widaryanto E 2013 Studi pemberian airdan tingkat naungan terhadap pertumbuhan bibit tanaman cabe jamu (*Piper retrofractum* Vahl*) J Prod Tan* **1** 325-332

[8] Setyowati N 2012 *Perbanyakan Garut Dari Bibit Cabutan Sisa Panen Dengan Aplikasi Berbagai Pupuk Kandan* (Bogor: Pusat Penelitian Biologi LIPI)

[9] Sitompul S M 2016 *Analisis pertumbuhan tanaman* (Malang: UB Press)

[10] Swadija O K, Padmanabhan V B and Vijayaraghavakumar 2013 Growth and yield of arrowroot intercroped in coconut garden as influence by organic management *J Rootcrops* **30** 67-72

[11] Taiz L and Eduardo Z 2010 *Plant Physiology* (London: Sinauer)

[12] Pate H S, SH Bodi S H, Dikwahal H D, Jibung G G, and Molchen M D 2014 Respons of Polynesian Arrowroot Tacca leontopetloides to Fertlizer and Intra-row Spacing at Garkawa *J Biol Agric Healthcare* **4** 97-101

IOP Conf. Series: Earth and Environmental Science **142** (2018) 012049 doi:10.1088/1755-1315/142/1/012049

Effect aquadest-extracted *Gloriosa superba* seed as mutagen on morphology of *Artemisia annua*

S I Rahmawati[1], A Susilowati[2], A Yunus[3] and Y Widyastuti[4]
[1]Department of Bioscience Graduate School, Universitas Sebelas Maret, Surakarta -57126, Central Java, Indonesia
[2]Department of Biology, Faculty of Mathematics and Science, Universitas Sebelas Maret, Surakarta -57126, Central Java, Indonesia
[3]Department of Agrotechnology, Faculty of Agriculture, Universitas Sebelas Maret, Surakarta - 57126, Central Java, Indonesia
[4]Research Center for Medicinal Plant and Traditional Drug, Tawangmangu, Indonesia

E-mail : yunus.uns7@yahoo.com

Abstract. *Gloriosa superba* is a plant that contains colchicine in all parts of organs, especially in the seeds. Its extract is as a mutagen to produce plants with polyploid cells. *Artemisia annua* is a plant that produces active ingredients artemisinin as malarial drugs, hemorrhoids therapy, aromatherapy, antiviral, anticancer, and anti-bacterial. The aims of this research was to determine the effect aquadest-extracted *Gloriosa superba* seed as a mutagen to *Artemisia annua* morphology. Extraction of *Gloriosa superba* seeds obtained from Sukoharjo using maceration method with aquadest solvent (1: 1). The extracts were diluted (0, 25, 50, 75 and 100%) for *Artemisia annua* sprinkling with different times (0, 30, 60 and 90 minutes). Observations of morphology *Artemisia annua* included height, stem circumference, number of branches, number of leaves, leaf width and leaf length. The treatments did not affect plant morphology observation included height, stem circumference, number of branches, number of leaves, leaf width, and leaf length. The EB treatment (100%, 30 minutes) was higher (120 cm) than other. In all treatments stem circumference about 2.5 cm, number of branches ranged between 40-50, leaves width ranged 9 -16c m, and leaf length ranged 8 -15 cm.

1. Introduction

Gloriosa superba is an herbaceous vine and consisted of tubers, stems, leaves, flowers and seeds containing colchicine. Previous research has shown that the *Gloriosa superba* seed has a higher content of colchicine than other organs of this plant. Colchicine is a chemical used as a polyploid induction in plants because it is effective and easily soluble in water [1]. *Gloriosa superba* seed extract can be utilized as a mutagen in diploid plants into polyploid plants. *Artemisia annua* is a medicinal plant from China and cultivated in Indonesia which produces artemisinin as malaria drug. Artemisinin is sesquiterpene lactone endoperoxide extracted and isolated from leaves *Artemisia annua* as anti malarial drug [2]. Pathogen *P. falciparum* has been resistant to quinine as a malaria drug [3], thus attempting another alternative plant as a malaria drug. Polyploidy plants often exhibit new phenotypes, such as broader leaves, excellent quality, high yield and increased resistance to environmental stresses and disease, and increase the secondary metabolite content of a plant [4]. The aim of this research is to know the effect aquadest-extracted *Gloriosa superba* seed obtained from Mulur, Sukoharjo on morphology of *Artemisia annua*.

2. Methods

This research used sampling method and experimental laboratory method. *Gloriosa superba* obtained by sampling method in Mulur, Sukoharjo during January-February 2017. Samples used in the form of seeds which are then dried using oven with temperature 70°C for 24 hours. Dry seeds extraction using maceration method with aquadest solvent (1:1) for 24 hours with 3 replications. Analysis of colchicine content and level on extract of *Gloriosa superba* seeds using TLC-Densitometry method. The aquadest-extracted *Gloriosa superba* seeds was divided into several concentrations (0, 25, 50, 75 and 100%) for the treatments. *Artemisia annua* seeds were obtained from B2P2TOOT, Tawangmangu. Seedling was done in B2P2TOOT Karangpandan garden, Karanganyar at the beginning of the rainy season. The seeding process of *Artemisia annua* seeds to be sprouts ± 2 weeks. At 2-week-old strong sprouts were then treated with aquadest-extracted *Gloriosa superba* seeds. Treatment of *Artemisia annua* sprouts was performed with above mentioned extract concentrations and soaking sprouts times (0, 30, 60 and 90 minutes) **(Table 1)**. Sprouts of treatment were grown on polybags. Observations of morphology *Artemisia annua* included height, stem circumference, number of branches, number of leaves, leaf width and leaf length. The parameters of morphological observation were done once every week until 50% of the plant population forms primordial flowers. Observation data of height, stem circumference, number of branches, number of leaves, leaf width and leaf length were recorded every week. Data were used as an analysis to compare the morphology between the treated plants.

Table 1. Design of the treatments

Times (minutes) / Concentration	A (0)	B (30)	C (60)	D (90)
A (0%)	AA	AB	AC	AD
B (25%)	BA	BB	BC	BD
C (50%)	CA	CB	CC	CD
D (75%)	DA	DB	DC	DD
E (100%)	EA	EB	EC	ED

3. Results and Discussion

Artemisia annua is a medicinal plant that produces artemisinin as a malaria drug. Increased biomass and artemisinin levels in these plants were performed by induction of ploidies using colchicine. Utilization of *Gloriosa superba* as a plant containing colchicine is used for induction of *Artemisia annua* sprouts. Ploid-induced plants produce not only high artemisinin, but also morphologically superior to diploid plants. In this research conducted a study on the effects of morphology *Artemisia annua* that the soaking spourts by aquadest-extract *Gloriosa superba* seeds. The results of this study show the morphology of *Artemisia annua* treatment results include height, stem circumference, number of branches, number of leaves, leaf width and leaf length **(Table 1)**.

Artemisia annua sprouts was used in soaking treatment aquadest-extracted in 2 weeks old with an average height of ± 1cm. Treatment of soaking sprouts using variable concentration of aquadest-extract *Gloriosa superba* seed and soaking time. Sprouts result of treatment planting on polybag with soil media, husk and manure. Morphological observation data were observed a week after sprouting soaking treatment until 50% of the plant population formed primordial flowers. The problems experienced during the growth process of *Artemisia annua* sprouts are the weather and caterpillar pests. These conditions affect the growth of sprouts with one another which is a repeat of the treatment. The observed data each week varies due to the condition until the plants form primordial flowers.

Table 2. Morphological data of *Artemisia annua*

Sample	Morphology					
	Height (cm)	Stem Circumference (cm)	Number of branches	Number of leaves	Leaf width (cm)	Leaf length (cm)
AA	100	2	55	45	10	11
BA	108	2.5	53	56	10	11
CA	110	2.5	47	43	11	10
DA	98	2	43	39	11	10
EA	98	2	23	28	9	11.5
AB	97	2	49	39	9	8
BB	110	2.5	40	38	12	11
CB	98	2	42	30	11	9
DB	104	2.5	53	33	11	11.5
EB	120	2.5	40	30	10	11
AC	105	2.5	44	37	10.5	9
BC	100	2.5	57	35	12	11
CC	100	2	51	42	13	12
DC	103	2.5	50	37	14	13
EC	100	2.5	50	36	12	12.5
AD	108	2.5	54	39	9	10
BD	98	2	48	41	11	10
CD	93	2	38	30	12.5	11
DD	106	2.5	51	46	13	12
ED	100	2.5	54	50	16	15

The plant height morphology was not different between treatments. The presence of caterpillar pests that feed on leaves early in *Artemisia annua* growth inhibits plant growth. In addition, entering the rainy season with a little more sunlight causes the plants easily die because it is too wet. The ability of plants with each other on the same treatment is also different is shown by the adaptation to seasonal changes. Plant height at the end of the observation is not much different from each other. This may be due to environmental factors during the growth of *Artemisia annua* which suffered some disruption.

Other morphological observations include circumference stem, number of branches, number of leaves, leaf width and leaf length no significant difference. The morphological measurements observed and recorded for all plants are similar. In this study cannot be concluded diploid and polyploid plants because morphologically all the plants were difference between each other. The leaves of *Artemisia annua* look greener and thicker than the control plants.

The morphological effect of *Artemisia annua* on treatment with aquadest-extracted *Gloriosa superba* seeds in this study cannot be clearly presented. The growth of this plant is strongly influenced by many environmental factors because basically this plant is still being studied for cultivation in Indonesia. In addition, plant morphology is also strongly influenced by growing environmental conditions, so morphological data is only used as supporting data to infer the effect of the treatment that has been done. To be able to know more clearly and precisely needs to be observed cytology such as stomata and plant chromosomes of plants treatment.

4. Conclusion

The aquadest-extracted *Gloriosa superba* seed showed no significant difference between the treatments. Height, stem circumference, number of branches, number of leaves, leaf width and leaf length in treated plants showed no significant different data. In other observation, *Artemisia annua* by treatment aquadest extract of *Gloriosa superba* seeds has thicker leaf and look greener than control plant. The morphology of treated plants is superior when compared to diploid plants cannot be proven in this study.

Acknowledgment
This research was supported by the Ministry of Research Technology and Higher Education and Dr. Ir. Yuli Widyastuti as field from Research Center for Medicinal Plant and Traditional Drug, Tawangmangu, Indonesia. *Artemisia annua* seeds in this research were obtained from B2P2TOOT, Tawangmangu, Karanganyar.

References

[1] Haryanti S, Hastuti R B, Setiari N, Banowo A 2009 Pengaruh kolkisin terhadap pertumbuhan, ukuran sel metafase dan kandungan protein biji tanaman kacang hijau (*Vigna radiata* (L) Wilczek) *J. Penelit. Sains Teknol.* **10** 112-120
[2] World Health Organization 2001 The use of antimalarial drugs: Report of a WHO Informal Consultation 13–17 November 2000 *Geneva*. (WHO/CDS/RBM/2001.33)
[3] World Health Organization 2004 More than 600 million people need effective malaria treatment to prevent unacceptably high death rates *Press release* WHO/29, 22 April
[4] Osborn TC, Pires J A, Auger DL, Chen Z J., Lee HS, Comai L, Madlung A, Doerge R W, Colot V and Martienssen R A 2003 *Trends. Genet.* **19** 141-147

IOP Conf. Series: Earth and Environmental Science **142** (2018) 012050 doi:10.1088/1755-1315/142/1/012050

Performance of Mentik Wangi rice (*Oryza sativa*, L.) M2 generation from gamma ray irradiation

A Yunus[1], Parjanto[1], Nandariyah[1] and S Wulandari[2]
[1] Lecturer of Agrotechnology Study Program, Agriculture Faculty, Universitas Sebelas Maret, Surakarta, Indonesia
[2] Student of Agrotechnology, Faculty of Agriculture, Universitas Sebelas Maret Surakarta, Indonesia

E-mail: yunus.uns7@yahoo.com

Abstract. The objective of this research is to produce Mentik Wangi rice with shorter flowering age, shorter rice stem and high yield. This research was conducted in Palur Village, Mojolaban Sub-district, Sukoharjo District, from April to August 2016. This research used descriptive method, performed by observing each individual and comparing to the control average. Observational variables included plant height, total number of tillers, number of productive tillers, panicle length, number of grain per panicle, panicle density index, 1,000 filled seed weight, seed weight of cluster, flowering age and M_2 mutant selection. The results showed that there were several plants indicated mutation, there were 7 plants indicate short stem mutation with height 85 cm to 97 cm. The plants which indicated short flowering age were 5 weeks after planting up to 6 weeks after planting. The mutated plants indicating of the highest number of productive tillers which were 27 up to 36. Each component result includes total number of tillers, number of productive tillers, number of grain of panicle as well as the weight of 1000 seeds in the plant indicated some mutation that has a high level of diversity in each treatment.

1. Introduction

Rice is a staple food for all Indonesian people. This increase in demand of rice per year can reach up to 2-3% which is considered to be a very high [1], thus, there is necessary to increase the production and quality improvement of rice by plant breeding on developed rice and local rice. Local rice plant breeding is now widely practiced, by developing local superior properties and removing the weakness of the rice.

Mentik Wangi rice varieties are local rice originating from Magelang, Central Java. Mentik Wangi has a characteristic that is fragrant aroma, besides the soft texture and white color that gain interest in society. Mentik Wangi variety also has a harvest age of 4 months. The trunk is not solid and it is easily collapsed by the wind [2]. One way can be done to improve the quality and reduce or eliminate the weakness of Mentik Wangi rice is trough breeding mutation. Breeding mutation uses gamma ray irradiation by irradiating the rice seeds that will be planted with gamma rays in several dosages. Nuclear Power Agency (BATAN) has produced 15 new varieties of rice planting. Experiments using gamma ray irradiation aimed to produce new varieties by improving the superior properties possessed by one of such Mentik Wangi rice.

IOP Conf. Series: Earth and Environmental Science **142** (2018) 012050 doi:10.1088/1755-1315/142/1/012050

2. Methods

2.1. Location and time research
The experiment was conducted at Experimental Field of Faculty of Agriculture, Sebelas Maret University, Palur, Mojolaban Sub-District, Sukoharjo District. The study was conducted from April to August 2016.

2.2. Materials
The material used in this research was the second generation of fragrant rice generated from the first generation that has been treated gamma ray radiation in BATAN.

2.3. Research design and data analysis
The study was conducted using descriptive design with observation treatment on each individual of each radiation dose by comparing the average value with control treatment to know the difference and the effect of irradiation on the growth of Mentik Wangi rice. The data obtained were analyzed descriptively by comparing each individual plant at each radiation dose with average control.

3. Result and Discussion

3.1. General condition of field
Based on the Central Statistics Agency data, the total area of Mojolaban sub-district is 3.554 Ha or about 7.65% of the total sukoharjo regency (46,666 Ha). The area consists of 2,169 Ha (61.03%) of paddy fields and 1,385 Ha (38.97%) instead of paddy fields. Non-wetland area used for the yard is 89.1% of the total area not the rice field. The percentage is not another wetland. Geographically, Sukoharjo District located in the position: 110^0 57 '33.7' E to 110^0 42' 06.79' E, 7° 32 ' 17.00" S, 7° 49' 32.00" S. The soil in this experimental field is included in alluvial soil types. Alluvial soil is one type of soil formed by the result of sediment. This land is formed in the area of rivers, lakes in lowland areas or basins that usually occur soil sediment

The altitude of the experimental field was 104 meters above sea level, so that this land has a sloping to hilly structure. The annual rain intensity is 1,052 mm with average annual rainfall intensity of 19 mm. Mojolaban area has a fairly high amount of rainfall, which every year reaches 1,422 mm and topographically Mojolaban is a low plain. The results of the laboratory analysis showed that the soil in the experimental field has a total N content of 0.28%, KPK of 18.05%, organic C 1.16%, organic matter content of 2.00%, pH of 5.70 and C/N ratio of 4.14%. Organic matter content was high where the soil concluded as a healthy soil. The pH of 5.70 indicated that the soil belongs to a neutral soil, since it has a pH greater than 5.5 and less than 8.0. The pH of a soil may be affected by soil temperature, climate or rainfall, vegetation, internal soil drainage and human activity.

3.2. Variable of plant growth

3.2.1. Plant height Based on Table 1, plants that have a low plant height is a plant with 300 Gray radiation with a height of 85 cm. Dose 300 Gray was a very high dose that make plants easily mutated and can even damage the DNA or genes, in order to avoid such high damage and undesirable mutations, the appropriate irradiation dose ranges from 150 Gray to 250 Gray [3]. Irradiation treatment can give effect inhibits plant heigt [4]. The lowest plant height is the expected, because the short rice contributes to productivity because the rather short stems are more resistant to collapse.

Table 1. Plant Height In Various Doses of Gamma Ray Irradiation

Dosage (Gray)	Lowest (cm)	Highest (cm)	Range	Average (cm)
0	118	124	118-124	120.80
100	90	162	90-162	121.62
200	97	150	97-150	119.85
300	85	130	85-130	117.72

3.2.2. Total Number of Tillers The total number of tillers is the total number of tillers produced by each clump. Through the calculation of the number of tillers obtained the results presented in Table 2.

Table 2. Total Numbers of Tillers In Various Doses of Gamma Ray Irradiation

Dosage (Gray)	Lowest (tillers)	Highest (tillers)	Range	Average (tillers)
0	14	36	14-36	22
100	6	45	6-45	25
200	4	41	4-41	22
300	3	38	3-38	19

The total number of tillers can be calculated when the plant has entered the maximum vegetative phase, which was 50-60 days after planting [5]. Based on the research result, that the lowest tiller number was at 300 Gray irradiation of 3 tillers and the highest total number of tiller shown in 100 Gray irradiation which amounts 45 tillers.

The result of the study of the diversity of the number of tillers per dose of irradiation was evident and has considerable differences due to mutations. The lowest seedlings on the research result were the total number of tillers i.e. 3 tillers at 300 Gray radiation, the least number of tillers according to previous observation that rice plants can be said little if each clump has number of tillers <10 and number of saplings is said to be a lot if every family has number of tillers> 20 [6]. In this study, the most tillers with the amount of 45 tillers one clump on plants with irradiation dose of 100 Gray.

3.2.3. Number of Productive Tillers The number of productive tillers determines the amount of grain production in each crop, as each tiller produces one grain panicle. The results of the research showed the results of productive tillers at each irradiation dose treatment can be seen in the Table 3.

Table 3. Number of Productive Tillers In Various Doses of Gamma Ray Irradiation

Dosage (Gray)	Lowest (tillers)	Highest (tillers)	Range	Average (tillers)
0	7	30	7-30	16
100	4	36	4-36	16
200	4	29	4-29	15
300	3	27	3-27	13

The plants with the highest average number of productive tillers were on the control treatment and irradiation of 100 Gray with an average of 16 tillers. The results showed that for average number of productive tillers between control and irradiation treatment 100 Gray had the same number of same productive tillers i.e. 16 tillers, while the average number of the lowest tillers on rice with 300 Gray irradiation was 13 tillers. Rice with irradiation 200 Gray had 15 productive tillers.

3.2.4. Panicle length The length of panicle can affect the yield of rice grain productivity both pine (fill) and empty grain. The length of panicle in various types of irradiation is presented in Table 4.

Table 4. Panicle Length In Various Doses of Gamma Ray Irradiation

Dosage (Gray)	Lowest (cm)	Highest (cm)	Range	Average (cm)
0	19.8	26.8	19.8-26.8	25.18
100	18.2	32.2	18.2-32.2	25.73
200	18.8	28.6	18.8-28.6	24.22
300	16.2	29.8	16.2-29.8	25,11

The treatment which had the highest panicle length of 32.2 cm was shown by 100 Gray irradiation treatment. Types of panicles are divided into two: panicles and heavy panicles. The type of heavy panicle has a slight panicle but is large and long, whereas many type panicles have many small and short panicles. Most of the high yielding varieties with high yields include many panicle types, while local varieties include heavy panicle types.

3.2.5. Number of seed per panicle The amount of grain content of each panicle of rice is an indicator of a rice production. The perfect grain indicates that the results obtained in rice farming is very good. Observation of the number of grains of this malai content as a comparison of empty grain and grain contents generated on every malai. The calculation of the number of grains per panicle can be seen in Table 5.

Table 5. Number of Seeds per Panicle of Mentik Wangi Rice In Various Doses of Gamma Ray Irradiation

Dosage (Gray)	Lowest (seeds)	Highest (seeds)	Range	Average (seeds)
0	98	170	98-170	140
100	73	186	73-186	140
200	98	167	98-167	137
300	43	217	43-217	136

Studies in cereals irradiated with gamma rays at doses of 100 Gray and 200 Gray have a high generative growth which is supposed to be in similar with another cerealia crops such as rice and sorghum [7]. Giving various doses to a seed affects the rate of mutation that occurs. If irradiation doses are too low, the resulting mutation is low, but if the irradiation dose is too high it can turn off the mutation material or become sterile.

3.2.6. Panicle density index Panicle is a generative part of rice, panicle is formed when rice starts flowering. The index of panicle indicated how solid the panicle is filled by the grains of grain. The observation of Malai Strain Index on Wangi Mentik rice research can be seen in Table 6.

Table 6. Panicle Density Index In Various Doses of Gamma Ray Irradiation

Dosage (Gray)	Lowest	Highest	Range	Average
0	3.98	7.83	3.98-7.83	5.63
100	3.05	6.81	3.05-6.81	5.40
200	4.11	7.88	4.11-7.88	5.67
300	2.65	7.27	2.65-7.27	5.38

The panicle density index is obtained from the number of grains per panicle divided by panicle length. The ideal rice has long panicle and dense seed grain. The greater the number of grains per panicle, the higher the index of panicle density.

3.2.7. The 1000 filled seed weight Pithy weight of 1000 seeds is the total weight prediction at harvest, because the weight of 1000 seeds is an indication that the seed weight is ideal. Weight of 1000 seeds is a physical quality test on a seed, while the physiological quality tests include, germination, strength and growth power [8]. Weighing results of 1000 Seeds on Mentik Wangi research can be seen in Table 7.

Table 7. Weight of 1000 Filled Seed In Various Doses of Gamma Ray Irradiation

Dosage (Gray)	Lowest (g)	Highes t (g)	Range	Average (g)
0	19.3	38.2	19.3-38-2	30.00
100	14.0	36.5	14.0-36.5	26.24
200	11.4	38.3	11.4-38.3	24.13
300	10. 8	42.2	10.8-42.2	23.87

Gamma-ray radiation has the potential to obtain mutants with morphological and desirable agronomic characters such as short plant, short life-span and high production [9]. Each of these generations is a valuable mutant source that must be selected according to the characteristics that approximate the desired properties [10].

3.2.8. Seed weight per cluster Grain is a fruit of rice covered lemma and palea. This fruit is formed after the pollination and fertilization of lemma and palea and other parts will form the skin of rice. Grains of grain are endosperm containing starch and partly occupied by embryos located in central or lemma. The weight of grain per plant clump is influenced by number of tillers, number of tillers, number of grain and percentage of pithy rice. The weight of Mentik Wangi is presented in Table 8.

Table 8. Seed Weight per Cluster In Various Doses of Gamma Ray Irradiation

Dosage (Gray)	Lowest (g)	Highest (g)	Range	Average (g)
0	18.24	98.74	18.24-98.74	51.47
100	6.49	126.40	6.49-126.40	53.14
200	10.85	86.11	10.85-86.11	48.51
300	4.82	86.51	4.82-86.51	39.50

The weight of grain per cluster is influenced by number of tillers, number of productive tillers, number of grain and percentage of filled rice seed. The weight of rice grain is the yield per hectare component. Based on the observation of the seed weight per cluster of this stock, it shows that the contour has the highest average weight shown in the irradiation treatment of 200 and 300 Gray with a weight of 51.47 g. The lowest average weight is on the 300 Gray irradiation treatment which is 39.50 g. Rice with irradiation 200 Gray has an average weight of 48.51 g, almost 10 g lower than than 300 Gray irradiation result. For irradiation control treatment, 100 Gray, 200 Gray and 300 Gray that has the highest average of rice is in the 100 Gray irradiation with weight of 53.14 g. Seed weight per cluster of 100 Gray irradiation has a weight ranged from 6.49 g to 126.40 g, for irradiation 200 Gray has a weight ranged from 10.85 g to 86.11 g. Irradiation 300 Gray has a weight range per cluster of 4.82 g to 86, 51 g.

3.2.9. Flowering age The plants flower during the maximum vegetative phase. Measurement of flowering age is done by observing the population of each flowering plant every week. Rice is said to flower overall if 50% of the population is already flowering. The observation result on the age of flowering of Mentik Wangi is presented in Table 9.

Table 9. Flowering Age In Various Doses of Gamma Ray Irradiation

Dosage (Gray)	Lowest (weeks)	Highest (weeks)	Average (weeks)
0	7	8	8
100	5	8	7
200	5	8	7
300	6	8	8

Flowering age is the age parameter of harvest. Where plant flowering age is mature then the age of harvest is also mature [11]. The longest one is in irradiation 100 Gray and 200 Gray, where the plants flowering early at 5 weeks after planting relative to other treatment. The results showed that the control had a flowering age of 7 weeks to 8 weeks, 2 or 3 weeks longer compared with irradiation treatment of 100 and 200 Gray.

3.2.10. M2 selection result The result of M2 mutant selection is obtained in accordance with the objective to obtain Mentik Wangi rice with short flowering age, short stems and high productive tillers that some plants will be replanted as M3.

Table 10. M2 Selection Result of Flowering Age Properties

Strain	Flowering Age (weeks after planting)
R1G61T11	5
R2G81T6	5
R2G80T7	5
R3G37T3	6
R1G11T8	6
R2G77T20	6
R3G16T6	6

Observation result shows that there are some plants that indicate mutations where the plant has a shorter flowering age. There are 3 plants from the 7 plants that have shorter flowering age at the age of 5 WAP (Week After Planting), at 100 Gray irradiation there is 1 plant of R1G61T11 strain and at irradiation 200 Gray there are 2 plants of R2G80T7 and R2G81T6 strain. In irradiation 300 Gray has plants indicated mutation with the shortest flowering age of 6 WAP, one of which is on R3G16T6 strain. Irradiation of 300 Gray produces plants that have a flowering age that is not considered short.

Table 11. M2 Selection Result of Short Plant Height Properties

Strain	Plant Height (cm)
R3G53T3	85
R3G27T20	89
R1G32T5	90
R1G66T20	91
R3G45T20	93
R2G19T5	97
R1G26T1	97

In accordance with the purpose of this study, one of the objectives is to get a plant that has a short stem, which we know from Rice Seed Center Mentik Wangi rice has a plant height of approximately 120 cm. The shortest plant is R3G53T3 strain, where R3 mean that the plant is in irradiation of 300 Gray. Irradiation 100 Gray has the shortest plant height of 90 cm in R1G32T5 strain, whereas in irradiation of 200 Gray has the shortest plant which 7 cm difference with 100 Gray irradiation which is 97 cm at R2G19T5 strain. Giving doses of irradiation received by plants not only changes the height of the plants but also changes the nature of other morphology and physiological properties [12].

Table12. M2 Selection Result of Productive Tillers Properties

Strain	Number of Productive Tillers
R1G33T1	30
R2G70T1	28
R1G1T20	27
R2G1T2	27
R2G89T11	27
R3G6T20	27
R3G39T6	27

The results showed that from all the existing plants, there are 7 plants of mutant selected based on the number of productive tillers. Results from the 7 results mutant selection based on the number of productive tillers, the highest number of productive tillers shown at 100 Gray irradiation of R1G33T1 strain of 30 tillers. On irradiation 200 Gray of all the selected plant crops that have a number of productive tillers as many as 28 tillers in R2G70T1 strain, for plant from irradiation of 300 Gray of mutant selection there is lowest productive tillers with 3 tillers, for 300 Gray irradiation is also have the lowest average with 13 tillers. 200 Gray irradiation has an average number of productive tillers of 15 tillers, 1 tillers difference with 100 gray control and other irradiation treatment. The highest number of productive tillers totaling 27 tillers showed on R3G6T20 and R3G39T6 strain.

4. Conclusion

Conclusion that can be taken from this research of Mentik Wangi Rice M2 Generation From Gamma Ray Irradiation are;

1. Mentik Wangi which has short stems properties are R3G53T3 with height 85 cm, R3G27T20 with 89 cm height, R1G32T5 with height 90 cm, R1G66T20 with height 91 cm, plant number R3G45T20 rod 93 cm, number R2G19T5 and R1G26T1 with a height of 97 cm.
2. Components of the results include the total number of tillers, the number of productive tillers, the number of grain per panicle and the weight of 1000 seeds in the plant occurs indication of mutations that have a high level of diversity in each treatment.

3. There are rice plants that have shorter flowering age compared to controls i.e with 5 weeks flowering age after planting on rice with irradiation of 100 Gray and 200 Gray is shown by R1G61T11, R2G81T16 and R2G80T7 strain, for flowering age 6 weeks after plant shown in R3G37T3, R1G11T8, R2G77T20 and R3G16T6 strain.
4. The result of mutant selection showed that some of the highest number of productive plants were shown by R1G33T1 strain of 36 tillers, R2G70T1 strain with 29 tillers, for number of 27 tillers are shown by R1G1T20, R2G1T2, R2G89T11, R3G6T20.

Acknowledgement
The authors are grateful to faculty of Agriculture Universitas Sebelas Maret for the financial support

References
[1] Deptan 2008 Neraca Bahan Makanan (Jakarta: Departemen Pertanian Republik Indonesia)
[2] Abdullah B, Prajitno Al Ks, Mudjisihono R 2006 Keragaan Beberapa Genotipe Padi Menuju Perbaikan Mutu Beras Subang (Id): Balai Besar Penelitian Tanaman Padi Sukamandi
[3] Albokari M M A, Alzahrani S M, Alsalman S M 2012 Radiosensitivity Of Some Local Cultivars Of Wheat (Triticum Aestivum L) To Gamma Irradiation *J Bot* **41** 1-5 Bangladesh
[4] Sarduie N S, Sirchi G R, Torabi M H 2010 Assessment of dissimilar gamma irradiations on barley (Hordeum vulgare sp) *J Plant Breeding Crop Sci* **2**
[5] Anditasari N 2009 Efisiensi Serapan S Dan Hasil Tanaman Padi (Oryza Sativa L) Dengan Pemberian Pupuk Kandang Puyuh Dan Pupuk Anorganik Di Lahan Sawah (Musim Tanam II) Skripsi Fakultas Pertanian UNS
[6] Las I, Widiarta I, Suprihatno B 2004 Perkembangan Varietas Dalam Perpadian Nasional Inovasi Pertanian Tanaman Pangan (Bogor: Puslitbang Tanaman Pangan)
[7] Khush G 2000 Strategies For Increasing The Yield Potential Of Rice In: Sheehy J *Redesigning Rice Photosynthesis To Increase Yield Journal Rice Research Institute* **7** 207-212
[8] Bewley J D, Bradford K J, Hilhorst H W M, Nonogaki H 2013 *Seed Physiology of Development Germination and Dormancy* 3rd Edition (New York: Springer)
[9] Haris A, Abdullah, Bakhtiar, Subaedah, Aminah, Jusoff K 2013 Gamma Ray Radiation Mutant Rice On Local Aged Dwarf Middle-East Journal Of Scientific Research **15** 1160-1164
[10] Martin A, Pippa J M, Carlos B, Tearall K 2009 Mutation discovery for crop improvement *J Exp Bot* **60**
[11] Lestari A D, Nugraha Y 2007 Keragaan Genetik Hasil dan Komponen Hasil Galur-galur Padi Kultur Anter *Penelitian Pertania Pangan* **26** 8-13
[12] Emrani A, Razavi A, Rahimi M F 2013 Assessment of Gamma Ray Irradiation Effects On Germination And Some Morphological Characters In Tow Corn Cultivars *Int J Agric Crop Sci*

Dynamic models of farmers adaptation to climate change (case of rice farmers in Cemoro Watershed, Central Java, Indonesia)

Sugihardjo, J Sutrisno, P Setyono and Suntoro
Graduate Program, Universitas Sebelas Maret, Surakarta, Indonesia

Email : giek_bb@yahoo.com

Abstract. Farming activities are generally very sensitive to climate change variations. Global climate change will result in changes of patterns and distribution of rainfall. The impact of changing patterns and distribution of rainfall is the occurrence of early season shifts and periods of planting. Therefore, farmers need to adapt to the occurrence of climate change to avoid the decrease productivity on the farm land. This study aims to examine the impacts of climate change adaptation that farmers practiced on the farming productivity. The analysis is conducted dynamically using the Powersim 2.5. The result of analysis shows that the use of Planting Calendar and Integrated Crops Management technology can increase the rice productivity of certain area unity. Both technologies are the alternatives for farmers to adapt to climate change. Both farmers who adapt to climate change and do not adapt to climate change, experience an increase in rice production, time after time. However, farmers who adapt to climate change, increase their production faster than farmers who do not adapt to climate change. The use of the Planting Calendar and Integrated Crops Management strategy together as a farmers' adaptation strategy is able to increase production compared to non-adaptive farmers.

1. Introduction

Farming activities is generally very sensitive to the vary of climate change. Agriculture sector is very affected with the existence of climate change [1]. The indicators of climate change phenomenon are the rising of temperature and uncertainty of rainfall which affects to the shift of early season and planting period. This will cause the lower productivity of land followed by disturbance of food security [2]. Some countries will face food insecurity caused by climate change [3]. Thus, farmers need to adapt to climate change phenomenon to avoid lower productivity on their farming land.

Adaptation is basically an adjustment conducted to human or natural system in reacting the impact of climate change that is being or going to occur. Adaptation against climate change may consider a responds against climate change phenomenon [4]. Meanwhile, for farmers, adaptation means to maintain income or at least to minimize loss possibility emerged due to the climate change. Thus, income may become an indicator of farmers' success in adapting against the climate change. Without adapting, climate change will cause bigger loss [5]. Farmers who able to remain their income or minimize their loss by conducting adaptation, it is said that they are success adapting.

Farmers at the Cemoro watershed of Bengawan Solo River, have particular customs and ways to maintain their success in farming, against the climate change effects. Farmers have conducted adaptation due to climate change effects to their farming land, according to the local condition they have experienced. Analysis related to the forms of farmers' adaptation against climate change, should be conducted to gain best adaptation model for farmers' profit.

2. Methods

This research is conducted at Cemoro watershed in Central Java, Indonesia. It includes parts of Semarang regency, Boyolali regency, Karanganyar regency and Sragen regency. This study is addressed for farmers with rice farming.

The organization of adaptation model by farmers against climate change, uses a dynamic model with support of Powersim 2.5. Dynamic model is a picture of complex system that strived for describing system with good output. There are three steps in organizing dynamic model:

a. First step is to arrange the black box, that shows relation between input, process and output.
b. Second stage is to arrange the causal relation between variables that are related to the farmers' adaptation to face the climate change phenomenon.
c. Third step is to arrange dynamic model of farmers' adaptation against the climate change.

3. Result and Discussion

Observation at the research area shows two strategies applied by the farmers in adapting to climate change. They are Planting Calendar utilization and Integrated Crops Management utilization. It is according to the study conducted by [6]. Planting Calendar is basically giving information related to the area's potential wide to the upcoming planting season: Planting Season I, Planting Season II, or Planting Season III. Integrated Crops Management is an effort to increase farmers' productivity and income, through land management, water, plant, organism and pest in integrated and continued way.

Both strategies applied by the farmers are form of adaptation, noticing balance and harmonious relation between components that support farming. Scenario of farmers' adaptation in facing climate change is formulated based on the strategies applied by the farmers. Scenario of farmers' adaptation model against climate change is seen on Table 1.

Table 1. Scenario of farmers' adaptation in facing climate change at Cemoro Watershed in Central Java, Indonesia

Scenario	Applied Technology	Description
Scenario 1	Not practicing adaptation	Average wide of land owned by farmers 0.35 ha of irrigation rice field and 0.36 ha of rain-fed rice field
Scenario 2	Practicing Planting Calendar and Integrated Crops Management	
Scenario 3	Practicing only Planting Calendar	
Scenario 4	Practicing only Integrated Crops Management	

Scenario 1 is a model in condition that farmers do not adapt to the climate change phenomenon. Scenario 2 is a model in condition that farmers adapt to climate change by applying both Planting Calendar and Integrated Crops Management. Scenario 3 is a model in condition that farmers adapt only the Planting Calendar to their farming activities. Meanwhile, scenario 4 is a model in condition that farmers adapt only Integrated Crops Management as an adaptation against climate change.

Black box may be formulated by paying attention to the arranged scenario, as a foundation to formulate dynamic model of farmers' adaptation in facing climate change. Black box shows several indicators that describe relation between input, process and output. Adaptation process to climate change is meant to gain expected output and to avoid unexpected output. The expected output for farmers means improvement in production and income. Meanwhile, the unexpected output is decline in production and income. Adaptation process may run well by keeping attention to existed-input. There are three inputs that influence adaptation process, such as controllable input, uncontrollable input and policy input. Amongst three, controllable input is the only input that might be adjusted according to the willing of how adaptation be processed. Uncontrollable and policy variables have 'given' character. Black box shows relation between input, process and output, as seen on Figure 1. Meanwhile, diagram of causality related to the adaptation against climate change is shown on Figure 2.

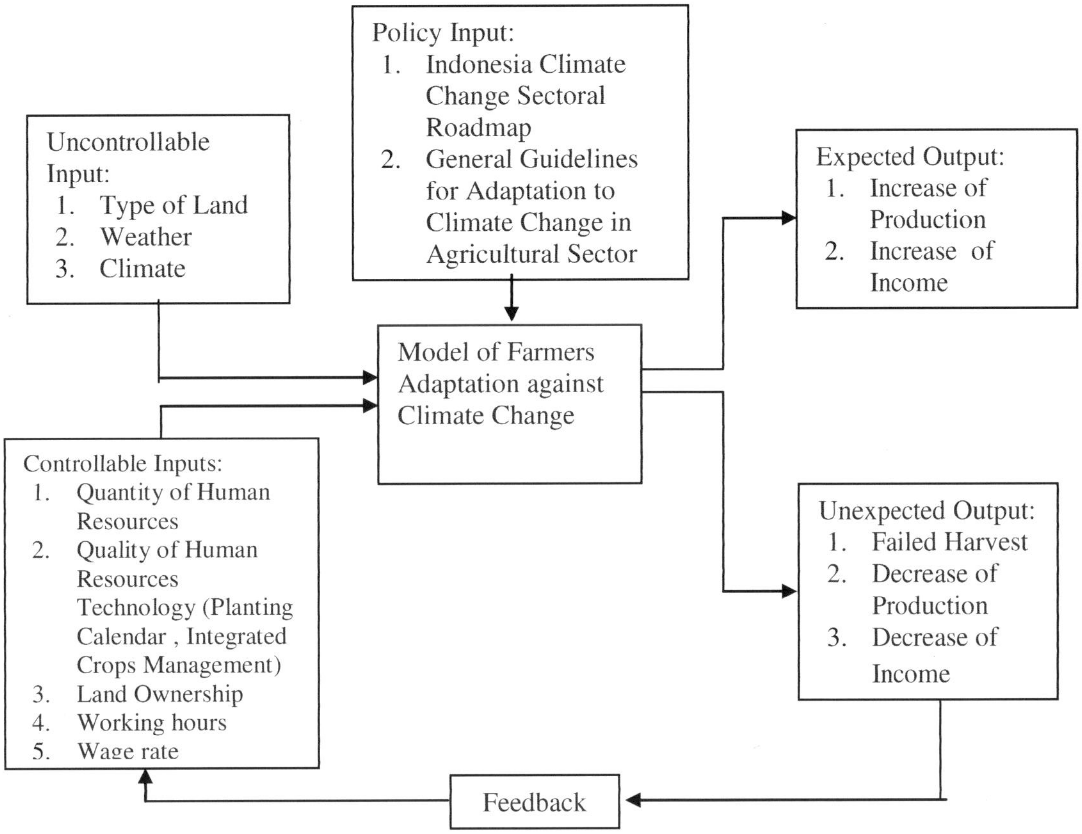

Figure 1. *Black Box*of Farmers' Adaptation Model to Climate Change

The analysis result of dynamic model by using Powersim 2.5 shows that farmers who use technology of Planting Calendar and Integrated Crops Management, are able to increase rice productivity. The use of Planting Calendar and Integrated Crops Management at once, is a scenario chosen for the farmers' adaptation dynamic model against climate change. Another scenario is that farmers only practice Planting Calendar or Integrated Crops Management. Comparing the two models, it is reviewed that the usage of both Planting Calendar and Integrated Crops Management at once, is able to increase productivity of farming and production of rice. The next impact is increase of farmers' income and farmers' welfare.

The increase of production due to farmers' adaptation against climate change, is not significant in the beginning. However, the big gap of production increase, will appear on the next period. Figure 3 shows difference of rice productivity between adaptive farmers and non-adaptive farmers. It is seen that farmers who do not conduct adaptation, produce fewer rice than ones who adapt to climate change.

IOP Conf. Series: Earth and Environmental Science **142** (2018) 012051 doi:10.1088/1755-1315/142/1/012051

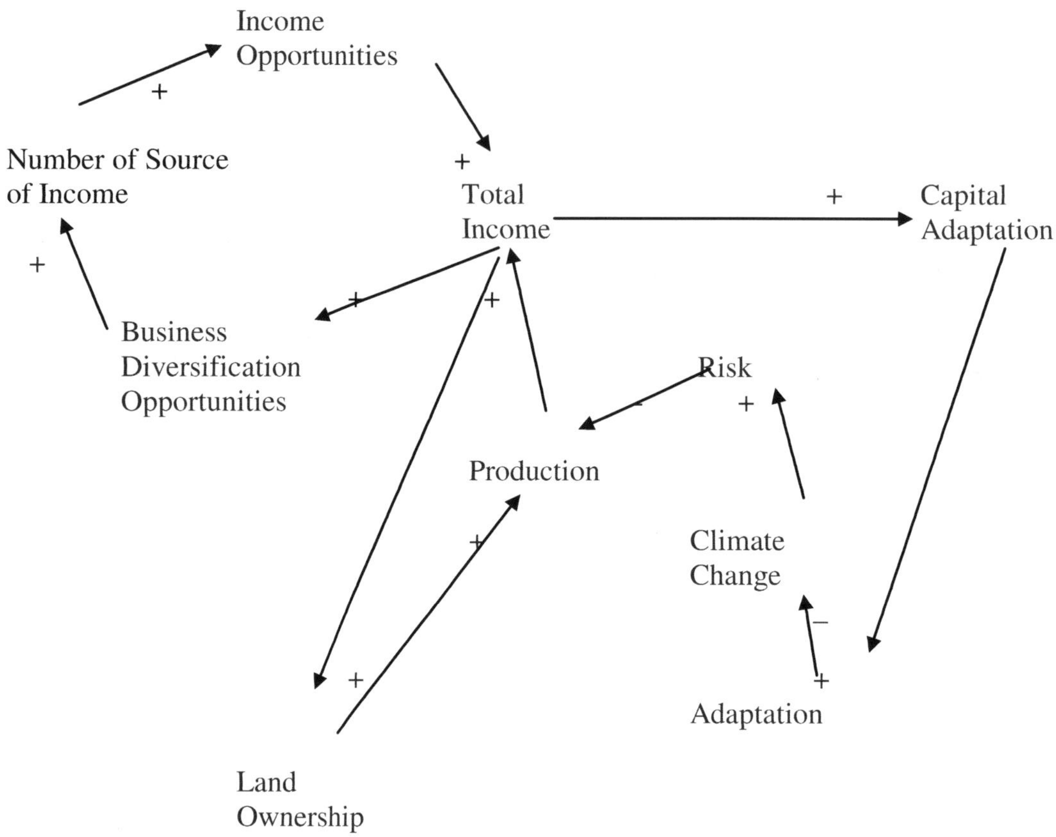

Figure 2. The causal relationship of dynamic models of adaptation against climate change

Scenario 1 describes the amount of rice production gained by farmers who do not adapt to climate change, while scenario 2 describes the amount of rice production gained by farmers who adapt to climate change by practicing Planting Calendar and Integrated Crops Management. Scenario 3 describes the amount of rice production gained by farmers who practice only the Planting Calendar, while scenario 4 describes the amount of rice production gained by farmers who practice only the Integrated Crops Management. Based on the Figure 3, it is seen that adaptive and non-adaptive farmers, have more amount of rice production time after time. However, adaptive farmers experience faster increase of rice production compared to non-adaptive ones.

In 2015, the gap of the amount of rice production between adaptive and non-adaptive farmers against climate change, is small. However, time has proven that the gap is growing increase year by year. Table 2 shows the rice production gap between adaptive and non-adaptive farmers against climate change.

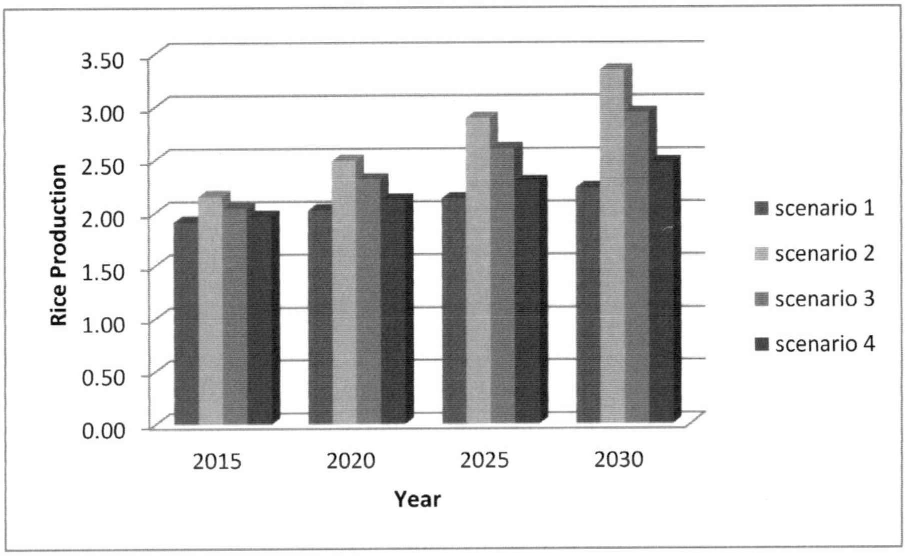

Figure 3. Gap of Rice Production between Adaptive and Non-Adaptive Farmers against Climate Change

As seen on Table 2, if farmers practice both Planting Calendar and Integrated Crops Management (scenario 2), then an increase in rice production is most significant compared to others. In 2015, the gap of rice production is 12.57%, increases to 49.8% in 2030. As well as with scenario 3 and scenario 4 where they have more significant increase of rice production, compared to scenario 1, but less than scenario 2. This proves that effective adaptation ways, will minimize the production decline due to climate change. This condition should be informed to the farmers, so that they are willing to adapt to climate change, in order to increase production and income.

Table 2. Gap of Rice Production between Adaptive and Non-Adaptive Farmers against Climate Change

Year	Production (ton)				Production Gap (%)		
	Scenario 1	Scenario 2	Scenario 3	Scenario 4	(3)-(2)	(4)-(2)	(5)-(2)
1	2	3	4	5	6	7	8
2015	1.91	2.15	2.04	1.96	12.57	6.81	2.62
2020	2.02	2.49	2.31	2.12	23.27	14.36	4.95
2025	2.13	2.89	2.60	2.29	35.68	22.07	7.51
2030	2.23	3.34	2.94	2.47	49.78	31.84	11.66

Description :
Scenario 1 : Not practicing adaptation
Scenario 2 : Adapting with Planting Calendar and Integrated Crops Management
Scenario 3 : Adapting with Planting Calendar
Scenario 4 : Adapting with Integrated Crops Management

4. Conclusion

The results of the analyses show that each scenario shows an increase in production. The highest amount of production is emerged at condition that the farmers apply the adaptation in the form of Planting Calendar and Integrated Crops Management. Farmers who do not adapt to climate change, have the fewest rice production

References

[1] Tripathi, Amarnath and Mishra, Ashok K 2017 Knowledge and passive adaptation to climate change: An example from Indian farmers, *Clim. Risk Manag.* **16** 195–207

[2] Maharjan S K, E R Sigdel, B R Sthapit, B R Regmi 2010 Tharu Community's Perception on Climate Change an Their Adaptive Initiation to Withstand its Impacts in Western Terai of Nepal *Int. NGO J.* **6** 035-042

[3] Food and Agriculture Organisation 2008 Climate change adaptation and mitigation in the food and agriculture sector *Technical Background Document From The Expert Consultation*

[4] Babatolu and Akinnubi 2016 Smallholder farmers' perception of climate change and variability impact and their adaptation strategies in the upper and lower Niger river basin development authority areas, *Nig. J. Petrol. Environ. Biotech.* **7** 3

[5] Gbetibouo 2009 Understanding farmers perceptions and adaptations to climate change and variability: the case of the limpopo basin South Africa *IFPRI Discussion Paper* No 0084 (Washington DC: IFPRI)

[6] Irawan 2012 Adaptasi perubahan iklim untuk mempertahankan produksi beras di pulau Jawa (sustain rice production in java trought climate change adaptation) (Bogor: Balai Penelitian Tanah)

Morphology, production, and chemical content performance of black rice Matesih accession with several comparisons

Nandariyah[1], E Purwanto[1] and A N Meidini[2]

[1]Study Program of Agrotechnology, Faculty of Agriculture, Universitas Sebelas Maret (UNS) Surakarta
[2] Student of Faculty of Agriculture, Universitas Sebelas Maret (UNS) Surakarta
Jl. Ir. Sutami 36A, Surakarta, Central Java, Indonesia 57126

E-mail: nandariyah@staff.uns.ac.id

Abstract. Rice (*Oryza sativa* L.) is an important food crop in Indonesia. In Matesih area, Karanganyar, recently found new varieties of black rice cultured by local farmers which morphology and chemical content have not yet identified. The purpose of this research was to obtain information of morphology, production, and chemical content of black rice matesih accession and to compare the appearance in order to detect the superiority of black rice matesih accession with the comparison of other accession of black rice. There were four accessions of black rice tested, namely Matesih Accession, Klaten Accession, Toraja, and Cempo. Research data were divided into qualitative data which processed by scoring, and quantitative data are processed with simple descriptive statistic. The kinship test was done by using NTSYSpc program with SIMQual and SIMInt function. The observation of qualitative properties indicates that accession matesih has a form that is relatively similar to other accessions. Qualitatively, accession matesih superior at leaf length, leaf width, plant height and culm diameter. Klaten accession has higher production than accession matesih. Matesih accession has the advantage of having shorter period on heading time and harvest time than other accessions. Matesih accession has the highest amylose content, lower protein content than klaten accession and lower content of anthocyanin than toraja accession. The kinship analysis showed that matesih accession and klaten accession has close kinship.

1. Introduction

Rice (*Oryza sativa* L.) is an important food commodity in Indonesia, South East Asia and some country. Rice for people need always increase from year to year similar with increasing of people number. There is some kind of rice consumed of people in Indonesia like white rice, red and black rice.

One kind of rice in Indonesia is black rice that has black colour of pericarp. Black rice is a localy variety that has cultured by farmers. Actually, the black colour mean of anthocyan contain in pericarp, aleurone and endosperm that have compound of red-blue and deep of purple colour. [1] reported that anthocyanin has functionally as antiocxidant and catch free radical then secure from aging, cancer and degenerative disease.

Black rice cultivar is less popular than white rice and so not much research about this varieties like identification of morphology and biochemistry aspect. Identification of locally black rice is important in order to know the specific characters need for breeding program later. One location of black rice developing and culturing is Matesih district of Karanganyar resident. Some farmer developing and culturing the locally black rice but they don't know about specifically and production potential also where the originally of this rice. Research is important and aim to know about performance of Matesih accession including the morphology and chemical content compared with some local black rice varieties to know but the superiority of Matesih accession has. Identification need to improve the quality of important characters and reduce the lack of characters in order to developing new variety that has a high productivity [2].

2. Methods

Research conducted in March-December 2016 in Ngadiluwih district of Matesih Karanganyar resident Central Java Indonesia. Experiment method in paddy land to know Matesih variety productitvity compared with other local varieties of Klaten, boyolali, Toraja and Cempo varieties. From the research can be determined about specific characteres of each variety. Analysis of data used simple statistics for quantitative data and descriptive qualitative. There were 43 variables observed for performance, morphology characters, production, chemical contain of grains and taste of rice. Cluster analysis used by SIMQUAL (Similarity Qualitative) and SIMINT (Similarity Interval.).

3.Results and Discussion

3.1.Plant Morphology

*3.1.1 Leaf Morphology*Morphology of leaves paddy observed showed that Matesih variety have the longest and widest then other varieties (Table 1). Size of leafs is determining the fungtion for light number, nutrition and water take and growth and development of paddy. physiologycaly process occurred in leaf related with resistence against disease, and response of plant against environment stress and yield depend on leaf size [3]. [4] reported that their positive correlation between leaf measure and paddy grain yield.

Table 1. Morphology of black rice leaf characters

Characters	**Matesih**	**Klaten**	**Toraja**	**Cempo**
Length	45.93 ± 6.66	38.67 ± 4.32	37.06 ± 4.98	37.56 ± 4.67
Width	1.1 ± 0.10	1 ± 0.09	1.07 ± 0.07	1.04 ± 0.06
Angle	Upright	Upright	Uprrigth	uprigth
Colour shape	Green	Green	Young green	Young green
Characters	Hairs	Hairs	Hairs	Hairs
Colour of neck	Green	Green	Young green	Young green
Colour of tongue	White	White	White	White
Form of tongue	*2-cleft*	*2-cleft*	*2-cleft*	*2-cleft*
Colour of ear	White	White	White	White
Leaf blade colour	Green	Green	Green	Green
Flag angle leaf formed	Straigth	Straight	Straight	Straight

Qualitative character leaf colour of Matesih variety was different with Toraja and Cempo variety but same with Klaten variety. Green colour mean with more chlorophyl content of leaf compared with young green colour. The higher chlorophyl content of Matesih very related with higher rate of photosynthesis becaure of more light quality absorbed [5] higher chlorophyll indicated of higher photosynthesis rate.Other characters same belong of four varieties were leaf anle, hair on leaf shapel colour of neck, colour of tongue, form of tongue, colour of ear leaf, colour of blade, and form of flag leaf angle.

3.1.2. Stem Morphology Matesih variety has plant height and heighest of plant diameter compared by other varieties. [6] said about paddy, decreasing of plant height become semi dwarf is a primary target in order to improve the resistence against plant collapsed, but the result of [7] showed that improving semi dwarf of paddy can limitate of canopy for photosynthesis and number of biomass and then decrease grain yield.

Table 2. Morphology of Black Rice Stem

Characters	**Matesih**	**Klaten**	**Toraja**	**Cempo**
Plant height	136.6 ± 7.02	126.5 ± 6.46	97.2 ±7.46	118. 43 ± 10
Stem Diameters	0.57 ± 0.11	0.57 ± 0.11	0.51 ± 0.08	0.52 ± 0.10
Stem Power	Strong	Strong	Strong	Strong
Stem angle	upright	Upright	Upright	Upright
Colour of stem segment	golden yellow	golden yellow	golden yellow	golden yellow

Matesih Variety has heighest diameter of stem it was same with Klaten variety. Result of research [8] showed that stem diameter is important to measure because it was related with photosynthesis rate and resistence against stem breakness. [5] reported that breackness in paddy devided of twoo category iecollaps of root and stem collaps. Root collaps because root not enough to cach to the soil then all stem breacked. Breackness of stem because croocked or cutting of segment based of stem by stress of up stem. Strongness of stem or thickness of bending stem important in supporting resistence against breacked mainly at up of stem part. Qualitative characters of Matesih variety was same with other varieties in thickness, angle and yellow colour of stem segment.

3.1.3Panicle Morphology Measurement morphology of panicles only for three varieties, that is matesih, klaten and Toraja accessions because one variety cempo was harvesting failure. Some factors according to [9] empty grains cause of not occur fertilization, less of water supply when generative phase, or environment not supported.

Table. 3. Morphology of Black Rice Panicle

Characters	**Matesih**	**Klaten**	**Toraja**	**Cempo**
Length of panicle	20.30 ± 1.92	20.31 ± 1.31	18.91 ± 1.37	-
Panicle Type	Compact, medium	Compact, medium	Compact, medium	-
Secunder panicle branch	Many	Many	Little	-

Longest of panicle owned by Klaten followed by Matesih. Panicle long take effect to how many grains in panicle, number and density of spikelet, rate of filled grains. and grain size made long of panicle as character that can influence rice yield [10, 11]. Panicle type at three varieties was same that is between compact and medium while panicle branch of Matesih and Klaten have many branch and the smallest is Toraja owned.

3.1.4 Morphology of Grain

Table 4. Morphology of Black Rice Grain

Character	Matesih	Klaten	Toraja	Cempo
Lemma and Palea colour	Brownish brown	Brown spotting	Brownish brown	-
Hair existence at Lemma and Palea	Short hairy	Short hairy	Short hairs	-
Grain hair	No hairy	No hairy	No hairy	-
Long Grain	9.74 ± 0.48	9.52 ± 0.56	9.89 ± 0.51	-
Wide Grain	2.97 ± 0.19	3.03 ± 0.14	3.16 ± 0.17	-
Thickness Grain	1.88 ± 0.10	1.78 ± 0.12	1.90 ± 0.08	-

Result of observation in lemma and palea showed the variation between variety. Lema and palea in Matesih and Toraja is brownish brown and Klaten has brown spoting lemma and palea at background of straw. Lemma and palea of three varieties showed short hairs. Unknown function of hair on lemma and palea but in according research result of [12] hairs and bumps at husk shape (lemma danpalea) contain of hight silica. Size of grain in according scale of [13] including very long (> 7.50 mm).Size and form of grain is stabil variety properties. Slim and long grain. Butir ramping usually has many damage of the short and thick grain made the specific tool in processing [14]. According to [15] slim grain usually break at processing.

3.2. Flowering and Harvesting Age

Matesih variety has early in flowering and harvesting age compared with other varieties. According to [16] long harvest age of rice if above 151 days after planting medium if 125 - 150 days after planting and short harvest age if harvest at 105 - 124 days after planting then very short harvest age if 90 - 104 days after planting then ultra short harvest age if below 90 days after planting.

Table 5. Flowering age and harvest age of Black rice

Character	Matesih	Klaten	Toraja	Cempo
Flowering age	52	55	59	98
Harvest age	117	121	126	-

3.3 Production

According to [17] higher number of clumps as characters need in rice production because number of clumps very related with panicle number. According to [14] 1000 grain weight gave information about size and density of grain. Matesih variety has grain weight many than Toraja own but smaller than Klaten.

Table 6. Production of Black rice

Character	Matesih	Klaten	Toraja	Cempo
Total clumps	12.27 ± 2.61	12.50 ± 3.22	11.86 ± 3.70	12.67 ± 3.98
Total productive clumps	10.82 ± 3.25	12.46 ± 2.89	11.25 ± 3.31	-
Weight 1000 seeds (gram)	32	30	28	-
Grain weight per clump (g)	17.01 ± 5.72	21.62 ± 7.50	14.97 ± 7.05	-
Production per Ha (ton)	2.136	2.715	1.88	-

3.4 Chemical Content

According to [18] rice protein more quality because contain of eight of ten amino acid essential compared with other plant own. According to [19] grain with little amilosa gave the sticky and clot if cool. Chemical contain in black rice very important to know is antosianin. Anthocyanin made black rice as healthy food [20] with antioxidant that can inhibit inflammation, anti cancer, increase blood sirculation, slow down damage and aging decresing cholesterol and sugar, and stomach acid [21].

Table 7. Chemical contents of grain

Character	Matesih	Klaten	Toraja	Cempo
Protein (%)	6.38	6.78	6.38	-
Amilosa (%)	23.61	23.44	8.92	-
Antosianin (ppm)	2250.96	1807.77	4347.56	-

3.5 Organoleptic Test

Rice grain colour of three accessions black rice actually purple is blackish. According to [22] colour is form of precipitation of pigment anthocyanin in large quantities on the surface of rice. Organoleptic test of three kind of rice showed difference among varieties. Matesih and Klaten has hard texture while Toraja accession has more lenient related with amilosa content. Fragrances of three accessions rice was same there is a little fragrant.

Table 8. Organoleptic Test of Rice

Character	Matesih	Klaten	Toraja	Cempo
Rice colour	Purple	Purple	Purple	-
Texstur	Sticky	Sticky	Thiny	-
Aroma	Fragrance slightly	Fragrance slightly	Fragrance slightly	-

3.6 Similarity Analisys

3.6.1 Qualitatif Characters

Table 9. Similarity coefficient of black rice based on qualitative character

	AksesiMatesih	AksesiKlaten	AksesiToraja
Matesih	1.000		
Klaten	0.870	1.000	
Toraja	0.739	0.739	1.000

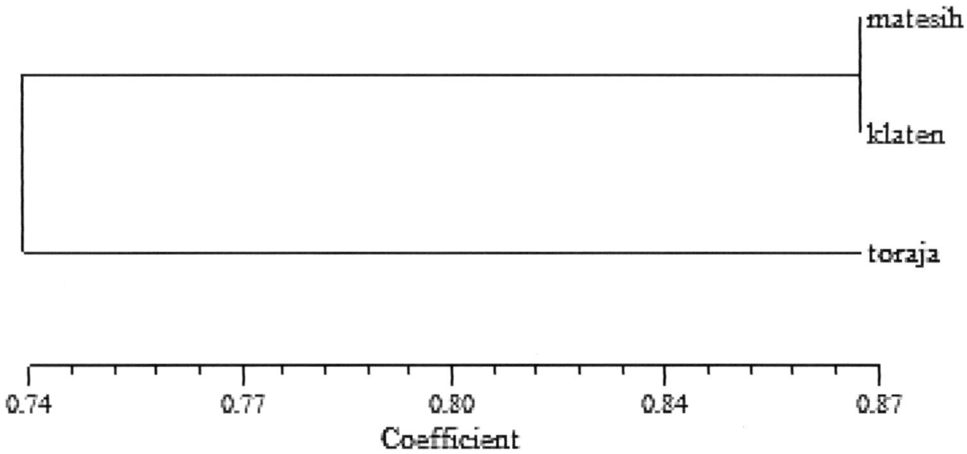

Figure 1. Dendogram based on qualitative morphology characters

There is Similarity analysis showed that Matesih has similar with Klaten accession 0.87 while Toraja and Klaten has similarity coefficient of 0.74. Even though in breeding selection genetic distance not the only thing to be reckoned but the other interesting characters need to be included. Correlation factor of vegetative and generative important to be calculated so that more focused and effective [23].

3.6.2. Quantitative Character

Table 10. Matrixs of similarity coefficient based on quantitatif characters

	Matesih	Klaten	Toraja
Matesih	0.000		
Klaten	110.882	0.000	
Toraja	524.282	635.006	0.000

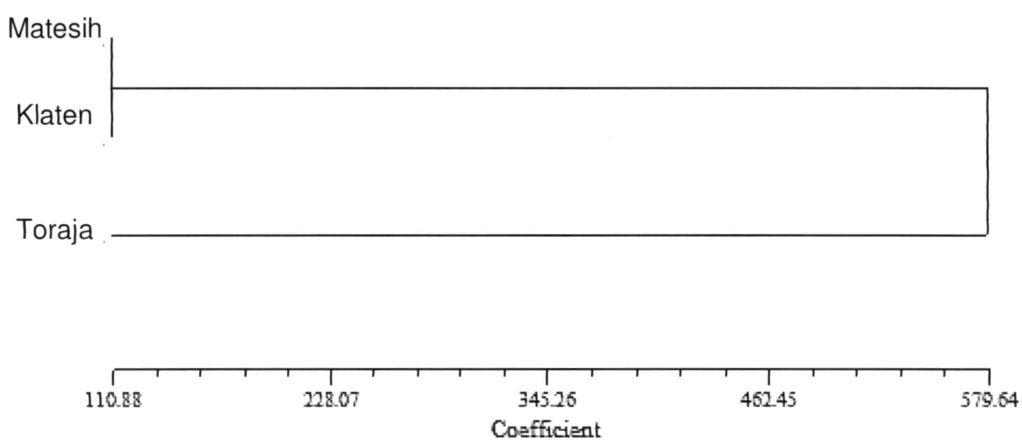

Figure 2. Dendogram of morphology based on quantitative characters

Dendogram of morphology characters based on quantitative characters showed that Matesih accession has similarity with Klaten accession compared with Toraja accession. On wide of leaf, stem diameter, long of panicle, number of clump, and amilosa content.

4. Conclusions

From this study it may be concluded that Matesih accession has qualitative characters relatively same among accessions with low variation like long of leaf, wide leaf, plant height, and stem diameter that Matesih has better than comparisons accessions. Klaten accession has more grain yield than Matesih accession. Matesih accession has superiority about early flowering and harvesting age compared with there comparisons.Matesih accession has high amilose content but lower protein content than Klaten accession and lower anthocyanin content than Toraja accession had highest amylose content. Matesih and Klaten accession has similarity coffecient compared with Toraja.

Acknowledgement

Acknowledgment to Mr Asih as chairman of the farmer group in Matesih Karanganyar Central Java who has provided facilities during this research took place.

References

[1] Jusuf M, Rahayuningsih S A, Ginting E 2008 Purple sweet potato *Res. Publ.develop. Agric.* **30** 13-14.

[2] Kristamtini, Taryono M, Bunanda P et al 2014 Genetic variability of local cultivar Black rice *J Agrobiogen* **10** 69-76

[3] Perez-Perez J M, Esteve-Bruna D, Micol J L 2010 QTL analysis of leaf architecture *J Plant Res* **123** 15-23

[4] Alvaro B 1979 Leaf area estimation, growth analysis, and yield evaluation in grain sorghum (Sorghum *bicolor* L. Moench.) (Iowa States University: Retrospective Theses and Dissertations)

[5] Sterling M, Baker C J, Berry P M et al. 2003.An expiremental investigation of the lodging of wheat *Agr Meteorol* **119** 149-165

[6] Khush G S 1999 Green revolution: Preparing for the 21st century *Genome* **42** 646-655 PMID: 10464789

[7] Kuroda E, Ookawa T, Ishihara K 1989 Analysis on difference of dry matter production between rice cultivars with plant height in relation to gas diffusion inside stands *Jap J Crop Sci* **58** 374-382 DOI:10.1626/JCS.58.374

[8] Li-Li W, Zhong-Li L, Jun-Min W et al. 2011 Morphological, anatomical, and physiological characteristics involved in development of the large culm trait in rice *Aus J Crop Sci* **5** 1356-1363.

[9] Polisafaris 2010 The cause of Empty Rice URL: *http://polisafaris.blogspot.com/2010/04/penyebab-padi-hampa-gabug.html*

[10] Erbao L, Yang L, Guocan W et al. 2016 Identification of candidate gene for panicle length in rice (*Oryza sativa* L.) via association and linkage analysis *Front Plant Sci* **7** 596

[11] Huang C, Wanneng Y, Lingfeng D et al. 2013 Rice panicle length measuring system based on dual-camera imaging *Comput. Electr Agric.* **98** 158-165

[12] Krishnarao R V, Jagadish K, Jandhyala S 2001 Studies on the formation of black particles in rice husk silica ash *J Eur Ceram Soc* **21** 99-104.

[13] BPS [International Rice Research Institute] 2005Rice quality training manual (Philippines: International Rice Research Institute)

[14] IRRI [International Rice Research Institute] 2006 Grain quality (Philippines: International Rice Research Institute)

[15] Susan R 2015 Rice Thickness Grading, World-Grain URL: *www.world-grain.com/articles/news_home/features/2015/Rice-thickness-Grading*

[16] BBPADI [Balai Besar Penelitian Tanaman Padi] 2015 Classification of harvest age of Rice. BBPADI

[17] Miller B C, Hill J E, Roberts S R 1991 Plant population effects on growth and yield in water-seeded rice *J Agron* **83** 291-297

[18] Huebner F R, BIetz J A, Juliano B O 1990 Rice cultivar identification by high-performance liquid chromatography of endosperm proteins *Cereal Chem* **67** 129-135.

[19] Damardjati D S 1995 Characteristcal for standardization grain quality as based of development agribusines and agroindustyi of ricei in Indonesia Oration inauguration of Ahli Peneliti Utama (Bogor: Balai Penelitian Bioteknologi Tanaman Pangan)

[20] Chutipaijit S, Chaum S, Sompornpailin K 2011 High contents of proline and anthocyanin increase protective response to salinity in *Oryza sativa* L. spp. Indica In Australian *J Crop Sci* **5** 1191-1198

[21] Ujjawal K 2016 Black rice: Research, history, and development (Switzerland: Springer International Publishing)

[22] Chaudhary R C 2003 Speciality rices of the world: Effect of WTO and IPR on its production trend and marketing *J Food Agric Environ* **1** 34-41

[23] Maulan Z, Kuswinan T, Sennang S R et al. 2014 Exsploration of germ plasm variability local rice origin of Toraja and Enrekang based on morphology characters

IOP Conf. Series: Earth and Environmental Science **142** (2018) 012053 doi:10.1088/1755-1315/142/1/012053

Determination colchicine content in aquadest-extracted *Gloriosa superba* seed from Sukoharjo and Gunung Kidul

S I Rahmawati[1], A Susilowati[2], A Yunus [3] and Y Widyastuti[4]
[1]Department of Bioscience Graduate School, Universitas Sebelas Maret, Surakarta -57126, Central Java, Indonesia
[2]Department of Biology, Faculty of Mathematics and Science, Universitas Sebelas Maret,
Surakarta -57126, Central Java, Indonesia
[3]Department of Agrotechnology, Faculty of Agriculture, Universitas Sebelas Maret, Surakarta - 57126, Central Java, Indonesia
[4]Research Center for Medicinal Plant and Traditional Drug, Tawangmangu, Indonesia

E-mail : yunus.uns7@yahoo.com

Abstract. Colchicine is a toxic alkaloid compound from the *Colchicum autumnale* plant. Colchicine can be used in agriculture as anti-mitosis in the procurement of plants with polyploid cells. *Gloriosa superba* is an herb contains colchicine as a mutagen (polyploid) potential in its parts, especially in the seeds. The aims of this research were to determine colchicine content from aquadest-extracted *Gloriosa superba* seed obtained from Sukoharjo and Gunung Kidul. *Gloriosa superba* seed samples were obtained naturally in January - April 2017. Samples were divided into immature seed samples from Mulur, Sukoharjo, Central Java and old seed samples from Krakal Beach, Gunung Kidul, Yogyakarta. Extraction seeds of *Gloriosa superba* using the maceration method with aquadest solvent (1: 1). Determination content of colchicine extract using TLC-Densitometry method. The results indicated that colchicine content of old seed *Gloriosa superba* samples 37,6 μg/μl (± 6,63) higher than colchicine content of immature seed *Gloriosa superba* samples 12,84 μg/μl (± 2,88)

1. Introduction

Gloriosa superba is an herbaceous vine plant found in the yard of Mulur, Sukoharjo and hill around Krakal Beach, Gunung Kidul. Usefulness and benefits of this plant is not widely known by the surrounding population. This plant is usually used as animal feed by the local people and allowed to grow on the sidelines of other plants. The abundance of this plant is found wildly and widely in Mulur, Sukoharjo and Krakal Beach, Gunung Kidul. *Gloriosa superba* L. is a perennial climber and is used as an ayurvedic medicinal herb to cure diseases in various parts of Africa and Southeast Asia. The plant was under threatened category due to its imprudent harvesting from wild as it is extensively used by medicinal industries for its colchicine content [1]. The high colchicine content accompanied by prospects of good availability from both wild and cultivated sources make the seeds of *Gloriosa superba* a potential commercial source of colchicine and colchicoside [2]. The colchicines content has been estimated to be 0.6% in seeds [3] and colchicosides 0.8%. The seeds are the best source of colchicines as their content is 2–5 times higher than in tubers. Thiocolchicoside (TCC), a semi-synthetic derivative of naturally occurring colchicoside from the seeds of *Gloriosa superba* [4] and *Colchicum autumnale*. It is chemically 2-demethoxy-2-glucosidoxythiocolchicine with a chemical formula $C_{27}H_{33}NO_{10}S$. TCC has been used clinically for humans as a centrally acting muscle relaxant agent for the treatment of orthopedic, traumatic and rheumatological disorders [5]. This compound also has anti-inflammatory and analgesic effects [6]. It is an analogue of colchicines, since they share the same benzo(alpha) heptalenic moiety [7]. The aim of this study was determine of colchicine content in aquadest-extracted of *Gloriosa superba* seeds obtained from Sukoharjo and Gunung Kidul.

2. Methods

The method used in this research is the sampling method. *Gloriosa superba* is a wild plant located in Mulur, Sukoharjo and Krakal Beach, Gunung Kidul. Sampling conducted in January-April 2017. Samples obtained in the form of seeds *Gloriosa superba* consisting of immature seeds and old seeds obtained at a different place. *Gloriosa superba* immature seed samples were obtained from Mulur, Sukoharjo, while the old seed sample was obtained from Krakal Beach, Gunung Kidul. Both seed samples immature and old seed were dried using oven at 70°C for 24 hours. The extraction of both *Gloriosa superba* seeds using the maceration method with aquadest solvent (1:1). Determination content of cholchicine on aquadest-extracted of *Gloriosa superba* seeds based on parameters of selectivity or specificity, accuracy, precision and linearity using TLC-Densitometry method [8]. Both samples of seed extract on silica plate was eluted using phase chloroform: diethylamine (9:1). The elution results were observed under UV light to detect the presence of a colchicine compound in a sample indicated by color and Rf value. The silica plate after eluted then sprayed using vanillin sulphate to see the color change of the compound. Colchicine area at selected wavelengths in the analysis using a densitometer.

3. Result and Discussion

Gloriosa superba seeds sample in this study consisted of immature seed samples obtained from Mulur, Sukoharjo and old seed samples obtained from Krakal Beach, Gunung Kidul. *Gloriosa superba* seed samples were dried using oven at 70°C for 24 hours. Both samples of dried seeds were extracted using the same method of maceration method with aquadest solvent (1:1) for 24 hours. The extraction result was aquadest-extracted of immature seed and old seed of *Gloriosa superba*. Aquadest-extracted seeds using TLC-Densitometry method for determination of colchicine content. Results of colchicine content in immature seeds and old seeds sample were presented in Table 1.

Table 1. Colchicine Content of *Gloriosa superba*

Samples	Colchicine Content
Immature seeds	12.84 μg/μl (± 2.88)
Old Seeds	37.65 μg/μl (± 6.63)

3.1. Immature Seeds Sample of Glorisa superba

Immature seeds of *Gloriosa superba* were obtained wildly around Mulur's yard, Sukoharjo. This plant is a wild plant that grows around the houses and residents in the gardens. Immature seeds of *Gloriosa superba* was conductued in January-February 2017 during the rainy season. Immature seeds drying using oven with temperature 70°C for 24 hours. Extraction of immature seed of *Gloriosa superba* using the method of maceration with water solvent (1:1) for 24 hours with 3 times repetition. The extraction result is aquadest-extracted immature seed which then analyzed using TLC-Densitometry to presence, determine and calculate the colchicine content contained in immature seed extract.

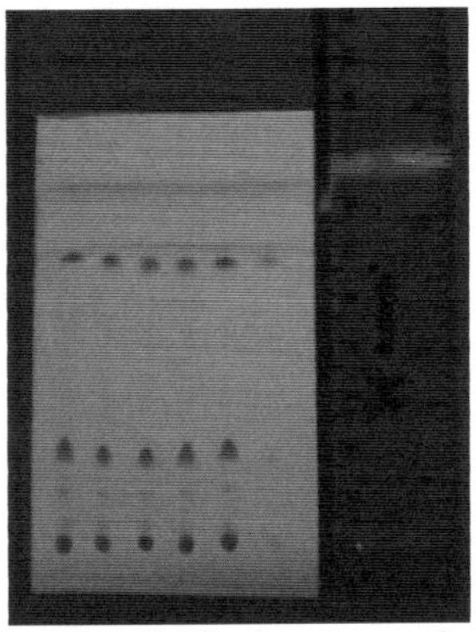

Figure 1. TLC Results of *Gloriosa superba* Immature Seeds

Aquadest-extracted immature seeds of *Gloriosa superba* eluted on solution chloroform:diethylamine (9:1). The elution results were then seen using UV light (Figure 1). The Rf values and the colors between the samples of the eluted compound appear to be the same as the standard colchicine (right). Detection of colchicine compounds was also done by spraying vanillin sulphate to determine the color change of elution compounds formed on silica plates. The spray results show no change in color after being sprayed. This shows that the compound in the extract is colchicine because it has the same color and Rf value as the standard colchicine (right). Determination of levels of colchicine on immature seeds extract using densitometric method. Colchicine content in immature seeds extract is 12.84 μg/μl (± 2.88).

3.2. Old Seeds Sample of Gloriosa superba

Gloriosa superba old seeds sample obtained in March-April 2017 at Krakal Beach, Gunung Kidul. This plant grows wildly and vine over the coastal hills. Old seed sampling is done around the hills of the Krakal Beach entering the summer. *Gloriosa superba* around the Krakal beach usually begins to grow seeds and flowering at the beginning of the rainy season, old after 2-3 months and after that it will dry and then die. The old seeds of *Gloriosa superba* are ripe brown and start to dry. Seed samples were dried using oven at 70°C for 24 hours. The dried samples was extracted using maceration method with aquadest solvent (1:1) for 24 hours with 3 replications. Determination levels of colchicine contained in the extract of old seeds using the TLC-Densitometry method.

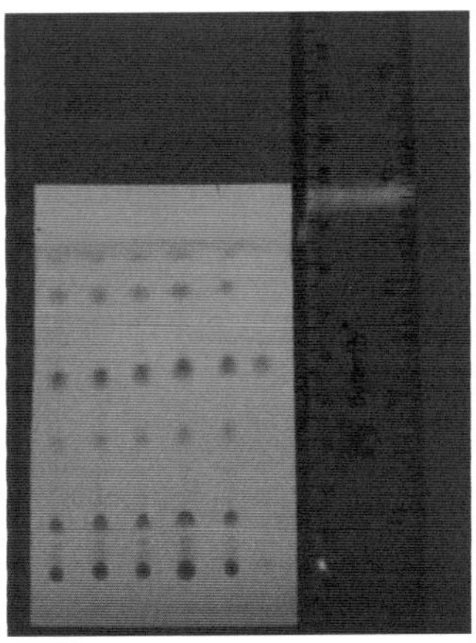

Figure 2. TLC Results of *Gloriosa superba* Old Seeds

The aquadest-extracted *Gloriosa superba* old seeds was analyzed by using TLC method to determine the presence or absence of colchicine compounds in the extract. The extract that has been on silica plate then elution using solution chloroform:diethylamine (9:1). Elution results are more clearly seen using UV light to know the compounds that are formed. The Rf value and the color of the formed compound look the same as the standard colchicine. Spraying silica plate elution using vanillin sulphate is also done to see clearly the color of extract samples and standard colchicine are formed. The results of the aquadest-extracted *Gloriosa superba* old seeds using TLC and spraying vanillin sulphate showed that the compound formed is a colchicine compound with yellow color and the same Rf value between samples with standard colchicine. Determination of the value of colchicine content in the aquadest-extracted samples of *Gloriosa superba* old seeds using densitometry method. The results showed the colchicine levels in aquadest-extracted old seeds was 37.65 μg/μl (± 6.63).

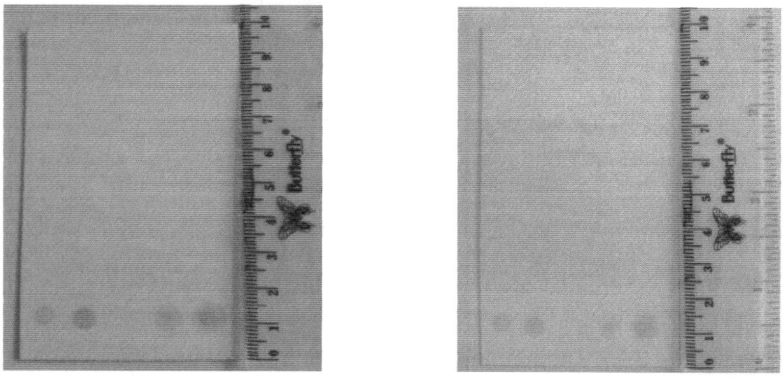

(a) (b)

Figure 3. (a) TLC result, (b) TLC result after spraying with vanillin sulphate

The research that has been done on determination levels of colchicine contained in the aquadest-extracted immature seed and old seed of *Gloriosa superba* showed different grade values. The colchicine level in immature seeds aquadest-extracted was (12.84 μg/μl (± 2.88)) lower than in old seed aquadest-extracted (37.65 μg/μl (± 6.63)) .This was possible that there are factor seed maturity, sampling time and sampling area. Other that may affect is the environmental factor where it grows from both plants. *Gloriosa superba* that grows and develops in Mulur, Sukoharjo is in the yard of the residents and around the garden of the residents. The plant grows creeping under the shade of large trees are not experiencing water shortage. *Gloriosa superba* plant that grows and develops on Krakal Beach, Gunung Kidul is located in the hills around the coast and creeps among other coastal vegetation. The plants grow wildly with the conditions of the hot beach environment and are on a hill that is possible away from the water source. The growth plants of the two different *Gloriosa superba* can be observed in the morphology of leaves, flowers, stems, and seeds. The plants grown in Mulur have greener leaves, striking flowers, short stems and large seeds. In plants grown on the Krakal Beach has a slightly yellowish leaves, smaller flowers, high and large creeping stems, and smaller seeds. In different growing conditions it is possible that the content and level of colchicine in the seeds in the two plants is different. Colchicine is a secondary metabolite compound produced by plants. In severe environmental conditions, secondary metabolites may be higher than plants grown in nutrient-rich environments.

4. Conclusion

Colchicine content on *Gloriosa superba* aquadest-extracted obtained from two different places has different colchicine values. Based on the research that has been done, the factor of time, place and environmental condition of *Gloriosa superba* seed sampling affect the value of colchicine content even though using the same method of extraction. The old seed sample has a higher content of colchicine than the immature seed sample of *Gloriosa superba* plant. Further research can be done by sampling old seeds and immature seeds at once in both places to compare clearly the content of colchicine.

Acknowledgment
This research was supported by the Ministry of Research Technology and Higher Education and Dr. Ir. Yuli Widyastuti as field from Research Center for Medicinal Plant and Traditional Drug, Tawangmangu, Indonesia. *Artemisia annua* seeds in this research were obtained from B2P2TOOT, Tawangmangu, Karanganyar.

References

[1] Jana S, and G S Shekhawat. 2010. Critical review on medicinally potent plant species : *Gloriosa superba*. *Journal Elsevier, Fitoterapia* **82** 293-301
[2] Sivakumar G, Krishnamurthy KV. *Gloriosa superba* L. — a very useful medicinal plant. In: Singh VK, *Govil* JN, Hashmi S, Singh G, editors. *Recent Progress in Medicinal Plants*, Vol. 7. Ethnomedicine and Pharmacognosy, Part II. Texas: Series Sci Tech Pub, Texas, USA; 2002. p. 465–82
[3] Sarin Y K, Jamwal P S, Gupta R K, Atal C K 1974 Colchicine from seeds of *G. superba*. *Curr. Sci.* **43** 87–90
[4] Carta M, Murru L, Botta P, Talani G, Sechi G P, De Riu P L 2006 The muscle relaxant thiocolchicoside in an antagonist of GABAA receptor function in the central nervous system. *Neuro Pharmacol* **51** 805–15
[5] Sandouk P, Bouvier d'Y voire M, Chretien P, Tillement J P, Scherrmann M 1994 Single-dose *bioavailability* of oral and intramuscular thiocolchicosidein healthy volunteers. *Biopharm. Drug Disp.* **15** 87–92
[6] Giavina-Bianchi P, Giavina-Bianchi M, Tanno LK, Ensina L F C, Motta A A, Kalil J 2009 Epileptic seizure after treatment with thiocolchicoside. *Ther. Clin. Risk. Manage.* **5** 635–7
[7] Janbroers J M 1987 Review of the toxicology, pharmacodynamics and pharmacokinetics of *thiocolchicoside*, a GABA-agonist muscle relaxant with anti-inflammatory and analgesic action *Acta Ther.* **13** 221–50
[8] Hilmi A, Sudjarwo, and Asri D 2013 Validasi *Metode* Kromatografi Lapis Tipis-Densitometri untuk Penetapan Kadar Kolkisin Dalam Infus Daun Kembang Sungsang (*Gloriosa superba* Linn.). Departemen Kimia Farmasi, Fakultas Farmasi, Universitas Airlangga. *Jurnal Berkala Ilmiah Kimia Farmasi* **2** 2

IOP Conf. Series: Earth and Environmental Science **142** (2018) 012054 doi:10.1088/1755-1315/142/1/012054

Farmers's perception and strategies for the development of sustainable livelihoods in disaster prone areas

S Anantanyu, Suwarto and Suminah
Faculty of Agriculture, Universitas Sebelas Maret, Surakarta, Indonesia

Email: sap_anan@yahoo.com

Abstract. Empowerment is a strategy to develop and build economy and community in both physically and mentally. One effort is to help generate a plan for community development in disaster-prone areas. For that, this study aims to develop farmers profile, to describe farmer perception towards farming effort and to formulate the empowerment strategies in the development of sustainable livelihoods in disaster prone areas. This study uses mixed methods. Farmers population use in this study were live in two villages of landslides prone areas, that are Beruk and Wonorejo which belongs Jatiyoso Subdistrict Karanganyar regency (Central Java). In depth structured interview was conducted to 150 farmers under Focus Group Discussion (FGD) followed with. data analysis using SWOT analysis. The results showed level of farm management is in anxiety level, perception of farmers toward the availability of agricultural inputs is at a reasonable level and the agricultural information becomes the reduction factors. The result of QSPM matrix calculation through SWOT analysis on livelihood of agricultural, resulting some strategy according to the priority level that are development of conservation farming, strengthening the farmers capacity in agricultural products processing, strengthening farmer groups and improving the performance of agriculture extensionist.

1. Introduction

Agriculture can be considered as an attempt to establish an artificial ecosystem which has function to provide food for human [1]. In order to establish a stable ecosystem, it means the plant must be maintained well on a steady land. This means that every unit of energy and nutrients which removed from the soil in an attempt to obtain agricultural products should be returned to the system in another format. Agricultural system refers to a specific arrangement of farming operations that are managed based on the physical, biological, and socio-economic basis and in accordance with farmers' goals, capabilities and resources owned by farmers [2]. Farming system is a unique agro-ecosystem, because it has a combination of different physical, biological, and human resources. It is further explained that the genetic resources, techniques and strategies that farmers that can be chosen to create, to maintain, and to develop a farming system that are largely determined by the characteristics of the ecosystem.

In developing the agricultural livelihood strategies, the structure of program directed the objectives and outputs for sustainable mechanisms of life with respect to biodiversity conservation [3]. Specifically the following themes are included into strategies for decreasing poverty reduction and sustainable livelihood promotion. The nature of poverty and vulnerability, including causes and indicators, differ from each site. The poor and vulnerable become poor in different ways, at different times, in different places. There are some strategies to overcome the poverty therefore they will be responsive and adaptive to the local situation. The strategies presented here are general approaches that will be addressed to the specific needs for each demonstration site. In Jatiyoso sub district, the condition of agricultural land mostly in the form of dry land with oblique contours so that it has potential landslides. In addition, the dependence of community on agricultural land, as a source of livelihood, has encouraged the exploitation of land so that it causes landslides. This condition leads to complex problems related to livelihoods where the majority of the society are farmers, such as the absence of harvests or agricultural products which cannot be taken evenly throughout the year and lack of farmers' income due to landslides. This condition has an impact on the availability and accessibility of local food. The efforts to restructure pattern of agricultural land use through sustainable livelihood strategy becomes an alternative solution to the problems of farming communities.

The objectives of this research are to arrange the farmers profile in disaster prone areas, to describe farmer perception on agricultural business and to formulate empowerment strategy in sustainable livelihood development in disaster prone areas.

2. Method

The research was conducted in Beruk and Wonorejo Village, Jatiyoso Sub-district, Karanganyar Regency, because most of the agricultural land is disaster prone area. This research using mixed method [4]. In order to obtain the farmer profile data, structured interviews with 15 farmer respondents were performed to get an overview of site condition, general agricultural impression, and in depth interview and farmer involvement in FGD. This combination of data collection techniques [5], will provide a better understanding of rural livelihoods. The quantitative data were analyzed using descriptive statistics, while the qualitative data were done by using SWOT Analysis, namely: First, the input stage by analyzing the external and internal environment then evaluated through IFE and EFE matrix; second, the strategy formulation phase using IE matrix and SWOT matrix [6].

3. Results and Discussion

3.1 Location Overview

Wonorejo Village is located in Jatiyoso District, Kayanganyar Regency. The location is a mountainous land with slope between 30-85%. This location is in 1000 - 2300 mdpl. Overall, the area of Wonorejo village is an area with its breadth 2,045,175 ha. The distance between Wonorejo Village and Jatiyoso Sub-district is about 12 km. Administratively, Wonorejo village has 17 hamlets. Beruk village is located in Jatiyoso sub-district of Kayang-anyar Regency. The location area has a mountainous land with slope between 23-85%. This location is in 850 - 2200 mdpl. Overall, the Beruk village has an area of 690,2740 hectares. The distance between Beruk Village and Jatiyoso Subdistrict is about10 km. Administratively Beruk village has 9 hamlets.

Based on the village monograph data in 2014, the population of Wonorejo Village is 7,076 inhabitants and the population of Beruk Village is 560 peoples. The population in Wonorejo village is higher than Beruk village due to the area of Wonorejo is wider than Beruk village.

3.2 Respondents and Perception to Agriculture

The characteristics of farmers provide an overview about farmers condition as individual and the general condition of farm household. These characteristics include farmer age, the farming duration, and number of family member. The summary of farmer profile (characteristic) is presented in Table 1.

Table 1. The farmer characteristics

No	Description	Wonorejo Village		Beruk Village	
		Frequency (orang)	Percentage (%)	Frequency (total person)	Percentage (%)
1.	Age				
	a. < 30 years old	1	1.33	2	2.67
	b. 30 – 50 years old	42	56.00	43	57.33
	c. > 50 years old	32	42.67	30	40.00
	Total	75	100.00	75	100.00
2.	Education (School level)				
	a. Elementary (6 years)	56	74.67	62	82.67
	b. Junior (9 years)	13	17.33	7	9.33
	c. High (12 years)	4	5.33	6	8.00
	d. University (>12 years)	2	2.67		
	Total	75	100.00	75	100.00
3.	Farming duration				
	a. < 5 years	2	2.67	0	0.00
	b. 5 – 10 years	12	16.00	2	2.67
	c. > 10 years	61	81.33	73	97.33
	Total	75	100.00	75	100
4.	The total of family member				
	a. Small (< 4 people)	39	52.00	25	33.33
	b. Medium (4 – 5 people)	34	45.33	34	45.33
	c. Large(>5 people)	2	2.67	16	21.33
	Total	75	100.00	75	100.00
5.	Number of working family member				
	a. 1 person	4	5.33	39	52.00
	b. 2 people	44	58.67	32	42.67
	c. 2 people or more	27	36.00	4	5.33
	Total	75	100.00	75	100.00

Source: Primary Data Analysis

In the Table 1 above, it can be seen that from 75 respondents in each village, most data which obtained show that the frequency of farmer's age were the age of 30-50 years, while the smallest frequency at age less than 30 years, this condition shows that the low interest of young generation who works in the agricultural sector. The level of formal education of respondents varies from elementary school to university. The frequency of education of farmers is dominated by elementary school graduates (SD) who were as many as 60 people in Beruk Village and 56 people in Wonorejo Village.

The experience of farming is one of the factors which influences production. The experience of farming of the respondents mostly has experience for more than ten years with the number of respondents 73 farmers (97.33%) in the Beruk village and the number of respondents of 61 farmers (81.33%) in Wonorejo Village. The number of family members of the farmer as respondents in Beruk and Wonorejo villages varies considerably. Most of the number of family members ranged from 4 to 5 people with a percentage of 45.33%, while the smallest is the number of family members more than 5 people.

Perception of farmers on agricultural business are reflected in the perception of farming adequacy which seen from farmers assessment toward the availability of production factors owned by farmers. The detailed information can be seen in the Table 2.

Table 2. The distribution of Farmers Based on the Farming Adequacy Perception

No.	Description	Areas							
		Beruk Village				Wonorejo Villlage			
		TT	KT	CT	ST	TT	KT	CT	ST
a.	Farm land for farming	-	22	32	21	-	15	35	25
b.	Family labor that helps farming	-	23	34	18	5	26	32	12
c.	Outsourcing workers who can help	1	4	41	29	2	31	33	9
d.	Quality seeds / seeds use	-	4	59	12	-	24	46	5
e.	Fertilizer factory used	-	9	59	7	-	11	63	1
f.	Organic fertilizer / the cage use	-		37	38	-	14	39	22
g.	Pesticides which are suitable for pest control and disease	-	10	61	4	-	19	49	7
h.	Water / irrigation for watering or irrigating	3	16	35	21	-	5	22	48
i.	Capital (money) to buy production facilities	-	32	42	1	-	36	36	3
j.	Useful agricultural techniques for managing crops and yields	-	32	43	-	3	47	23	2
k.	Agricultural tools needed to manage crops and the result of farming	-	5	68	2	2	32	41	-
l.	Collaborative management	33	27	3	12	1	42	29	3
m.	Agricultural information	61	9	4	1	4	60	9	2

Source: Primary Data Analysis
Description: TT = Not Available, KT = Less Available, CT = Simply Available, ST = Very Available

There are thirteen factors of production which related to the distribution of farmers' perceptions based on the basis of whether it is adequate, available, unavailable or not. From the factors above, the agricultural information are still not enough according to the farmers in Beruk and Wonorejo Villages. The organic fertilizer or cage is the most widely used by farmers in Beruk Village, while water / irrigation is the most available in Wonorejo Village. Farming land for farming in both villages is sufficient, the labor availability, factory fertilizer, pesticides, capital, and agricultural tools are enough too. From the description above about the perception of farmers toward the agricultural business, It shows condition that:

1. Land as the main capital in agricultural availability which is very diverse. There are some farmers with limited land and other with adequate land. The level of land adequacy is strongly influenced by the availability of family labor. The use of land with a relatively high slope for farming has become common. Farmers' awareness for not planting seasonal crops on slopes is still low. Actually some farmers have obtained information about the danger of landslides, but all this time their knowledge has not been able to encourage the farmers to stop doing farming activities which have been done.
2. According to most farmers, they consider that the availability of water is still very adequate, but there are some of them who say that it is less adequate. The damage of environment slowly but surely causes the decreased of water supply. Absorption areas which are on the top used to have some plants and now they decrease their function as agricultural land.
3. The availability of labor, both family and outside workers, is very relative. The agricultural labor is dominantly done by farmers aged 30 years up to over 50 years. Young people (under 30 years old) are mostly not interested in agriculture, prefer to work in the city. Farmers' working experience is more than 10 years old and they acquired it through generation.
4. It is easy to obtain some production facilities, such as seeds, fertilizers, pesticides, and agricultural equipment. This condition causes the perception of farmers toward the adequacy of this production facility. The opening of transportation access (road) facilitates the distribution / sale of production facilities.
5. Two things that are considered inadequate are agricultural information and group cooperation in the management of farming. Agricultural information meant by farmers is the necessary knowledge when farmers face problems, for example: pests or diseases. From the existing farmer groups, they do not show the cooperation between farmers to carry out the joint activities.

Table 3. The SWOT Matrix of Agriculture Livelihoods in Beruk Village and Wonorejo Village Jatiyoso Sub-District Karanganyar

INTERNAL FACTORS / EXTERNAL FACTORS	POWER (S)	WEAKNESS (W)
	1. The land is fertile 2. Road access is available 3. Water sources are available 4. Experience in horticultural cultivation 5. Producing horticultural products mainly vegetables and fruits 6. Production facilities are available 7. Available agricultural cooperatives	1. Have a high slope 2. Young workers happy to wander 3. Lack of farm management cooperation 4. Irrigation channels are inadequate 5. Cultivation technology is not yet available 6. Ability to process agricultural products is not adequate
Opportunity (O)	**Strategy (SO)**	**Strategy (WO)**
1. Agricultural technology is growing rapidly 2. Demand for qualified agricultural products increase 3. Farm location is suitable for tourism object 4. A lot of time to do business outside the farming. 5. There is a farmer group organization	1. Production of pilot farm (S1, S4, O5) 2. Development of partnership cooperation (S7, O2, O3, O5) 3. Improvement of Marketing Cooperation (S5, S6, O2, O1,) 4. Optimizing the availability of facilities in supporting the improvement of product quality (S1, S2, S6, O1, O2) 5. Growing institutional and partnership functions (S7, O4, O5) 6. Strengthening of Farmer Group (S4, S7, O5) (S2, S5,S8, O5) 7. Establishing the cooperation between the cooperative commission with farmer group in managing agriculture productivity so that farmers can easily sell their yields. (S2,S5,S8,05)	1. Strengthening the capacity of farmers in the processing of agricultural products (W6, W3, O2, O5) 2. Strengthening of farmer groups (W3, O5) 3. Procurement of agricultural cultivation technology (W5,W6, O1, O5) 4. Increasing non-agricultural business opportunities in the local area (W2, O4) 5. Improve irrigation channels as agricultural technology develops (W4, O1) 6. Utilization of location as a tourist village (W1, O3)
Threat (T)	**Strategy (ST)**	**Strategy (WT)**
1. Many farmers do not care about landslides 2. The open land is easy to cause the landslide 3. The selling price of the product fluctuates 4. Performance of agricultural extension workers is still low	1. Counseling on conservation farming (S1,S4, T1, T2, T4) 2. Increasing the cooperation among group members in selling crops to obtain stable prices (S5, S8, T3) 3. There is Evaluation on extension worker performance and farmer group (S7, T4).	1. Development of conservation farming (W1, T1, T2) 2. Improvement of Agricultural Extension Performance (W6, T4) 3. The need to foster youth gene-rasi in agriculture so that younger generation is encouraged in advancing their farming in their village (W2, T4) 4. Improvement of irrigation channels to prevent erosion (W1, W4, T1, T2)

Source: Data Analysis Results

3.3 Development Strategy

3.3.1 SWOT analysis To obtain an overview of agriculture as a livelihood, the SWOT analysis which was used in this research include strengths, weaknesses, opportunities, and thread. The data required in the SWOT analysis is obtained through observation, interview, and FGD, which are re-confirmed. Table 3 presents the SWOT Matrix of this research.

Table 4. The Identification of QSPM Matrix of Agriculture Livelihood in Beruk Village and Wonorejo Village, Jatiyoso Sub-District, Karanganyar Regency

Alternative Strategies		1 Development of farming conservation		2 Strengthening Farmers Capacity in the process of agricultural products		**3** Farmers Group Strengthening		**4** Improvement of Agricultural Extension Performance	
Main Factors	**Value**	**AS**	**TAS**	**AS**	**TAS**	**AS**	**TAS**	**AS**	**TAS**
Opportunity									
1. Agricultural technology is growing rapidly	0.1	4	0.4	3	0.3	1	0.1	2	0.2
2. The demand for qualified agricultural products increases	0.09	4	0.36	1	0.09	3	0.27	2	0.18
3. The farm location is suitable for tourism	0.09	4	0.36	3	0.09	1	0.09	2	0.18
4. A lot of time to do business outside the farm.	0.13	3	0.39	4	0.52	2	0.26	1	0.13
5. There is a farmer group organization	0.1	1	0.1	2	0.2	4	0.4	3	0.3
Threat									
1. Many farmers do not care about the danger of landslides	0.13	4	0.52	1	0.13	2	0.26	3	0.39
2. Open land can cause landslide	0.14	4	0.56	1	0.14	2	0.28	3	0.42
3. The selling price of the product is fluctuate	0.07	2	0.14	1	0.07	4	0.28	3	0.21
4. the work performance of agricultural extension is still low	0.1	2	0.2	1	0.1	3	0.3	4	0.4
	1.00								
Power									
1.The land is fertile	0.07	4	0.28	1	0.07	2	0.14	3	0.21
2. Road access is available	0.06	1	0.06	4	0.24	3	0.18	2	0.12
3. Water sources are available	0.09	4	0.36	1	0.09	2	0.18	3	0.27
4. Horticultural cultivation experience	0.08	4	0.32	1	0.08	2	0.16	3	0.24
5. Producing the horticultural products mainly vegetables and fruits	0.1	1	0.1	4	0.4	3	0.3	2	0.2
6. Production facilities are available	0.08	4	0.32	1	0.08	3	0.24	2	0.16
7. Agricultural cooperative commission is available	0.06	1	0.06	2	0.12	4	0.24	3	0.18
Weakness									
1. Has a high slope	0.1	4	0.4	1	0.1	2	0.2	3	0.3
2. Young workers are happy to wander	0.09	3	0.27	4	0.36	1	0.09	2	0.18
3. Lack of farm management cooperation	0.05	2	0.1	1	0.05	4	0.2	3	0.15
4. Irrigation channels are inadequate									
5. Cultivation technology is not yet available	0.05	4	0.2	1	0.05	3	0.15	2	0.1
	0.08	4	0.32	1	0.08	2	0.16	3	0.24
6. The ability to process agricultural products is not adequate	0.09	1	0.09	4	0.36	3	0.27	2	0.18
	1								
TOTAL			5.91		3.2		4.75		4.94

Source: Data Analysis Results

3.3.2 QSPM analysis The QSPM matrix analysis is used to select the main strategies where can be applied among several alternative options of existing strategies. Some strategies are analyzed based on the internal factors (strengths and weaknesses), external factors (opportunities and threats), and the

IOP Conf. Series: Earth and Environmental Science **142** (2018) 012054 doi:10.1088/1755-1315/142/1/012054

weight and total value provided. Analysis of QSPM matrix of agricultural livelihoods in Beruk Village and Wonorejo Village Jatiyoso Sub-district of Karanganyar Regency can be seen in Table 4 as follows:

In Table 4, there are four alternative strategies which are considered as important. Those strategies are development of farming conservation, strengthening the capacity of farmers in the process of agricultural products, strengthening of Farmer Groups, increasing the agricultural extension performance. After the calculation above, the total score for the first strategy is 5.91, second is 3.2, third is 4.75 and fourth is 4.94.

4. Conclusions

The conclusions and recommendations that can be taken are the description of farmers' profiles in both research villages shows that the management level is at the anxiety level, the perception of farmers on the availability of agricultural production facilities, which include: land, labor, seeds / seedlings, fertilizers, pesticides, irrigation, capital, technology, agricultural tools, management, and agricultural information are at sufficient level or they are available, agricultural information becomes factor which less sufficient. According to SWOT analysis resulted some strategies of livelihood development effort (livelihood) of farmer of Beruk Village and Wonorejo Village, Jatiyoso Sub-district, Karanganyar Regency. The results of the QSPM matrix calculation on the SWOT analysis of livelihood of agriculture, it produces strategies according to the main priorities, namely development of farming conservation, strengthening the capacity of farmers in the processing of agricultural products, strengthening farmer groups and improving the performance of agricultural extensionist.

References

[1] Nasution A H 2008 *Pengantar Ilmu-ilmu Pertanian* PT Pustaka Litera Antar Nusa Jakarta
[2] Reijntjes C, Haverkort B and Bayer A W 1999 *Pertanian Masa Depan: Pengantar untuk Pertanian Berkelanjutan dengan Input Luar Rendah* Terjemahan oleh Y Sukoco Kanisius Yogyakarta
[3] Frankenberger T R, Luther K, Becht J and McCaston M K 2002 *Household Livelihood Security Assessments: A Toolkit for Practitioners* TANGO International Inc Arizona
[4] Tashakkori A and Teddlie C (Ed) 2010) *Handbook of Mixed Methods: In Social and Behavioral Research* Terjemahan oleh Daryatno Pustaka Pelajar
[5] Murray C 2001 *Livelihoods Research: Some Conceptual and Methodological Issues* Chronic Poverty Research Centre Department of Sociology University of Manchester
[6] David F R 2012 *Manajemen Strategis: Konsep Terjemahan* Salemba Empat Jakarta

IOP Conf. Series: Earth and Environmental Science **142** (2018) 012055 doi:10.1088/1755-1315/142/1/012055

The effect of shade and vermicompost application on yield and flavonoid levels of Tempuyung (*Sonchus arvensis*)

D P Putri[1], Y Widyastuti [2], W S Dewi[2] and A Yunus[3]
[1]Student of Agrotechnology Studi Program, Agriculture Faculty, Universitas Sebelas Maret, Surakarta, Indonesia
[2]Lecturer of Agrotechnology Studi Program, Agriculture Faculty, Universitas Sebelas Maret, Surakarta, Indonesia
[3]Research center for Medicinal Plant and Traditional Drug, Tawangmangu, Indonesia

E-mail: yunus.uns7@yahoo.com

Abstract. This study aims to determine the effect of shade and vermicompost on yield and level of flavonoids in Tempuyung (*Sonchus arvensis*). The study was conducted in May-August 2016, in Tegal Gede Village, Kab. Karanganyar. Secondary metabolite content analysis was performed in B2P2TOOT laboratory, Tawangmangu. This study used a Completely Randomized Design (CRD) arranged in a split plot. Treatment consists of shade as main plot and the dose of vermicompost as sub plot. Shade treatments are 0% (without shade), 50% and 75%. The dose of vermicompost are 0 g / polybag (control), 250 g / polybag, 500 g / polybag and 750 g / polybag. Each treatment was repeated 3 times and each in experimental unit consist of 2 sample plants. The results showed that the higher levels of shade tend to decrease vegetative growth of plants. Moreover, vermicompost give significant effect on leaf weight, l number, wet and dry weight. The combination of 0% shade and 750 gram vermicompost gave the highest value ($P < 0.05$) to leaf weight ± 186.68 g, wet weight ± 252.08 g and dry weight ± 30,76 gTempuyung. The combination of 50% shade with 0 g vermicompost and 75% shade with 250 g vermicompost show to increasing the content of flavonoid compounds.

1. Introduction

Indonesia has many kinds of medicinal plants that have been developed well or still grows wild. Tempuyung is one of the potential medicinal plants in Indonesia and has long been utilized by the public as herbal treatment [1]. Tempuyung has benefits to treat gout, diuretics, kidney stones, bladder stones, gallstones, swelling drugs, tranquilizers, cough, asthma and bronchitis. Tempuyung grows in wild in the open exposed area to sunlight or less protected and the soil slightly damp, like the edge of the ditch, roadside, between the stones, cliffs and sloping walls [2]. Tempuyung is a pioneer plant that can live in marginal soil that has minimal nutrient or event heavy metal. Tempuyung included in the category of nonnitrophilous species that can grow on low soil N content [3].

Simplicia Tempuyung leaves are commonly useas drug, therefore needs a cultivation techniques to improve the quantity and quality of crude content of TTempuyung leaves. Vermicompost application has a significant effect on root length, leaf fresh weight and dry weight of Tempuyung [1] at dosages of 0.25 kg/ polybag, 0.5 kg / polybag and 0.75 kg / polybag. Effect of shade treatment is expected to

IOP Conf. Series: Earth and Environmental Science **142** (2018) 012055 doi:10.1088/1755-1315/142/1/012055

increase secondary metabolite in Tempuyung, that is flavonoid. The shade treatment of Tempuyung will affect the intensity of the light obtained.

Previous study [4] showed the highest increment of flavonoids (11,42%) is in the shade level of 50% [5]. Tempuyung which receive stress shade plants tend to have higher number of branches and stems than in the absence of shade. Therefore, to study the effect of shade on medicinal plants is very important for cultivation management of medicinal to achieve higher biomass and active ingredient.

2. Methods

2.1. Material and tools

The materials used in this study are seed Tempuyung, media Latosol soil, vermicompost, ethanol, methanol, hexan and ethyl acetate. The tools used in this study include hoe, meter, an oven, a knife or scissors, analytical balance, weighing a kilogram, ruler, camera, plastic tubs, stationery, grinder, plates silica gel 254, chamber, saucer porcelain beaker 50 ml, waterbath, sonicator, a pipette, tweezers and the filter paper.

2.2 Treatment Design

This study uses a completely randomized design (CRD) which are arranged in a split plot and arranged randomly. The treatments consisted of shade as the main plot and doses of vermicompost as a sub plot. Shade consists of 0% (no shade), 50% and 75%. Vermicompost dose consisting of 0 g / polybag (control), 250 g / polybag, 500 g / polybag and 750 g / polybag. Each treatment were repeated 3 times, and each experimental unit consisted of two plant samples.

2.3 Observation of growth

The growth were observed at every week and after harvesting time. The observed factors were number of leaves, leaf weight, leaf length, crown diameter, root length, root weight, dry leaf weight, chlorophyll, wet weight, and dry weight

2.4. Flavonoid analysis

Flavonoid analysis was done by TLC (Thin Layer Chromatography). Thin layer chromatography (TLC) is a method of chemical component separation and development is strongly influenced by the type and polarity of separated chemicals. The common and widely used silicone phase is silica gel mixed with CaSO4 to increase the adhesion of silica gel particles to other widely used absorbent (alumina) auxiliaries are alumina, cellulose powder, starch and sephadex. Parameters in thin layer chromatochoses are retention factor (Rf), a comparison of the distance traveled by the solute to the distance traveled by the mobile phase. The distance traveled by the phase of motion (cm): Distance traveled solute (cm) Rf = (1) Rf price is generally smaller than 1, whereas if multiplied by 100 will be worth 1-100, so this parameter can be used for qualitative calculations in sample testing with thin layer chromato- gos [6].

2.5. Statistic analysis

Data were statistically treated by ANOVA, Duncan's multiple ranges test at probability level 0.05 was performed to compare mean.

3. Results and discussion

3.1. General Conditions Research

This research was conducted at B2P2TOOT Medicinal Plant Garden, Tanjungsari Village, Tegal Gede Village, Karanganyar. Geographical location of the research site at 7036,687 'S and 110057,766' E. The study site is altitude of 215 meters above sea level. The temperature at the study site ranged from

27^{0}C-30^{0}C. Type of soil in this area is a red latosol land, textured clay and with the content of nutrients that are.

3.2. Weight leaves

The leaves are a very important part of the plant, where the photosynthesis take place to obtain energy. There is no interaction (P <0.05) between the shade and vermicompost. The highest interaction generated by N0V3 (without shade and vermicompost 750 g) of ± 186.68 g. N2V0 combination treatment (75% shade and without vermicompost) provides the lowest leaf weight values of ± 12.25 grams.

Table 1. Interaction Shade and Vermicompost on Leaf Weight (g)

Treatment	Without Vermicompost (V0)	Vermicompost 250 g (V1)	Vermicompost 500 g (V2)	Vermicompost 750 g (V3)
Without Shade (N0)	76.78 ± 35.59	168.66 ± 72.42 b	154.42 ± 84.05 b	186.68 ± 89.14 b
50% shade (N1)	16.13 ± 3.88	26.34 ± 12.1 a	34.80 ± 10.26 a	33.09 ± 22.83 a
Shade 75% (N2)	12.25 ± 3.8	16.33 ± 14.7 a	12.81 ± 16.4 a	16.63 ± 5.12 a

Previous study [1] explains that vermicompost contain cytokines that can increase endogenous cytokines, either directly or indirectly to regulate the production of hormones to stimulate cell division and form new shoots, which can affect the leaf area and leaf dry weight. While the greater level of shade, the leaves lose weight, plants growing under shade will have leaves that are wider and thinner than the leaves of plants that live in the open [7]. Giving light shade that minimize the entry of plants to cause the slow process of photosynthesis and growth Tempuyung. Variable observation leaf weight was positively correlated with a variable number of leaves *($r = 0.985$ **)*.

Table 2. Interaction Shade and Vermicompost on the Number of leaves (leaf)

Treatment	Without Vermicompost (V0)	Vermicompost 250 g (V1)	Vermicompost 500 g (V2)	Vermicompost 750 g (V3)
Without Shade (N0)	25 ± 9.98 b	46.33 ± 2.46 c	42.16 ± 3.40 c	45.5 ± 20.3 c
50% shade (N1)	8.83 ± 3.12 a	10 ± 1.8 ab	13 ± 6.4 ab	10 ± 2.17 ab
Shade 75% (N2)	7.33 ± 6.08 a	8.5 ± 13.93 a	6 ± 5.89 a	10.5 ± 2.17 ab

Based on the interaction table shows the combined treatment of N0V3 (without shade and vermicompost 750 grams) shows the average value of the highest number of leaves 45.5 ± strands. Combination treatment N2V2 give an average value that is the lowest number leaves ± 6 strands. [8], shelter reducing solar light energy needed by plants to grow and thrive. This will lead to reduced photosynthetic production resulting in a reduction in the formation of the leaves. According to [9] the fertilization of vermicompost causes the plants to flourish, the stems are higher, the leaves are dark green and the leaf texture is good compared to the plants without the vermicompost. According to the previous study improvement of growth and yield as a result of vermicompost application is due to the improvement of soil quality and the availability of micro and macro nutrients [10], as well as increased soil microbial activity. Occurs a positive correlation between the number of leaves and leaf length *($r = 0.825$ **)*.

Table 3. Effect of the Long Leaf Shade (cm)

Shade (%)	Long Leaf (cm)
0	31.067 ± 3.9 c
50	25.125 ± 6.11 b
75	18.500 ± 6.22 a

The application of shade showed the value of a long decline leaves on each level of shade. The application vermicompost did not show the effect of significantly different ($P > 0.05$) to the length of the leaf. This is reinforced by the research [11] that light passing or emanating from adjacent foliage makes plants grow to support the specific genetic and physiological processes of the leaves. Changes in gene expression underlie the adaptation, or the avoidance of light compromised by the environment or it can be said that the growth response in the environment on the light manipulated in direct sun avoidance responses, can physiologically inhibit the growth of leaves. The leaves are positively correlated with crown diameter ($r = 0.928$ **), if the value of the leaf length increases, the diameter of the leaf canopy is also increasing. The longer the canopy showing very good plant growth, because growth header that lengthening the more leaves to photosynthesize.

Table 4. Effect of the Diameter header Shade Leaf Total (cm)

Shade (%)	The longest canopy (cm)
0	39.7448 ± 9.33 c
50	29.2057 ± 4.06 b
75	23.2813 ± 9.30 a

Shade treatment results were significantly different ($P < 0.05$), treatment without shade has an average value of the highest crown diameter ± 39.7448 cm. Fifty percent of shade treatment had an average value ± 29.2057 cm, while 75% shade treatment had the lowest average value that is equal to ± 23.2813 cm. Vermicompost treatment to variable diameter canopy provides results that are not significantly different *($p > 0.05$).*

3.3. Chlorophyll levels

Chlorophyll is the green substance found in all green plants. Shade treatment showed shade intensity significantly *($p < 0,05$)*, whereas vermicompost treatment did not significantly affect chlorophyll content ($p > 0.05$). There is no interaction between shade intensity and vermicompost dose to chlorophyll content of Tempuyung.

Table 5. Effect of Shade to Chlorophyll levels

Shade (%)	Chlorophyll levels (%)
0	38.10 ± 1.35 b
50	34.50 ± 2.19 a
75	32.98 ± 2.19 a

Based on Table 5, the highest chlorophyll content indicated by the plant with no shade that is equal to ± 38.10%, whereas treatment with shade intensity of 50% and 75% that is equal to 33.74%. This because the higher the intensity of irradiation will further increase the chlorophyll content of leaves that make the process of photosynthesis increases. This study is in line with [12], which examined the effect of light as a single factor on Myrtus communis, which resulted in the treatment without shade giving higher levels of chlorophyll compared to the 70% shade treatment. Vermicompost treatment did not significantly affect the levels of chlorophyll. Results of correlation analysis showed that the leaf chlorophyll positively related *($r = 0.754$ **)* with a weight Tempuyung leaves. Increased chlorophyll will increase the rate of photosynthesis which accelerate the formation of the leaves that are part of the plant fresh weight [13].

3.4. Roots of Tempuyung

Root as the entry point of mineral (translocate nutrients) from the soil to get to all parts of the plant. Shade treatment against root length variable result significantly different ($P < 0.05$). Table 6 shows the treatment without shade give the highest root length of 44.52 cm, while the length of the lowest root level of 75% was obtained auspices of 15.12 cm. Effect of vermicompost did not give significantly different results against Tempuyung root length ($P > 0.05$).

Table 6. Effect of Shade to Root Length (cm)

Shade (%)	Root Length (cm)
0	44.52 ± 10.75 c
50	22.38 ± 8.67 b
75	15.12 ± 6.31 a

This is presumably due to the small light that can penetrate paranet so stunted plant growth. In the previous study [14] states the auspices cause absorption of light for photosynthesis is very low and cause growth to slow down. The analysis results show a root length correlated positively *(r = 0.969 **)* with root weight. This can be understood the longer the root, it will increase the weight of plant roots Tempuyung. The shade treatment was significantly different ($P < 0.05$) against the weight of root Tempuyung.

Table 7. Effect of Shade on Root Weight (g)

Shade (%)	Root weight (g)
0	39.64 ± 12.8 b
50	7.05 ± 4.41 a
75	3.91 ± 3.35 a

Treatment without shade giving an average of the highest root weight of ± 39. 64 g. While the shade 50% and 75% was significantly different to the weight of the roots with an average of ± 5.48 g. According to recent study [15] when plants lack light then the process of photosynthesis becomes low, consequently the photosynthesis results can be distorted by the process of respiration, food reserves are reduced so that plant growth is inhibited.

Table 8. Effect of Vermicompost on Root Weight (g)

Vermicompost (g)	Root weight (g)
0	11.28 ± 12.12 a
250	19.89 ± 24.40 b
500	17.45 ± 16.21 b
750	18.84 ± 19.66 b

Treatment without vermicompost gives the average of the lowest root weight of ± 11.28 g. While the provision of vermicompost significantly different to the weight of the roots with an average of ± 18.72 g. [16] stated that vermicompost contains plant-growing hormones. The hormone not only stimulates rooting in the graft, but also spurs the growth of plant roots in the soil, spurring new branch twigs on the stems and branches of trees, and spurring leaf growth.

3.5. Wet weight

Wet weight is the size of the resulting pile of photosynthesis in plants in the form of heavy plant that still contains moisture is not constant. [17] explains that wet weight and dry weight of plants play an important role in plant growth, because the plant can grow strongly if the wet weight and dry weight of the plant is high. Vermicompost shade treatment and significantly different ($P < 0.05$) on the wet weight. There is no interaction between the two treatments.

Table 9. Interaction Shade and Vermicompost on wet weight (g) at Tempuyung

Treatment	Without Vermicompost (V0)	Vermicompost 250g (V1)	Vermicompost 500g (V2)	Vermicompost 750g (V3)
Without Shade (N0)	111.52 ± 48.1	235.19 ± 72.4 b	235.27 ± 121.7 b	252.08 ± 100.22 b
50% shade (N1)	22.8 ± 6,87	35.86 ± 17.1 a	50.60 ± 14.52 a	44.23 ± 31.27 a
Shade 75% (N2)	17.83 ± 6.32	18.63 ± 20.3 a	22.22 ± 25 a	22.26 ± 7.88 a

The results of the wet weight of the land without a shade higher than in land shaded. This condition is due to the growth of wet weight, Tempuyung plants need the help of good photosynthesis process, shade treatment can accumulate photosynthate products at light level required in the process of photosynthesis in plants [18]. Wet weight gain in this study caused also by vermicompost which has the ability to increase the stability of aggregates and increase total porosity of the soil.

The input of organic matter contributes nutrients to the soil [19], the higher the nutrient content in the soil can increase crop yields. Nutrients will be used to accelerate the process of photosynthesis. Based on the results of the correlation, so that the weight of high light intensity is needed. With the wet plant Tempuyung positively correlated *($P < 0.05$)* with all variables except observation at variable levels of correlation and the extract did not occur at lower levels of flavonoids are negatively correlated.

3.6. Dry weight

Dry weight measurement is part of plant biomass measurements. There is interaction and vermicompost treatment giving shade to dry weight Tempuyung. The value obtained N0V3 treatment that is equal to 31.36 grams, while the value of the lowest dry weight obtained N2V1 treatment with an average value of 2.98 g. According to the table 13 can be seen that the higher the intensity the lower the auspices of the dry weight of the resulting, contrary to the provision of higher vermicompost will raise Tempuyung plant dry weight.

Table 10. Interaction Shade and Vermicompost on Dry Weight (g) on Tempuyung (Sonchus arvensis)

Treatment	Without Vermicompost (V0)	Vermicompost 250 g (V1)	Vermicompost 500 g (V2)	Vermicompost 750 g (V3)
Without Shade (N0)	10.81 ± 4.67	31.36 ± 10.65 c	20.93 ± 2.06 b	30.76 ± 7.91 c
50% shade (N1)	3.46 ± 0.7	4.72 ± 1.86 a	6.21 ± 1.59 a	4.85 ± 3.48 a
Shade 75% (N2)	3.16 ± 0.94	2.98 ± 1.92 a	3.94 ± 5.17 a	3.46 ± 1 a

This is in line with the statement [20] which states that the granting of enhanced shade intensity by 75% causing the dry weight of various types of plants is lower. This is due in addition to the total leaf area is reduced, as well as the depletion of the leaves, which can cause leaves to shade each other and result in a decrease in the rate of photosynthesis. Also plant dry weight is closely associated with the availability of nutrients in growing media. In addition, vermicompost can also improve soil physical properties such as soil structure associated with the development of plant roots. Based on the results of the correlation, Tempuyung plant dry weight were positively correlated ($P < 0.05$) with all variables except observations on the extract did not happen correlation levels and negatively correlated with the levels of flavonoids.

3.7. Analysis of Flavonoids

The content of the extract is a quantitative calculation by comparing the extract obtained with the weight of the starting material. The picture shows the extract content of Tempuyung, which is the highest result obtained by treatment of N2V2 (75% shade and 0.5 kg of vermicompost fertilizer) by 22%, while the N0V1 treatment obtained the lowest percentage rendement of 14%.

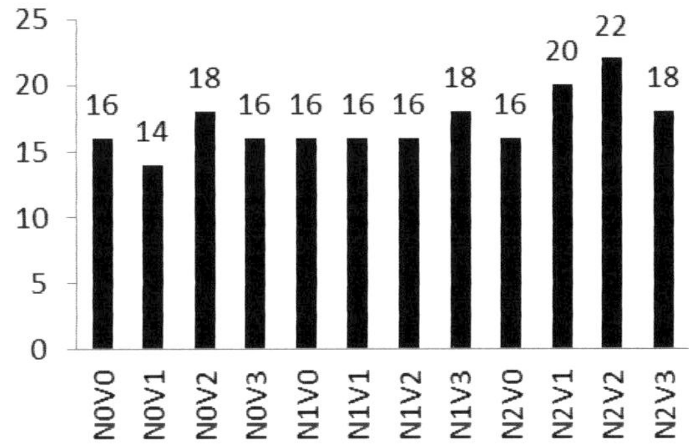

[a] N0= shade 0%; N1= shade 50%; N2= shade 75%; V0= 0 g; V1= 250 g; V2= 500 g; V3= 750 g

Figure 1. Graph Extract yield levels on Tempuyung

Levels of this extract show the content of active compounds or secondary metabolites present in the plant. The greater the level of extract, the content of active compounds in the plant more and more, while the fewer levels of extract the active compound content in the plant the less. It can be understood that in this extraction method, using ethanol which serves to dissolve the resin, fat, oil, fatty acid compounds which then this solution is vaporized in the water bath and leaving the residue [21]. Based on the results of the correlation between extract levels with flavonoids in Tempuyung, it is known that there is no correlation between the two variables. It can be understood that the value of the extract is not necessarily directly proportional to the level of flavonoids Tempuyung, because in an extract contains a variety of other secondary metabolites.

Based on a qualitative analysis was conducted using TLC (Thin Layer Chromatography) dispersed in the form of pigment compounds, it appears that each treatment shows the results of size and thickness of different colors. Identification of flavonoid compounds in this study using Thin Layer Chromatography [6]. TLC used in the identification of flavonoids is made of silica gel with a size of 9 cm x 13 cm. Plate TLC silica gel is activated by means of the oven at 100° C for 1 hour. The aim is to remove the water contained in the TLC plate. Condensed extract extraction results then diluted with hexane as much as ± 5 ml. Hexane is used to dissolve fat and fatty acids. Then spotted along the plate by using a micro pipette at a distance of 1 cm from the bottom line and each treatment was given within 1 cm. Furthermore, in the elution using eluent gives the best results on the TLC separation, namely n-hexane - ethyl acetate - methanol ratio (5: 5: 1) [6].

In Figure 2. the chromatogram detection at 366 nm UV light showed a few patches of color that indicates the presence of a compound. The resulting number of spot colors signify the amount of active compound contained in a plant. The thicker the spot colors produced by the thicker the active compound content in plants. The test results are read on a flavonoid UV lamps generate Rf 0.5 (white column picture 3) with a red stain that is suspected of flavonoids and previous study[22] states that if the form of orange, red, purple or yellow, positive means flavonoids.

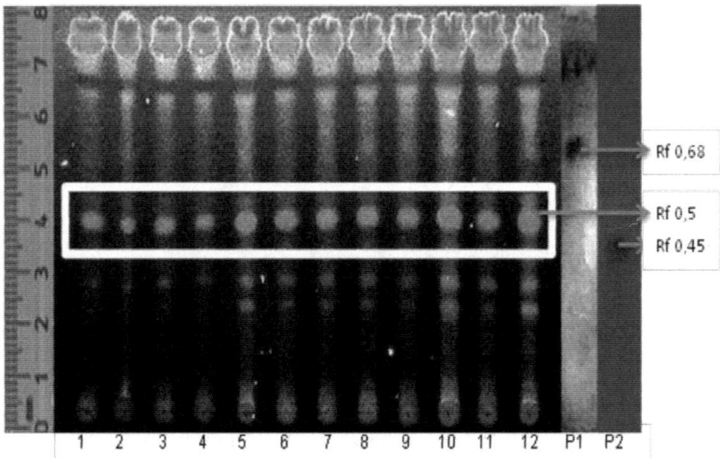

Information :
The white box is a flavonoid compound with Rf 0.5
P1: is a comparison flavonoids isodeoxifelantopin with Rf 0.68
P2: is a comparison flavonoid quercetin with Rf 0.45
Figure 2. Cross-section on Chromatography Silica Gel Thin Layer Chromatography

Plants evolved in various ways to survive from stress, one of them by producing secondary toxic metabolites. Poisonous metabolites generally accumulate in vacuoles, extracellular trichomes, or extracellular cells. Glycosylation is an important modification that occurs in various secondary metabolite compounds. Based on the origin of the biosynthesis, the plant's natural metabolite products can be divided into three main groups, namely terpenoids, alkaloids, and phenylpropanoids as well as groups of antioxidant phenolic compounds [23]. There are two groups of sources of antioxidants, namely synthetic antioxidants (antioxidants obtained from the synthesis of chemical reactions) and natural antioxidants (antioxidants extracted natural ingredients or contained in natural ingredients). Natural antioxidants come from phenolic compounds such as flavonoids. Flavonoids are secondary metabolites produced by plants [24]. These compounds can be toxic to other organisms, which work by disrupting the function of cell proteins. Some metabolites interact with molecules that have basic cellular functions, such as DNA and proteins involved in cell division [25]. [26] mentions that the formation of secondary metabolites governed by nutrition, decreased growth rate, feedback control, enzyme inactivation, and enzyme induction.

According to the table 11, it is known that the treatment has the greatest scoring flavonoids contained in N2V1 treatment (shade level of 75% and vermicompost 250 g) and mistreatment N2V3 (shade level of 75% and 750 g of vermicompost). Other researchs support this finding was carried out by [27], namely that examines the effects of vermicompost fertilizer and other organic fertilizers on the tuber timber. The content of phenolics and flavonoids were significantly higher ($p < 0.01$) in vermicompost than mineral fertilizers and compost. Metabolite secondary is produced more when the plant in a state gripped, but if the plant is too stressed or otherwise the plant will produce fewer secondary metabolites [28].

Based on the correlation results showed that levels of flavonoids correlated negatively on all variables observation and did not correlate with the levels of the extract. Although fertilizer able to improve secondary metabolites, but according to the previous finding [29] the expression of secondary metabolites depends not only on the increase in plant biomass but also depends on the activity of the enzyme. Because secondary metabolites is the result of the synthesis and degradation of enzymes that occur during the process of plant metabolism. Differences content of plant secondary metabolites depending on how plants adapt to the environment, such as light factor and how plants absorb nutrients [26]. The light shade conditions contain UV-A rays, blue, green and red light slightly, but rich in infrared light [30]. This affects on the production of active ingredients contained in the plant.

Table 11. Result and Scoring Calculation Spot Size Color Flavonoids in Tempuyung (%)

Code	Treatment	Spot Size (%)	Scoring Color (%)	scoring Flavonoids	Harkat
P1	Standards	4	3	7	moderate
1	N0V0	2	3	5	Low
2	N0V1	2	2	4	Low
3	N0V2	3	3	6	moderate
4	N0V3	2	2	4	Low
5	N1V0	4	4	8	Many
6	N1V1	4	3	7	moderate
7	N1V2	4	3	7	moderate
8	N1V3	4	3	7	moderate
9	N2V0	4	3	7	moderate
10	N2V1	4	4	8	Many
11	N2V2	4	3	7	moderate
12	N2V3	4	3	7	moderate

Information :

Scoring diameter spot	Scoring color:	Scoring flavonoids:
1: diameter 0-24%	1: very pale	2-3: very low
2: diameter of 25-54%	2: pale	4-5: Low
3: diameter of 55-74%	3: light	6-7: medium
4: diameter 75-100%	4: very light	8: lots

4. Conclusion

The higher the level of shade giving up to 75% tends to lower the growth of plants, including ± 14.51 g of leaf weight, leaf number ± 8 strands, leaf length ± 18.5 cm ± 23.28 cm crown diameter, root length ± 15 12 cm, wet weight ± 20.34 g, and ± 3.39 g dry weight and chlorophyll content ± 32.98%. Giving vermicompost at a dose of 750 g / plant gives the highest yield to the weight of leaf ± 78.80 grams, number of leaves ± 22 strands, Gross weight ±106.33 g. Combination treatment without shade and vermicompost 750 g provides the highest scores ($P < 0.05$) to the weight of leaf ± 186.68 g, ± 252.08 g wet weight and dry weight of 30.76 g± Tempuyung. The combination treatment of 50% shade with 0 g vermicompost, and 75% shade with 250 g vermicompost, increasing the content of flavonoid compounds.

Acknowledgement
This research was supported by the Medicinal Plant and Traditional Drug, Tawangmangu, Indonesia.

References

[1] Arifin H, Anggraini N, Handayani D et al 2006 Standarisasi ekstrak etanol daun *Eugenia cumini* Merr *J Sains Teknik Farmasi* **11**(2):88-93

[2] Awad M A, Patricia S W, Anton de Jager 2001 Effects of light on flavonoid and chlorogenic acid concentrations in the skin of 'Jonagold' apples *Scientia Horticulturae* 2001 ; 88: 289-298

[3] Dalimartha, S 2005 Tanaman Obat di Lingkungan Sekitar Cet 1 Puspa Swara Jakarta

[4] Dalimunthe, Isral, Meiriani 2015 Respons pertumbuhan bibit kakao (Theobroma cacao L) terhadap pemberian pupuk organik *J Online Agroekoteknologi* **3**(1):188-197

[5] Departemen Kesehatan Republik Indonesia 2008 Farmakope herbal Indonesia Indonesia

[6] Gatut W A S, Sundari, 2011 Perubahan karakter agronomi aksesi plasma nutfah kedelai di lingkungan ternaungi *J Agron* **39**:1-6

[7] Ghulamahdi M, Aziz S, Nirwan 2008 Peningkatan laju pertumbuhan dan kandungan flavonoid klon daun dewa (Gynura pseudochina (L) DC) melalui periode pencahayaan *Bul Agron* **36** (1) : 40-48
[8] Kurniawati A, Darusman, Rani 2005 Pertumbuhan,produksi dan kandungan riterpenoid dua jenis pegagan (Centella asiatica L (Urban)) sebagai bahan obat pada berbagai tingkat naungan *Bul Agron* **33** (3): 62-67
[9] Manivannan S, Balamurugan, Parthasarathi, Gunasekharan and Ranganathan, 2009 Effect of vermicompost on soil fertility and crop productivity beans (Phaseolus vulgaris) *J Environ* **30** (2): 275-281
[10] Mariana PR, Sipayung M, Sinuraya 2012 Pertumbuhan dan pengaruh produksi bawang merah (Allium ascalonicum L) dengan pemberian vermikompos dan urine domba *J Online Agroekoteknologi* **1**(1)
[11] Octaviani I, Noly Z A, Suwirmen 2016 Pertumbuhan dan kadar artemisinin cina baru(Artemisia Vulgaris L) pada intensitas cahaya dan komposisi media tanam yang berbeda *J Biocelebes* **10**(2):43-51
[12] Omar N F, Hassan S A, Yussof U K et al 2012 Phenolics, flavonoids, antioxidant activity and cyanogenic glycosides of organic and mineral-base fertilized cassava tubers *J Molecules* **17**, 2378-2387
[13] Pantilu L I, Feky R M, Nio S A et al 2012 Respons morfologi dan anatomi kecambah kacang kedelai (Glycine max (L) Merill) terhadap intensitas cahaya yang berbeda *J Bioslogos* **2** (2): 79-87
[14] Salisbury F B, Ross 1991 Fisiologi Tumbuhan Bandung: ITB
[15] Siswanto U, Sukarjo E, Risnaily 2004 Respon tanaman Tempuyung (Sonchus Arvensis L) pada berbagai takaran dan aplikasi vermikompos *J Ilmu-Ilmu Pertanian Indonesia* **6** (2): 83-90
[16] Suharja, Sutarno2009 Biomassa, kandungan klorofil dan nitrogen daun dua varietas cabai (Capsicum annum) pada berbagai perlakuan pemupukan Nusantara Bioscience 1: 9-16
[17] Sukarjo E I 2004 Toleransi beberapa jenis Curcumaspp Terhadap intensitas naungan *J Ilmu-Ilmu Pengetahuan Indonesia* **4**(2): 97-105
[18] Tripatmasari M, Wasonowati C, Aliyanti VR 2010 Pemanfaatan naungan dan pupuk kotoran sapi terhadap pertumbuhan dan kandungan triterpenoid pegagan (Centella asiatica L) *J Agrovigor* **3**(2):137-145
[19] Wattimena, G A 1992 Bioteknologi Tanaman Bogor: PAU Institut Pertanian Bogor Bogor
[20] Widiastoety D, Bahar FA 2000 Pengaruh intensitas cahaya terhadap pertumbuhan anggrek dendrobium *J Hort* **5** (4):72-75
[21] Sinha R K, Agarwal S, Chauhan K et al 2010 The wonders of earthworms & its vermicompost in farm production: Charles Darwin's 'friends of farmers', with potential to replace destructive chemical fertilizers *Agricultural Sciences* **1**, 76-94 doi:104236/as201012011
[22] Wang Y, Folta K M 2013 Contributions of green light to plant growth and development *American Journal of Botany* **100**(1): 70–78
[23] Mendes M M, L C Gazarini, ML Rodrigues 2001 Acclimation of Myrtus communis to contrasting Mediterranean light environments: effects on structure and chemical composition of foliage and plant water relations *Environ Exp Bot*, **45** pp 165-178
[24] Mathivanan S, Chidabaram A, Sundaramoorthy P, Bakiyaraj R 2012 Effect of vermicompost on growth and yield of groundnut (Arachis hypogaea L) *Inter J Environt Biol* **2**(1):7-11
[25] Pandey S, Katiyar 2010 Determination and Comparison of The Curcuminoid Pigments in Turmeric Genotypes (Curcuma Domestica Val) High performance Liquid Chromatography *J Pharm Pharm Sci* **2** (4) : 125-127
[26] Riesinger P, Hyvönen T 2006 Impact of management on weed species composition in organically cropped spring cereals *J Biological Agriculture & Horticulture* **24** (3) : 257-274
[27] Croteau R, Kutchan T M, and Lewis N G 2000 Natural products (secondary metabolites) *Biochemistry & Molecular Biology of Plants* **24**:1250-1318

[28] Saija A, Scalese M, Lanza M, Marzullo D, Bonina F, Castelli F 1995 Flavonoids as antioxidant agents: importance of their interaction with biomembranes *Free Radic Biol & Med* **19**(4):481-486
[29] Sirikantaramas S, Yamazaki M, Saito K 2008 Mechanisms of resistance to self-produced toxic secondary metabolites in plants *Phytochem Rev* **7**:467-477
[30] Nofiani R 2008 Artikel ulas balik: Urgensi dan mekanisme biosintesis metabolit sekunder mikroba laut *J Natur Indonesia* **10**(2):120-125

IOP Conf. Series: Earth and Environmental Science **142** (2018) 012056 doi:10.1088/1755-1315/142/1/012056

The effect of cutting origin and organic plant growth regulator on the growth of Daun Ungu (*Graptophyllum pictum*) through stem cutting method

S P Pratama[1], A Yunus [2], E Purwanto [2] and Y Widyastuti[3]
[1]Student of Agrotechnology Studi Program, Agriculture Faculty, Universitas Sebelas Maret, Surakarta, Indonesia
[2]Lecturer of Agrotechnology Studi Program, Agriculture Faculty, Universitas Sebelas Maret, Surakarta, Indonesia
[3]Research center for Medicinal Plant and Traditional Drug, Tawangmangu, Indonesia

E-mail: yunus.uns7@yahoo.com

Abstract. *Graptophyllum pictum* is one of medical plants which has important chemical content to treat diseases. Leaf, bark and flower can be used to facilitate menstruation, treat hemorrhoid, constipation, ulcers, ulcers, swelling, and earache. *G. pictum* is difficult to propagated by seedling due to the long duration of seed formation, thusvegetative propagation is done by stem cutting. The aims of this study are to obtain optimum combination of cutting origin and organic plant growth regulator in various consentration for the growth of Daun Ungu through stem cutting method. This research was conducted at Research center for Medicinal Plant and Traditional DrugTanjungsari, Tegal Gede, Karanganyar in June to August 2016. Origin of cuttings and organic plant growth regulator were used as treatments factor. A completely randomized design (RAL) is used and data were analyzed by F test (ANOVA) with a confidence level of 95%. Any significant differences among treatment followed with Duncan test at α = 5%. The research indicates that longest root was resulted from the treatment of 0,5 ml/l of organic plant growth regulator. The treatment of 1 ml/l is able to increase the fresh and dry weight of root, treatment of 1,5 ml/l of organic plant growth regulator was able to increase the percentage of growing shoots. Treatment of base part as origin of cuttings increases the length, fresh weight and and dry weight of shoot, increase the number of leaves. Interaction treatment between 1 ml/l consentration of organic plant growth regulator and central part origin of cuttings is capable of increasing the leaf area, whereas treatment without organic plant growth regulator and base part as planting material affects the smallest leaf area.

1. Introduction

Indonesia has many plant species that contain high efficacy for natural medicines that can be use to heal various diseases that are free from side effects. Pharmacological studies on crude extracts of Daun Ungu plant are reported for anti-inflammatory, oxytocic and anti implantation activities [1]. Medicinal plants have important benefits in the world of health and have been used by humans since ancient times, one of them is Daun Ungu (*G. Pictum*). Daun Ungu is one of the traditional herbs that usually grows in Indonesia as a bush that has a chemical and nutritious content to treat the disease [2]. Leaves can be used to treat wounds, swelling, ulcers, skin diseases, liver and ears [3]. Daun Ungu containing saponins, tannins, and flavonoids glycosides [4]. The presence of flavonoid extract cause a decrease in blood sugar levels [5].

Daun Ungu is difficult to propagate with seeds because no seeds formation and require long time to obtain the seed, therefore it is done by vegetative propagation using stem cutting. The cutting is important because of the ease of reproduction [6]. Vegetative propagation is generally chosen due to the difficulty of obtaining seeds in continuity according to irregular season and seed keeping period. The benefits of propagation using cutting methodsare capable of produce similar plant to its parent in a relatively short time and simple [7]. The origin of cuttings were taken from the shoots, middle and the base. Those origin have a different C/N ratio and chance of growing cuttings as well. The origin of cuttings for multiply vegetatively should come from the parent tree which its growth rate has been known, as well as the quality and quantity of its production.

Plant growth regulator called ZPT, are the organic non-nutrient compounds that affect to the physiological processes of a plant. Organic ZPT is more environmental friendly, easy to obtain, safe to use and cheaper. In this study organic ZPT containing cytokines (zeatin and kinetin), auxin (IAA), and gibberellins was used. Cytokinin works for the growth of side shoots, auxin serves to trigger cell elongation in the end meristem and roots. Moreover, Giberelin trigger the elongation of stems and accelerate cell division.

2. Methods

The research was conducted in garden of medicinal plants B2P2TOOT, Tanjungsari, Tegal Gede, Karanganyar. Research in the field was done for three months in June until August 2016. The geographic location of the research location is at -7 0 36.687 'S and 110 0 57.766' E. The altitude of this research location is about 215 meters above sea level. Temperature is ranging from 27°C - 30°C.

Factorial experimental design with combination of 2 factors of treatments that are organic plant growth regulator and origin of cuttings treatments. The first factor is organic ZPT (S) consisting of four levels S0 (Without organic ZPT), S1 (organic ZPT 0.5 ml / L), S2 (organic ZPT 1 ml / L) and S3 (organic ZPT1.5ml / L). The second factor is the origin of cuttings (B) consisting of three levels that are B1 (Shoots) B2 (Middle Section) B3 (The base). The origin of this cuttings comes from the parent plants of Daun Ungu that have a lifespan of approximately 3 years. The origin of this cuttings each section has a number of different nodes. The top part has 8 nodes, the middle has 6 nodes, and the base has 4 nodes. Observed data will be analyzed based on F (ANOVA) test with 95% confidence level. Any significant different will be proceed to DMRT test at 5%.

3. Results and Discussion

3.1. Shoot Length

Shoots are newly grown part of plants to form a new part that can be stems, juvenile leaves, prospective flowers and fruit candidates. Length of shoot is used as parameter to determine whether the plant is growing or not.Calculations are done from the base to the tip of the shoot. Observation of shoot length is done in the last week which is the 8th week or when cuttings are ready to be moved. The treatment of cuttings origin gives a significantly different result to shoot length variables. The top of the shoot has the lowest average shoot length than the other parts, which is 2.05 cm. Secondly, the middle section has an average shoot length of 6.61 cm and the base has the highest average shoot length with a value of 9.35 cm (Figure 1).

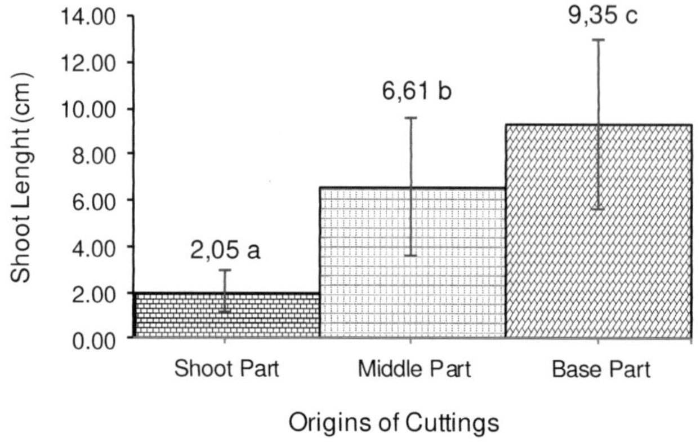

Figure 1. Histogram of the effect of origin of cuttings to the length of shoot

Based on these results then the best origin of cutting to support the long shoots is the base of shoot. This happens because the base has larger diameter of the stem than the tip and the center of shoot. Large trunk diameter has food storage more than the smaller diameter [8]. The diameter of the base of the stem has the largest size so the food or energy reserves to support the growth of shoot length is greater so that the cells in the area of elongation will grow longer. In Figure 1 the length is smaller as the stem goes to upper part, this is due to the food storage in upper part is getting smaller so that the growth of shoot is low as well.

3.2. Percentage of Growing Shoots

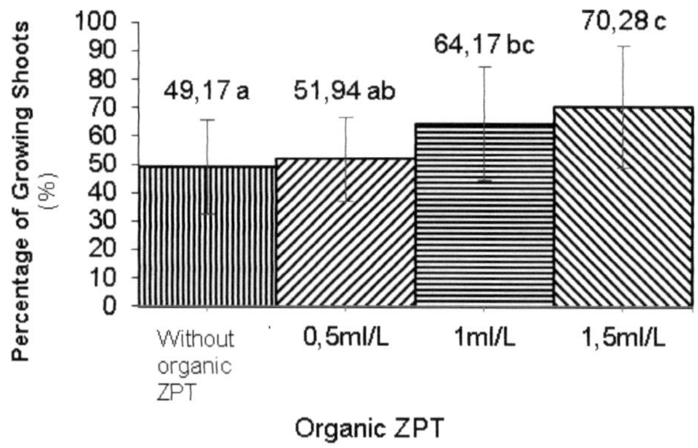

Figure 2. Histogram of the effect of organic ZPT towards the percentage of growing shoots

The lowest percentage of shoot growth occurred without organic ZPT treatment as indicated by the mean of 49.17%. The percentage of growing shoots that appear to be low in the treatment without organic ZPT is due to unavailibility of growth hormone, thus, stem cuttings become less able to trigger the emergence of new shoot buds. The organic ZPT treatment of 1.5 ml/l has the highest percentage of shoots with an average value of 70.28% because the amount of organic ZPT given is suitable to trigger shoot growth, whereas in the organic ZPT treatment of 1 ml/l the shoot growth percentage tends to decrease with average score of 64.17%. The decrease of shoot percentage growth in organic ZPT treatment 1 ml/l is suspected because the amount of hormone given not enough so buds grow less optimal. Applying plant growth regulators needs to consider the accuracy of the dosage, because if the dose applied is too high or too low it will not trigger growth but inhibit the growth of plants instead and cause toxicity in all plant tissues.

3.3 Shoots Weight

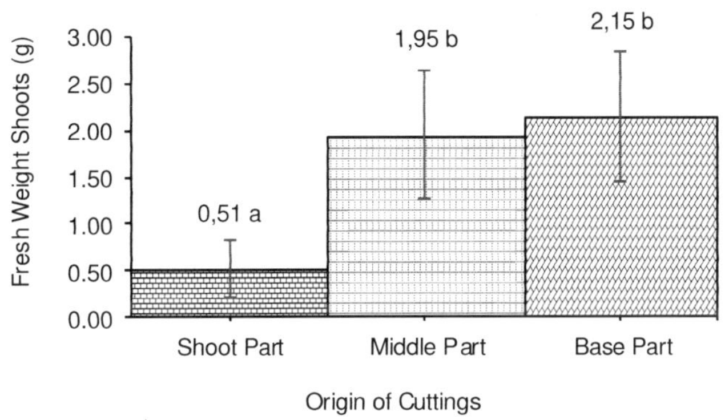

Figure 3. Histogram of origin cuttings effect on fresh weight shoots

The appearance of shoots is an important part in determining the success of cuttings. The presence of shoots on cuttings will help the process of formation of natural growth regulator which is then circulated to the bottom or basal to form the roots [9]. The weight of these shoots was measured in the last week of observation measured in terms of fresh weight buds and dry weight of buds.The treatment of shoot part has the lowest fresh weight of shoots compared to the center and the base. This is assumed because the shoots only produce buds that have a small size and length. Treatment of the base has the highest fresh weight of shoots because it has a larger shoot length value than other parts so that the average value of fresh weight shoots are larger as well. Another factor that affects the fresh weight of buds is the presence of water content in the bud. Wet or fresh weight is strongly influenced by water content in plant tissue. Treatment of the base has a larger size and longer shoot length so it can store more water which can affect the fresh weight of the shoots.

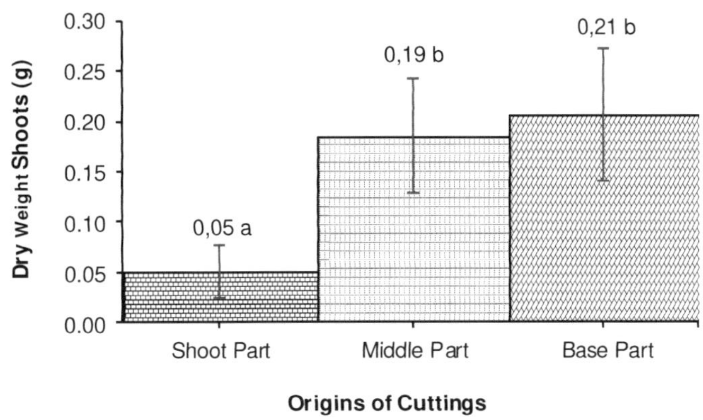

Figure 4. Histogram of origin cuttings effect on dry weight shoots

Based on the 5% duncan test the shoots treatment has the lowest dry shoot weight value compared to the middle and the base part. This is directly proportional to the fresh weight of buds, the greater the value of fresh weight the dry weight of the buds is also greater. The shoot part treatment has the lowest dry shoot weight is assumed because the shoots have only a small amount of food or energy reserve to form the shoot so it only produces shoots that have small size and length. The treatment of the base has the highest dry weight of the shoots because it has enough food or energy reserves to trigger the growth of the size and length of the larger shoots so that it affects the shoot dry weight compared to other parts. Another factor is the impact of the photosynthesis process. Photosynthesis results in increased dry weight of plants due to CO_2 retrieval, whereas the process of catabolism of respiration causes CO2 exposure and reduces dry weight [10].

3.4 Root Lenght

Based on the analysis of the variety of zeatin treatment and the origin of cuttings gives no significant difference on root length variables. Treatment without zeatin has a root length with an average of 24.19 cm. The zeatin treatment of 0.5 ml/l has an average root length of 30.24 cm. Zeatin treatment of 1 ml/l has an average root length value of 25.97 cm. Treatment of 1.5 ml/l has an average value of root length of 27.27 cm. This suggests that with the little addition of zeatin may affect root growth but with a certain amount. Cytokinin has the ability to encourage the occurrence of cell division and tissue differentiation on certain shoot bud formation and root growth [11]. The shoot treatment has the highest root length with an average value of 27.95 cm. The middle part has the second longest root length value after the shoot, which is 27.33 cm while the base has the lowest root length with an average value of 25.48 cm. It is suspected that the base of the stem has hardened so that the root is difficult to grow. The cutting material with high C/N ratio will be easier and faster to form the root but because the cell wall begins to harden so the root primordia is difficult to penetrate, thus, the roots are rather difficult to form [12].

3.5 Root Weight

The cutting position and age of origin cuttings important role in rooting and plant growth [13]. Root weight is one indicator of success in the cuttings. The weight of the roots is closely related to the number of roots in which the greater the number of roots, the greater the weight of the roots. The occurrence of roots on the multiplication of stem cuttings is very important because at the time the roots have not appeared then the cuttings grow with the food or energy reserves that exist at the origin of the cuttings itself.

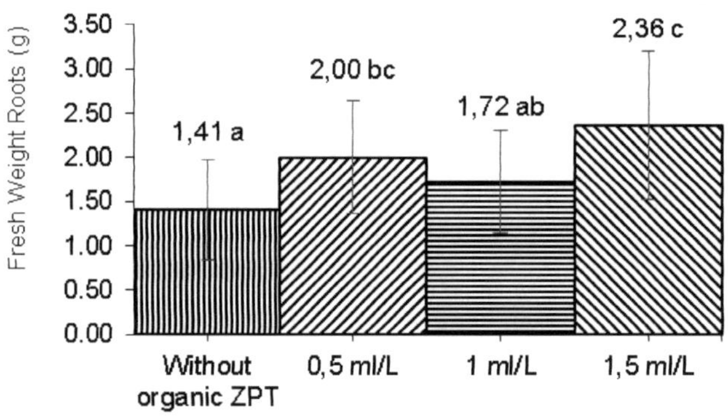

Figure 5. Histogram of the Effect of Organic ZPT towards Fresh Weight Roots

Growth can be shown by increasing plant height, length, width, and leaf area, as well as the dry weight of each organ including roots, stems, leaves and fruits. The organic ZPT treatment has the lowest fresh root weight compared to the others, whereas the 1.5 ml/l organic ZPT treatment gave the highest yield of fresh root weight. Treatment of organic ZPT 1.5 ml/l gives the most effective effect to increase root weight. Cytokinin is a plant hormone that affects many developmental processes, one of which is able to increase the shoots of plants, such as buds [14]. Shoots and roots have a close relationship in plants. The good growth of shoots will lead to the good formation of the leaves, so the process of photosynthesis increases, thus resulting more carbohydrates and may be used to form roots [15].

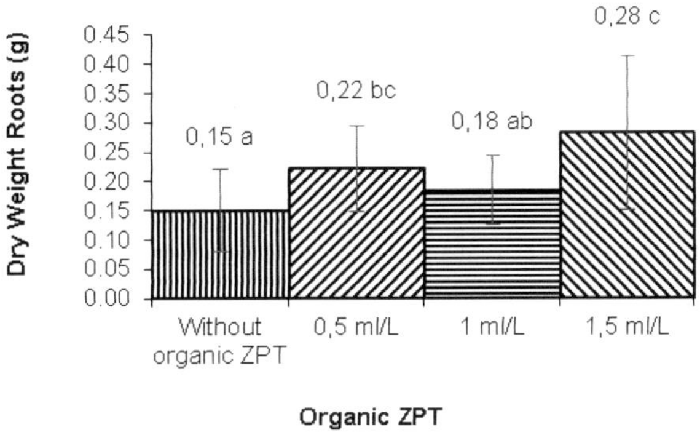

Figure 6. Histogram of the Effect of Organic ZPT towards Dry Weight Roots

The organic ZPT treatment of 1.5 ml / l has the highest dry weight value. This is directly proportional to the fresh weight of the roots, which organic ZPT 1.5 ml / l has the highest dry weight of plant roots that is affected by the administration of organic ZPT. Good root growth allows plants to produce a lot of energy for the purposes of metabolic processes and for further growth process, so indirectly dry weight also increases [15].

3.6 Leaf Amount

Leaves are one of the most important plant organs [16]. The leaves become the site of the process of photosynthesis that will produce food for plants and make up the dry matter of plants. Food products from this photosynthesis process will be distributed to all parts of plant body

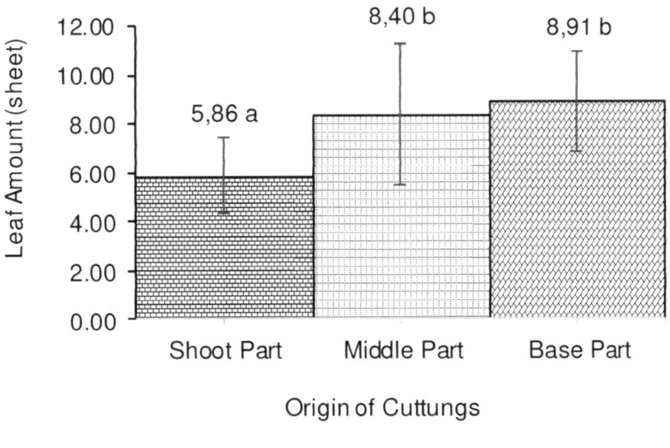

Figure 7. Histogram of the effect of origin cuttings on the number of leaves

According to the analysis, the variety treatment of origin of cuttings gives significantly different result to the variable of the number of leaves. The top of the shoot has the greatest number of leaves compared to the center and the base, while the middle and the base part has the same effect. The top of the shoot has the largest number of leaves with an average value of 5.86 pieces. The middle has the second greatest number of leaves with an average value of 8.40 strands. The base has an average number of shoots of 8.91 strands (Figure 5). This is directly proportional to the percentage of growing shoots, the higher the percentage of shoots growth, the more the average leaves grow in each cut.

3.7 Leaf Area

Leaf is the place where the process of photosynthesis to construct the plant dry matter is occur. Leaf area is the result of leaf length, leaf width and leaf constant multiplication [10]. Leaf area on a plant is important to be counted because it can affect the process photosynthesis in plants. An important factor to note in measuring leaf area is the accuracy of measurement results and measurement speed [16].

Table 1. Effect of organic PGR and origin of cuttings towards Leaf area (sheets) on Daun Ungu done by Stem cuttings (*G. pictum*)

Organic ZPT	The Origin of Cuttings			Average
	Shoot Part	Middle Part	Base Part	
Without Organic ZPT	13,94 [a]	16,86 [a]	13,08 [a]	14,63
0,5 ml/l	27,20 [c]	22,88 [b]	23,18 [b]	24,42
1 ml/l	27,76 [c]	29,76 [c]	21,54 [b]	26,35
1,5 ml/l	27,20 [c]	28,08 [c]	20,94 [b]	25,41
Average	24,03	24,40	19,69	22,70

The combination of treatment of cuttings with organic ZPT on the wide varieties of these leaves gives significantly different results. The treatment of the origin of the central cuttings with organic ZPT 1 ml / L has the largest average leaf width, which is 29.76 cm2, whereas the lowest average value is on the treatment of origin of the base cuttings with no organic ZPT, which is 13.08 cm2 . This is presumably because the middle treatment of 1 ml / L has an active cell that still divides and has enough food reserves to enlarge the leaf area, other than that, addition of organic ZPT 1 ml/l can trigger the leaf area growth to become more optimal. Treatment of the base without organic ZPT only has the smallest average value of the leaf area, because this section is old so that there is only a few cells that divide, but in the absence of application of organic ZPT make the leaf area only grow small. The use of exogenous cytokines through the leaves can increase the growth of leaf area and delayed leaf senesen [17].

4. Conclusion

Treatment of organic ZPT of 0.5 ml / L is able to increase root length; 1 ml / L treatment is able to increase fresh and dry root weight and treatment of 1.5 ml / L is able to increase the percentage of growing shoots. Treatment of the origin of the base cuttings can increase the length of shoots, fresh weight, dried buds, and the number of leaves. The combination of treatments between organic ZPT 1 ml / L and the central cuttings were able to increase the leaf area, whereas the treatment without organic ZPT and the origin of the cuttings gives the smallest leaf area effect.

Acknowledgement

This research was supported by the Medicinal Plant and Traditional Drug, Tawangmangu, Indonesia.

References

[1] Keloth K S, Jessy E M, Kerryn J 2011 Effect of ethanol extract of Graptophyllum pictum (L) Griff on cisplatin induced nephrotoxicity in rats *J Department of Pharmaceutical Chemistry* **57** (2) : 52-65

IOP Conf. Series: Earth and Environmental Science **142** (2018) 012056 doi:10.1088/1755-1315/142/1/012056

[2] Wahyuningtyas E 2005 The Graptophyllum pictum extract effect on acrylic resin complete denture plaque growth *Dentist Journal* **38** 201–204

[3] Lestari P, Khumaida N, Sartiami D and Mardiningsih TL 2015 Selection criteria of Graptophyllum pictum resistance to doleschallia bisaltide cramer (lep: nymphalidae) attack based on insect feeding preference *Sabrao Journal of Breeding and Genetics* **47** 172-184

[4] Olagbende-Dada SO, Ogbonnia SO, Coker HAB and Ukpo GE 2010 Blood glucose lowering effect of aqueous extract of Graptophyllum pictum (linn) griff on alloxan-induced diabetic rats and its acute toxicity in mice *African Journal of Biotechnology* **10** 1039-1043 ISSN 1684–5315

[5] Aulia W, Rahmawati, Dessy S M 2015 Activity of ethanol extract of purpleleaves (Graptophyllum pictum(Linn)Griff) onalloxan-induced diabetes mice *International Journal of PharmTech Research* **3** 497-501

[6] Deng X, Bagping C C, Jian Wen D, MuLin C 2000 Cutting propagation techniques for Quisqualis indica Journal of Zhejiang Forestry College **17** 38-488http://encnkicomcn

[7] Hartmann H T, Kester D E 1983 Plant Propagation Principles and Principles 4 th edition New York (US) : Pentee hall nine, Englewood

[8] Suwasono H 1989 Hormon tumbuhan Jakarta (ID): Rajawali

[9] Dwijoseputro 1980 Fisiologi tumbuhan Jakarta (ID) : PT Gramedia

[10] Gardner F P 1991 Fisiologi, Tanaman Budidaya Susilo Herawati (Translator) Jakarta (ID) Translating from: Physiology of crop plant

[11] Werner T, Motyka V, Strnad M, Schmulling T 2001 Regulation of plant growth by cytokinin USA

[12] Kastono D, Sawitri H, Siswandono 2005 Pengaruh nomor ruas setek dan dosis pupuk urea terhadap pertumbuhan dan hasil kumis kucing *J Ilmu Pertanian* **12** 56 – 64

[13] Santoso B B, Parwata A 2014 Seedling growth from stem cutting with different physiological ages of jatropha curcas l of west nusa tenggara genotypes *International Journal of Applied Science and Technology* **4** 6-16

[14] Gajdošová S, Spíchal L, Kamínek M, Hoyerová K, Novák O, Dobrev P I, Galuszka P, Klíma P, Gaudinová A, Žižková E, Hanuš J, Dančák M, Trávniček B, Pešek B, Krupicka M, Vanková R, Strnad M and Motyka V 2011 Distribution, biological activities, metabolism, and the conceivable function of cis-zeatin-type cytokinins in plants *J of Experimental Botany* **8** 2827-2840

[15] Purwanto 2004 Pengaruh isomer sodium nitrofenol terhadap pertunasan dan pertumbuhan bibit tanaman pisang Jurnal Penelitian UNIB **10** 105-108 http://ejurnaltripodcom/dilp04html

[16] Haryanti S 2010 Pengaruh naungan yang berbeda terhadap jumlah stomata dan ukuran porus stomata daun Zephyranthes rosea Lindl *J Anatomi Fisiologi* **18** 100 - 130

[17] Januwati M 1992 Faktor-faktor ekologi yang mempengaruhi pertumbuhan tanaman sirih (Piper betle Linn) Warta Tumbuhan Obat Indonesia

[18] Beltrano J, Caldiz DO, Barreyro R, Vallduri GS and Besus R 1994 Effect of foliar applied gibberelic acid and benzyladenin upon yield component in sunflower (Helianthusannus L) *J Plant Growth Regulation* **15** 101-106

[19] Munarti, Kurniasih S 2014 Pengaruh konsentrasi IAA dan BAP terhadap pertumbuhan setek mikro kentang secara in vitro *J Pendidikan Biologi, FKIP, Universitas Pakuan* **1**(1) : April 2014

IOP Conf. Series: Earth and Environmental Science **142** (2018) 012057 doi:10.1088/1755-1315/142/1/012057

The impact of seasonal rice price changes on rice self-consumption in farm household of rural Java

S W Ani and E Antriyandarti
Study Program of Agribusiness, Faculty of Agriculture, Universitas Sebelas Maret, Surakarta, Indonesia

E-mail: susi_wuri@staff.uns.ac.id

Abstract. Seasonal rice price changes are very volatile and not predictable. This price changes have a heterogeneous impact on public consumption. The problem of seasonal rice price changes is not only experienced by consumers, but also in the farmers side as producers. The objective of this study is to provide a detail overview and description of the changing seasonal rice self-consumption of farm households in rural Java in response to seasonal rice price changes and income shocks to anticipate seasonal scarcity. This paper constructs a theoretical model to address such seasonality of food deprivation by using one year of seasonally farm household panel data, empirically tests the extent to which farmers in rural Java can smooth their rice self-consumption from season to season in response to income shocks. The result shows that rice farmers increase their rice self-consumption when prices are high.

1. Introduction

Seasonal rice price changes are very volatile and not predictable, including in Java, the main producer of rice in Indonesia [1]. This price changes have a heterogeneous impact on public consumption. The problem of seasonal rice price changes is not only experienced by consumers, but also farmers as producers. This study will focus on the impact of seasonal rice price changes on the rice-self consumption of farm household in rural Java. Java Island produces the highest amount of paddy in Indonesia more or less53% of total production. Specifically, in Java Island, 16% of total production comes from Central Java. The total harvested area in Central Java is 17.474.67 km^2 or 14.41% of the total harvested area in Indonesia [2].

Most of population in rural Java live in households and they are cultivate paddy tp produce rice. The majority of households that consumed home-produced rice produce enough rice to meet the needs at some point through the year. In the typical agricultural sector in rural Java, the rice harvest season begins in rainy season or season I. Some areas in rural Java experienced with the lowest incidence of insufficient stocks occurred following the peak of the harvest. The change of rice price will affect to the household income. Previous study [3] found the rice price will influence the farm household welfare in the long-run. Farmers as producers face the problem of seasonal rice price change, not only experienced by consumers in the urban area.

The objective of this study is to provide a detailed overview and description of the changing seasonal rice self-consumption of farm households in rural Java in response to seasonal rice price changes and income shocks to anticipate seasonal scarcity.

2. Theoretical Framework

In this study, the farm household is assumed to maximize its utility subject to its constraints, namely a production function, a time and a budget constraint. Let equation (1) be the utility function, which is quasi-concave with positive partial derivatives.

$$\underset{C,X,l}{Max}\, U(C, X, l, z) \tag{1}$$

The factors are household consumption of commodity rice, with I denoting income, l is leisure and z is household characteristics (e.g. household age, number of family member, number of family laborers). Income equals to the self-sufficiency ratio, such that

$$I(\text{remittance}) = \Pr.C + Px.X \tag{2}$$

With Pr denoting price of rice, C is food consumption (rice), Px is price of non-food commodity, and X is non-food commodity. Here, total income spends for $Pr.C$ + $Px.X$, so, no saving for household. The production of rice is assumed to be influenced by level of food (rice) and non-food commodity consumed, total labour for farming L^f, current input M (e.g. seed, fertilizer, chemical and tractor rent), fixed capital $\overline{K}$, land T (operated land or planted land) and parameter condition which affect production δ (e.g. cropping pattern, irrigation system).

Although the total income of household equations derived from the household production model, as mentioned above, by doing the contain of all predictions of conventional utility maximizing models, we can get the optimal total income of household and partial derivative of optimal total income with respect to rice price.

$$\frac{\partial I^*}{\partial \Pr} \gtrless 0 \tag{3}$$

The effect of a rice price change on total income will thus depend on the income effect and the substitution effect. When rice price decreases, the farmer need to decrease income in order to keep purchasing power constant. Then for normal good, a decrease in income will lead to a decrease in demand. An increase in rice price means that demand of rice will go down due to substitution effect. If the rice price goes up, it is like a decrease in income, which for normal good means a decrease in demand. The optimal total income of household may increase or decrease due to the change of rice price. As a consequence, the farmer may decrease or increase the marketed surplus of food (rice).

According to the previous finding [5], when poor farmers produce a marketed surplus of food, it may happen that this surplus fall when the price of food rises. For this reason, this study examines the effect of self-consumption as well as a marketed surplus of food. The marketed surplus of food (ms_a) is defined as follows.

$$ms_r = F - C \tag{4}$$

The response of marketed surplus to price is:

$$\frac{\partial ms_r}{\partial \Pr} = \frac{\partial F}{\partial \Pr} - \left.\frac{\partial C}{\partial \Pr}\right|_{u=constant} - ms_r \frac{\partial C}{\partial I^*} \tag{5}$$

where
Pr = rice price
ms_r = surplus of food (rice)
F = production of rice
C = consumption of rice

The first term on the right-hand side is the supply response in production and is positive. The second term is unambiguously positive. The third term, for a normal good, is negative. In this study, rice is assumed as a normal good [6]. While the net will generally be positive, poor households may have such a high income elasticity of food consumption that the entire expression is negative. This effect will be reinforced by low elasticity of supply response and a low substitution effect between food and other goods. All responses to the rice price change will affect to their rice self-consumption from season to season in response to income shocks.

3. Method

3.1. Empirical Model

Based on the theoretical framework, to estimate the changing seasonal rice self-consumption of farm households in rural Java in response to seasonal rice price changes and income shocks to anticipate seasonal scarcity, this study will use the empirical model as follows:

$$RSC = \alpha_0 + \alpha_1 Educ + \alpha_2 Age + \alpha_3 Land + \alpha_4 Yield + \alpha_5 Income + \alpha_6 Riceprice + \alpha_7 Occupation + \alpha_8 Season + \alpha_9 \mathrm{Re}mit\tan ce + u \quad (6)$$

Where:
RSC : Rice self-consumption
Educ : Number of successive years in school of household head
Age : Age of household head
Land : Total hectares of agricultural land (farm land) in the given year
Yield : Total production in kg/season
Income : Total income of household
Rice price : Price of a kilogram of rice
Occupation : Dummy occupation of household
If the member of family have off-farm job = 1, otherwise = 0
Season : Dummy of season
If the household planted paddy in season 1 = 1, otherwise = 0
Remittance : Total money and in kind remitted by household members and relatives to the farm household

3.2. Analysis Methods

The panel data analysis is applied to estimate the changing seasonal rice self-consumption of farm households in rural Java in response to seasonal rice price changes and income shocks to anticipate seasonal scarcity. The null hypotheses of individual effect are rejected. Consequently, the Ordinary Least Square (OLS) method is not used for analysing seasonal rice self-consumption. After applying the Hausman test to investigate the kind of effect that exists (fixed effect or random effect) on rice self-consumption function, the result indicated that we should use random effect model to estimate the changing seasonal rice self-consumption ([7];[8]).

4. Result and Discussion

This study conducted in Cilacap and Grobogan which is the largest area for paddy cultivation in Central Java. Table 1 illustrated the farm household characteristic in both districts. The farm size in Grobogan district is larger than in Cilacap, therefore the agricultural income of farm household in Grobogan is higher as well. However, the rice self-consumption in Cilacap is larger than in Grobogan.

The results of estimating equation (6) are summarized in Table 2. The estimated parameters of age, are significantly positive for equation. These findings suggest that human capital of household contribute to increase of rice self-consumption. The older household head tend to keep the production for rice self-consumption.

The estimated parameters of land, yield, income, and remittance are significantly negative to the rice self-consumption. This result indicates that an increase in land, yield, income and remittance lead a decrease in rice self-consumption. The rice farmers prefer to sell their product to the market rather than to keep for their consumption when they have larger land, yield, income and remittance. In contrast, the estimated parameter of rice price is significantly positive for rice self-consumption. This implies that the increase of rice price has a negative effect on marketed surplus of rice production. Rice farm household would keep their production when rice price increases. It is suggested that rice farmer in the study areas are net rice consumers [9,10]. The variable of occupation and season do not have any effect to rice self-consumption.

Table 1. Household Characteristic

Characteristics	Cilacap District	Grobogan District
Number of farm households	51	50
Number of family labor (person/hh)	5.07	4.02
Area of agriculture owned land (ha/hh):	0.26	0.55
Paddy field	0.22	0.37
Farm size (ha/hh)	0.26	0.55
Yield of paddy (ton/planting)		
Session I	1.3	3.7
Session II	1.1	2.2
Session III	-	8.6
Average Farm Household Income (IDR/year):		
Agricultural income	5.449.902	28.193.300
a. Rice income	5.449.902	19.135.600
b. Non-rice income	0	9.057.700
Off-farm income	24.883.529	68.171.700
Remittance	1.587.255	410.000
Farm household income	31.920.686	96.775.000
Farm assets (excluding land):		
Cow and buffalo (head/hh)	0.29	0.86
Tractor owned (number/hh)	0.02	0.088
Sprayer (number/hh)	0.71	0.88
Thresher (number/hh)	0	0.04
Rice self-consumption (kg/sesion) :		
Session I	602	227
Session II	470	395
Session III	-	0

Source : Farm Household Survey, 2017
Note : Exchange rate in 2017 was US$1= IDR 13.382

Table 2. The Changing of Rice Self-Consumption

Independent variable	Coefficient	Standard Error
Constant	-695.6886**	(320.0284)
Education	17.4787	(14.0382)
Age	8.0478**	(3.972812)
Land	-0.1075***	(0.0299)
Yield	-0.1438***	(0.049)
Income	-9.27e-07**	(4.44e-07)
Rice price	0.0739**	(0.0328)
Occupation	133.8937	(112.0421)
Season	115.2349	(82.5940)
Remittance	-.00002**	(.00001)
R^2	0.9555	
Wald chi^2	38.75***	
Hausman Test	0.8749	
Number of observations	178	

Source: Farm Household Survey, 2017

5. Conclusion

The estimation results of rice self-consumption functions show that rice price had positive influence on rice self-consumption or in other words rice price had negative impact on marketed surplus of rice. This indicates that if price increase, rice farmer will increase their rice self-consumption. The result shows that some farmers increase their rice self-consumption when prices are high. However, we still need to examine carefully whether the increase of rice production off sets the increase of rice consumption or not, rice price fluctuation does not affect the household welfare or not, and Indonesian government do not need to import rice more or not.

References

[1] Istiqomah, Zeller and Taubadel. 2005. Volatility and Integration of Rice Markets in Java, Indonesia: A Comparative Analysis before and After Trade Liberalisation. Conference on International Agricultural Research for Development.

[2] BPS [Badan Pusat Statistik (Central Bureau of Statistics)], https://www.bps.go.id/linkTableDinamis/view/id/895, 2017/6/23

[3] Sobrevinas, A.B. and Barrios, E.B. 2009. Impact of the Rice Trade Policy Reforms on Household Welfare in the Philippines.UPSS Working Paper No. 2009-07 July 2009. School of Statistics, University of the Philippines Diliman.

[4] BPS [Badan Pusat Statistik (Central Bureau of Statistics)], https://www.bps.go.id/LinkTableDinamis/view/id/963, 2017/7/11

[5] Sadoulet and Janvry. 1995. Quantitative Development Policy Analysis. The Johns Hopkins University Press.

[6] Cristovao, Hartono, Slamet and Mulyo, JH. 2007. AnalisisKonsumsiBerasPada Tingkat RumahTangga di Propinsi Daerah Istimewa Yogyakarta.*Agroekonomi.* 14 (1), pp. 49-60.

[7] Maddala, GS. 2001. *Introduction to Econometrics. Ohio State University*.Prentice-Hall. New York

[8] Baum, C. F. (2006) *An Introduction to Modern Econometrics Using Stata*. Texas: Stata Press.

[9] Mulwanyi, A., Hutagaol, P., and Sinaga, B. M. (2011) "Impact of rice purchasing policy on welfare of both producers and consumers in Indonesia," *Journal of International Society for Southeast Asian Agricultural Sciences*, **17**, 2011, pp. 48–57.

[10] McCulloch, Neil. 2008. Rice Prices and Poverty in Indonesia. Bulletin of Indonesia Economic Studies. Vol. 44 (Issue 1), pp. 45-64.

Drought resistant of bacteria producing exopolysaccharide and IAA in rhizosphere of soybean plant (*Glycine max*) in Wonogiri Regency Central Java Indonesia

A Susilowati, A A Puspita and A Yunus
Bioscience Study Program, Graduate Program, Universitas Sebelas Maret, Surakarta, Indonesia

E-mail: arisusilowati@staff.uns.ac.id

Abstract. Drought is one of the main problem which limitating the agriculture productivity in most arid region such as in district Eromoko, Wuryantro and SelogiriWonogiri Central Java Indonesia. Bacteria are able to survive under stress condition by producte exopolysaccharide. This study aims to determine the presence of exopolysaccharide-producing drought-resistant bacteria on rhizosphere of soybean (*Glycine max*) and to determine the species of bacteria based on 16S rRNA gene. Isolation of bacteria carried out by the spread plate method. The decreased of osmotic potential for screening drought tolerant bacteria according to the previous equation [12]. Selection of exopolysaccharide-producing bacteria on solid media ATCC 14 followed by staining the capsule. 16S rRNA gene amplification performed by PCR using primers of 63f and 1387r. The identificationof the bacteria is determined by comparing the results of DNA sequence similarity with bacteria databank in NCBI database. The results showed 11 isolates were exopolysaccharide-producing drought tolerant bacteria. The identity of the bacteria which found are *Bacillus sp*, *Bacillus licheniformis*, *Bacillus megaterium* and *Bacillus pumilus*.

1. Introduction

Drought is one of the main problem whichlimiting agricultural productivity in the arid region [1]. Wonogiri is an area that has low rainfall levels. In 2013, between Augusts to September in three districts in Wonogiri namely, Eromoko, Wuryantro, and Selogiri encounter drought because there is no rainfall. Limited water availability in the soil can inhibit the absorption of nutrients by the roots of plants, especially legumes.

Drought can affect on the growth of microorganisms and plants as the water becomes a limiting factor for plants and microorganisms to survive. Microorganism adaptation to drought is to secrete exopolysaccharide (EPS) in high quantities [2]. Bacteria are able to survive under stressful conditions due to the production of EPS [1]. EPS is a structural component of the extracellular matrix in biofilms that are synthesized by cells of microorganisms in response to physiological stress on the environment [3]. EPS have important role for bacteria to protect against a broad range of environmental stress. Such as the bacterium Pseudomonas sp. which increases the production of EPS during the dry season [4]. Production of EPS of Rhizobium sullae KYGT207 strains isolated from dry land in Southern Algeria is known to have a role in the absorption of water and nutrients. EPS in sandy soils can protect plants from stress, lack of water and contribute to the formation of soil aggregates [2].

Bacteria from a broad range of genus are an important component of the soil. The bacteria involved in the various biotic activity of the soil ecosystem to be dynamic for the exchange of nutrients and support agricultural production [5]. A group of bacteria that can be found in the rhizosphere, which is advantageous in improving the growth of plants, can be classified as plant growth-promoting rhizobacteria (PGPR) [6, 7]. Pseudomonas, Azospirillum, Azotobacter, Klebsiella, Enterobacter, Alcaligenes, Arthrobacter, Burkholderia, Bacillus, and Serratia have been reported as PGPR able to promote plant growth [7].

Bacteria belonging to the PGPR are one of high producing IAA. One of the PGPR is one of potential IAA-producing bacteria. IAA-producing bacteria touch some physiological processes of plants by inserting the IAA that it generates to the plant. The effect on crops is changing its IAA concentration thus helps in the formation of lateral roots, adventitious roots and primary root elongation [8]. Condensed root can increase the nutrients quickly, thus increasing plant growth [9]. Also, rhizosphere bacteria are able to secrete siderophores, dissolving phosphate, and act as a biocontrol that protect crops from fungi, pathogens, and pests [10].

Soybean (*Glycine max*) is one of the primary sources of protein crops in Indonesia. Soybean becomes one of the primary food commodities other than rice and corn [11]. Soybean is one of the agricultural products commonly grown in upland areas in the district Eromoko, Wuryantro, and Selogiri, Wonogiri. To develop sustainable agriculture, it is needed to know the potentially exopolysaccharide producing bacteria as producers of IAA in the rhizosphere of soybean crops that grow under drought stress conditions.

2. Methods

2.1. Isolation of Rhizosphere Bacteria

Bacteria isolated from the dry soil around the roots of soybean (*Glycine max*) using a dilution series (Platting Method). In the rhizosphere soil of soybean plants as much as 1 gram dissolved in 10 mL of normal saline (0.85% NaCl), vortex, and heated in a water bath at a temperature of 80°C for 20 minutes, then carried serial dilutions, to obtain 10-1 dilution, Dilution performed until dilution to 10-2. Each series of dilutions taken 100 mL put in a petri dish that already contains media NA and propagated using stem L to rub it on the surface so that droplets of the suspension evenly. The cup and its contents were incubated upside down in an incubator at 38°C for 24-48 hours. Furthermore, the different morphology of each colony that grows on the medium NA selected and grown on agar slant NA as the pure culture.

2.2. Preparation of Medium and Osmotic Pressure

Nutrient Broth medium are sterilized, added with polyethylen glycol (PEG) 6000 while still hot for lowering osmotic potential to follow the equation according to [12]:

Ψs = - (1:18 x 10-2) C - (1:18 x 10-4) C2 + (2.67 x 10-4) CT + (8:39 x 10-7) C2T

Description:

Ψs = osmotic potential (Mpa)

C = concentration

T = ambient temperature (oC)

Osmotic potential that is set is -1 MPa, -1.5 MPa, -2 MPa.

2.3. Screening Drought Tolerant Bacteria

1000 mL of a suspension of bacteria grown on NB medium with a population of around 1,5x108 CFU / mL was inoculated into 20 ml NB medium that has been coupled with PEG 6000. These cultures were incubated in an incubator shaker at a speed of 80 rpm at a temperature of 38°C for 24 hours, Optical Density measurement using a spectrophotometer at a wavelength (λ) 570 nm. Value of drought tolerance to Optical Density (OD) was determined as: Highly Sensitive OD <0.3; Sensitive OD = 0.3-0.4; Tolerant OD = 0.4-0.5; Highly Tolerant OD> 0.5 [13].

2.4. Test of exopolysaccharide production
A selection of potential bacteria producing-exopolysaccharide conducted by growing a loop of bacterial isolates on solid media in a Petri dish ATCC 14 for seven days [4] at a temperature of 38ºC. Colonies of bacteria that form slime or mucus thick exopolysaccharide show that it contains.

2.5. Capsules Staining
Glass objects cleaned using 70% alcohol. Then at the bottom of the glass object is labeled with a marker. As much as one drop of physiological saline dripped on the surface of the glass object. As much as a loop of bacterial culture was swapped on the surface of the glass object and homogenized and then fixed over the Bunsen burner until visible dry evenly. Results of preservation of bacteria dropped with a drop of crystal violet and allowed to stand for 3 minutes. Then rinsed with water and dried. Mixture drops with a Chinese ink at the edges of the glass objects. Mixture smear results are then dried. Furthermore, preparations added immersion oil for observation with a microscope.

2.6. IAA production
IAA produced by the bacteria is measured by a colorimetric method with the addition of Salkowski reagent [14]. Bacterial isolates were inoculated into 10 ml of nutrient broth (NB) medium that has been added 0.2 Mm L-tryptophan and NB media without the addition of L-tryptophan. The cultures were incubated and shaken (80 rpm) at room temperature for 48 hours. 3 ml cultures of each treatment are divided into two microtubes and then centrifuged (10,000 rpm) for 15 minutes. A total of 2 ml of the filtrate obtained is inserted into a sterile test tube and add 2 ml of Salkowski reagent. Then, the suspension was incubated for 60 minutes in the dark at room temperature. After that, IAA levels were measured using a spectrophotometer at a wavelength of 530 nm.

2.7. Amplification of 16S rRNA encoding-gene
16S rRNA gene was amplified using the Polymerase Chain Reaction (Perkin Elmer PCR system geneAmp 2400). PCR mix done by mixing a universal primer for the group of bacteria, which is 63F (5'-CAG GCC CAC ATG TAA CAA GTC-3 ') and 1387r (5'-GGG CGG WGT CAA GGC GTA-3'), 1.25µl, 2X KAPA 2G Fast Ready Mix 12.5 mL, 2 mL total DNA template, as many as 8 mL ddH2O. Cycle used consisted of pre-denaturasi at a temperature of 95ºC for 3 minutes, denaturation at a 95ºC for 15 seconds, annealing at 55ºC for 15 seconds, elongation at a temperature of 72ºC for 5 seconds for 30 cycles and finalizing at a temperature of 72ºC for 1 minute. The process of denaturation, annealing, and elongation each consisting of 35 cycles [15]. The PCR products then sequenced at 1st Base Singapore.

2.8. Data Analysis
Data were analyzed descriptively includes the exopolysaccharide capsule which is produced as the bacteria tolerant to drought, especially PGPR bacteria capabilities as the production of IAA. 16S rRNA gene sequences were analyzed by bioinformatics at GenBank data center by using the program Basic Local Alignment Search Tool (BLAST) on the website National Center for Biotechnology Information (NCBI), (http://www.ncbi.nlm.gov/BLAST/).

3. Results And Discussion

3.1. Drought Tolerant Rhizosphere Bacteria
Drought stress conditions were made by the addition of polyethylene glycol 6000 (PEG-6000) to the culture medium to adjust the osmotic pressure [12]. PEG-6000 can be used to mimic the magnitude of potential ground water or drought stress level [16]. A selection of bacteria that are resistant to drought observed begin at -1 MPa pressure. Bacteria that have a value of Optical Density (OD) greater than 0.5 at the osmotic pressure of certain bacteria are categorized as a genuinely tolerant [13]. The selected bacteria can survive and grow well under conditions of drought stress.

Tolerance to drought stress conditions of 34 isolates was selected 11 best isolates with OD values at pressures above 0.5 MPa -1. At 11 isolates contained 9 isolates were highly tolerant of drought, drought-tolerant 1 isolates, and 1 isolates were highly sensitive to drought (Table 1). Isolates E2, E4, E5, E11, S3, W6, W11, W13, W19 grouped into categories that are very tolerant because it has the OD> 0.5 in the third osmotic pressure, ie -1,0MPa, -1,5MPa, -2, 0MPa. E13 isolates included in the class tolerant because of the decrease in osmotic pressure OD value fell to 0.4879 at the osmotic pressure -2,0MPa. Isolates S6 listed in the class is very sensitive because of the decrease in osmotic pressure OD value decreased to 0.2745 at the osmotic pressure -2,0MPa. The high osmotic pressure outside bacterial cells causes the fluid inside the cell will diffuse out of the cell, so that the destruction of microbial cell walls will cause lysis. Tolerance is a condition where bacteria can survive and persist in the environment is less favorable, namely drought. Meanwhile, the sensitive is a state in which bacteria cannot survive because the environment does not support life.

Table 1. Tolerance levels of rhizosphere bacteria of soybean roots (*G. max*) on NB media added PEG 6000

Bacterial Isolate Code	Optical Density (OD) Value in Osmotic Pressure (MPa)			Category
	-1,0	-1,5	-2,0	
E2	1.7596	0.7493	0.5946	highly tolerant
E4	1.3652	1.1079	0.7866	highly tolerant
E5	0.8516	0.7227	0.6322	highly tolerant
E11	1.6334	1.4092	1.1641	highly tolerant
E13	1.5300	0.8929	0.4879	Tolerant
S3	1.2897	0.7952	0.7357	highly tolerant
S6	0.9746	0.3361	0.2745	very sensitive
W6	1.6741	1.5578	1.3743	highly tolerant
W11	1.7177	1.3042	1.1583	highly tolerant
W13	1.5870	1.4436	1.2403	highly tolerant
W19	1.5428	0.9903	0.5989	highly tolerant

Populations of soil bacteria decreased with stress or stress on the water, but not entirely the same, and certain soil bacteria can withstand extreme drought conditions [13]. Bacteria can survive in drought conditions through several mechanisms such as the production of exopolysaccharide (EPS), biofilm formation, and production osmolit to avoid the shortage of water in the cell [17].

3.2. Producing Bacteria exopolysaccharide

Source sucrose contained in ATCC medium can be used by bacteria to form exopolysaccharide. Colonies of bacteria that form slime or mucus thick exopolysaccharide show that it provides exopolysaccharide. Bacteria then examined the physical structure of the capsule staining. Capsule of the bacteria can not be colored with Indian ink [18]. Exopolysaccharide often found around the outside structure of prokaryotic and eukaryotic microbial cell. The physical structure exopolysaccharide in the form of a capsule to the cell wall of the massive slime formed outside the bacterial cell membrane [19].

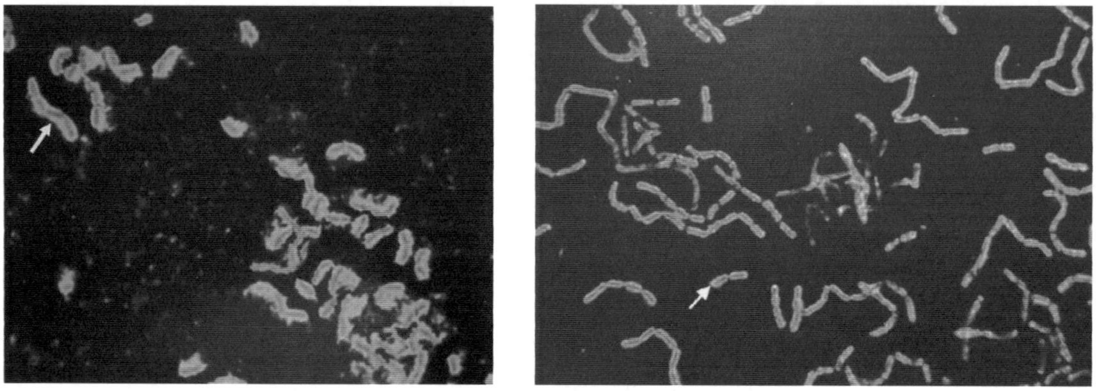

Figure 1. The bacterial capsule (arrow) is seen as a white exopolysaccharide layer not colored by ink dye

3.3. Producing Bacteria Indole Acetic Acid (IAA)

The ability of bacteria to produce IAA was tested by using colorimetric method with the addition of reagent salkowski according to [14]. Salkowski addition of reagents used to detect the presence of the compounds synthesis of IAA. The data (Table 4) show that each bacterial isolate produce IAA less and tend to be low. IAA production has decreased in drought stress conditions [20]. Based on research [17]. All the strains showed the ability to produce IAA with the L-tryptophan as a precursor. Some species of bacteria have been known as a producer of IAA, such as Pseudomonas sp., Bacillus sp., Azotobacter sp. and others [21]. However, some strains show little variation synthesis, and even strains of the same genus (Bacillus) produced a different amount of IAA in liquid media. This can happen because the production of IAA can be affected by conditions of culture, the phase of growth and the availability of substrate.

Table 2. IAA-producing bacteria from soybean plant rhizosphere

Code of bacterial isolate	IAA concentration (ppm)		color reaction
	Without L-triptophan	with the addition of L-triptophan	
E2	-0.5928	1.0652	++
E4	-0.4802	0.5474	+
E5	-0.3221	0.9249	+
E11	-0.6462	0.0573	+
S3	-0.8932	0.5296	+
W6	-0.8557	0.1660	+
W11	-0.5632	0.1996	+
W13	-0.7628	0.1264	+
W19	-0.2707	0.2411	+
E13	-0.7885	0.2984	+
S6	-0.2845	0.4664	+

Annotation: IAA= <1 $\mu g\ mL^{-1}$ (+), 1–10 $\mu g\ mL^{-1}$ (++), 11–50 $\mu g\ mL^{-1}$ (+++), >51 $\mu g\ mL^{-1}$ (++++) color: low (+), medium (++), high (+++).

3.4. Bacterial Identity based on 16S rRNA gene sequence

Isolates E5 has a similarity 97% with *Bacillus licheniformis*, isolates E11 has a similarity 97% with *Bacillus licheniformis*, isolate S3 has a similarity of 96% with *Bacillus megaterium*, isolate W6 has a similarity of 98% with *Bacillus pumilus*, isolates W11 has a similarity 97% with Bacillus sp., isolates W13 has a 97% similarity with *Bacillus pumilus* isolate E13 has a 98% similarity with *Bacillus licheniformis*, isolates S6 has a 98% similarity with *Bacillus pumilus*. BLAST analysis results which have sequences with ≥99% similarity showed the same species. Sequences with ≥97% similarity included in the same genus [22].

Table 3. Species identical to bacterial isolates found in rhizosphere of soybean by the gene encoding of 16S rRNA

Isolate name	The closest relatives	No. Access	% similarity
E5	*Bacillus licheniformis*	JQ965662.1	97%
E11	*Bacillus licheniformis*	JX237861.1	97%
S3	*Bacillus megaterium*	KM279706.1	96%
W6	*Bacillus pumilus*	KC845305.1	98%
W11	*Bacillus sp.*	HQ704718.1	97%
W13	*Bacillus pumilus*	KJ195695.1	97%
E13	*Bacillus licheniformis*	KJ831075.1	98%
S6	*Bacillus pumilus*	KC845305.1	98%

Bacillus are found as PGPR, as in some of the following research. *Bacillus pumilus* and *Bacillus licheniformis* isolated from the rhizosphere alder glutinosa [L.] Gaertn. has an activity as a strong growth driver [23]. *Bacillus megaterium* reportedly included in the solvent phosphate bacteria that help the growth of Tectona grandis and Chukrasia tubularis on plantations [24]. Based on research Kavamura [17] was selected as the strain of bacteria PGPR for *Zea mays* L. under water stress is *Bacillus* sp. and *Pantoea* sp.

4. Conclusion

It is concluded in this study there are 11 isolates were exopolysaccharide-producing drought tolerant bacteria. The bacteria identified as *Bacillus sp*, *Bacillus licheniformis*, *Bacillus megaterium* and *Bacillus pumilus*.

References

[1] Ali S Z, Vardharajula S, Linga V R 2013 Isolation and Characterization of drought-tolerant ACC deaminase and exopolysaccharide-producing fluorescent *Pseudomonas* sp *International Crops Research Institute for Semi-and Tropics*: 1-10
[2] Putrie R F W, A T Wahyudi, A A Nawangsih, E Husen 2013 Screening of rhizobacteria for plant growth promotion and their tolerance to drought stress *Microbiology*, 7(3): 94-104
[3] Marvasi M, Visscher PT, Martinez LC 2010 Exopolymeric substance (EPS) from *Bacillus subtilis*: polymers and genes encoding their synthesis *FEMS Microbiol Lett* 313 (1):1-9
[4] Santi L P, Ai D, Didiek H G 2008 Peningkatan Kemantapan Agregat Tanah Mineral oleh Bakteri Penghasil Eksopolisakarida *Menara Perkebunan*, 76(2): 93-103
[5] Ahemad, M, and M Kibret 2014 Review, Mechanisms and applications of plant growth promoting rhizobacteria: Current perspective *Journal of King Saud University-Science*, 26: 1-20
[6] Aryanth, I N P, Dian P L, Nurmi P D P 2004 Potensi Isolat Bakteri Penghasil IAA dalam Peningkatan Pertumbuhan Kecambahan Kacang Hijau pada Kondisi Hidroponik *Jurnal Mikrobiologi Indonesia,* 9(2): 43-46
[7] Yadav J, J P Verma, K N Tiwari 2010 Effect of plant growth promoting Rhizobacteria on seed germination and plant growth Chickpea (*Cicer arietinum* L) under in vitro condition *Biological Forum – An International Journal*, 2(2): 15-18
[8] Leveau J H, S E Lindow 2005 Utilization of Plant Hormone Indole-3-Acetic Acid for Growth by *Pseudomonas putida* strain 1290 *American Society for Microbiology*, 71(5): 2365-2371
[9] Lambrecht M, Y Okon, A V Broek, Jos V 2000 Indole-3-acetic acid: a reciprocal signaling molecule in bacteria-plant interaction *Trends in Microbiology*, 8(7): 298-300

[10] Ma Y, M N V Prasad, M Rajkumar, H Freitas 2011 Plant growth promoting rhizobacteria and endophytes accelerate phytoremediation of metalliferous soils *Biotechnology Advance*, 29: 248-258

[11] Ekawati I 2009 Prospek Budidaya Kedelai Sistem Organik di Kabupaten Sumenep *Agritek*, 10(2): 80-87

[12] Michel B E and M R Kaufmann 1973 The Osmotic Potential of Polyethylene Glycol 6000 *Plant Physiol*, 51: 914-916

[13] Alikhani,H A and L Mohamadi 2010 Assessing tolerance of rhizobial lentil symbiosis isolates to salinity and drought in dry land farming condition 19th World Congress of Soil Science, Soil Solutions for a Changing World 13-16

[14] Gordon S A and R P Weber 1997 Colorimetric Estimation of Indolacetic acid *Plant Physiol*, 26: 192-195

[15] Marchesi J R, Sato, T, Weightman, A J, Martin, T A, Fry, J C, Hiom, S J, Wade, W G 1998 Design and evaluation of useful bacterium specific pcr primers that amplify genes coding for bacterial 16S rRNA *Appl Environ Microbiol*, 64: 795-764

[16] Efendi R, Sudarsono, Satriyas I, Eko S 2009 Seleksi Dini Toleransi Genotip Jagung terhadap Kekeringan *Penelitian Pertanian Tanaman Pangan*, 28(2): 63-68

[17] Kavamura V N, S N Santos, J L da Silva, M M Parma, L A Avila, A Visconti, T D Zucchi, R G Taketani, F D Andreote, I S de Melo 2013 Screening of Brazilian cacti rhizobacteria for plant growth promotion under drought *Microbiological Research*, 168: 183-191

[18] Vandevivere P, and P Baveye 1992 Improved preservation of bacterial exopolymers for scanning electron microscopy *J Microscopy*, 167: 323-330

[19] Steinmetz I, M Rohde, B Brenneke 1995 Purification and Characterization of Exopolysaccharides of *Burkholderia* (*Pseudomonas*) *pseudomallei Infection Immunity*, 63(10): 3959-3965

[20] Sandhya V, S K Ali, M Grover, G Reddy, B Venkatcswarlu 2009 Alleviation of drought stress effect in sunflower seedlings by the exopolysaccharides producing *Pseudomonas putida* strain GAP-P45 *Biol Fertile Soil*, 46: 17-26

[21] Ahmad F, Ahmad, I, Khan, MS (2005) Indole acetic acid production by the indigenous isolates of azotobacter and fluorescent pseudomonas in the presence and absence of tryptophan 29, 29-34 *Turk J Biol* 29 (2005) 29-34

[22] Drancourt M, C Bollet, A Carlioz, R Martelin, J Gayral, D Raoult 2000 16S rDNA Sequence Analysis of a Large Collection of Environmental and Clinical Unidentifiable Bacterial Isolates *Journal of Clinical Microbiology*, 38(10): 3623-3630

[23] Manero G, F Javier, R Solano, Beatriz, Probanza, Agusti'n, Mehouachi, Jalel, R Tadeo, Francisco, Talon, Manuel 2001 The plant-growth-promoting rhizobacteria *Bacillus pumilus* and *Bacillus licheniformis* produce high amounts of physiologically active gibberellins *Physiologia Plantarum*, 111(2): 206-211

[24] Mohan V, A Radhakrishnan 2012 Screening of phosphate solubilizing bacterial isolates for the growth improvement of *Tectona grandis* Linn *Research Journal of Microbiology*, 7(2): 101-113

IOP Conf. Series: Earth and Environmental Science **142** (2018) 012059 doi:10.1088/1755-1315/142/1/012059

Analysis of soybean production and import trends and its import factors in Indonesia

I H Ningrum, H Irianto and E W Riptanti
Department of Agribusiness, Faculty of Agriculture, Sebelas Maret University, Surakarta, Indonesia

E-mail: Intanhusna@ymail.com

Abstract. This study aims to analyze the factors affecting soybean imports in Indonesia and to know the trend and projection of Indonesian soybean production as well as the import in 2016-2020. The basic method used in this research is the description analysis method. The data used are secondary data in the form of time series data from 1979-2015. Methods of data analysis using simultaneous equations model with 2SLS (Two Stage Least Square) method and Trend analysis. The results showed that the factors affecting soybean imports in Indonesia are consumption and production. Consumption has positive effect while production is negatively affected. The percentage changed in soybean imports is greater than the percentage change in consumption and production of soybeans. Consumption is positively influenced by imports and production, while production is influenced positively by consumption and negative by imports. The production trend of soybean in 2016-2020 has a tendency to increase with a percentage of 11.18% per year. Production in 2016 is projected at 1.110.537 tons while in 2020 it will increase to 1,721,350 tons. The import trend in 2016-2020 has a tendency to increase with an average percentage of 4.13% per year. Import in 2016 is projected at 2.224.188 tons while in 2020 it will increase to 2.611.270 tons.

1. Introduction

The population and food are the two things that are related to one another. The more number of people than the need for food is also increasing, including soybean demand. Soybean is one of the food sources of protein that favored by Indonesian society. The study [1] in the event of a 1% population growth, the demand for soybeans will increase 10.6% which means the increase of population in Indonesia resulted in increased consumption of soy so that soybean demand is also increasing.

National soybean production in Indonesia decreased from 843.153 tons in 2012 to 779.992 tons in 2013. In 2014 and 2015, the national soybean production increased to 954.997 tons and 963.183 tons [2]. Meanwhile, soybean harvested area tends to decrease during the last 16 years an average of about 0.97% per year while the soybean crop productivity tends to increase by an average of 1.67% per year. As a result of the increase in productivity that is greater than the decrease in harvested area, the average production increased by 0.8% per year with an average production of 819.442 tons per year.

Indonesian soybean consumption is 2.2 million tons per year, this figure is far from the average rate of production that only 819.442 tons per year, as a result, the government should import soybeans. Domestic soybean production is currently only able to meet about 35% of consumption, and more than 65% of Indonesian soy needs to be met from imports. Indonesian soybean imports continue to rise each year with an average percentage of 7.73%. In 2013 the import volume of soybeans only reached 1.788.962 and it increased to 1.965.811 tons in 2014 and 2.256.932 tons in 2015 [3]. According to previous study [4], the supply of soybeans both production and the import of 70% was used for tempeh and tofu while the remaining 30% to meet the needs of other food.

Dependence on imported soybean will be analyzed to determine the development in the coming years. Based on the previous year's data, it will be seen how the development of Indonesia's soybean production and imports in the next few years. This study was conducted to determine trends and projections of production and import soybean Indonesia in 2016-2020 and also to determine what factors affecting soybean imports in Indonesia. Analytical results from this study were expected to be a material consideration in decisions relating to the import of soybean and others soybean policy.

2. Methods

The basic method used in this research was the description analysis method. The data used are secondary data in the form of time series data from 1979-2015 are sourced from the publication of the BPS, the Center for Agricultural Data and Information System (Pusdatin), Food and Agriculture Organization (FAO), World Bank, and other agencies. Data used include import of soybean, production and consumption of Indonesian soybean, per capita income, domestic soybean prices, international soybean prices and the value of the real exchange rate of Rupiah against the US Dollar. Analysis of the factors affecting soybean imports in Indonesia will be done using simultaneous equations. However, in this study we do not conduct endogeneity test due to the unavailability data. The analyzed models were consist of three endogenous variables, namely import of soybean (IM), soybean consumption (K) and soybean production (PD), which will be analyzed by Two Stage Least Square (2SLS). The equation used to analyze soybean imports in Indonesia are as follows:

$$LnIM = \alpha_{10} + \beta_{11}LnK + \beta_{12}LnPD + \beta_{13}LnKR + \beta_{14}LnHI + \mu \quad (1)$$
$$LnK = \alpha_{20} + \beta_{21} LnIM + \beta_{22} LnPD + \beta_{23} LnY + \mu \quad (2)$$
$$LnPD = \alpha_{30} + \beta_{31} LnK + \beta_{32} LnHD + \beta_{33} LnIM + \mu \quad (3)$$

Description:

A	:	constant
βi	:	the degree of slope of regression to i
IM	:	import of soybean (tons)
K	:	soybean consumption (tons)
PD	:	soybean production (tons)
KR	:	exchange rate (rupiah / US $)
HI	:	the price of International soybeans (US$/tons)
Y	:	income per capita (Rupiah)
HD	:	the price of domestic soybean (rupiah/tons)
μ	:	error

There are several steps to analyze the factors affecting soybean imports in Indonesia from the third equation. Among them is the identification test and rank condition, econometric and statistics test. Identification test and rank condition is done by considering the rules of identification (order condition) as follows:

K - k < m-1: the equation is *under-identified*
K - k = m-1: it is *exactly-identified*
K – k > m-1: it is *over-identified*

Where: K is number of variables predetermined (exogenous predetermined) in the model, k is number of variables predetermined (exogenous predetermined) in a given equations, M is number of endogenous variables in the models and m is number of endogenous variables in a given equations.

2SLS analysis can be done if a model can be identified (exactly-identified or over-identified). While in the adequacy requirement (rank condition), the equation model is identified if and only if at least one determinant is not equal to zero in order (M-1) [5]. The next step is the econometric tests, done to get a model that qualifies BLUE (Best Linear Unbiased Estimator) with normality, multicollinearity, heteroscedasticity, and autocorrelation test. The next is the statistical test. Tests conducted with test statistics R^2, F test and t test.

Analysis of trends and projections of production and import of soybean is done by using a model of trends in accordance with the data analyzed, to select the most suitable trend model for forecasting can evaluate trend model [6]. The most suitable model will give the approximate value close to its actual value. The difference between actual value and allegations referred to as an error or mistake. Trends matches will provide the most minimal error, for it can be used to select the criteria that match the trend partly by the standard error of estimation (SEE) or R-square or square Adj. R. Based on the trend found, it can be used to project soybean production and import during 2016-2020. Trend value is already known, often referred normal forecasts, but better forecast would be obtained if the effect of cyclic (S) and random (R) are included, given the components of a data *time series* annual consist of trend (Y'), the variation of cyclic (S) and random variation (R). The equation will be used to project the production and import of soybean Indonesia in 2016-2020 are:

$$SR = Y.\ Y'^{-1}.\ 100\% \qquad \text{Eq.(4)}$$
$$Yt = Y'\ SR$$

Where Yt is the actual value in year t, Y' is a trend and SR values are the combined effect of cyclical variation value and random/residue [7].

3. Results and Discussion

3.1. Analysis of suspected Factors Affecting Imports Soybean in Indonesia

Based on the of identification with order condition the result that all the equations are over identified while testing the adequacy requirement (rank condition) indicates that the determinant of all the equations is not equal to zero in order (3-1), which means that this equation is identified. Based on the identification test with the order condition and rank condition it can be concluded that the model can be estimated by the Two Stage Least Square (2SLS) method. All test econometrics to test the model by using the classical assumption states that the model qualifies BLUE (Best Linear Unbiased Estimator) with eligibility normality, no multicollinearity symptoms, homoscedasticity and no autocorrelation so that the estimated models can be used as a model equation.

Based on Table 1, results of the analysis with the simultaneous equations by 2SLS method show that, imports of soybean in Indonesia jointly influenced by consumption, soybean production, exchange rates and international soybean prices. Indonesian soybean imports partially influenced by the consumption and production of soybeans. While the exchange rate and international soybean prices partially no effect on soybean imports in Indonesia as shown in eq.I.

Consumption has a positive effect on soybean imports. The increase of soybean consumption will cause the imports of soybean will also increase. This equation shows that the percentage change in soybean imports is greater than the percentage change in consumption. This condition illustrates that a slight increase in consumption will result in the addition of imported proportions more. The proportion of imports increase more than the increase of consumption may be caused by the condition of Indonesian soybean imports were not restricted by the quota. Since 1979 restrictions on soybean imports through quotas no longer applied after previously applied from 1973, so this condition has triggered importers to import more. Soybean imports now are made by Registered Importers (IT) and Importers Manufacturers (IP), as well as the addition of state-owned enterprises as the importer of soybeans that participated in the Soybean Prices Stabilization Program, therefore, opened the opportunity for another state other state-owned enterprises than the Company BULOG to participate.

The number of parties that can import and unlimited imported soybeans makes import increased by a greater proportion than the increase in soybean consumption.

Table 1. Regression Analysis 2SLS

Equation	Variable	Coefficient	Standard Error	t-statistic	Prob
I	Constant	-2.651	3.145	-0.843	0.406
	LnK	2.662***	0.341	7.801	0.000
	LnPD	-1.450 ***	0.351	-4.134	0.000
	LnKR	-0.930	0.092	-1.010	0.320
	LnHI	-0.227	0.167	-1.358	0.184
	R Square		0.955		
	Adjusted R Square		0.949		
	F-statistic		168.027 ***		
II	Constant	1.830	0.973	1.880	0.096
	LnIM	0.304***	0.100	3.049	0.005
	LnPD	0.542***	0.092	5.864	0.000
	LnY	0.062	0.045	1.361	0.183
	RSquare		0.962		
	Adjusted R Square		0.958		
	F-statistic		276.571 ***		
III	Constant	-3.791	2.369	-1.600	0.119
	LnK	2.007***	0.281	7.146	0.000
	LnHD	-0.132	0.085	-1.553	0.130
	LnIM	-0.691***	0.193	-3.574	0.001
	R Square		0.652		
	Adjusted R Square		0.620		
	F-statistic		20.586 ***		

Source: Secondary Data analysis, 2017
Note: *** Significant at $\alpha = 1\%$

Consumption of soybean jointly influenced by imports of soybean, soybean production, and income per capita, while partially influenced by imports and soybean production as shown in eq.II. Consumption is positively influenced by imports and production. The change of soybean consumption is smaller than of import and production for soybeans. Although soybean imports have increased, people tend to increase their consumption is not too much because people have the alternative option to consume other foodstuffs. Similarly, the industry, their own production capacity with the availability of capital and the need for raw materials that have been adjusted. So, although soybean imports more they tend to increase the production of not more. Moreover, the addition of soy production that occurs is the result of efforts to substitute imported soybeans, so if the production increase, soybean imports will tend to be reduced. This is because the availability of soybeans in the market tends not to increase much.

Production has a negative effect on soybean imports, if the soybean production increases, soybean imports will decline. Based on the eq.I show that the percentage change in soybean imports is greater than the percentage change in production of soybeans, the increased production will result in a decrease in the proportion of imports that much more. This happens because one focus of the government is to achieve self-sufficiency soybeans so expect the level of import dependence can be reduced.

Soybean production is jointly influenced by consumption, the domestic soybean price, and soybean imports, while partial, production positive influenced by consumption and negative by imports, as presented in Eq.III. The change of soybean production is greater than that of soybean consumption, meanwhile, the change of soybean production is smaller than that of soybean imports. A lower soybean production makes cheaper prices at the farm level, as imports of soybeans occur constantly to meet domestic demand. The Increasing of soybean imports gives the effect to the decreasing of domestic soybean prices, but it does not have any effect on reducing the price of processed soybean products and animal food products from soybeans. The higher imports, the market power will be lower in determining the input prices and if import activity continues to rise, there is a possibility that the people who get in charge in the market can be functioned as the price recipients [8].

The exchange rate has no effect on soybean imports is partial because soy is the consumption needs that must be met. In addition, soy is a commodity that is highly maintained a stable supply and price, as it will greatly affect the economic conditions of consumer goods other given soy is one kind of strategic commodities, therefore strengthening or weakening rupiah would not affect the number of imported soy. Although the rupiah is down, soybean imports will remain to be done in order to meet consumption and maintaining economic stability in the country.

International soybean prices also had no effect on soybean imports partially. According to research previous study [9] the dependence on imported soybeans is one of the factors causing no effect of soybean import prices on soybean demand. The quantity of imported soybean exceeds the quantity of locally made imported soybean price has its share large enough so that the amount of soybean demand is not affected by the price of soybean itself. The most effective solution to meet the domestic soybean shortage in the short term is to import soybeans. So regardless of the prevailing price of imported soybean, the government will continue to buy soybean imports. This is done to ensure the availability of soy for households to be consumed directly and also for the industry which is used as raw materials. Therefore, it is high or low prices prevailing soybean imports have no effect on soybean demand in Indonesia.

3.2. Trends and Projections of Indonesia Soybean Production in 2016-2020

According to the analysis of production trends (in appendix), obtained the best production trends equation model is *Cubic*. This model will be used to look at the trend of soybean production during the years 2016-2020. Indonesian production trend chart can be seen in the following figure:

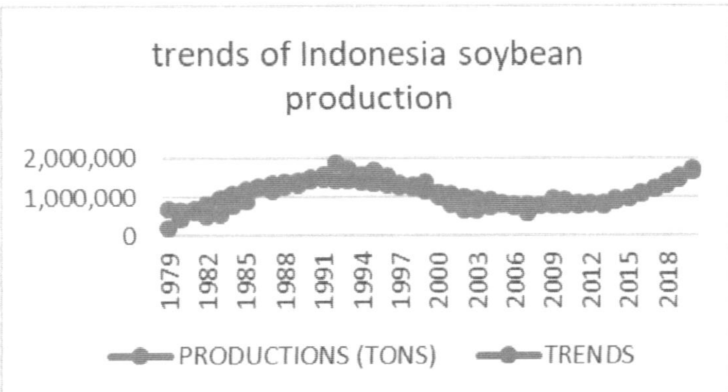

Figure 1. Trends of Indonesia Soybean Production

Based on Figure 1, it can be seen that the trend of Indonesia soybean production is to follow the non-linear patterns that tend to increase then decrease. The trend of soybean production in 2016-2020 has a tendency to increase with a percentage of 11.18% per year. If the consumption figures are not much increased the imports can be reduced. However, if the consumption figures also increase, could not be ascertained if the number of imports will be reduced.

About 50 countries around the world grow soybean and the total soybean production was estimated at 269 million Mg in 2012/2013. Among the 20 highest producing countries, the USA is currently the largest producer of soybean, followed by Brazil, Argentina, the People's Republic of China and India [10]. Although as one of the most soybean producing countries in the world, the Chinese state is also the largest importer of soybeans from the United States. The increase in consumption of vegetable oils in China, resulting in dependence on soybean imports to be very high. In addition, there is competition between cereal crops such as corn to soybeans. The decision to focus more cereal crops make Chinese soybean production tends to decline [11]. Over the past five years, the trend of production in China decreased by 5.05% per year [12].

Indonesia soybean production in 2016-2020 can be sought from the trends that have gained value. Here are the results of the calculation of the projected soybean production in 2016-2020.

Table 2. Projection of Indonesia Soybean Production in 2016-2020

Year	Trends Production (Y')	Average SR	Yt = Y'SR
2016	1,062,744.85	104.5	1,110,537
2017	1,175,598.62	104.5	1,228,466
2018	1,309,786.21	104.5	1,368,688
2019	1,466,584.63	104.5	1,532,538
2020	1,647,270.95	104.5	1,721,350
Average			1,392,316

Source: Secondary Data Analysis, 2017

Based on projected results showed that Indonesia soybean production in 2016-2020 continued to increase. Production in 2016 is projected at 1,110,537 tons, while in 2020 it will increase to 1,721,350 tons. Increased production is affected by the consumption is increasing as well. Increased consumption of an impact on the growing demand to encourage farmers to increase production. Nevertheless, the production target of 2019 to achieve self-sufficiency of 3 million tons of soybean is not expected to be sustainable. Based on the projection, soybean production in 2019 was only able to achieve 51% of its production targets.

Based on the agriculture center of research and development, to increase soybean production in Indonesia needs to be developed five sources, namely expansion of soybean growth, increased productivity, increased yield stability, suppression, and suppression gap is the result of yield loss. Of the five sources of soybean growth is then that has the largest contribution to the increase in production is the expansion of the area [14]. Availability of farmers who want to cultivate soybeans is important, besides the availability of land [15]. Farmers were not willing to plant soybeans if they can't profit from the farm. farmers will suffer losses if the community prefers soybean imports because it has a cheaper price. The establishment of the purchase price the government does not help if the price of soybean imports is still cheaper than the price of local soybean. To overcome this problem, it is necessary to keep the price of imported soybeans remains under control.

3.3. Trends and Projections of Indonesian Soybean Imports in 2016-2020

Based on the analysis of import trends (in appendix), the model obtained the best import trend equation model is quadratic. The trend of Indonesia soybean imports chart can be seen in the following figure:

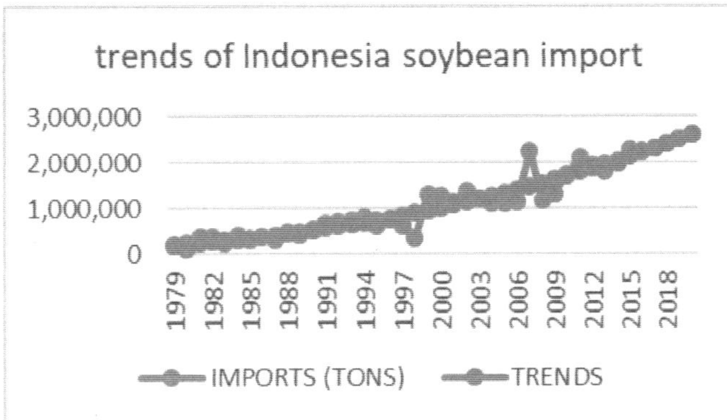

Figure 2. Trends of Indonesia Soybean Import

Referring to Figure 2 it can be seen that the trend of Indonesian soybean imports is to follow the nonlinear pattern that is likely to increase. The import trend in 2016-2020 has a tendency to increase with an average percentage of 4.13% per year. If you look at the trend of the previous production also tends to rise, increasing import trend may mean that soy consumption in Indonesia is also increased so that despite increased production, import figures have not been able to decline if the increase in consumption is still higher.

The development of world import volumes tends to increase in the period 2009 - 2013, with an average of 6.59% per year. The increase in import volume is quite high in China with an average increase of 10.48% per year. Indonesia's import volume average increased by 9.33% per year. Two countries tend to decrease, namely, Japan fell by an average 4.63% per year on average and Taiwan fell 2.25% per year [12]. China, Japan, and Taiwan are included in the three soybean importing countries in East Asia, where the fourth position is South Korea. Contributions fourth are as much as 70% of the international soybean trade. Countries in East Asia has reduced soybean planting, resulting in the production of fewer, consequently, these countries have to import soybean [15].

The result of trends that have gained value can then be used to project the value of Indonesia's soybean imports in 2016-2020. Here are the results of the calculation of the projected soybean production in 2016-2020.

Table 3. Projection of Indonesia Soybean Imports in 2016-2020

Year	Trends import (Y')	Average SR	Yt = Y'SR
2016	2,226,245.398	99.91	2,224,188
2017	2,320,012.762	99.91	2,317,869
2018	2,415,841.769	99.91	2,413,610
2019	2,513,732.419	99.91	2,511,410
2020	2,613,684.713	99.91	2,611,270
Average			**2,415,669**

Source: Secondary Data Analysis, 2017

Based on projected results can be seen that the Indonesian soybean imports in 2016-2020 continue increasing. Based on Eq.I, the increase in consumption is likely to result in import increase with a higher percentage of the increased production while potentially lead to a reduction of imports with the higher percentage, but the effect of consumption greater than production. It seems that the next five years consumption increase is very large and exceeds the increase in production, so even though production increased, imports have not been able to decline.

IOP Conf. Series: Earth and Environmental Science **142** (2018) 012059 doi:10.1088/1755-1315/142/1/012059

4. Conclusions

Factors affecting soybean imports in Indonesia is the consumption and production of soybeans. If consumption increases, import and production will also increase and if production increases, imports will be decline while consumption is increasing. The opposite relationship, if the imports increased, then consumption will also increase, but production will be decline. The production trend of soybean in 2016-2020 has a tendency to increase with a percentage of 11.18% per year. Production in 2016 is projected at 1.110.537 tons while in 2020 it will increase to 1.721.350 tons. The import trend in 2016-2020 has a tendency to increase with an average percentage of 4.13% per year. Imports in 2016 are projected at 2.224.188 tons while in 2020 it will increase to 2.611.270 tons.

References

[1] Adetama DS 2011. Thesis Department of Planning and Public Policy, Faculty of Economics, University of Indonesia.
[2] Badan Pusat Statistik 2016 Tanaman Pangan, Badan Pusat Statistik Republik Indonesia, Jakarta.
[3] Badan Pusat Statistik 2016 Statistik Perdagangan Luar Negeri: Impor 2015 Jilid I, Badan Pusat Statistik Republik Indonesia, Jakarta.
[4] Susilowati E, Oktaviani R, Arifin B, Arkeman Y 2014. The Evaluation of Cooperative's Role in Soybean Supply Chain: A Case Study of Tofu and Tempeh Cooperatives in Indonesia. *International Journal of Administrative Science & Organization* **21** 121-127.
[5] D. Gujarati (Third Edition: International Editions) 1995. Basic Econometrics. McGraw-Hill Book Co., Singapore p. 665-669.
[6] B. Juanda and Junaidi 2012. *Ekonometrika Deret Waktu: Teori dan Aplikasi,* IPB Press, Bogor, P. 45.
[7] Sudjana 1975. *Statistika untuk Ekonomi dan Niaga.* Tarsito, Bandung.
[8] Arnawa IK, Tamba IM, Anindita R 2015. The Impact of Market Power on Soybean Price in Indonesia. *Asia Pacific Journal of Sustainable Agriculture Food and Energy (APJSAFE)* **3** 1-6
[9] Anjani SR, Darwanto DH, Mulyo JH 2015. Analisis Permintaan Kedelai di Indonesia. *SEPA* **12** 42-47.
[10] Liu XB, Sheng CL, Herbert SJ, Chin KL, Qi L 2015. Mapping soybean physiology research based on the web of science. *International Journal of Plant Production* **9** 261-280.
[11] Jamet JP and Chaumet JM 2016. Soybean in China: adaptating to the liberalization. *OCL (Oilseeds and fats, Crops and Lipids)* **23** 1-9.
[12] PUSDATIN 2016. Outlook Komoditas Pertanian Tanaman Pangan Kedelai, Pusat Data dan Sistem Informasi Pertanian, Jakarta.
[13] Amang B, Sawit H dan Rachman A 1996. *Ekonomi Kedelai di Indonesia*, IPB Press, Bogor.
[14] Hasan N, Nuryani E, Hendrawan R 2015. Analysis of Soybean Production and Demand to Develop Strategic Policy of Food Self Sufficiency: A System Dynamics Framework. *ScienceDirect Procedia Computer Science* **72** 605 – 612.
[15] Kim CM, Min-Chang, Lee, EunSu 2014. Soybean Import Demand Analysis in East Asia: Korea, China, and Japan. *Management Reviewed. An International Journal* **9**

Paddy farmer households' participation and food security level in special effort program in Seputih Raman sub-district of Central Lampung Regency

K K Rangga and Y A Syarief
Department of Agricultural Social Economic, Lampung University, Indonesia

E-mail: korrangga@yahoo.com

Abstract. The objectives of this study are to study the participation of paddy farmers in the Special Effort program to increase paddy production, to study the level of household food security of paddy farmers, and to analyze the correlation between farmer participation and food security level of paddy farmers. The location was chosen purposively in Seputih Raman sub-district. The data were collected from December 2016 to February 2017. The population of this study was paddy farmers who participating in Special Effort program. The hypothesis was tested by using Spearman's Rank correlation test. Farmer household's food security was measured objectively based on the share of household's food expenditure and subjectively based on the opinion, views, and attitudes or farmers' opinions on food availability, food distribution, and household food consumption. This research showed that farmers' participation in Special Effort program in Seputih Raman Sub Ditrict, Central Lampung Regency belonged to medium classification, household food security either objectively or subjectively was in food resistant condition of medium classification, and there was significant correlation between farmers' participation and food security level of paddy farmer household.

1. Introduction

Society participation in Special Effort program activity of increasing paddy production is a form of active society use in development. As effort to realize national food sovereignty and resilience, governor through Indonesia Republic Ministry of Agriculture compiles and implements Special Effort Program of increasing Soy, Corn, and Paddy Production. This Special Effort Program of increasing Soy, Corn, and Paddy Production is a national program done by, from, and for farmer, also helped by overseeing and supervision from instructor, college student/ alumni, also Babinsa (Village Coaching Non Commissioned Officer) to do goodness together in order to hasten paddy corn, soy production increase, in all Indonesia territory [1]. Lampung province is one of province which implements Special Effort Soy, Corn, and Paddy Production Increase program. In order to reach the paddy production increasing target of one ton dried milled grain in Lampung province, Lampung Governor has submitted letter number 520/0154/III.1/2015 by 21st January 2015 to Minister of Agriculture, subject: Paddy Production Increasing Target in Lampung Province year 2015 – 2016; and followed by its description per regency/ city corresponded by letter number: 5210.21/0245/III.12/2015 by 28th January 2015 [2].

The success of Soy, Corn, and Paddy Production Increasing activity requires synergy and coordination of the whole intercross significance official from the planning, execution, monitoring, and evaluation also reporting through Soy, Corn, and Paddy Production Increase Coaching Facilitation which implemented by team stage by stage (central-village) also formed and set which fit the valid condition. At village level, society role and participation, in this case paddy farmers, is really to be expected. Participation is society's voluntary involvement based on society's determinant and self-awareness in development program [3]. Participation is willingness to build the success of every program suiting every individual skill without sacrificing his/ her own interest [4]. Participation is the participation in development, in the activities, and in utilization, also enjoying the development results [5]. The existence of society participation in Soy, Corn, and Paddy Production Increasing program expectedly increases production, enables the society to realize food security, an fulfills the society's food and nutrition need at once. Based on the description, then the research objective is to study: (1) paddy farmer participation level in Special Effort program of increasing paddy production, (2) paddy farmer household's food security level, and (3) connection of farmer participation to farmer household food security level in Seputih Raman sub-district, Central Lampung Regency.

2. Methods

This research was taken place in Seputih Raman sub-district, Lampung province, Indonesia, selected purposively considering that it is the biggest paddy production field in Central Lampung regency, and most of the farmer group member (75%) is on advanced, medium and high class. Data collection was conducted from August to September 2016. The population of this research was 303 members of farmer group implementing the paddy production increasing Special Effort program. From the population, the total sample was 30 respondents [6]. To review the questionnaire accuracy and precision, validity and reliability tests were conducted.

The first and second objectives were answered using descriptive analysis, while the third objective, the answer of the hypothesis, was addressed using non parametric statistic analysis with Spearman's Rank [7] correlation test with SPSS 21 help. Especially, the second objective was to find out farmer household's food security in objective and subjective analyses. The objective measurement was done by comparing the amount of food expenditure with the household total income. If household expenditure share $<60\%$, the household can be said as food resilient; and if the share $\geq 60\%$, the household can be said as not food resilient [8]. Subjective measurement according to [9] is a food security measure based on farmer household opinion, viewpoint, demeanor or opinion to his food situation regarding the household food availability (regarding sufficiency and stability of food availability), household food distribution (regarding food's accessibility or affordability), and farmer household's foodstuff consumption (regarding farmer household member's knowledge on food concerning the quality of food consumed and the ability of choosing food fitting the farmer household's need or sufficiency so that it's safe to be consumed).

3. Result and Discussion

3.1. Respondent Characteristics

The average age for the respondents is 43 years, which is productive age. 13 respondents (43.33%) have formal education level at elementary school, and 12 respondents (40%) have experienced junior and senior high school education level. Respondent family dependent is 3 persons in total. Average experience in paddy farming is 16 years. Average size of the owned farming field is 0.56 ha.

3.2.Farmer Participation Level in Paddy Production Increasing Special Effort Program

The farmer participation in the paddy production increasing Special Effort program viewed from some indicators referring to [4, 10] theory is as follows: (1) Participation in planning phase, (2) participation in program implementation, (3) participation in assessment, and (4) participation in utilization. Based on research result, it can be seen that 14 correspondents (46.67%) have high participation level in paddy production increasing Special Effort program, and average respondents belong to medium level. It means that respondent has had high will to participate in the paddy production increasing Special Effort program.

Viewed from each participation phase, the farmer participation in program planning process belongs to average classification, that can be seen from the farmers which follow the meeting of program planning and farmer willingness to help in preparation and data collection of Group Definitive Plan or Group Needs Definitive Plan, and farmer willingness in giving input and identifying problem happening in the paddy production increasing Special Effort program. The participation level in the paddy production increasing Special Effort program implementation is put into medium classification. Farmers have participated actively in implementing the paddy production increasing Special Effort program activity. The implementation of paddy production increasing Special Effort program includes tertiary irrigation system rehabilitation, farmer machine and tool provision assistance by governor and superior seed provision, Integrated Crop Management Implementation Movement activity, and field optimization enhancement. Farmer participation level in program assessment or evaluation of Special Effort stands in medium classification. This thing can be seen from farmers that have high willingness and skill to assess or evaluate the paddy production increase Special Effort program activity. The Assessment or evaluation includes the suitability of superior seed number given by governor to farmer, the suitability farming machine and tool which borrowed from governor to farmer, also the accompaniment done by several parties. Farmer participation level in utilizing the paddy production increasing Special Effort program is in medium classification. It means that farmers have benefitted from the paddy production increasing Special Effort program conducted. The participation level in this stage can be seen from whether or not there is profit either economically or physically which can be obtained after all the program implementation stages have been done [11]; another benefit which received by farmer is assistance benefit from governor including farmer machine and tool, and superior seed provision to support farmer farming activity, counseling activity, also cooperation accompaniment from several party.

3.3. Paddy Farmer's Food security Level in Seputih Raman sub-district

3.3.1. Objective Measure Objective food security level in this research is measured based on household food expenditure share. According to [12], the higher the household food expenditure share, the lower will be the household food security level. The result of research shows 24 respondent households (80.00%) with food expenditure share <60%, and 6 respondent households (20.00%) with food expenditure share ≥60%. High food expenditure share shows that household income is still low and food is the primary priority for household expenditure. It is in line with [13] research stating that the higher the food expenditure proportion, the lower household is the food security, in this case household prioritizes its food needs fulfilment focusing on cheap and useful food to overcome hunger.

3.3.2. Subjective Measure Based on result of research and observation on the paddy farmer household's food availability, food distribution and foodstuff consumption, it can be seen that paddy farmer household's food security level in Seputih Raman sub-district subjectively tends to fall into medium category with average value of 40.86. It means that paddy farmer households in Seputih Raman sub-district have sufficient food security. This is due to the paddy farmer household's food availability and food distribution aspects belonging to medium category with average value 12.65 and 10.39, food consumption aspect still belonging to low category with average value 17.82.

Based on each aspect, household's food availability aspect is included into medium category with average value of 12.65, meaning that farmer household has fulfilled its household food needs adequately. Some of farmer household food availability comes from self production and buying from market and stores around. There is plant production especially vegetable cultivated by farmer for own consumption, and partially sold in raw, but there is also the processed one sold later. For primary food production like paddy, farmers prioritize more the harvest saved as stock, because rice is the only primary food in the area and farmers will sell the harvest if it exceeds stock capacity. Food availability stability at household level is based on the food availability sufficiency and household member's eating frequency a day. Farmer households commonly do not prepare food reserve for one month in the near future, but they prepare food for just couple of days. The type of primary food usually consumed by farmer household is rice without another primary food mixed, like corn or yam, while the farmer household's food eating frequency is 3 times a day.

From farmer household's food distribution aspect including accessibility/ and affordability, the farmer household's food is included into medium category with average value of 10.39. It means that food distribution in paddy farmer household is good enough. The field condition shows that farmer household's food accessibility/ affordability to obtain foodstuff is easy enough, since they produced themselves the foodstuff they need. Regarding the foodstuff they buy, the way to obtain it is also quite easy, because there is market in respondent residence area with pretty close distance (500 m) and affordable price, and there are stores also merchant selling around every morning.

Viewed from consumption aspect, farmer household is included into low category with average value of 17.82, meaning that the knowledge on food consumed by farmer household is still low. The field condition shows the respondent's lack of knowledge on: food which needed by body in right composition, all types of food useful to body health and in the consumption of food, farmer households do not consider the quality of food consumed. Farmer households only consumed the food as it is and what they want suiting their taste is the only one matter without specific considerations regarding the food quality. In similar vein, [13] suggest that household with low income like farmer focuses its spending on fulfilling the food needs in quantity, while the quality aspect is less considered. [14] say that the high farming revenue income actually doesn't affect the farmer household's food consumption habit, because the most important thing from food is the foodstuff's ability of filling in the stomach, so the food doesn't need to be expensive or luxurious.

3.4. Hypothesis Testing

The result of hypothesis testing between farmer participant and farmer household food security level obtains correlation coefficient value of 0.923** at significance level of 0.000 (table 1). That value is smaller compared with value α 0.1 or at confidence interval of 90 percent, meaning that H_1 is supported, there is a very significant correlation between farmer participation and paddy farmer household's food security level variables in Seputih Raman sub-district. The higher the farmer participation in Special Effort program, the higher is the household's food security. In similar vein, [9] says that farmer group participation affects the household's food security in *Desa Mandiri Pangan* (Food Self-Sufficient Village).

Table 1. Result of Correlation Analysis on Farmer Participation and Paddy Farmer Household's Resilience Level

Variable X	Variable Y	Correlation Coefficient ($r_{s)}$)	Sig (1-tailed)
Participation	Paddy farmer household food security level	.973**	.000

Description :
r_s:*Rank Spearman*
** :Very significant at Confidence interval of 90% (α=0,1)

4. Conclusion

Considering the results it can be concluded that: (1) Paddy farmer participation level in the paddy production increasing Special Effort program in Seputih Raman sub-district is included into medium classification; (2) paddy farmer household's food security level in Seputih Raman sub-district objectively and subjectively belongs to food security condition with medium classification; and (3) there is a very significant correlation between farmer participation and paddy farmer household's food security in Seputih Raman sub-district and Central Lampung regency.

References

[1] Kementrian Pertanian (Agricultural Ministry) 2015 Agricultural development policy in 2015-2016 RI's Agricultural Ministry Jakarta

[2] Dinas Pertanian Tanaman Pangan dan Hortikultura Provinsi Lampung (Food Plant and Horticultural Agricultural Service of Lampung Province) 2015 manual instruction of special effort building to improve paddy, corn, and soy production (UP2PJK) of lampung province in 2015 Lampung

[3] Soetomo 2006 Strategies of community development Pustaka Pelajar Yogyakarta

[4] Ndraha 1990 Community development: preparing the taking-off community Jakarta

[5] Murtiyanto N 2011 Public participation Downloaded *http://bagasaskara.wordpress.com/2011/10/12/partispasi-masyarakat-teori-ringkas* Accessed on February 12, 2016

[6] Sugiarto D, Siagian L T, Sunaryanto 2003 Sampling technique PT Gramedia Pustaka Utama Jakarta

[7] Siegel S 2011 Non-parametric statistics for social studies PT Gramedia Jakarta

[8] Purwaningsih Y 2010 Household expenditure pattern according to food security level in central java province (Yogyakarta, Indonesia: Faculty of Agriculture, Gadjah Mada University of Yogyakarta)

[9] Rangga K K 2014 The effectiveness of micro-scale enterprise affinity group in realizing the household food security in food self-sufficient village in lampung province Dissertation Sebelas Maret University Surakarta

[10] Dwiningrum S I A 2011 Decentralization and public participation in education Pustaka Pelajar Yogyakarta

[11] Batubara R H 2015 Public participation in gsmk program in gedung aji sub district of tulang bawang regency. *Thesis*. Lampung University

[12] Ilham N and Sinaga B M 2007 The use of food expenditure as composite indicator of food security. *Journal SOCA* **7** (3) : 213-328.

[13] Amaliyah H and Handayani S M 2011 An analysis on the relationship between food expenditure and consumption proportion and the paddy farmer household's food security in klaten regency. *SEPA* **7**(2) 110-118. *http://agribisnis.fp.uns.ac.id/wpcontent/uploads/2013/-10/06Husnul-Amaliyah-Sugiharti-Mulya-Analisis-Hubungan-Proporsi-Pengeluaran-dan-Konsumsi-Pangan-Dengan-Ketahanan-Pangan-1.pdf* Accessed on November 1, 2014

[14] Hernanda T AP, Indriani Y and Listiana I 2013 Income of corn farming and food security of farmer households in simpang sub district of ogan komering ulu (oku) selatan regency JIIA 1(4)

IOP Conf. Series: Earth and Environmental Science **142** (2018) 012061 doi:10.1088/1755-1315/142/1/012061

The analysis of export commodity competitiveness in Central Java Province at period 2011-2015

M G Elysi[1], Darsono[2] and E W Riptanti[2]
[1]Student Department of Agribusiness, Faculty of Agriculture, Universitas Sebelas Maret, Surakarta, Indonesia
[2]Staff Department of Agribusiness, Faculty of Agriculture, Universitas Sebelas Maret, Surakarta, Indonesia

E-mail : meiviana67@gmail.com

Abstract. This study aims to determine the competitiveness of furniture export in Central Java Province Indonesia in terms of comparative and competitive advantages and to formulate the strategies to enhance the competitiveness of furniture export in Central Java Province. Descriptive analytic method was used for this research. Data used in this research are primary and secondary data. Data were analyzed using RCA (*Revealed Comparative Advantage*), TSI (*Trade Specialization Index*) and SOAR (*Strength, Opportunity, Aspiration, Result*). The results showed that furniture commodity in Central Java Province had strong comparative competitiveness with the value of RCA>1 and had strong competitiveness with positive index values in the range of 0 to 1. Based on SOAR analysis, strategy measures can be formulated, namely maintaining the existing and expanding targeted markets, improving product designs (innovations) and improving raw materials efficiency.

1. Introduction

International trading is an activity which done by a resident of a country with the population of other countries on the basis of mutual agreement. The existence of international trade has contributed to the industrialization, transportation advances, globalization and the presence of multinational companies [1]. One of the international trade activity has an important role is export. Indonesia's export structure divided into two parts, namely the export of oil and gas (oil) and non-oil exports [2].

Table 1. Values of oil and gas and non oil exports Indonesia years 2011-2015 (US$ million)

Tahun	Non Migas	Migas
2011	162.019.60	41.477.04
2012	153.043.00	36.977.30
2013	149.918.76	32.633.03
2014	145.961.21	30.018.78
2015	131.730.33	18.551.93

Based on Table 1, indicating that non-oil exports is much larger than the value of its exports of oil and gas exports, so the government needs to pay attention to non-oil exports over again to be developed, given its role with considerable value. In the non-oil sector are the flagship products of wood, the furniture. Furniture is one of the products of wood that have high economic value and profitable. Central Java province is one of the provinces that have the potential and have the advantage in producing furniture. The province can locate its position as the largest furniture export support compared to other provinces, so that the province of Central Java can be considered as one of the biggest furniture center in Indonesia.

Table 2. Commodity export value furniture Central Java province against total non-oil exports Central Java province years 2011-2016 (US$)

Year	Export Value of Furniture	Total Exports of Non Oil and Gas	Share (%)
2011	237.959.026	4.259.587.721	5.59
2012	258.839.029	4.513.026.249	5.74
2013	341.867.270	4.871.878.841	7.02
2014	395.680.140	5.297.193.878	7.47
2015	451.513.101	5.261.656.280	8.58

Commodity export value of furniture Central Java province Years 2011-2015 shows the increase from year to year, so the furniture industry in Central Java is very potential to be developed considering the market increasingly attractive furniture. The contribution of commodity exports of furniture in Central Java province in affect total value of non-oil exports, efforts are needed to stabilize and improve the competitiveness of commodity exports of furniture, with competitiveness increases expected commodity export furniture Central Java province is able to be an added value to the regional income. The export activities to various destination countries of course the furniture industry in Central Java province get competition from other furniture centers in Indonesia. Therefore, there is a need for an analysis to determine the extent of the furniture export competitiveness in Central Java province based on comparative advantage and competitive advantage, in order to support the strategy in furniture export competitiveness in Central Java province. So the purpose of this study: (1) determine the competitiveness of export commodities furniture Central Java province Years 2011-2015 in terms of comparative and competitive advantages, and (2) formulating the strategy effort in improving the competitiveness of commodity exports furniture Central Java province.

2. Methods

2.1. Research Basic Method

The basic method of research used in this research is descriptive analysis method using periodic data (*timeseries*).

2.2. Location Determination Method

Determining the location of research conducted intentionally (*purposive*) that is in Central Java Province.

2.3. Key Informant Determination Method

Determination *of key informants* in the study conducted in purposive *stakeholders* involved and the effect on commodity export activity furniture in Central Java province, as well as determine the condition of the commodity exports of furniture in Central Java province as a whole.

2.4. Types and Sources of Data

The data used in this study are primary data and secondary data.

2.5. Analysis Method

2.5.1. Revealed Comparative Advantage (RCA) RCA method is one method to determine the competitiveness of export commodities of the country or the industry in terms of comparative advantage. This study did not use the RCA generally comparing the export of a commodity specific countries with export commodity world countries, but in a smaller scope, which is the ratio of commodity exports furniture Central Java province with the export commodities of Indonesia's furniture, systematically value of RCA used in this study are as follows:

$$RCA = \frac{Xij/Xt}{Wij/Wt} \quad \ldots\ldots\ldots\ldots\ldots\ldots\ldots (1)$$

Where the RCA is the competitiveness of commodity exports furniture Central Java province in terms of comparative advantage, Xij is the value of commodity exports furniture Central Java province, Xt is the value of total exports of Central Java province, Wij is the value of commodity exports of Indonesian furniture, and Wt is the total value of Indonesian exports. There are two possible values of competitiveness obtained from the calculation, that if RCA> 1, means shows that commodity export furniture Central Java province has a strong comparative competitiveness. If the RCA <1, meaning shows that commodity export furniture Central Java province has a weak comparative competitiveness [3].

2.5.2. Trade Specialization Index (TSI) Trade Specialization Index (ISP) is used to calculate the competitive advantages of commodity furniture Central Java in Indonesia. Analysis of these TSI can describe the position of Central Java for commodity furniture, which tends to be the exporter or importer and put the furniture industry on a stage (competitive position) industrialization that can describe the capabilities gained through the performance of the maximum gain in terms of value over the technology or devices wisdom government. The formula used is as follows:

$$TSI = \frac{Nx' - Nm'}{Nx' + Nm'} \quad \ldots\ldots\ldots\ldots\ldots\ldots (2)$$

Where the TSI is Trade Specialization Index, Nx 'is the value of commodity exports furniture Central Java, and Nm' is the value of commodity imports furniture Central Java. Values of this index has a range between -1 to +1. If the positive value above 0 to 1, then the furniture commodity is said to have a strong competitive competitiveness or Central Java tends to be as an exporter of commodities furniture. Instead, competitiveness on weak or inclined as the importer, if the negative value below the 0 to -1.

TSI can also be used to identify the position of the competitiveness of a commodity in trade which is divided into phases, as follows: (1) Phase introduction, in this phase, the value index of industrial TSI *latercomer* is -1.00 to -0.50; (2) Phase Import Substitution TSI is the index value rose between -0.51 to 0.00; (3) Stage Growth is the ISP index value rose from 0.01 to 0.80; (4) Phase Maturity is the index value is in the range 0.81 to 1.00 [4].

2.5.3. Matrix SOAR (Stregth, Opportunity, Aspiration, Result) SOAR analysis is one tool to the strategic planning approach that focuses on strength and trying to understand the whole system by incorporating opinions from *stakeholders.* the relevant SOAR framework focuses on the formulation and implementation of positive strategies by identifying strengths, build creativity in the form of opportunities, encourage stakeholders to share their aspirations and determine the size and outcome means [5]. Result or results indicate efforts competitiveness strategy of increasing commodity exports furniture Central Java province obtained by looking at the aspect of power, opportunities and aspirations.

3. Ressults and Discussion

3.1. Export Competitiveness Commodity Furniture Central Java province in terms of the Comparative Advantage

Revealed Comparative Advantage (RCA) in this study was defined that if the share of commodity export of furniture in the total export of commodities from Central Java province is greater than the market share of commodity export furniture inside total commodity exports in Indonesia, Central Java province is expected to have a comparative advantage in producing specialized furniture commodity based efficiency compared with other countries. RCA is also an index to find out how the performance of commodity exports furniture based on the pattern of trade.

Table 3. Development of RCA value commodity furniture Central Java Province years 2011-2015

Tahun	RCA Value
2011	12.51
2012	10.83
2013	11.71
2014	11.76
2015	12.05
Total	**58.86**
Average	**11.77**

Based on Table 4, it can be seen that in the period 2011-2015, shows that the average value of RCA commodity furniture Central Java province has a value above 1, which amounted to 11.77 with RCA index ratio of 1.00. This means that during this period furniture commodity Central Java province has a comparative advantage strong competitiveness in the international market compared with other provinces to commodity furniture.

The value of comparative competitiveness of commodity exports furniture Central Java province relies heavily on changes in the exchange rate against the US dollar. Besides natural resources as well as a determining factor in the analysis of comparative competitiveness, which can be seen from the availability of land resources are used for the development of wood *supply all* raw material furniture. Central Java province has its own forests to produce wood for furniture raw material. The forest is located in Perum Perhutani Unit I Central Java with a total area of 635.43 thousand hectares, which is then used as a function according to forest production as a raw material in the manufacture of furniture [6].

The main problems faced by the wooden furniture industry in Central Java province is the imbalance between supply and demand for wood. According to the Industry and Trade [7], the demand of wood each year to reach 3-3.5 million m3, while the ability of Central Java province to supply the raw materials are still about 400 thousand m3 per year, derived from forestry are the main suppliers. Attempts to cover the shortage of raw material needs is done by bringing raw materials from outside of Central Java province as from West Java, East Java and Kalimantan.

Despite these difficulties the wood raw material, furniture production in Central Java province still able to grow and grow fast due to the *image* well-known of furniture products to the international market due to increased demand. The raw materials of furniture more varied since the raw material is no longer solely teak and mahogany, but also started a lot of use, Mindi wood, wood bayur, combined with sea grass, water hyacinth, banana trees, even with metal or brass. In addition to the raw materials that vary in furniture manufacture is supported by the human resources or furniture craftsmen that have been experienced on the field and carried out by the hands of skilled craftsmen who get the skills and expertise of the patrimony. Skills are also supported by the various facilities available are courses that specialize school field development and engineering expertise timber (PIKA). Therefore, the quality of the product can be trusted by consumers, so the furniture industry in Central Java province comparative able to compete in international markets.

3.2. Commodity export competitiveness furniture Central Java province in terms of competitive advantage

Analysis used to measure competitive advantage of the export competitiveness of furniture Central Java province is using Trade Specialization Index (TSI). This competitive advantage illustrates the position of Central Java for commodity furniture tends to be the exporter or importer and put the furniture industry in a competitive position industrialization stage which can represent a capability

IOP Conf. Series: Earth and Environmental Science **142** (2018) 012061 doi:10.1088/1755-1315/142/1/012061

gained through the performance of maximum benefit in terms of technology or device value of government policies.

Table 4. Development of ISP value commodity furniture Central Java Province years 2011-2015

Year	ISP Value
2011	0.97
2012	0.99
2013	0.99
2014	0.99
2015	1.00
Total	**4.94**
Average	**0.99**

Table 4 shows the average value of the commodity TSI furniture Central Java province in the period 2011 to 2015 amounted to 0.99 where the value indicates the positive value of the index is in the range of 0 to 1, so that the said commodity furniture competitiveness has a strong competitive or Central Java tends to be as an exporter of commodities furniture. TSI can also be used to identify the competitive position of a commodity in trade. Furniture commodity competitiveness position of Central Java province showed the average value of the TSI that is in the range of 0.81 up to 1.00, the average value can be inferred that the commodity furniture Central Java province at the stage of *maturation.*

Furniture products in Central Java province is able to qualify for export qualify. Meanwhile, in order to do export, the furniture industry to be recognized as a Registered Exporter of Forestry Industry Products (ETPIK), which is an industrial company that has had industrial permit issued on the basis of legislation in force and meet the requirements under the rules of the Ministry of Trade No. 20 / M -DAG / PER / 5/2008 concerning the export of forestry industry products [8].

So far the government continues to encourage and develop policies to improve the competitiveness of furniture with running a promotional strategy through the Minister of Industry No. 90 / M-IND / PER / 11/2011 [9]. Promotion strategies carried out by constantly strive to facilitate the participation of a furniture company at the exhibition. One of the exhibition is the Indonesian International Furniture Expo (IFEX). Indonesian International Furniture Expo (IFEX) is one of the medium used to promote innovation and excellence furniture and handicraft products to a global audience, while helping increase the value of exports [10].

The increase in export value is also strengthened by the issuance of regulations regarding the legality of the timber, which aims to ensure Indonesian timber supply on the world market are not obtained illegally. The Government has established the Timber Legality Verification System (SVLK) for forest management and wood raw material processing companies to implement these policies through Decree No. 64 / M-DAG-PER / 10/12 [11], concerning the provision of export products of the forestry industry through, that exports of forestry industry products shall be equipped with a V-Legal Documents [12].

4. Efforts Strategy to Increase Export Competitiveness Furniture

Efforts strategy to increase competitiveness of commodity exports furniture Central Java using SOAR analysis. SOAR matrix can be seen in Table 5. Following the formulation of strategies for improving efforts commodity export competitiveness furniture Central Java Province, namely (1) Maintain the existing export markets and expand the target market; (2) Improve the product design (innovation); (3) Creating a raw material efficiency.

Table 5. Matrix SOAR competitiveness improvement furniture export commodities Central Java Province

STRENGTH	OPPORTUNITY
1. Experienced Human Resources 2. Image products are well known to the international market 3. The selling price of competitive products so as to compete in the export market 4. Quality of products trusted consumers 5. On time in completing order so as to meet the demand of export market 6. The existence of an effective work process, craftsmen work according to order. Not producing a failed product	1. The market share of furniture export is wide open 2. Capital support from banks is very helpful for furniture production activities to meet the demand of export markets 3. Central Java has its own forest area in producing wood for furniture raw materials
ASPIRATION	**RESULT**
1. Expect new product innovations to increase 2. Expect there is easy capital assistance 3. Expect to be able to participate in various prestigious furniture exhibition events inside and outside the country 4. Expect the raw materials eficience 5. Expect a balance between the needs and supply of raw materials	1. Maintaining existing export markets and expanding target markets (S1, S2, S3, S4, S5, S6, O1, O2. A1, A4) 2. Improve product design (innovation) (S1, O1, A1, A3) 3. Create raw material terminal (O3, A2, A5, A6)

5. Conclusion

Based on the analysis of Revealed Comparative and the analysis *of Revealed Comparative Advantage* (RCA) in the data period 2011 to 2015, furniture commodity in Central Java province has an average value of 11.77 or more than one. That's means during that time, furniture commodity in Central Java province has a comparative advantage and strong competitiveness in the international market compared to other provinces. Moreover, according to the analysis of Trade Specialization Index (ISP) for furniture commodity in Central Java province has an average value of 0.99 which is in the range of 0 to 1, so the furniture commodity is said to have strong competitiveness means Central Java tends to be an exporter. To the position result, the competitiveness of furniture commodity has average value of at 0.81 up to 1.00, so that the furniture commodity is at a stage of maturation exports. Efforts strategies can be formulated for improving the competitiveness of commodity exports furniture Central Java Province based on analysis of SOAR include: (1) Maintaining the export market existing and expanding target markets, (2) Improve the product design (innovation), (3) Efficiency material raw.

References

[1] Rahimah, A. 2012. *Administrasi Kepabeanan & Ekspor Impor*. Malang.
[2] Departemen Perindustrian dan Perdagangan. 2008. *Pengembangan Industri Pengolahan Rotan Indonesia. Biro Umum dan Humas*. Jakarta.
[3] Balassa, B. A. (1965). Trade liberalization and revealed comparative advantage. *Manchester School of Economic Studies*. **32**(2) : 99-122.
[4] Tambunan, T. (2004). *Globalisasi dan perdagangan internasional*. Ghalia Indonesia. Bogor.
[5] Stavros Jacqueline, Cooperrider David, Kelly D. Lynn. 2003. Strategic Inquiry, Appreciative Intent: Inspiration to SOAR. A New Framework for Strategic Planning. *Journal Ai Practitioner*, Volume November 2003 page 1-21.
[6] BPS. 2016. *Jawa Tengah dalam Angka 2016*. Badan Pusat Statistik, Semarang.
[7] Dinas Perindustrian dan Perdagangan Jawa Tengah. 2014. *Perkembangan Ekspor*. E-Paper Info INDAG. Semarang.
[8] Menteri Perindustrian dan Perdagangan. 2008. *Ketentuan Umum di Bidang Ekspor*. Keputusan Menteri Perindustrian dan Perdagangan No. No. 20 / M -DAG / PER / 5/2008. Jakarta.

IOP Conf. Series: Earth and Environmental Science **142** (2018) 012061 doi:10.1088/1755-1315/142/1/012061

[9] Menteri Perindustrian dan Perdagangan. 2011. *Peta Paduan (Road Map) Pengembanan Kluster Industri Furniture*. Keputusan Menteri Perindustrian dan Perdagangan No. 64 / M-DAG-PER / 10/12. Jakarta.

[10] Indonesia Tourism Forum, 2016. IFEX Sebagai Pendorong Pertumbuhan Industri Mebel dan Kerajinan Indonesia. *http://www.indonesia-tourism.com.* Diakses 13 April 2017.

[11] Menteri Perindustrian dan Perdagangan. 2012. *Ketentuan Ekspor Produk Kehutanan*. Keputusan Menteri Perindustrian dan Perdagangan No. 64 / M-DAG-PER / 10/12. Jakarta.

[12] Wibowo, B. 2016. *Terobosan Mengangkat Daya Saing*. Indonesian Barcore Association. Jakarta.

IOP Conf. Series: Earth and Environmental Science **142** (2018) 012062 doi:10.1088/1755-1315/142/1/012062

Building agribusiness model of LEISA to achieve sustainable agriculture in Surian Subdistrict of Sumedang Regency West Java Indonesia

E Djuwendah, T Priyatna, K Kusno, Y Deliana and E Wulandari
Agribusiness Study Program, Faculty of Agriculture, Universitas Padjadjaran, Indonesia

E-mail: endah_djuwendah@yahoo.com

Abstract. Building agribusiness model of LEISA is needed as a prototype of sustainable regional and economic development (SRRED) in the watersheds (DAS) of West Java Province. Agribusiness model of LEISA is a sustainable agribusiness system applying low external input. The system was developed in the framework of optimizing local-based productive resources including soil, water, vegetation, microclimate, renewable energy, appropriate technology, social capital, environment and human resources by combining various subsystems including integrated production subsystems of crops, livestock and fish to provide a maximum synergy effect, post-harvest subsystem and processing of results, marketing subsystems and supporting subsystems. In this study, the ecological boundary of Cipunegara sub-watershed ecosystem, administrative boundaries are Surian Subdistricts in Sumedang. The purpose of this study are to identify the potency of natural resources and local agricultural technologies that could support the LEISA model in Surian and to identify the potency of internal and external inputs in the LEISA model. The research used qualitative descriptive method and technical action research. Data were obtained through interviews, documentation, and observation. The results showed that natural resources in the form of agricultural land, water resources, livestock resources, and human labor are sufficient to support agribusiness model of LEISA. LEISA agribusiness model that has been applied in the research location is the integration of beef cattle, agroforestry, and agrosilvopasture. By building LEISA model, agribusiness can optimize the utilization of locally based productive resources, reduce dependence on external resources, and support sustainable food security.

1. Introduction

Building LEISA agribusiness model is needed as a prototype of sustainable rural and regional economic development (SRRED) on watersheds (DAS) in West Java Province. Therefore, this agribusiness system is developed in the agribusiness development region that pays attention to ecological, social and functional boundaries.

Sustainable agricultural development is essentially producing food and conserving agricultural resource base. The term used to describe the terminology refers to ecological agriculture. One of the sustainable agricultural systems is Low External Input Sustainable Agriculture (LEISA), which emphasizes the efficient use of internal production factors to create sustainable agriculture. The five principles of sustainable agriculture are ecological stability, economic sustainability, just, humane, and flexible [1]. LEISA system combines the components of plants, animals, soil, water, climate, and humans in the production system in order to complement each other and have a sense of ecologically and economically proper integrated agricultural system [2].

Ecologically, Surian Subdistrict is located in sub-watershed of Cikandung, Cipunagara, which is one of rural areas that have great potential of natural resources for the development of LEISA system based on agriculture, livestock, and forestry. However, the development of Surian Subdistrict is slow, as can be seen in the lower economic growth rate compared to the economic growth rate of Sumedang Regency, and in the poor road infrastructure. This indicates that the synergy among the subsectors of agriculture, industry, and services is still weak. Therefore, the agricultural sector needs to be encouraged to be the regional economic base in the Surian Subdistrict.

The potential of agricultural resources existing in this region has not been fully utilized in supporting sustainable agriculture. This is evident from the fact that many farmers still rely on chemical fertilizers and drugs in the practice of plant cultivation, and that animal manure has not been optimally used as fertilizer or biogas in agricultural activities [3].

This study aimed to identify the potential agricultural resources in Surian Subdistrict of Sumedang Regency to support the sustainable agriculture of LEISA model, and to obtain a study of the potential of internal and external inputs in the agribusiness model of LEISA.

2. Methods

This research used qualitative method using case study technique. Case study is defined as a research strategy in which researchers carefully investigate a program, event, activity, process, or group of individuals [4]. The ecological boundary of the research location was Cikandung sub-watershed ecosystem, and is administratively located in the Surian Subdistrict of Sumedang Regency of West Java. The data used consisted of primary and secondary data. Primary data were obtained through interview with informants, participatory observation, and documentation. Secondary data were obtained through literature review of some research results. The object of the research was the various integrated farming of LEISA model.

3. Research and Discussion

3.1. The Potential of Surian Subdistrict in the Agricultural Development of LEISA System

Low external input and sustainable agriculture (LEISA) is agriculture that optimizes the utilization of natural resources (land, water, plants, animals, and humans) available somewhere that is economically proper, ecologically sound, and socially adapted to the local culture in that location [1].

Table 1 shows that the condition of agricultural resources of Surian Subdistrict has potency to support sustainable agriculture applying LEISA system. Most of the land in Surian Subdistrict is the land utilized by the community for the plantation of some crops such as Albasia, coffee, pepper, and banana. The sources of livelihood of the inhabitants are agriculture and livestock. The availability of local inputs for the development of the LEISA system in the Surian Subdistrict encompasses considerable agricultural land using rotary cropping pattern of rice–rice–non rice/*palawija* (soybean or corn), wetland irrigation, livestock manure, availability of crop residues, use of organic fertilizers, and integrated pest control.

In farming, farmers in Surian Subdistrict joined 5 combined farmer groups (*gapoktan*), namely Ranggamukti Jayasari, Giri Mukti, Suriatani, and Tunas Harapan. Cultivated commodities in the wetland and heterogeneous gardens were wetland paddy (6,324 quintals), soybean (1,791 quintals), tree yam (140 quintals), corn (357.43 quintals), sweet potato (120 quintals), fruits, and various timbers. The livestock farming included beef cattle (1,209 heads), goats (1,773 heads), sheep (1,766 heads), chicken (28,600 heads), and duck (1,805 heads). In addition, in the field of fisheries, types of fish that were mostly cultivated were carp (16,544 kg) and parrotfish (15,540 kg) [5]. Some agroindustry activities that were growing in Surian Subdistrict included tofu and tempeh business, and cassava chips and banana chips industries.

Table 1. The Potential of Agricultural Resources of Surian Subdistrict in 2016

Agricultural Resource	Information
Land resource (Ha)	
- Wetland	951.07
- Dry land	455.45
- Forest	3,114.65
- Height	170-500 m ASL
Agro climate condition:	
Rainfall	2075 mm per year
Rainy days	172
Humidity	25
Average temperature	23-27 °C
Water availability:	9-12 months coming from Cikandung River,
Water resource	Cipunagara Watershed of Sumedang Regency
Wetland cultivation techniques	Ip-200, *jajar legowo* 2 : 1 intercropping
Irrigation	Half-technical self-help
Agricultural and plantation commodities:	Agricultural Waste
paddy, soybean, corn, sweet potato, peanut, banana, and coffee	Straw, husk leaves of corn, cassava leaves, sweet potato leaves, banana leaves, coffee skins
Agroindustry type:	Agroindustry waste
Manufacture of tofu, tempeh, banana and cassava chips	Soy skins, banana skins, cassava skins, tofu dregs, and *onggok* of tofu

Local inputs were in the form of animal waste that can be used as manure to cultivate rice crops. Animal waste was obtained from beef cattle groups found in almost all villages in the Surian Subdistrict. Based on the information of interviews conducted to breeders in the Ranggajaya cattle breeder group located in the village of Ranggasari, the animal waste produced about 4 kilograms per cow and 2 kg per sheep or goat. Thus, livestock manure produced in the Surian Subdistrict per day was predicted to reach 4.8 tons of cow dung and 7.08 tons of goat and sheep dung. Similarly, the urine of goat, cow, and sheep can be used as liquid organic fertilizer and local microorganism (MOL) to help fertilize the soil. However, only about 40 percent of the livestock manure potential was utilized as organic fertilizer. Other local inputs in the form of agricultural and agroindustrial waste can be used as the raw material for composting, or for animal feed.

According to farmers, the harvest or paddy straw residue reached 4 tons per hectare. Surian Subdistrict had a harvest area of wetland paddy of 1,840 hectares, thus the potential of paddy straw that can be processed into animal feed or returned to the soil as organic fertilizer was 7,360 tons.

Livestock maintenance is one component in farming that will integrate with other commodities cultivated by farmers. Thus, if small-scale farming oriented to the family business, then the development program is based on an integrated agricultural system. Several potential integrated agricultural technologies to support sustainable agriculture system in Surian Subdistrict included intercropping, agroforestry, and integration of laying duck maintenance with paddy rice.

3.1.1. *Multi Cropping Pattern in the Wetland Ecosystem and Intercropping Pattern in the Dry Land Ecosystem* The commodities cultivated in the wetland of Surian Subdistrict were rice, corn, soybeans, cassava, green beans, and vegetables. Wetland was generally planted twice with rice and once with palawija as the alternative crops (rice–rice–non rice/palawija), both of which were planted in monoculture. Meanwhile, dry land like moor was planted with vegetable crops. There were also farmers who implemented intercropping pattern with paddy-fish integrative farming (mina padi) pattern, that is, (1) fish as the interval between rice planting seasons, which is usually done while waiting for rice seedlings to be ready to plant, which is between 20 and 30 days; and (2) the maintenance of fish with paddy by using jajar legowo paddy planting system.

3.1.2. *Jajar Legowo Planting System* Legowo planting system is the way of planting wetland paddy with the pattern of several rows of plants interspersed with empty rows. Crop that should be planted in the empty rows are moved to be as insertion plants in the rows. Previously, jajar legowo planting was commonly applied to areas with many pests and disease attacks. On each empty row between every legowo unit, a shallow ditch can be made. The trenches can serve to collect the golden snail, reduce iron-poisoning level in paddy plants, or work for the maintenance of small fish [6]. The implementation of this system can reduce, even stop, the population of plant hopper pests in the wetland without the use of synthetic pesticides, so the results from time to time will remain, and the results do not contain residues of synthetic pesticides. Another advantage, this model can improve soil fertility because the fish manure and food waste serve as fertilizer.

According to information from agricultural extension officers, only about 30 percent of farmers in Surian Subdistrict applied this legowo-farming pattern. This is because the large majority of farmers were still farming with conventional system.

3.1.3. *Agroforestry* Agroforestry is a farming system that combines forest trees with agricultural crops, livestock, and/or fish on the same land to form a pattern of cultivation with the aim to optimize the social, economic, and ecological benefits [7].

Pepper and cardamom are plantation crops that are excellent for the community in Surian through the intercropping cultivation pattern with banana plants, animal feed, moluccan albizia (Falcataria moluccana Miq.) and magnolia blumei (Manglietia glauca Bl). By looking at the pattern of utilization of plantation land in Surian Subdistrict, it can be said that this region has developed agrosilvopasture. Agrosilvopasture is a combination of forest plants or woods and grasses of forage in order that the conservation is guaranteed and the demand for forage is sufficient without damaging the environment and the weight of the beef livestock and reducing feed conversion.

3.1.4. *Integration of laying duck maintenance with paddy rice* In order to utilize agricultural waste and plant pest control in Surian Sub-district, there are many farmers who apply the maintenance business of laying duck with traditionally maintained rice paddy. The form of management is grazing on the move. In this system of integration, duck farms feed on the remnants of rice that are lost or left behind during harvest, the grass growing in paddy fields, insects, snails, stone crabs, small frogs and so on [8]. Meanwhile, rice paddy will benefit from the reduction of pest attacks such as weeds, insects and others because it has been eaten by the duck livestock. In addition, organic fertilizer is obtained in the form of manure from ducks while grazing.

Integration of ducks and paddy fields provides benefits in terms of egg production and better rice production, supported by the reciprocal benefit (interaction) of integrated cohesive integration between ducks and rice [9]. The results of the previous [10] in North Sumatra showed that ducklings integrated with irrigated rice paddy fields could control the golden snail pest and the implications of duck and rice egg production increased significantly.

3.2. Obstacles of the Potential of Agriculture in Surian Subdistrict

In order to develop the potential of Surian Subdistrict, there were still several obstacles, including that the available human resources had not been able to be developed well yet both in terms of knowledge and skills in the processing of manure into compost, liquid fertilizer or biogas, in the technologies for processing straw or animal feed, and in *jajar legowo* farming pattern technology and paddy-fish integrative farming. Some had already been accustomed to use outside input because they found it more practical as it does not require much labour. There are limitations of the field extension workers (*PPL*), where not all villages have *PPL*, as well as poor infrastructure and transportation, and lack of water ponds to store water during the water crisis in dry seasons.

3.3. Development of Agribusiness Model LEISA

Agribusiness model of LEISA is one of Green Business Model, which is a business model that does not cause externalities to the environment physically, socially, or economically. Agribusiness Model of LEISA is an agribusiness system that applies sustainable agribusiness model with low external input. The system is developed in order to optimize the utilization of locally-based productive resources, including land, water, vegetation, microclimate, renewable energy, appropriate technology, social capital, environment, and human resources (including youth and women), by combining various subsystems that include:

(a) integrated and sustainable production subsystem of crops, livestock, and fish, so that they complement each other and provide maximum synergistic effect;
(b) processing subsystem of yields and post-harvest;
(c) marketing subsystem; and
(d) supporting subsystem.

Sustainable agricultural system must be built on the foundation of renewable resource that comes from the farm and the surrounding environment. The classification of internal and external resources will be helpful in understanding and developing agribusiness with LEISA model. Inputs as productive resources used differently in each agribusiness subsystem are divided into 2 (two) kinds: internal inputs and external inputs. In developing the agribusiness model of LEISA, the inputs or productive resources used are based on the internal inputs by minimizing external inputs.

Table 2. Internal and External Inputs in LEISA Agribusiness Model

Internal Inputs	**External Inputs**
A. Food Production Subsystem	
Agro-polis as a unit of integrated farming area based on watershed ecosystem	Unit of farming area outside the agro-polis area
The sun as the renewable source of energy in the process of photosynthesis of plants	Artificial light is used in greenhouses for production
Water: coming from rain or local irrigation networks	Water: comes from large reservoirs, centralized distribution/deep wells
Nutrition (nitrogen from air fixation) coming from soil and recycled organic matter	Nutrition coming from chemical fertilizers; and imported from abroad
Performing weed control and pest diseases in biological, cultural, and mechanical way	Control of pests, weeds and diseases: with chemical herbicides and insecticides
Seed: produced from self/local farming	Seed: hybrids or other varieties that are traded
Agricultural machinery: assembled and managed by the farmers and local communities	Farm machinery: bought and used frequently
Labor: coming from the family and/or those around the farming	Labor: hired or outsourcer labor force
Capital focusing on the utilization of social capital	Capital: loans from a money-lending institution on credit
Production management initiated by the farmers and local communities	Management: from input traders, field extension workers, and so on
B. Subsystem of Post-Harvest and Crop Processing	
Locally based raw materials and other materials	Raw materials and other materials purchased from outside the area
The raw materials processed using renewable energy from local resources	Raw materials are processed using non-renewable energy
Labor coming from the local area	Labor coming from outside the area

Tools and machinery produced and developed by the local residents	Tools and machinery purchased from the other outside the area
Waste recycled	Waste not recycled
C. Crop Marketing Subsystem	
Residents and local government are proactive in making agro-eco-system potential for agro-tourism.	Government and private sectors are proactive in making the local agro-ecosystem zone potential for the development of tourism.
Food and non-food distribution agencies are initiated by farmers and/or partner institutions, such as farmer groups, food barns, village-owned enterprises (*BUMDes*), and cooperatives	Distribution agencies of food and non-food agricultural products are not initiated by farmers and/or partner institutions that favor farmers.
Village markets, local retails, and retails in the outside area market local food and non-food production.	Rural markets and retails market non-local food and non-food products.
The marketing management of the crops is arranged by local farmers and entrepreneurs	The management of marketing the crops is initiated by outsiders.
D. Supporting Subsystem	
Information and communication system is based on local specific agribusiness.	Agribusiness information and communication system is general.
Agribusiness research and development is conducted by local people and local government.	Agribusiness research and development is organized by government, private parties, and external capital owners.
Micro-Economic Institution includes Joint Business Groups, Cooperatives, Barns, and Village-Owned Enterprises	Financial institution: Bank
The effort to improve ecosystem services is initiated by local communities and local	Effort to improve ecosysterm services is initiated by external parties.

Source :[11]

In the production subsystem, LEISA's method refers to the efforts of (1) optimizing the utilization of existing local resources by combining various components of farming systems, i.e. crops, livestock, fish, soil, water, climate, and humans in order to complement and give the greatest synergistic effect; and (2) the use of external inputs, which is only made when necessary, to supplement the less elements in the agro-ecosystem and increase biological, physical, and human resources. In utilizing external inputs, the main concern is given to the mechanism of recycling and minimization of environmental damage.

Agribusiness system of LEISA model, which integrates agriculture, livestock, fisheries, and other locally available resources, will support sustainable agriculture. This is in line with the results of research conducted by [12-14], which found that the combination model of crops, livestock and fish has been strongly recommended to ensure the sustainability of intensification, and support the development of rural areas.

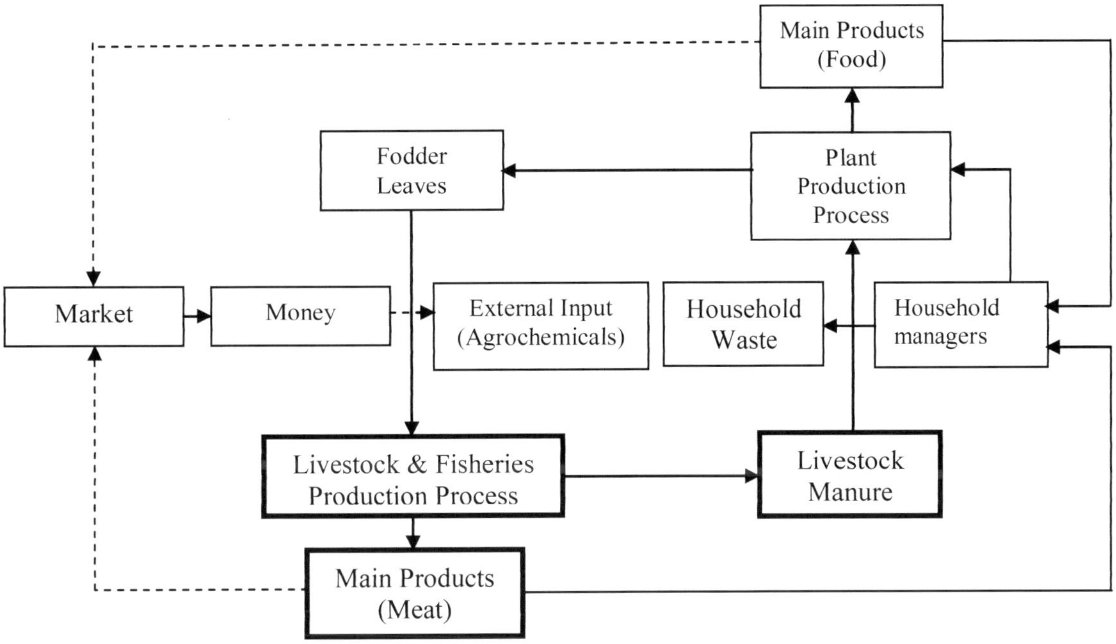

Figure. 1. Energy Cycle in Agribusiness Model of LEISA [12]

4. Conclusion

Ecologically, the agricultural resources of Surian Subdistrict have supported sustainable agriculture of LEISA model, but technically and socially, constraints still exist especially in the limited human resources (skill) and infrastructure (market and street). Agribusiness model of LEISA seeks to optimize the utilization of local resources through the subsystems of production, post-harvest and processing, processing and utilization of wastes, widespread distribution of local food through community empowerment and the provision of ecosystem services conducted by local communities involving micro-economic institutions. Thus, the agribusiness model of LEISA can develop regional economic activity.

References

[1] Reintjes *et al* 1999 *Pertanian Masa Depan: Pengantar Untuk Pertanian Berkelanjutan dengan Masukan Luar Rendah* Edisi Indonesia Terjemahan Sukoco Kanisius Yogyakarta
[2] Das A 2013 *Integrated farming: an approach to boost up family farming* LEISA India **15**(4)
[3] Wikarta K 2015 *IbM Membangun Ekobisnis Model LEISA* Laporan Akhir Ipteks Bagi Masyarakat (IbM) DRPMI Unpad Bandung
[4] Cresswell J W 1998 *Research Design:Qualitative and Quantitative Approaches* London: SAGE Publications
[5] BPS Kabupaten Sumedang 2016 *Kecamatan Surian dalam Angka*
[6] Sarlan A *et al* 2013 *Sistem Tanam Legowo* Badan Penelitian dan Pengembangan Pertanian Kementerian Pertanian Jakarta
[7] Mayrowati H and Ashari 2011 Pengembangan agroforestry untuk mendukung ketahanan pangan dan pemberdayaan petani sekitar hutan *Jurnal Forum Penelitian Agro Ekonomi* Pusat Sosial Ekomi dan Kebijakan Pertanian Bogor
[8] Evans A J and Setioko A R 1985 Tradisional System of Layer Flock Management in Indonesia. In Duck Production Science and World Practise. D J Farrell and P Stapleton (Eds) University od New England pp 306-322

[9] Abduh U *et al* 2003 *Integrasi ternak itik dengan sistem usahatani berbasis padi di Kabupaten Sidrap Sulawesi Selatan (Duck-rice integration in farming system in Sidrap Regency, South Sulawesi)* Balai Pengkajian Teknologi Pertanian http://peternakan.litbang.pertanian.go.id/fullteks/lokakarya/loli04-28.pdf

[10] Wasito and Khairiah 2004 *Peranan itik diintegrasikan dengan padi lahan sawah irigasi untuk mengendalikan keong emas di Sumatera Utara* Proc. Seminar dan Ekspose Nasional Sistem Integrasi Tanaman-Ternak. Denpasar, Bali. 20-22 Juli 2004 Puslitbang Peternakan, BPTP Bali dan CASREN. pp. 186-194

[11] Tuhpawana *et al* 2016 *Pembangunan Ekonomi Perdesaan Berkelanjutan di Kabupaten Sumedang* Laporan Academic Leadership Program (ALG) DRPMI Universitas Padjadjaran Bandung

[12] Surahman M and Sudrajat 2009 *Sistem Pertanian Terpadu* Naskah Akademis: Pengembangan Model Ecovillage, pp 63-82, LPPM dan Fakultas Ekologi Manusia IPB Bogor

[13] Priyanta and Djadjanegara A 2004 *Pengembangan usaha sapi potong pola integrasi* Prosiding Lokakarya Nasional Sapi Potong Menuju 2020 Strategi Pengembangan Sapi Potong dengan Pendekatan Agribisnis dan Berkelanjutan. pp. 63-82,Yogyakarta, 8-9 Oktober 2004 Badan Litbang Pertanian Puslitbang Peternakan Bogor

[14] Davendra C 2007 Perspectives on Animal Production Systems in Asia *Livestock Science* **106**

IOP Conf. Series: Earth and Environmental Science **142** (2018) 012063 doi:10.1088/1755-1315/142/1/012063

The performance of M2 generation of *Mentik Wangi Susu* rice resulted from gamma ray iradiation

A Yunus[1], Parjanto[1] and I Y Pratama[2]
[1]Lecturer of Agrotechnology Studi Program, Agriculture Faculty, Universitas Sebelas Maret, Surakarta, Indonesia
[2]Student of Agrotechnology Studi Program, Agriculture Faculty, Universitas Sebelas Maret, Surakarta, Indonesia

Email : yunus.uns7@yahoo.com

Abstract. This research is about the morphology of *Mentik Wangi Susu* rice M2 generation rice after gamma ray irradiation. The research objective is to create second generation with new and better properties of plant growth and lifespan trough mutation. The study was conducted in farming land of Klumprit Village, Mojolaban Sub-District, Karanganyar District at April to July in 2016. The dosage of gamma ray used in this study were 0,1 kGy; 0,2 kGy; 0,3 kGy and control (without irradiation). The result showed that mutation occured in the gamma-irradiated plant as indicated by shorter plant and harvesting time compared to without irradiation.

1. Introduction

The innovation of new rice varietiy was expected to increase the productivity of rice production. The innovative approach to improve the rice farming efficiency is trough combination of technology components that have synergistic effects on is the mutual supportive and gives better effect on growth as well as productivity.

Mentik Wangi Susu is one of local rice variety that has some good properties. Previous study [1] was stated that *Mentik Wangi Susu* rice is a local variety originated from Magelang Central Java Indonesia. *Mentik susu* rice has a resistentcy to collapse and tasty rice. The rice is characterized with white and milk looks like color. *Mentik Wangi Susu* rice often called the Indonesian's Japan rice due to its good taste and sticky but not as same as sticky rice. However, *Mentik Wangi Susu* rice is similiar with other local rices which have some disadvantages. The rice has long cultivation period, and low rice yield, in contrast with national rice varieties that have short harvesting time and high productivity [2]. To geta local rice variety could be served by plant mutation using gamma ray radiation. This method has been frequently used to induce plants to mutate. This gamma ray has the ability to penetrate in deep to plant tissue. Mutation induction was directed to alter one or more important characters while other original characters are retained [3].

2. Methods

The research on the performance of *Mentik Wangi Susu* M2 generation produced by gamma ray radiation was held in agricultural land of Klumprit village, Mojolaban sub-district, Sukoharjo regency. The study was conducted for four months from April to July 2016. New variety of *Mentik Wangi Susu* rice was generated through gamma-ray radiation in collaboration with the National Atomic Energy

Agency (BATAN). The materials were used in this study including of *Mentik Wangi Susu* M2 seeds of gamma ray radiation results with a dose of 0.1 kGy; 0.2 kGy; 0.3 kGy and *Mentik Wangi Susu* rice without gamma-ray radiation as a control.

This research was conducted experimentally by planting *Mentik Wangi Susu* rice seeds M2 (M1 yield) of gamma ray radiation result. The planting was consisted of 4 levels of radiation dose, namely: 0.1 kGy; 0.2 kGy; 0.3 kGy and no radiation. The variables were observed including of physiological properties of plants considered as mutant, including of plant height (cm), productive number, average number of productive grain, panicle length, weight of 1000 Grains Asset, Grain per panicle amount, potential result and harvest age (day). The study was used descriptive analysis by comparing rice plants of various doses of gamma irradiation treatment with irradiated control treatment, since the research purpose was selected each individual suspected mutation.

3. Results and Discussion

3.1 General Condition of Research

The height of the experimental field of *metik susu* rice is 104 meters above sea level, so that this land has a sloping land structure to hilly. The one-year rain intensity is 1,052 mm with average of one-year rainfall intensity is 19 mm.

Rice crop is a plant that has the nature of life in a humid place, this plant is not a water plant but a living plant needs a lot of water. This characteristic that makes rice crops planted on soil that can hold water longer, soil that has low pH and high Al affecting rice plant growth. Thus, rice plants grown under these conditions will suffer plant stress resulting in stunted growth of rice crops and results in the first season experiencing drought during the vegetative period. The second season of the plant was an explosion of rice fungus attack

3.2 Growth Modifiers

3.2.1 Plant height Short rice crops were contributed to productivity because the shorter stems are more resistant to collapse [4]. The measurements results of high rice cultivated of *Mentik Wangi Susu* rice are presented in Table 1 and 2 below.

Table 1. Plant Height on Maximum Vegetative Growth of *Mentik Wangi Susu* Rice Based on Mutant Selection of Short Plant

Irradiation Dose (kGy)	Lowest Plants Height (cm)	Highest Plants Height (cm)	Range	Average (cm)
0	105	144	105-144	130.87
0.1	94	141	94-141	116.90
0.2	81	133	81-133	103.88
0.3	81	107	81-107	93.33

Based on measurement result in TTable 1, the highest of plant height average is control group (without irradiation) that is 130.87 cm. The height average of irradiated plant was shorter than the control group that is of dose of 0.1 kGy with a height of 116.90 cm and a dose of 0.2 kGy with a height of 103.88 cm, while the lowest of plant height average was a dose of 0.3 kGy with height of 93.33 cm. *Mentik Wangi Susu* rice were irradiated with gamma rays is more collapse resist to the *Mentik Wangi Susu* rice without irradiation (control).

The measurement results also shown the highest plant is the control group of 144 cm. The height of the irradiated plant was shorter than control that is dose of 0.1 kGy with a height of 141 cm and a dose of 0.2 kGy with a height of 133 cm, while the lowest of plant height average was a dose of 0.3 kGy with 107 cm height. *Mentik Wangi Susu* rice was irradiated with gamma rays is more collapse resistant than *Mentik Wangi Susu* rice without irradiation (control).

Table 2. Plant Height of Maximum Vegetative Growth of *Mentik Wangi Susu* Rice Based on Mutant Selection of Tiller Number

Irradiation Dose (kGy)	Lowest Plants Height (cm)	Highest Plants Height (cm)	Range	Average (cm)
0	105	144	105-144	130.87
0.1	91	163	91-163	107.83
0.2	69	142	69-142	106.67
0.3	81	107	81-107	93.33

Based on measurement result of Table 2, the highest plant height average is control (without irradiation) of 130.87 cm. The average height of the irradiated plant was shorter than the control, i.e. the dose of 0.1 kGy with height of 107.83 cm and the dose of 0.2 kGy with the height of 106.67cm, while the lowest plant height average is 0.3 kGy with height of 93.33 cm. *Susu Mentik* rice that in irradiated with gamma rays is more collapse resistant than *Mentik Wangi Susu* rice that is not in irradiation (control).

Measurement results also showed the highest plant height is the dose of 0.1 kGy with a height of 163 cm. Control with height of 144 cm. The height of the irradiated plant was shorter than the control, i.e. the dose of 0.2 kGy with height of 133 cm, while the lowest plant height average is 0.3 kGy with height of 107 cm. *Mentik Wangi Susu* rice that in irradiated with gamma rays is more collapse resistant than *Mentik Wangi Susu* rice that is not in irradiation (control). This suggests that the administration of gamma ray irradiation in the *Mentik Wangi Susu* rice seedlings occurred gene mutation that controls the height of the plant.

3.2.2 *Total Tiller Number* Tiller is an important agronomic trait that ultimately determined the number of panicles, seeds and grains in a certain area. The number of tiller productive is the number of panicles that appear in a clump of rice plants and one of the parameters to measure the crop productivity. Through the calculation of the tiller number was obtained the results presented in Table 3 and 4 below.

Table 3. Total Tiller Number of Mentik wangi susu Rice based on mutant selection of short plant

Irradiation Dose (kGy)	Lowest Tiller Number	Highest Tiller Number	Range (plant)	Average
0	9	20	9 - 20	11
0.1	8	38	8 - 38	17
0.2	6	24	6 - 24	15
0.3	5	25	5 - 25	15

Mentik wangi susu rice that has been studied has the highest tillers number total is dose of 0.1kGy with 17 tillers with a range of 8-38 productive tillers of each plant. Dose of 0.2 kGy with the tillers number average is 15. Dose of 0.3 kGy with the average of tillers number is 15 not different with the tillers number of dose of 0.2 kGy. This was influenced by the plants number observed that is control group only 30 plants.

Mutation induction is one of way to improve the plants diversity, the tillers number was formed in a single clump of plants were determined by plant genetic [2]. This means that rice with irradiation gives better total tiller number than irradiated rice, then based on the productive tillers number was selected the best 30 plants at each dose that exceeds the control presented in the Table 4.

Table 4. Total Number of Tillers of *Mentik Wangi Susu* Rice based on Mutant Selection of Tillers Number

Irradiation Dose (kGy)	Lowest Tiller Number	Highest Tiller Number	Range (plant)	Average
0	9	20	9 – 20	11
0.1	9	26	9 – 26	14
0.2	7	23	7 – 23	14
0.3	5	25	5 – 25	15

Mentik wangi susu rice that has been studied based on the observation of the tillers number has the highest tillers number is dose of 0.2 kGy with 15 tillers with range of 5-25 tillers productive each plant. Dose of 0.2 kGy with the average of tiller number is 14. Dose of 0.3 kGy with the average tillers number is 14 is the lowest. This was influenced by the number of plants observed that is control only 30 plants and dose 0.3 kGy only 24 plants. [5] argued that the tillers number productive depends on the number of formed tillers. Another factor that influenced the formation of productive tillers is the factors of soil, soil with high moisture level can stimulated the plant roots development as in accordance to [6] reported the development of tillers number will be high if the condition of aerobic soil and roots grow well. The dried soil can push the roots deeper and increase the compactness of the soil surface and the strength of the soil that produces more growing plants.

3.2.3 *Average of Panicles Productive Number* The number of panicles productive was calculated based on the number of tillers that produce panicles and rice grains. The calculation of the number of grains per panicle can be seen in the following Table 5 and 6.

Table 5. Average of Panicles of *Mentik Wangi Susu* Rice based on Mutant Selection of Short Plant

Irradiation Dose (kGy)	Lowest of Panicle Number	Highest of Panicle Number	Range	Average
0	5	21	5 - 21	40
0.1	10	72	10 – 72	21
0.2	9	42	9 - 42	19
0.3	11	62	11 - 62	28

An important factor for obtaining high grain yield is dependent on the formation of tillers number that determined the panicles number. Based on the observation results on Table 5, the lowest of productive grain number average of *Mentik Wangi Susu* rice is in the dose of 0.2 kGy with 19 panicles whereas the highest is the dose of 0.3 kGy with 28 panicles. The average of productive grain number on doses of 0.1 kGy, 0.2 k Gy and 0.3 kGy are lower than the control group so that the irradiation given an effect to the average of productive grains number of *Mentik Wangi Susu* rice.

Table 6. The average of Panicles Number of *Mentik Wangi Susu* Rice based on Mutant Selection of Tillers Numbers

Irradiation Dose (kGy)	Lowest of Panicle Number	Highest of Panicle Number	Range	Average
0	5	21	5 - 21	40
0,1	9	39	9 - 39	18
0,2	5	32	5-32	17
0,3	11	62	11 - 62	28

The productive panicles number based on the observation of mutant selection of tillers numbers is the number of panicles that appear in a clump of rice plants and is one of the parameters to measure the crop productivity. Based on the results of observation on Table 7 the lowest of productive grain number average on *Mentik Wangi Susu* rice is at the dose of 0.2 kGy with 17 panicles number while the highest is the dose of 0.3 kGy with 28 panicles numbers. The average of productive grain number on doses of 0.1 kGy, 0.2 kGy and 0.3 kGy are lower than the control group so that the irradiation given effect on the average of productive grains number of *Mentik Wangi Susu* rice.

3.2.4 *Weight of 1000 Field out Grain* The weight of 1000 field out grains is one of the parameters were used in research to determine the crop quantitative character. Based on observation result of weight of 1000 field out grains can be seen in Table 7 and 8 below.

Table 7. Weight of 1000 filled grains per panicle of*Mentik Wangi Susu* Rice Based on Mutant Selection of Short Plant

Irradiation Dose (kGy)	Lowest of 1000 grains weight	Highest of 1000 grains weight	Range	Average
0	18.5	23.8	18.5 -23.8	22.223
0.1	19.9	28.8	19.9 -28.8	22.057
0.2	20	31.1	20 – 31.1	24.992
0.3	12.6	26.7	12.6 -26.7	21.917

The lowest of 1000 grain weight is 0.3 kGy dose of 12.6 grams and the highest is 0.2 kGy dose of 31.1 grams. The of 1000 grains weight of 0.2 kGy doses is heavier than the weight of the highest of 1000 grains weight of the control group, i.e. 23.8 grams, this proven that irradiation has a good effect on the grain weight because grain weight was related to rice production. Gamma ray radiation has the potential to create diversity to obtain mutants with morphological characteristics and desirable agronomic characteristics such as low plants, short life and high production.

Table 8. The Weight of 1000 Filled Grainsper Panicle of *Mentik Wangi Susu* Rice based on Mutant Selection of Tiller Number

Irradiation Dose (kGy)	Lowest of 1000 grains weight	Highest of 1000 grains weight	Range	Average
0	18,5	23,8	18,5 - 23,8	22,22
0,1	19	27,90	19 - 27,90	23,69
0,2	19,60	27,20	19,6-27,20	23,27
0,3	12,6	26,7	12,6 - 26,7	21,92

The lowest of 1000 grain weight is 0.3 kGy dose of 12.6 grams and the highest is 0.1 kGy dose of 27.90 grams. The 1000 grains weight of 0.1 kGy doses is heavier than the highest of 1000 grain weight of the controls group, i.e. 23.8 gram. It was proven that irradiation has a good effect on grain weight because grain weight was related to rice production. Rice productivity was highly dependent on the number of productive tillers and panicle length. Crop productivity on a certain land was expected to determine the varieties results that can adapt to the environment [7].

3.2.5 *Number of Grain per Panicle* Number of grains per panicle is one of components to determine the potential of rice crops [6]. The grains number per panicle can be seen in Tables 9 and 10 below.

Table 9. Average number of grains per panicle of *Mentik Wangi Susu* Rice Based on Mutant Selection of Short Plant

Irradiation Dose (kGy)	Lowest average	Highest average	Range	Average
0	87	235	87 – 235	181
0.1	78	236	78 – 236	150
0.2	77	184	77 – 184	125
0.3	69	209	69 - 209	110

The lowest number of productive grain of *Mentik Wangi Susu* rice is at the doses of 0.3 kGy which is equal to 69,667 while the highest is dose of 0,1 kGy which is 236. The number of productive grain of doses of 0.1 kGy and 0.2 kGy is higher than control group so that irradiation given effect to the number of productive grains of *Mentik Wangi Susu* rice. Cereals given doses of gamma rays at doses of 0.1 kGy and 0.2 kGy have a high generative growth, one of the cereals yield produced such as rice and sorghum.

The grains number of per panicle was affected by the panicles length. The ability of plants to express the panicles length was strongly influenced by the period of panicle initiation that was included in the critical period of the plant. Lack of nutrients and water in the panicle initiation period can caused the panicles formation is not maximal so that the affected to the seed will be formed.

Table 10. The average of grains number per panicle on Various Dose of Gamma Rays Irradiation of *Mentik Wangi Susu* Rice Based on Mutant Selection of Tillers Number

Irradiation Dose (kGy)	Lowest number of grains per panicle	Highest number of grains per panicle	Range	Average
0	87	235	87- 235	181
0.1	35	238	35 – 238	133
0.2	76	197	76 -197	131
0.3	69	209	69-209	110

The lowest number of productive grain is Mentik wangi susu rice at the doses of 0,1 kGy that is 35,33 and the highest is doses of 0,1 kGy that is 238. The number of productive grain at doses of f 0,1 kGy and 0,2 kGy is higher than control group so irradiation was given effect to the amount of productive grain of *Mentik Wangi Susu*.

3.2.6 *Potential Results* The potential results in weight of 1000 field out grains is one of the parameters used to measuremen the crop productivity. Based on panicles number, the average of grains number per panicle can be seen in Table 11 and 12.

Table 11. Yield Potential of *Mentik Wangi Susu* Rice per 1000 Seeds Based on mutant selection of short plant

Dose Irradiation (kGy)	Lowest Yield Potential (Gram /Clump)	Highest Yield Potential (Gram /Clump)	Range	Average (Gram / Clump)
0	13.402	105.525	13.402-105.525	42.066
0.1	27.760	180.432	27.760-180.432	66.186
0.2	16.484	102.844	16.484-102.844	51.749
0.3	25.016	280.785	25.016-280.785	71.261

The lowest yield potential is the control dose of 13,402.80 grams and the highest is the dose of 0.3 kGy i.e. 280.785.60 grams. The highest potential yield of 0.3 kGy dose has more potential than control of 42.066.99. This was proven that irradiation has a good effect to the yield potential of *Mentik Wangi Susu* rice of control group because this yield potential was related to rice production. Gamma ray radiation has the potential to create diversity so obtained mutants with morphological characteristics and desirable agronomic characteristics such as low plants, short life and high production.

Table 12. Yield Potential of *Mentik Wangi Susu* Rice per 1000 Seeds based on mutant selection of tillers number

Irradiation Dosage (kGy)	Lowest Yield Potential (Gram/clump)	Highest Yield Potential (Gram/clump)	Range	Average (Gram/ Clump)
0	13.402	105.525	13.402-105.525	42.066
0.1	15.390	124.616	15.390-124.616	56.730
0.2	16.484	102.844	16.484-102.844	51.749
0.3	25.016	280.785.60	25.016-280.785	71.261

The highest yield potential of observations based on mutant selection of tillers number is 0.3kGy dose of 71.261.63 grams and the lowest yield potential is control dose of 42.066.99 gram. Potential result of dose 0.1 kGy; 02kGy; and 0.3kGy is better than control group of 42.066.99 gram. It was proven that irradiation has a good effect on grain weight because grain weight was related to production.

3.2.7 *Mutant Selection Results of Short Trunk and Tillers Number* Based on observations of various doses of gamma ray irradiation, the treatment based on observations of the short trunk, was presented in the following Table.

Table 13. Mutant Selection Results of Short Plant of Mentik wangi susu Rice on 0.1 kGy irradiation dosage that have High Yield Potential

Code	High Plant (Cm)	Thillers Number	Panicles Number	Weight of 1000 Seeds (g)	Grain Number Per Panicles	Yield Potential (Gram/Clump)
G14T5	141	11	12	21.5	215.33	55.556
G23T3	135	12	14	19.2	171.33	46.054
G63T11	134	18	20	19.9	107.67	42.85
G15 T5	134	11	11	21.4	222.67	52.415
G39T17	132	8	14	27.2	88.67	33.764
G38T6	132	16	21	21.4	216.00	97.070
G39T10	132	13	21	20.4	114.00	48.837

Table 14. Mutant Selection Results of Short Plant of Mentik wangi susu Rice on 0.2 kGy irradiation dosage that have High Yield Potential

Code	High Plant (Cm)	Thillers Number	Panicles Number	Weight of 1000 Seeds (g)	Grain Number Per Panicles	Yield Potential (Gram/Clump)
G13T2	133	17	31	2	G13T2	133
G3T18	132	14	15	2.38	G3T18	132
G16T5	130	12	20	2.27	G16T5	130
G5T7	127	14	14	2.31	G5T7	127
G3T17	121	13	10	2.65	G3T17	121
G15T4	120	18	15	3.11	G15T4	120
G15T3	118	18	24	2.5	G15T3	118

Table 15. Mutant Selection Results of Short Plant of Mentik wangi susu Rice on 0.3 kGy irradiation dosage that have High Yield Potential

Code	High Plant (Cm)	Thillers Number	Panicles Number	Weight of 1000 Seeds (g)	Grain Number Per Panicles	Yield Potential (Gram/ Clump)
G7T8	107	17	33	19	126.00	79.002
G31T20	107	25	62	21.6	209.67	280.785
G41T4	107	14	37	20.1	108.00	80.319
G17T17	105	19	38	24.6	122.33	114.357
G6T1	103	18	27	19.7	177.00	94.146
G9T6	100	15	21	20.8	103.67	45.28
G26T13	100	20	40	20.3	166.33	135.062

Table 13 to 15 were contained studies based on mutant-selection of short plant selection treated with irradiated doses of 0.1 kGy have a shorter height with other dose treatments. Irradiation of 0.1 kGy dose has 36.862% more superior that irradiation of 0.2 kGy dose of 34.549% and dose of 0.3 kGy with presentasae of 28.588%

Based on observations of various doses of gamma ray irradiation treatment observations based on mutant selection of thiller number, presented in the following Table.

Table 16. Mutant Selection of Thiller Number of Mentik wangi susu Rice of 0.1 kGy Irradiated Dosage that Have High Yield Potential.

Code	Thiller numbers	Panicles Number	Weight of 1000 Seeds (g)	Grain Number Per Panicles	Yield Potential (Gram/ Clump)
G24T31	26	39	23.3	84.00	2.896
G28T6	25	39	20.5	97.67	2.570
G15T11	22	26	23.4	99.33	3.274
G17T13	22	13	23.1	206.67	6.2
G33T16	19	22	23.8	238.00	7.437
G28T4	18	19	22.5	112.67	3.557
G37T22	18	29	26.2	112.67	3.673

Table 17. Mutant Selection of Thiller Number of Mentik wangi susu Rice of 0.2 kGy Irradiated Dosage that Have High Yield Potential.

Code	Thiller Numbers	Panicles Number	Weight of 1000 Seeds(g)	Grain Number Per Panicles	Yield Potential (Gram/ Clump)
G13T20	23	21	19.6	197.67	81.359.60
G9T5	22	19	23.2	195.33	86.102.93
G3T13	21	28	21.7	155.00	94.178.00
G3T20	21	21	22.6	144.67	68.658.80
G17T17	21	14	27.2	101.67	38.714.67
G18T7	20	23	27.1	165.00	102.844.50
G8T6	20	27	22.5	106.67	64.800.00

Table 18. Mutant Selection of Thiller Number of Mentik wangi susu Rice of 0.3 kGy Irradiated Dosage that Have High Yield Potential.

Code	Thiller numbers	Panicles Number	Weight of 1000 Seeds (g)	Grain Number Per Panicles	Yield Potential (Gram/ Clump)
G7T8	17	33	19	126.00	79.002
G31T20	25	62	21.6	209.67	280.785
G41T4	14	37	20.1	108.00	80.319
G17T17	19	38	24.6	122.33	114.357
G6T1	18	27	19.7	177.00	94.146
G9T6	15	21	20.8	103.67	45.281
G26T13	20	40	20.3	166.33	135.062

Table 16 to 18 are the selection result of second-generation of Mentik wangi susu rice that allegedly have mutations with lower plant and large number of tillers. The lower plant was expected the plant will be sturdy and not easily exposed to wind. Number of productive tillers is the number of rice plants that produce panicles. The higher number of productive tillers will have a positive impact on rice productivity. The research results were presented in the above based on the tillers number at many variation of irradiation doses. The dose of 0.3 kGy have the highest impact with the percentage of 36.21%, followed by 0.2 kGy irradiation dosage with percentage of 34,74% and the irradiation dose of 0.1 kGy has the lowest percentage of 30,05%. Table 16 to 18 is the selection result of 30 plants that have more tillers number of than control. The rice of M2 selection results are the next planting material for M3 was expected to have better individual properties that M2 generation.

3.2.8 *Harvest Age* Harvesting was done when the plant is 110 days after planting and when the seeds are ripe physiologically, which is about 90-95% panicles have yellow. Harvest was done by cutting the straw about 20-25 cm above the soil surface, harvested rice was placed or stacked on the tarpaulin, then done threshing rice used pedal treasure. The study began on April 6th with seeding. Seeds of *Mentik Wangi Susu* rice seedlings for 21 days were planted on Tuesday April 26th, 2016 and harvested on August 23, 2016 for controlled treatment, which means that the plant life is 140 days. Irradiation treatment of 0.1 kGy at 135 days harvest age, 0.2 kGy at 134 days harvest age and 0.3 kGy at 132 Days harvest age. The plant was harvested simultaneously per treatment due to the occurrence of pest infestations. It was done to reduce the loss of grain due to pest infestations affected the level of data validity.

IOP Conf. Series: Earth and Environmental Science **142** (2018) 012063 doi:10.1088/1755-1315/142/1/012063

4. Conclusion

Based on the research performed, it can be concluded that there are variety of M2 rice plants that is in the strain of short-aged plant from irradiation of 0.1 kGy, which are: G15 T5; G38 T6; G23 T3; G39 T10; G63 T11; G39 T17; G14 T5 strains, and in the treatment with irradiated dose of 0.2 kGy are G5 T7; G3 T17; G13 T2; G15 T3; G15 T4; G16 T5; G3 T18 strains; and in the tillers number properties from irradiation 0.1 kGy are: T4; T22; T31; T16; T6; T11; T13 strain, from the irradiation dose of 0.2 kGy are G18T7; G3T13; G9T5; G8T6; G13T20; G3T20; G17T17 strain and at a dose of 0.3 kGy are T4; T17; T1; T20; T13; T6; T2 strain. Obtained individual plants that allegedly have mutations for the properties of short plants and plentiful tillers

References

[1] Nurmayetti, Atman 2013 Keunggulan kompetitif padi sawah varietas lokal disumatra barat. *Jurnal pengkajian dan pengembangan teknologi pertanian.* **16** (2) : 100-108

[2] Makarim A K, Suhartati E. Morfologi dan Fisiologi Tanaman Padi. Balai Besar Tanaman Padi

[3] Yulianti F C, Martasari, Karsinah dan T. Hartanto 2010 Variasi Genetik Jeruk Keprok (*Citrus reticulata* Blanco) Hasil Radiasi Sinar Gamma Menggunakan Penanda ISSR *Buletin Plasma Nutfah* **16** (2)

[4] Sobrizal, Sanjaya S, Carkum et al 2004 Mutan Padi Pendek Hasil Lradiasi Sinar Gamma 0,2 Kgy Pada Varietas Atomita 4 Puslitbang Technologi Isotop dan Radiasi BATAN

[5] Yusnaweti, Sevindajuta 2012 Uji Beberapa Varietas Dan Jumlah Benih Terhadap Pertumbuhan Dan Produksi Padi Gogo Pada Ultisol

[6] Zhang W, Si-Tu S 1989 Influence of drainage practice on rice yield Pp 65-84 in *Proceeding of the 7 th Afro-Asean egional Conference* vol. D.7 International Commisionon Irrigation and Drainge Tokyo

[7] Kuswanto H 1997*Analisis Benih* (Yoyakarta: Andi)

IOP Conf. Series: Earth and Environmental Science **142** (2018) 012064 doi:10.1088/1755-1315/142/1/012064

Antagonism of rice phylloplane fungi against *Cercospora oryzae*

A Mardani and Hadiwiyono
Agrotechnology Department, Universitas Sebelas Maret, Indonesia

Email: hadi_hpt@yahoo.com

Abstract. Narrow brown leaf spot (NBLS) caused by *Cercospora oryzae* Miyake is one of the important obstacle in rice cultivation that can decrease the productivity up to 40%. It has been known well that some phylloplane fungi are antagonistic to some leaf diseases. Phylloplane fungi of rice however haven't been studied much and poorly understood as biological control agent of rice pathogen such *C. oryzae*. The research aimed to study the antagonism of some phylloplane fungi of rice against *C. oryzae*. At least 14 isolates of phylloplane fungi were collected which consisted of six pathogenic and eight nonpathogenic variants. All of nonpathogenic isolates were antagonistic against *C. oryzae* both in vitro and only one isolate could not inhibit the infection of the pathogen in vivo. Some isolates were identified as *Aspergillus, Mucor, Penicillium, Fusarium*, and *Trichoderma*. The isolate of *Mucor* and *Fusarium* could inhibit the highest growth of pathogen on potato dextrose medium that were at 36.0% and 35.5% respectively. Whereas on artificial inoculation on rice, some isolates such *Penicillium* and *Fusarium* could inhibit most effectively and were significantly different to Mencozeb application with dosage 5g L-1.

1. Introduction

Rice is one of important cereal plants. National rice consumption of Indonesia in 2011 was 79.1 Kg capita. It was higher than consumptions of cassavas and sweet potatoes, that were 3.0 Kg capita and 0.1 Kg capita respectively [1]. According to BPS [2], rice consumption for a week in 2013 was about 1.626 Kg, and then it increased in the next year.

NBLS caused by *C. oryzae* is one of rice cultivation obstacle. The symptoms start with lesion 2-10 x 1 mm. On the resistant varieties, spot will be narrower, shorter, and dark brown. If the disease occures in the susceptible varieties, the symptoms will be larger and has grey necrosis in the center [3]. The disease could decrease yield until 40% [4].

Planting resistant varieties is alternative way to control NBLS. However, resistant varieties are just able to stand for some years until the new virulent fungi will be developing [5]. Fungicides spraying are also able to control NBLS, but it has a lot of negative impact to human health [6]. Based on the reasons above, it needs other alternative to control the disease.

Phylloplane fungi is a part of microorganism living in the leaf surface [7, 8]. The fungi inhibited the growth of *Alternaria brassicae*. Evueh and Ogbebor [9] reported that phylloplane *Trichocladium* sp. and *Trichophyton* sp. had capability to inhibit *Colletotrichum*. Although it is effective as an antagonist, using phylloplane fungi to control *C. oryzae*is not discover yet. This research aimed to find antagonistic phylloplane fungi and their potency as antagonist against *C. oryae*.

2. Materials and methods

The researches were conducted at laboratory and greenhouse of Pests and Diseases Observation, Palur Surakarta from August 2016 until May 2017. Tool needed for laboratory research was inchase, while the materials were isolate of *C. oryzae*, isolates of phylloplane fungi, PDA medium (*Potato Dextrose Agar*), and PDB medium (*Potato Dextrose Broth*). In the greenhouse, materials needed were soil from rice field, polibag, and IR64 seed.

The researchers used Completely Randomized Design (CRD), three replications with one factor that was phylloplane fungi application. Antagonism in vitro test had 2 treatments that were control and application of phylloplane fungi + *C. oryzae*. In the antagonism in vivo test there were 4 treatments that were fungicide, without pathogen or phylloplane fungi application/no inoculation, application of *C. oryzae*, and application of phylloplane fungi+pathogen.

Observation variables were percentage of growth inhibition, spot area, inhibition percentage in antagonism in vitro test, incubation period, and disease inhibition effectiveness. Percentage of growth inhibition was measured with formula:

$$H\ (\%) = \frac{R2 - R1}{R2} \text{ x } 100\%$$

H = inhibition percentage
R_1 = the distance between pathogen colony center to antagonist colony
R_2 = the distance of pathogen colony away from antagonist colony

Incubation period was measured from day of symptoms appearance. it was observed from day 1st after C. *oryzae* inoculation. spot area were measured on day-7 after *C.* b*oryzae* inoculation. Disease growth effectivity was measured with formula:

$$\frac{A - B}{A} \text{x } 100\%$$

A = spot area of *C. oryzae* inoculation
B = spot area of no inoculation treatments

3. Results and discussion

3.1. Phylloplane fungi identification

Phylloplane fungi isolated from 3 parts, that were leaf (D), shealth (P), and stem (B). Different fungus were found in each parts of rice. Five fungus were found in the leaf, 7 fungus in shealth, and 2 fungus in the stem. Identification using morphological is the common way to identify. However, morphological identification is not accurate, microscopic observation is needed [10].

Identification results showed in Table 1. nine out of 14 fungus can be identified, while 5 fungi cannot. Fungi that is not have particular characters, cannot be classified to certain genus.

Identification of microscopic characters showed that there were *Aspergillus* sp., *Penicillium* sp., *Mucor* sp., *Curvularia* sp., *Botrytis* sp., *Fusarium* sp. and *Trichoderma* sp. Pathogenicity test resulted that 6 fungi were pathogen, that were *Botryosphaeria* sp., *Botrytis* sp., 2 fungus *Curvularia* sp., D2 and P7. D were fungi isolated from leaf, P were fungi isolated from sheath, B were fungi isolated from stem. "+" were pathogen fungi, and "-" were nonpathogenic fungi

Table 1. Identification and Pathogenicity of Rice Phylloplane Fungi Isolates

Isolates	Species	pathogenicity
D 1	*Aspergillus* sp.	-
D 2	-	+
D 3	*Mucor* sp.	
D 4	*Botrytis* sp.	+
D 5	*Penicillium* sp.	-
P 1	*Fusarium* sp.	-
P 2	-	-
P 3	-	-
P 4	*Curvularia* sp.	+
P 5	*Curvularia* sp.	+
P 6	*Botryosphaeria* sp.	+
P 7	-	+
B 1	*Trichoderma* sp.	-
B 2	-	-

3.2. Antagonism in vitro test of phylloplane fungi against C. oryzae

Phylloplane fungi impacted significantly to C. *oryzae* growth on PDA. Picture 1 about growth inhibition of *C.oryzae* by rice phylloplane fungi show the highest inhibition is *Mucor* sp. (36%). *Fusarium* sp. has 35.5% inhabitation that not significant different with *Mucor* sp. Phylloplane fungi with the lowest inhabitation is B3.2 (4.4%).

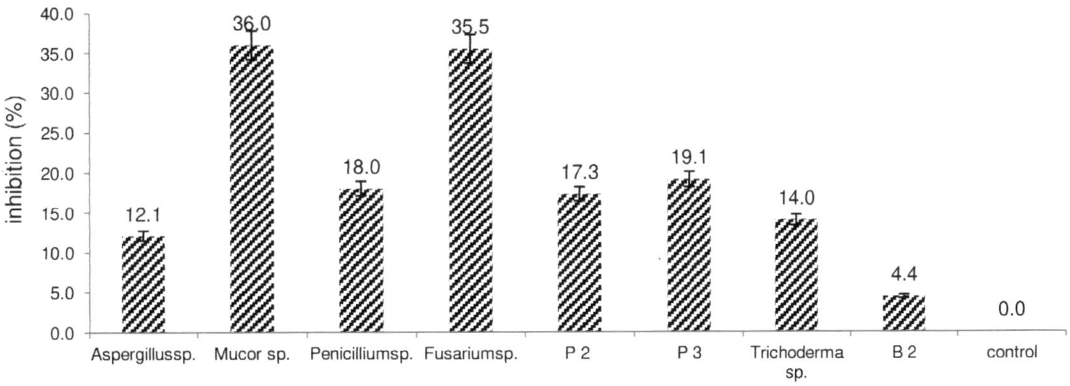

Figure 1. Growth Inhibition of *C. oryzae* by Rice Phylloplane Fungi on PDA Medium

Mechanisms of phylloplane fungi to inhibit pathogen are parasitism [9] antibiotics production, and volatile metabolites [11]. *Mucor* sp. has capability to grow rapidly [12], it is possible that the fungi inhibit *C. oryzae* through nutrition competition.

Fusarium sp. has been reported, can inhibit pathogen by nutrition competition, antibiotic production, and induced resistant [13]. Figure 2A indicate that *Fusarium* sp. grows rapidly and surrounds *C. oryzae*. so that, the growth of *C. oryzae* was limited.

Antagonistic *Trichoderma* sp. grew faster than pathogen [14]. However, *Trichodema* sp. colony has zone that not overgrown by both colonies (Figure 2B). It indicates that the zone is caused by certain compounds produced by *Trichoderma* sp. [15], revealed that *Trichoderma* sp. was able to inhibit pathogen growth by producing antibiotics and secondery metabolites.

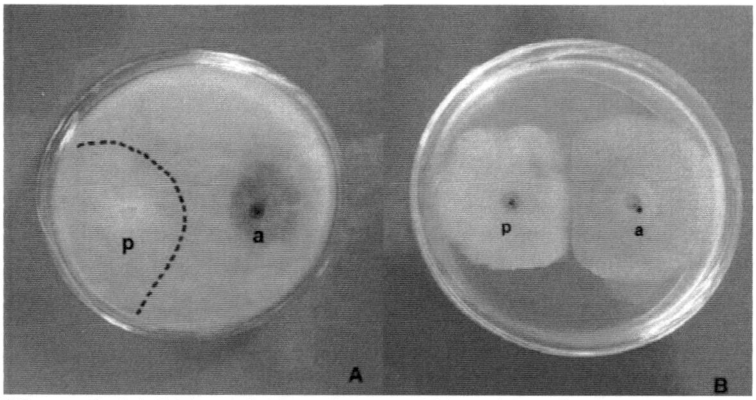

Figure 2. Inhibition of *Fusarium* sp. (A) and *Trichoderma* sp. (B) to *C. oryzae* on PDA Medium. P = pathogen, a = antagonist

3.3. Spot area appear on antagonism in vivo test of phylloplane fungi against C.oryzae

Application phylloplane fungi effected significantly to *C. oryzae* symptom. Control negative (C-), *Penicillium* sp., dan *Fusarium*sp. formed no spot (spot large=0.00 mm^2). *C. oryzae* inoculation has 2.34 mm^2, Mancozeb 2.11 mm^2. The largest spot was in P3, 7.00 mm^2.

The highest disease inhabition effectiveness were *Penicillium* sp. and *Fusarium* sp. The results showed that both fungi inhibited *C. oryze*non and were significantly more effective then Mancozeb 5 g L^{-1}. P3 had the lowest disease inhibition effectiveness that was 8.82%.

Table 2. Spot area, incubation Period, and Disease Inhibition Effectiveness of *C. oryzae* by Rice Phylloplane Fungi

Isolates of Phylloplane Fungi	Spot Area (mm^2)*	Incubation Period (DAI)	Disease Inhibition Effectiveness (%)
Aspergillus sp.	2.89±2.46 bc	3	48.28±42.81 abc
Mucor sp.	2.56±1.90 bc	3	53.02±34.84 abcd
Penicillium sp.	0.00±0.00 a	-	100.00±0.00 d
Fusarium sp.	0.00±0.00 a	-	100.00±0.00 d
P2	3.78±2.91 bcd	3	40.11±37.96 abc
P3	7.00±2.60 d	3	8.82±15.28 a
Trichoderma sp.	5.33±1.73 cd	3	13.56±11.74 ab
B2	1.89±2.14 ab	3	65.27±39.39 cd
Mancozeb**	2.11±0.38 bc	3	61.19±7.07 bcd
No inoculation	0.00±0.00 a	-	100.00±0.00 d
C. oryzae	5.444±2.34cd	3	-

The values followed by different letter are non-different based on Duncan's test at level of 5%. *data transformed into $\sqrt{(x + 0.5)}$. **application with dosage 5g L^{-1}

Penicillium sp. had capability to inhibit growth of *Phytophthora capsici* [16]. The fungi have mechanism of induced resistant [17]. It is also parasitism to the pathogen [18] and produce antibiotics [19, 20].

Treatment B2 had the lowest inhibition in antagonism in vivo test. However, B2 formed the lowest spot area in antagonism in vivo test (1.89 mm^2). This result was not significant compared with *Penicillium* sp., *Fusarium* sp., and no inoculation. It was assumed that the fungi induced resistant to *C. oryzae*.

Spot area of *Trichoderma* sp. was at about 5.33 mm^2. This is not significant compare with treatment of *C.oryzae*. P3 had 7.00 mm^2 in spot area. The opposite result is possibly caused by different condition during antagonism in vitro and in vivo test. Uncontrolled environmental during antagonism in vivo made *Trichoderma* sp. and P3 could not grow optimally on the rice plant.

Spot appear in day-3 after inoculation, 3 treatments not form symptom, that were no inoculation, *Penicillium* sp., and *Fusarium*sp. The fungus effective for controlling *C. oryzae*. However, it can assumed caused by the loss of pathogen inoculum after inoculation.

4. Conclusions

There are 8 fungus isolated from rice, that are *Aspergillus, Mucor, Penicillium, Fusarium, Trichoderma*. P2, P3, and B2. Phylloplane fungus effective against *C. oryzae*. *Mucor* and Fusarium have the highest inhibition, that were 36% and 35.5%. in the artificial inoculation, *Penicillium* sp. And *Fusarium* sp. could inhibit most effectively.

After conducting the research, the recommendations to give are the need for research about antifungal compounds produced by phylloplane fungus so that inhibit *C. oryzae*. There is also need for research about phylloplane fungi application in the field.

References

[1] Kemendag Kementerian Perdagangan. 2013. *Laporan akhir analisis dinamika konsumsi pangan masyarakat Indonesia*. (Jakarta (ID):Kementerian Perdagangan).

[2] BPS Badan Pusat Statistika. 2016. *Statistika Indonesia* 2016. (Jakarta (ID): Central Statistics Agency of Indonesia).

[3] Ou SH. 1985. *Rice Diseases*. (Surey (UK): Commonwealth Mycological institute).

[4] Hollier, C., 1992. Narrow brown leaf spot. P. 18 in *Compendium of Rice Diseases*. (St. Paul (CAN): APS Press).

[5] Sah DN, Rush MC. 1985. Pathogenic race of *Cercospora oryzae* in the Southern United States. *Phytopathol.* 75:1354.

[6] Mahyuni EL. 2015. Faktor risiko dalam penggunaan pestisida terhadap keluhan kesehatan pada petani di kecamatan berastagi kabupaten karo 2014. *Kesmas* 9(1):79-89.

[7] Elkhateeb WA, Abdel Nasser AZ, Mohamed BM, Mohamed H, Ghoson MD. 2016. Investigation of diversity of endophytic, phylloplane and phyllosphere mycobiota isolated from different cultivated plants in new reclaimed soil, Upper Egypt with potential biological applications. Inter. *J Med. Pharm. Rese* 2 (1):23-31.

[8] Fonseca A, Inacio J. 2006. Phylloplane yeasts. Rosa C, Peter G. 2005. Yeast Handbook. (New York (NY): Springer-Verlag. Berlin).

[9] Evueh GA, Ogbebor NO. 2008. Use of phylloplane fungi as biocontrol agent against *Colletotrichum* leaf disease of rubber (*Hevea brasiliensis* Muell. Arg). *J. Afr. Biotech.* 7(15): 2569-2572.

[10] Gautam AK, Bhadauria R. 2012. Characterization of *Aspergillus* species associated with commercially stored triphala powder. *Afric. J. Biotech.* 11(104):16814-16823.

[11] Yadav SL, AK Mishra, PN Dongre et al. 2011. Assessment of fungitoxicity of phylloplane fungi against *Alternaria brassicae* causing leaf spot of mustard. *J. Agric. Tech.* 7(6):1823-1831.

[12] Kaur R, Kaur J, Rama SS. 2010. Nonphogenic *Fusarium* as a biological control agent. *Plant Pathol. J.* 9(3):79-91.

[13] Bormer MH, Michael S, Christian HA, et al. 2017. Competition assays and physiological experiments of soil and phyllosphere yeasts identify *Candida subhashii* as a novel antagonist of filamentous fungi. Microbiol 17(4):1-12. DOI 10.1186/s12866-016-0908-z.

[14] Herliyana EN, Ratna J,Darmono T, et al. 2013. uji in-vitro pengendalian hayati oleh *Trichoderma* spp. terhadap *Ganoderma* yang Menyerang Sengon. *J. Silvikul. Trop.* 4(3):190-195.

[15] *Lelana* NE, Illa A, Nina M. 2015. Antagonisme test of *Aspergillus* sp. dan *Trichoderma* spp. to *Fusarium* sp., the causal agent of dumping off on Alazia. *J. Pen. Hut. Tan.* 12(1):23-28.

IOP Conf. Series: Earth and Environmental Science **142** (2018) 012064 doi:10.1088/1755-1315/142/1/012064

[16] Suharna N. 2003. Interaction of *Trichoderma harzianum*, *Penicillium* sp. and *Pseudomonas* sp. and the antagonistic capasity to *Phytophthora capsici* in vitro. *Berita Biol.* 6(6): 747 (in Indonesia).

[17] Madi L, Katan J. 1998. *Penicillium janczewskii* and its metabolites, applied to leaves, elicit systemic acquired resistance stem rot caused by *Rhizoctonia solani*. *Physiol. Mol. Plant. Pathol.* 53:163-175.

[18] Yang L, Xie J, Jiang D, et al. 2008. Antifungal substances produce by *Penicillium oxalicum* strain PY-1 potential antibioties against plant pathogenic fungi. *Wold J. Microb. Biotech.* 24:909-915.

[19] Amaria W, Efi T, Rita H. 2013. Seleksi dan identifikasi jamur antagonis sebagai agens hayati jamur akar putih (*Rigidoporus microporus*) pada tanaman karet. *Bul. RISTRI* 4(1):55-64.

[20] Nicoletti R, De SM, De SS, et al. 2004. Identification of funitoxic metabolites produced by some *Penicillium* isolates antagonistic to *Rhizoctonia solani*. *Mycopathol.* 61:3-18.

IOP Conf. Series: Earth and Environmental Science **142** (2018) 012065 doi:10.1088/1755-1315/142/1/012065

Effectiveness of cow manure and mycorrhiza on the growth of soybean

M N A Muktiyanta[1], Samanhudi[2], A Yunus[2], B Pujiasmanto[2] and S Minardi[3]
[1] Faculty of Agriculture, Universitas Sebelas Maret (UNS) Surakarta, Indonesia
[2] Study Program of Agrotechnology, Faculty of Agriculture, Universitas Sebelas Maret (UNS) Surakarta, Indonesia
[3] Study Program of Soil Science, Faculty of Agriculture, Universitas Sebelas Maret (UNS) Surakarta, Indonesia

E-mail : samanhudiuns@gmail.com

Abstract. Soybean is one of the major food crop commodities in Indonesia. The needs of soybean each year is always increasing, but the the production rate is low. The research aimed to know the influence of treatment doses of cow manure and mycorrhiza towards growth and yield of soybeans. This research was conducted using Randomized Complete Block Design with two factors. The first factor is the dose of cow manure: S_0 (0 g/plot), S_1 (781.25 g/plot), S_2 (1562.5 g/plot), and S_3 (2343.75 g/plot). The second factor is the dose of mycorrhiza: M_0 (0 g/plot), M_1 (100 g/plot), and M_2 (200 g/plot). The observed parameters is plant height, the number of productive branches, weight of 100 seeds, root length, fresh weight of biomass, dry weight of biomass, conversion calculation results of soybeans per hectacre and the percentage of roots infected with mycorrhiza. Data were analyzed with ANOVA at 5% significance level, continued with Duncan test at 5% confidence level. The results showed that no interaction between the two treatments. Doses of cow manure provides significant influence to plant height and the length of the root. Whereas, the doses of mycorrhiza provides significant effect to the number of productive branch, weight of 100 seeds, dry weight of biomass, and the conversion of soybean yield per hectare.

1. Introduction

Soybean is one of the main food crops in Indonesia after paddy and maize. National need for soybeans reaches 2.54 million tons every year. Soybean production in 2015 amounted to 998.87 thousand tons of dry beans or increased by 43.87 thousand tons (4.59 percent) compared to 2014 [1]. The constraint in increasing soybean productivity today is the lack of productive land supporting capacity.

One option to increase soybean productivity is by fertilizing efficiently [2]. Fertilization to overcome the low productivity of soybean and the decrease of soil fertility can be done by means of manure and biological fertilizer (mycorrhiza). Provision of organic fertilizer into the soil, in addition to increasing the pH and soil organic matter content, can also increase mycorrhizal activity in infecting plant roots. Thus, it is necessary to conduct research on the application of cow manure and mycorrhiza in effect on increasing soybean productivity.The purpose of this research is to know the effect of dosage of cow manure and mycorrhizal dose on growth and yield of soybean crop, to know the effect of interaction between dose of cow manure and mikoriza on growth of soybean crop.

2. Methods

2.1. Location and Time Research
Field trials conducted in February 2016 until April 2016 in the BPP of Selogiri District, Wonogiri. Analysis of the results carried out in the Laboratory of Ecology and Management of Plant Production, analysis of mycorrhiza took place in the Laboratory of Biological Soil and soil analysis took place in the Laboratory of Chemistry and Soil Fertility in Faculty of Agriculture, Universitas Sebelas Maret.

2.2. Tools and Materials
The tools used in this study include field analysis equipment are hoes, ruler, cameras, stationery, logbook and sprinkler hose; equipment for laboratory analysis include an analytical balance, oven, scissors, waterbath, etc. Materials used in this study include soil samples and chemical for laboratory analysis covering 70% alcohol, distilled water, 10% KOH, 1 N HCl, 0.05 % trypanblue, 1 N ammonium acetate, 10% NaCl, 45% NaOH, 0.1N HCl, 2% boric acid and other chemical materials.

2.3. Research Design and Data Analysis
This research used a Randomized Complete Block Design (RCBD) with two factors. The treatment used was dose of cow manure (S0 = 0 g / plot, S1 = 781.25 g / plot, S2 = 1562.5 g/plot, and S3 = 2343.75 g/plot). The second factors is the dose of mycorrhiza: M0 (0 g/plot), M1 (100 g/plot), and M2 (200 g/plot). Data was analyzed with ANOVA at 5% level, continued by test of Duncan at 5% level.

3.Results and Discussion

3.1. Plant height
The result of variance analysis showed that the dosage of cow manure had significant effect on soybean plant height. Giving mycorrhiza had no significant effect and there was no interaction between the two treatments. The results of this study are in line with [3] stated that organic fertilizers can create better soil conditions, such as available nutrients, oxygen, and water needed by soybean plants in optimal and balanced quantities, so as to increase soybean growth.

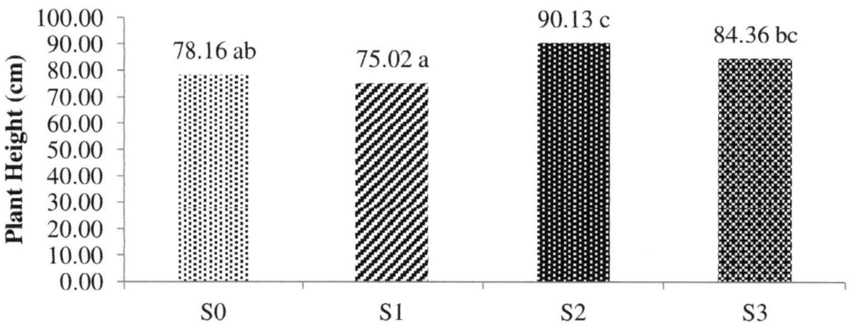

Inter figures followed by the same letter are not significantly different in the Duncan Multiple Range Test of 5%

Figure 1. Effect of doses cow manure on soybean plant height

Treatment of dosage of cow manure 1562.5 g / plot was significantly different with dosage of cow manure 0 g / plot.[4] states that manure derived from cows contains many decompost microorganisms that are beneficial to the increasing of the type and population of soil microorganisms. Microorganisms can improve the physical properties so that the drainage and aerase of the soil become good.

3.2 Productive Branch

The result of variance analysis showed that the dosage of mycorrhizal dose gave a real effect to the productive branch of soybean crop. This result is in line with the study of [5] which states that mycorrhizal affects the number of branches because mycorrhiza is able to provide nutrients for plants. [6]) states that mycorrhiza plays a role in the absorption of phosphorus nutrients. This nutrient can help in the formation of flowers and soybean pods.

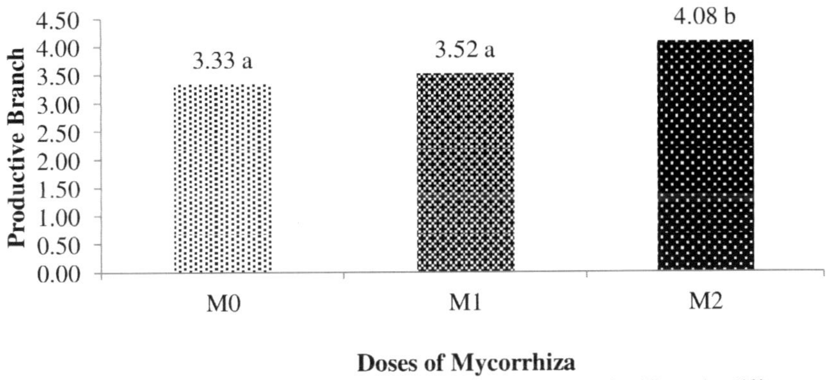

Inter figures followed by the same letter are not significantly different in the Duncan Multiple Range Test of 5%

Figure 2. Effect of doses mycorrhiza on productive branch of soybean

Treatment of mycorrhizal doses of 200 g / plot has the highest mean value of productive branches of 4.08 branches. Treatment of mycorrhizal dose of 200 g / plot was significantly different from the dose of mycorrhizal 0 g / plot and mycorrhizal dose of 100 g / plot. [7] states that mycorrhiza can help the absorption of water and nutrients so that plant growth can increase along with the addition of mycorrhiza.

3.3 Weight of 100 seeds

Variation analysis results obtained that the dosage of mycorrhizal dose gave a significant effect on the weight of 100 seeds. The dosage of manure and the interaction between the two treatments did not significantly affect the weight of 100 seeds. [8] state that giving different doses of mycorrhizae increasing the weight of soybean seeds because mycorrhiza can help nutrient uptake and improved nutrient uptake status especially phosphorus.

The picture 3 shows that the effect of mycorrhizal administration of 200 grams / plot has the highest average to the weight of 100 seeds is 17.95 grams. The dosage of mycorrhizal doses of 200 grams / plot was significantly different from the 0 gram / plot dosage treatment, while the dosage of mycorrhizal 100 gram / plot was not significantly different from the dosage of mycorrhizal 0 gram / plot. [9] state that mycorrhiza can nourish plants, especially phosphorus from the soil.

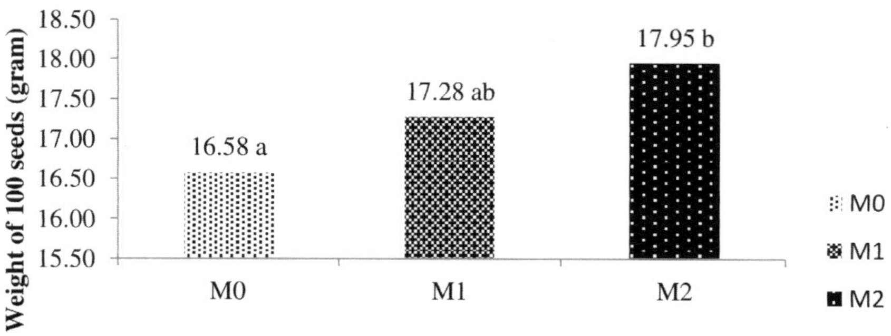

Doses of Mycorrhiza
Inter figures followed by the same letter are not significantly different in the Duncan Multiple Range Test of 5%

Figure 3. Effect of doses mycorrhiza on weight of 100 seeds

3.4. Length of Roots
Variation analysis results obtained that the treatment of doses of cow manure has a real effect. This result is in line with [10] it states that the addition of organic material from manure into the soil will improve the physical properties of the soil, it facilitates the roots in its growth. Manure is able to improve the physical and chemical conditions of the soil, and encourage the life (development) of microorganisms.

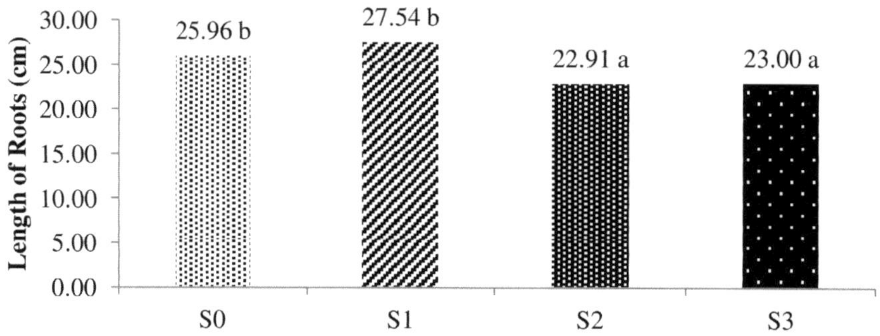

Doses of Cow Manure
Inter figures followed by the same letter are not significantly different in the Duncan Multiple Range Test of 5%

Figure 4. Effect of doses mycorrhiza on length of roots

Figure 4 shows that the average dose of cow manure is highest to root length that is at dose 781.25 gram / plot with average root length 27.54 cm. Dosage of cow manure 0 g / plot has no significant effect on dosage of cow manure 781.25 g / plot. The dose of cow manure 1562.5 g / plot and 2343.75 g / plot has a significant effect on dosage of cow manure 0 g / hectare.

3.5. Dry Weight of Biomass
Variation analysis result showed that dosage of mycorrhizal dose gave a significant effect on dry weight of biomass. This result is in line with [11] study that the increase in dry weight is due to an association between mycorrhiza and plants. Mycorrhizas symbiotic with plant roots can increase water and phosphorus uptake to be exploited by leaves and stems.

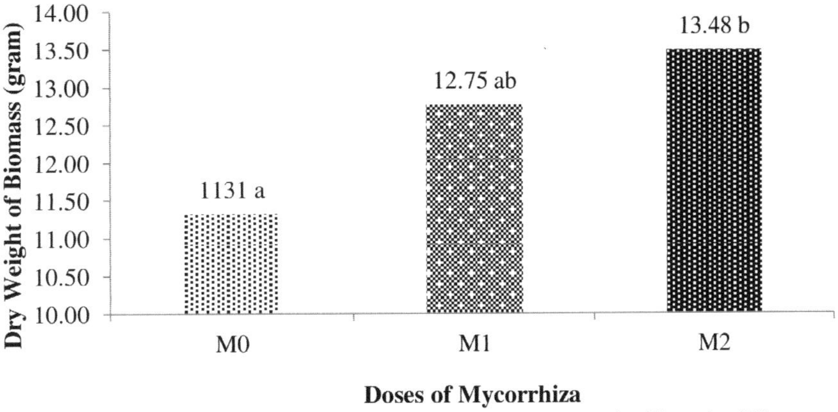

Inter figures followed by the same letter are not significantly different in the Duncan Multiple Range Test of 5%

Figure 5. Effect of doses mychorrhiza on dry weight of biomass

Figure 5 shows that the highest dry weight is found in the dose of mycorrhizal 200 gram / plot with an average weight of 13.48 grams. Treatment dose mycorrhizal 0 gram / plot has a real difference to the treatment dose of mikoriza 200 gram / plot. Treatment of 0 gram / plot of mycorrhizal doses was not significantly different from the treatment of mycorrhizal doses of 100 grams / plot, and mycorrhizal treatment of 100 grams / plot was also not significantly different from the treatment of mycorrhizal doses of 200 grams / plot. Mycorrhizal doses may increase the dry weight of the plant. Plants which are given mycorrhiza will grow more fertile because of the larger root surface area to absorb nutrients and the number of leaves more to support the process of photosynthesis so as to produce more dry matter [12].

3.6. The Conversion of Soybean Yield Per Hectare

The result of variance analysis showed that the dosage of mycorrhizal dose gave a significant effect on the conversion of soybean yield per hectare. The dose of cow manure and the interaction between the two treatments did not significantly affect the conversion of soybean yield per hectare. Mycorrhizae can increase the ability of plants to utilize the available resources on the land so that the plants are not lack of nutrients and can increase production yield [13].

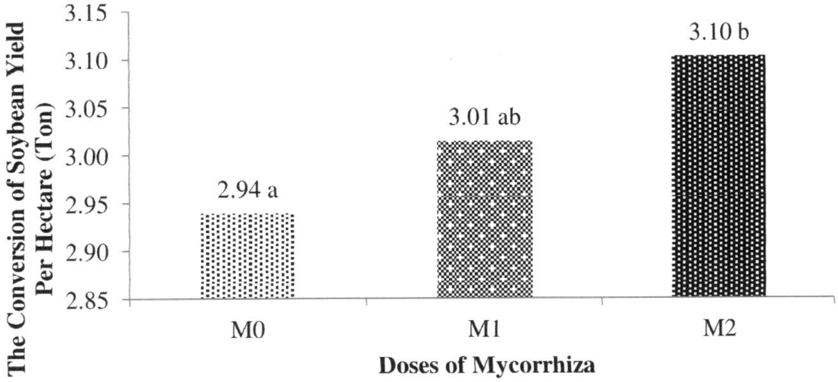

Inter figures followed by the same letter are not significantly different in the Duncan Multiple Range Test of 5%

Figure 6. Effect of doses mychorrhiza on the conversion of soybean yield per hectare

Figure 6 shows that the highest average calculation of soybean yield per hectare is found in the treatment of 200 gram / hectare of mycorrhizal dose with an average of 3.10 ton / hectare. Treatment of mycorrhizal 0 grams / hectare was not significantly different from the treatment of mycorrhizal doses of 100 grams / hectare and the treatment of 100 grams / hectare was also not significantly different from the treatment of mycorrhizal dose of 200 grams / hectare. Treatment of 0 gram / hectare mycorrhizal dose was significantly different from the treatment of mycorrhizal dose of 200 gram / hectare. Plants which are given mycorrhizal applications will have higher yields than non-mycorrhizable plants [14].

3.7 The percentage of infections Mycorrhiza

Mycorrhiza as a biological fertilizer can give benefit plants in the supply of nutrients, especially Phosphor for plants. Mycorrhizal infecting the root system of plants can increase the capacity to absorb nutrients. Infected mycorrhizal roots will produce internal hyphae, external hyphae, and spores that are effective sources of inoculum [15].

Table 1 The effect of fertilizer dose and mycorrhizal fertilizer on percentage of fertilizer efficiency (%)

Doses of mycorrhiza	Doses of Cow Manure				Average
	S0	S1	S2	S3	
M0	0.00	0.00	0.00	0.00	0.00
M1	1.00	1.00	1.00	1.00	1.00
M2	1.00	1.00	1.00	1.00	1.00
Average	0.67	0.67	0.67	0.67	0.67

Based on Table 1 shows that the percentage of soybean plants infected with mycorrhiza is 67%. According [16] .states that giving mycorrhiza can increase the percentage of infected roots and form symbiotic associations that can increase root growth. Symbiotic mycorrhizas with host plants will result in external hyphae that can increase the ability of roots to absorb water and nutrients in the soil.

4. Conclusions

Based on the results of research Effectiveness Of Cow Manure And Mycorrhiza On The Growth Of Soybean showed no interaction between the two treatments. Doses of cow manure provides significantly influence to the high of plants and the length of the root. Whereas, the doses of mycorrhiza provides significantly influence to the number of productive branch, weight of 100 seeds, dry weightof biomass, and the conversion of soybean yield per hectare.

Acknowledgement

The authors are grateful to the Ministry of Research, Technology and Higher Education of the Republic of Indonesia that has funded this research through the scheme of Masterplan Percepatan dan Perluasan Pembangunan Ekonomi Indonesia (MP3EI) for fiscal year 2015.

References

[1] BPS (Badan Pusat Statistik) 2015 Production of Paddy, Maize, and Soybean (forecast I in 2015) Berita Resmi Statistik (Badan Pusat Statistik) **16** (1) :1-12
[2] Martodireso and Suryanto 2001 Pemupukan Organik Hayat Kanisius Yogyakarta
[3] Efendi 2010 Increase of Growth and Yield of Soybean Using Green Organic Matter and Manure *J Floratek* **1** (5): 65 -73
[4] Novizan 2007 *Sawi dan Selada* Penebar Swadaya Jakarta
[5] Samanhudi, Sudadi, Pujiasmanto B, Mahasti W 2017 The Influence of Kascing and Mycorrhiza on the Growth and Results of *Tribulus terrestris*. *Jurnal Agrotek Indonesia* **2** (1): 1-6. ISSN

: 2477-8494
[6] Leovini H, Kastono and Widada J 2014 The Effect of Arbuscula Mycorrhizal Fungi Application, Phosphate Fertilizer and Compost Dosage to Seedling Growth of Sugarcane (*Saccharum officinarum* L.) on Coast Sand. *Jurnal vegetalika* **3**(1): 102-115
[7] Nurhayati 2010 Effects of Inoculation Time of Mycorrhiza Vesicular Arbuscular on Fusarium Oxysporum Infection and Tomatoes Growth. *Jurnal Agrivigor* **9** (3): 280-284. ISSN 1412-2286
[8] Zuhry E and Puspita F 2008 Giving of Arbuscular Mycorrhizal Fungi in Red Yellow Podzolik to Soybean Growth and Production (*Glyci'ne max*(L) Merill) *Jurnal SAGU* **7** (2): 25-29. ISSN 1412-4
[9] Yamawaki K, Matsumura A, Hattori1 R, Tarui1 A, Hossain MA, Ohashi Y and Daimon H 2013 Effect of Inoculation With Arbuscular Mycorrhizal Fungi On Growth, Nutrient Uptake and Curcumin Production Of Turmeric (Curcuma Longa L.) *Journal Agricultural Sciences* **4** (2): 66-71. DOI : 10.4236/as.2013.42011
[10] Prasetyo A, Listyorini E and Utomo WH 2014 Relationship of Soil Physical Character, Rooting and Cassava Yield Second Year At Alfisol Jatikerto Due to Organic and Inorganic Fertilizer *Jurnal Tanah dan Sumberdaya Lahan* **1** (1): 27-37
[11] Rengganis RD, Hasanah Y and Rahmawati N 2014 Role of arbuscular mychorrhizal fungi and rock phosphate fertilizer on growth and production of soybean (*Glycine max* (L.) Merrill) *Jurnal Online Agroekoteknologi* **2** (3): 1087- 1093. ISSN: 2337- 6597
[12] Muis R, Ghulamahdi M, Melati M, Purwono and Mansur I 2016 Compatibility of Arbuscular Mycorrhizae Fungi Inoculants with Soybean Plants in Saturated Soil Culture *Penelitian Pertanian Tanaman Pangan* **35** (3): 229-238
[13] Ferry Y and Rusli 2014 The Effects of Mycorrhizal and NPK Fertilization Dosage for Growth and Production of Robusta Coffe among Productive Coconut Plantation *J Littri* **20** (1) : 27-34. ISSN 0853-8212
[14] Ahmed FE, Yagoub SO and Elsheikh AE 2000 Effects of Mycorrhizal Inoculation and Phosphorus Application on the Nodulation, Mycorrhizal Infection and Yield Components of Faba Bean Grown Under Two Different Watering Regimes *J of Agriculture Science* **8**(2): 107-116. DOI: 41.67.20.41:8080/handle/123456789/1761
[15] Anas I and Tampubolon JLO 2004 Soil-Sand Misture Growth Medium and Inorganic Fertilizers for Arbuscula Mycorrhizal Fungus (AMF) Inoculant Production *Jurnal Bulletin Agronomi* **32** (1) : 26-31
[16] Laksono AB, Dewi IR, Suherman C and Santoso J 2013 Effect of Arbuscular Mycorrhizal Fungi on Growth of Cinchona (*Cinchona ledgeriana Moens*) Shoot Cutting Cib5 and QRC Clones. *Jurnal Penelitian Teh dan Kina* **16** (2): 83-90

IOP Conf. Series: Earth and Environmental Science **142** (2018) 012066 doi:10.1088/1755-1315/142/1/012066

Aplication of giberelins on flowering and yield of two varieties of shallot in lowland

E Triharyanto, S Nyoto and I Yusrifani
Agrotechnology Department, Agriculture Faculty, Universitas Sebelas Maret Surakarta, Indonesia

E-mail: eddytriharyanto@yahoo.co.id

Abstract. Shallot is one of horticultural commodities which has difficulties in flowering and producing seeds. Flowering of shallot generally occurs in highlands at 9-12° C. Flowering in lowland can be supported with vernalization or replace cold temperature with gibberelin (GA_3). This research is aimed to determine the effect of varieties, the concentration of GA_3 applied and their interaction on flowering and yield of shallots grown in the lowlands, 98 m altitude with Vertisol-type soil. The research was conducted in a randomized complete block design (RCBD) with two factors, which were varieties (V): Bima and Mentes and concentration of GA_3 (C) 0 ppm, 50 ppm, 100 ppm, 150 ppm and 200 ppm and repeated three times. Varieties have significant effect ($P<0.05$) on plant height, leaf number, the number of bulbs per clumps, weight of fresh bulbs per clumps, and percentage of small and large bulbs produced. Bima varieties were able to flower and produced seeds, while Mentes varieties could not produce flowers and seeds. GA_3 concentration have no significant effect in all of the observed component. GA_3 can't replace cold temperature for supported flowering in varieties Mentes which were planted in lowlands. There was interaction between varieties and GA_3 concentration occurs in variable percentage of small and large bulbs.

1. Introduction

Shallot is one of the horticultural commodities whose needs are always increasing every year, but not always followed by increased productivity. According to data from BPS [1], there is a decrease in shallot productivity in Indonesia by 0.39% in 2015 compared to 2014. One of the causes of decreasing shallot production is the low quality of seed used [2] decline in the long term as seeds so as to increase the likelihood of decreasing virus like *Shallot latens virus* from the previous generation to the next generation [3].

Seeds can be used as alternatives in these problems. The use of seeds as planting materials will minimize transmission of the virus from the previous generation [4] and more efficient use because it only takes approximately 1.5 kg per Ha of shallot seeds. However, the production of seeds in Indonesia is still low due to the difficult flowering occurs especially in the lowlands. Not all varieties can flower in lowlands. The varieties used in this research are varieties of Bima and Mentes because both of these varieties are superior varieties of shallot which are of great interest to the public. Bima varieties of shallot have been able to adapt to flowering in lowland but with low yield, while Mentes variety to enter the generative phase requires cold temperatures (high land).

Environmental factors such as temperature, photoperiode and rainfall will affect the process of forming shallot flowers [5]. Shallot plant is a biannual and plant long days in which to flowering takes cold temperatures (<18° C) and photoperiode >12 hours [6], while in the lowlands of relatively high temperatures (>29° C) with a photoperiod 12 hours or less. The cold-temperature function can be replaced by a vernalization treatment (cold temperature) or the administration of a growth regulator in the form of gibberellin which can specifically replace the cold temperature function to induce flowering induction.

Shallot basically have gibberellin hormones that are inactive and in limited quantities depending on the variety used so that the concentration of gibberellin given will have different effects depending on the needs of each variety. Giberelin used in this research is the gibberellic acid (GA_3) which applied by soaking the seed bulbs before planting (30 minutes) and spraying the seedlings age 2-5 WAP at a concentration of 0 ppm, 50 ppm, 100 ppm, 150 ppm and 200 ppm.

According to Barani et al [7], concentrations below 100 ppm have not had any effect on plant growth because they have not been effectively used, while concentrations between 100 ppm and 200 ppm have a significant effect on the number of flowering plants per plot and seed weight per plant umbel [8]. This research was conducted with the aim to find out the response of varieties of Bima and Mentes planted in the lowlands in terms of flowering and yield of seeds and bulbs produced, knowing what the appropriate concentration to encourage flowering and improvement of crop production and to know the interaction between the varieties tested and the concentration of GA_3 was applied.

2. Methods

This research was conducted from May to August 2016 in Gunung Mijil village, Jaten sub-district, Karanganyar district.Materials used include shallot crop land, the seed bulbs and Mentes varieties, GA_3, SP36, ZA, KCl and Urea. The tools used to support this research include hoes, analytical scales, meters, measuring cylinders, sprayers, slide length, cameras, logbooks and stationery.

This research was conducted with randomized complete block design (RCBD) with two factors, such as the variety and concentration of GA_3 Varieties consists of two levels such as Bima variety and Mentes variety whereas the concentration of GA_3 consists of 5 levels such as 0 ppm, 50 ppm, 100 ppm, 150 ppm and 200 ppm.The two factors are then combined and repeated three times.

The observed variables were divided into 3 main components ie growth components consisting of plant height and number of leaves, flowering and seed forming components consisting of the time of flower emergence, percentage of flowering plants per plot, number of flower stalks per Ha, number of flowers per flower stalk, number of seeds per flower stalk and the weight of the seeds per flower stalk as well as the yield component consisting of the number of bulbs per plant clump, the weight of fresh bulbs per plant clump, the weight of fresh bulbs per Ha, the weight of dry bulbs after drying per Ha, the percentage of small, medium and large per plot.

3. Result and Discussion

3.1 General condition of research location

This research was conducted in Gunung Mijil village, Jaten sub-district, Karanganyar district at altitude of 89 mdpl (lowland).The study took place from May to August 2016 with relatively high rainfall of 168 mm in May, 242.5 mm in June, 255.5 mm in July and low in August of 36.5 mm and relative humidity 78 , 6% in May, 78.75% in June, 79% in July and 78.1% in August.The air temperature is relatively high at 26.9 to 38.4 ° C in May, from 26.9 to 38.6 ° C in June, from 26.8 to 38.6 ° C in July and 26.7 to 38, 7 ° C in August.The temperature is too high for the induction and initiation of flowering (> 26 ° C), but optimal for bulb formation and ripening seeds of shallot (9).

3.2 Growth components

The growth component consists of plant height and number of leaves obtained by measuring at ages 2 WAP to 7 WAP.Plant height is obtained by measuring from the base of the plant (ground level) to the longest leaf end of the sample plant while the number of plant leaves is obtained by counting the number of leaves that can form on each sample plant clump.Plant height and number of leaves of the plant are presented in Table 1.

Table 1. Average plant height (cm) and number of leaves (leaf) shallot plants aged 7 WAP in a wide variety, the concentration of GA $_3$ and the interaction between varieties and concentrations of GA_3

Concentration of GA_3 (ppm)	Plant height (cm)		Average	Number of leaves (strands)		Average
	Variety			Variety		
	Bima	Mentes		Bima	Mentes	
0	41.73	34.86	38.56	36	42	38.97
50	38.37	35.50	36.93	36	36	35.80
100	39.17	33.20	36.46	34	36	34.57
150	37.43	32.47	34.95	25	36	30.37
200	39.60	33.57	36.58	31	43	36.90
Average	39.43 b	37.96 a	(-)	32.09 a	33.44 b	(-)

Description: The numbers followed by the same letter are not significantly different based on the DMRT test of 5%.

The result of variance analysis showed that varieties had significant effect on plant height and number of plant leaves ($p<0.05$). Bima variety has higher plant height (39.43 cm) than Mentes variety (37.96 cm) but has fewer number of leaves (32.09 sheets) than Mentes varieties (33.44 leaf).This difference is because the two varieties have different genetics that will produce different phenotypes [10].Abdissa et al [11] revealed that vegetative growth of plants is influenced by the ability of plants to synthesize some of the protein components available and needed.

The concentration of GA_3 did not show any real effect on plant height and number of leaves of shallot plants ($p>0.05$).This is thought to be due to improper application (immersion in seed bulbs before planting and spraying on plants).Excessive application will cause gibberellin to be given gibberellin glucoside which is inactive so that can not be used by plants.Table 1 shows that the plant without the application of GA_3generating plant height and number of leaves were higher than in plants by application of GA_3. Application of GA_3 tends to decrease the plant height and number of leaves produced.This result is contrary to the opinion and Benkeblia Williamson [12], which revealed that GA_3 applied capable of increasing the length of the leaf by influencing the degradation of chlorophyll.There is no interaction between the varieties used and the concentration of $GA_3$was applied ($p>0.05$).

3.3 Purification and forming seeds components

The flowering component consists of the time of the flower emergence, the percentage of flowering plants per plot, the number of flower stalks per Ha and the number of flowers per stalk while the seeding component consists of the number of seeds per stalk and the weight of the seeds per stalk.The flowering component is presented in Table 2 and the seeding component is presented in Table 3.

Table 2. Average of the emergence of interest (DAP), percentage of flowering plants (%), the number of flower stalks per hectare and the number of flowers per stalk on a wide variety, the concentration of GA_3 and the interaction between varieties and concentrations of GA_3

Concertation of GA_3 (ppm)	Time of the flower emergence (DAP)		Percentage of flowering plants per plot (%)		Number of flower stalk/Ha		Number of flowers per stalk	
	Variety							
	Bima	Mentes	Bima	Mentes	Bima	Mentes	Bima	Mentes
0	32	-	31.67	0	40,740	0	48.27	0
50	32	-	7.04	0	25,925	0	43.90	0
100	30	-	15.56	0	66,666	0	48.17	0
150	30	-	14.07	0	59,259	0	32.85	0
200	29	-	29.63	0	134,567	0	32.65	0
Average	30,47	-	19.57	0	65,432	0	41.17	0

Description: Mentes varieties of shallot can not flower and produce seeds, so can not be analyzed

Flowering occurs only on the Bima varieties of shallot, while the varieties of Mentes can not be flowered so it can not produce seeds.No occurrence of shallot varieties of flowering in suspected Test the temperature is too high because that is an average of 26.7 to 38.7° C at the time of the study.Triharyanto et al. [5] revealed that the temperature is above 29° C will impede the flowering process.Mentes varieties are allegedly unable to adapt well to enter the generative phase when planted in the lowlands, while the varieties of Bima have been able to adapt so that it can flower if planted in lowland even with low yield.

Shallot is a biannual plant in which the plant can flower after two seasons (spring and summer).Jasmi et al [13] revealed that flowering induction is influenced by the presence of stimuli in the leaves that are affected by cold temperature induction (vernalization). Flowering induction is the initial stage of flower formation because at this stage there is a change in the biochemical response of the apex layer which is a signal of change from the vegetative to the generative phase [14]. Plant which induced will continue initiation and form the flower buds, while the induced failed plants will return to the vegetative phase [15].

Bima varieties grown in the lowlands can be flowering and produce seeds but with low yields.Shallot of Bima varieties are flowering at the age of 30 DAP.The timing of this flower is faster than the description of the varieties (50 DAP). Giving GA_3 speed up the emergence of interest shallot varieties of Bima but was unable to push flowering Mentes varieties.This is contrary to the opinion of Sorensen et al.[15], which revealed that GA_3 can be used to replace the function of cold temperatures to stimulate cell elongation umbel shallots from the growing point.

Shallot varieties Bima applied GA_3 with a concentration of 200 ppm can be flowering at the age of 29 DAP, while at a concentration of 100 ppm and 150 ppm flowering at the age of 30 DAP as well as the concentration of 0 ppm and 50 ppm flowering at the age of 32 DAP.This result is basically the Bima varieties grown in lowland flowering and administration of GA_3 could accelerate the emergence of interest in one plot although the percentage of flowering plants per plot are low (average 19.57%).

Application of GA_3 can not increase the percentage of flowering plants per plot generated.Bima shallot varieties without the application of GA_3 as a greater percentage of flowering plants per plot were higher than the Bima shallot varieties with the application of GA_3. The concentration of GA_3 50 ppm, 100 ppm and 150 ppm resulted in a low percentage of flowering plants, respectively 7.04%, 15.56% and 14.07%. GA_3 was applied at a concentration of 200 ppm resulted in 29.63% of flowering plants per plot. This result is lower than in Bima shallot crop varieties without the application of GA_3 (31.67%).

Flower stalks produced per Ha Bima varieties averaged 65,432 stalks. Each plant produces 2-5 flower stalks per hill plant. According to Patra et al.[16] GA_3 applied will increase the synthesis and translocation of flowering hormones that will increase the size of the flowers and the maximum number of flowers per plant. The flower stalk comes from the core buds that are in the center of the stem and grow as well as the leaves in general. Varieties Bima applied GA_3 with a concentration of 200 ppm produced 36 stalks of flowers/2.7 m^{2}, equivalent to 134 567 flower stalks/ha while that of the untreated GA_3 resulted in the number of flower stalks are lower at 11 peduncle/2.7 m^2 or equivalent to 40,740 stalks / ha.

The number of flowers per flower stalk produced varies each individual.The interest rate per flower stalk is calculated after the interest is harvested. According Rahayu and Berlian [17] each flower stalk can produce 50 - 200 flower buds depending on varieties used.Plants that have undergone induction and flowering initiation will then proceed to the stage of forming the flower parts [14].Flower buds that are formed then will do pollination.Successful flowers in their pollination will then develop and produce seeds, while pollination that fails will cause the flowers to dry and fall.Bima shallot crop varieties without the application of GA_3 generates average number of flowers per stalk a higher interest rate than plants that applied GA_3. This suggests that GA_3 is not able to increase the number of flowers per stalk is produced and tends to decrease the number of flowers per stalk produced. Bima shallot crop varieties without the application of GA_3 resulted in the number of flowers per stalk around 48.27 flowers while at a concentration of 50 ppm produced an average 43.90 flower / stem, the concentration of 100 ppm produced an average 48.17 flower / stem, the concentration 150 ppm yields an average of 32.86 flowers / stalks and a concentration of 200 ppm yields an average of 32.65 flowers / stalks.

Table 3. Average number of seeds per stalk and seed weight per stalk (gram) in a wide variety, the concentration of GA $_3$ and the interaction between varieties and concentrations of GA_3

Concertation of GA_3 (ppm)	Number of seeds per stalk		Weight of seeds per stalk (gram)	
	Variety		Variety	
	Bima	Mentes	Bima	Mentes
0	65.93	0	0.29	0
50	60.33	0	0.30	0
100	65.50	0	0.50	0
150	71.00	0	0.41	0
200	62.18	0	0.32	0
Average	64.99	0	0.36	0

Description : Mentes varieties of shallot can not flower and produce seeds, so can not be analyzed

The number of seeds per stalk produced is influenced by the success of pollination on each flower. Reproductive organs that have been ripe will continue the process of pollination and fertilization.Pollination shallot can occur alone or cross with the help of polinator. Palupi et al. [18], revealed that the seeds produced in the lowlands are still relatively low at below 30%. The number of seeds per stalk and seed weight per stalk are presented in Table 3. According Sumarni et al. [6] seed formation occurs at high temperature is 35° C and the temperature in this study about 26.7 to 38.7^{o} C which is the temperature optimum for seed formation.Table 3 shows that the number of seeds produced in the shallot crop without GA_3 is on average 65.93 seeds, the concentration of 50 ppm produced an average 60.33 seeds, a concentration of 100 ppm produced an average 65.50 seeds, the concentration of 150 ppm yields an average of 71 seeds and a concentration of 200 ppm yields an average of 62.18 seeds. According Helaly et al. [19] revealed that the administration of GA $_3$ is able to increase the fertility rate of shallot, so that the beans are produced more and more. The formation of seeds only occurs in the varieties of Bima while the varieties of Mentes can not be flowered so it can not produce seeds anyway.

The weight of the seeds per flower stalk is obtained by weighing the weight of the seeds after drying in the sun for 2 days. The high number of seeds is not always followed by the high seed weight. The resulting beans have different sizes, so that every seed produced will also vary. The results showed that administration of GA $_3$ can increase the weight of seeds produced. Treatment without GA_3 yields 0.29 g/stalk whereas the application of GA $_3$ with a concentration of 50 ppm, 100 ppm, 150 ppm and 200 ppm respectively produce seed weight per stalk on average 0.30 grams, 0.50 Grams, 0.41 grams and 0.32 grams.

3.4 Yield components

The yield components include the number of bulbs per plant family, the weight of fresh bulbs per plant clump (table 4), the weight of fresh bulbs per Ha, the weight of dry bulb per Ha (table 5) and the percentage of small, medium and large bulbs produced per plant clump (table 6). The yield component is calculated after the bulbs are harvested. Shallot is a bulb with a layer that is quite a lot and quite thick.

Table 4. Average number of bulbs per plant and fresh weight of bulbs per plant (g) on a wide variety, the concentration of GA $_3$ and the interaction between varieties and concentrations of GA_3

Concertation of GA_3 (ppm)	Number of bulbs per plants		Average	Weight of bulbs per plants (gram)		Average
	Variety			Variety		
	Bima	Mentes		Bima	Mentes	
0	11.27	11.13	11.20	60.01	64.73	62.37
50	8.87	12.60	10.73	49.15	54.15	51.65
100	9.13	12.80	10.97	42.94	53.82	48.38
150	9.00	13.67	11.33	44.05	59.52	51.78
200	8.47	15.00	11.73	36.22	62.82	49.52
Average	9.35 a	13.04 b		46.47 a	59.01 b	

Description: The numbers followed by the same letter are not significantly different based on the DMRT test of 5%

The number of bulbs produced is obtained by counting the number of bulbs in the sample plants after harvesting. The result of variance analysis showed that varieties had significant effect ($p < 0.05$) on the number of bulbs per plant family.According to Hailu et al. [20] varieties have different attributes and levels of adaptation to the environment, this is what causes their productivity to vary. Bima varieties produce bulbs with a lower number than the varieties of Mentes.Bulbs produced on the average Bima variety 9.35 bulbs per hill while Mentes varieties produce 13.04 bulbs per plant clump.

Based on the results of analysis of variance concentration of GA_3 did not significantly affect the number of bulbs per plant ($p>0.05$). There is no significant difference in the number of bulbs produced in plants that applied GA_3 and plants that are not applied GA_3.This suggests that without the application of GA_3 will produce a pretty good number of bulbs due to the variety used is able to adapt well in low and productivity is high enough.According to Gautam et al. [21] bulb yields are more influenced by climatic conditions such as temperature, nutrition and spacing and the condition of seeds used as planting material. The temperature in this study is quite high (26-36° C) which is the optimum temperature for bulb formation. This is as revealed by Dinarti et al. [22] which revealed that high temperatures during the bulb-forming phase will lead to an increase in carbohydrate accumulation to bulb parts as well as enzymatic activity that increases the process of sucrose translocation to the bulb. No interaction between varieties with the concentration of GA_3 was applied ($p>0.05$).

The number of bulbs per hill produced will affect the weight of fresh bulbs per plant clump. Based on the result of correlation test, the number of bulbs per plant was positively correlated (62.4%) with the weight of fresh bulbs per plant clump, which means that the greater the number of bulbs per hill the weight of fresh bulbs per hill is also heavier. Based on the analysis of varieties, varieties affect the weight of fresh bulbs per plant grown ($p<0.05$). Bima varieties produce fresh bulb weight per plant clump 46,47 gram and lower than with varieties of Mentes yielding fresh bulb weight per plant of 59,01 gram. This is directly proportional to the number of bulb produced.

Based on the results of analysis of variance, the concentration of GA_3 did not significantly affect the weight of fresh bulbs per family of plants. Shallot without the application of GA_3 is able to generate fresh bulb weight higher than the shallot is applied GA_3.Asra [23] revealed that the administration of GA_3 could lead to an increase in the enzyme proteinase that converts the protein into amino acids and the enzyme lipase which converts fat into fatty acids and glycerol so the food reserves stored in the bulbs will become more mobile and can be transported evenly to all parts of the plant , this led to heavy fresh bulbs per family of plants that applied GA_3 tends to be lower than that of the untreated GA_3. No interaction between varieties used at a concentration of GA_3 was applied.

Table 5. Average weight of fresh bulb per ha (tons) and weight of dry bulb per ha (ton) on a wide variety, the concentration of GA_3 and the interaction between varieties and concentrations of GA_3

Concertation of GA_3 (ppm)	Weight of fresh bulb per Ha (tons) Variety		Average	Weight of dry bulb per Ha (tons) Variety		Average
	Bima	Mentes		Bima	Mentes	
0 ppm	18.0	16.6	17.3	15.6	14.0	14.8
50 ppm	18.0	13,9	15.9	15.1	10.5	12.8
100 ppm	18.4	18.1	18.3	14.8	12.5	13.6
150 ppm	13.6	16.4	15.0	10.2	12.5	11.4
200 ppm	10.2	17.0	13.6	8.9	13.7	11.4
Average	15.7	16.4		12.9	12.7	

Description: The numbers followed by the same letter are not significantly different based on the DMRT test of 5%

Based on the results of analysis of variance, variety and concentration of GA_3 did not significantly affect the fresh and dry weights per ha ($p>0.05$). The weight of fresh bulbs per high plant family is not always followed by the weight of fresh bulbs per Ha, as each individual plant produces different fresh bulbs. The weight of the dry bulb is used as the shallot productivity parameter. Bima varieties produce dried bulbs higher than the description. According Balitsa (2015) shallot varieties Bima produce 9.9 tons / ha dry bulbs, while this study varieties Bima able to produce an average of 12.9 tons / ha. Production of dried varieties of Mentes varieties according to descriptions of varieties of 12.7 tons / ha (range 7.10 - 27.58 tons / ha). According to Getahun [24] bulb yield is also affected by the type of soil and the availability of water both irrigation water and rain water. The availability of water at the time of the research is quite good, so the bulb productivity is also good. The concentration of GA_3 did not increase the weight of fresh and dried bulbs per hectare produced. This is presumably because the GA_3 applied too much (applied by immersion and spraying) so that the impact is likely to decrease the production of shallot. No interaction between varieties GA_3 were used at a concentration of GA_3 was applied.

The result of variance analysis showed that the varieties had significant effect on the percentage of small and large bulbs produced. Bima varieties produce more bulb (70.02%) than Mentes variety (56.61%), but lower yield small bulb (10.55%) than Mentes variety (20.79%). Bulb size is determined by the number of layers of bulbs that wrap the bulbs and the amount of moisture and nutrients that can be stored in the bulbs. The more layers of bulbs the larger the size. Each variety has its own differences in absorbing nutrients and producing photosynthesis. This difference is because each variety has a different metabolic processes so that the size of the bulbs will be different generated. Differences in response to this environment causes the difference in response to GA_3 was applied. Table 6 shows that the Mentes varieties applied GA_3 with a concentration of 50 ppm produced the highest percentage of large bulbs but lowest percentage of small bulbs and not significantly different from the Mentes varieties without the application of GA_3. Application of GA_3 concentration of 100 ppm, 150 ppm and 200 ppm tends to lower the percentage of large bulbs in the Mentes varieties and increase the percentage of small bulbs. This shows that the administration of GA_3 on the variety seems inappropriate Mentes so will reduce plant growth and development.

Tabel 6. Average percentage bulb size produced at multiple concentrations, a wide variety and interaction between varieties and concentrations of GA_3

Concertation of GA_3 (ppm)	Percentage bulb size (%)								
	Bulb size								
	Small		Average	Medium		Average	Large		Average
	Variety			Variety			Variety		
	Bima	Mentes		Bima	Mentes		Bima	Mentes	
0 ppm	3.4 a*	12.2 ab*	7.8	19.1	18.0	18.5	77.5 cd	69.8 bcd	73.7
50 ppm	3.4 a*	34.3 c*	18.9	14.5	29.4	21.9	82.1 cd	36.2 a	59.2
100 ppm	9.0 ab*	21.4 bc*	15.2	24.1	26.3	25.2	66.9 bcd	52.3 ab	59.6
150 ppm	16.5 ab*	21.5 bc*	19.0	18.0	23.7	20.8	65.6 bcd	54.8 abc	60.2
200 ppm	20.5 bc*	14.6 ab*	17.5	21.5	15.5	18.5	58.0 abcd	69.9 bcd	63.9
Average	10.55 a	20.79 b		19.42	22.60		70.02 b	56.61 a	

Description: The number followed by the same letter has no real effect based on the 5% DMRT test and the 10% level (*)

GA_3 were applied to Mentes variety is different result with GA_3 were applied to Bima variety. Table 6 shows that the administration of GA_3 at concentrations of 50 ppm produces large bulbs lowest and highest small bulbs on Mentes varieties. They show a real difference to Mentes varieties without application of GA_3. Application of GA_3 with a concentration of 100 ppm, 150 ppm and 200 ppm (Table 6) showed a tendency to increase the percentage of large bulbs and lower percentage of small bulbs although not significantly different to Test the varieties without the application of GA_3. This shows that the administration of GA_3 on Test the variety does not really give a higher impact compared to Test the varieties without the application of GA_3. Large bulb are better when used as seeds, but it is recommended to use medium bulbs because it is more economical [20], whereas for consumption is preferred for large bulbs because it stores more photosynthate and more layers that surround it so it is not easy to shrink.

4. Conclusion

Bima variety of shallot which planted in lowland can be flowering on average at 30 DAP with low percentage of flowering plants per plot (average 19.57%) and low seed yield (0.36 gram / stalk).Mentes varieties can not bloom or produce seed if planted in lowland. GA3 applied to the varieties of Bima concentration of 200 ppm able to produce high flower stalk per ha (134,567 stalks / Ha) and accelerate flowering (29 DAP).On Mentes varieties, GA3 application has no effect on flowering and seed formation. Interaction between varieties with a concentration of GA_3 shown on bulbs quality. Application of GA_3 in Bima varieties tend to lower the percentage of large bulbs and increase the percentage of small bulbs, while the application of GA_3 on Mentes varieties tends to increase the percentage of large bulbs and reduced the percentage of small bulbs produced per clump.

References

[1] Badan Pusat Statistik 2016 Berita konsumsi dan pengeluaran Jakarta Badan Pusat Statistik Republik Indonesia
[2] Triharyanto E, Samanhudi, Pujiasmanto B and Purnomo D 2013 Kajian pembibitan dan budidaya bawang merah (*Allium ascalonicum* L.) melalui biji botani (*True Shallot Seed*) *Seminar Nasional Fakultas Pertanian UNS Surakarta*
[3] Kurniawan A dan Suastika G 2013 Deteksi dan Identifikasi Virus pada Umbi Bawang Merah *J Fitopatologi Indonesia* **9**(2): 47-52
[4] Triharyanto E dan Purnomo D 2014 Study of Viability and Seed Structure of Shallot *J. Agricultural Science and Technology* B **4**: 121-125

[5] Triharyanto E, Purnomo D, Yunus A and Samanhudi 2017 Study of planting season to flowering potency, fruits and seed yield *Global J Engineering Science and Research Management* **4**(2): 78-84 D 10.5281/zenedo. 344632

[6] Sumarni N, Suwandi, Gunaeni N and Putrasamedja S 2013 Pengaruh variety dan cara aplikasi GA_3 terhadap pembungaan dan hasil biji bawang merah di dataran tinggi sulawesi selatan *J Hortikultura* **23**(2): 153-163

[7] Barani M, Akbari N and Ahmadi H 2013 The effect of gibberelic acid (GA_3) on seed and spouting of potato bulbs (*Solanum bulbosum* L.) *African J. Agricultural Research* **8**(29): 3898-3903 D 10.5897/AJAR09.419

[8] Berson, Mariati and Rosita 2015 Produksi biji bawang merah samosir aksesi simanindo terhadap konsentrasi GA_3 dan lama perendaman di dataran tinggi samosir *J Online Agroteknologi* **3**(3): 1147-1151

[9] Idhan A, Syam'un E, Zakaria B and Riyadi M 2015 Potential selection of flowering and bulb production in fourteen onion variety (*Allium ascalonicum* L.) at lowland and upland. *Int J Curr Res Biosci Plant Biol* **2**(7): 63-67

[10] Tambak DGP, Siregar LAM and Rosmayati 2013 Respons pertumbuhan dan produksi beberapa variety bawang merah (*Allium ascalonicum* L.) dengan pemberian kompos limbas kakao pada tanah inseptisol *J Online Agroekoteknologi* **2**(1): 95-102

[11] Abdissa Y, Tekalign T and Pant LM 2011 Growth, bulb yield and quality of onion (*Allium cepa* L.) as influenced by nitrogen and phosphorus fertilization on vertisol I. growth attributes, biomass production and bulb yield *African J. Agricultural Research* **6**(14): 3252 – 3258 D 10.5897/AJAR10.1024

[12] Williamson DD and Benkeblia N 2015 Effects of gibberelic acid and calcium chloride on colour, phenolic compounds, carotenoids and quality attributes of white cabbage (*Brassica oleracea*) during storage under refrigeration *Int J Food Sci Nutr Diet* **4**(7): 239-245 D 10.19070/2326-3350-1500043.

[13] Jasmi, Sulistyaningsih E and Indradewa D 2013 Pengaruh vernalisasi umbi terhadap pertumbuhan, hasil dan pembungaan bawang merah (*Allium cepa* L. Aggregatum group) di dataran rendah *Ilmu Pertanian* **16**(1): 42-57

[14] Fahrianty D 2013 Peran vernalisasi dan zat pengatur tumbuh dalam peningkatan pembungaan dan produksi biji bawang merah di dataran rendah dan dataran tinggi *Tesis Sekolah Pascasarjana Institut Pertanian Bogor*

[15] Sorensen A, Mariati and Siregar LAM 2015 Tanggap pertumbuhan vegetatif dan generatif bawang merah terhadap konsentrasi dan lama perendaman GA_3 di dataran rendah *J Online Agroteknologi* **3**(1): 310-319

[16] Patra SK, Beura S and Shasani T 2015 Efficience of GA_3 on growth and flowering regulation of in vitro raised hybrid gerbera under shade net *Agricultural Science Digest* **35**(3): 173-177

[17] Rahayu E and Berlian N 2004 Bawang merah (Jakarta: Penebar Swadaya) pp 11-12

[18] Palupi ER, Rosliani R and Hilman Y 2015 Peningkatan produksi dan mutu benih botani bawang merah (*True Shallot Seed*) dengan introduksi serangga penyerbuk *J Hort* **25**(1): 26-36

[19] Helaly AA, Abdelghafar MS, Al-Abd MT and Alkharpotly AA 2016 Effect of soaked *Allium cepa* L. bulbs in growth regulators on their growth and seed production. *Adv Plants and Agriculture Research* **4**(3): 00139 D 10.15406/apar.2016. 04.00139

[20] Hailu G, Tibebu S and Melese T 2014 Participatory evaluation of improved shallot (*Allium cepa* var. Aggregatum) varieties and their bulb size effect on yield traits in Wolaita Zone, Southern Ethiopia *J. Biology, Agriculture and Healthcare* **4**(13): 103-111

[21] Gautam IP, Khatri B and Paudel GP 2006 Evaluation of different varieties of onion and their transplanting times for off season production in mid hills of nepal *J. Nepal Agric Res* **7** : 21-26

[22] Dinarti D, Purwoko BS, Purwito and Susila AD 2011 Perbanyakan tunas mikro pada beberapa umur simpan umbi dan pembentukan umbi mikro bawang merah pada dua suhu ruang kultur *J Agronomi Indonesia* **39**: 97-102
[23] Asra R 2014 Pengaruh hormon giberelin (GA_3) terhadap daya kecambah dan vigoritas *Calopogonium caeruleum. Biospesies* **7**(1): 29-33
[24] Getahun D 2016 Bulb placement effects on growth performance and yield of shallot (*Allium cepa* L. Aggregatum group) *Internasional J. Research Studies in Agricultural Sciences* **2**(8): 33.38 D 10.20431/2454-6224.0209005

IOP Conf. Series: Earth and Environmental Science **142** (2018) 012067 doi:10.1088/1755-1315/142/1/012067

Adaptation of six shallots varieties to phosphate solubilizing bacteria on the flower formation, seeds fromation, and yields on the lowland

E Triharyanto, Sudadi and S Rawandari
Agrotechnology Department, Agriculture Faculty, Universitas Sebelas Maret Surakarta, Indonesia

Email: eddytriharyanto@yahoo.co.id

Abstract. Using seeds as planting materials is a solution to improve the quality and quantity of shallot. This study aims to determine the interaction between shallot varieties and Phosphate-Solubilizing Bacteria (PSB) on the flowering and shallot yield on the lowlands. The research was conducted in Mijil Village, Jaten, Karanganyar, 98 m altitude with Vertisol-type soil order in June to December 2016, using Randomized Complete Block Design (RCBD) with two factors. Shallot varieties used as factors are Bima, Manjung, Ilokos, Bima (bulb seeds), Mentes and Rubaru. PSB factors are control and with PSB inoculation. Observed variables included plant height, number of leaves, flowering percentage, seed formation and shallot bulb yield. Results showed that there was no interaction between varieties and PSB inoculation on all observed variables. However, PSB inoculation were able to increase the number of flowering plants and seed weight per plot. Bima variety have the highest average yield compared to other varieties in terms of number of leaves, number of bulbs per plant and bulb weight per plot (fresh harvest weight 317.74 g equivalent to 17.65 ton per hectare and dry weight 288.16 g consumption equivalent to 16 ton per hectare).

1. Introduction

Shallot is a horticulture commodity with a lot of utilities, high economic value and good market prospect. The demand for shallot in the Indonesia is always increasing but not balanced with the production. The need for shallot consumption in 2014 was 1.35 million tons, while the production was 1,234 million tons resulting in a deficit of shallot [1]. Shallot cultivation can be done with vegetative (bulbs) or generative (seeds) manners.

Generally, farmers use bulbs seedlings because it has advantages such as planting ease, seedless and shorter harvest time when compared to using seeding materials. However, the weakness of bulb seedling is prone to infectious diseases from fungus such as *Fusarium* sp and viruses such as *Onion Yellow Drawf Virus* (OYDV) (ranging from 53-93% crop, correspondently) thereby decreasing the quality and quantity of shallot [2]. Meanwhile, seeding material or True Shallot Seed (TSS) has the advantage that the less usage of planting materials, transport and easy storage [3], and relatively safe from disease transmission. However, the use of TSS has not developed because flowering of shallot is still difficult and the seeds produced have low viability [4]. Flowering shallot is difficult because the shallot is biannual plant (two seasons region) [5].

Flowering problems can be overcame by inoculation of Phosphate-Solubilizing Bacteria (PSB). The role of PSB in this study was to increase the absorption of P element so as to increase the flower formation [6], as biocontrol [7] and to produce phosphatase enzymes to hydrolyze organic P into inorganic P [8]. The objective of this research is to study the ability of shallot varieties, PSB inoculation and their interaction in yielding flowers and seeds in lowland.

2. Methods

The experiment was conducted in the paddy field of Gunung Mijil Village, Jaten Sub-district, Karanganyar District with 98 m altitude and Vertisol-type soil. The experiment was carried out from June to December 2016. The material included are shallot seed, PSB, SP 36 250 kg/ha, Urea 50 kg/ha, ZA 100 kg/ha and KCl 200 kg/ha. The tools used are hoe, trowel, ruler, hemacytometer, microscope, scales and term slide.

The experiment used Randomized Complete Block Design (RCBD) with 2 treatment factors. The first factor is shallot varieties include Bima variety, Manjung, Ilokos, Bima (bulb from seed yield), Mentes and Rubaru. The second factor was PSB $1{,}10^6$ cells/g of soil and without PSB, in total there were 12 treatment combinations and 3 replication. Implementation includes soil preparation, preparation of PSB, application of PSB, preparation of planting materials, planting, maintenance and harvesting. The data were analyzed using analysis of variance (ANOVA) based on F test with 5% confidence level, if there are significant differences, further analysis using Duncan Duncan (DMRT) test with 5% confidence level were implemented.

3. Result and disscusion

3.1 General condition of research location

The soil used in the study was classified as Vertisol soil. The soil analysis used for the study showed pH of 7.43 (slightly alkaline), organic matter was 1.99% (low), total N was 0.29% (medium), available P is 2.39 ppm (very low) and available K is 0.124 me/100g (low). Based on these results, this soil can be included in the low soil fertility category. Environmental conditions during the period are 26.7-38.7 0C of air temperature, 78.1-79% of air humidity (RH) and 36.5-255.5 mm/month rainfall. Based on these weather conditions, conclude that this are not a proper condition for shallot cultivation. Shallot plants grow well in the dry season with rainfall of 25-125 mm/month [9].

3.2 Plant vegetative growth

Plant height were significantly differ among varieties ($P<0.05$), whereas PSB inoculation has no significant effect on plant height. Figure 1 shows that Rubaru varieties have the shortest plants that this does not match the description which it can reach 35-44 cm. This might be due to Rubaru varieties derived from the highlands and have not been able to adapt in the lowlands. In variables of leaf number, shallot varieties gave significant effect while PSB inoculation had no significant effect. Bima variety had the highest number of leaves of 32.17 leaflets (Figure 2), according to the product description potentially 15-50 leaf were formed. The number of leaves was not only influenced by PSB but also by nutrients such as N. Nitrogen main role is to stimulate overall growth, especially stems, branches and leaves [10].

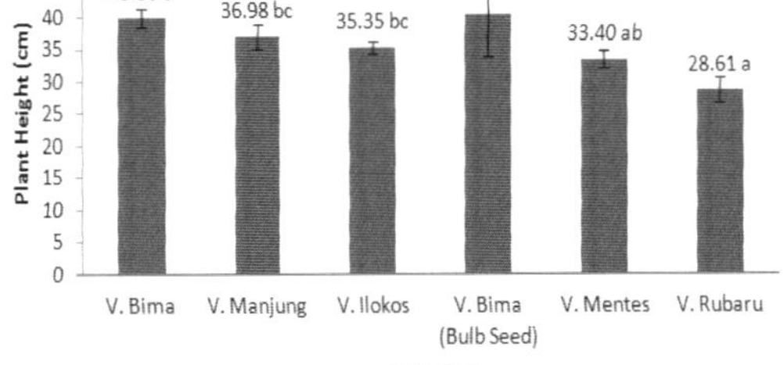

Figure 1. Shallot varieties influence toward plant high on lowland.

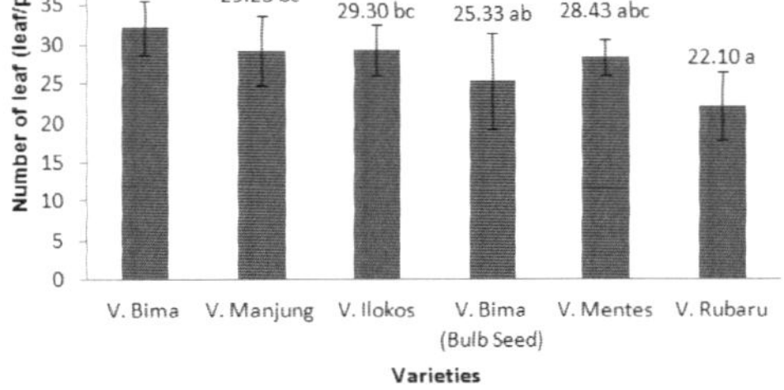

Figure 2. Shallot varieties influence toward the number of leaf of shallot on lowland.

3.3Flowering

3.3.1 Flower induction periode. Flowering of shallot varieties is particularly difficult in Indonesia, especially in the lowlands. Based on Table 1, flower induction periode of Manjung variety inoculatied with PSB was 49 days whereas without PSB inoculation was not flowering. Flower induction periode of Ilokos variety without bacteria inoculation was 48 days while using PSB inocultation was 56 days. It is assumed that PSB was able to increase the P element, but its ability is different [11]. In addition, flowering is influenced by the ability of each variety to flower and correspondent to weather conditions during planting.

3.3.2 The number of plant per plot. Based on Table 1 the percentage of flowering plants per plot of Manjung variety without inoculation of PSB was 0%, where as with inoculation PSB was 1.23%. The percentage of Ilokos variety without PSB inoculation was 1.23%, where as with PSB inoculation was 2.47%. Inoculation of PSB tends to increase the percentage of flowering, it is assumed that PSB can increase uptake of P. Sumarni et al. [12] states that to produce flower, shallot plants require more P fertilizer.

3.3.3 Number of flower stalks per plant cluster. Table 1 shows the number of flower stalks per cluster was one. The number of flower stalks is categorized as good if there are 2-4 flower stalks per cluster. The low number of flower stalks per plant was thought to be due to unsuitable environmental factors, genetic factors (varieties) and the availability of nutrients P which was still low. This is aligned with the results of Nkaa et al. [13] study that the role of P in plants can stimulate the formation of flowers.

3.3.4 The number of flower per stalk. Bima variety without PSB had flower per stalk rate of 46.8, while with PSB treatment was 46.4. Ilokos variety without PSB inoculation resulted in the number of flowers per stalk of 44.5 whereas with inoculation PSB was 42.75 (Table 1). PSB inoculation tends to decrease the number of flowers per stalk, it is assumed that PSB was not optimum in increasing P so that P availability was still lacking. According to Purba [14] the ability of PSB in improving the uptake of P elements vary.

Table 1. Effect of interaction of varieties and Phosphate-Solubilizing Bacteria on flowering of shallot

Treatment	Flower induction periode (days)	Number of plant/plot (%)	Number of flower stalks per cluster	Number of flower per stalk
Bima – Bacteria	45.80	4.32	1.40	46.80
Bima + Bacteria	45.80	3.09	1.20	46.40
Manjung – Bacteria	0.00	0.00	0.00	0.00
Manjung + Bacteria	49.00	1.23	1.00	47.00
Ilokos – Bacteria	48.00	1.23	1.00	44.50
Ilocos + Bacteria	56.00	2.47	1.00	42.75
Bima (Bulb Seed) – Bacteria	0.00	0.00	0.00	0.00
Bima (Bulb Seed) + Bacteria	0.00	0.00	0.00	0.00
Mentes – Bacteria	0.00	0.00	0.00	0.00
Mentes + Bacteria	0.00	0.00	0.00	0.00
Rubaru – Bacteria	0.00	0.00	0.00	0.00
Rubaru + Bacteria	0.00	0.00	0.00	0.00

3.4 Seed formation

3.4.1 The number of seed per stalk. The number of seeds per flower stalk ranged from 0 to 116 seeds. Based on Table 2, the highest number of seeds comes from Ilokos variety without PSB inoculation of

116 seeds, while the variety which not produce any seeds are Bima (bulb seed), Rubaru and Mentes varieties. The number of seeds per stalk is influenced by the number of pollinated flower stalks.

3.4.2 Seed weight per flower stalk. Table 2 shows the highest seed weight per stalk produced by Ilokos variety without PSB of 0.253 g. The weight of the seeds per stalk is influenced by the number of flowers that produce the seed, the more seeds are formed and the size of the seeds, the weight of the seeds per stalk is heavier. The formation of flowers and seeds is influenced by environmental factors, varieties and nutritional needs in early stage of seed filling.

3.4.3 Seed weight per plot. The highest seed weight per plot produced by Bima variety with PSB is 0.409 g / plot (Table 2). This is because the variety of Bima has the largest number of flower stalks. PSB inoculation tends to increase seed weight per plot, because it can increase the number of flowering plants per plot so as to increase the weight of seeds per plot.

Table 2. Effect of interaction of shallot varieties and Phopshate-Solubilizing Bacteria inoculation toward shallot seed formation

Treatment	Number of seed per stalk (gram)	Seed weight per flower stalk (gram)	Seed weight per plot (gram)
Bima - Bacteria	93.60	0.23	0.38
Bima + Bacteria	94.80	0.22	0.41
Manjung - Bacteria	0.00	0.00	0.00
Manjung + Bacteria	91.00	0.23	0.23
Ilokos - Bacteria	116.00	0.25	0.17
Ilocos + Bacteria	107.75	0.23	0.33
Bima (Bulb Seed) - Bacteria	0.00	0.00	0.00
Bima (Bulb Seed) + Bacteria	0.00	0.00	0.00
Mentes - Bacteria	0.00	0.00	0.00
Mentes + Bacteria	0.00	0.00	0.00
Rubaru - Bacteria	0.00	0.00	0.00
Rubaru + Bacteria	0.00	0.00	0.00

3.5 Plants yield

3.5.1 Bulb weight per plant. The highest bulb weight per plant was produced by Bima variety (bulb from seed yield) without PSB inoculation (Figure 3) of 62.63 g/plants cluster. Shallot varieties treatments have no significant effect on bulb weight per plant cluster variable, as well as inoculation of PSB treatment. Bulb weight can be influenced by the size of bulb plant material and the need of K elements [12]. Role of K element is increasing bulb size and bulb yield per plant cluster. Deficiency of K element will inhibit the process of leaf growth, so the process of photosynthesis is hampered causing small bulb size [14].

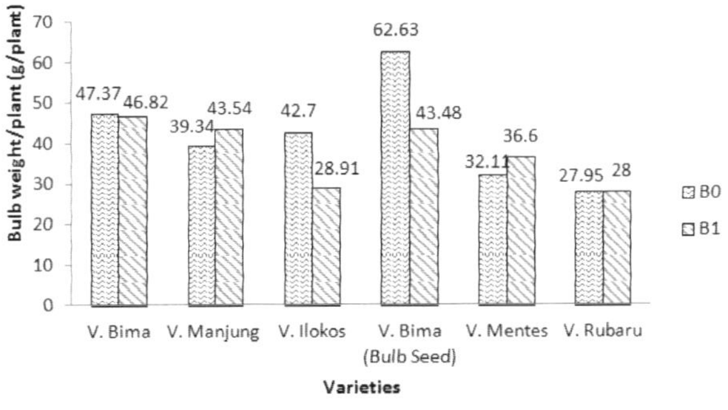

Figure 3. Effect of shallot varieties on bulb weight per plant on lowland.

3.5.2 The number of bulb per plant cluster. Table 3 shows that shallot varieties significantly affect the number of bulbs per plant cluster. The largest number of bulbs is Bima variety of 8.4 bulbs. According to Sofiari et al. [15] the number of bulbs of Bima shallot varieties ranged from 1-9 bulbs. Bulb is as a place to store food. The more leaves are, the more leaf midrib, so the number of bulbs will be plentiful [16].

3.5.3 Bulb weight per plot. Bulb weight per plot includes fresh weight and dry weight. Bulb weight per plot is related to the number of bulb, the weight, and size of the bulb per plant. Bima varieties produce fresh weight equivalent to 17.65 ton / ha fresh weight (Table 3). Fresh weight is influenced by the abundance of water absorption of photosynthesis resulting abundance photosyntate translocation to bulbs, thus affecting fresh weight of bulbs.

Shallot varieties have a significant effect on bulb weight, while PSB inoculation has no significant effect. Table 3 shows the Bima variety have dry weight of bulb equivalent to 16 ton / ha. This is due to the interaction between genetic and environmental factors, so that the shallot is able to grow well.

Table 3. Effect of shallot varieties on the number of bulb per plants cluster and weight of shallot bulb grown on lowland vertisol.

Varieties	Number of bulb / plant	Fresh weight of bulb/plot (g/0.18 m^2)	Fresh weight of bulb/ha (ton/ha)	Dry weight of bulb/plot (g/0.18 m^2)	Dry weight of bulb/ha (ton/ha)
Bima	8.40 c	317.74 b	17.65 b	288.16 c	16.00 c
Manjung	8.10 bc	256.73 ab	14.26 ab	227.56 bc	12.64 bc
Ilokos	7.53 bc	214.91 a	11.94 a	186.00 ab	10.33 ab
Bima (Bulb Seed)	6.37 ab	241.71 ab	13.43 ab	206.39 abc	11.47 abc
Mentes	8.30 bc	209.32 a	11.63 a	174.20 ab	9.68 ab
Rubaru	5.23 a	163.65 a	9.09 a	121.93 a	6.77 a

3.5.4 Large bulb diameter. Based on figure 4, Ilokos variety have the highest percentage in large bulb which is 83.61%. The diameter of the shallot bulb is influenced by the K element. According to Ispandi [17] K element is needed in the formation, enlargement and lengthening of the bulbs. According to Prajapatu and Mohidi [18], the role of element K is to increase the yield on shallot crops. The characteristics of shallot bulbs that are preferred by farmers are round bulb, dark red, 2 cm in diameter and have strong aroma [19].

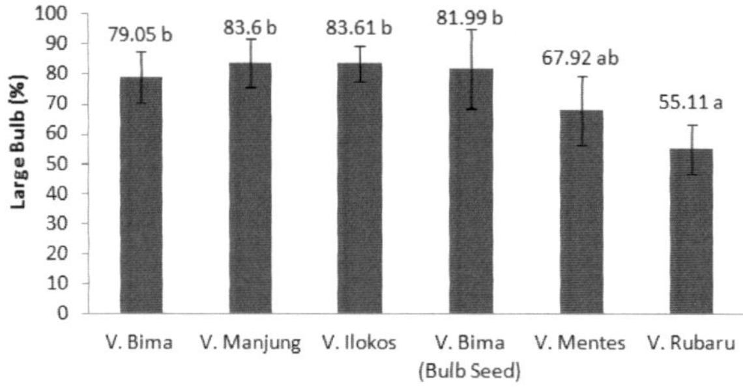

Figure 4. Effect of shallot varieties on % of large bulb diameter on lowland.

3.5.5 Small bulb diameter. Figure 5 shows that the varieties of Bima (bulb from seed yield) without PSB produce 0% of small bulbs, that this is in accordance with the description that shows Bima variety has a large bulb size. This is because the seeds of Bima variety bulb (bulb from seed yield) larger than other varieties. Large bulb has a relatively larger bulb layer and has a larger cross-sectional area so it can improve the nutrient and water absorption capacity [20]. This is in line with Purwanto [21] research, large bulbs have relatively more food reserves for the formation of plant growth energy. Aklilu and Lemma [22], stated that bulbs grown from seed have large bulb size, with ranging from 1 to 3 bulb per plant cluster.

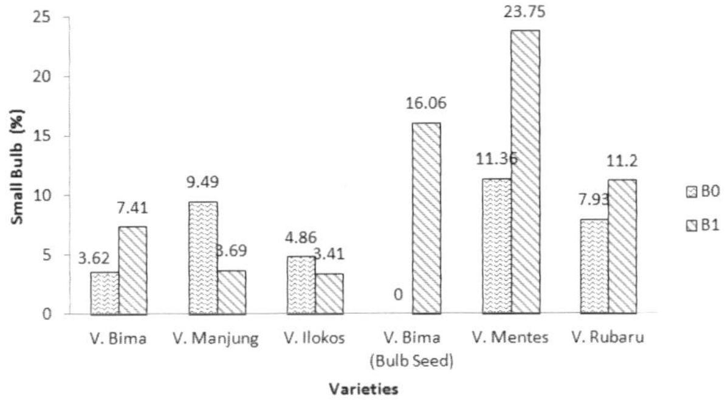

Figure 5. Effect of shallot varieties on % of small bulb diameter of shallot on lowland.

4. Conclusion

From this research of Adaption of some shallot variety with phosphate-solubilizing bacteria on the flower formation, seeds fromation, and yields on the lowland, can be conclude that Bima variety was able to produces dry bulb per plot of 288.16 g equal to 16 ton/ha and percent of big bulb 86.34%. PSB inoculation could trigger the increasing rate of the number of plants flowering per plot, increasing the number of seeds per stalk and seed weight per plot in flowering plants. Bima variety with PSB inoculation was able to produce the highest seed weight per plot of 0.409g.

References

[1] Badan Pusat Statistik 2016 Berita konsumsi dan pengeluaran Jakarta Badan Pusat Statistik Republik Indonesia

[2] Smekalova K, Helena S and Karel D 2017 Distribution of viruses in the shallot germplasm collection of the Czech Republic – Short Communication. *Hort. Sci* (Prague) **44** (1) 49–52 doi: 10.17221/271/2015-HORTSCI

[3] Jasmi, Endang S and Didik I 2013 Pengaruh vernalisasi umbi terhadap pertumbuhan, hasil, dan pembungaan bawang merah (*Allium Cepa* L. Aggregatum Group) di dataran rendah. *J Ilmu Pertanian* **16** (1) 42 – 57

[4] Triharyanto E, Samanhudi ET and Pujiasmanto B 2013 Kajian pembenihan dan budidaya bawang merah (*Allium ascalonicum L.*) melalui biji botani (true seed). Makalah disampaikan pada *Seminar Nasional* Fak. Pertanian UNS Surakarta dalam Rangkan Dies Natalis Tahun 2013

[5] Triharyanto E and Purnomo D 2014 Study of viability and seed structure of shallot. *Journal of Agricultural Science and Technology* B **4** 121-125

[6] Schoebitz M, Ceballos C and Ciampi L 2013 Effect of immobilized phosphate solubilizing bacteria on wheat growth and phosphate uptake *J of Soil Science and Plant Nutrition* **13**(1)1-10

[7] Trujillo ME, Vela´zquez S and Migue´lez MS 2007 Characterization of a strain of *Pseudomonas fluorescens* that solubilizes phosphates in vitro and produces high antibiotic activity against several microorganisms. First International Meeting on Microbial Phosphate Solubilization, 265–268. Springer The Netherlands.

[8] Zhongqi H, Thimothy S G and Wayne H 2004 Enzymatic hydrolysis of organic phosphorus in swine manure and soil *J Environ. Qual.* **33** 367-372

[9] Perdana S N, Wiwin S D Y and Mudji S 2015 Pengaruh aplikasi biourin dan pupuk terhadap pertumbuhan dan hasil tanaman bawang merah (*Allium Ascalonicum* L.) *J Produksi Tanaman* **3**(6) 457 – 463

[10] Hirel B,Thierry T and Peter J 2011 Improving nitrogen use efficiency in crops for sustainable agriculture *J Sustainability* **3**:1452-1485 DOI:10.3390/su3091452. ISSN 2071-1050

[11] Atekan, Nuraini Y and Handayanto E 2014 The potential of phosphate solubilizing bacteria isolated from sugarcane wastes for solubilizing phosphate. *J of degraded and mining lands management* **1**(4) 175-182 ISSN: 2339-076x, DOI:10.15243/jdmlm.2014.014.175

[12] Sumarni N, Rosliani R and Suwandi 2012 Jarak tanam dan dosis pupuk NPK untuk produksi bawang merah dari benih umbi mini di dataran tinggi *J Hort* **22**(2):148-155

[13] Nkaa FA, Nwokeocha OW and Ihuoma O 2014 Effect of phosphorus fertilizer on growth and yield of cowpea (*Vigna unguiculata*). *J of Pharmacy and Biological Sciences* **9**(5) 74-82 ISSN: 2278-3008, p-ISSN:2319-7676

[14] Purba 2014 Applications of NPK phonska and KCL fertilizer for the growth and yield of shallots (*Allium ascalonicum*) in Serang, Banten *J Application Science*. **4** (3): 197 – 203

[15] Sofiari, Kusmana E and Basuki RS 2009 Evaluasi daya hasil kultivar lokal bawang merah di Brebes *J Hort*. **19**(3) 257-280

[16] Nana SA and Salamah Z 2014 Pertumbuhan tanaman bawang merah (*Allium cepa* l.) dengan penyiraman air kelapa (*Cocos nucifera* l.) sebagai sumber belajar biologisma kelas XII. Jupemasi-Pbio **1**(1): 82-86 ISSN: 2407-1269

[17] Ispandi A 2003 Pemupukan P dan K dan waktu pemberian pupuk pada tanaman ubi kayu di lahan kering vertisol *J Ilmu pertanian* **10**(2): 35-50

[18] Prajapati K and Modi HA 2012 The importance of potassium in plant growth – a review. Indian *J of Plant Sciences* **1**(02-03): 177-186 ISSN: 2319-3824

[19] Basuki RS 2009 Analisis tingkat preferensi petani brebes terhadap karakteristik hasil dan kualitas bawang merah varietas local asal dataran medium dan tinggi *J Hort* **19**(4): 475-483

[20] Sufiyati Y, Imron SAK and Fikrinda 2006 Pengaruh ukuran fisik dan jumlah umbi per lubang terhadap pertumbuhan dan hasil bawang merah (*Allium ascalonicum* L.) *J Floratek* **2**:43-54

[21] Purwanto AM 2013 Pengaruh ukuran benih terhadap pembentukan terhadap biomassa tanaman bawang merah pada tingkat pemberian pupuk nitrogen yang berbeda *Agritech* **15**(1):23-31

[22] Aklilu S and Lemma D 2015 The performance of true seed shallot lines under two methods of planting at different environments of Ethiopia *J of Agriculture and Environmental Management*. **4**(3): 174-179. ISSN 2315 – 8719.

IOP Conf. Series: Earth and Environmental Science **142** (2018) 012068 doi:10.1088/1755-1315/142/1/012068

Defense reactions of bean genotypes to bacterial pathogens in controlled conditions

B Uysal[1] and K K Bastas[2]
[1]Igdır University, Faculty of Agriculture, Department of Plant Protection Igdır, Turkey
[2]Selcuk University, Faculty of Agriculture, Department of Plant Protection, 42031 Konya, Turkey

E mail: kbastas@selcuk.edu.tr

Abstract. This study was focused on the role of antioxidant enzymes and total protein in imparting resistance against common bacterial blight caused by *Xanthomonas axonopodis* pv. *phaseoli* (Xap) and halo blight caused by *Pseudomonas syringae* pv. *phaseolicola* (Psp) in bean. Activities of Ascorbate peroxidase (APX), Catalase (CAT) and total protein were studied in resistant and susceptible bean genotypes. Five-day-old seedlings were inoculated with a bacterial suspension (10^8 CFU ml^{-1}) and harvested at different time intervals (0, 12, 24 and 36 up to 72 h) under controlled growing conditions and assayed for antioxidant enzymes and total protein. Temporal increase of CAT, APX enzymes activities showed maximum activity at 12 h after both pathogens inoculation (hpi) in resistant cultivar, whereas in susceptible it increased at 72 h after both pathogens inoculation for CAT and 12, 24 h for APX enzymes. Maximum total protein activities were observed at 12 h and 24 h respectively after Xap, Psp inoculation (hpi) in resistant and maximum activities were observed at 24 h and 72 h respectively after Xap, Psp inoculation (hpi) in susceptible. Increase of antioxidant enzyme and total protein activities might be an important component in the defense strategy of resistance and susceptible bean genotypes against the bacterial infection. These findings suggest that disease protection is proportional to the amount of enhanced CAT, APX enzyme and total protein activity.

1. Introduction

Common bean (*Phaseolus vulgaris* L.) is a crop of major societal importance and is a major source of protein and essential nutrients. Worldwide, common bean is the most consumed legume, providing up to 15% of total daily calories and 36% of total daily protein in parts of Africa and the America. Turkey is an important bean producer. Bean production of 638.496 tons, is achieved by growing on 501.767 hectares of land in Turkey [1]. Beans may be affected by a number of distinct diseases, including bacterial diseases. Common bacterial blight caused by *Xanthomonas axonopodis* pv. *phaseoli* (Xap), halo blight caused by *Pseudomonas syringae* pv. *phaseolicola* (Psp), are the most important bacterial diseases in bean. These bacterial diseases are observed in all the fields of bean production in Turkey. There is no specific pesticide to control bacterial spot disease of bean caused by *Xanthomonas axonopodis* pv. *phaseoli*. Therefore, genetic resistance is very important to control these bacterial diseases in bean plant. Plants have a variety of endogenous biochemical defense mechanisms that can be induced in response to attack by insects and pathogens [2]. Enzymes play an important role in plant defense mechanisms. For many years, the role of oxidative enzymes and of their metabolic products in the defense mechanisms of infected plants has been studied [3]. There is worldwide interest in identifying safe and effective novel materials to induced resistance for the control of plant pathogenic bacteria [4]. The aim of this present study was to test activities of some plant defense related enzymes and proteins on susceptible and resistant bean genotypes.

2. Materials and methods

2.1 Planting material and bacterial inoculation preparation
Two different bean cultivars (resistance and susceptible) were obtained from Dept. of Field Crop, Ataturk University and Eastern Anatolia Agricultural Research Institute. All seed samples were surface sterilized with 3% (v/v) sodium hypochlorite solution for 4 min and were thoroughly washed with distilled water three times and surface dried. Plants were grown in 10 cm pots in a soil mix containing sand, perlite, and peat compost under 25◦C with a 14 h/10 h light/dark conditions. *Xanthomonas axonopodis* pv. *phaseoli* (Xap) and halo blight caused by *Pseudomonas syringae* pv. *phaseolicola* culture was provided by Dr. M. Figen Donmez, Dept. of Plant Protection, Iğdır University. Two days old cultures of Xap and Psp were harvested from NA plates, suspended in sterile deionized water, and adjusted to a concentration of 10^8 CFU mL^{-1} as determined by a spectrophotometer at 660 nm.

2.2 Enzyme Assay
Sample collection: Resistance and susceptible bean plants were harvested 0, 12th, 24th, 48th, 72nd hours post inoculation (hpi) and stored at -80 ^{0}C for subsequent analysis [5].

2.3 Catalase assay
Bean seedlings (1 g) were homogenized in 5 ml of 50 mM potassium phosphate buffer pH 7.0 in a pre-chilled mortar and pestle on ice. The homogenate was centrifuged at 10,000 rpm for 20 min at 4 ^{o}C and the supernatant served as enzyme source. All experiments were carried out at 4 ^{o}C. Catalase was estimated by following the procedure of [6]. The 900 μL reaction mixture contained 50 mM potassium phosphate buffer and 30 mM H_2O_2. The reaction was initiated by adding 100 μL of enzyme extract and the decrease in A_{240} nm was measured for 1 min. The enzyme activity was calculated by the absorbance values (mmole/g^{-1}FW/da^{-1}).

2.4 Ascorbate peroxidase assay
Bean seedling (1 g) was homogenized in 5 ml of 50 mM potassium phosphate buffer pH 7.0, containing a pre-chilled mortar and pestle on ice. The homogenate was centrifuged at 10,000 rpm for 20 min at 4°C and the supernatant served as enzyme source. All experiments were carried out 4°C. Ascorbate peroxidase activity was determined according to [7], by monitoring the decrease in the absorbance at 290 nm over 2 min. The reaction mixture consisted of 50 mM potassium phosphate buffer (pH 7) with 0.1 mM EDTA, 0.5 mM ascorbic acid and 100 μL of enzyme extract. The enzyme activity was calculated by the absorbance values (mmole/g^{-1}FW/da^{-1}).

2.5 Total Protein Assay
One gram of leaf tissue was homogenized with 50 mM potassium phosphate buffer (pH 7.0), and centrifuged at 10, 000 rpm for 20 min at 4 ^{o}C and the supernatant was used as crude extract. Protein concentration was measured according to [8] serum albumin served as a standard.

3. Results and discussion

Seedlings of two different bean genotypes (resistant and susceptible) were analyzed for CAT and APX and Total Protein with or without pathogen inoculation. All the cultivars showed an increased level of enzyme activity after pathogen inoculation. Highest CAT and APX enzymes activities showed at 12 h after both pathogens inoculation (hpi) in resistant cultivar (Figure 1,2,5,6), whereas in susceptible it increased at 72 h after both pathogens inoculation for CAT and 12, 24 h for APX enzymes (Figure 3,4,7,8).

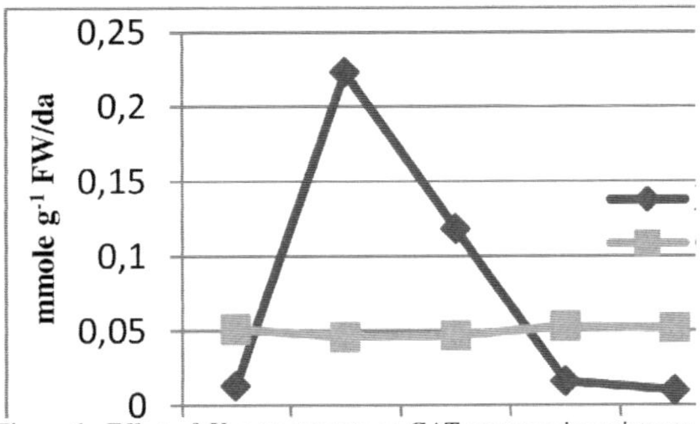

Figure 1. Effect of Xap treatments on CAT enzymes in resistance bean plants, inoculated with pathogen and control

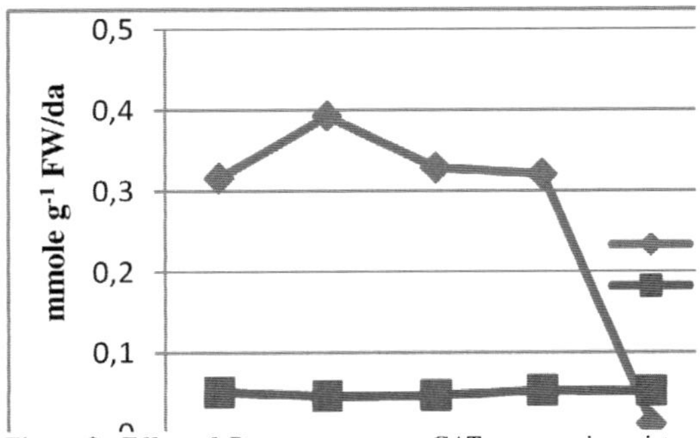

Figure 2. Effect of Psp treatments on CAT enzymes in resistance bean plants, inoculated with pathogen and control

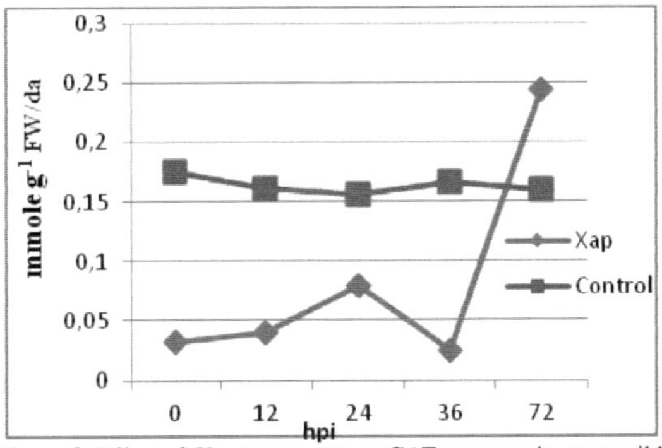

Figure 3. Effect of Xap treatments on CAT enzymes in susceptible bean plants, inoculated with pathogen and control

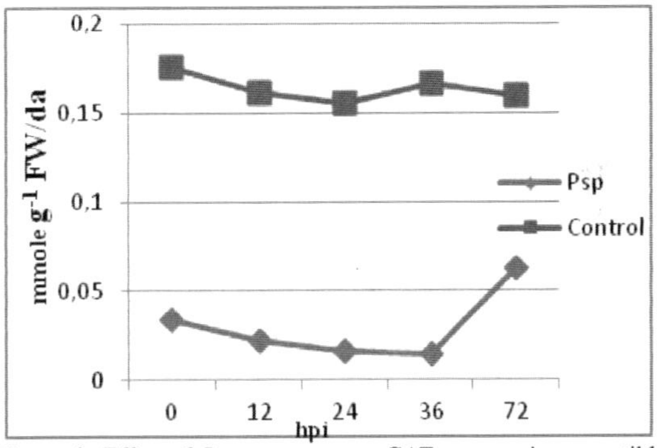

Figure 4. Effect of Psp treatments on CAT enzymes in susceptible bean plants, inoculated with pathogen and control

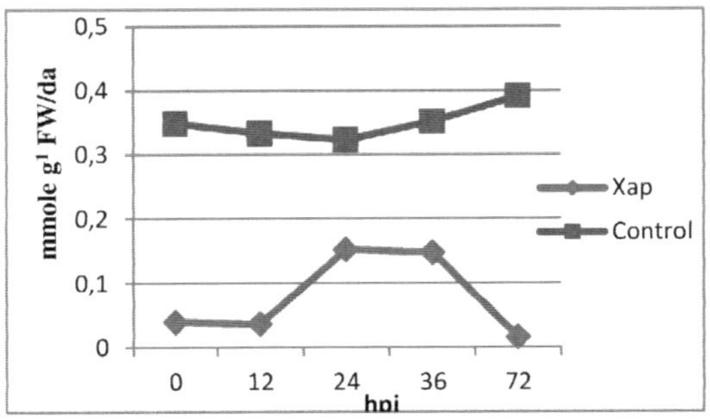

Figure 5. Effect of Xap treatments on APX enzymes in resistance bean plants, inoculated with pathogen and control

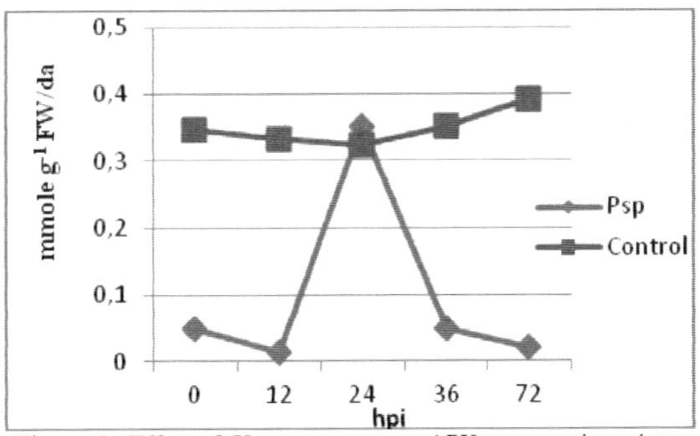

Figure 6. Effect of Xap treatments on APX enzymes in resistance bean plants, inoculated with pathogen and control

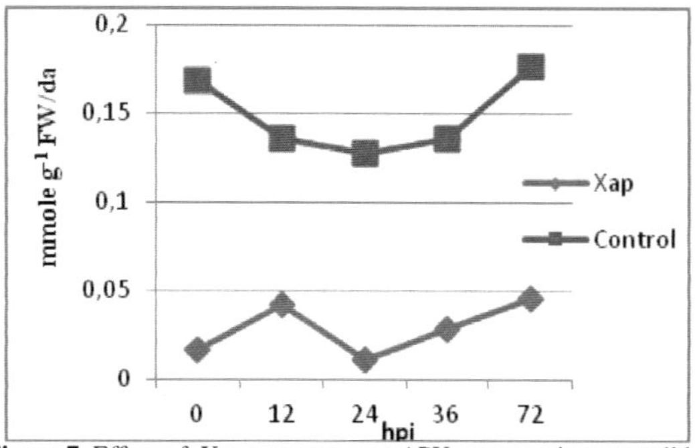

Figure 7. Effect of Xap treatments on APX enzymes in susceptible bean plants, inoculated with pathogen and control

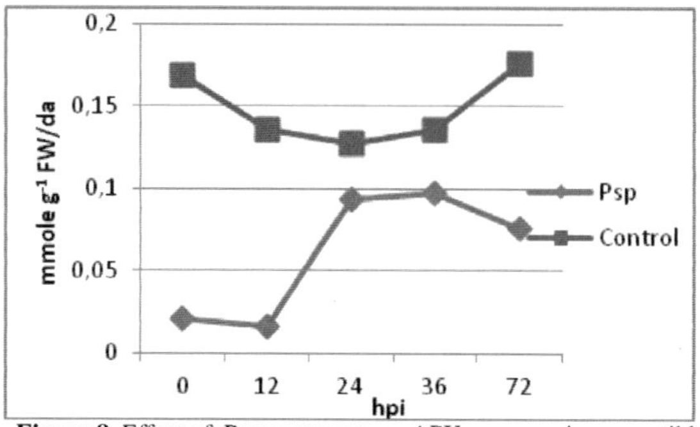

Figure 8. Effect of Psp treatments on APX enzymes in susceptible bean plants, inoculated with pathogen and control

The highest maximum total protein activities were observed at 12 h and 24 h respectively after Xap and Psp inoculation (hpi) in resistant (Figure 9,10) and also maximum activities were observed at 24 h and 72 h respectively after Xap, Psp inoculation (hpi) in susceptible (Figure 11,12).

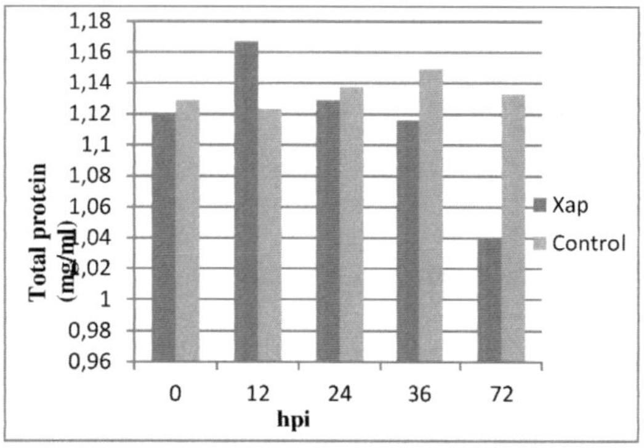

Figure 9. Effect of Xap treatments on Total Proteinin in resistance bean plants, inoculated with pathogen and control

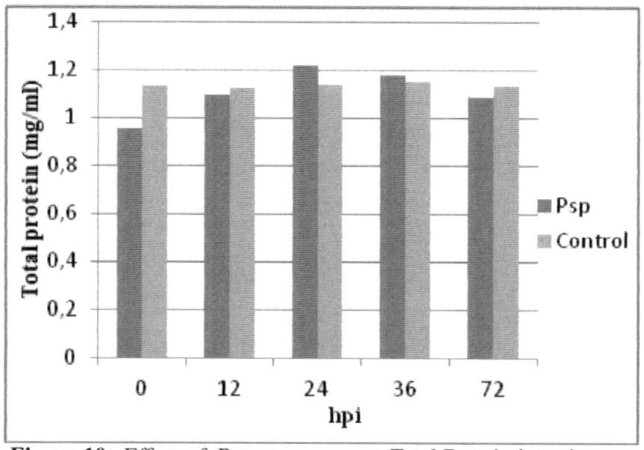

Figure 10. Effect of Psp treatments on Total Protein in resistance bean plants, inoculated with pathogen and control

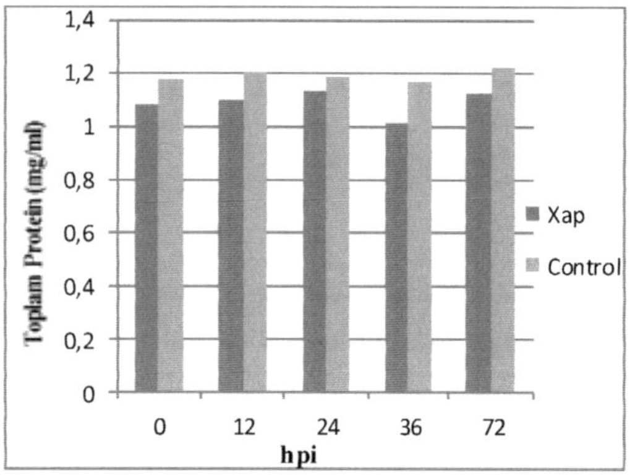

Figure 11. Effect of Xap treatments on Total Proteinin susceptible bean plants, inoculated with pathogen and control

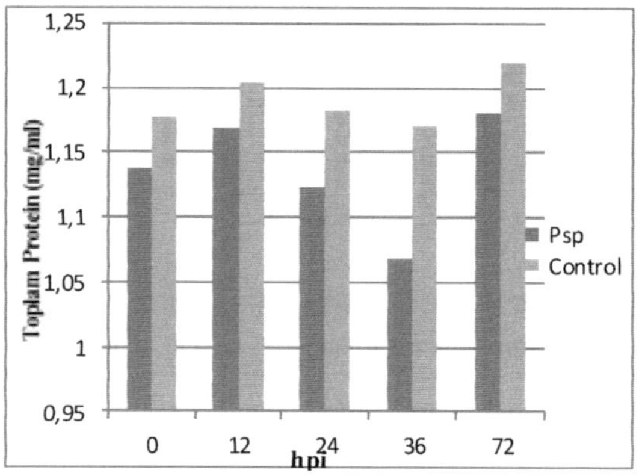

Figure 12. Effect of Psp treatments on Total Protein in susceptible bean plants, inoculated with pathogen and control

Our studies indicated that CAT and APX enzyme plays an important role in the defense system of the infected resistant bean plants pathogen inoculation compared to uninoculated control and also the susceptible under controlled growing conditions. Similar studies were observed increased APX and CAT enzymes [9].

IOP Conf. Series: Earth and Environmental Science **142** (2018) 012068 doi:10.1088/1755-1315/142/1/012068

References

[1] Anonymous 2014 FAO Statistics Food and Agriculture Organization of United Nations Agriculture Data (FAO); http://wwwfaostatcom

[2] Bastock RM, Karban R, Thaler JS, Weyman PD, Gikhrist D 2001 Signal interaction in induced resistance to pathogens and insect herbivores *Eur J Plant Pathol* **107**:103-111

[3] Waheed A, Tehmina A 2011 Quantitative Changes in Defense System of Tomato Induced By Two Strains of *Bacillus* against *Fusarium* Wilt *Ind J Fund Appl Life Sci* **1**: 2231-6345

[4] Reddy M S 1991 Biological control of plant diseases In: McClay AS, editor Biological control of pests in Canada Canada: Alberta pp33-42

[5] Teranishi Y, Tanaka A, Osumi M and Fukui S 1974 Catalas activity of hydrocarbon utilizing candida yeast *Agr Biol Chem* **38**:1213-1216

[6] Aebi H 1984 Catalase in Vitro *Method Enzyme* **105**: 121-126

[7] Nakano Y, Asada K 1981 Hydrogen peroxide is scavenged by ascorbate specific peroxidase in spinach chloroplasts *Plant Cell Physiol* **22**: 867-880

[8] Bradford M M 1976 A rapid and sensitive method for the quantitation of microgram quantities of protein utilizing the principle of protein-dye binding *Anal Biochem* **72**, 248–254

[9] Sumayo MS, Kwon D and Ghim S 2014 Linoleic acid-induced expression of defense genes and enzymes in tobacco *Journal of Plant Physiology* **171**:1757–1762

IOP Conf. Series: Earth and Environmental Science **142** (2018) 012069 doi:10.1088/1755-1315/142/1/012069

Farmer readiness for adopting stevia cultivation (a case study at District of Pasir Jambu, Regency of Bandung)

D Supyandi, Y Sukayat and H Hapsari
Department of Social Economics of Agriculture, Universitas Padjadjaran

E-mail: dika.supyandi@unpad.ac.id

Abstract. Recognized as a complementary for conventional sugars made from cane, coconut, corn, and palm, as well as a substitute for synthetic sweetener, recently stevia has accepted significant attention in order to fulfill increasing demand for sweeteners in Indonesia. Stevia has several advantages, among other is having 200-300 times sweetness level compared to cane sugar with low-calorie level. In Indonesia, stevia was introduced from Japan, Korea and China, and has been cultivated in several areas, among other is in West Java, particularly at District of Cikajang (Garut), District of Pangalengan (Bandung) and District of Ciwidey/Pasir Jambu (Bandung). Introducing new commodity and/or technology has usually faced constraints and sometimes rejection. However, considering the potentials and increasing demand for it, stevia cultivation widespread need to be stimulated. This paper describes several conditions of farmer community at District of Pasir Jambu in terms of their readiness to adopt stevia cultivation in their land. Community readiness model was used to guide the structure of thinking in data collection process at farmer level in order to compose possible best intervention based on farmer aspiration and condition. In addition, several references from previous research reports, journal articles as well as government reports were used to sharpen analysis of data and information collected from the field.

1. Introduction

Demand towards natural sugars has increased in line with the growth of food and beverage industries and increasing household consumption. Even though Indonesian natural sugar production tends to increase in the period of 2012-2016 [1], hitherto its availability has not been able to fulfill national sweetener needs. This situation fostered food and beverage industries to use synthetic sweetener, which is not only to meet the demand but also in order to cut their cost of production.

People apprehension towards the use of synthetic sweeteners, in addition to increasing cases of diseases caused by excessive sugar consumption have stimulated the search for low-calorie natural sweetener sources. One of the sources is stevia. Stevia has potential to be a low-calorie natural sweetener, a complementary for sugar from cane, coconut, corn, and palm, as well as a substitute for synthetic sweeteners. It has 200-300 times sweetness level compared to cane sugar [2]. In addition, stevia has other advantages, such as strong root for protecting land from erosion and able to cope with drought. From health viewpoint, several benefits of consuming stevia are reducing hypertension, will not causing caries, rich with vitamins that are not available in synthetic sweeteners, contain antioxidants, reducing diabetes, and curing stomach ache [3].

In Regency of Bandung, a farmer group that constantly cultivates stevia since 2010 is Mulyasari Farmer Group, located in Village of Cibodas, District of Pasir Jambu. Recently about 2 hectares of farmers' land cultivated stevia, produced about 500-600 kg stevia leaves in every two weeks after 40 days planting. The demand of stevia to this farmer group increased, but they could not fulfill it. Farmer hesitancy to cultivate stevia is presumed as the main factor of this situation. The financial aspect is also a "classical" problem faced by farmers. In addition, less support from local government to develop this commodity worsen farmer (group) inability.

Notwithstanding, as the main actor in the development of stevia, the farmer is also having the most important role in the success of this process. This role must be started by farmer readiness to develop, which represents several aspects, from an ability to make a decision, to cultivate on the farm, to organize themselves in terms of farm management and marketing, and other capacities. According to Behar and Hydaker [8], the concept of ''community readiness'' offers a vital influence to improve the planning and implementation process for communities. Knowing important factors to the successful implementation should "help communities assess their own strengths and weaknesses". Furthermore, they go on to say that this understanding could "support technical assistance efforts by helping to determine areas of focus and strengthen areas of weakness".

Considering the previous statements, inquiries toward farmer readiness are important prior to implementing activities. High farmer readiness level is expected to be more successful to achieve activity's goals. In this context, the goal is to cultivate stevia successfully at farmer level.

2. Method

The research implemented descriptive qualitative design, used a case study at Mulyasari Farmer Group, which cultivate stevia, located at Village of Cibodas, District of Pasir Jambu, Regency of Bandung. Qualitative method is frequently an interpretative method, because the result data usually resulted from field data interpretation [5], while case study is a descriptive technique, the object is now, only one case (community, family, or individual) and deep explorative in nature [6].

Primary and secondary data were used in this research. Primary data were obtained by interview with farmer group members and informants and through observations. Purposive sampling to group patron, management, and members was implemented. Secondary data were obtained from report documentations, journals, and related government institutions. Descriptive analysis data was used; refer to the identification of characteristics of human groups, things or events. In short, qualitative descriptive involves conceptualization process and results in the formation of classification schemes [4].

In order to conceptualize and guide structure of thinking in data collection, community readiness model was used. Community readiness model integrates communities' culture, resources and level of readiness to address the issues more effectively [11]; developed to provide communities with a theoretical framework, a process, and specific tools to facilitate readiness [8]; it also can be used for specific and certain situations of a community [9].

The process of community readiness model development is (1) identify the issue, (2) define the community, (3) conduct key respondent interviews, (4) score interviews to determine level of readiness and (5) develop strategies based on level of readiness and conduct workshops or trainings [10]. Moreover, Plested et al. [11] categorized stage of community readiness (from low to high) into (1) No Awareness, (2) Denial/Resistance, (3) Vague Awareness, (4) Preplanning, (5) Preparation, (6) Initiation, (7) Stabilization, (8) Confirmation/Expansion, and (9) High Level of Community Ownership. In order to assess farmer readiness, four deep interviews with farmer group leader, a farmer group member, a cooperative management, and a local government leader have been conducted.

3. Results and Discussion

3.1. *Stevia cultivation in Regency of Bandung*

Stevia was introduced in Regency of Bandung in 2005. It used stevia seeds originated from Vietnam, involving four demonstration plots at Districts of Pasir Jambu, Pangalengan, Banjaran, and Cikalong. Since 2010, stevia has been developed intensively at District of Pasir Jambu, particularly at Village of

Cibodas. Initially, it was planted in 15-hectares of land, with hundreds of farmers involved. However, it decreased significantly into only two hectares, which is used for plasma stock, and only 25 farmers involved. According to farmer group leader, this plasma stock is different from its first introduced. It is a new variety of stevia seed, developed independently by farmers. They named the variety "Cibodas Manis 3" (CM3), named after the village.

At District of Pasir Jambu, development of stevia is facing several obstacles, among others first, hitherto stevia development has not been received sufficient attentions and responses from policymakers. It is indicated by a very limited amount of government based research and budget on stevia development. Even, at the national level, only due to demands from large industries, such as Sidomuncul, Martha Tilaar, and Jamu Jago, stevia has been cultivated in several places in Java. Second, in Regency of Bandung, particularly at District of Pasir Jambu, stevia was developed in communities that familiar with and common to a tea plantation. Without intensive and serious approaches and assistances, this could result in a difficulty to disseminate technology and to raise culture for stevia cultivation. Third, as large industries processing stevia do not exist, stevia cultivation will face a challenging situation. Hitherto, stevia produced by Pasir Jambu farmers has only been sold to certain communities (herbal and organic consumers), even though so far "local" demand always exceeds supply, and to a cooperative that processed it into several processed foods. Developing new consumers has also faced challenges as stevia taste is different from other sweeteners such as cane sugar or palm sugar. Hence, for several reasons, stevia has been mixed with other synthetic sweeteners such as xylitol, which is actually unhealthy to consume. With several potential processed product diversifications in further industries, such as for food, cosmetics, and pharmacy, so far farmers still only produced stevia-based food raw materials. Fourth, at farmer level cultivating stevia actually requires a large amount of money, in particular for seed provision. Therefore, existing stevia farmers saw this situation as an opportunity and recently they also started to produce stevia seeds instead of final stevia products.

Since 2016, Indonesian government released a regulation stating stevia can be used for food. Since then, demand towards stevia increased, it is even getting stronger since a commercial sweetener product has been released a sweetener with stevia on national television. In addition, increasing demand of stevia overseas, such as from Japan, the US and European Union [12] also plays an important role of farmer attractiveness to cultivate stevia. Further information stated that next year, the Regency of Bandung administration through Office of Agriculture, Plantation, and Food Plant promotes the development of stevia in broader Regency of Bandung areas, in particular at Districts of Kertasari and Ciparay.

Further development of stevia in Pasir Jambu will require several requirements, namely, first, processed industries based on stevia product must be available and ready for accepting stevia produced by farmers. So far, only one cooperative received stevia and processed it into sweet tea products with several tastes. Second, in order to keep stevia plasma for next cultivation, the government should help the farmer to register farmer stevia variety for intellectual property rights (IPRs) and plant variety protection. This breeding process could be served as a type of participation plant breeding (PPB). Third, the government should provide regulation and support stevia to develop. The dominance of other sources of sugar, such as cane, coconut and palm should be complemented by "new" other sources such as stevia. It is political in addition to technical and economic considerations. Hence, the government indeed plays a very important role.

3.2. *Farmer readiness to cultivate stevia*

Since has been introduced at District of Pasir Jambu, stevia cultivation has never been refused and rejected by farmers. Almost all members of farmer groups located at District of Pasir Jambu recognize stevia and have the ability to cultivate. However, at the location, from about 600 members of farmer groups, only about 25 farmers cultivating stevia and coffee, the others cultivate coffee and other commodities. Decreasing cultivated land and production in the past were mainly resulted from the disability of market to absorb production as well as high cost needed to cultivate.

Recently, there is a plan from Regency of Bandung administration to develop District of Pasir Jambu, particularly in Village of Cibodas into a tourist destination. It is planned to develop forestry educational tourism. The government has improved several infrastructures such as road, accommodation and sanitation facilities, as well as supports from certain offices in Regency of Bandung. Moreover, this district is going to be developed as a center for several community-based agriculture developments in Regency of Bandung, which will conduct training, provide apprenticeship and carry out researches, which are not only for farmers from Bandung but also opened for farmers from all over Indonesia. Among several commodities to be developed, stevia is one of the advantage commodities which receives special attention.

Based on community readiness assessment to adopt stevia cultivation at District of Pasir Jambu, Regency of Bandung, overall farmer readiness score is 6.46. This score shows that community is on "initiation" stage, which means that "enough information is available to justify efforts, activities are underway".

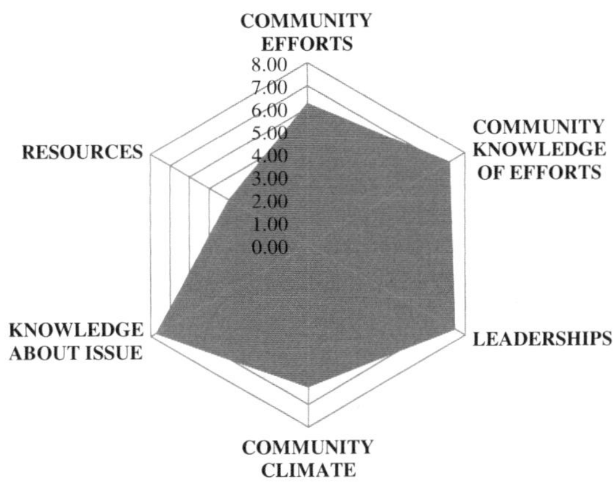

Figure 1. Farmer readiness assessment results to adopt stevia cultivation at District of Pasir Jambu

Moreover, based on a model developed by Plested et al. [11] detail score of every assessment component can be depicted in Table 1.

Table 1. Farmer readiness assessment components to adopt stevia cultivation at District of Pasir Jambu

Components	Score	Definition
Community Efforts	6.25	Efforts (programs/activities) have been implemented
Community Knowledge of Efforts	7.25	There is evidence that the community has specific knowledge of local efforts, including contact persons, clients involved, etc.
Leadership	7.25	Leaders are supportive of continuing basic efforts and are considering resources available for self-sufficiency
Community Climate	6.25	The attitude in the community is "this is our responsibility" and is now beginning to reflect modest involvement in efforts
Knowledge about Issue	7.75	Community members have knowledge of, and access to, detailed information about local prevalence
Resources	4.00	The community has individuals, organizations, and/or space available that could be used as resources
Average	6.46	Enough information is available to justify efforts. Activities are underway

Table 1 shows that farmer knowledge about the issue is the highest score in this assessment. This score supports the issue that actually, stevia has been recognized and pretty familiar with a farmer in the location. It is also supported by a local leader who gives serious attention to cultivating stevia, which is indicated by a high score as well. Farmers also identified that several efforts have been conducted. Even though still at a very beginning involvement, farmers put stevia development as an important part of their life. Due to a small development of stevia, so far only small amount of resources has been used to cultivate stevia. However, the farmer is actually willing to provide their resources if stevia development will be extended.

3.3. *Development strategies*

Based on community readiness assessment to adopt stevia cultivation at District of Pasir Jambu Regency of Bandung, three weaknesses components identified, namely resources, community efforts and community climate. Hence, importunate strategies to implement are to increase quality and quantity of these three components. Several strategies to overcome these weaknesses can be formulated as can be seen in Table 2.

Table 2. Development strategies to cultivate stevia at District of Pasir Jambu

Components	**Stevia Cultivation**	
	Score	**Strategies**
Goals of Community Development	6.46	Provide community-specific information
Corrective Actions		• Plan publicity efforts associated with start-up development of stevia cultivation in community • Attend meetings to provide updates on progress of the potential and development of stevia in community • Search for additional resources and potential funding to develop stevia in location • Begin some basic evaluation efforts, in particular, related to farmer involvement in developing stevia • Insist government (local, regional) give further attentions and responses related to stevia development in the location • Collect information related to the gap between farmer and buyer expectation related to stevia (processed) products

Table 2 shows that actually, development stevia at District of Pasir Jambu has significant potential to deliver. Internally, farmer's community is adequately ready to adopt stevia development. Better organization and management will easily attract farmers to involve. Government attentions and responses (at local as well as regional/national) through supportive policies (including financial source provisions) will help the development rate very much. Politically, the government should encourage alternative sweetener sources development, in addition to "conventional" sources. Similarly, certain industries involvement and collaboration with farmers will also increase the adoption process significantly.

4. Conclusion

Farmer readiness to cultivate stevia at District of Pasir Jambu, Regency of Bandung is at "initiation" stage, which is actually having sufficient information to be further developed. In order to increase farmer (group) capacity (quantity and quality), several strategies can be implemented, in particular, related to management and organization of farmers, government policy supports, and involvement of related industries.

Acknowledgment

We would like to thank Universitas Padjadjaran for providing financial support to conduct this research through "Hibah Internal UNPAD" scheme.

References

[1] Hairani R I, Aji J M M and Januar J 2013 Analisis Trend Produksi dan Impor Gula serta Faktor-Faktor yang Mempengaruhi Impor Gula Indonesia *Berkala Ilmiah Pertanian* **1** 77-85
[2] Maudy E 1992 Peluang Bisnis Stevia *Tahun XXIII* **274** 18-19
[3] Donna G, Bonvie B, Bonvie L 2000 *The Stevia Story: A Tale of Incredible Sweetness & Intrigue* B E D Publications
[4] Silalahi U 2012 *Metode Penelitian Sosial* (Bandung: PT Rafika Aditama)
[5] Sugiyono 2012 *Metode Penelitian Kualitatif dan Kuantitatif dan R&D* (Bandung: Alfabeta)
[6] Rusidi 1983 *Metode Ilmiah* Laboratorium Sosiologi Penyuluhan Fakultas Pertanian Unpad
[7] Plested B A, Edwards R W, and Jumper T P 2006 *Community Readiness: A handbook for successful change* (CO: Tri-Ethnic Center for Prevention Research)
[8] Behar L B and Hydaker W M 2009 Defining Community Readiness for the Implementation of a System of Care *Adm. Policy Ment. Health.* **36** 381–392
[9] Trautman K, Burkhart K C, and Rice T 2012 *Accessing a Community Readiness* (Institute of Agriculture and Natural Resources: The University of Nebraska-Lincoln Extension)
[10] National Institute on Drug Abuse 1997 *Community readiness for drug abuse prevention: Issues, tips, and tools* (Rockville MD: National Institute on Drug Abuse)
[11] Plested BA, Jumper T P, and Edwards R W 2006 *Community Readiness: Advancing Suicide Prevention in Native Communities (Community Readiness Model Handbook*). Center for Applied Studies in American Ethnicity, Colorado State University, Fort Collins, CO
[12] EUFIC (European Food Information Council) 2009 *Stevia: a natural sweetener with potential* http://www.eufic.org/article/en/nutrition/Sweeteners/artid/stevia-natural-sweetener-with-potential.

Utilizing Bacillus to inhibit the growth and infection by sheath blight pathogen, *Rhizoctoniasolani* in rice

R Margani, Hadiwiyono and S Widadi
Agrotechnology Department, Agricultural Faculty, Sebelas Maret University Surakarta, Indonesia

Email author: hadi_hpt@yahoo.com

Abstract. *Rhizoctonia solani* Kuhn is a common pathogen of rice. The pathogen causes sheath blight of rice. The pathogen can cause loss in the production of rice up to 45%. So far, the disease however is still poorly taken care of by the farmers and researchers, so the control measures is nearly never practiced by the farmers in the fields. It due to the unavailability of effective control method of the disease. Therefore, development to control the disease is important. *Bacillus* is one of popular bacteria which is effective as biological control agent of a lot of pathogens in plants, but it has not been used for control sheath blight in rice yet. The current researches were aimed to study the potential of *Bacillus* collected from healthy rice as candidates of biological control agent of the disease. The results showed that some isolates showed indications to inhibit significantly the growth and infection of the pathogen. We obtained at least five isolates of *Bacillus* collected from leaves, sheath, and stem of healthy rice fields. All of the isolates could effectively inhibit the growth of *R. solani* in vitro on potato dextrose medium at range 30.33-58.00%, whereas in vivo B05 isolate was the most effective in inhibiting the infection of pathogen at 30.43%. It was not significantly different ($P \geq 0.05$) to application of hexaconazol with dosage of 2 ml L^{-1}.

1. Introduction

Rice is one of the main commodities in Indonesia. The demand of rice will continue to increase along with the increase in population [1]. Based on data from [2], Indonesia rice production is 44.72 million tons in 2013, and was decreased to 44.45 million tons in 2014. Efforts to supply the demand for rice production must be continued, but the challenge in increasing production is also increasing due to climate change, threat of pests and plant pathogens [3].

Rhizoctonia solani Kuhn is one of the pathogens that attacks rice. *R. solani* is a pathogen that causes rice sheath blight disease that becomes one of the main problems in the attempt to increase rice production, especially in intensive cultivation system [4]. The pathogen can cause yield loss up to 45% [5]. This pathogen can survive in soil, plant debris, and has a wide range of hosts [6]. *R. solani* can survive as sclerotia and sclerotia produced by *R. solani* can survive in soil for more than one year [7].

The control measures were nearly never practiced by the farmers in the fields. Therefore, the development to control the disease is needed. Some examples of biological agents that have been developed as biological pesticides were *Bacillus*, *Pseudomonas fluorescent* group, *Actinomycetes*, and *Trichoderma* [8]. Antagonistic bacteria, particularly of the genus *Bacillus*, are most widely used as biological agents to control many plant pathogens [9] including *Rhizoctonia* [10]. This study aims to evaluate the inhibitory capacity of *Bacillus* obtained from the isolation of healthy rice plants from the field to *R. solani* in vitro and in vivo.

2. Methods

The experiment was conducted in laboratory and green house of the Laboratory of Plant Pests and Diseases Control, Faculty of Agriculture, Universitas Sebelas Maret from August 2016 to April 2017. Materials used in the lab includes alcohol 70%, NA medium, PDA medium, LB medium, soil, rice seeds, *Bacillus* isolates and *R. solani* isolates. The tools used in this research are Petri's dish, test tube, measuring cylinder, plastic wrap, vortex, ose needle, Bunsen's lamp, loop, microscope, Erlenmeyer's tube, autoclave, shaker, camera, blight plant of blight, healthy rice plant and Polybags.

The treatments were arranged by Completely Randomized Design (CRD) consisted of 8 treatments and repeated three times. The treatments were control (no biological agents and no pathogen (K-)), negative control in the presence of pathogens (K+), fungicides chemicals (P0), *Bacillus* 01 (B01) isolates, *Bacillus* B02 (B02) isolates, *Bacillus* B03 (B03) isolates, *Bacillus* B04 (B04) isolates and *Bacillus* B05 (B05) isolates. Observated data were analyzed using *F* test followed by Duncan's Multiple Ranged Test (DMRT) with 5% confidence level.

Variables observed in this study were growth inhibition, onset of symptoms, disease incidence, and spotting area. The inhibition of pathogenic fungi was performed during in vitro antagonism test in the laboratory. Inhibition measurements were performed when the inhibition zone was formed between *Bacillus* and *R. solani* isolates. The inhibition measurement of the fungus with the formula of the colony radius away from the antagonist is reduced by the radius of the colony approaching the antagonist divided by the colony radius away from the antagonist multiplied by 100%. Observation of *R. solani* symptoms was performed on treated rice plants, which were observed daily with direct observation when there was appearance of spots on the rice stem. The observation of disease incidence was done by the number of affected plant samples divided by the total number of plants multiplied by one hundred percent. The area of spots was observed using millimeters of grid. Observations were performed on all spots that appeared from the treatment.

3. Result and Discussion

3.1 Characteristics of *Bacillus* isolates

Bacillus isolates were obtained from healthy rice plants. The isolates obtained were 9 with the origin of different parts of the rice plant(leaf, stem and sheath). The characterization of the isolates obtained was done by observing the colony morphology directly on the Nutrient Agar (NA) medium on Petri's dish (Table 1). Observation was done to morphological types of bacterial colonies include shape, edges, elevation, color and surface [11].

Table 1. Character morphology of colonies and gram testing of bacterial isolates from rice plant

No	Source	Character Morphology					Gram
		Shape	Edge	Elevation	Surface	Color	
1	Leaf	Circular	Entire	Flat	Smooth	White	+
2	Leaf	Irregular	Entire	Flat	Smooth	Cream	-
3	Leaf	Irregular	Serrate	Flat	Smooth	White	-
4	Sheath	Irregular	Lobate	Flat	Smooth	White	-
5	Sheath	Irregular	Lobate	Flat	Smooth	White	-
6	Sheath	Irregular	Entire	Flat	Smooth	White	+
7	Stem	Circular	Undulate	Flat	Smooth	White	+
8	Stem	Irregular	Entire	Flat	Smooth	White	+
9	Stem	Irregular	Undulate	Flat	Smooth	White	+

Based on the observations, the character of colonies of bacterial isolates obtained, were varied with circular (rounded), irregular (rounded irregular) colonies, with flat elevation, with edge variation of the entire (smooth), undulate (curvy), and lobate (Wavy) and smooth surface (smooth). The colonies of bacterial isolates obtained varied, some were white and cream. Colonies of bacteria were dominated by the irregular edge shapewith flat elevation, Smooth surface and white color. Bacterial isolates showed colony morphological characteristics which tend to vary in culture medium [12].

Bacterial morphological characters obtained from observations in nutrient media to conform to a statement from Tay et al. [13] that *Bacillus cereus* can grow well on NA medium incubated at an average temperature of 28.8 ^{o}C. The most common form of bacterial colonies morphology is irregular, this is in accordance with the results of research [14] that bacteria class *B. cereus* has an irregular shape. Similarly, the colony elevations are flat in all bacterial colonies observed. The observations indicated that the morphology were quite diverse; one of the causes is the bacterial density. Several factors affecting the population density of bacteria are plant species, plant age, tissue type, habitat and resilience formed by plants and environmental factors [15-16]. Isolates of purified bacteria, then characterized by morphology, Gram testing using KOH 3% [17]. According to [18] *Bacillus* is a Gram positive and rod-shaped bacterium. Nine bacterial isolates obtained, showed 5 Gram positive and 4 gram negative isolates.

3.2 Characteristics of *R. solani* isolate

The isolate of *R. solani* was identified visually and microscopically. The pure cultures and hypha of *R. solani* used, are presented in Figure A. The microscopic appearance of *R. solani* is shown in Fig. B where there are arrows. Without the arrows it shows the *R. solani* hypha that has almost engle branching at 90 degree,, 6-10 μm wide, and hyaline color [19].

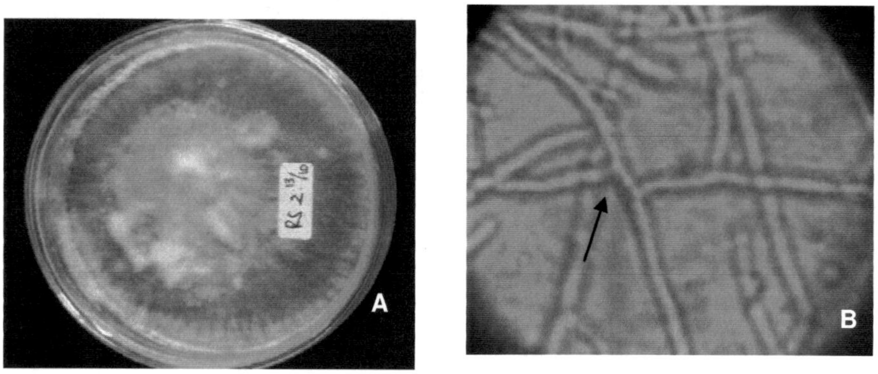

Figure 1. Characteristics of *R. Solani*: (A) *R. Solani* culture on PDA medium, (B) Hypha of *R. Solani* that makes angle of 90^0

3.3 Bacillus antagonism against R. solani in in vitro

The results of this study showed that the growth of pathogenic fungi could be inhibited by *Bacillus* isolates with varying inhibitory power (Figure 2). The result of observation of antagonism tested, showed that the treatment of isolate B05 has the highest inhibitory, which is 58.00%. *Bacillus* B02 isolate was ranked second in inhibiting *R. solani* at 48.67%. The next *Bacillus* isolate inhibiting *R. solani* was B01 at 47.67%. *Bacillus* B04 isolate was at 38.67% inhibitory capacity and *Bacillus* isolate having the lowest inhibition to *R. solani* was B03 at 30.33%. This is in line with statements from [20] that the use of *B. cereus* isolate can suppress *R. solani* and *S. rolfsii* by 68.9% and 33% and were able to slow down the growth rate of *R. solani*on PDA media. According to [21] strong inhibition of pathogenic fungi caused by the presence of fungicidal compounds produced bacterial cultures.

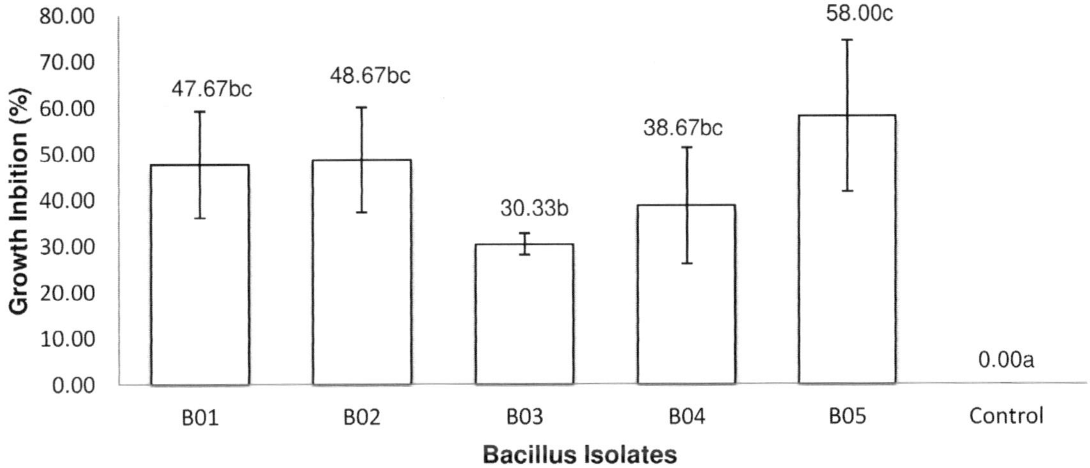

Figure 2. Growth inhibition of *R. solani* by *Bacillus* on PDA medium

Bacillus isolates having ability to inhibit growth of *R. solani* on PDA medium indicates that the isolates produce some antibiotic compounds involved in antagonistic mechanism. The result supports previous research done by [22] that *Bacillus* sp. has ability to produce antibiotic compounds. The formation of inhibitory zones indicated a mechanism of antibiosis by *Bacillus* isolates [23]. This is in accordance with the research of [24] where endophytic bacteria have antibiotic ability to form highly variable inhibitory zones. Inhibition of *R. solani* is indicated by the distance between pathogenic fungi and *Bacillus* bacteria (Figure 3).

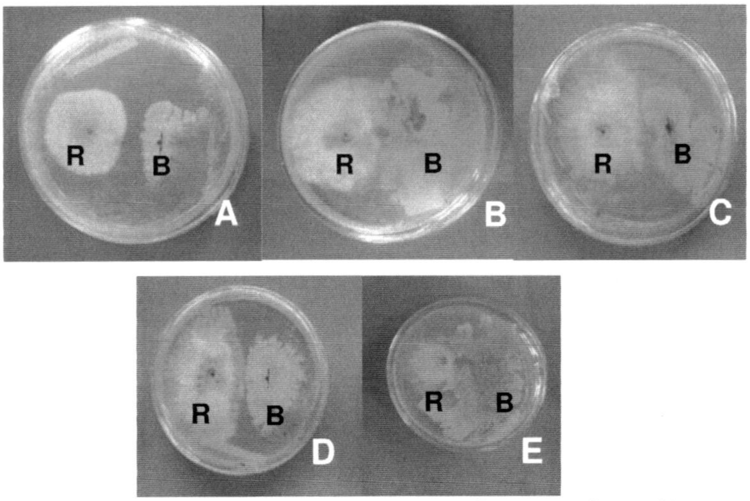

Figure 3. *Bacillus* antagonism against *R. solani* on PDA media: (A) Isolate B01 to *R. solani*; (B) Isolate B02 to *R. solani*; (C) Isolate B03 to *R. solani*; (D) Isolate B04 to *R. solani*; (E) Isolate B05 to *R. solani*; R: isolate *R. solani*; B: isolate *Bacillus*

The inhibition of pathogen growth is characterized by the formation of clear zones in PDA medium, indicating that *B. cereus* produces certain metabolite compounds that can inhibit *R. solani* [20]. Metabolite compounds were suspected as antifungal; thereby inhibiting the growth of pathogenic fungi. The results of [25] showed inhibition by class *B. subtilis* up to 50% of *R solani* in vitro and suspected *B. subtilis* produced antibiotic compounds such as inturin A and surfactin.

3.4 *Bacillus* antagonism against *R. solani* in in vivo

Results of *Bacillus* test showed that the appearance of symptoms between the treatments was not significantly different from each other. The appearance of symptoms was emerged at 3.39 to 10.00 days after inoculation. The disease incidences were at the ranges 33.33% to 100%. The effectiveness in inhibiting of disease was at 0% to 54.35% (Table 2).

Table 2. Bacillus isolate test on *R. solani* infection at green house

Bacillus Isolates	Appearance of symptoms	Disease Incidence (%)	Effectiveness in Inhibiting of Disease (%)
B01	8.33 ± 1.53 c	55.57±19.24 bc	14.49±18.10 abc
B02	10.00 ± 0.00 c	66.67±00.00 cd	09.42±16.31 ab
B03	8.67 ± 0.58 c	100.00±0.00 d	00.00±00.00 a
B04	8.67 ± 1.53 c	66.67±33.34 cd	11.59±20.08 ab
B05	8.67 ± 1.15 c	55.57±19.24 bc	30.43±13.57 bcd
Heksaconazol**	8.00 ± 1.73 c	33.33±00.00 b	54.35±28.42 d
No *Bacillus*	3.39 ± 0.35 b	88.89±19.24 cd	-

The value followed by the same letters in the same column are non-significantly different based on Duncan's test at 5%.* Based on data transformed in to $\sqrt{x+1}$; ** Dosage 2 mL L^{-1}.

Tests at the in vivo level of 5 isolated *Bacillus* in isolateB01 showed the earliest symptoms when compared with isolate B02. When the symptoms appear more slowly, *Bacillus* isolates slow down the growth rate of *R. solani*, according to [20] statements that the *B. cereus* treatment allows to slow or delay the timing of symptoms in plants. However, B01 isolate showed a lower incidence of disease and spotted area compared to B02 isolate. For the highest incidence of disease occurs in the treatment of B03 that is 100%, where rice plants inoculated *R. solani* all show symptoms. We suspected that B03 isolates cannot colonize the plants. This is in accordance with the statement of [26] that emphasis on the intensity of disease attacks is related to the ability of bacteria to colonize the plants and produce the same metabolism as secondary to protect the plants from pathogenic infection.

Bacillus isolates based on the results of the study showed the effectiveness of various disease suppressors. The highest *Bacillus* isolate in suppressing *R. solani* pathogen was B05 treatment with 30.43%. Isolates B01 has 14.49% of disease suppression efficacy, B02 isolate is 9.42%. Isolate B03 effectiveness of disease suppression is 0%, this is due to incident disease that happened 100%, so all plants infected by *R. solani*. The use of hexaconazole fungicide has the highest disease suppression efficacy of 54.35%. Isolate B03 shows that it is in line with in vitro test. Isolate B03 in in vitro test showed inhibitory resistance to *R. solani* but in vivo test could not inhibit. According to [27] it is possible that the isolate has no or less ability to produce antifungal compounds, or less adaptive so the antagonism role disappears.

Isolate B04 shows a somewhat weak resistor to *R. solani* growth at the in vitro level, but has a strong inhibitory at the in vivo level. The isolate has a weak antibiotic mechanism to inhibit *R. solani* growth directly, but indirectly has other mechanisms that are strong enough to suppress symptoms when applied to seeds or plants. In other words, B04 isolates do not directly suppress the growth or development of pathogens, but rather increase plant inhibition to pathogens. The mechanisms of plant inhibition that are affected by antagonistic bacteria are known as the induced systemic resistance [28].

4. Conclusion

The conclusions can be drawn from this study is that all of the isolates could effectively inhibit the growth of *R. solani*in vitro on PDA at 30.33-58.00%, and isolate B05 was the most effective. It also occurred in vivo that the isolate was most effective inhibiting the infection of pathogen at 30.43% that was non-significantly different different ($P \geq 0.05$) from the application of hexaconazol with dosage 2 mL L^{-1}.

References

[1] Sinar Tani 2011 *The rice innovation for facing climate change* Jakarta

[2] Badan Pusat Statistik *2015 Agricultural and mining data* Jakarta Central Statistics Agency of Indonesia

[3] Susanti E,Ramadhani F, Runtunuwu E and Amien I 2012 The impacts of climate change on the attack of plant pest organisms (ppo) and anticipation and adaptation strategies (Dampak perubahan iklim terhadap serangan organism pengganggu tanaman (opt) serta strategi antisipasi dan adaptasi) (Bogor: Agro-climate and Hydrology Research Center)

[4] Jaya prakashvel M andMathivanan N 2012 Morphological and pathological variations of rice sheath blight inciting south Indian *Rhizoctonia solani* isolates *Arch. Phytopathol. Plant Prot.* **45**(4): 455-467

[5] Kumar K V K, Reddy M S, Kloepper J W, Lawrence K S, Groth D E and Miller M E 2009. Sheath blight disease of rice (*Oryza sativa* L.) an overview *Biosci. Biotechnol. Res. Asia.* **6**:465–480

[6] Groth D E 2008 Effect of cultivar resistance and single fungicide application on rice sheath blight, yield, and quality *Crop Protect.* **27**: 1125-1130

[7] Guo Q, Kamio A, Sharma B S, Sagara Y, Arkawa M and Inagaki K 2006 Survival and subsequent of rice sclerotial diseases fungi, *Rhizoctonia oryzae* and *Rhizoctonia oryzae-sativae*, in paddy fields *Plant Dis* **90**: 615-622

[8] Supriadi 2006 Analysis of the risk of biological agents for the control of pathogens in plants *J Lit Pert* **25**: 75-80 (in Indonesia)

[9] Elhamshary O I M and Khattab A A 2008 Evaluation of antimicrobial activity of *Bacillus subtilis* and *Bacillus cereus* and their fusants against *Fusarium solani. Res. J. Cell Mol. Biol.* **2**: 24-29

[10] Mizumoto S, Hirai M and Shoda M 2007 Enhanced iturin A production by *Bacillus subtilis* and its effects on suppression of the plant pathogen *Rhizoctonia solani* Appl Microbiol Biot **75**: 1267-1274

[11] Habazar T and F Rivai 2004 *Plant pathogenic bacteria*. (Padang: Andalas University Press (in Indonesia)

[12] Hatmanti A 2000 Introduction of *Bacillus* spp. Oseana **25**(1): 31- 41 (in Indonesia)

[13] Tay L, Goh K T and Tan S E 2008 An Outbreak of *Bacillus cereus* food poisoning *J Singapore Medical*. **23**(4): 214- 217

[14] Salaki L 2011 Isolation and characterization of indigenous bacteria (*Bacillus cereus* Frank.) as biological controlling agent of cabbage pest *Eugenial* **7**(1): 10-15 (in Indonesia)

[15] Marwan H, Sinaga M S, Giyanto and Nawangsih A A 2011 Isolation and selection of endophytic bacteria for the control of blood diseases on banana *J HPT Tropika* **11**(2): 112–119 (in Indonesia)

[16] Melnick R L, Zidack N K, Bailey B A, Maximova S N, Guiltinan M and Backman P A 2008 Bacterial endophytes: *Bacillus* spp. from annual crops as potential biological control agents of black pod rot of cacao *Biol. Control* **46**(1): 46–56

[17] Schaad N W, Jones J B and Lacy G H 2001 *Xanthomonas* in Laboratory guide for identification of plant pathology bacteria, third edition (St. Paul Minnesota: APS Press)

[18] Arwiyanto T, Asfanudin R, Wibowo A, Martoroedjo and Dalmadiyo G. 2007. The use of *Bacillus* local isolates to suppress Temanggung tobacco linger disease *Berkala Penelitian Hayati* **13**: 79-84(in Indonesia)

[19] Semangun H 1993 Diseases of food crops in Indonesia (Penyakit-penyakit tanaman pangan di Indonesia) GadjahMada University Press Yogyakarta

[20] Hidayah N and Yulianti T 2015 *Bacillus* cereus antagonism test against *Rhizoctonia solani* and *Sclerotium rolfsii Buletin TanamanTembakau, Serat dan Minyak Industri* **7**(1): 1-8. (in Indonesia)

[21] Liu C H, Chen X, Liu TT, Lian B, Gu Y, Caer V, Xue Y R and Wang B T 2007 Study of the antifungal activity of *Acinetobacter baumannii* of its antifungal components *Appl. Microbiol. Biotechnol.* **76**: 459-466

[22] Abidin Z, Aini L Q and Abadi A L 2015 Effect of *Bacillus* sp. and *Pseudomonas* sp. against the growth of *Sclerotium rolfsii* Sacc pathogens causes of seedling fall in seedlings (Pengaruh bakteri *Bacillus* sp. Dan *Pseudomonas* sp. Terhadap pertumbuhan jamur pathogen *Sclerotium rolfsii* Sacc. Penyebab penyakit rebah semai pada tanaman kedelai) *J Hama Penyakit Tumbuhan* **3**(1):1-10

[23] Saputra R, Arwiyanto T and Wibowo A 2015 Activity test of several isolates of *Bacillus* spp. against bacterial wilt disease (*Ralstonia solanacearum*) in several varieties of tomatoes and their identification *Proc. of National Seminar Masy Biodiv Indon* **1**(5): 1116-1122 Doi: 10.13057/psnmbi/m010525 (In Indonesia)

[24] Munif A, Wiyono S and Suwarno 2012 Isolation of endophytic bacteria from dry land rice and its potential are biocontrol agents and growth promoters (Isolasi bakteri endofit asal padi gogo dan potensinya sebagaai agens biokontrol dan pemacu pertumbuhan) *J Fitopatol.Indones.* **8**(3): 57-64

[25] Seema M and Devaki N S 2012 In vitro evaluation of biological control agents against*Rhizoctoniasolani. J Agricultur Technol* **8**(1): 233-240

[26] Jatnika W, Abadi A L and Aini L Q 2013 Influence of *Bacillus* sp. and *Pseudomonas* sp. on the progression of the dull disease caused by the pathogenic fungus *Peronosclerosporamaydis* in corn plants (Pengaruh aplikasi *Bacillus* sp. dan *Pseudomonas* sp. terhadap perkembangan penyakit bulai yang disebabkan oleh jamur pathogen *Peronosclerospora maydis* pada tanaman jagung) *J HPT* **1**(4): 19-29

[27] Rustam, Giyanto, Wiyono S, Santoso D A and Susanto S 2011 Selection and identification of antagonistic bacteria as biological control agent of bred rice blight disease *J Penelitian Pertanian Tanaman Pangan* **30**(3): 164-171 (In Indonesia)

[28] Van Loon L C, Bakker P A H M and Pieterse M J 1998 Systemic resistance induced by rhizhosphere bacteria *Ann. Rev. Phytopathol.* **36**:453-483

Sabuk Janur: tools to move community participation in reducing natural disasters and environment (case study at Lawu mount slope in Indonesia)

E Lestari, S H Purnomo, Komariah and A Wibowo
Universitas Sebelas Maret
Jl. Ir. Sutami 36 A Surakarta 57126 Jawa Tengah Indonesia

E-mail: agungbersahaja@gmail.com

Abstract. Globalization in general both traditional values and local knowledge that exist in a society, but this is different in the community of mountain slope lawu on the Java Island precisely in Karanganyar regency of Central Java is still maintained local cultural values to maintain sustainability Springs. This is a foothold for cheaper the benefits of community culture that play an important role in environmental conservation and social solidarity to cope with natural disasters and the environment. The research method used is qualitative with explanatory case study approach. The results showed that the Myth of Sabuk Janur proved able to build public awareness in preserving the environment and overcoming natural disasters and the environment. The high level of community participation in every development is evident from every stage of community participation in a program. Where the solidarity of social visible in all conditions, namely social solidarity in the nature of the in-celebration and the nature of overcoming the disaster that befall. The preservation of local cultural values is still done because of the existence of the identity of local communities. This is because the integration of national value and local potential becomes a prerequisite in every development plan. Thus the cultural approach is an effective strategy in solving the problems exist in an area, especially in overcoming natural disasters and the environment.

1. Introduction

Basically human ecology sees a mutual relationship between man and his environment. Human ecology is seen as the relationship between man and God Almighty, with his neighbor and with the environment (nature and artificial) in the vicinity. Man has two traits that underlie his relationship to his environment: first, his biological nature and his instincts as biological living beings; need for food, reproduce, remove dirt, require air and space. Second, have values, norms, minds, attitudes and perceptions about the world around and place itself in it [1].

Development based on environmental in principle strives for sustainable development. Development based on environmental is an effort to develop local communities with their ecosystems towards higher productivity and higher levels of fulfillment that are sustainably implemented in both ecological and social terms. Natural resource management should be done as efficiently as possible and by always maintaining an ecological system to meet the basic needs of its. Thus to achieve a balance between production activities and environmental carrying capacity required a change in production mode and consumption style. The community life on the slopes of Lawu Mount is modest, utilizing what they have is very relevant to support environmentally conscious development.

Globalization in general abandon traditional values as well as local knowledge that exist in a society, however this differs in the community of lawu mountain slopes on the island of Java precisely in Karanganyar regency of Central Java which still maintain the local cultural values to preserve the springs . This becomes a foothold for further exploration of local cultural values that play an important role in influencing people's participation in environmental conservation and social solidarity to cope with natural and environmental disasters.

According to [1], environmental sustainability will be achieved when the use of existing resources is not drained out without adequate compensation. In addition, the community must also be prepared to utilize resources wisely. Conflicts that occur in the utilization of resources by various parties involved as actors in the actual development process is a process of interaction between people, technology and resources, and humans as the driving force.

Meanwhile, according to [2] related to resource utilization, the need for programs: (1) maintaining and improving land productivity by utilizing local resources (such as potential crop species, local superior seeds, green manure, compost and biofertilizer) ; (2) increasing the diversification of food crops; (3) optimization of integrated farming system through plant livestock systems, plant-based fisheries combined with efforts to increase crop production; (4) cultivating and planting the yard with useful plant species (such as family medicinal plants) and adapted to local conditions, and utilizing technologies that farmers can easily adopt and implement; (5) enhancing mentoring, guidance and training for farmers and farmer groups to encourage and enhance farmer self-reliance; (6) increasing the role of women in farming; (7) conducting surveys and mapping of food availability and vulnerability aimed at obtaining a true picture of the food conditions in Indonesia.

Conserving biodiversity means ensuring the continuity of sustainable development. There are four things that stand out that biodiversity has important economic prospects; (1) biodiversity is a potential source of genetic wealth; (2) biodiversity in the forests is the only hope of human life because there are scientific medicines; (3) having biodiversity means having a great choice to develop science and technology; (4) have biodiversity of varying species richness [3].

[4] defines local wisdom or indigenous knowledge as follows: "Indigenous knowledge is local knowledge that is unique to a given culture or society. That is important as a basis for a society which facilitates communication and decision-making ". Local people have many local wisdom / local knowledge in managing water resources in order to remain sustainable, development will be effective when there is synergy with the knowledge possessed by the local community.

UU no. 22 Tahun 1999 and PP no. 20 Tahun 2000, stipulates that local governments have full responsibility for strengthening local cultures. Local governments should create an atmosphere conducive to the growth of social groups, encouraging the growth of values and norms that enable them to enrich the social life of local communities and provide a driving infrastructure for the growth of civil society with high social capital. Thus, the management of water resources must be integrated which is implemented in a cross-sectoral manner by taking into account the function of water as an economic, ecological and social function.

Some of the reasons for the disturbance of water sources are: (1) the decreasing of land as water catchment area due to the development of residential and industrial areas; (2) declining water quality as a result of disposing of various wastes into rivers or water sources; (3) the decreasing of environmental carrying capacity to the sustainability of water resource function and benefit due to uncontrolled upland land use behavior; And (4) disruption of water resources sustainability and threatened sustainability of irrigation waterworks functions as a result of uncontrollable return of excavated materials for development [5].

Integrated and sustainable development needs to be realized in the optimal management of water resources fairly and maintain ecological balance. Therefore, it is necessary to manage water resources based on local culture and local wisdom for effective management. This research will explore further how local cultural values and local wisdom play an important role in influencing community participation in environmental conservation and social solidarity to overcome natural and environmental disasters.

2. Methods

This research is focused on the community on the slope of Mount Lawu in Karanganyar Regency. There are a variety of reasons that researchers deem to set a very urgent research setting to do. First, the region has an important role in conservation to reduce the sedimentation rate of the Bengawan Solo River. Secondly, in that area local wisdom plays an important role in maintaining harmony and maintaining environmental balance. Based on the problems posed in this research, which is more emphasis on the process and meaning of the problem, then the right type of research is descriptive qualitative research with phenomenology approach[6]. Because the research will be conducted in one district and focus on a particular problem, the analytical technique used is a single case analysis[7]. According to [8], in a single case analysis in each case the analysis process was performed using an interactive analysis model. In this analytical model, the three components of the analysis are: data reduction, data presentation and drawing conclusions or verification.

3. Results and Discussion

3.1. Local Culture of Lawu Mount Slope

Karanganyar District, Central Java, is indeed a very rich location with culture and tradition. One of the areas that contribute a lot to the diversity of traditions is the area at the foot of Lawu Mount. In the area of Kalisoro Village, Tawangmangu, which is famous for its clean tradition of Monsodiyo tradition. Before we see more about what is '*Mondosiyo*' traditional ceremony, now we first check the origin of this traditional ceremony. It is said that there is a very cruel and cruel king named '*Prabu Baka'*. He caused the society can not live peacefully because he is the king of human eaters and to stop the arbitrary acts perpetrated by the tyrant, then '*Patut Tetuka'* is also present to kill and break the chain of misery of the people. He comes from the hermitage '*Pringgodani'* then named with '*Eyang Kancanegara*.' In an attempt to kill '*Prabu Baka*,' Tetuka slams the king's head on a rock (now called '*Watu Gilang'*). After that, the garlic plant grows from the former lying fangs '*Prabu Baka* 'teeth, while in other places grow onion on the ground where the molars are scattered on the ground. Both plants are also a major commodity of the residents of Pancot and Blumbang. However, before King of Baka died, he also ordered that every '*Wuku Mondosiyo*,' on Tuesday '*Kliwon'* clean village '*Mondosiyo.'*

This traditional ceremony is a tradition of chicken catching throw which is a clean village tradition as a form of gratitude after passing the previous trials. Chickens in this case are the votive media for some residents of Dusun Pancot. Mondosiyo ceremony is one of the cultural tourism wealth of Karanganyar. Mondosiyo tradition itself takes place once in seven months (Javanese calendar) and fall on Tuesday Kliwon Wuku Mondosiyo. While the peak of the implementation of this traditional ceremony is at Pancot Village Market Hall. The procession of clean tradition of Dusun Pancot village begins two days before Tuesday Kliwon is on Sunday Pon. On that day, residents of Dusun Pancot will collect rice to be processed into gandhik and also other types of cuisine as complementary offerings. The procession continued on the following day on Monday Wage where the notice was made by hitting bende in the evening. When the peak activity comes, the activity starts from the morning. The tradition of Mondosiyo begins with slaughter of goats and chickens at Punden Bale Patokan in the morning around 7 pm. The ceremony then continues in the 10 o'clock range when all offerings are prepared in punden and continue at 1 pm where Manyar Sewu gending begins to be played. The peak of the event is a chicken catching start at 4 pm. Before the event began, traditional entertainment like the usual reog was held while thousands of residents filled the Pancot Village Market Hall. Not to forget also before the start of the water event that was taken from Bale Patokan first splashed in Batu Gilang contained in the building in the market area. Shortly after, the peak of Mondosiyo tradition began. Chickens brought by voting citizens began to be thrown into the roof of the hall. Meanwhile, thousands of residents in attendance will scramble and try to catch the chicken. For citizens who throw this thing is a form of gratitude for the trials that have been passed.

Meanwhile, for citizens who catch this is a form of business to get luck and blessing when they get a chicken.

The next culture is '*Srawung Seni Sukuh*,' is a big pray packed in the form of art performances, discussion and distribution of plant seeds. The event was held to derive the noble values of Sukuh Temple, the preservation of nature around the slopes of Lawu Mount, and the culture of the people around the temple and the slopes of the mountain. '*Srawung Seni Sukuh'* event closed with the distribution of plant or tree seedlings. Visitors who are present are welcome to take tree seeds for free to take home and planted in the yard. This is held regularly every year as an effort to preserve nature and respond to global warming issues that are currently being discussed a lot. '*Srawung Seni Sukuh'* event became the tourism magnet of Karanganyar district when commemorating the turn of the year. Art performances displayed by local, overseas, classical, and modern artists provide color and diversity in the performing arts.

3.2. Myths That Developed in Maintaining Environmental Sustainability

Behind the '*Sabuk Janur'* dance stored myth very strong for the people on the slope of Lawu Mount precisely in Plawan Village Hamlet Girimulyo Karanganyar District. Folklore whose main character '*Kyai Sabuk Janur"* stems from the presence of an abundant source of water in the village of Girimulyo. In the first time, source of springs used for the purposes of the population for drinking water, bathing and irrigating rice fields, once the water does not flow again. There are village elders concerned about the condition of people who lack water for daily necessities and even the planted rice also can not grow and dry so that the population fails to harvest.

Based on incident '*Kyai Sabuk Janur'* suspicious with existing springs. The Kyai then along the water source location and it turns out the flow of water for the residents is blocked by a large rock. '*Kyai Sabuk Janur"* ask for help to God Almighty by doing a meditation ritual on the big rock that became the water barrier and it turns behind the big rock inhabited a very large centipede. The gulf centipede is disturbed to go out and attack the Kyai. With the '*Janur Kuning'* weapon Sang '*Kyai Sabuk Janur'* is successful mencabuknya hinga injured But with the policy of '*Kyai Sabuk Janur'* centipede is not in the kill and even healed from the wound. Finally, the big centipede helped remove large stones with the help of other branch clubs. Residents of the community finally get a source of clean water and lancer back.

The '*Sabuk Janur'* Myth is finally developed in a '*Sabuk Janur'* Dance. This dance is finally able as a tool to move the society to participate in every development, especially the community participation in maintaining environmental sustainability and social solidarity in overcoming natural disasters and limits that befall.

According to the story of '*Prabu Baladewa'* puppets at the time of the war Westernayudha, told '*Kresna'* to meditate in Grojogansewu. This is to avoid Baladewa fought in the battlefield, because kesaktiaannya without enemies who can match it. To the east of Grojogan Sewu there is a similar waterfall called Pringgodani. This place is an antagonistic setting of King Anom Gatotkaca his son Bima. To get there through the narrow and steep streets. Here there is a hermitage also there is a grave which supposedly is the grave of Gatotkaca. The grave was sacred and many pilgrims came. Above there is Pringgosepi forest.

In addition to the puppet myth, there are also other myths trusted by the community around for generations. Dikisahakan there is a pair of young people who are making love in the lake Grojogan Sewu. They bathe together under the waterfall. Because it is a prohibition, then the two young people are cursed to be the stones scattered under the waterfall.

There is also a myth that until now still held firm and trusted. This is related to the existence of a bridge called "kretek pegat". Who called the bridge "pegat". The name is given based on empirical experience, that the bridge often leads to pegatan or divorce. At some point in the evening, it is often seen that a grandfather passes through the mist covering the clogs of pegat. They believe that the grandfather was a supernatural creature andhahan (subordinate) ruler in Grojogan Sewu. Namely Kyai Baladewa.

3.3. The Local Wisdom of The Community to Do Conservation

Geographical conditions in mountain areas that have steep slopes will certainly result in the presence of disaster-prone areas, especially landslides. Landslides often occur in areas with steep slopes. This disaster happens because of the greedy human error itself. The condition of sloping soil is still desperate to be used for real agricultural land for land conservation areas. But to overcome these disasters, people in Blurmbang Village have their own way. This method has been done by ancestors first and inherited to the present generation and still used. That way is to regulate the flow of water in agricultural areas so as not to stagnate. So create a channel that the water continues to flow arela land by means of twisting winding. In making this channel farmers rely on their long-used instincts from past work experience. This method is good enough to overcome the problem of landslide disaster that usually occurs in the agricultural area. This is evidenced by the absence of such longter land disasters to date. Farmers argue that landslide disaster occurs because there is a puddle in the field so that in the soil also contains a lot of water. This condition makes the deep soil become loose and makes the ground easy to move. By using the method then the loose soil only on the surface only so it will not potentially landslide disaster.

Geographical condition in Girimulyo village is very supportive to overcome the existing landslide disaster that usually haunts the people who live on the slopes. The area in Girimulyo village is very safe from the disaster. This is evidenced by the disaster never happened in Girimulyo Village. This is due to the natural condition of Girimulyo Village area which consists of rocks. Most of the villages in Girimulyo Village consisting of rocks make this a rare disaster. Even if it happens but the scale is small and only occurs in the area of agricultural land. The area in Girimulyo village only has a layer of soil as deep as 3-4 meters and then in it all the rocks. But 5 years ago the area in Girimulyo Village ever happened tornado disaster. This disaster has resulted in some houses damaged. The community does not know the origin and cause of the disaster. The area had never experienced this disaster before. That way the people do not know and able to overcome them. For a flood disaster that might attack the community in Girimulyo Village, the community has entered into an unwritten agreement. The agreement is a prohibition to take stones in the river. These stones serve as a brace of water that may occur during the rainy season comes. Until now residents in Girimulyo village no one wants to take stones in the river.

3.4. The Local Wisdom of the community to do water management

The environment is a basic need for human life. To maintain the availability of those basic needs, the environment requires sustainability for its functions, such as the availability of water, fertile soil and air that sustains for plants to live. Farmers' efforts and behavior in managing water resources are closely related to their perception of the environment. In the village of Wonorejo most are farmers, where in everyday life is always dealing with nature. In this activity they are guided by the experiences and knowledge they capture in managing the sustainability of water resources. There are various local wisdom of the community in managing the water source in order to remain sustainable.

'Mamayu hayuning bawana' is a slogan that remains firmly held by most Javanese people, especially the people at the foot of Mount Lawu. This is a guide in conducting activities that are closely related to the natural environment. "Memayu hayuning bawana" means protecting the safety and prosperity of the world. Their behavior in understanding the motto is shown by maintaining themselves and the physical environment in symbolic-ritual. Local people still perform rituals to obtain supernatural powers, also aims to maintain the balance between humans and the natural environment. They argued that the destruction of nature brought a very disastrous disaster. Floods, landslides, and droughts are caused by human hands that do not conserve natural resources. They hope that the values in the slogan are included in education in schools since primary education, in the hope that all human beings are aware of the importance of nature sustainability.

'Nandur Kebecikan, ndheder kautaman. Nandur will be slogan ngunduh null '. It means to plant virtue, to train virtue. Planting the good will reap the good. For that everyone is obliged to do good to others and do something for the common good. For example, planting trees in the forest is an activity that is routinely done by the community of Gunung Lawu Mountain together with PERHUTANI to maintain the sustainability of the water source. This activity is done with the hope that the land is not avalanched, and the water source remains sustainable. This is evidence that communities around forests have high participation in maintaining the ecological balance of forests without expecting economic returns from the results of such activities. They also planted trees on their own land. People have local wisdom in planting trees that are expected to keep the water source sustainable. They plant trees (avocado, banyan tree, dadap, aren) around the stream of the spring water. The plants according to their observations so far are very important in maintaining the water source. With their collective consciousness, they multiply tree planting together with local community leaders independently.

'Dadio wong kang nandur wiji keli'. Meaning to Be the one who planted the beans carried by the current (drift), ie berbatlah unconditionally; Beramalah for a long term. Life is not only in the world because after the life of the world there is life akherat. Local wisdom of the community is reflected in maintaining the source of water supplied to the PDAM, various actions taken include: (1) checking the flow of water, if one is blocked immediately removed; (2) if any damaged pipeline is repaired immediately, even funds for repairs come from the residents' dues, without waiting for funding for maintenance from the PDAM; (3) provide a water filter to prevent clogging; (4) provide lids or beds to water sources from leaf litter; (5) clearing the trenches around the river; (6) creates a small river for a small stream of water sources in a particular sub-village rather than the water being disposed of; And (6) does not dispose of trashes of agricultural produce to the stream of spring water.

'Alam paringane Gusti, mulo kito jogo alam iki'. The point is that man and nature are God's creations, so humans must nurture, utilize and preserve them well. Thus farmers act strongly influenced by local cultural values. They assume that water resources are used for the common good. This means that the sentence reflects the local wisdom of the community in looking at the water source. The public recognizes that the spring is a common property that must be nurtured, guarded and tuned for the next generation. It also reflects in the practice of Psal 33 ayat 3 UUD 1945 namely "Earth, water, and natural resources contained underneath are controlled by the state and used as much as possible for the welfare of the people". They let some of the water resources in their area be taken for the benefit of PDAM Karanganyar District

3.5. The Dynamics of Local Communities in the face of outside Intervention

The dynamics of modern life today, the color of life dominated by attitudes and behaviors that are the fruit of western cultural transformation as a direct influence of globalization. The western society considers that matter is the reality of life so that every activity must produce material even though the path taken is not in accordance with the values of humanity as adopted by the people of the eastern world. The life of the world is the responsibility of everyone is the executor of such responsibilities as well as in groups. In any activity utilizing nature, it should not be excessive or with no damage to the ecosystem or nature itself, environmental sustainability is maintained.

The inhibiting factor in the environmental conservation process in Blumbang Village is the same ie economic factor. They are already thinking about the profit they will receive from their actions. For example, in the recommendation to plant perennials on slope sites with a slope of 45-60 degrees they do not do. This is because if they plant these crops then the agricultural land for the cultivation of horticulture crops will be reduced. So that will directly reduce the income of these farmers. In addition, gotong royong in working on agricultural land has begun to disappear. In the past they worked on their fields helping each other. For example, in cultivating the land could have been done by 15 people who performed voluntarily. But the activity has begun to fade and disappear. For reasons of increasing demand and need of money to meet the needs of the main reason why the culture gotong royong now began to fade.

The environmental conservation factor in Girimulyo Village is the same as in Blumbang Village which is economic factor. People in Girimulyo Village have started to think of profit in every activity. For example, there are rules for planting sengon on the sidelines of the land of pertania. This is to keep the soil condition from avalanche. Initially people want and implement these recommendations. But lately they kapok and do not want anymore. This is because the large sengon trees will block the sunlight into the main plant. So that the main plant gets less light so that will affect its production. Planted shade of sengon trees also result in humid conditions that will be susceptible to disease.

The level of village accessibility will affect the mobility of the population. A good level of accessibility will support the vast mobility of its population. This condition makes the residents can relate to the outside world so that it will increase their knowledge and insight. Accessibility to a good village will also affect access to education and health for the community. Ease of access to health and education services is very influential from accessibility to the village. Accessibility level of Blumbang Village and Girimulyo Village is easy, this is because the road conditions are well paved. A good infrastructure will also facilitate the transportation of agricultural products to the market. The ease of transportation will have an impact on the smoothness of the agricultural marketing process. That way, the products from the agricultural products in both villages can compete in the market.

The location of Blumbang Village itself is quite strategic. Blumbang Village is on the main jalu of Central Java and East Java. Favorable conditions for the development of agriculture in the region. This is because the harvest can be sold to Central Java or East Java. In addition, the distance to the market is also not too far away. Close distance allows transportation costs to be reduced.

The location of Girimulyo Village itself lies the main route to the direction of the gentlemen. The location of Girimulyo Village is strategic for the route to the location of the tea garden destination. People use this geothermal condition to sell vegetables in polybags. Communities in Girimulyo Village through KWT with food self-sufficiency program have made vegetable planting program in polybag. This program requires each house to have vegetable plants of 20 polybags. Apparently this program is very beneficial to the community. This is because tourists who pass this village to the tourist attractions are interested in vegetable crops in the polybag and intend to buy. With the desire to buy plants in the polybags provide their own benefits to the community. For tomato planting sold for 15 thousand and strawbery sold 20 thousand per polybag. Apparently this program is very beneficial to the community. This is because tourists who pass this village to the tourist attractions are interested in vegetable crops in the polybag and intend to buy. With the desire to buy plants in the polybags provide their own benefits to the community. For tomato planting sells for 15 thousand and strawbery sold 20 thousand each polobag.

The supporting factor for conservation in Blumbang Village is the value of togetherness among the people who are still thick. This togetherness value becomes the main key for environmental conservation in Blumbang Village. The value of togetherness is reflected from the pleasure of their inter-relationship bersilatmimi. They are always active and come to RT meetings. RT meetings are the right forum for PPL to socialize programs that lead to environmental conservation. For example, the balancing in the use of manure with chemical fertilizers. With these rules that have been agreed with them simultaneously perform these recommendations. This is because these suggestions have been mutually agreed upon so that they have the obligation to do so as well. This is a joint decision.

Supporting factors for conservation in Girimulyo Village are mutual assistance and togetherness among the people who are still awake. People in Girimulyo Village still uphold the value of solidarity and mutual cooperation among the people. The value is still well preserved despite the oncoming globalization that is threatening. The good cooperation of the people in Girimulyo Village is evident from the community's participation in the work of devotion. In addition, if there are people who need to build a house then they will volunteer to help. They will work from starting to build the foundation until the house is erect. They work without being paid money.

The value of togetherness also awakened quite well in the community in Girimulyo Village. This is reflected in the unity of the community in sustainability existing culture. The culture is Tarian Sabuk Janur that describes the history of the fertile Girimulyo region. This dance consists of 50 dancers and

30 gamelan musicians. The high enthusiasm of the people who want to take part in the preservation of this culture illustrates the unity in society in Girimulyo Village.

4. Conclusion

Community life at the foot of Mount Lawu always faced with nature, Efforts and behavior of farmers in managing water resources is closely related to the values of local culture that exist, one of the most entrenched and become the paetisipasi community to conserve the environment and social solidarity in overcoming disaster Nature and environment is the myth of the Sabuk Janur. Villagers in mountainous areas are predominantly farmers, who in their daily lives are guided by the experiences and knowledge they capture in managing the sustainability of water resources. Based on existing myths, there are various local wisdom of society that still survive in managing water resources in order to remain sustainable.

'Mamayu hayuning bawana' is a slogan that remains firmly held by most Javanese people, especially the people at the foot of Lawu Mountain. This is a guide in conducting activities that are closely related to the natural environment. "Memayu hayuning bawana" means protecting the safety and prosperity of the world. 'Nandur Kebecikan, ndheder kautaman. Nandur will be slogan ngunduh null '. It means to plant virtue, to train virtue or to plant good will be good. 'Dadio wong kang nandur wiji keli'. It means to be the one who planted the beans carried by the current (drift), ie berbatlah unconditionally; Beramalah for a long term. Life is not only in the world because after the life of the world there is life akherat. 'Alam parigane Gusti, mulo kito jogo alam iki'. The point is that man and nature are God's creations, so humans must nurture, utilize and preserve them well. 'Revenge son putu ben komanan'. That means that in utilizing natural resources always remember the content it contains for the next generation. Local cultural values proved to have a high role in managing conservation farming. Therefore, every process of planning of development program should be based on local culture.

References

[1] Poerbo 1999 *Lingkungan Binaan Untuk Rakyat* (Bandung: AKATIGA)

[2] Sutanto R 2005 *Tantangan Dlobal Menghadapi Kerawanan Pangan dan Peranan Pengetahuan Tradisional dalam Pembangunan Pertanian* dalam Wahono, Widyanta dan Kusumajati (Penyunting) “Pangan, Kearifan Lokal Dan Keanekaragaman Hayati : Pertaruhan Bangsa Yang Terlupakan” Cindelaras Pustaka Rakyat Cerdas Yogyakarta

[3] Mangunjaya F 2006 *Hidup Harmonis dengan Alam : Esai-Esai Pembangunan Lingkungan Konservasi dan Keanekaragaman hayati Indonesia* (Jakarta: Yayasan Obor Indonesia)

[4] Warren D M Slikkerveer L J and Brokensha D 1995 Introduction In D M Warren, L J Slikkerveer, and D Brokensha (Eds) *The Cultural Dimension of Development : Indigenous Knowledge Systems* (London: Intermediate Technology Publication)

[5] Kodoatie dkk 2002 *Pengelolaan Sumber Daya Air dalam Otonomi Daerah* (Yogyakarta :ANDI)

[6] Bogdan dan Biklen 1982 *Qualitative Research for Education* (United States of America: Mc Graw-Hill, Inc)

[7] Yin R K *Case Study Research: Design and Methods Beverly Hills* (California: Sage Publication)

[8] Miles, Matthew B Huberman and A Michael 1992 *Analisis Data Kualitatif* (Terj Tjetjep Rohendi Rohidi) (Jakarta: Universitas Indonesia)

IOP Conf. Series: Earth and Environmental Science **142** (2018) 012072 doi:10.1088/1755-1315/142/1/012072

The land use potential of flood-prone rice fields using floating rice system in Bojonegoro regency in East Java

H Irianto[1], Mujiyo[2], E W Riptanti[1] and A Qonita[1]
[1] Department of Agribusiness, the Faculty of Agriculture, Universitas Sebelas Maret.
[2] Department of Soil Science, the Faculty of Agriculture, Universitas Sebelas Maret.

Email: erlynawida@staff.uns.ac.id

Abstract : Bojonegoro regency occupies the largest flood-prone rice fields of about 14,198 hectares, in East Java province. Floods commonly occur due to Bengawan Solo river over-burst, particularly in rainy season. The fields are potential for cultivating rice, but floods lasting for months causing these areas to be unproductive. The objective of this article is to examine the potential land use of flood prone rice fields in Bojonegoro regency using floating rice system as an effort to maintain productivity in rainy season. The method of this study is referential study about the rice production using floating cultivation system in other regions, which are later compared with the physical condition of the fields in Bojonegoro. The results of analysis show that rice cultivation using floating system can maintain rice production in flood prone areas during rainy season. The potential production of rice is 5-6 tons/ha. However, technical problems for cultivating rice cannot be ignored since farmers are not familiar with cultivating flooded fields. This article also explains alternatives of floating rice cultivation technique, which can be implemented effectively and efficiently. Pioneer work of developing floating rice in Bojonegoro that has been done by the Team of Faculty of Agriculture of UNS, Surakarta, is expected to serve as a medium for accelerating the adoption of cultivation technology innovation to farmers.

1. Introduction
Food buffer stock, particularly rice, is the main priority in national development. Besides as main staple food for more than 95% of Indonesian people, rice also provides job opportunities for approximately 20 million of farmer households in the village. According to Indonesian Central Bureau of Statistics or [1], in the last five years, rice production in Indonesia increase 3.61% each year. However, rice production rate is not able to meet rice consumption, which reaches 98.63 kg per capita in a year.

Diversity of rice production in Indonesia is influenced by the climate. Extended harvest failure is often attributable to extreme climate, like flood and drought. Results of some studies reveal that failure in harvest resulted from extreme climate, taking forms of flood and drought were tend to increase [2]. In addition, flood potential level also increases in low land areas because of environmental damage in rainwater catchment areas that increases river water discharge in the lower course and the increase of seawater surface [3].

For farmer households living in flood prone areas, flood will cause a loss of agricultural production. In order to cope with that problem, technology of floating rice cultivation is developed. This is a rice cultivation technique that uses rafts as cultivating media as an effort of adapt with flood.

Therefore, the aim of this article is to examine the potential land use of flood prone rice fields in Bojonegoro regency using rice-floating system as an effort to maintain productivity in rainy season.

2. Methods

The basic method of this study is exploratory research, which is a research aiming at seeking for explanation of observed phenomena, problems, or behaviors on potential land use of flood prone rice fields in Bojonegoro using floating rice system [4]. The method of this study is referential study of products of rice cultivation using floating system in other regions, which are later compared with the physical condition of the fields in Bojonegoro. The research location, Bojonegoro regency, was purposively selected by considering that Bojonegoro occupies the largest flood-prone rice fields, 14,198 hectares, in East Java province.

3. Results and Discussion

3.1. The Potential for Developing Floating Rice in Bojonegoro Regency

Bojonegoro is a district implementing regional autonomy whose one of the development missions is "empowering society and optimizing regional potential". Bojonegoro regency occupies human and natural resources to develop agricultural sector. The resources face some threats from external factors of disasters, like flood. This condition happens nearly every year [5] and [6]. Bojonegoro has the largest flood-prone rice fields, 14,198 hectares, in East Java province [7], parts of flood-prone areas in Bojonegoro, reaching 65,989.13 hectares (Figure 1). Flood is caused by discharge of Bengawan

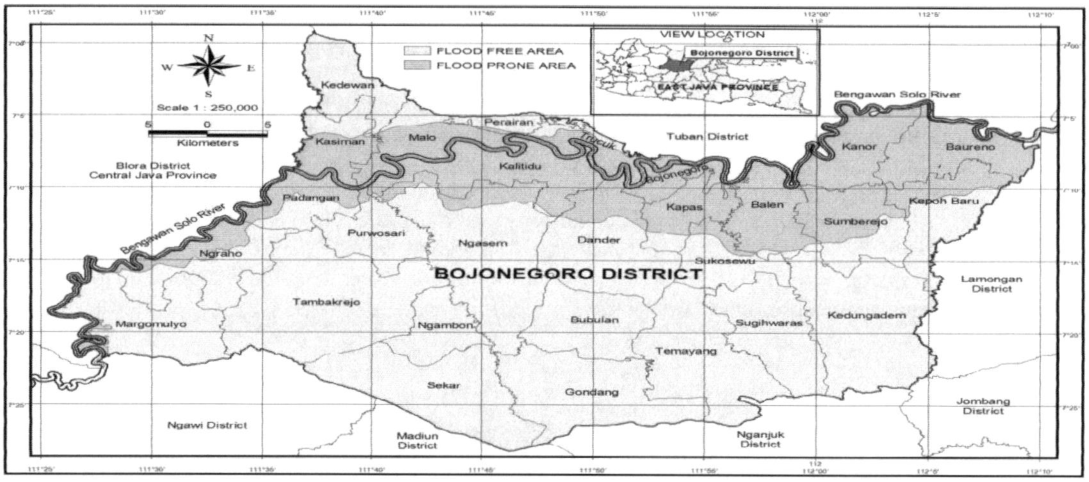

Figure 1. Map of Flood-Prone Areas in Bojonegoro Regency

Figure 2. Flooded Agricultural Fields in Sedayu Village, Baureno Sub-district, Bojonegoro Regency

One of possible efforts applied to unproductive rice fields in rainy season due to flood is adopting technology of floating rice cultivation. Floating rice system enables farmers to keep nurturing rice even though flood happens. On the other hands, farmers face limitations that their roles as either humans, farming expert, or manager, require a process of accepting or refusing new technology. Farmers' decision to decline or agree to take an innovation of technology is attributable to their social and economic factors [10]. To ensure that innovation resulted from research by certain institution or research institution can be delivered to target group or farmers needs a process of transferring agricultural information, while the speed of adopting agricultural innovation is influenced by internal and external factors [11]. Thus, it is important to examine more deeply and comprehensively the potential land use of flood prone rice fields using floating rice system, so that farmers can implement technology for cultivating floating rice and be able to improve food security at household and regional levels.

3.2. Alternative of Effective and Efficient Floating Rice Cultivating Technique

Floating rice is a type of plants that can be raised in some regions in the South and Southeast Asia. The viability of rice depends on their ability to cope with water when they are submerged and the duration flooding. Floating rice cultivation technique can also increase production [12]. It is a rice growing technique using the media of rafts to adapt with flood. Rafts function to support rice so that rice will not fall whenever wind blows and will not sink whenever flood happens. Rafts are made of bamboo to make them floating, the central part of it is made of bamboo cut in two and arranged like a fence, which later is filled with straw waste and coconut fiber mixed with organic compost, and the upper part is covered with net. Raft cultivating media can be used up to six growing seasons (3years) [3].

Floating rice cultivation uses System Rice Intensification (SRI) method, which is a method to improve rice productivity with integrated natural resources use and management (plant, soil, water, biota, and nutrition) to improve organic rice productivity [13]. Plants are moved to cultivating media (rafts) after rice seeds have been sown for 10 days and fertilizer is spread to stems and leaves to work more effectively. The organic fertilizers used include is Complementary Liquid Fertilizer and MOL (Micro Organism Local) [3].

The areas for cultivating floating rice are flooded fields. Planting media of rafts are prepared on land. Rafts are made of three stalks of bamboo with minimum diameter of 7 cm and length of 5 m. The planting media include straw waste, coconut fiber waste, and organic compost. Nets are used to keep planting media not scattered during planting process. Floating rice fields in Bojonegoro apply SRI (System of Rice Intensification) method with single rice seeds previously sown in 14 days (14 days after sowing).

Fertilizing floating rice fields is done by observing the development and the age of the plants. During the first cycle or vegetative growth (0-60 days), rice requires fertilizer containing 5% of nitrogen (N) and phosphorus (P). Transition from vegetative to generative growth (55-70 days) of rice requires fertilizers that contain phosphorus (P), and during the last cycle, generative growth (last 2 months), rice requires fertilizers containing potassium (K). Organic liquid fertilizers are obligatory in floating rice fields. Fertilizers can be effectively sprayed directly to stems and leaves.

In general, the steps for cultivating floating rice fields are as follows: (1) ensure that the water height of fields for floating rice media exceeds 20 cm during one planting season, at minimum; (2) prepare organic compost media mixed with soil with ratio of 2:1, processed with decomposer; (3) a. prepare bamboo with minimum diameter of 7 cm and cut 5 m (3 stalks), b. prepare bamboo and cut, 2 meter in length each (5-6 stalks), and then cut into small pieces until they shape like mats, and match the bamboo, as well as tie them at wider parts, (4) bring rafts to flooded areas in which floating rice fields will be located, (5) separated with SRI pattern with seedling period of 10-14 days after sowing, the seeds will be pulled out and moved to floating rice fields, (6) bamboo mats are attached on rafts, in which the softer parts face up, (7) spread organic planting media with 3-5 cm thickness, (8) before the floating rice fields are ready, create lines to help manage planting distance (25 cm x 25 cm), (9) plant rice that have been sowed with L pattern (between root and stem), 0.5-1 cm width, and one seed in one hole, and then prepare 20-30 seeds and plant them at the edge of rafts for refilling, (10) prepare organic liquid composts containing elements required as the media of plant growth regulator (PGR) and complementary liquid fertilizer, (11) spray plant growth regulator (PGR) and complementary liquid fertilizer 3-4 times in a season, (12) observe the growth of pests and control with organic pattern, including refilling when needed, (13) after rice is ripening, harvest them and prepare sacks and boats to bring the crops to land, and then thresh and dry the rice, just like treatments when cultivating conventional rice fields [14].

This floating rice-based agro-ecology system can exploit flood for the life of villagers. It is more important that this system enables farmers to adapt well with flood and they will not encounter scarcity because floating rice can grow well in flood [15]. This proves that floating rice is potential to be developed, particularly on the fields facing annual flooding. Floating rice cultivation technique has both drawbacks and strengths. The strengths of this technique include: (1) it does not require watering and irrigation canals, (2) it does not need tractors to cultivate the soil, (3) it does not require chemical fertilizers and organic pesticides, (4) it does not need certain treatments to clean grass or weed, (5) it reduces straw waste and coconut fiber, (6) it optimizes the use of abandoned/unproductive fields due to prolonged flood (one growing season), and (7) it is free from drought in rainy season, particularly for areas vulnerable to annual flood. Meanwhile, problems in applying floating rice cultivation are among others: (1) technical problems for cultivating rice cannot be ignored since farmers are not familiar with cultivating water-logged fields and (2) making rafts requires a large sum of money, especially in the beginning of planting.

Floating seedling method offers some benefits: (1) preventing plants to be soaked, (2) optimizing growing plants in rainy season so that the following planting period will not be late, (3) saving fields, which are usually cultivated by farmers, and (4) facilitating seed treatment during fertilization and pests and weeds (organism) control. Floating media can be repeatedly applied for the next planting periods, because the sources of water plants do not go up and are sustainable [16]. If all flooded areas are optimized with floating rice, it is possible to improve rice productivity and economic condition of lower-middle class people, farmers can adapt with environmental and climate changes. In addition to floating rice, flooded fields can also be combined with fish cattle [17].

The followings are the results of research conducted in various countries and other regions, including Vietnam, Myanmar, India and Indonesia. In general, the yield of floating rice varies depending on the land elevation. In deeper water, yield is slightly higher. Average yield is about 1 ton/ha. Of all the existing floating rice varieties, the most popular ones are Nang Tay C, Tau Binh, Nang Dum, Nang Kieu, Tau Nut, Trung Hung [17]. In Myanmar, it is reported that the total net benefit of floating rice-leeks shows the highest net benefit (VND 24.8 mil./1000 m2), followed by floating rice-sweet corn- two baby corn crops and cattle systems (VND 18.5 mil./1000m2), and floating rice-chili (VND 16.7-17.7 mil./1000m2). If farmers cultivate monoculture of rice either two or three crops, the net benefits are ranging from VND 2.2-4.8 mil./1000m2 respectively. Alternatively, farmers integrate vegetables; the net benefit was ranging from VND 5.6 to 11.7 mil./1000m2. The net benefit of the two rice crops is the lowest (VND 1.3-2.3 mil./1000m2). Surprisingly, if farmers convert from two intensive rice crops to two maize crops, the net benefit can reach to VND 21 million/1000m2. The results support the argument that the floating rice-based systems allows farmers diversified into profitable upland crops, which can help farmers to improve their total income [18]. The results of other research show that the red floating rice varieties as a low-volume but high-value crop is the development of markets. Floating rice has high protein, vitamin E and anthocyanin compared with high-yield white rice, but has a different texture and flavour [19]. In India, the results of research by [20] reported that the production of rice with floating system ranges from 1,628 to 3,000 kg/ha. Meanwhile, members of Taruna Mekar Baru farmer group in Ciganjeng village, Ciamis Regency, West Java, Indonesia successfully harvested rice in flooded fields. IR 64 rice is planted on water of 1 meter in depth, resulting 6.2 tons of rice per hectare [21].

Figure 3. Experiment of Floating Rice Cultivation in Bojonegoro Regency by Team of Faculty of Agriculture of UNS Surakarta

Pioneer work of floating rice development in Bojonegoro that was conducted by the Team of Faculty of Agriculture of UNS, Surakarta [22]. The development was initiated by conducting experimental study of floating rice cultivation using growing media and floating materials obtained from Sedayu Village, Baureno Sub-district, Bojonegoro Regency (Figure 3). The design of experiment was Randomized Complete Block Design (RCBD) with 18 types of treatments, each of which is repeated to three blocks. Each experiment unit was 10 m^2 (2 m x 5 m) in width. The types of treatments were combinations of floating material uses (drum and *petung* bamboo), mixtures of planting materials (soil : organic fertilizer = 1 : 1, soil : chaff ash = 1 : 1 and soil : organic fertilizer : chaff ash = 1 : 1 : 1), and planting distance (15 cm x 15 cm, 20 cm x 20 cm and 25 cm x 25 cm). The variety of planted rice was IR-64 because farmers are familiar enough to plant this variety and it has high productivity and short harvest time. Rice grew quite well in 25 days after planting, despite of the fact that the effect of treatments of the growth and development of plants cannot be reported yet in this paper.

4. Conclusion

Floating rice cultivation is a technique of nurturing rice with the media of rafts to adapt with flood. This is feasible to be further developed on the fields flooded in years because this can maintain rice productivity during rainy season, with potential production of 5-6 tons/ha. However, technical problems for cultivating rice cannot be ignored since farmers are not familiar with cultivating flooded fields. Thus, based on the research results, farmers require development and counseling on floating rice cultivation in order to improve their skills and to yield optimum crops. Pioneer work of developing floating rice in Bojonegoro done by the Team of Faculty of Agriculture of UNS, Surakarta, is expected to serve as a medium for accelerating the adoption of cultivation technology innovation to farmers.

Acknowledgements

We would like to express our gratitude to Universitas Sebelas Maret for funding this research skim Penelitian Unggulan UNS in 2017. Our sincere appreciation is extended to all parties who have provided information to complete this article.

References

[1] Badan Pusat Statitik 2016 *Statistik Indonesia* 2016 (Jakarta: Badan Pusat Statistik)

[2] Boer R and Suharnoto Y 2012 Climate change and its impact on Indonesia's food crop sector, The Sixth Forum on Natural Resource Management: Water and Food in a Changing Environtment on 11-13 April 2012 at SEARCA headquarters, Los Banos, Philippines,

[3] Adinata K 2012 *Petunjuk teknis padi apung* (Cilacap: IPPHTI)

[4] Nueuman and Lawrence W 1994 Social research methods: *Qualitative and quantitative approach*. Allyn and Bacon

[5] Agustono 2011 Peran sektor pertanian dalam pertumbuhan dan stabilitas produk domestik regional bruto di Kabupaten Bojonegoro *J SEPA* **8** (7) 14 – 21
[6] Rositha A and Taufik M 2011 Studi analisa banjir dengan menggunakan teknologi sig di Kabupaten Bojonegoro *National Seminar VII* Civil Engineering of ITS Surabaya
[7] Aziz M L, 2012 Undergraduate Thesis Universitas Negeri Yogyakarta, Indonesia
[8] Suyoto, 2016 Membangun kemandirian pangan dari Bojonegoro untuk Indonesia, Sidang Regional Dewan Ketahanan Pangan Wilayah Tengah Tahun 2016 (Se Jawa dan Kalimantan) Palembang,
[9] Riptanti E W, Supardi S and Qonita A, Sustainable food security on farmer households in flood prone areas, International *J of Applied Environmental Sciences* **9** (6) 1435-1450
[10] EM Rogers & FFShoemaker 1971 Communication of innovation, New York: Free Press,
[11] A Sukmawati, (2011)Tingkat adopsi dan difusi inovasi beternak ayam broiler di Kota Kendari, **J of Agrisep 10** (1)
[12] RKUpadhyay, 2016 How rice (Oryza sativa L), a semi-aquatic plant adapt to natural flood or submerged condition? A physiological perspective **J Sains Malaysiana 44** (6) 879-882
[13] Berkelaar D, 2001 The system of rice intensification: Less can be more Bulletin ECHO Development Notes, Issue 70
[14] Saputra D W, 2015 Analisis pendapatan kelayakan usahatani sawah apung di Desa Ciganjeng Kecamatan Padaherang Pangandaran Jawa Barat, The Faculty of Economics and Management of Institut Pertanian Bogor, Bogor,
[15] Kien N V, 2016 The values and recovery progress of floating rice-based agro-ecological systems for adaptation to climate change in the Vietnamese Mekong Delta *J Earth Sci Clim Change* **7** 9 (Suppl)
[16] Sulaiman F E dan Sodikin 2016 *Persemaian padi terapung di areal sawah Lebak Sumsel*, Departement of Agronomy, Faculty of Agriculture, University of Sriwijaya, Palembang
[17] Vo-Tong Xuan, 1975 Rice cultivation in the Mekong Delta, *J South East Asian Studies* **8** (1) 88-111
[17] Nugroho R A 2013 *Adaptasi iklim: Petani Ciamis kembngkan padi terapung* Kompas, Jakarta,
[18] Nguyen V, Vo O and Huyuh D, 2015 Comparing the cost and benefits of floating rice-based and intensive rice-based farming system in the Mekong Delta *J Asian Economic and Social Society* **5** (9) 202-207
[19] Pittock J and K Nguyen, 2017 *The role of traditional floating rice system (*Australian: Australian National University)
[20] Prasanta N, Sarma P K, Ravindra, Dutta S, Rajbongshi R, Baruah S, Sarma D, Sarma M K, Borah P, Srinivas Ch and Hazarika G N 2016 Building climate resilient agriculture throught tradisional floating rice in flash flood affected areas of the North Bank Plains Zone of Assam *India J of Traditional Knowledge* **15** (4) 632-638
[21] Djuni 2013 *Adaptasi iklim: Petani Ciamis kembangkan padi terapung*, Kompas dated 15 March
[22] H Irianto, Mujiyo, EW Riptanti and A Qonita, 2017 *Adopsi teknologi budidaya padi apung sebagai strategi menjaga ketahanan pangan di daerah rawan banjir Kabupaten Bojonegoro* Progress Report of Research, Hibah PNBP UNS TA 2017, Universitas Sebelas Maret, Surakarta

IOP Conf. Series: Earth and Environmental Science **142** (2018) 012073 doi:10.1088/1755-1315/142/1/012073

The fficiency of Mycorrhiza biofertilizer treatment to the growth and yield of soybean

Samanhudi[1], B Pujiasmanto [1], Sudadi [2], I H Putra [3] and H M Mumtazah [3]
[1]Study Program of Agrotechnology, Faculty of Agriculture, Universitas Sebelas Maret (UNS) Surakarta, Indonesia
[2]Study Program of Soil Science, Faculty of Agriculture, Universitas Sebelas Maret (UNS) Surakarta, Indonesia
[3]Undergraduate Student of Faculty of Agriculture, Universitas Sebelas Maret (UNS) Surakarta, Indonesia

E-mail: samanhudiuns@gmail.com

Abstract. Soybean is one of the major commodities in Indonesia. Due to its high demand, its requires an effort to increase the production. Soybeans are generally cultivated in dry land, for that its need a special management to increase the yield. The association between Mycorrhiza and roots help the plant to get water and nutrients. In this regard Mycorrhiza expected to increase soybean yield and efficiency. This research aim is to study the dose of Mycorrhiza on the growth and yield of soybean efficiently. The experiment was conducted in Selogiri District, Wonogiri, while the analysis of Mycorrhiza and soil was inFaculty of Agriculture, Universitas Sebelas Maret Surakarta from February to April 2016. Randomized Complete Block Design (RCBD) with two factors was emplyed for this experiment. The treatments are compost dose (derived from Waste Management Faculty of Agriculture UNS) and Mycorrhizal dose (obtained from BPPT Serpong). The result showed that the Mycorrhiza treatmentwas able to improved the growth and yield of soybean. The most efficient dose of is Mycorrhiza treatment at 0.64 ton ha^{-1}.

1. Introduction

Soybean is one of important commodities in Indonesia. Soybean production was increased by 4.59% from 2014 to 2015 (in total 998.87 thousand tons) however it does not the national market demand [1]. For that, Indonesian government set an effort in form of strategies by land and science intensification to improve the soybean production in attempt to meet the demand. One of strategy is to use Mycorrhiza as biofertilizer.

Soybeans are generally cultivated in dry land, therefore requires special management to increase its production. Mycorrhiza is a symbiotic association between fungus and plant roots which form a complex interaction. Mycorrhiza associations help the plan to obtain water and nutrients in dry conditions and poor nutrients [2]. *Arbuscular Mycorrhiza* can increase growth and yield of soybean. To improve Soybean yields its need the right application of Mycorrhiza dose, in order to get the fertilizer efficiency. Right amount of fertilizer would be give a better result of Soybean growth and production [3], for that this things need to be studied

IOP Conf. Series: Earth and Environmental Science **142** (2018) 012073 doi:10.1088/1755-1315/142/1/012073

2. Methods

2.1. Location and Time Research
Field trials conducted in February 2016 until April 2016 in the BPP of Selogiri District, Wonogiri. Analysis of the results carried out in the Laboratory of Ecology and Management of Plant Production, analysis of mycorrhiza took place in the Laboratory of Biological Soil and soil analysis took place in the Laboratory of Chemistry and Soil Fertility in Faculty of Agriculture, Universitas Sebelas Maret.

2.2. Tools and Materials
The tools used in this study include field analysis equipment are hoes, ruler, cameras, stationery, logbook and sprinkler hose; equipment for laboratory analysis include an analytical balance, oven, scissors, waterbath, etc. Materials used in this study include soil samples and chemical for laboratory analysis covering 70% alcohol, distilled water, 10% KOH, 1 N HCl, 0.05 % trypanblue, 1 N ammonium acetate, 10% NaCl, 45% NaOH, 0.1N HCl, 2% boric acid and other chemical materials.

2.3. Research Design and Data Analysis
This research used a Complete Block Randomized Design (RAKL) with factorial pattern arranged in a single dose treatment factor that is mycorrhiza inoculants (BPPT Serpong production). Mycorrhiza dosage is 0 ton ha^{-1}, 0.64 ton ha^{-1} and 1.28 ton ha^{-1}.

3.Results and Discussion

3.1.General Conditions Research Sites
This research was conducted in Kaliancar village in Selogiri District of Wonogiri which has 106 meters above sea level altitude. Land which used for soybean cultivation is the BPP Selogiri office yard with large of the yard is 312.625 m^2. Environmental conditions (temperature) the annual average in 2015 around the area of cultivation range is 26.5°C. The desired temperature of soybean crop is between 21-34°C, but the optimum temperature for growth of the soybean crop is 23-27°C.

Table 1. Some Vertisol soil chemical properties were used in the experiments

No	Analyzed properties	Analysis	Results	Explanation
1	pH	Elektrometrys	6,58	Neutral
2	Total N	Kjeldhal	0,19%	Low
3	P provided	Brays	5,78 ppm	Low
4	K provided	Flame	0,25 me%	Low
5	Organic materials	Walkley& Black	0,53%	Very Low

Laboratory analysis of vertisol soil in BPP Selogiri shows the pH value is 6.58 which is classified in the category of neutral. Based on international standards, criteria total N of vertisol soil has a low percentage is 0.19% [4]. The loss of N from the soil for used by plants or microorganisms and N in nitrate form easily washed away by rainwater. The result of the laboratory analysis of available P is 5.78 ppm and K provided is 0.25 me%. [5] results of P provided in a preliminary analysis of the land included low criteria when value shows<6.5 ppm.

3.2. Variable of Plant Growth

3.2.1. Plant height Plant height is used as growth parameter of a plant in vegetative phase. Obstacles that arise during the vegetative phase are the caterpillars eat leaves until plant shoots. Prevention is performed by spraying a systemic pesticide with active ingredient in form fipronil to the entire body of the soybean crop at a dose of 1 ml/ l.

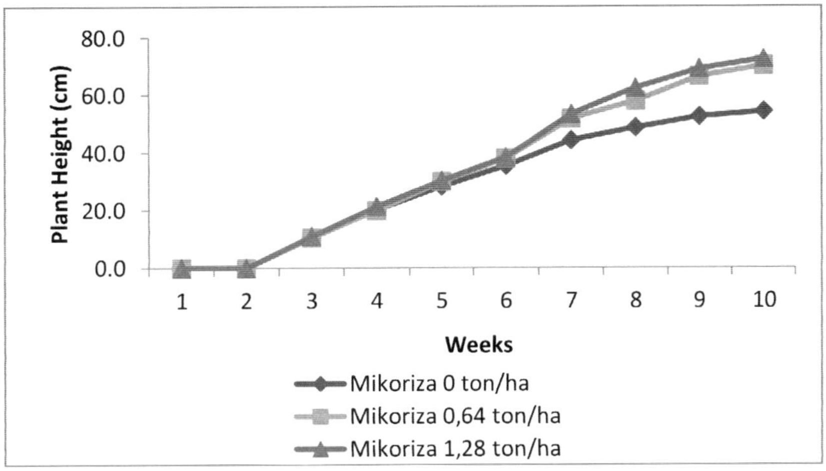

Figure1. Effect of arbuscular mycorrhiza on plant height of soybeans in vertisol soil

Figure 1 shows the growth of plants from first week to 10 weeks after planting (MST). Week 6 shows reduction in the rate of growth because it started to get generative period. The generative period indicated by the emergence of interest. Result of variance analysis showed that dose of mycorrhiza increases soybean height.

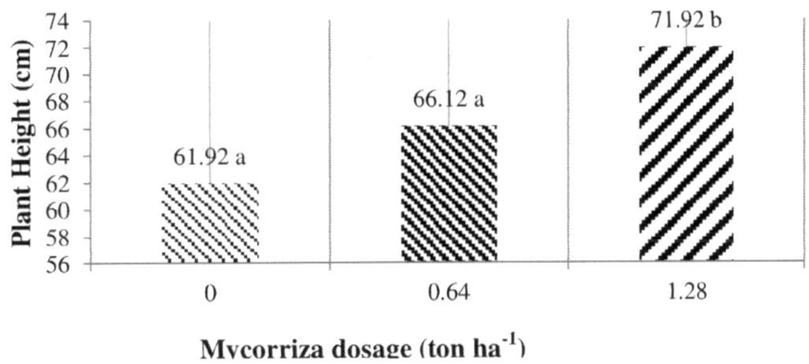

Figure 2. Effect of arbuscular mycorrhiza dose on soybeans height in vertisol soil

Effect of doses of mycorrhiza on soybean height in figure 2 shows the highest plant height due to the effect of mycorrhiza 1.28 ton ha^{-1}. Dose of mycorrhiza 0 ton/ha with an average plant height of 61.92 cm has the same effect with dose of mycorrhiza 0.64 tonnes/ha with average height of 66.12 cm plants. [6] that mycorrhiza can help absorption of water, so stem lenght and number of leaves can grow optimally and increase in line with the addition of mycorrhiza dose.

3.2.2 Productive Branch Dose of mycorrhiza increase the number of productive branches. The more productive branches, the amount of fruit will increase [7]. inoculation FMA is an effort that can be done to improve availability, uptake and help phosphor transfer [8]. Phosphorus has an important role in the body such as plant cell division, formation of flowers, fruits and seeds, and stimulating root development.

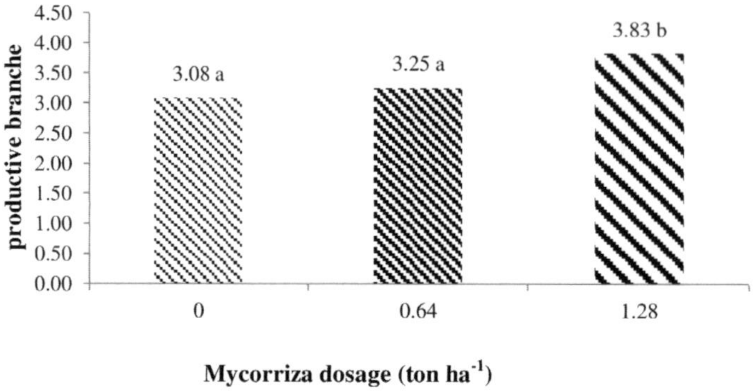

Figure 3. Effect of arbuscular mycorrhiza dose on the number of soybeans productive branches in vertisol soil

Referred to Figure 3, the dose of mycorrhiza 0.64 ton ha^{-1} generates an average of 3.25 productive branches, whereas a dose of 0 ton/ha produce 3.08 but both have the same effect on the productive branches. Arbuscular mycorrhiza dose of 1.28 ton ha^{-1} generates average of 3.83 productive branches. Dosage of 1.28 tonnes/ha generates more productive branches and significantly different compared to the dose of mycorrhiza 0 ton ha^{-1} and 0.64 ton ha^{-1}.

3.2.3 The Filling Pods per Plant Dose of 0 ton ha^{-1} generates an average number of pods 43.72; dose of 0.64 ton ha^{-1} generates average of pods 49.30 and mycorrhiza dose of 1.28 ton ha^{-1} generates average number of pods 54.38. Number of filling pods increased with increasing doses of mycorrhiza but no difference in the effect of three doses. [9] explains that the number of fillinf pods per plant was not affected by fertilizer, but influenced by varieties or genetic factors.

[10] the large number of filling pods depending on the number of pods pithy and the number of empty pods. [11] adds that the genetic properties of soybean plants greater role in determining the age of flowering. The sooner plant entered the flowering phase can add a variety of opportunities to form more pods.

3.2.4. Weight of 100 Seeds Seeds weight carried by weighing seeds in random and repeated three times. Giving dose arbuscular mycorrhiza of 0 ton ha^{-1} generates 17.14 g weight and increased in a dose of 0.64 tonnes/ha that is 17.36 g. Dose mycorrhiza of 1.28 ton ha^{-1} also showed an increase to 17.89 g. The increase seed weight has the same effect with increasing doses of mycorrhiza, because of variations in seed weight depends on a variety of genetic factors [12]. According to Conservation et al. (2013) each variety has a genetic advantage that different so that each variety has different seed production also vary. In addition, seed development is also a factor formation of seeds.

3.2.5. Dry Stover Weight Dry plant weight reflects the growth of plants and the amount of nutrients absorbed. The increase of dry stover weight with the addition of mycorrhiza show that increasing doses of mycorrhiza, the nutrients are absorbed more and more. In accordance [13] highest dry weight at FMA administration which showed that plant growth is affected by the FMA.

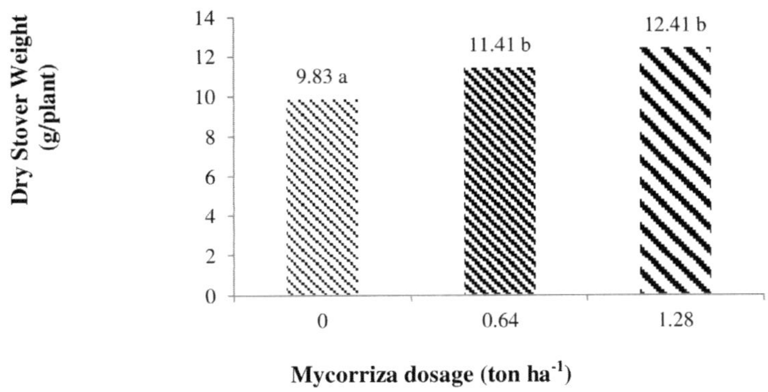

Figure 4. Effect of arbuscular mycorrhiza dose on the weight of soybeans dry stoverin vertisol soil

Figure 4 shows the dose mycorrhiza of 0 ton ha^{-1} significantly different from the 0.64 ton ha^{-1} and 1.28 ton ha^{-1}. Dose mycorrhiza of 0.64 ton ha^{-1} generates 11.41 grams weight and increase after the dose was increased to 1.28 ton ha^{-1} that is 12.41 g weight. FMA symbiotically to plant roots can increase water uptake and phosphorus that can be utilized to leaves and stems to grow [13].

3.2.6. Dry Pods Weight per Hectare Dry pod is soybean pods are dried until the color is yellow and hard. Giving mycorrhiza significantly affect to dry pods of soybeans weight gain. Add that the use of arbuscular mycorrhiza inoculation on soybean plants in addition to increasing germ vigor, also increase grain yield and protein content.

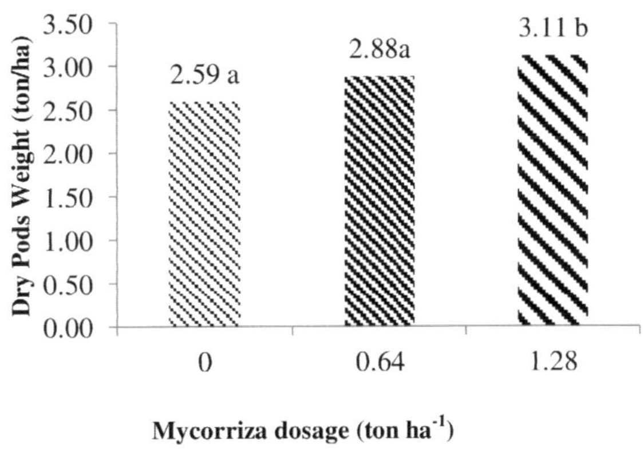

Figure 5. Effect of arbuscular mycorrhiza dose on the weight of soybeans dry pods per hectare in vertisol soil

Each additional of mycorrhiza dose has effect to each of the dry pod yield. Figure 5 shows that each individual dose of mycorrhiza 0 ton ha^{-1}, 0.64 ton ha^{-1} and 1.28 ton ha^{-1} have a different effect on the weight of dry pod and its weight increased from 2.59 tons; 2.88 ton and 3.11 ton. Dose of mycorrhiza 1.28 ton/ha had the highest influence on the weight of dry pods of soybean of 3.11 ton.

3.2.7 The percentage of infections Mycorrhiza Mycorrhiza used is an arbuscular mycorrhiza fungi (AMF) with zeolite medium. Based on laboratory results, each 10 gram equals 25 mycorrhiza spores. [15] mycorrhiza inoculation of 7.5g or 30 spores per plant showed mycorrhiza inoculation in sterile soil gives the best results in the increase of crop growth. Based on the results of laboratory analysis of mycorrhiza infection present in all dose mycorrhiza. Corresponding calculation results showed that the highest percentage of mycorrhiza is 83 %.

3.2.8. Fertilizer Use Efficiency The different doses of fertilizer gain different influence on the growth and yield. According to [16] use of fertilizer too little can lead to a lack of nutrients in the soil and if too much is also a negative impact. The effect of a dose of arbuscular mycorrhiza on fertilizer use efficiency did not show improvement.

Dosage of mycorrhiza 0.64 tonnes/ha is 6.14 %. The next increase mycorrhiza dose is 1.28 tonnes/ha showed a lower percentage that is 5.60 %. The percentage decrease shows that increasing the percentage of mycorrhiza did not increase efficiency. Based on these data showed the most efficient outcome is dose of mycorrhiza 0.64 ton ha^{-1}.

4. Conclusions

From this study it is concluded that dose of *Arbuscular Mycorrhiza* is able to increases the growth and yield of soybean, however, the Mycorrhiza dose does not improve the fertilizer efficiency. The most efficient dose of Mycorrhiza according to the growth and yield of soybean was 0.64 ton ha^{-1}.

Acknowledgement

The authors are grateful to the Ministry of Research, Technology,and Higher Education of the Republic of Indonesiathat has funded this research through the scheme of Masterplan Percepatan dan Perluasan Pembangunan Ekonomi Indonesia(MP3EI) for fiscal year 2015.

References

[1] BPS [Badan Pusat Statistik] 2011 Production of rice, corn and soybeans (provisional figures of 2010 and forecast figures I 2011) Jakarta: Press Release 14: 1-9-

[2] Hapsoh 2003 Compatibility MVA and some soybean genotypes at different levels of soil drought stress ultisol: morfofisiologis responsive and results (dissertation) Bogor: Agricultural University graduate

[3] Moelyohadi Y, Harun MU, Munandar, Hayati R and Gofar N 2013 Pengaruh kombinasi pupuk organik dan hayati terhadap pertumbuhan dan produksi galur jagung (*Zea mays.* L) hasil seleksi efisien hara pada lahan kering marginal *J Lahan Suboptimal* **2** (2): 100-110. [4] Balittanah 2005 Instructions analysis of soil, water, fertilizer, and crop. Research and Development of Soil and Agro-climate. Bogor

[5] Helmi 2013 Assessment and preparation of P-available phosphorus fertilizer recommendations on a rice field in the district of Pidie district Way *J Sains Riset* **3** (1)

[6] Nurhayati 2010 Effect of timing of vesicular arbuscular mycorrhizal tomato growth *J Agrivigor* **9**(3): 280-284

[7] Surtinah 2007 Studies on the vegetative growth of the relationship with the production of tomato plants (Lycopersicum esculentum, Mill) *A Scientific Pert* **4**(1)

[8] Nasr AH, Zare M, Alizadeh O and Naderi NM. 2013 Improving effects of mycorrhiza symbiosis on sorghum bicolor under four levels of drought stress *J African of Agric Res* **8**(43)

[9] Efendi 2010 Increased growth and soybean production through a combination of organic fertilizers with manure lamtoro gung *A Floratek* **5**: 65 -73

[10] Krisnawati A and Adie MM 2015 F5 populations soy based agronomic characters. *Proc of the National Seminar on Biodiversity Indonesia Society* **1**(3): 434-437. DOI: 10.13057/psnmbi/m010309

[11] Jusniarti, Marni Y and Haryoko W 2014 Growth and yield varieties of soybean (Glycine max L.) under the auspices of the peat *J Embrio* **7** (2) (90-102). ISSN No.2085-403X

[12] Hayati E, Mahmud T and Fazil R 2012. Effects of types and varieties of organic fertilizers on growth and yield of pepper plants (Capsicum annuum L.) *A Floratek* **7**: 173-181 [13] Rengganis RD, Hasanah Y dan Rahmawati N. 2014 The role of arbuscular mycorrhizal fungi and phosphate rock fertilizer on the growth and production of soybean (*Glycine max* (L.) Merrill). *J Online Agrotek* **2**(3) : 1087- 1093. ISSN: 2337- 6597

[14] Jannah H 2011 Response of soybean to associations arbuscular mycorrhizal fungi on dry land *GaneÇSwara* **5**(2)
[15] Syam'un 2006 Production of potato (*Solanum tuberosum* L.) varieties of granola that applied organic fertilizer vermicompost and arbuscular mycorrhizal inoculation. *Bulletin of the Faculty of Agriculture Forestry Research* Hasanuddin University.
[16] Xiang Y, Jiyun, J, Ping HE and Ming-zao L 2008 Recent advance on the technologies to increase fertilizer use efficiency *J AgricSci China* **7**(4)

The influence of ihsan attitude and economic condition to farmer household food security

H Hendrarini[1], E S Rahayu[1], Kusnandar[1], R J Sunarsono[2] and T Soedarto[3]
[1]Departement of Agriculture, Faculty of Agriculture, Sebelas Maret University, Jl. Ir. Sutami Number 36A Jebres, Surakarta city, 57126, Indonesia
[2]Departement of Management, Faculty of Economics and Business, University of Pembangunan Nasional Veteran East Java, Jl. Raya Rungkut Madya, Gunung Anyar, Surabaya city, 60294, Indonesia
[3]Departement of Agriculture, Faculty of Agriculture, University of Pembangunan Nasional Veteran East Java, Jl. Raya Rungkut Madya, Gunung Anyar, Surabaya city, 60294, Indonesia

Email: hamidahrini1@gmail.com

Abstract. Ihsan attitude is one unique attitude that was owned by farmers in Bangkalan Madura, Indonesia. This study was focused to examining the relationship of ihsan attitude and economic condition on household food security of farmers in Bangkalan. The ihsan attitude will be one a new view regarding to the human resource perspective in agribusiness. For that this study was aimed to analyse the influence and the effect of that attitude to food security of farmer household. Data were collected from 360 respondents, followed with descriptive analysis. The social ecological model was the basis concept in this study, moreover its also used to test the three variables. In this study, a structural equation model tools call WarpPLS was employed. Results from this study would be provide ihsan attitude picture to economic conditions on household food security of farmers indicated, that would be important for formulation of the farmer empowerment policy in the future.

1. Introduction

One element that becomes a prosperity variable in a country is food security. Food security becomes a very crucial variable in almost countries in many parts of the world, including Indonesia. In the 1970s food security began to be an international issue [1]. In the early stages, the focus on of food security concept lies in the availability of food at the national and international levels. The concept was based on the world food crisis from 1972 to 1974. The concept of food security has been discussed in depth by Food and Agriculture Organization, food security as the starting point of world food situation evaluation, based on the thoughts that emerged at the UN world food conference in 1974 [2].

Food security for a country is very crucial, especially for countries with very large populations such as Indonesia, where by 2020 the Indonesian population is expected to reach 271 million people and estimated to 305 million people by 2035. Ministry of Agriculture (2004) states that agricultural development is one of the backbone of national development and its implementation should be synergistic with other development sector. Food security according to the requires simultaneous fulfilment of two sides: (a) the availability of adequate food for the entire population in the amount, quality, safety and affordability, of domestic products, and (b) the consumption side which is the ability of every household to access enough food for each member to grow healthy and productive from time to time.

2. Methods

A quantitative approach will be used in this study where the researcher will involve the administration of a structured set of questions or statements with predefined response options aimed at a large number of respondents [3]. The data and source format used is clearly identified and well defined. Compiling and formatting the data collected follow a neat procedure that is mostly numerical [3]. This research is designed to explain the influence between variables or relationships affect (causality) between one variable to another variable through hypothesis testing or confirm the relationship influence between variables or constructs where the relationship between variables that are built or tested is a relationship that has been researched and tested before by researchers either conceptually or through a data analysis technique.

2.1. Analytical Tool

WarpPLS is an analytical tool used in this study. The flow of use of this analytical tool will begin by designing a structural model (inner model) involving several variables involved in this study. Then the next step is to design the measurement model (outer model) followed by constructing the path diagram and convert the path diagram to the system of equations. Further measurements were made to determine the value of path coefficient, loading factor, and weight. Which is continued by determining the value of goodness of fit and the last is hypothesis testing.

3. Result and Discussion

3.1. Model Fit And Quality Indices

In the WarpPLS analysis there is a fit model size and quality index to be met, here are the results of the analysis presented in the following table.

Table 1. Model Fit And Quality Indices In Farmer Behavior Model to Meet Food Needs

MF &QI	Fit Criteria	Result	Note
APC	P < 0.05	0,411 P < 0,001	good
ARS	P < 0.05	0,495 P<0,001	good
AARS	P < 0.05	0,492 P<0,001	good
AVIF	Acceptable if ≤ 5, ideally ≤ 3.3	1,001	good
AFVIF	Acceptable if ≤ 5, ideally ≤ 3.3	1,496	good
GoF	Small >= 0.1, medium ≥ 0.25, large ≥ 0.36	0,505	good
SPR	Acceptable if ≥ 0.7, ideally = 1	1	ideal
RSCR	Acceptable if ≥ 0.9, ideally = 1	1	ideal
SSR	Acceptable if ≥ 0.7	1	good
NLBCDR	Acceptable if ≥ 0.7	0,5	enough

Note : MF&QI is Model Fit and Quality Indices; APC is Average path coefficient; ARS is Average R-squared; AARS is Average adjusted R-squared; AVIF is Average block VIF; AFVIF is Average full collinearity VIF; GoF is Tenenhaus Goodness of Fit; SPR is Sympson's paradox ratio; RSCR is R-squared contribution ratio; SSR is Statistical suppression ratio; NLBCDR is Nonlinear bivariate causality direction ratio.

From the table above, it shows that the overall model is fit. After the testing is done, we can identify that all criteria in the fit and quality indices model are met, Thus the model is said to be good and can be used to explain the phenomenon (system) that is studied and can be used for hypothesis testing. In table 1 above we can see that the value of APC and ARS each has a number of 0.411 and 0.495. This shows that Average Path Coefficient (APC) and Average R-Squared (ARS) meet the fit and quality indices model. Then the AVIF (Averaged block VIF) value also shows a number smaller than 5 that is 1,001 where this shows that there is no multicollinearity in the model under study. For the value of other fit and quality indices model indicators such as AARS, AFVIF, GoF, SPR, RSCR, SSR, and NLBCDR generate numbers that match the predefined criteria to achieve the fit model.

3.2. Goodness of Fit Outer Model

3.2.1. Validity Test. The required load value for a construct to be said as valid is at least 0.533. A model is also said to be capable of having sufficient discriminant validity if the AVE root for each construct is greater than the correlation between kostruk and other constructs. Here are the results of the tests performed.

Table 2. Validity Test

Variable	X1	X2	Y1
Ihsan Attitude (X1)	0,594	-0,228	0,581
Farmer Economic Condition (X2)	-0,228	0,749	0,085
Household Food Security of Farmers (Y1)	0,581	0,086	0,796

From the Table 2, we can identify that the root value of the average variances extracted (AVE) in the numbers with bold print on the main diagonal of the table, while the number outside the main diagonal is the correlation coefficient between variables. When the AVE root value is greater than the correlation coefficient of that variable with other variables, then it is said that the instrument is valid. The table above shows that the research instrument in the form of questionnaires for all variables is said to be valid

3.2.2. *Reliability Test* The loading value required for a construct to be reliable and satisfy the rule of thumb is a minimum of 0.733. Here are the results of analysis in identifying the reliability value of a construct.

Table 3. Reliability Test

Variable	Composite Reliability	Cronbach Alpha
Ihsan Attitude (X1)	0,683	0,640
Farmer Economic Condition (X2)	0,777	0,570
Household Food Security of Farmers (Y1)	0,938	0,924

Based on the above table, we could see that the whole variable indicates good reliability figures. Although there are some variables that have values below 0.7 but the numbers are not too far from the specified limit. Thus, we can conclude that all indicators of each variable is valid and also reliable as a measure of research variables.

3.2.3. Factor Loading The result of factor loading shows that the strongest indicator (dominant) as a measure or formator of Attitude Ihsan (X1) is X1.6 with a factor load of 0.907 close to 100 percent. Then the identified indicators are able to perform the measurements properly are X1.4 and X1.5 with a factor loading of 0.886 and 0.878. While other variables are identified able to make measurements with the amount of load factor is X1.2 and X1.3 with the factor load of 0.536 and 0.550. Meanwhile, other indicators such as X1.1, X1.7, X1.8, and X1.9 were identified as the indicators are able to measure with the lowest ability like X1.1 with value 0.254 and also indicator X1.7, X1 .8, and X1.9, each of them which has negative values of -0.058, -0.375, and -0.040. Then the table above also shows that the strongest indicator (dominant) as a measure (reflect) variable Farmer Economic Conditions (X2) is indicator X2.2 with a factor loading of 0.885, Then followed by indicators that also have strong enough ability to measure the variable that is X2.3 indicator with a factor loading of 0.864. Then the identified indicator has the weakest ability to measure the variable is the indicator X2.1 with a factor loading of 0.392. The strongest indicator (dominant) as a measure (reflective) household food security of farmers (Y1) is Y1.8 with a factor load of 0.919, and also Y1.9 with a factor load of 0.917. Then followed by other indicators such as Y1.1, Y1.2 and Y1.4 which also have a good factor load in measuring the variables which are respectively are 0.830, 0.859 and 0.869. The overall indicator used to measure the variable of Farmer's Household Food Security is identified as being overall positive. The last indicators to be gauges such as Y1.3, Y1.5, Y1.6 and Y1.7 are indicators that have a good enough value to reflect these variables with factor load values of 0.699, 0.778, 0.636 and 0.576, respectively.

4. Conclussion

Household food security of farmers is identified as an important variable in the welfare of the people in a country in which this research is tested together with some influencing variables such as ihsan attitude (X1) and also farmer economic condition (X2). In this study it was identified that the attitude of ihsan (X1) and also the farmers economic condition (X2) had an effect with a high enough significance value, especially for the ihsan attitude which each of is 0.687 and 0.135. This result indicates that the attitude of ihsan owned by farmers in Bangkalan, Madura island gives a strong influence on food security of farm households. The better or increased attitude ihsan owned by farmers, it will also increase the food security of farm households. By knowing this, of course the related parties such as the government can strive and empower the ihsan attitude owned by the farmers so that the food security of farm households which become one of the indicators of the welfare of a country can be realized. Through this research, it is identified that the farmers in the Bangkalan Madura area have a good ihsan attitude, where the ihsan attitude is the attitude that is able to provide food security owned by the farmers.

One of the characteristics possessed by farmers is the economic condition owned by farmers. In this study, we identify that the farmer economic condition has positive relationship to household food security of farmers. The economic condition in this study is the economic condition of the family in terms of the status or position of the family economy either in terms of income or livelihood of farmers in Bangkalan Madura in meeting their families needs. We also identify that the better the economic condition owned by the farmers, the better the food security in the farmer's household.

Based on the results above, the researchers expect the parties which have responsibility on the preservation of household food security of farmers such as the government is expected to contribute through the implementation of policies or programs that can be implemented related to household food security of farmers. In addition, the government is also expected to formulate further policies on improving economic conditions owned by farmers. Through the implementation of these policies, it is expected to realize food security owned by farmers' households.

Reviewing the results of this study, it is hoped that further research can identify deeper about the classification of economic conditions that are divided into several levels of economic conditions, so that it can be known exactly at a level where economic conditions will actually be able to affect the food security of farm households. In addition, in-depth research on the ihsan attitude variable also need to be done back in the realm of agribusiness because of this attitude that has its own value for the realm of human development.

References

[1] Maxwell S and Frankenberger T 1992 *Household food security: concepts, indicators, and measurements: a technical review* New York, NY, USA and Rome, UNICEF and IFAD

[2] Food and Agriculture Organization 1996 *The State Of Food and Agriculture*

IOP Conf. Series: Earth and Environmental Science **142** (2018) 012074 doi:10.1088/1755-1315/142/1/012074

[3] Burns Alfin C and Ronald F Bush 2014 *Marketing Research Seventh Edition. International Edition* England: Pearson Education Limited

Effect of CMC and arabic gum in the manufacture of jackfruit velva (*Artocarpus heterophyllus*)

B Yudhistira , N H Riyadi , A D Pangestika , S R Pertiwi

Food Science and Technology, Faculty of Agriculture, University Sebelas Maret, Jl. Ir. Sutami 36 A, Kentingan, Surakarta 57126, Indonesia

Email: atikapangestika@gmail.com

Abstract Velva is one type of frozen dessert which is made from fruit/vegetable with ice cream maker, low fat and high fiber content. Jackfruit is a raw material for the manufacture of velva because of the high fiber content of 2.31 gr. The use of a stabilizers combination of CMC and arabic gum in the manufacture of velva will provide a better gel mix than single use. The purpose of this research is to know the influence of variation of CMC and arabic gum stabilizer on the characteristics (physical, chemical, and sensory) of jackfruit velva (*Artocarpus heterophyllus*) and determine variations in the most appropriate combinations of stabilizers to produce jackfruit velva with the best quality. This research applied Completely Randomized Design consist of one factor which is the combination of CMC and arabic gum levels in the making of jackfruit velva with two replicates and two replications of the analysis. The data obtained then analyzed statistically using one way analysis of variance (ANOVA), when there is a significant difference, then followed by Duncan's Multiple Range Test (DMRT) at significance level of 0.05. The results of this study concluded that the jackfruit velva with the addition of various concentrations of CMC and arabic gum is significantly affecting the taste, texture and overall parameters, but no significant difference on the color and flavor parameters of jackfruit velva. Based on the results of physical characteristics, chemical and sensory jackfruit velva with the addition of a stabilizing concentration of CMC and arabic gum 1: 1 result in best jackfruit velva. The best jackfruit velva with stabilizing the concentration of CMC and arabic gum 1: 1 contains a water content of 61.95%, dietary fiber 2.231%, total dissolved solids 20.38 ^{o}Brix, overrun 19.709%, meltdown 28.215 minutes. As for the color attribute score 3.72; Taste 4; flavor 3.60; Texture 3.68, and overall 3.88.

1. Introduction

Velva is one type of frozen dessert which is made from fruit/vegetable with ice cream maker, low fat and high fiber content. This product is made from a puree of fruits, sugar, sucrose and stabilizer which are frozen so as to obtain a smooth texture resembling ice cream [1].

Jackfruit is a raw material for the manufacture of velva because of the high fiber content of 2.31gr. Jackfruit production in Indonesia in 2014 according to the Ministry of Agriculture amounted to 644,291 tons. Jackfruit has a sweet taste and a spesific flavor, that has the potential to be developed as a processed food product.

The use of stabilizers is important in the production of velva fruit. According to [2], the main function of the stabilizer is to bind water in the mixture, so that the formation of large ice crystals can be avoided and also to maintain the shape and texture of the product during storage. There are different types of stabilizers used in the manufacturing of velva fruit. The common stabilizers used in the manufacture of ice cream and frozen desserts are CMC, gelatin, sodium alginate, carrageenan, arabic gum and pectin. Selection of the stabilizer on its effect on overrun and texture [1].

CMC (Carboxyl Methil Celullose) is the most important cellulose derivative in food applications. Its main function is to bind water, give viscosity in aqueous phase and stabilize other components or prevent sineresis [3]. According to [4], the use of CMC as a stabilizer does not require aging time to produce perfect stability so as to shorten production process time.

Arabic gum is the sap of the acacia tree (*Acasia senegal*). The sap is dried, sated and then process to be powdered. Arabic gum is a heteropolysaccharide complex, consisting of D-galactopranose, L-ramnopiranosa, L-Arabinopiranosa, L-Arabinofuranose and D-glucoronic acid and a small amount of protein (2%) [5]. Arabic gum easily soluble in hot or cold water but not soluble in alcohol or other organic solvents. Gum Arabic is used as a stabilizer for desserts such as ice cream and sherbet for its ability to absorb water.

The addition of arabic gum aims to prevent the formation of larger ice crystals by binding large quantities of water. Ice cream used arabic gum as a stabilizer generally does not melt quickly [6]. The use of a stabilizers combination of CMC and arabic gum in the manufacture of velva will provide a better gel mix than single use, it can produce good quality Velva. The purpose of this research is to study the influence of variation of CMC and arabic gum stabilizer on the characteristics (physical, chemical, and sensory) of jackfruit velva (*Artocarpus heterophyllus*) and determine variations in the most appropriate combinations of stabilizers to produce jackfruit velva with the best quality.

2. Methods

2.1 Materials

The ingredients used in this work including Jackfruit, water, sugar, citric acid. Food grade hydrocolloids; CMC (Carboxyl Methyl Cellulose) and arabic Gum. The ingredients used to produce jackfruit velva.

The chemical test materials used for analysis in the study, among others, the test dietary fiber by using buffer Na-phosphate 0.1M pH 7, alpha amylase enzyme, 1% pepsin enzyme, 1N hydrochloric acid (HCl), sodium hydroxide (NaOH) 1 N , 95% ethanol, acetone (CH3) ¬2¬CO, beta amylase enzyme, Test water content using toluene

2.2 Velva Preparation

The first stage is Stripping jackfruit and jackfruit seeds disposal. Then the jackfruit is washed and cut into pieces. The jackfruit pieces then blancing at 90°C for 5 minutes. The jackfruit pieces are then crushed with a blender until it becomes a fruit pulp. The subsequent addition of citric acid is 0.1% by weight of puree which has been dissolved in water. The mixture is stirred for approximately 15 minutes. The homogeneous mixture is then frozen with a 4°C refrigerator for 24 hours to allow the stabilizer to bind the free water present. The next stage is freezing by using ice cream maker for 30 minutes and stored in the freezer temperature -12°C. According to the formulations shown in Table 1

Table 1 Jackfruit Velva Formulation

Ingredients (g)	Comparison of CMC and Arab Gum (%)				
	F1 0:1	F2 1:3	F3 1:1	F4 3:1	F5 1:0
Jackfruit	200	200	200	200	200
Water	100	100	100	100	100
Sucrose	50	50	50	50	50
Citrit acid	0,3	0,3	0,3	0,3	0,3
CMC	0	0,75	1,5	2,25	3
Arabic Gum	3	2,25	1,5	0,75	0

2.3 Physical and Chemical Analysis

According to Masrshal and Arbuckle (1986), overrun was estimated using a standard 50 ml cup, according to the equation as follows:

$$\% \text{ Overrun} = \frac{\text{Volume of ice cream} - \text{volume of mix}}{\text{volume of mix}} \text{ x } 100\%$$

Melting property was analyzed at 25 $\pm 2^{o}$C. Measurements were made by taking one spoon velva (± 5 gram) -20°C on the plate. Velva allowed to melt at room temperature (27°C) and measurement using stopwatch [7].

Jackfruit velva were allowed to melt at room temperature at 25°C before they were subjected to total soluble solid analysis. Total soluble solid was determined using a refractometri. Dietary fiber analysis was estimated according to the method described by multienzim, while water content according to the method by thermovolumetri.

2.4 Sensory Evaluation

Sensory evaluation was carried out using 5-point hedonic scale [8]. The panelist suggests personal likes or dislikes, in addition to his favorite level.

2.5 Weighting Evaluation

Determination of the best formula using compensatory method [9]. The decision making process of the best formula is done by weighting technique of additive. The principle of testing on the method is by giving weight on each characteristic in accordance with the importance of the characteristics of the resulting formula

2.6 Statistic Analysis

The data obtained then analyzed statistically using one way analysis of variance (ANOVA), when there is a real difference, then followed by Duncan's Multiple Range Test (DMRT) at significance level of $\alpha = 0.05$.

3. Results and Discussion

3.1 Physico-chemical analysis of jackfruit velva.

3.1.1 Overrun

The table 1 show the use of a combination of CMC and arabic gum in jackfruit velva production, gives the highest average overrun of 25.015% in the formulation of F2 (1: 3), because CMC and arabic gum are polysaccharides. According to [10], Hydrocoloid polysaccharides will bind water molecules, which causes increased water phase viscosity in both cream and foam, thus protecting the whole system against sineresis and ultimately resulting in foam stabilization. The high foam on the overrun shows that more air is trapped, giving higher foam expansion [11].

The value of carrots velva overrun is 11.05% - 19.79% using a combination of CMC and gum arab stabilizers [12]. The value of the jackfruit velva overrun with the addition of a combination of CMC and gum arab stabilizer averaged 16.15% - 25.02%, the results obtained are not much different from the overrun value of the reference velva.

3.1.2 Meltdown

The table 1 show The highest average meltdown was produced by F3 using a combination of CMC and arabic gum stabilizer (1: 1) of 28.22 minutes while the lowest average meltdown was obtained by F5 using a 1% CMC stabilizer of 21.13 min. According to [13] the combination of stabilizers CMC and gum arab can optimize the water binding power so that the dough viscosity increases and the melting power becomes slower. The ability of the stabilizer to bind the water causes the water molecules trapped in the gel structure formed by the stabilizer so that the dough viscosity increases and the meldown of the ice cream increases [7].

The value of the reference carrot velva meltdown is between 20.25 -27.00 minutes using a combination of CMC and gum arab stabilizers [12]. The value of the meltdown of the jackfruit velva with the addition of a combination of CMC and arabic gum stabilizer averaged 21.13 - 28.22 minutes. The results obtained were not much different from the meltdown value of the reference velva.

3.1.3 Total Disolved Solids

An increase in the value of dissolved solids is also associated with a decrease in water content (Vail ea al., 1978). The average total of the highest dissolved solids was produced by F2 using a combination of stabilizer CMC and arabic gum (1: 3) of 22.00 °Brix while the average total of the lowest dissolved solids was produced by F5 using a 1% CMC stabilizer of 19.00 °Brix . The use of a combination of CMC and arabic gum stabilizers in the making of jackfruit velva higher total soluble solids than single stabilizer use, this is because the stabilizer used contributes to the increase in total soluble solids in the jackfruit jackfruit.

3.1.4 Water Content

Based on testing of water content of jackfruit velva, all formulations (F1, F2, F3, F4 and F5) showed significant differences between treatments. The combination of CMC and arabic gum 0: 1 has the highest moisture content compared with other formulations of 65.31%. While the formulation of F5 (comparison of combination of CMC and arabic gum 1: 0) yielded the lowest value of 60.12%. The increased use of CMC concentrations will provide a low moisture value, according to [14], the increase in CMC concentration in solution can lead to a rise in viscosity. The high viscosity is due to the available water being mostly trapped in the material matrix due to the influence of the stabilizer, the stabilizer having to bind the free water so that the measured water content is lower.

3.1.5 Dietary Fiber

Based on the testing of dietary fiber against jackfruit velva all formulations (F1, F2, F3, F4 and F5) showed significant differences between treatments. Showed that the jackfruit velva in the formulation F1 (comparison of combination of CMC and gum arabic 0: 1) had the lowest value of dietary fiber compared with the other formulation of 0.9785%. While the formulation of F4 (comparison of combination of CMC and gum arab 3: 1) yielded the highest value of 2.74%. The fiber in the jackfruit velvet comes from puree, sugar and stabilizer.

Table 2. Effect of combination concentration variation of CMC and arabic gum stabilizer on physicochemical characteristics of jackfruit velva

Characteristics	Formulation				
	F1	F2	F3	F4	F5
water content (%)	65.31[c]	63.4[b]	61.95[b]	61.67[ab]	60.12[a]
dietary fiber (%)	0,979[a]	1,825 [c]	2,231 [d]	2,743 [e]	1,56 [b]
total dissolved solids (°Brix)	21,33[cd]	22,00[d]	20,38[b]	20,83[bc]	19,0[a]
Overrun (%)	16,476[a]	25,015[b]	19,709[a]	20,147[a]	16,15[a]
Meltdown (Minutes)	25,75 [c]	25,967[c]	28,215[d]	22,172[b]	21,13[a]

Information:
*: F1: Jackfruit velva with stabilizer concentration CMC and arabic gum 0:1
F2: Jackfruit velva with stabilizer concentration CMC and arabic gum 1:3
F3: Jackfruit velva with stabilizer concentration CMC and arabic gum 1:1
F4: Jackfruit velva with stabilizer concentration CMC and arabic gum 3:1
F5: Jackfruit velva with stabilizer concentration CMC and arabic gum 1:0
**: The numbers followed by different letters in the same column show a real difference in the siq a = 0.05 level

3.2 Sensory Evaluation

Sensory testing is very important in product development because it determines consumer acceptance of the product. A score test is also called scoring. The scoring is to score the value or to place the sensory miti value on the material tested at the quality level or hedonic scale level. Test scoring can be done on research of very specific sensory properties, such as, texture, color, taste, flavor [15].

The result of the analysis of variance in Table 3 shows that for the jackfruit velva all formulations (F1, F2, F3, F4 and F5) were not significantly different between the treatments on the color parameters. The results of the analysis of jackfruit velvet with color parameters in research on the Japanese pumpkin velva products with the addition of stabilizers CMC and arabic gum at various levels of concentration does not cause different effects on the color of products in various treatments.

The result of variance analysis in Table 3 shows that for the jackfruit velva all formulations (F1, F2 F3, F4 and F5) are significantly different between the treatments on the taste parameter. The tendency of adding a stabilizer to the soursop velva, with a certain concentration can improve the taste, but the addition of more stabilizers will decrease the value of organoleptic flavor. According to [14], CMC affects the threshold of flavor. At a certain concentration there is a decrease in sweet taste threshold and a rise in sour and bitter taste. The results of the analysis of variance in Table 3 show that for the jackfruit velva all formulations (F1, F2 F3, F4 and F5) did not differ significantly between treatments of flavor parameters.

Table 3. Influence of Variation of Concentration of Addition of CMC and Gum Arab to Sensory Character of Velva Jackfruit

Formulation	Parameters				
	Color	Taste	Flavor	Texture	Overall
F1	3.76[a]	3.64[bc]	3.52[a]	3.20 [abc]	3.76 [c]
F2	3.52[a]	3.20[ab]	3.36[a]	2.64 [a]	3.16 [ab]
F3	3.72[a]	4.00[c]	3.60[a]	3.68 [c]	3.88 [c]
F4	3.36[a]	3.28[ab]	3.56[a]	3,48 [cb]	3.40 [bc]
F5	3.32[a]	2.84[a]	3.24[a]	3.00 [ab]	2.76 [a]

*:F1: Jackfruit velva with stabilizer concentration CMC and arabic gum 0:1
F2: Jackfruit velva with stabilizer concentration CMC and arabic gum 1:3
F3: Jackfruit velva with stabilizer concentration CMC and arabic gum 1:1
F4: Jackfruit velva with stabilizer concentration CMC and arabic gum 3:1
F5: Jackfruit velva with stabilizer concentration CMC and arabic gum 1:0
**: The numbers followed by different letters in the same column show a real difference in the siq a = 0.05 level
***: Score 5 = really likes, 4 = likes, 3 = neutral, 2 = dislikes, 1 = very dislike

substances that can cause flavor and color in food, but hydrocolloids can have a synergistic effect on adding flavors to the emulsion [6].

The result of the analysis of variance in Table 3 shows that for jackfruit velva all formulations (F1, F2 F3, F4 and F5) are significantly different between treatments on Texture parameters. An increase in CMC concentration in solution can also lead to an increase in the viscosity of the dough. With increasing viscosity, more water is bound. So that the formation of rough ice crystals more can be avoided and the resulting texture is soft. According to [4], the stabilizer is effective for producing a soft texture through its ability to bind water in a mixture of velvety batter. Soft texture can also be obtained when much air is trapped in the dough during freezing, resulting in a high overrun.

The result of the analysis of variance in Table 3 shows that for the jackfruit velva all formulations (F1, F2 F3, F4 and F5) are significantly different between treatments against the Overall parameter. Overall, the combination of CMC and gum arab pentasbil in jackfruit velvet products was significantly different from taste, texture and overall parameters, whereas on color and aroma parameters.

3.3 Weighting Evaluation

Weighting is a decision-making technique in a process that involves multiple factors simultaneously by weighting each of these factors. Weighting can be done objectively statistically or subjectively by setting it based on certain considerations. The weighted sum is a compensatory method, which can mean that the worst criterion score is compensated for the best score. This method is a special form of Multi Attribute Value Theory (MAVT) [16].

Based on Table 4 the results of the assessment given on each test parameter of jackfruit fruit velvet characteristic, the highest final score is found on the addition of CMC concentration and 1: 1 gum arabic with the score of 0.7389. This indicates that the concentration of CMC and arabic gum 1: 1 produced the physical, chemical and sensorial characteristics of the best jackfruit velvet and can be accepted by panelists. Jackfruit Velva with CMC and arabic gum concentration 1: 1 (F3) has moisture content of 61.95%, food fiber 2.231%, Total soluble solid 20.38 obrix, overrun 19.709%, melt 28.215 minutes . As for the color attribute score 3.72; Taste 4; Aroma 3.60; Texture 3.68, and overall 3.88.

Table 4. Weighted Jackfruit Velva Test Value of each Formulation

Karakteristik	BV	BN	Formulation									
			F1		F2		F3		F4		F5	
			DV	NH	DV	NH	DV	NH	DV	NH	DV	NH
Water content (%)	1	0.100	0.0	0.000	0.368 0154	0.037	0.647	0.0647	0.701	0.070	1.000	0.100
Dietary fiber (%)	1	0.100	0	0.000	0.480	0.048	0.710	0.071	1	0.100	0.330	0.033
Total Disolve °Brix	1	0.100	0.777	0.078	1	0.100	0.460	0.046	0.61	0.061	0.0	0.0
Overrun (%)	1	0.100	0.037	0.004	1	0.100	0.402	0.040	0.451	0.045	0.0	0.0
Meltdown	1	0.100	0.652	0.065	0.683	0.068	1	0.100	0.147	0.015	0.0	0.0
Color	1	0.100	1	0.100	0.455	0.045	0.909	0.091	0.091	0.009	0.0	0.0
Taste	1	0.100	0.690	0.069	0.310	0.031	1	0.100	0.379	0.038	0.0	0.0
Flavor	1	0.100	0.778	0.078	0.333	0.033	1	0.100	0.889	0.089	0.0	0.0
Texture	1	0.100	0.538	0.054	0	0.000	1	0.100	0.808	0.081	0.346	0.035
Overall	1	0.100	0.893	0.089	0.357	0.036	1	0.100	0.571	0.057	0.0	0.000
Sum	10			0.536		0.499		**0.8128**		0.565		0.168

Description: BV = Variable Weight BN = Normal Weight, DV = Dimensionless Value, NH = Value Result

4. Conclucion

Based on physicochemical testing of jackfruit velva with addition of various concentrations of CMC and arabic gum, significantly different to the parameters of overrun, total dissolved solids, meltdown, dietary fiber and water content of jackfruit velva. Based on sensorial testing of jackfruit velva with the addition of various concentrations of CMC and arabic gum are significantly different to taste, texture and overall parameters. But no significant difference to the color parameters and aroma of jackfruit velva. Based on the results of physical, chemical and sensory characteristics of jackfruit velva with addition of CMC and arabic gum concentration 1:1 yields the best jackfruit velva and acceptable to consumers.

Reference

[1] Sommer, HH. 1947. *The Teory and Practice of Ice cream Making*. Published by the author, Madison, Wisconsin.
[2] Herschdoerfer, SM. 1972. *Quality Control in The Food Industry*. Vol. III. Academic Press, New York dan London
[3] Pomeranz, Y. 1991. *Functional Properties of Food Components*. Academic Press, London
[4] Arbuckle, W S. 1968. *Ice Cream*. AVI Publishing Company, Inc., Wesport, Connecticut
[5] Carriedo, Nisperos MO. 1994. *Edible Coatings and Films Based on Polysaccharides*. technomic Publising Company,Inc. Bazel
[6] Glicksman, M. 1983. *Food Hydrocolloids Vol II. CRC Press*. Boca Raton, Florida
[7] Bodyfelt. F W., J. Tobias, dan G. Trout. 1998. *The Sensory Evaluation of Dairy Product*. AVI Publishing Company, Westport, connecticut
[8] Larmond E .1977. Laboratory Methods for Sensory Evaluation of Food. Research Branch, Canada, Department of Agriculture, Ottawa Publication
[9] Sullivan, William G, Elin M Wicks dan C Patrick Koelling. *Engineering Economy: Sixteenth Edition*. Pearson. Virginia
[10] Sikora, M., Badrie, N., Deisingh, N. K., & Kowalski, S. 2008. *Sauces and dressings: a review of properties and applications. Critical Reviews in Food Science and Nutrition, **48***: 50-77
[11] Pasban, A., Mohebbi, M., Pourazarang, H. & Varidi, M. (2014a). *Effects of endemic hydrocolloids and xanthan gum on foaming properties of white button mushroom puree*

studied by cluster analysis: A comparative study. Journal of Taibah University for Science, **8**: 31-38

[12] Rini, K A., Ishartani D., Basito. 2012. *Pengaruh Kombinasi Bahan Penstabil CMC dan Guma Arab terhadap Mutu Velva Wortel (Daucus Carota L.) Varietas Selo dan Varietas Tawangmangu*. Jurnal Teknologi Pangan. Vol. **1** (1) 86-94. ISSN: 2302-0733

[13] Setianawati, Hijrah Nora. 2002. *Penggunaan Kombinasi bahan Penstabil pada Pembuatan Velva Kweni (Mangifera odorata Griftt).* Skripsi Jurusan Gizi Masyarakat dan Sumber Daya Keluarga, Fakultas Pertanian, Institut Pertanian Bogor

[14] Ganz AJ.1997. *Cellulosa Hydrocolloid*. Avi Publishing Co. Inc. Westport, connectiont
Glickmasn, M. 1969. *Gum Technology in The Food Industries*. Academic Press. New York

[15] Soekarto, S. T. 1985. *Penilaian organoleptik*. Bharatara Aksara. Jakarta

[16] Keeney, R dan Raiffa H. 1976. *Decisions with multiple objective: preferences and value trade-offs*. Willey.New York

IOP Conf. Series: Earth and Environmental Science **142** (2018) 012076 doi:10.1088/1755-1315/142/1/012076

Sustainable development of smallholder crop-livestock farming in developing countries

S Ates[1], H Cicek[2], L W Bell[3], H C Norman[4], D E Mayberry[5], S Kassam[6], D B. Hannaway[7] and M Louhaichi[1, 8]

[1)] Department of Animal and Rangeland Sciences, Oregon State University, Corvallis 97333, USA
[2)] FiBL, Department of International Cooperation, Switzerland
[3)] CSIRO Sustainable Agriculture Flagship, PO Box 102, Toowoomba, Qld 4350, Australia
[4)] CSIRO Agriculture and Food, Private Bag 5, Wembley, WA 6913, Australia
[5)] CSIRO Agriculture & Food, St Lucia QLD 4067, Australia
[6)] International Center for Agricultural Research in the Dry Areas (ICARDA), Cairo, Egypt
[7)] Department of Crop and Soil Science, Oregon State University, Corvallis 97333
[8)] International Center for Agricultural Research in the Dry Areas (ICARDA), Amman, Jordan
e-mail: serkan.ates@oregonstate.edu

Abstract Meeting the growing demand for animal-sourced food, prompted by population growth and increases in average per-capita income in low-income countries, is a major challenge. Yet, it also presents significant potential for agricultural growth, economic development, and reduction of poverty in rural areas. The main constraints to livestock producers taking advantage of growing markets include; lack of forage and feed gaps, communal land tenure, limited access to land and water resources, weak institutions, poor infrastructure and environmental degradation. To improve rural livelihood and food security in smallholder crop-livestock farming systems, concurrent work is required to address issues regarding efficiency of production, risk within systems and development of whole value chain systems. This paper provides a review of several forage based-studies in tropical and non-tropical dry areas of the developing countries. A central tenet of this paper is that forages have an essential role in agricultural productivity, environmental sustainability and livestock nutrition in smallholder mixed farming systems.

1. Introduction

A large proportion of rural poor populations across the world derive a major share of their livelihood from mixed farming systems. Smallholder farming enterprises have a critical role in rural development and supporting local food systems through integrating multiple agricultural value chains from vegetables and fruits to animal products. Additionally, they provide a number of social and agro-ecological benefits such as increased land use efficiency, improved biodiversity and access to affordable animal-source food for rural communities. These systems are threatened by population pressure, climate change, recurrent droughts and degradation of natural resources. These pressures are more acute in areas with climates that are marginal for existing practices [1]. In developing countries, the negative trends of unsustainable use of resources are often compounded by mismanagement, poor infrastructure, political instability, and ineffective socioeconomic policies. Addressing these issues will lead to improved sustainability, food security and rural livelihoods.

Forage plants sown for direct grazing, conservation or cover crops have a central role in agronomic sustainability and animal production in mixed crop-livestock farming systems. Benefits of forages include improved ruminant production, enhanced soil health and fertility, increased carbon sequestration, root disease management in cropping systems, increased biodiversity, and reduced economic risk through diversification [2] [3] . While the opportunities to expand the area of forage crops is limited due to the competition with field crops for land and water resources, increases in forage production may be possible through intercropping, alley cropping, or integration of legumes via crop rotation. Alternative fodder plants that are adapted to harsh climatic conditions and can be grown on lands unsuitable for cropping have significant potential for reducing the feed gap for livestock, improving animal health and contributing to food security and diversifying the income source of resource-poor farmers [4][5]. Revegetation of marginal land with perennials can lead to substantial increases in environmental health, including reduced soil salinity, increased carbon sequestration and reduced methane emissions from ruminants[6]. Developing varieties and management techniques for dual purpose (food-feed) legumes and cereals, cover crop grazing techniques, and intensifying cropping systems (by replacing fallows with feed and forage legumes) are promising options[7] [2][8]. Furthermore, increased efficiency in farming practices, value-added approaches, sound agricultural policies and specific value chain development are needed to improve access to inputs and markets for smallholders.

2. Integration of forage legumes into cropping systems

2.1 Forage legumes in crop-livestock farming in non-tropical dry areas

Forage legumes are essential to agricultural sustainability in dryland cereal-growing areas of the world where grazing livestock is a dominant enterprise. In dry areas where livestock are an integral part of the production system, forage legumes have increased crop yield and animal production [9] [2][8][10]. Legumes can be inserted into crop rotations in various ways depending on the production system, site-specific needs, farmer capabilities, and the amount of precipitation received. As conservation agriculture gains momentum in the dryland areas, there is potential for better integration of forage legumes into cereal based-cropping systems [11]. Based on one of three principles of conservation agriculture, diverse rotations, farmers are encouraged to grow forages legumes to avoid pest and weed pressures and improve soil quality. In addition, cereal–legume rotations extend the time during which the soil remains covered, and thus less prone to erosion and water loss.

In the dryland farming system of the Mediterranean region, several long-term barley- and wheat-based rotation experiments evaluated the value of forage legumes [11] [2]. Results demonstrated that forage and grain legumes are excellent alternatives to continuous cereal cultivation or cereal–fallow rotations in Mediterranean-type climate areas, leading to more efficient and sustainable cropping systems. The increased efficiency was also highlighted in a long-term study that compared yields and profitability of wheat (*Triticum aestivum* L.) when grown after wheat, fallow, a grazed mixture of medic species (*Medicago* spp.), and common vetch (*Vicia sativa* L.) cut for hay, over ten growing seasons in Syria [2]. It was reported that wheat after lentils were generally lower (2.22 t ha^{-1}) than after vetch (mean 2.56 t ha^{-1}) and after medic (2.40 t ha^{-1}). Inclusion of grain legumes in the rotations boosted profits significantly because of their high grain prices and valuable straw. Replacing fallow with vetch for hay production increased the average gross margin by US$126 ha^{-1} $year^{-1}$, and growing vetch for hay in rotation with wheat produced greater profit than continuous wheat, by $254 ha^{-1} $year^{-1}$. Similarly, on-farm trials in Tunisia have shown that replacing weedy fallows with forage legumes like vetch or sulla (*Hedysarum coronarium* L.) increased productivity under conservation agriculture [12]. The dry matter production (DM) of sulla averaged 6.2 t DM ha^{-1} with a crude protein content greater than 17%. Similarly, DM production of vetch and vetch/cereal mixtures ranged from 3.0 to 5.0 t DM ha^{-1}, as compared to weedy fallows that produced only 1.0 t DM ha^{-1}, illustrating the clear advantage of forage [12].

2.2 Forage legumes in crop-livestock farming in the semi-arid tropics

Research is also demonstrating the potential of using forage legumes in crop-livestock systems in the semi-arid regions of Eastern Indonesia. Livestock production in this region is constrained by low quality and unreliable availability of forage during the dry season, while yields of stapled cereal crops (rice and maize) can be increased by improved nitrogen supply. The herbaceous forage legumes, *Clitoria ternatea, Lablab purpureus* and *Centrosema pascuorum,* have shown to be well adapted and can be grown in either rotation or relay systems following either wet-season maize, or rice crops where irrigation water is limited or unreliable. Grain yields of a following maize crop were increased by 50% (1.4-1.6 t ha^{-1}) where legume was cut and removed, and by 90% (2.6-2.8 t ha^{-1}) where legume biomass was retained; the persistence of these benefits in subsequent years is being assessed [13]. This demonstrates that the transfer of nitrogen to subsequent crops is significantly diminished where legume material is cut and removed, as is often the case in smallholder cut and carry systems. Devising ways to enable livestock to graze the legumes directly is likely to enhance the N benefits provided by legumes to subsequent crops.

In addition to these herbaceous legumes, leguminous shrubs such as Leucaena can provide higher quality forage during the dry season in a variety of ways to improve cattle production. They can be used to enhance cattle fattening growth rates, but may also have value in other cattle systems. In some regions cattle are free-grazing, but a penned at night, where supplementing with small amounts of legume hay (1% of body weight) during the dry season enables cows to maintain body weight and condition, while unsupplemented cows lost 5% of their body weight. This is likely to have impacts on their subsequent fertility and calving interval, though this has not yet been shown. High mortality rates and low growth rates of Bali calves have been found in Eastern Indonesia, but this can be improved with grass hay and concentrate (18% CP), but this is expensive and complicated [14] [15]. Forage legumes have the potential as a cheaper alternative to a concentrate ration and preliminary studies have shown similar benefits for calf growth rates.

A series of experiments in Indonesia demonstrated that Bali (*Bos javanicus*), Ongole and Brahman (both *Bos indicus*) cows can all maintain weight on rice-straw based diets with small levels of supplementation [16] [17] [18]. Tree legumes such as *Gliricidia sepium*, *Leucaena leucocephela* and *Sesbania grandiflora* fed at approximately 0.3% liveweight (3-4 kg fresh weight) provided sufficient energy for maintenance, although the amount of supplement fed would need to be increased for cows in the late stages of pregnancy, lactating or being used for draught [19] . Even at low levels of inclusion in the diet, tree legumes comfortably meet the rumen degradable N requirements of cattle, without the risk or cost of feeding urea. Tree legumes provide a cheap and locally supplement for livestock throughout the year, including the dry season, and can be planted as living fences around the farmer's house or animal pens, where they can be harvested easily with little labour cost.

3. Utilizing trees and shrubs in forage systems

Alley cropping is a type of agroforestry in which trees or shrubs are planted simultaneously in rows with agricultural crops. It can be applied to improve forage production, mitigate drought, serve as a windbreak and capture windblown sediments in dry areas. The incorporation of trees or shrubs into farming systems may also have environmental advantages such as the lowering of the water tables causing secondary salinity, increasing functional diversity and improving soil stability[6][20][21]. Shrubs can be directly grazed or palatable tree fodder can be cut during autumn or winter-feed gaps. This can reduce supplementary feeding costs and increase paddock carrying capacity, enabling farmers to increase animal and crop production from the same land area[22][23]. Once mature, trees and shrubs can be pruned after grazing and used as a green manure or mulch to improve soil organic matter and improve crop growth. Grazing can enhance the production of woody material and proper pruning can increase the growth of green forage, extend plant longevity and improve fodder quality [23].

3.1 Alley cropping in productive soils

Saltbushes (*Atriplex nummularia* Lindl., *Atriplex halimus* and *Salsola vermiculata*) are commonly planted in arid areas of the southern Mediterranean region. They have many advantages because of their wide adaptability to harsh agro-climatic conditions and ability to grow for a longer period than many other crops and forage species. As they require little care after establishment, their production cost is low [24]. Atriplex species are used to fill the summer/autumn or early winter-feed gap typical of Mediterranean-type climates in southern Europe [25], southern Mediterranean [22] [26][27]. *Acacia cyanophylla* is often used in North Africa as it fixes nitrogen and provides edible leaves for livestock. In Central Asia, *Morus alba* L. and *Morus nigra* L. are commonly planted as hedgerows and their leaves can be fed to livestock. Another well-adapted species to the cold temperature of Central Asia is *Haloxylon ammodendron* that can tolerate direct grazing [28].

In south Mediterranean, studies on alley cropping systems showed that barley grain and straw yields had significant (20-100%) increases when cropped with *Atriplex halimus* L. in alley cropping systems as compared to mono-cropped barley [23]. In another study, the agronomic benefits of the cactus-barley alley cropping system were evaluated in Tunisia[29]. Similar to the results reported by [23], total biomass (straw plus grain) of barley cultivated between the rows of spineless cactus increased from 4.24 to 6.65 t ha^{-1} and the grain from 0.82 to 2.32 t ha^{-1} as compared to barley alone. These results were attributed to the change of the micro-environment created by alley-cropping with cactus, which created a beneficial 'wind breaking' role that reduced water loss and increased soil moisture. The barley crop stimulated an increase in the number of cactus cladodes and fruits, while the cactus increased the amount of root material, contributing to the soil organic matter.

The alley-cropping system with *Atriplex nummularia* L. also proved efficient in the semi-arid regions of Morocco (annual rainfall 200-350 mm). Barley was cropped using a seeding rate of 160 kg ha^{-1} between *Atriplex* rows planted at 333 plants ha^{-1}. Compared to the farmers' mono-cropping system, dry matter consumable biomass yield of *Atriplex* was significantly higher in the alley-cropping system and was more profitable than mono-cropping. [30] determined the net benefit from *Atriplex* monocropping and barley-*Atriplex* alley cropping over 15 years were 732 \$ ha^{-1} and 3,343 \$ ha^{-1}, respectively. The assessment of alley cropping system indicated that this technology is agronomically feasible and economically profitable. Thus, system should be extended on a larger scale in the agro-pastoral areas of North Africa and West Asia.

3.2 Use of shrubs in saline and arid areas

The area of human induced salinity is increasing rapidly and a major constraint to livestock production in many areas. Shallow water tables that are associated with poor irrigation practice is another major constraint to forage production. Only few plants can tolerate the combined stresses of salinity and waterlogging [31]. It is estimated that 5% of the world's land surface is cultivated salt-affected land, which includes 19.5% of irrigated agricultural land [32] [33]. A large proportion of salt-induced soil degradation occurs on land used by smallholder farmers, who are dependent on this resource for food and feed needs [34]. Considering the reliance of smallholder communities on saline waters and salt-affected soils for their livelihoods, direct and indirect impacts of salinity are often borne by the most vulnerable sectors of the population. In the past 15 years, there has been a concerted effort in to identify plants that can persist in saline, waterlogged and/or arid areas while providing nutrients for ruminants during critical feed gaps [35] . This work has largely focused on halophytic shrubs such from the Chenopodiaceae family as *Atriplex, Mariana and Rhagodia* species as they are naturally found in arid areas and are tolerant of salinity and aridity [36].

Halophytic shrubs are not ideal as a sole diet due to low to moderate digestibility of the organic matter, excessive salt and/or sulphur accumulation and production of plant secondary compounds such as oxalate or saponin [37] [38]. They tend to be used as a drought reserve or to fill annual feed shortages within grazing systems [39]. It is therefore recommended that farmers supplement ruminants grazing these shrubs with crop residues, hay or grain [21]. Despite these limitations, shrubs offer a valuable source of crude protein, sulphur, vitamin E and minerals within meat and wool production systems [40] [41]. There is ongoing work in Australia to select genotypes of *Atriplex nummularia* and *Rhagodia preisii* with higher nutritional value and improved 'palatability' to livestock [42]. Elite genotypes have been commercialized in Australia and testing of this material is being undertaken in Afghanistan, Tibet and Pakistan. Preliminary results from Afghanistan indicated that these elite genotypes can be successfully grown in marginal areas and have high potential to reduce the widespread feed shortage in Afghanistan [43].

4. Dual-purpose ('graze and grain') management of cereal crops

One area that has received little evaluation in smallholder production systems is the dual-purpose use of grain crops, where they are grazed during their early vegetative stage and later harvesting the grain and straw at maturity. Within a systems context, grazing crops can enable other pastures to be rested, thus increasing productivity of the whole-system[44] . This has been a common farmer practice in mixed farming systems in many developed countries [45] [7]. [7] summarize a range of reported studies and find consistently greater economic returns for dual-purpose cereal and canola crops compared with a grain-only option, particularly in dry seasons. However, the bioeconomic efficiency of dual-purpose management of cereal crops has not been widely assessed in developing world. Especially in the southern Mediterranean, where soil degradation, poverty and food insecurity are prevalent, adoption of dual-purpose management can provide desirable livelihoods outcomes.

Depending on the environment, grazing intensity and timing, and the resilience of plant species to grazing, satisfactory grain production can be obtained from dual-purpose cereal crops. Grazing at the tillering development stage does not necessarily reduce grain or fodder production [46] [47] and in some cases may even increase crop yields. For example, in a field trial in Tibet, [48] reported that cutting barley during the season to simulate grazing resulted in an increased grain yield of 0.9 t ha^{-1}. They hypothesized that the reduced plant height caused by cutting reduced lodging, which is a big problem for farmers in this region due to variety, management and seasonal conditions. Studies conducted in north-west China, have shown that cutting timing and height greatly influence the capacity of a wheat crop to respond after defoliation, with significant trade-offs between forage provision and grain yield recovery [49]. This seems to be particularly severe where crops have only a short period in which to recover while in longer season environments, the trade-off is less. Grazing of immature cereal crops (before stem elongation) may also have a positive impact on the nutritive value of the regrown forage material [50][51][52]. Given the widespread use of cereal straws in the diets of livestock in the West Asia-North Africa region, any improvement in the feeding value of these crop residues would have substantial effect on ruminant productivity.

The high nutritive value of immature winter annuals is aligned with the most productive period of growing lambs[44][53]. For example, [54] reported that crossbred lambs that grazed winter wheat had impressive liveweight gains (320-360 g/day) in Australia. Similarly, early spring liveweight gains of weaned merino lambs that consumed wheat (227 g head/d) and oat (221 g head/d) from tillering to booting stage were similar to the liveweight gains of lambs fed on a concentrate feed diet (256 g head/d) in Turkey [53]. At the vegetative phase, the energy content of dual-purpose crops is consistently measured at greater than 12 MJ ME/kg DM [35]. An issue when grazing crops is spatial variation in grazing intensity. Recent work in Australia found that preferred grazing areas could be manipulated through the use of a mineral supplement to attract sheep from over- to under-utilised areas of a wheat crop [55]. This could be a useful and practical grazing management practice in developing world where fencing to control grazing is not used in small segmented farms.

5. Crop residues and straw-based feeding

5.1 Stubble grazing, management and land tenure in WANA

Crop residues are a valuable livestock feed resource through the developing world. They are widely available, cheap to purchase, and can be stored for long periods to meet periods of feed shortages or reduce daily labour requirements for cutting feed for livestock. Over the last three decades, the feed gap has become more acute because of increased sheep numbers and expansion of cereal mono-cropping at the expense of food and forage legumes and rangeland resources in the south Mediterranean[9][56]. Consequently, cereal stubbles and straw have become the primary source of feed for small ruminants in the region [40][8]. Stubble grazing practices in the Mediterranean basin often lean towards intensive, high stocking rate methods, leaving soil barren. These intensive grazing practices also damage soil structure, create compaction, and lower the soils capacity to capture rainwater [57]. Although crop residues are perceived to be the essential nutrient source during the summer months, the dry matter intake and nutritional quality of the stubble decreases linearly with an increased number of grazing days and stocking rates [58][59]. Immediately after harvest, higher quality plant parts (leaves and some grain) are inadvertently left on the field. These higher quality materials have been shown to benefit livestock in liveweight gain, but animals consume these materials in a few weeks depending on the stocking rate [39]. Thus, the feeding value of stubbles declines over time and there is a point where the high-value components have been eaten and high fibre content of remaining stems will constrain intake to sub-maintenance levels. Provision of energy, protein and mineral supplements is required to utilize the remaining stubble.

5.2 Straw-based feeding in in the semi-arid tropics

Crop residues such as rice straw and maize stover often provide the bulk of livestock diets. However, they have low digestibility, and do not contain sufficient metabolisable energy (ME) or crude protein (CP) to maintain animal liveweight. Large amounts of research has been conducted to try to improve the feeding value of cereal straws through treatment with additives such as urea, ammonia and enzymes. These treatments have not been widely adopted by smallholder farmers because they are considered costly, labour intensive, technically difficult, or potentially dangerous to animal health [60]. Researchers in India have identified cultivars of maize, millet and sorghum that have higher quality stover without compromising grain yields [61] [62][63]. Supplements can be also used to increase the energy and protein content of livestock diets, but they need to be cost effective, easily accessible and practical to be adopted by farmers.

Crop by-products as such onggok (residue after starch extraction from cassava) or rice bran can also be used to provide additional energy to cattle fed rice-straw based diets. However[16] found that in Ongole cattle, feeding rice bran did not provide additional ME intake or liveweight gain compared to feeding gliricidia. It is important to note that while crop-residue based diets are sufficient for maintenance of livestock with low energy requirements, even with supplementation, they will not support high growth rates. However, by feeding rice straw to cows with low maintenance requirements, higher quality feed can be directed to high-value animals such as growing calves.

6. Socioeconomic and policy issues

Policies to promote the adoption of productivity-enhancing measures that will help improve the management of natural resources should be developed and enforced in developing countries. Improving the feed base through increased forage availability requires knowledge management, an understanding of the behavior of local communities, and the ability to communicate and reduce gender inequalities. Improved productivity and equitable access to inputs for crop-livestock production are needed to help increase household incomes as well as asset wealth and well-being through healthy livestock holdings. Recognizing that women play important roles within integrated crop and livestock production systems, greater attention must be focused on equitable access to knowledge and public and private services in the development of socially sustainable forage production systems. It is crucial to enhance national adoption of research-based recommendations through more effective linkages among national and international research systems, marketing and service providers, and farmers and consumers. Without proper technical backstopping and improved efficiency in crop-livestock production, producers in developing countries will not be able to compete with the more efficient foreign producers in either the domestic or export markets.

Understanding the whole-of-system implications of proposed technologies is also required. For example, studies have demonstrated that intensifying livestock production systems that require increased forage resources can have positive economic outcomes but can have negative impacts on food self-sufficiency and labour requirements [64]. The impact of particular technologies on both market and production risk is also often unexplored, this often requires whole-farm bio-economic modelling approaches. Furthermore, there is often a wide diversity of resource availabilities, capital access, skills and knowledge and production emphasis amongst small-holder producers. Hence, the impacts of particular technologies is likely to differ amongst them. For example, analysis of integrating forage legumes in Eastern Indonesia shows that farmers with access to underutilised land and more livestock are likely to benefit more [65].

Currently, lack of quality seed of improved forage varieties, poorly functioning seed and fodder markets, and biased national policy towards the production of strategic food grains and crops constrain the economic incentives for sustainable forage production and marketing. Redirecting subsidies on forage crops has provided successful outcomes for smallholder crop-livestock producers in various Mediterranean countries [66] [67]. Thus, public funds should be invested in supporting forage-based crop-livestock farming and research, including seed systems, rather than cereal mono-cropping or industrial-scale meat and milk production.

Reference

[1] World Bank 2008 World Development Report The World Bank Washington DC USA
[2] Entz M H, Baron V S, Carr P M, Meyer D W, Smith S and McCaughey W P 2002 Potential of forages to diversify cropping systems in the Northern Great Plains *Agronomy Journal* 94(2) 240-250
[3] Christiansen S, Ryan J, Singh M, Ates S, Bahhady F, Mohamed K, Youssef O and Loss S 2015 Potential legume alternatives to fallow and wheat monoculture for Mediterranean environments Crop and Pasture Science **66(2)** 113-121

[4] Masters D G, Revell D K and Norman H C 2010. Managing livestock in degrading environments In Sustainable improvement of animal production and health (N. E. Odongo, M. Garcia and G. J. Viljoen, eds.) pp. 255-267 Food and Agriculture Organization of the United Nations (FAO) Rome Italy
[5] Nefzaoui A, Louhaichi M and Ben S H 2014 Cactus as a tool to mitigate drought and to combat desertification *Journal of Arid Land Studies* **13** 121-124
[6] Barrett L E G, George R J, Hamilton G, Norman H C and Masters D G 2006 Multi-disciplinary approaches suggest profitable and sustainable farming systems for valley floors at risk of salinity *Australian Journal of Experimental Agriculture* ***45***(11) 1415-1424
[7] Bell L W, Moore A D, and Kirkegaard J A 2014 Evolution in crop–livestock integration systems that improve farm productivity and environmental performance in Australia *European Journal of Agronomy* **57** 10-20
[8] Porqueddu C, Ates S, Louhaichi M, Kyriazopoulos A P, Moreno G, del Pozo A, Ovalle C, Ewing M A and Nichols P G H 2016 Grasslands in 'Old World' and 'New World' Mediterranean climate zones: past trends, current status and future research priorities Grass and Forage Science **71**1-35
[9] Ryan J, Singh M and Pala M 2008 Long term cereal based rotation trials in the Mediterranean region: implications for cropping sustainability *Advances in agronomy* **97** 273-319
[10]Ates S, Keles G, Inal F, Gunes A and Dhehibi B 2015 Performance of indigenous and exotic× indigenous sheep breeds fed different diets in spring and the efficiency of feeding system in crop–livestock farming *The Journal of Agricultural Science* **153**(3) 554-569
[11] Loss S, Haddad A, Khalil Y, Alrijabo A, Feindel D and Piggin C 2015 Evolution and adoption of conservation agriculture in the middle east In Conservation Agriculture (pp. 197-224) Springer International Publishing
[12] Yau S K, Bounejmate M, Ryan J, Baalbaki R, Nassar A and Maacaroun R 2003 Barley–legumes rotations for semi-arid areas of Lebanon *European Journal of Agronomy* **19**(4), 599-610
[13] Hosang E J, Nulik J, Kanahau D, Abi Y and Bell L W 2016 Nitrogen contribution from forage legumes in maize farming system in West Timor Indonesia *Proceedings of the 2016 International Nitrogen Initiative Conference* Solutions to improve nitrogen use efficiency for the world 4 – 8 December 2016 Melbourne Australia www.ini2016.com
[14] Jelantik I G N, Mullik M L, Leo-Penu C, Jeremias J, Copland R 2008 Improving calf survival and performance by supplementation in Bali cattle *Australian Journal of Experimental Agriculture* **48** 954-956
[15] Jelantik I G N, Mullik M L, Leo-Penu C and Copland R 2010 Factors affecting the response of Bali cattle (*Bos sondaicus*) calves to supplementation prior to weaning Animal Production Science **50**(6) 493-496
[16] Antari R, Syahniar T M, Mayberry D E, Pamungkas D, Anderson S T and Poppi D P 2014 Evaluation of village-based diets for increasing the weight and condition of Ongole (Bos indicus) and Bali (Bos javanicus) cows in Indonesia *Animal Production Science* **54**(9) 1368-1373
[17] Antari R, Ningrum G P, Mayberry D E, Pamungka D, Quigley S P and Poppi P D 2014 Rice straw, cassava by-products and tree legumes provide enough energy and nitrogen for liveweight maintenance of Brahman (Bos indicus) cows in Indonesia *Animal Production Science* **54**(9) 1228-1232
[18] Syahniar T M, Antari R, Pamungkas D, Mayberry D E and Poppi D P 2012 The level of tree legumes required to meet the maintenance energy requirements of Ongole (Bos indicus) cows fed rice straw in Indonesia *Animal Production Science* **52**(7) 641-646
[19] Mayberry D E, Cowley F, Cramb R, Poppi D, Quigley S, McCosker K and Priyanti A 2016 Final Report 2016-16: Improving the reproductive performance of cows and performance of fattening cattle in low input systems of Indonesia and northern Australia Australian Centre for International Agricultural Research
[20] Le Houérou H N 1992 The role of saltbushes (Atriplex spp.) in arid land rehabilitation in the Mediterranean Basin: a review Agroforestry Systems **18** 107-148

[21] Norman H C, Silberstein R P and Byrne F 2008 Achieving profitable and environmentally beneficial grazing systems for saline land in Australia 12th meeting of the FAO-CIHEAM sub-network on Mediterranean pastures and fodder crops, "Sustainable Mediterranean Grasslands and their Multi-Functions", Elvas, Portugal; 01/2008

[22] Osman A E, Bahhady F, Hassan N, Ghassali F and Ibrahim T A 2006 Livestock production and economic implications from augmenting degraded rangeland with Atriplex halimus and Salsola vermiculata in northwest Syria *Journal of arid environments* **65**(3) 474-490

[23] Ghassali , Osman A E, Singh M, Norton B, Louhaichi M and Tiedeman J 2011 Potential use of Mediterranean saltbush (Atriplex halimus L.) in alley cropping in the low rainfall-cropping zone of northwest Syria. *Range Management and Agroforestry* **32**(1) 1-8

[24] Nefzaoui A, Ben Salem H and El Mourid M 2011 Innovations is small ruminants feeding systems in arid Mediterranean areas. In: "R. bouche, A. Derkimba, F. Casabianca (eds) New trends for innovation in the Mediterranean animal production" EAAP publication No 129 Wageningen Academic Publishers, ISBN 978-90-8686-170-5, ISSN 0071-2477. pp. 99-116

[25] Papanastasis V P, Yiakoulaki M D, Decandia M and Dini-Papanastasi O 2008 Integrating woody species into livestock feeding in the Mediterranean areas of Europe *Animal Feed Science and Technology*, **140**(1) 1-17

[26] Guevara J C, Silva Colmer J H, Estevez O R and Paez J A 2003 Simulation of the economic feasibility of fodder scrub plantations as a supplement for goat production in the north-eastern plain of Mendoza, Argentina *J. of Arid Environment* **53** 85-98

[27] Malcolm C V and Pol J E 1986 Grazing and management of saltland shrubs *J. of Agricultural* Western Australia **27** 59-63

[28] Clifton K and M Louhaichi 2015 Land Tenure, climate change and livestock mobility in Central and Southern Asian Grasslands Book chapter In P.K Ghosh, S.K. Mhanta, J.B. Singh, P.S. Pathak, Grasslands: A Global Resource Perspective (pgs. 347-362) New Delhi, India: International Grasslands Congress

[29] Alary V, Nefzaoui A and El Mourid M 2007 How risk influences the adoption of new technologies by farmers in low rainfall areas of North Africa? In: El-Beltagy, A. M.C Saxena and Tao Wang (eds) Human and Nature – Working together for sustainable Development of drylands. Proceedings of the 8th International Conference on Development of Drylands, 25-28 February 2006 Beijing China ICARDA, Aleppo, Syria. pp. 802-810

[30] Laamari A, Boughlala M and Chriyaa A 2005 Adoption and impact studies in Morocco In: H. Shideed Kamil and M El Mourid (eds) Adoption and impact assessment of improved technologies in crop and livestock production systems in the WANA region The development of integrated Crop/Livestock Production in Low Rainfall Areas of Mashreq and Maghreb Regions (Mashreq/Maghreb Project). ICARDA, Aleppo, Syria, 160 pp En pp 107-118

[31] Barrett-Lennard E G, Malcolm C V, Bathgate A 2003 Saltland Pastures in Australia – a Practical Guide, 2nd edition Sustainable Grazing on Saline Lands (a sub-program of Land, Water and Wool), 176 pp

[32] Flowers T J and Yeo A R 1995 Breeding for salinity resistance in crop plants: where next? Australian Journal of Plant Physiology. 22, 875-884

[33] Ghassemi F, Jakeman A J, Nix H A 1995 Salinisation of Land and Water Resources. CAB International, Wallingford, UK, 526 pp

[34] Qadir M, Noble A D, Schubert S, Thomas R J and Arslan A 2006 Sodicity induced land degradation and its sustainable management: Problems and prospects *Land Degradation & Development* **17**(6) 661-676

[35] Masters D G, Thompson A N 2016 Grazing crops: implications for reproducing sheep Animal Production Science **56** 655–668

[36] Norman H C, Masters D G and Barrett-Lennard E G 2013 Halophytes as forages in saline landscapes: interactions between plant genotype and environment change their feeding value to ruminants *Environmental and Experimental Botany* **92** 96-109.

[37] Norman H, Freind C, Masters D, Rintoul A, Dynes R, Williams I 2004 Variation within and between two saltbush species in plant composition and subsequent selection by sheep Australian *J. of Agricultural Research* **55** 999-1007

[38] Al Daini H, Norman H C, Young P and Barrett-Lennard E G 2013) The source of nitrogen (NH4+ or NO3–) affects the concentration of oxalate in the shoots and the growth of Atriplex nummularia (oldman saltbush) *Functional Plant Biology* **40**(10), 1057-1064.

[39] Ben Salem H, Norman H C, Nefzaoui A, Mayberry D E, Pearce K L and Revell D K 2010 Potential use of oldman saltbush (Atriplex nummularia Lindl.) in sheep and goat feeding *Small Ruminant Research* **91**(1)13-28

[40] Ben Salem H and Smith T 2008 Feeding strategies to increase small ruminant production in dry environments. *Small Ruminant Research* **77**(2) 174-194

[41] Pearce K L, Norman H C and Hopkins D L 2010 The role of saltbush-based pasture systems for the production of high quality sheep and goat meat *Small Ruminant Research* **91**(1) 29-38

[42] Norman H C, Hulm E and Wilmot M G 2015 Improving the feeding value of old man saltbush for saline production systems in Australia – a review Halophytic and Salt Tolerant Feedstuffs: Impacts on Nutrition, Physiology and Reproduction of Livestock. Eds Hassan M. El Shaer, Victor Roy Squires CRC press Taylor & Francis ISBN 9781498709200 - CAT# K24977

[43] ACIAR 2016 Forage options for smallholder livestock in water-scarce environments of Afghanistan http://aciar.gov.au/project/ah/2012/021

[44] Thomas D T, Moore A D, Norman H C and Revell C K 2015 Small effects of deferment of annual pastures through grazing spring wheat crops in Western Australia can benefit livestock productivity. *Crop and Pasture Science* **66** (4), 410-417

[45] Anderson W K 1985 Production of green feed and grain from grazed barley in Northern Syria. *Field Crops Research* **10** 57-75

[46]Fieser B G, Horn G W, Edwards J T and Krenzer E G 2006 Timing of Grazing Termination in Dual-Purpose Winter Wheat Enterprises1. *The Professional Animal Scientist* **22**(3) 210-216

[47] Cazzato E, Laudadi V and Tufarelli V 2012 Effects of harvest period, nitrogen fertilization and mycorrhizal fungus inoculation on triticale (× Triticosecale Wittmack) forage yield and quality *Renewable Agriculture and Food Systems* **27**(4), 278-286

[48] McNeill A, Wilkins J, Piltz J, Tashi N, Zhu Se, Tao J, Tsamyu H T, Brown C, Waldron S, Cummins J, Rose C, Coventry D and Paltridge N 2014 Integrated crop and dairy systems in Tibet Autonomous Region, PR China FR2014-04 Australian Centre for International Agricultural Research http://aciar.gov.au/publication/FR2014-04

[49] Tian L H, Bell L W, Shen Y Y and Whish J P M 2012 Dual-purpose use of winter wheat in western China: cutting time and nitrogen application effects on phenology, forage production and grain yield. *Crop and Pasture Science* **63**(6) 520-528.

[50] Francia E Pecchioni N, Li Destri N O, Paoletta G, Taibi L, Franco V, Odoardi M, Stanca A M and Delogu G 2006 Dual-purpose barley and oat in a Mediterranean environment. Field Crops Res. **99** 158-166

[51] Jacobs J L, Hill J and Jenkin T 2009 Effect of different grazing strategies on dry matter yields and nutritive characteristics of whole crop cereals *Animal Production Science* **49**(7) 608-618

[52] Keles G, Ates S, Coskun B and Koc S 2013 Re-growth yield and nutritive value of winter cereals. In Revitalising Grasslands to Sustain our Communities: Proceedings, 22nd International Grassland Congress, 15-19 September 2013 Sydney Australia (pp. 951-953) New South Wales Department of Primary Industry

[53] Keles G, Ates S, Coskun B, Alatas M S and Isik S 2016 Forage yields and feeding value of small grain winter cereals for lambs. Journal of the Science of Food and Agriculture

[54] Dove H, Holst P J, Stanley D F and Flint P W 2002 Grazing value of dual-purpose winter wheats for young sheep. *Animal Production in Australia* **24** 53-56

[55] Thomas D T, Moir B, Mata G, Toovey A and Flower K 2014 Use of supplementary feeding stations in a grazed wheat crop to increase the dispersion of sheep and their utilisation of the crop.

Proceedings of the Spatially Enabled Livestock Management Symposium 2014 18 November 2014 Hamilton New Zealand p 10

[56] Ates S, Feindel D, El Moneim A and Ryan J 2014 Annual forage legumes in dryland agricultural systems of the West Asia and North Africa Regions: research achievements and future perspective *Grass and forage science* **69**(1) 17-31

[57] Bell L, Kirkegaard J A, Swan A, Hunt J R, Huth N I, Fettell N A 2011 Impacts of soil damage by grazing livestock on crop productivity *Soil and Tillage Research* **113** 19–29

[58] Treacher T T, Rihawi S and Owen E 1996 Stubble grazing by sheep. In: Proceedings of the 2nd FAO Electronic Conference on Tropical Feeds Livestock Feed Resources within Integrated Farming Systems FAO Rome

[59] Moujahed N, Abidi S, Youssef S B, Darej C, Chakroun M and Salem H B 2015 Effect of stocking rate on biomass variation and lamb performances for barley stubble in Tunisian semi arid region and under conservation agriculture conditions *African Journal of Agricultural Research* **10**(50), 4584-4590

[60] Doyle P T, Devendra C and Pearce G R 1986 Rice straw as a feed for ruminants (International Development Program of Australian Universities and Colleges: Canberra)

[61] Anandan S, Khan A A, Ravi D, Rao M S B, Reddy Y R and Blümmel M 2013 Identification of a superior dual purpose maize hybrid among widely grown hybrids in South Asia and value addition to its stover through feed supplementation and feed processing *Field crops research* **153** 52-57

[62]Bidinger F R and Blümmel M 2007 Determinants of ruminant nutritional quality of pearl millet [Pennisetum glaucum (L.) R. Br.] stover: I. Effects of management alternatives on stover quality and productivity *Field crops research* **103**(2) 119-128

[63] Blummel M and Rao P P 2006 Economic value of sorghum stover traded as fodder for urban and peri-urban dairy production in Hyderabad India *International Sorghum and Millets Newsletter* **47** 97-100

[64] Komarek A M, Bell W, Whish J P, Robertson M J and Bellotti W D 2015 Whole-farm economic, risk and resource-use trade-offs associated with integrating forages into crop–livestock systems in western China. *Agricultural Systems* **133** 63-72

[65] Gabb S R, Bell L W, Basuno E, Prestwidge D, Prior J and Guppy C 2017 Whole-farm impacts of forage legumes in small-holder crop-livestock systems In Proceedings of the 2017 Agronomy Australia Conference, 24–28 September 2017 Ballarat Australia. www.agronomyconference.com

[66] Lloveras J, Santiveri P, Vendrell A, Torrent D and Ballesta A 2004 Varieties of vetch (Vicia sativa L.) for forage and grain production in Mediterranean areas. Cahiers Options Méditerranéennes **62** 103-106

[67] Demir N, and Yavuz F 2010 An analysis on factors effective in benefiting from forage crops support. *Scientific Research and Essays* **5**(15) 2022-2026

IOP Conf. Series: Earth and Environmental Science **142** (2018) 012077 doi:10.1088/1755-1315/142/1/012077

Economic activities results and problems of the farm engaged in production of stick tomatoes in Erdemli (Mersin) Province

M Direk and S Topkara
Selçuk University, Agriculture Faculty, Department of Agricultural Economics, Konya-Turkey

E-mail: mdirek@selcuk.edu.tr

Abstract. This study was carried out to investigate the socio-economic status of the stick-grown tomatoes located in Erdemli district of Mersin province and to examine the annual activity results and to reveal the problems. The data used in the research were obtained from questionnaires conducted with 44 selected farms by stratified random sampling method. The data obtained is for the 2016 production period. In the farmer surveyed, the operating area per farm was 7.01 da. In the enterprises, the average active capital per farm (69,916 $), 89.62% of the land capital and 2,40% of the operating capital. The ratio of own capital in the passive capital is 85.22%. In the examined farm, the average net yield was 3,150 $, the agricultural income was 5,483 $, the financial profitability was 4.29% and the economic profitability was 4.39%. As a result of the pie tomato cost analysis, the cost of tomato was determined to be below the tomato sales price. In the study, the cost of 1 kg product was calculated as 0,23 $. The surplus supply in the time of the intensive production of tomatoes in the region of the study causes the prices to decrease. For this reason, it is necessary to try to ensure price stability by ensuring balanced distribution of the accrual during the season.

1. Introduction

Turkey is a country which has great potential for vegetable species and production with its suitable ecology. Tomato has a significant place amongst vegetables; it has good nutrition facts, food industry and good product in both domestic and foreign trade. Tomato is one of the most important income sources for the farmers in the areas it has been being produced. Tomato production is common in especially Marmara, Mediterranean, Aegean and Black Sea region. Climate in Turkey is highly sufficient for tomato production. The industry to process the vegetable has widely been built since 1970s. For these reasons, Turkey has rapidly been lifted in the lists of tomato producer countries and placed amongst those who has vast production volume such as China, USA, and India.

Tomato is one of the most common of all the vegetables being produced, being consumed and been a subject to trade. Together with being produced in many countries in the world, thanks to its suitable climate conditions, Turkey is one of the most important countries in tomato production in the world. Tomato is being used freshly, in all kinds of meals, with other foods, as resistant tomato juice, as concentrated tomato juice, as pickle, as sauce, as smashed or as dried [1].

Tomato consists of 94,1% water. It is 23 calories when eaten. Tomato is cheap and source of lots of vitamins. Because it is both nutritious and delicious, in many of the countries it is commonly consumed. Since it can be grown out of season, it is being consumed all four seasons. In the tomato there are vitamins A, B1, B, B6, C, K, niacin, protein, fat, carbohydrate, potassium, calcium, iron and lycopene in high volumes. Vitamins A, B6, B1, C and lycopene in the tomato as antioxidant have protective effects against various types of cancer and heart diseases. In the studies, it has been found out that the people who consume tomato at least twice a week are 34% less at risk of having prostate cancer than those who do not consume. Also it was seen that tomato strengthens the immune system, slows the aging of brain cells, and beautifies hairs and skin [2].

In tomato production and marketing, since there is not an association and producers cannot get organized, the efficiency in marketing cannot be sustained. Producers do not have power in negotiating with merchants and they cannot take part in pricing the product. In this issue, existing organizations and associations do not have the necessary authorization and they are not strong enough to strengthen producers' hand and to come up with solutions to producers' problems. Thus, existing legal legislations should be reconsidered and by doing so producers' competitive capacity should be increased and producers' associations should take more active part in pricing. Surplus supply during the intensive production months causes prices to critically decrease. For this reason, price consistency should be ensured by conserving the balance in distributing the harvest during the season [3].

Erdemli county geographically located in the middle of the Mersin province, besides its agricultural economy which depends on greenhousing is a rapidly growing tourism and summer resort city with modern structuring. It has a distinctive importance with its cultural and historical environment. The area of Erdemli is 2.078 km². 62% of this area is forest, 17% of this area is agricultural areas, 21% of this area is forages, stony and rocky areas. Total agricultural area in the count is 33080 da, vegetable production is being made in the 5845 da of this area. There are 2460 agricultural farms in Erdemli. 1315 of these farms produce vine tomatoes. Vine tomato production is widely being done in the areas whose altitude ranges between 300 and 1100 from sea level. These places are also called and characterized as "yayla". In 61,6% of the vegetable production area of Erdemli, tomato production is being done. The yield rate of vine tomato in the county is considerably high (9100 kg/da). It is a significant problem that the important income source of the region, vine tomato production, is being done by small and disorganized farms. Encouraging vegetable farming in bigger farms will enable the usage of advanced technologies. By doing this, the cost of production will reduce and safer production will be executed and related to all these factors marketing will be easier.

The subject of this study is to investigate the socio-economic status of vine tomato producers in Erdemli, Mersin. Besides this it has been done to examine the annual activity results and to discover problems. This study aims to put light to the existing condition by analyzing the agricultural who produce vine tomato. By doing so it will be possible to make suggestions to improve vine tomato production. It will be possible to determine on what scale a more beneficial production can be made between agriculture. Also it is possible to determine existing problems and developing possible solutions for these problems, and to detect the physical and financial quantities of the entries used in production. Determining the cost of unit area and per kg, discovering producers' socio-economic and cultural status, determining producers' expectations have been set as the goal of the study

2. Material and Method

A large part of the material consists of the data obtained from agriculture through questionnaires. A number of neighborhood, in which there are intensive amounts of vine tomato producers, who can represent the region as means of vine tomato production and economic conditions has been intentionally chosen. Tomato producer all considered as a whole and the land size of these calculated with the formula given below according to stratified random sampling [4].

$$n = \frac{\Sigma\,(\mathrm{Nh.Sh})^2}{\mathrm{N}^2.\mathrm{D}^2 \;+\; \Sigma\,(\mathrm{Nh.}Sh^2)} \qquad \mathrm{D}^2 = \mathrm{d}^2/\,\mathrm{z}^2$$

In the formula; N: Company number in population, Nh: Company number in h stratum, Sh^2: Variance of h startum, n: Number of sample, d: Allowed tolerance from population average. Z: z value in standart normal distribution according to error rate.

After sampling 44 company has been found sufficient. 35 of these small, 9 of these are large scaled, these investigated in two groups as small and large scaled. The questionnaires applied to the coincidentally. The population at present in the are examined as means of age, gender and education status; family labor presence converted into male labor unit (EIB). To determine the presence of revenue and working animal in farms converted into the common cattle unit (BBHB) and to be able to compare this with the business expanse group's reputations. The area presence of the farms and status of partition, population and labor status, education status, amount of capital and components are determined. Labor of farms at presence converted into (EIB). While calculating the total male labor in the area of study, annual work time was considered 251 days and daily work time was considered 10 hours. Farms' annual activity results were calculated and they are compared according to expanse groups. Gross output (GSUD), gross proceeds (GSH), business expense, gross income, avail, agricultural income were calculated as a result of annual activity. GSUD was reached by multiplying obtained herbal and animal product quantity with the product prices given to the farmer. Non-operating agricultural income and housing rent were added to the GSUD and GSH was reached. Non-operating agricultural income consisted of using man and machine power of the company on the agricultural businesses outside of the company and direct income support payments by national registry of farmers. Life cycle cost was made up of constant cost and variable expenses. Gross income was calculated by subtracting variable expenses from GSUD. Avail was reached by subtracting business expenses from GSH. Constant cost was subtracted from gross income and added to non-operating agricultural income to find agricultural income.

3. Result and Discussion

Population presence per company is 1,90 and women population (52,54%) is more than men population(47,76%). 65,11% of the population consist of people aging between 15-49. Women population in the group whose ages ranging between 15-49 is also more than that of men. 15-49 age group's population density is important for labor potential of the farms. It was determined in the farms that 95,8% of the population older than 6 years of age are literate and 4,2% are not. It was found out that 23,72% of the literate population are primary school, 17,23% of them are middle school, 25,66% of them are high school, 6,16% of them are college and university graduates. On the farms investigated, it was detected that there are average 3,12 EIB labor per company. 52,25% of this amount are men, 47,75% of this amount are women. There are differences in labor presence between the farms of different size groups. Labor presence of small scaled farms was found to be greater (3,23 EIB) than that of large scaled farms. Labor presence per company is 116 EIB, and 5,02% of this number is the family's own labor presence and 94,98% is foreign labor presence. Women labor ratio is more than men labor ratio in total labor presence. This situation is because there is a significant need for women labor in harvesting and other labor requiring processes in field farming, in hoe plants' labor and especially in vine tomato production which is being done intensively in the region. Total labor potential in farms calculated as 1128 EIG. 89,28% of this amount is family labor which can be used in business, 10,72% of this is the family labor that is being used outside of the business. This ratio might show large differences from business groups to business groups. It was detected that family labor which is used in business was calculated as 248,55 EIG and idle labor was calculated as 642,16 EIG. According to this approximately 61,3% of the labor is idle.

In the study, land area per company was determined as 7,01. 72,07% of this value is property areas, 27,97% of it is rented areas. In the study, no partnership in businesses was encountered. In the large scaled farms average area of the company is 8,43 and 56,58% of this is property areas and 43,41 of this is rented areas. In the small scaled farms average area of the company is 5,63 and 87,56% of this is property areas and 12,44% of this is rented areas. In the large scaled farms, property area's share in overall company area is significant and it is less than that of small scaled farms.

Total land capital per company was detected as 32.664$, 50,66% of this value is property areas and 49,34% of it is rented areas. In the small scaled farms, capital for property areas are greater than that of large scaled farms. The capital for buildings per company is 19.388$, 96,16% of this is housing,

3,84% of this are storehouses. Average capital for herbs was determined as 15.587$. 24,36% of the herb capital is for orchard areas and 75,64% of it is for fields. In the small scaled farms this value is 14.757$ while it is 16.418 in large scaled farms. There were no hunting and/or fishing capitals encountered in the investigated farms. Total land capital per company was determined as 62.655$, 47,15% of this value is land capital, 4,94% of it is land improvement capital, 27,99% of it is building capital and 10,37% of it is herb capital.

Table 1. The distribution of the active capital in investigated .

Capital Groups		Farm Groups ($)					
		1. Layer (1-3 da)		2. Layer (4-5)		average	
		$	%	$	%	$	%
Land investment	Soil land	26,888	45.08	43,391	53.77	32,664	46.72
	Land reclamation capital	2,799	4.69	4,043	5.01	3,421	4.89
	Building investment	16,015	26.85	19,598	24.29	19,388	27.73
	Plant resources	7,325	12.28	7,037	8.72	7,181	10.27
	Total	53,029	88.90	74,070	91.78	62,655	89.62
Fixed Capital	Tooling machine capital	5,249	8.80	5,916	7.33	5,583	7.99
	Total	4,635	7.77	5,262	6.52	5,583	7.99
Revolving Operating Capital	Money capital	1,984	3.33	1,369	1.70	1,677	2.40
	Total	1,984	3.33	1,369	1.70	1,677	2.40
Total business capital		6,619	11.1	6,632	8.22	7,260	10.38
Total active capital		59,648	100	80,702	100	69,916	100

Approximate tool-machine capital per company was determined as 5.583$. 62,33% of this value are tractors, trailers (8,68%), plough (1,53%), pulverizator (2,82%), cultivator (0,91%), other tool-machine capital (0,91%), rototiller (3,63), motopomp (0,98%). Money capital per company is 1.677$, in small scaled farms this is determined as 1.984$, in large scaled farms this is determined as 1.369$. In the farms investigated there has been no spare material capital encountered. Business capital per company was determined as 6.625$. 25,20% of this value is circulating capital, 74,78% of it is fixed capital. Fixed capital is 5.583$ and since there is no animal capital, all of this value is tool-machine capital. Circulating capital is 1.677$, all of this value is money capital.

Total active capital is 69.916$. 2,40% of this value is business capital, 89,62% of it is land capital. Land capital in large scaled farms (91,78%) are greater than that of small scaled farms(88,90%). Foreign capital per company was determined as 12.170$. 15,13% of this value are debts 84,87% of this is the value of rented areas. The amount of foreign capital shows differences according to company sizes. In the large scaled farms the amount of foreign capital is greater than that of other business groups. Foreign capital is 18.817$ per company. The main reason of the debts the farms have to the banks is because in the recent years the presence of tool-machine that farms have increased.

In the investigated farms, passive capital per company was determined as 82.346$. 85,22% of this is owner's equity and 14,78% of this is foreign capital. The status of passive capital may vary between business groups. In small scaled businesses, passive capital per company was determined as 65.172$. 91,52% of this value is owner's equity and 8,48% of this is foreign capital. In the small scaled farms, the reason for the owner's equity ratio to be more in the passive capital is because small scaled farms are operating less rent areas compared to large scaled farms. In the large scaled farms, total passive per company was determined as 99.520$. 81,09% of this is owner's equity and 18,91% of this is foreign capital.

Table 2. Distribution of passive capital in investigated .

	1-3 da				4-5 da	
	$	%	$	%	$	%
Foreign capital	5.524	8,48	18.817	18,91	43.815	14,78
Own capital	59.648	91,52	80.702	81,09	252.633	85,22
Total passive capital	65.172	100	99.520	100	296.448	100

In the investigated , profitability factor per company was determined approximately as 19,21%. This ratio shows differences according to business groups. In small scaled farms this ratio was determined as 27,28% while it was 11,93% in the large scaled farms. In the investigated farms, return on equity was detected as 4,29%. In the large scaled farms this was 1,87% and in the small scaled farms it was 6,72%. In the investigated farms, average economic profitability was detected as 4,39%. This varies according to the business groups. In the large scaled farms economic profitability was calculated as 2,07%, while it was 6,51% in small scaled farms. As a result of calculating return on equity and economic profitability, it was seen that in the large scaled farms the value is lower than that of small scaled farms. This is because; in the large scaled farms, although the owner's equity is in large amounts the net profit levels are way too low. In the investigated farms, capital turnover ratio was 19,50% and capital turnover time was calculated as 5,24 years. This varies between business groups. In the small scaled farms, capital turnover ratio was 22,6% while the capital turnover time was calculated as 4,42 years. In the large scaled farms; however, capital turnover ratio 16,4% while capital turnover time was detected as 6,06 years. In the investigated farms, financial leverage ratio was detected as 2,30%. This value in the small scaled farms was detected as 2,63% and in the large scaled farms it was 1,98%.

In the investigated farms, GSH per company was detected as 16.400$. 95,05% of this value was GSUD, 4,95% of this was non-operating agriculture income. As well as this, the rate of GSH to the active capital was calculated and noted as 4,37% per company. In small scaled farms, GSH per company was determined as 15.550$. Adding to this, the ratio of GSH to the active capital was detected as 26,07%. In the large scaled farms, GSH per company was 17.249$. Concurrently, the ratio of the GSH to the active capital was determined as 21,37%.

In the investigated farms, variable cost was calculated as 8.623$ per company. Variable costs amount increase as the size of a company increases. In the small scaled farms, total variable cost was calculated as 6.945$. In the large scaled farms, variable cost per company was calculated as 10.301$. In the investigated farms, fixed cost per company was calculated 4.625$. 35,35% of this value was depreciation expenses, 2,71% of it was building maintenance expenses and 58,90% of it was family labor fee. Fixed costs shows differences as the company sizes grow. In the small scaled farms, fixed cost per company was calculated as 4.362$. 52,17% of this value is depreciation expense, 36,02% of this is family labor fee and 11,81% of it is building maintenance expenses. In the large scaled farms, fixed cost per company was calculated as 4.889$. 34,01% of this value is depreciation expenses, 5,46% of it is building maintenance expenses and 60,53% of it is family labor fee. In the investigated farms, avail per company was calculated 3.150$. Avail shows differences as company sizes increase. While in the large scaled farms it was 2.058$, in the small scaled farms it was 4.242$. Avail per decare of the business area in the small scaled farms was 859$, in the large scaled farms it was 378$ and company average was calculated as 618$. The ratio of avail to the active capital was calculated and in the small scaled farms it was 26,07%, and in the large scaled farms it was 21,37% and company average was 23,72%. In the investigated farms, gross income per company was calculated 6.964$. While it was 7.812$ in the small scaled farms, it was 6.116$ in the large scaled farms. In the investigated farms, agriculture income per company was calculated 5.482$. This value varies according to the business groups. In the small scaled farms, agricultural income was 6.495$, in the large scaled farms it was 4.470$. Also vine tomato production per company was determined 8.756,19 kgs, in the small scaled farms it was 9.038,69 kgs, and in the large scaled farms it was 8.473,68 kgs.

In the investigated farms, gross vine tomato production value per company was calculated 3.048$. All of this value consists of main product. Because vine tomato harvest period starts with the 3rd week of June and goes until mid-August, there are no side product incomes. In the small scaled farms, vine tomato production value per company is 3.055$, and in the large scaled farms it is 3.040$. In the investigated farms, vine tomato production cost per company is 2.001$, in small scaled farms it is 2.248$, and in large scaled farms it is 1.860$. Company average unit tomato cost is 0,23$, in small scaled farms it is 0,25$, in the large scaled farms it is 0,22$. Vine tomato price in 2016 production period was 0,35$. According to agreed price of it, net profit average for farms per decare is 1.046$. In the small scaled farms vine tomato price was 0,34TL and producers made 807$ profit out of it. Large scaled farms made 1107$ profit and their apparent price for vine tomato was 0.36$. According to this information, small scaled farms made the best efficiency out of vine tomato per decare. One of the elements which decides the vine tomato production value; price is variable because price follows a variable course.

Existing problems of vine tomato production were detected. Main problem of it is high values of input price in research area, no alternative market exist so producers can't sell their products in the desired price and some problems occurring in the marketing. In the production phase of vine tomato, when compared to other vegetables it requires more input and labor, thus it has a high production cost. Not only that, but also quality is an important factor on vine tomato. The reason why it can't be in the desired quality is because tomato is a very sensitive plant against diseases and pests. Product quality may be effected dearly when the disease and pest amounts are high.

Vine tomato producers in the region, get their seedlings by ordering from the seedling producers as it requires expertise. Seedling supplying is one of the most important problems of producers as it is expensive. In the research area, it was detected that most of the vine tomato planting farms do not do organized alternation. For the farms that do alternation it was found out that they are doing alternation with green pepper and string beans. When the data of two years which is obtained from summer vegetables evaluated together, in tomato species on efficiency values pre-plant has significant effect and especially after plants of cabbage, broccoli and cauliflower had good efficiency values. Considering efficiency values obtain, for both years, cabbage parcel has found more effective. In tomato production, especially effect on efficiency, findings of both years was interpreted in such a way that alternation has a positive effects [5].

In the last half of the century, in agriculture activities there have been important changes. In the new applications, inorganic fertilizer usage has increased, in the battle against pests and diseases chemicals started to be used widely, monoculture production systems have gained significance and earth processing has been much more than ever. Chemical drugs and fertilizers pollute earth, underground and ground waters. Waste products corrupts earth, plants and eatables in a way that they could be harmful for living. Because of these reasons, ecologic agriculture is more important than the conventional agriculture systems which is widely used in many countries.

In the investigated farms, in the earth preparation process various agriculture tool and machine has been being used other than substitution tool and machines. In vine tomato production, it was detected that hard labor is being done more than machine usage. In earth preparation, it was found that in different times, with various processing tools in various depths the earth had to be processed. The aim is to prepare a better seed bed, struggle with the foreign weeds, reconfiguring the earth's physical structure to make it keep and transmit more water [6].

In the investigated farms, all of the farms has been using drip irrigation. Main reasons as to why producers have been using drip irrigation are reducing the labor, fertilization convenience and insufficient water resources. In the investigated farms, the biggest problem that they face is the insufficient water resources and being unable to water their plants in time related to this. Drip irrigation is a method in which sterilized water is being given to the roots of a plant by inserting the system underground or it is being given to the surface of earth through emitters by dripping [7]. This method has been widely used and each day users of this method increase in number, because it makes a good water economy and it enables the user to keep the earth humidity in the desired levels and it even enables the user to give the nutrients to the plant with water. Agricultural pest control has been being done frequently in the investigated farms, and no hormones are being used. Producers indicated that they often encounter diseases and pests. Mostly they encounter total absolute and green worm pest. Most of the vine tomato producers generally take technical assistance from pesticide dealers. Because total absolute pest's population was at high levels in 2016 production period and agricultural pest control was not strong enough against this; most of the producers indicated that they could not get the efficiency they desired to have.

In inorganic and organic conditions, it was reported that in the study that researchers did to investigate the effects of efficiency and fruit quality in tomato producing on brix applications' effect, just like the findings obtained, was insignificant as means of statistics [8].

Vine tomato is one of the agricultural products that needs frequent fertilization. This process must be done very carefully, it has to be in time and efficient. In the investigated farms, it was detected that generally they are doing the fertilization after each harvest. Vine tomato is one of the plants which requires foreign labor. Especially reasons such as increase in the area it was produced and not having enough family labor levels makes the foreign labor usage in large scaled farms needed. Generally, producers fulfill their foreign labor need from the farmer families around the region. Vine tomato producing areas are enlarged in the recent years with the increase of irrigation areas and cultivation areas. And thus labor requirement increased and because the labor presence in the region is not enough, it was supplied by neighborhoods near. Foreign labor was supplied from outside the county and thus this increased the fee paid to the foreign labor.

Vine tomato harvest frequency ranged between 25-30 in 2016 production period. When climate conditions are normal, a harvest was being done per 2-3 days. In the investigated farms, most of the farms are not happy with the vine tomato prices and it was detected that merchants determined the price of the vine tomato. Because producers have no alternative they got to sell their products to the merchants. Vine tomato producers in the region are in large numbers, although there are many producers in the region there are no association, and this is one of the other important problems. Producers in the mutual agreement that they need an association especially for the vine tomato producers. This situation shows that producers are having hard times in getting organized and that they are in need of a leader. In the research region, socio-economic conditions of environment, having a irrigation association, the tendency to have an association in farmers are the signs that promising associations might be founded. After required preparations for the desired association and its goals are set for sure, it will enable the region to develop further. If an association founded and headed towards activities such as production and marketing and if it finds information and foreign resources when farmers are in need, it would be beneficial for the farmers. It was detected that producers want an association to be formed to solve their problems with marketing and especially in the pricing. They at least want a base price for the products. Investigated farms has indicated that if a base price is to be set they would even consider their cultivation areas. Main reason for this increase would be the efficiency of vine tomato in the area is at desired levels and profit margin is high. Sarıaydın dam, which is being planned to be built on Silifke County's Lamas creek, which supplies irrigation for the research area, there will be a rise in the irrigated areas and there will be more production done in the area. In the investigated farms it was found that producers need information on growing vine tomato.

4. Conclusion

The findings after the research are; farmers are not very careful on choosing the seedling and producing vine tomato. However, choosing the appropriate tomato seedling is one of the most important elements which decides the amount and quality of the production. And related to this, high quality, more efficient and disease-resistant seed usage will increase efficiency and quality in the region. Before planting the seedling of vine tomato, the physical and chemical characteristics of the earth should be determined and proper fertilization must be done. When producing tomato, farms are facing lots of pests and diseases, and there is an apparent inexperience they have to fight against these challenges. Wrong applications in agricultural pest control, insensible and over usage of drugs are one of the most important problems; and this situation in domestic and foreign trade causes the products to be sent back and thus affecting the foreign trade badly. For human health and safe food, unnecessary use and over usage of the drugs must be prevented. It was detected that farms were not getting earth analysis done, are not being sensible and conscious about the amount and type of fertilizer. Producers are doing the applications of such with their experiences. To make the earth analysis common and to make the producers more conscious about the situation, concerned corporation must do the necessary education activities and try to spread awareness. By doing so, over usage of the fertilizer will be prevented and the producers will be made sure to use the fertilizer in a conscious manner and one of the most important input cost will be reduced. If most important input materials; drug and fertilizer choice are to be done by the help of an expert it would also reduce the input cost dearly in the region. After the harvest, grading, packaging and offering to market will sustain the profit and continuity. As a result, production must be done according to the techniques and be more conscious to have more efficient and profitable vine tomato in the region. Surplus supply in the intensive tomato production periods causes prices to decrease. For this reason, price consistency should be ensured by conserving the balance in distributing the harvest during the season. In the investigation region, the small farms which are not large enough for mechanization, should use these machines partnered up, this will help their economy. By partnering up in buying machines, small farms could have the costly machines and use them on a more comfortable way since their time to use these machines are limited in the season. Thus, their cost for machine per production unit will be reduced since it will be working one a larger area [9].

References

[1] Gazozcuzade N 2011 *Silifke Yayla Köylerinde Domates Üretiminde Hastalık Yönetimi* (Adana: Yüksek Lisans Tezi).

[2] MEGEP (Mesleki Eğitim ve Öğretim Sisteminin Güçlendirilmesi Projesi) Bahçecilik Domates Yetiştiriciliği 2008 Ankara.

[3] Ok Ü 2009 Tokat İl Merkez Köylerinde Sırık Domates'in Üretim ve Pazarlaması Y (Tokat: Lisans Tezi).

[4] Yamane T 1967 Elementary Sampling Theory Prentice Inc (NS, USA: Englewood Cliffs).

[5] Algan N, ve İ Duman 1996 Ekolojik Tarımda Rotasyon (Ekim Nöbeti), Ekolojik (Organik, Biyolojik) Tarım edt U Aksoy, A Altındişli Ekolojik Tarım Organizasyonu Derneği (Eto) s 17-22 Bornova-İzmir.

[6] Kösem S 2001 Baklagil Tarımı ve Geliştirilen Çeşitler.

[7] Kanber R 1999 Sulama Çukurova Üniversitesi Ziraat Fakültesi Genel Yayın No: 174 Ders Kitapları No: A-52 Adana.

[8] Karataş A, Padem H, Ünlü H 2005 Sera ve Tarla Koşullarında Yetiştirilen Bazı Sırık Domates Çeşitlerinin Verim ve Kalite Özelliklerini Karşılaştırılması *Süleyman Demirel Üniversitesi Fen Bilimleri Enstitüsü Dergisi* **9**(2) 42-49.

[9] Direk M, Bayramoğlu Z, ve Paksoy M 2002 Konya ilinde fasulye üretiminde karşılaşılan sorunlar ve çözüm önerileri *SÜ Ziraat Fakültesi Dergisi* **16**(30) 21-27.

IOP Conf. Series: Earth and Environmental Science **142** (2018) 012078 doi:10.1088/1755-1315/142/1/012078

The effect of mixed liming and NPK fertilizer to yield of some rice varieties on new openings of acid sulfate tidal swamp land

A Akhmad[1,2], W S Dewi[3], S Sagiman[2] and Suntoro[3]
[1]Doctoral Program of Agricultural Science, Graduate School,
Universitas Sebelas Maret (UNS). Jl. Ir. Sutami 36A, Kentingan, Surakarta, Indonesia
Email: asmahanakhmad@gmail.com
[2]Department of Soil Science, Faculty of Agriculture, Universitas Tanjungpura, Pontianak, Indonesia
[3]Soil Science Department, Agriculture Faculty, Universitas Sebelas Maret (UNS), Surakarta, Indonesia

E-mail: asmahanakhmad@gmail.com

Abstract. The strategies to meet the staple food needs in Indonesia is to open new paddy fields in the sub-optimal land. The research aims to get adaptive rice varieties with the highest yield on new openings of the acid sulfate tidal swamp applying mixed liming and NPK fertilizer. The experiment was conducted in a greenhouse at the Faculty of Agriculture, Tanjungpura University, Pontianak. The trials used a factorial completely randomized block design consisting of two factors. The first factor is a mixture of dolomite with NPK fertilizer, consisting of 3 levels (1 ton/ha dolomite and 60 kg/ha NPK; 2 ton/ha dolomite and 90 kg/ha of NPK, and 3 ton/ha dolomite and 120 kg/ha NPK). The second factor is rice varieties, consisting of 6 levels (Ciherang, Situ Bagendit, Inpara, Mira, Si Randah and Ringkak Janggut). Each treatment replicated four times. The results showed that the application of a mixture of 3 ton/ha dolomite and 120 kg/ha of NPK fertilizer showed the best results to improve rice yield on new opening of the acid sulfate tidal swap. Local rice varieties, Ringkak Janggut, applied 3 ton/ha dolomite and 120 kg/ha NPK fertilizer showed the best result of 1000 seed weight, i.e., 28.19 g, and total grain amount per panicle is 110.75 grains, with the lowest number of empty grains. Local rice varieties Ringkak Janggut potential to be developed as superior varieties on new opening acid sulfate tidal swamps by applying liming and fertilizer.

1. Introduction

Rice (*Oryza sativa* L.) is a staple food for most of the world population, especially in Indonesia. Rice crop is cultivated in wetland or dry land. Some obstacles of rice cultivation in Indonesia are the limited availability of land and majority of paddy fields are degraded. One strategy to overcome this problem is to open new paddy fields in the submarginal soil, such as tidal swamp. The new openings of the acid sulfate tidal swamp land have limiting factors, i.e., low pH and nutrient availability. Therefore, it is necessary to apply lime and NPK fertilizer. Besides that, it is required to find rice varieties that are tolerant of acidity and high production. Some rice varieties that can adapt to low pH and potentially high yields include Ciherang, Situ Bagendit, Inpara, Mira, Si Randah and Ringkak Janggut. This study aims to determine the effect of NPK fertilization and liming on the growth and yield of several varieties of rice, to obtain paddy that is adaptive and high production.

2. Methods

The experiment conducted at the greenhouse using randomized completely block design (RCBD) comprise two factors. The first factor was Rice Varieties (V), consists of six varieties, namely: Ciherang (V1), Situ Bagendit (V2), Inpara 3 (V3), Mira 1 (V4), Si Rendah (V5) and Ringkak Janggut (V6). The second factor was the dosage of mixed dolomite lime, and NPK fertilizer (P) consists of three levels. Those treatments are 1-ton dolomite + 60 kg NPK per ha (P1), 2-ton dolomite + 90 kg of NPK per ha (P2), and 3-ton dolomite + 120 kg NPK per hectare (P3). Each treatment replicates four times.

3. Results and Discussion

3.1. Total Grain Content Per Penicle

Treatment varieties and calcification of dolomite-NPK fertilization happens real influence on the treatment of liming dolomite-NPK fertilizer, to determine which treatment is different, then the Duncan Range Test (Table 1).

Table 1. Effect of Liming Dolomite-NPK fertilization Duncan Range Test to the Grain Total Content Per Penicle

Rank Mean	Name	Mean
1	P_3	79.42 a
2	P_2	73.21 ab
3	P_1	68.50 b

Table 1 shows that treatment P_3, P_1 significantly different from the treatment, but no significant effect of treatment with P_2, while P_2 treatment, no significant effect on the treatment P_1.

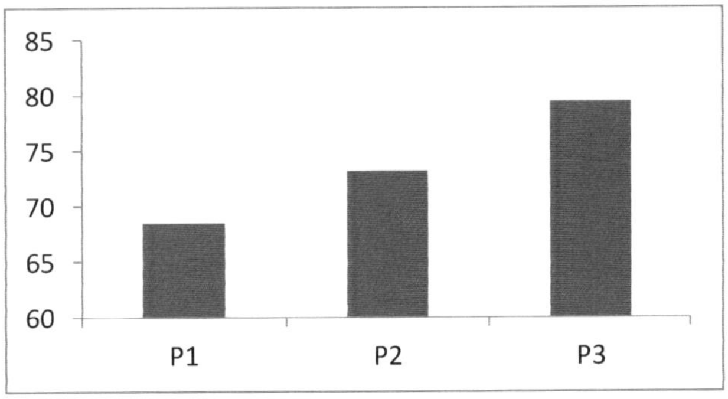

Figure 1. Histogram Total Grain Content Per Penicle

In Figure 1, can be explained that the P_3 is a treatment highs and lows P_1. Number of filled grain per panicle grain meant is truly filled with grain rather shiny color. NPK fertilizer can affect rice yields (number of grains per panicle and the number of filled grain per panicle) were associated with increased availability of nitrogen in the soil and nitrogen uptake by plants, but it is also the availability of Nitrogen and Phosphorus in the soil. The third element of this macro is a very important nutrient required by plants, where the interaction of these three elements will be able to support the growth and yield of rice is better. Nitrogen can increase the number of grains per panicle and the number of filled grain per panicle [1].

3.2. Empty Grain Per Penicle

In empty grain unfilled grain flour for charging failure caused by various factors, such as nutritional deficiencies while charging seeds. Many factors affect the emptiness grains. Such factors as of lodging, lack the intensity of sunlight, leaf-leaves dry as well as pests and diseases that cause low.

Analysis of diversity of varieties and liming treatment dolomite NPK-fertilizer to the number of grains per panicle empty, it turns out real effect on the treatment of varieties, to determine which treatment is different, then the Duncan Range Test (Table 2).

Table 2. Effect of Varieties Duncan Range Test to Total Empty Grain Per Penicle

Rank Mean	Name	Mean
1	V_4	5.64 a
2	V_3	4.91 ab
3	V_1	4.10 b
4	V_5	3.99 b
5	V_2	3.44 b
6	V_6	3.32 b

From Table 2 above shows that no significant V_4 to V_3, but significantly different from the V_1, V_2, V_5 and V_6, while the V_3 was not significantly different from the treatment V_1, P_2, P_5 and V_6. Among treatment V_1, P_2, P_3, V_5dan V_6 not real diffrent.

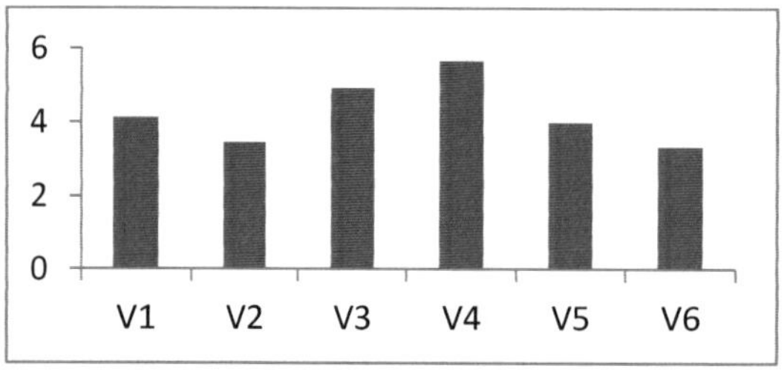

Figure 2. Histogram Total Empty Grain Per Penicle

In Figure 2, it can be explained that V_4 and V_6 is the highest treatment lows. ability to replenish grain rice-grain. Another opinion said that the large number of empty grain caused by lack of Nitrogen nutrients while charging seeds [2]. Nitrogen deficiency rice plants, will be the least amount of seedling and stunted growth, grains of rice produced would be an empty lot.

3.3. Number of Grain Per Penicle.

In a variable number of grains per panicle, it showed an obvious effect on the interaction between varieties and dolomite liming-NPK fertilizer. Where different treatment combinations, Duncan Range Test performed (Table 3).

Table 3. Duncan Range Test Interaction Effect to Total Grain Per Penicle

Name	Mean n Non-significant ranges		
Varieties/ Fertilizer	P_1	P_2	P_3
V_1	106.75 b	86.50 a	97.00 a
V_2	84.50 a	78.75 a	75.00 a
V_3	99.25 a	109.00 b	98.00 a
V_4	96.50 a	92.00 a	103.75 a
V_5	83.75 a	95.00 a	95.50 a
V_6	65.50 a	99.75 a	110.75 b

From Table 3 above can be explained, that the treatment V_6 and P_3, gives the number of grains per panicle best when compared with other treatments interactions.

IOP Conf. Series: Earth and Environmental Science **142** (2018) 012078 doi:10.1088/1755-1315/142/1/012078

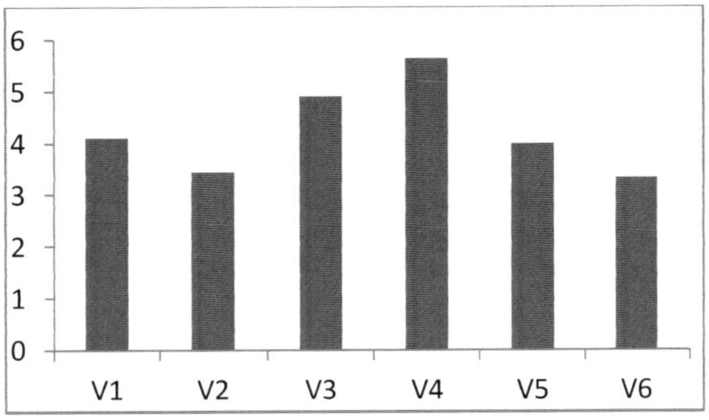

Figure 3. Histogram Number of Grain Per Penicle

In Figure 3, it can be explained that a combination treatment V_6P_3 highest and lowest V_6P_1

One of the components of rice yields is the number of grains per panicle. NPK fertilizer can affect rice yields. NPK fertilizer independently can increase the available nitrogen generative growth, including number of grains per panicle. Besides nutrients Phosphorus and Potassium together with the nitrogen will be able to support the rice yield and better. The nitrogen may increase the number of grains per panicle [1].

3.4. Weight 1,000 Grain Rice

In 1000 grain weight variable, it appears that the combined treatment of liming varieties and dolomite-NPK fertilization, occurs very real effect on the interaction. Which treatment combination significantly different, then the Test Distance Duncan, as presented in Table 4.

Table 4. Influence Interaction Duncan Range Test to Weights 1,000 Grain Rice

Name	Mean n Non-significant ranges		
Varieties/ Fertilizer	P_1	P_2	P_3
V_1	28.65 b	25.23 a	25.25 a
V_2	25.98 b	26.20 b	28.00 b
V_3	25.30 a	26.85 b	28.33 b
V_4	26.48 b	28.10 b	28.58 b
V_5	25.80 b	27.93 b	25.00 a
V_6	21.65 a	26.15 b	28.19 b

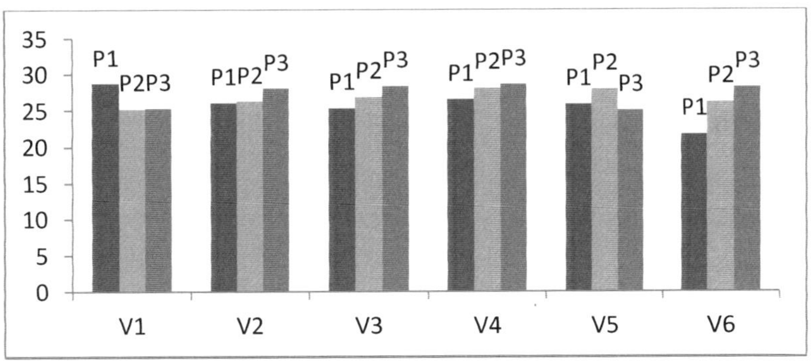

Figure 4. Histogram Weights 1,000 Grain Rice

In Figure 4, it can be explained that the combination treatment is the treatment V_1P_1 V_6P_1 highs and lows.

1000 grain weights of each variety are affected by the condition after flowering, such as the availability of nutrients, good or bad weather and the number of leaves. This condition affects the amount of carbohydrates produced by photosynthesis and will then determine the size of the grain.

From Table 4 is known, that each variety has an average weight of 1,000 grains different. 1000 grain weight is influenced by genetic and environmental factors and environmental conditions experienced during the time of planting seed filling. Weight of 1,000 grains of seeds is affected by the shape of grains as well as the size of the grain. The larger the grain size, has the potential weight of 1,000 grains also increase further, while the grain size is influenced by genetic factors plant itself.

Potassium can increase the weight of 1000 grain and rice crops increased tolerance to adverse climatic conditions and pest attack [3]. Nutrients Potassium helps the enzyme activity in the opening and closing of stomata and Potassium deficiency can inhibit the translocation of carbohydrate and nitrogen metabolism.

3.5. Grain Weight Per Clump and Converted Into Per Hectare

In variable grain weight per clump, it showed an obvious effect on the treatment of varieties and dolomite liming-NPK fertilization, between the treatment of varieties and dolomite liming-NPK fertilizer. Which treatment was significantly different, do Duncan Range Test (Tables 5 and 6).

Table 5. Effect of Varieties Duncan Range Test to Grain Weight Per Clump

Rank Mean	Name	Mean
1	V_6	12.75 a
2	V_5	12.34 ab
3	V_2	11.7 ab
4	V_3	11.48 ab
5	V_1	10.54 b
6	V_4	9.89 b

From Table 5 can be explained, that the V_6 no significant effect on the V_1, V_2, V_3 and V_5, but significantly different with V_1 and V_4. Among treatment V_1, V_2, V_3, V_4 and V_5 is not significant.

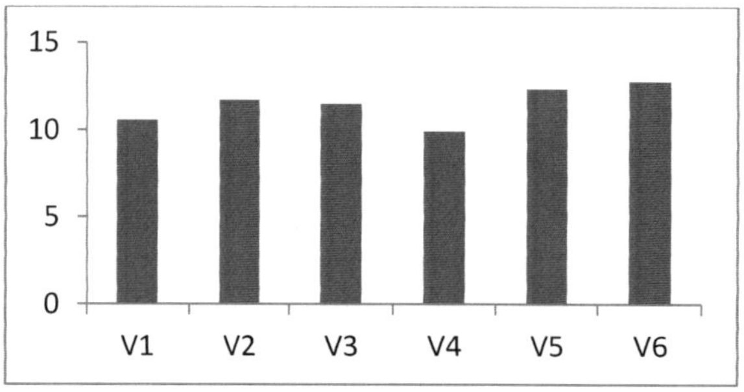

Figure 5. Histogram Grain Weight Per Clump

In Figure 5, it can be explained that the V6 is the highest treatment and lowest V4.

Table 6. Test Distance Duncan Effect of Liming Dolomite and NPK fertilization to Heavy Grain Per Clump

Rank Mean	Name	Mean
1	P_3	12.10 a
2	P_2	11.10 a
3	P_1	10.30 b

From Table 6, it can be explained, that no significant P_3 to P_2, P_3 and P_2 but significantly different with P_1.

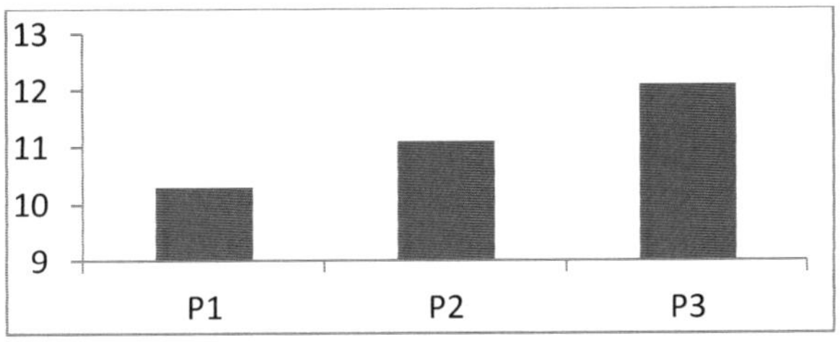

Figure 6. Histogram Grain Weight Per Clump

In Figure 6, can be explained that P_3 is the treatment of the highest and P_1 lows. Of variable grain production per hectare, it showed an obvious effect on the treatment of liming dolomite-NPK fertilization, liming of treatment varieties and dolomite-NPK fertilization. Which treatment was significantly different, then the Test Distance Duncan (Table 7).

Table 7. Effect of Liming Duncan Range Test Dolomite and NPK fertilization to the Grain Production Per Hectare

Rank Mean	Name	Mean
1	P_3	1.94 a
2	P_2	1.91 ab
3	P_1	1.64 b

Table 7 can be explained, that no significant P_3 to P_2, P_3 and P_2 but significantly different with P_1

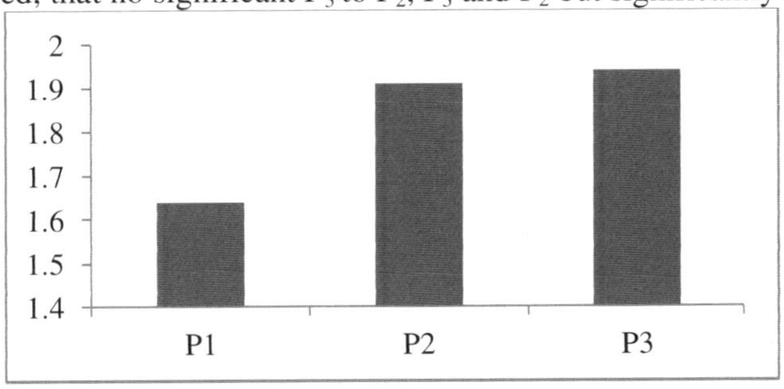

Figure 7. Histogram Grain Weight Per Hectare

In Figure 7, it can be explained that the P_3 is the highest treatment dan P_1 lowest. One of the considerations in selecting varieties of rice farmers to be cultured is the large potential for the results to be obtained. The yield per clump (per hectare) is determine how large the results of these varieties, so it is feasible to cultivated varieties [2].

As is well known, that the factors affecting the production of rice plants are: the number of plants per land area, number of tillers per hill, number of grains per panicle and grain weight of 1,000 grains. So the production of rice plants is the resultant of the photosynthetic process, assimilates decline due to respiration and translocation of dry ingredients into the crop. Increased production is directly proportional to the increase in the relative growth and yield net photosynthesis. Further explained that the components of the results is influenced by genetic factors, environmental factors such as liming and fertilizing, plant spacing and solar radiation.

So the highest production on V6 (Ringkak Janggut Variety). Ringkak Janggut is a local variety that is the result of a cross between the Ringkak Kuning (Sambas Local) with Rojolele (National Superior). The crosses result in a slightly broader leaf shape, slightly erect and slightly tapered at the end. The stem is not very high 106.08 cm, the maximum number of tillows is 10.67 stems and the number of productive tillers is 9.48 stems. Number of grain content per panicle 79.83 grains, empty grain per panicle 12.17 grains, grain total 110.75 grains and weight of 1,000 grains 28.19 grams.

IOP Conf. Series: Earth and Environmental Science **142** (2018) 012078 doi:10.1088/1755-1315/142/1/012078

4. Conclusion

Rice varieties adaftif and the highest production in the swamp tidal acid sulphate new openings: From research in Plastichouse, all variables reproductive (number of filled grain per panicle, number of grain hollow per panicle, number of grains per panicle, weight of 1,000 grains and production per clumps / per plot and conversion to a per hectare, significantly.

Varieties discovery Ringkak Janggut as a variety of the most adaptive and highest production, with a dose of dolomite liming 3 Ton-120 kg NPK / ha in Plastichouse, with a production of 1.94 tons / hectare.

References

[1] Fairhurst T, Witt C, Buresh R and Dobermann A 2007 *Rice A Practical Guide to Nutrient Management International Rice Research Institute*

[2]. Siregar A and Marzuki I 2011 The Efficiency of Urea Fertilization on N uptake and Yield of Lowland Rice (Oryza sativa, L.) *Jurnal Budidaya Pertanian* **7** (2): 107-112

[3] A. Dobermann T, Fairhurst 2000 *Rice, Nutrient Disorders and Nutrient Management. Potash & Phosphate Institute (PPI), Potash & Phosphate* Institute of Canada (PPIC) and International Rice Research Institute (IRRI)

Organic pollutant removal from edible oil process wastewater using electrocoagulation

Swati Sharma[1], Orhan Taner Can[2], Monsur Hammed[1], Dharmakeerthi Nawarathna[3], Halis Simsek[1]*

[1)] Department of Agricultural and Biosystems Engineering, North Dakota State University, Fargo, ND, USA
[2)] Department of Environmental Engineering, Bursa Technical University, Bursa, Turkey
[3)] Department of Electrical and Computer Engineering, North Dakota State University, Fargo, ND, USA

Email:halis.simsek@ndsu.edu

Abstract Wastewaters generated from vegetable oil industries contain a high concentration of organic pollutants that are detrimental to the aquatic ecosystem. Electrochemical processes are gaining importance in the treatment of inorganic and resistant organic pollutants in wastewaters. In this study, electrocoagulation (EC) was applied to remove organic pollutants and oil and grease from canola oil wastewater using aluminum (Al) and iron (Fe) electrodes. The application of EC in the wastewater achieved more than 80% removal of organic carbon and nearly 100% removal of suspended solids (SS). The effectiveness of EC is influenced mainly by current density, pH, electrolyte (NaCl), electrode contact time and electrode type. It was observed that Al electrode combination yielded better removal at a lesser time compared to that of Fe electrodes. However, varying current densities had its significance in terms of coagulation time only. Increase in current density achieved decrease in coagulation time. Both Al and Fe could remove between 52-59% of oil and grease from canola oil wastewater

1. Introduction

The amount of wastewaters discharged from edible oil industries is high because of a large volume of water required during the processing steps. The compositions and characteristics of wastewaters from vegetable oil refinery facilities vary depending on the type of crop used to produce oil [1]. The biological oxygen demand (BOD)/chemical oxygen demand (COD) ratio of 0.2 – typical of vegetable oil effluent – is appropriate to destroy microbes required for biodegradation. Most of the times, the vegetable oil wastewater discharge, although supposedly treated, does not meet the environmentally acceptable level. A number of methods such as absorption, coagulation, anaerobic treatment, reverse osmosis, and ultrafiltration have been found useful for treatment of vegetable oil wastewaters [1].

Electrocoagulation (EC) has some advantages on the treatment of edible oil industry wastewaters compare to the currently adopted wastewater treatment methods. Biological treatment methods require a vast land and facility combined with long treatment duration [2,3]. An appropriate treatment system should provide efficient pollutant reduction, uses less treatment time, requires simple installation, maintenance, and operation with the addition of little or no chemical compounds [3].

EC method has proved to be promising and thus attracted the attention of researchers around the world. EC techniques involve metallic dissolution through anodic reaction for creation of metallic hydroxide [4]. The positive metallic ion react with negative charged particles in the wastewater forming complex molecules – flocs – that coagulate without the addition of chemical [1,5]. Many reports have shown that EC can reduce heavy metals [4], nutrients [2], natural organic matters [6], dissolved organic carbons [6], COD [3,7,8], dyes [9-11] and microorganisms [12,13] from various types of wastewaters. EC has been successfully applied to treat vegetable oil refinery [1], olive mill [9], slaughterhouse [14-16], baker's yeast [7,17,18], dairy and tannery [19-21], and restaurant wastewaters [22]. Aluminum (Al) and iron (Fe) electrodes have found extensive applications for the EC treatment process because of their abundant availability and low cost. The main objective of this study is to investigate the treatment performance of EC processes on canola oil process wastewater sources in the removal of total organic carbon, oil and grease and suspended solids.

2. Materials and Methods

2.1. Sample collection and preparation

Canola oil process raw wastewater samples were collected using grab sampling method from Cargill West Fargo Oilseed Facility, West Fargo, ND. About one hour after collection, the samples were filtered through 1.2 μm glass microfiber filter paper followed by 0.45 μm pore size filter paper prior to determine initial parameters in the sample. The pollutant passing through a 0.45 μm filter paper was considered to be soluble components. About 50 ml of raw wastewater sample was filtered and used to determine initial concentrations of soluble COD (sCOD), dissolved organic carbon (DOC), total suspended solids (TSS), and oil and grease (O&G). The sCOD was analyzed using Hach testing kits TNT 821 and TNT 822 with the detection limits of 0-150 mg/l and 2-1500 mg/L, respectively. The DOC was analyzed using Shimadzu TOC-L analyzer. TSS was measured using the standard methods [32].

The O&G was analyzed using an Infracal 2 Analyer (Wilks Enterprise, Norwalk, Connecticut, USA). This method complies with EPA methods of 413.1 and 418.1 correlating to EPA 1664A (EPA/821-R-00-003). The method is designed to measure oil and grease by infrared determination in wastewater using S-316 solvent for the extraction procedure.

2.2. EC application of the raw wastewater sample

Raw wastewater sample was poured into a glass reactor for each set of batch EC experiments. The remaining samples were preserved at 4°C for the later usage. Every 3 or 4 days, any remaining wastewater samples were discarded and the fresh samples were collected from the facility. The working volume of 1 liter of raw wastewater sample was poured into glass reactor for the EC process. Al and Fe electrodes were used in two different combinations as Al|Al and

Fe|Fe. The electrodes were placed at 1 cm apart, connected to a DC power supply and immersed into the sample in the reactor (Figure 1). The experiments were carried out at different current densities of 0.002, 0.01, 0.019, 0.029 A/cm^2. The conductivity of the wastewater sample was 143 µS/cm which was too low to initiate EC process. Two g/L sodium chloride (NaCl) was added to the sample to increase the conductivity to about 4.93 mS/cm [23]. The surface area of each electrode used in the experiments was 12 cm x 7 cm where the immersed effective surface area of 8 cm x 6.5 cm. The efficiency of the process also depends on the sample pH. The pH in the reactor was maintained between 6.0-6.5 for Al-Al electrodes and 8.0-8.5 for Fe-Fe electrodes [23]. Sample solution of 50 ml from each run were collected at contact times of 10, 20, 30,40,50,60 and 70 minutes and tested for parametric analysis. The sample supernatant was filtered and analyzed for the presence of COD, DOC, TSS and O&G.

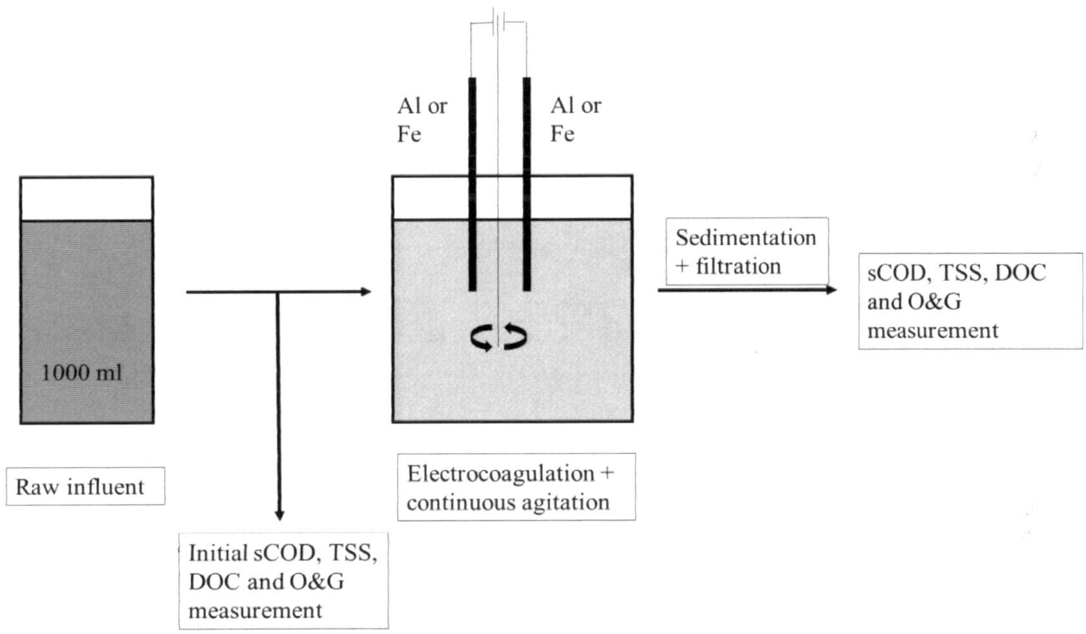

Figure 1. Schematic diagram of the electrocoagulation process in canola oil process wastewater.

3. Results and Discussion

3.1. Effect of current density on Al-Al and Fe-Fe electrodes

During the EC process, the samples were introduced to various current densities ranging between 0.002 to 0.029 A/cm^2. Figure 2 shows that effect of current density to coagulate colloidal particles depended on the electrode combination. It was observed that higher the current density, faster the process of coagulation. However, Al-Al electrode combination achieved coagulation at a time lesser than Fe-Fe combination. It can be deduced that the amount of Al^{3+} and Fe^{2+} ions produced for the applied current density are not sufficient to form enough flocs for complete removal of pollutants. El-Taweel et al., (2015) studied the removal of hexavalent chromium ions from wastewater through EC using iron electrodes. The study found that as the initial concentration of Cr(VI) ions increased , removal of these ions required increase in current density as well as electrocoagulation time for complete removal [4].

In our study, the reactor was run for 70 mins and samples were collected every 10 mins. The reactor was run even after coagulation was completed and flocs formed at the surface. At current density of 0.002 A/cm^2, coagulation was formed at Al|Al electrodes at a time interval of 25 mins, whereas, for iron electrodes, no coagulation was observed until 70 minutes. At higher current densities of 0.019 and 0.029 A/cm^2, time required to form coagulation was as low as 6 and 10 mins/L respectively.

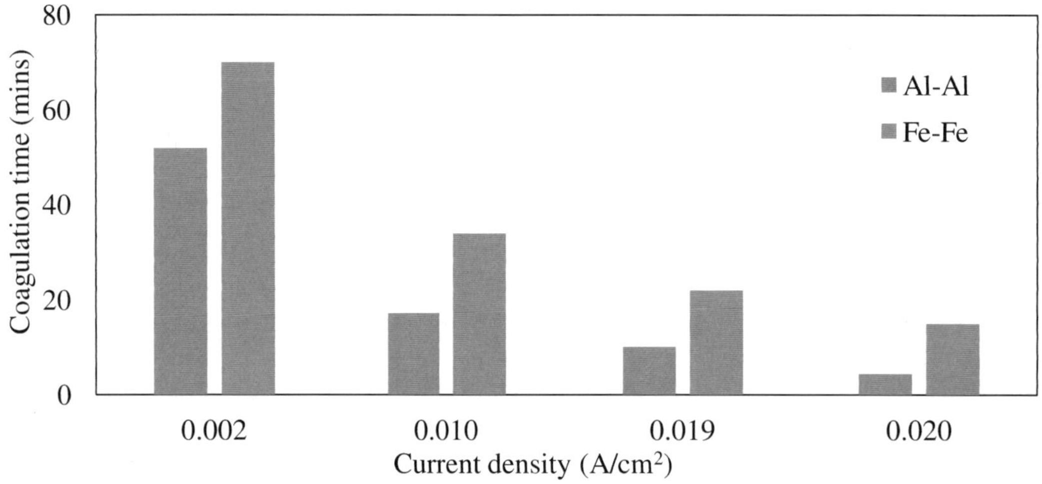

Figure 2. Coagulation time needed for electro-coagulation using aluminum and iron electrodes

3.2. Effect of electrode combination on removal of pollutants

3.2.1. Removal of sCOD

The impact of Al|Al and Fe|Fe electrode combinations with various current densities on the removal of sCOD was investigated (Figure 3 and 4). Samples were collected every 10 mins during the operation and initial and final concentration of sCOD was measured. The analysis was carried out both before and after coagulation was achieved in the reactor. Al|Al electrode combination could remove 75% at 0.002 A/cm^2 and upto 78% at 0.029 A/cm^2 (Figure 3a and 3b). However, at 0.002 A/cm^2, time required to achieve 75% removal was 70 mins compared to that of 10 mins at 0.029 A/cm^2. Fe|Fe achieved a removal efficiency of maximum 78% at the current density 0.029 A/cm^2 after 60 mins (Figure 4a and 4b). Removal efficiency depended on the contact time as well. Aluminum electrodes achieved higher removal efficiency at lower contact time than that of iron electrodes. It was also observed that iron electrodes took longer time of operation than aluminum electrode to attain the same amount of removal. From Figure 3 and 4, it can be concluded that aluminum electrodes achieved optimal removal of 78% at contact time of 10 mins at current density of 029 A/cm^2 and achieved steady state condition once the coagulation was formed. It was established in various studies that the current determines the dosage rate and size of flocs which in turn determines the removal efficiency of coagulation. Previous studies also stated that after 30 mins of coagulation at various current densities, the removal efficiencies tend to constant values in industrial and paper mill wastewater [24,25]. The removal of sCOD achieved stable concentration once coagulation was formed. The percentage removal of COD after EC was calculated as:

$$\text{Removal (\%)} = \frac{C_0 - C_f}{C_o} \times 100 \quad (1)$$

Where C_o and C_f are initial and final concentration measured in mg/L, respectively.

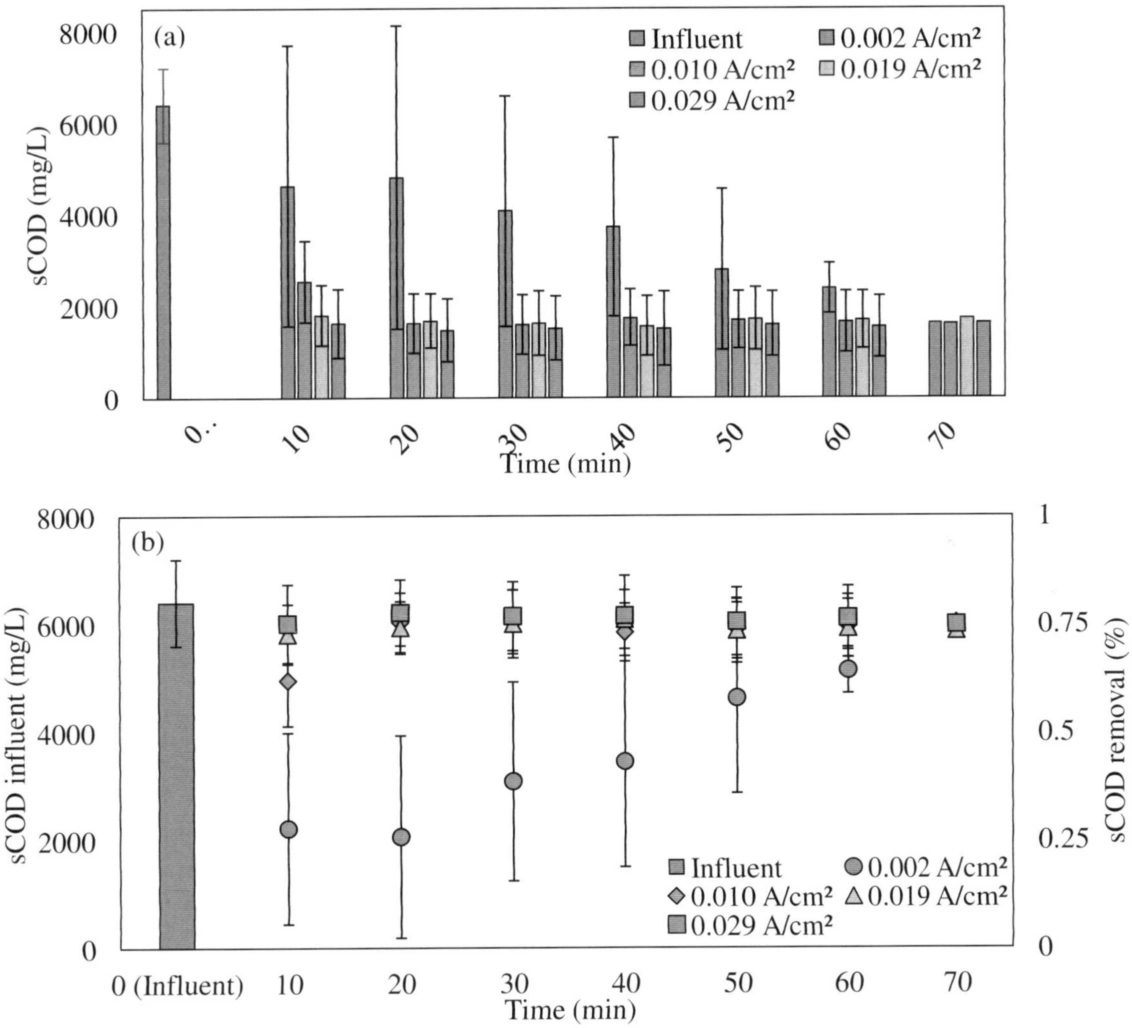

Figure 3. sCOD removal using electrocoagulation on canola oil wastewater using aluminum electrodes at various current densities a) concentration of sCOD in mg/L b) sCOD removal in percentage (%).

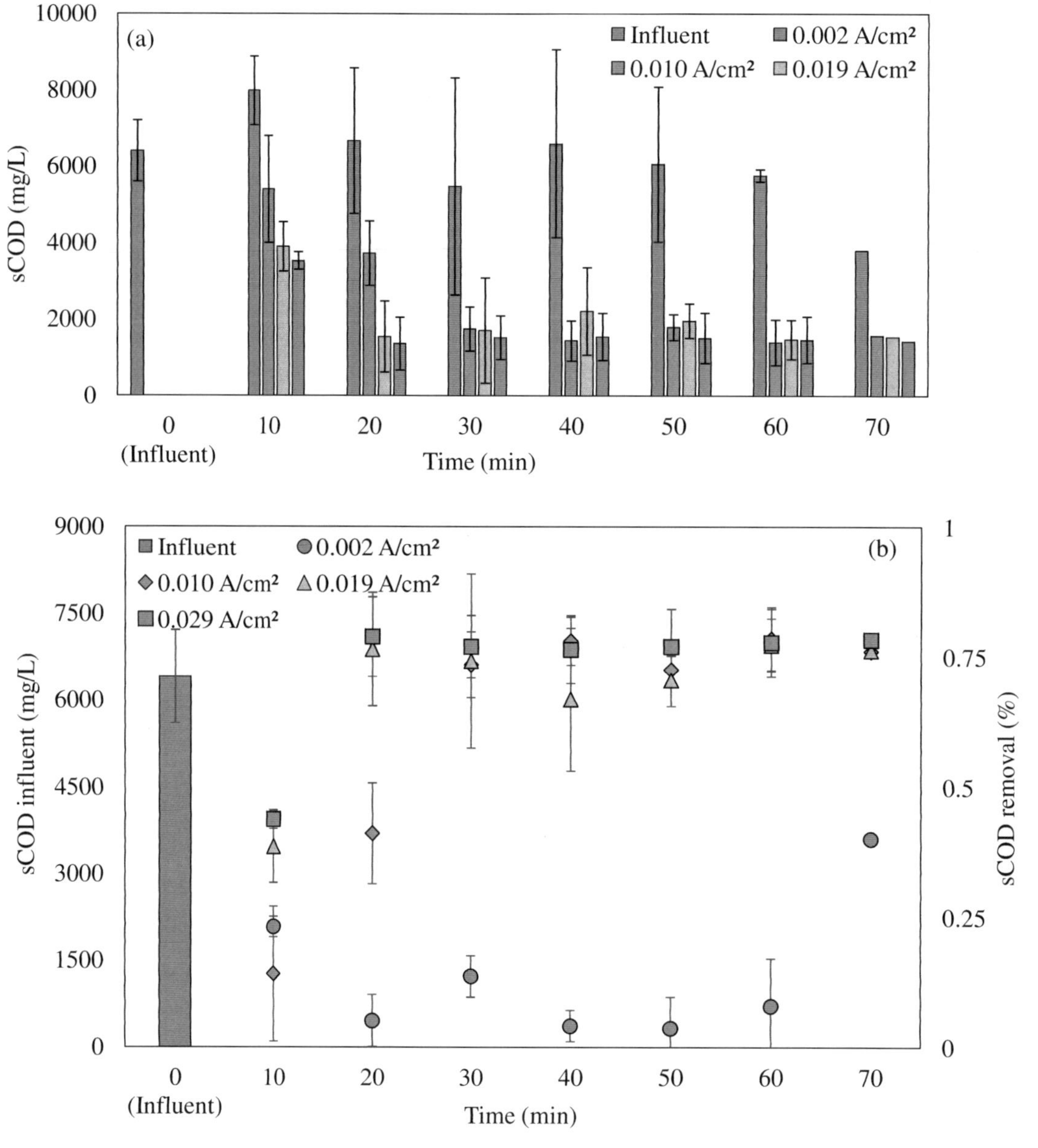

Figure 4. sCOD removal using electrocoagulation on canola oil wastewater using iron electrodes at various current densities a) concentration of sCOD in mg/L b) sCOD removal in percentage (%).

3.2.2. Removal of DOC

Removal of DOC using aluminum electrodes has yielded better performance compared to that of iron electrodes (Figure 5 and 6). Al-Al electrode combination was able to achieve more than 75% of DOC at all current densities that were applied (Figure 5a and 5b). On the other hand, Fe-Fe electrodes achieved maximum removal of 75% at 0.029 A/cm^2 at the end of 40 mins. Fe|Fe electrodes achieved not more than 16% removal in the concentration of DOC at current density of 0.002 A/cm^2 (Figure 6a and 6b). For Fe|Fe electrodes, removal efficiency was slower

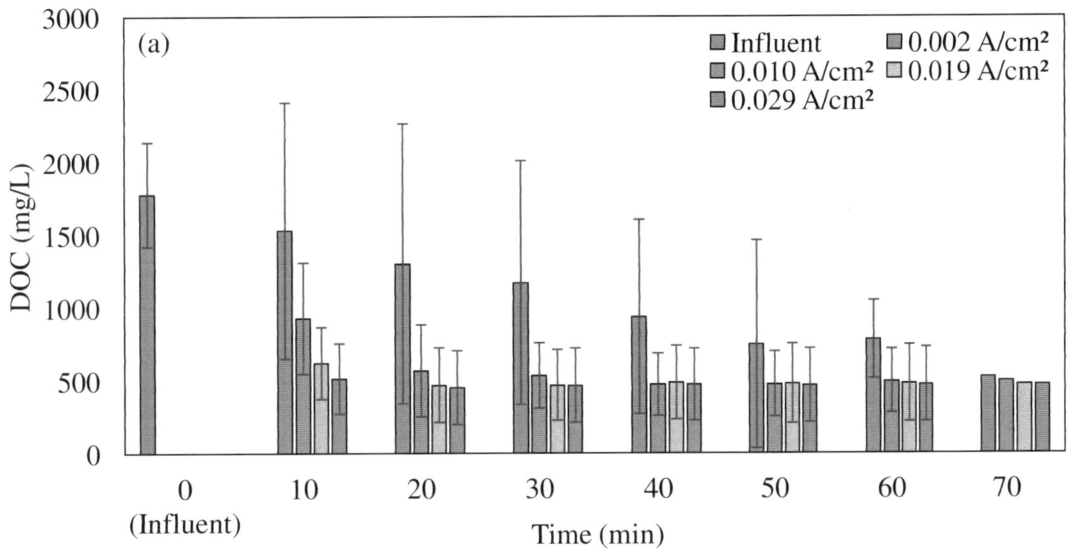

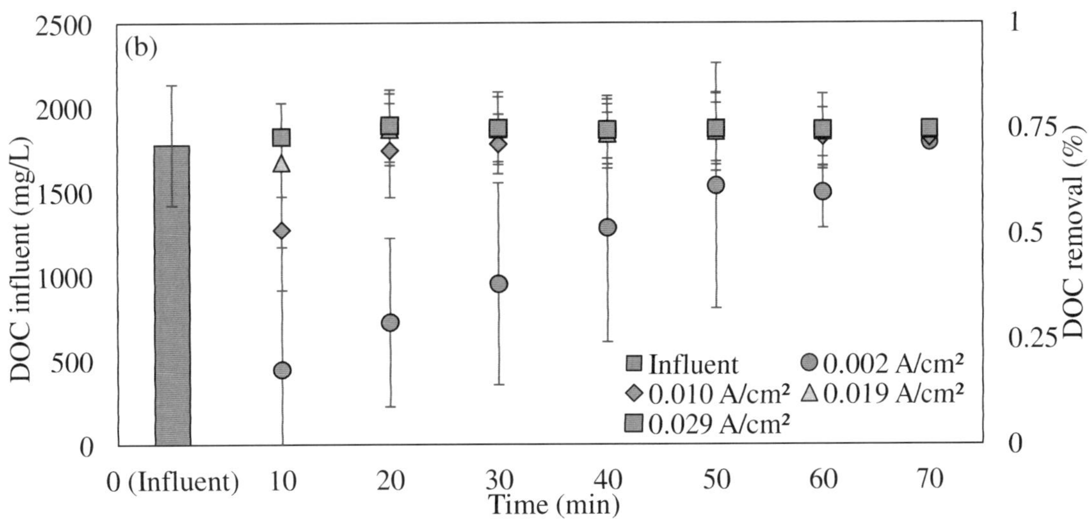

Figure 5. DOC removal using electrocoagulation on canola oil wastewater using aluminum electrodes at various current densities a) concentration of DOC in mg/L b) DOC removal in percentage (%).

compared to Al|Al electrodes at all current densities applied. Similar results were obtained in the study conducted by Un et al., (2009a) on vegetable oil refinery wastewater using EC in a batch mode by adding poly aluminum chloride. The study was conducted to investigate the effect of pH and current density on wastewater using aluminum electrodes, which was proved to be the superior combination compared to that of iron electrodes. El-Nas et al (2009) evaluated the removal efficiency of sulfate and COD on petroleum refinery wastewater using three types of electrode materials, which were stainless steel, aluminum, and iron. The study concluded that aluminum had better removal performance compared to the other two electrode materials.

3.2.3. Removal of oil and grease (O&G)

There was no significant difference in the removal efficiency of O&G (Figure 5) between the two electrode combinations (aluminum and iron). Both the electrodes could remove O&G between 52% and 59% from the canola oil wastewaters. Previous studies have also concluded that EC can remove more than 80% of O&G in vegetable oil wastewater sample [26-29]. This study observed that oil removal efficiency in both electrodes was nearly equal. However, at current density of 0.002 A/cm^2, iron electrodes could remove not more than 26% of O&G.

3.2.4. Removal of suspended solids

The results (Figure 8) showed that aluminum electrodes achieved almost 100% removal of total suspended solids at the end of each operation time compared to that of iron electrodes. Both electrode combinations achieved 100% removal of TSS and the results were significant in terms of contact time. Iron electrodes consumed longer time to achieved 100% removal compared to aluminum electrodes. However, it is important to mention that maintaining the pH of the electrolyte solution determines the time required to coagulate. Other studies that was conducted in textile, dairy and tannery wastewaters found similar results in the removal of suspended solids using EC [5,30,31].

Thus, comparing the removal rates and capability to remove organic carbon and suspended solids along with O&G, Al-Al electrodes combination was chosen to be superior compared to that of iron electrodes combination.

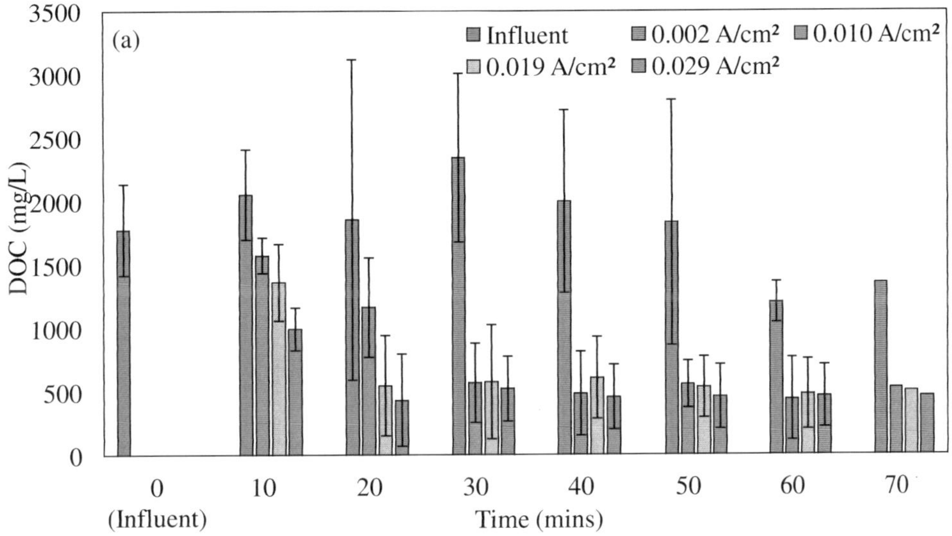

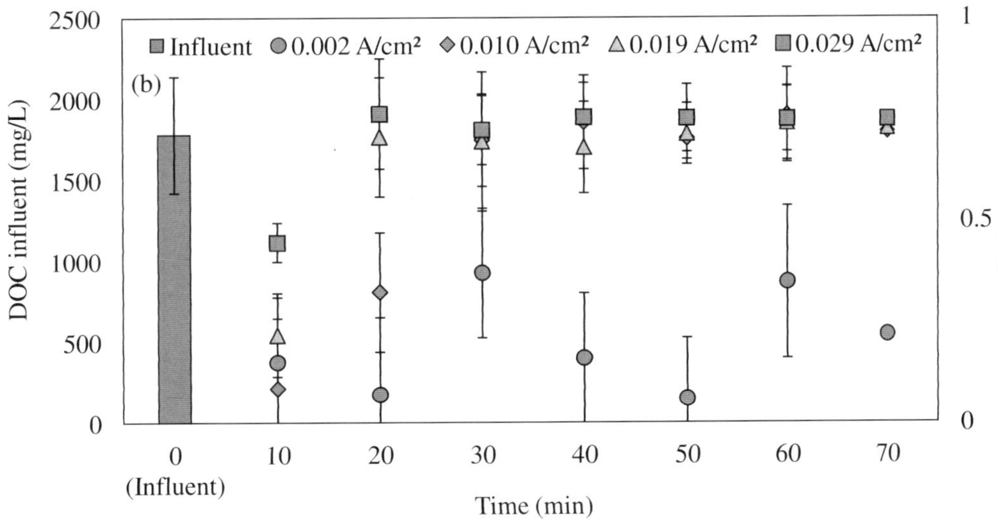

Figure 6. DOC removal using electrocoagulation on canola oil wastewater using iron electrodes at various current densities a) concentration of DOC in mg/L b) DOC removal in percentage (%).

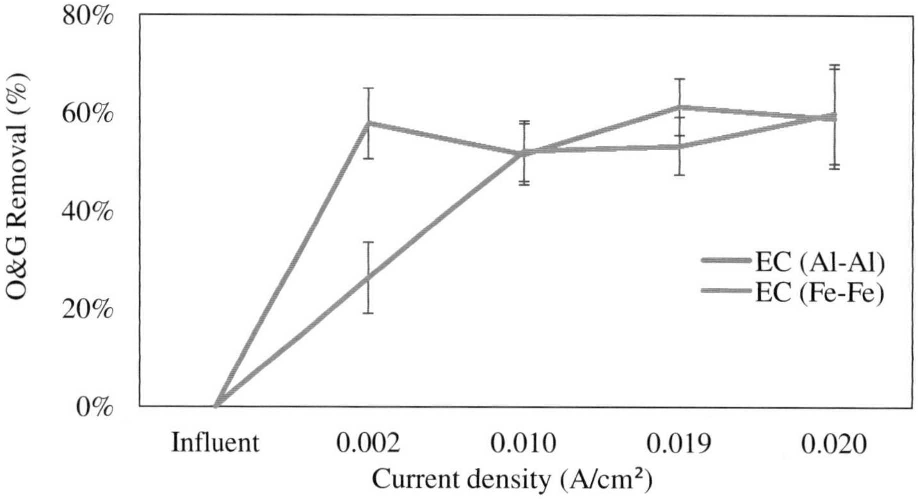

Figure 7. Oil and grease removal using combined electrocoagulation in canola oil wastewater

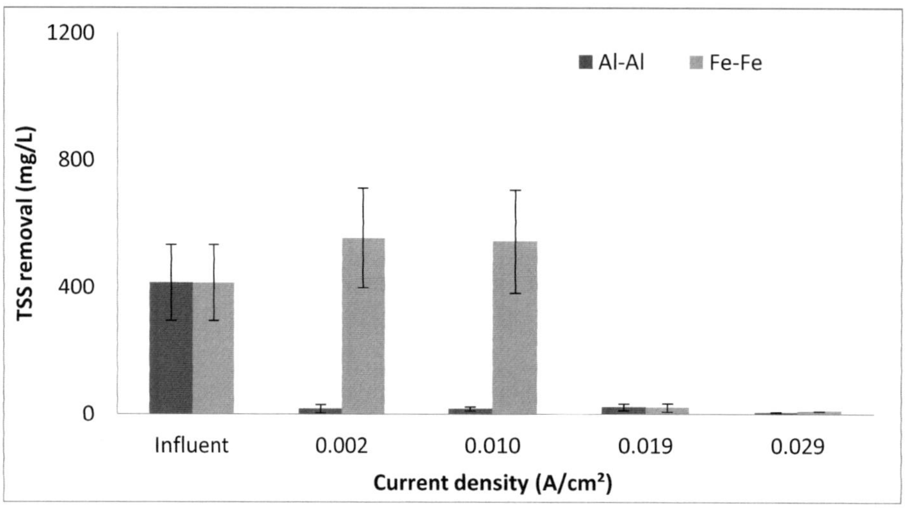

Figure 8. TSS removal by EC using Al-Al and Fe-Fe and EF process on Canola oil wastewater

4. Conclusion

The main conclusions that can be drawn from this study are, varying current density effects the coagulation time in edible oil wastewater. The higher the current density, lower time it takes to coagulate. Al-Al electrode combination achieved higher removal efficiency of organic pollutants at a lesser time compared to that of Fe-Fe electrode combination. The aluminum electrodes achieved more than 80% removal of DOC at all current densities under observation. Whereas, at low current densities, iron electrode combination could not yield more than 16% removal of DOC. Both Al-Al and Fe-Fe electrode combination could remove O&G between 52% and 59% showing that the removal efficiency was statistically insignificant for both electrode combinations. The EC process successfully removes nearly 100% of suspended solids using both Al-Al and Fe-Fe electrodes.

References

[1] Un U T, Koparal A S, Ogutveren U B 2009 Electrocoagulation of vegetable oil refinery wastewater using aluminum electrodes *J. of Environmental Management* **90** 428-433

[2] Nguyen D D, Ngo H H, Guo W et al 2016 Can electrocoagulation process be an appropriate technology for phosphorus removal from municipal wastewater? *Science of the Total Environmen*t **563** 549-556

[3] Tak B-y, Tak B-s, Kim Y-j et al 2015 Optimization of color and COD removal from livestock wastewater by electrocoagulation process: application of Box–Behnken design (BBD) *J. of Industrial and Engineering Chemistry **28** 307-315*

[4] El-Taweel Y A, Nassef E M, Elkheriany I et al 2015Removal of Cr(VI) ions from waste water by electrocoagulation using iron electrode. *Egyp J. of Petroleum* **24** 183-192

[5] Un U T, Koparal A S, Ogutveren U B 2009 Hybrid processes for the treatment of cattle-slaughterhouse wastewater using aluminum and iron electrodes. *J. of Hazardous Materials 164: 580-586*

[6] Feride U, Barıscı S, Kobya M et al 2014 Batch Electrocoagulation Reactor for Natural Organic Matter Removal from Synthetic and Real Water Samples. *J. of Selcuk University Natural and Applied Science* 369-379

[7] Al-Shannag M, Al-Qodah Z, Alananbeh K et al 2014 COD reduction of baker's yeast wastewater using batch electrocoagulation *Environmental Engineering and Management J.* **13** 3153-3160

[8] Al-Shannag M, Al-Qodah Z, Bani-Melhem K et al 2015 Heavy metal ions removal from metal plating wastewater using electrocoagulation Kinetic study and process performance *Chemical* Engineering *J.* **260** 749-756

[9] Hanafi F, Assobhei O, Mountadar M 2010 Detoxification and discoloration of Moroccan olive mill wastewater by electrocoagulation *J. of hazardous materials* **174** 807-812

[10] Alinsafi A, Khemis M, Pons M et al 2005 Electro-coagulation of reactive textile dyes and textile wastewater *Chemical engineering and processing :Process intensification* **44**: 461-470

[11] Kim T-H, Park C, Shin E-B et al 2002 *Decolorization of disperse and reactive dyes by continuous electrocoagulation process* *Desalination* **150** 165-175

[12] Cotillas S, Llanos J, Cañizares P et al 2013 Optimization of an integrated electrodisinfection/electrocoagulation process with Al bipolar electrodes for urban wastewater reclamation *Water research* **47** 1741-1750

[13] Massoudinejad M, Khashij M, Soltanian M 2014 Survey of electrocoagulation process in the removal of pathogen bacteria from wastewater before discharge in the acceptor water

[14] Asselin M, Drogui P, Benmoussa H et al 2008 Effectiveness of electrocoagulation process in removing organic compounds from slaughterhouse wastewater using monopolar and bipolar electrolytic cells. Chemosphere **72** 1727-1733
[15] Bayar S, Yıldız Y S, Yılmaz A E et al 2011The effect of stirring speed and current density on removal efficiency of poultry slaughterhouse wastewater by electrocoagulation method. Desalination **280** 103-107
[16] Ün ÜT, Koparal AS, Öğütveren ÜB 2009 Hybrid processes for the treatment of cattle-slaughterhouse wastewater using aluminum and iron electrodes *J of Hazardous Materials***164** 580-586
[17] Kobya M, Delipinar S 2008 Treatment of the baker's yeast wastewater by electrocoagulation *J. of hazardous* materials **154** 1133-1140
[18] Gengec E, Kobya M, Demirbas E et al 2012 Optimization of baker's yeast wastewater using response surface methodology by electrocoagulation Desalination **286** 200-209
[19] Feng J-w, Sun Y-b, Zheng Z et al 2007 Treatment of tannery wastewater by electrocoagulation *J. of Environmental Sciences* **19** 1409-1415
[20] Şengil İA 2006 Treatment of dairy wastewaters by electrocoagulation using mild steel electrodes *J. of hazardous materials* **137** 1197-1205
[21] Tchamango S, Nanseu-Njiki CP, Ngameni E et al 2010 Treatment of dairy effluents by electrocoagulation using aluminium electrodes *Science of the total environment* **408** 947-952
[22] Chen X, Chen G, Yue P L 2000 Separation of pollutants from restaurant wastewater by electrocoagulation. *Separation and* Purification *Technology* **19** 65-76
[23] Can O T 2014 COD removal from fruit-juice production wastewater by electrooxidation electrocoagulation and electro-Fenton processes *Desalination and Water Treatment* **52** 65-73
[24] Katal R, Pahlavanzadeh H 2011 Influence of different combinations of aluminum and iron electrode on electrocoagulation efficiency: Application to the treatment of paper mill wastewater. *Desalination* **265**: 199-205
[25] Garcia-Garcia A, Martinez-Miranda V, Martinez-Cienfuegos IG et al 2015 Industrial wastewater treatment by electrocoagulation-electrooxidation processes powered by solar cells Fuel **149** 46-54
[26] Xu X H, Zhu X F2004 Treatment of refectory oily wastewater by electro-coagulation process. Chemosphere **56** 889-894
[27] Canizares P, Lobato J, Paz R et al 2007Advanced oxidation processes for the treatment of olive-oil mills wastewater Chemosphere **67** 832-838
[28] Tir M, Moulai-Mostefa N 2008 Optimization of oil removal from oily wastewater by electrocoagulation using response surface method. Journal of Hazardous Materials **158** 107-115
[29] An CJ, Huang G, Yao Y et al 2017 Emerging usage of electrocoagulation technology for oil removal from wastewater: A review Science of the Total Environment **579** 537-556
[30] Khandegar V, Saroha A K 2013Electrocoagulation for the treatment of textile industry effluent - A review. *J. of Environmental Management* **128** 949-963

[31] Mollah M Y A, Schennach R, Parga J R et al 2001 Electrocoagulation (EC) - science and applications *J. of Hazardous Materials* **84** 29-41
[32] APHA, AWWA, WEF 2005 Standard methods for the examination of water and Wastewater 21st ed American Public Health Association Washington DC

Institute of Physics
Dirac House, Temple Back
Bristol BS1 6BE UK

ISSN: 1755-1307
ISBN 978-1-5108-6109-1

CURRAN ASSOCIATES INC.
proceedings.com